3D Printing of Foods

3D Printing of Foods

C. Anandharamakrishnan
National Institute of Food Technology, Entrepreneurship and Management - Thanjavur (an Institute of National Importance; formerly Indian Institute of Food Processing Technology - IIFPT), Ministry of Food Processing Industries, Government of India, Thanjavur, Tamil Nadu, India.

Jeyan A. Moses
National Institute of Food Technology, Entrepreneurship and Management - Thanjavur (an Institute of National Importance; formerly Indian Institute of Food Processing Technology - IIFPT), Ministry of Food Processing Industries, Government of India, Thanjavur, Tamil Nadu, India.

T. Anukiruthika
National Institute of Food Technology, Entrepreneurship and Management - Thanjavur (an Institute of National Importance; formerly Indian Institute of Food Processing Technology - IIFPT), Ministry of Food Processing Industries, Government of India, Thanjavur, Tamil Nadu, India.

WILEY

This edition first published 2022

Registered Offices
John Wiley & Sons, Inc., 111 River Street, Hoboken, NJ 07030, USA
John Wiley & Sons Ltd, The Atrium, Southern Gate, Chichester, West Sussex, PO19 8SQ, UK

Editorial Office
The Atrium, Southern Gate, Chichester, West Sussex, PO19 8SQ, UK

For details of our global editorial offices, customer services, and more information about Wiley products visit us at www.wiley.com.

Wiley also publishes its books in a variety of electronic formats and by print-on-demand. Some content that appears in standard print versions of this book may not be available in other formats.

Library of Congress Cataloging-in-Publication Data
Names: Anandharamakrishnan, C., author. | Moses, Jeyan A., author. | Anukiruthika, T., author.
Title: 3D printing of foods / C. Anandharamakrishnan, Indian Institute of Food Processing Technology (IIFPT), Thanjavur, India, Jeyan A. Moses, Indian Institute of Food Processing Technology (IIFPT), Thanjavur, India, T. Anukiruthika, Indian Institute of Food Processing Technology (IIFPT), Thanjavur, India.
Other titles: Three dimensional printing of foods
Description: First edition. | Hoboken, NJ : Wiley, [2021] | Includes bibliographical references and index.
Identifiers: LCCN 2021040694 (print) | LCCN 2021040695 (ebook) | ISBN 9781119669821 (hardback) | ISBN 9781119669753 (adobe pdf) | ISBN 9781119671800 (epub)
Subjects: LCSH: Three-dimensional food printing.
Classification: LCC TP372.85 .A53 2021 (print) | LCC TP372.85 (ebook) | DDC 664–dc23/eng/20211105
LC record available at https://lccn.loc.gov/2021040694
LC ebook record available at https://lccn.loc.gov/2021040695

Cover Design: Wiley
Cover Image: © Alex_Traksel/Shutterstock

Set in 9.5/12.5pt STIXTwoText by Straive, Pondicherry, India
Printed and bound by CPI Group (UK) Ltd, Croydon, CR0 4YY

C9781119669821_110322

Contents

Preface

Gone are the days when food was consumed by the local population within a short span after the cultivation/harvest season. Today, food processing and preservation strategies have made major strides in striving to ensure food and nutritional security, at large. In recent years, with emerging applications in a range of other fields, 3D printing of foods has gained interest in research and technology development. Food processing is a complex and interdisciplinary field, relating agriculture, a series of postharvest unit operations, an intricate food supply chain, and stringent quality/safety parameters, all interlinked with consumer perception and acceptance. With the rising need for food customization, and providing personalized foods in terms of taste, mouthfeel, appearance, and nutrition, 3D printing of foods is gaining importance. 3D printing is an additive manufacturing approach involving layer-by-layer fabrication, using robotic controls. Advantages in terms of bulk customization, rapid prototyping, reduced wastages, and increasing health care applications have been the major strengths of the technology. Aerospace, automotive, and medical industries are the biggest takers, and technology is revolutionizing the manufacturing sector. The present scenario of the rising global population has increased the demand for the production of foods. In this regard, 3D printing allows us to produce foods sustainably even utilizing underexplored food sources and waste/by-products. 3D printing of foods remains to offer a promising solution in satisfying global hunger, importantly, the 'hidden hunger': malnourishment. 3D food printing revolutionizes to customize foods based on genomic and biometric data, which are sensorially acceptable, making 'imagination into reality.' Eventually, food printing is poised to have a great impact on the economy, as it paves way for bridging the gap between small- and large-scale entrepreneurs, thereby providing a great degree of flexibility to consumers in choosing the food in the way they want to eat.

Food printing is a rapidly emerging field and has the attractiveness of a range of advantages focusing on product customization and better utilization of resources. With growing consumer demands, we can be assured that this field would take a key segment in the food processing sector. However, the R&D needs are high and need to be better addressed before widespread commercialization and adoption of the technology. The speed of production must match/supplement other food processing technologies; it remains a challenge to bring the perfect blend between customization and mass production. Very importantly, the cost of a food 3D printer needs to be majorly relooked, considering prospects of widespread commercialization and usage. Addressing such concerns can for sure bring about the best with this technology in the food industry. While it continues to remain a debatable concern if 3D printed food materials are categorized as 'processed foods,' we envisage that 3D food printing in combination with food processing techniques can bring about a range of synergistic applications. This article focuses on the foundation of 3D food printing, the aspect of 'food printability.' We explain the complexity of defining this term and the huge research gap that needs to be addressed. Hence, we can hope that with a strong understanding of the intricacies of food properties and 3D printing mechanisms, we would one day see food 3D printers in every kitchen!

C. Anandharamakrishnan
Jeyan A. Moses
T. Anukiruthika

1

Introduction to 3D Printing Technology

CHAPTER MENU

1.1 Introduction

Three-dimensional (3D) printing appears to be a revolutionizing solid-free fabrication (SFF) technique that grabs attention in recent years because of its inherent potential to transform virtual ideas into reality. Any manufacturing process involves a series of steps in converting the raw material into finished products. Manufacturing processes are classified as the additive process, subtractive process, formative process, and joining process (Bandyopadhyay and Heer 2018). As the name implies additive process involves the formation of an object by the addition of the material in a layered manner one above the other. On the other hand,

3D Printing of Foods, First Edition. C. Anandharamakrishnan, Jeyan A. Moses, and T. Anukiruthika.

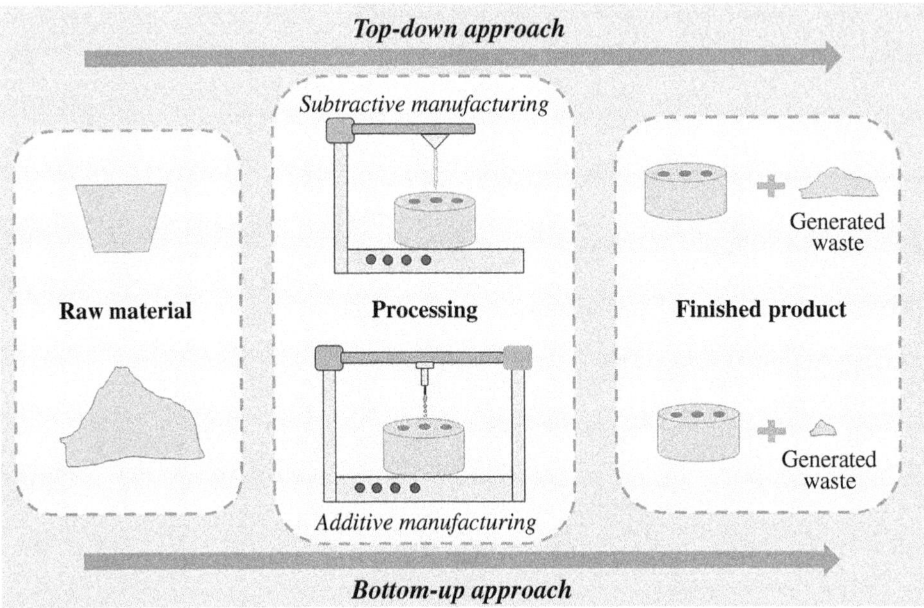

Figure 1.1 Schematic representation of subtractive and additive manufacturing.

the removal of the material by sculpturing an object out of the solid raw material is referred to as a subtractive process. In the formative process, the finished products are formed out of molten raw material as in the case of casting and forging. The latter is the joining process that combines the pieces of raw material either temporarily or permanently through fastening or welding. All the above processes other than additive manufacturing (AM) are grouped as traditional processing methods that involve a top-down approach (Tofail et al. 2018). In contrast, AM represents a bottom-up approach and is termed as a rapid prototyping (RP) layer-based technique that involves the direct fabrication of physical objects from raw material (Figure 1.1) (Hon 2007). The International Organization for Standardization (ISO)/ American Society for Testing and Materials (ASTM) defined AM as the process of creating a 3D object out of a computer-designed 3D model through deposition and fusion of material in a layerbylayer manner (Jiang et al. 2019). Several other terms that were used synonymously in place of RP and AM are free-form fabrication, ingress manufacturing, layered manufacturing, and digital manufacturing. 3D printing is one such technique of AM that allows for layer-by-layer construction of 3D objects with minimal processing and less wastage of raw materials (McClements 2017).

The ISO/(ASTM) 52900:2015 standard had classified AM processes into seven different categories based on its working mechanism as material extrusion (ME), binder jetting (BJ), material jetting (MJ), powder bed fusion (PBF), sheet lamination (SL), directed energy deposition (DED), and vat photopolymerization (VP) (Tofail et al. 2018). Among these 3D printing techniques, extrusion-based 3D printing is the most commonly used technique because of its simplicity and low cost (Jiang et al. 2019). Before food printing, 3D printing technology has widely applied for printing polymers (plastics, resins, and photopolymers), ceramics, metals, biomaterials, etc., with the assistance of external (thermal or mechanical) energy. The feed supply of these materials can be either in liquid or solid material in the form of powders/sheets/filaments. The process involves the deposition of feed materials in their fluidic state and gets fused and bonded together through appropriate chemical interactions. Raw materials such as polymers can be easily melted and bonded together due to their low melting point and glass transition temperatures while metals and ceramics require higher temperatures that employ an external heat source either laser or electron beams for post-deposition (Ligon et al. 2017). Each of the AM technologies has its advantages and limitations and the specific printing technique can be selected based on its applicability and end-use requirements. With the present scenario, 3D printing is expected to reach a peak of inflation in the coming years and is predicted to receive the main focus during 2019 and 2024 (Jayaprakash et al. 2019). This chapter covers the digital advancements of printing technology and printers, potential advantage, and applications of 3D food printing over the traditional food processing techniques.

1.2 Digital Manufacturing: From Rapid Prototyping to Rapid Manufacturing

Digital manufacturing involves the direct fabrication of objects without setting pre-tools or workpiece requirements. Although the terms RP and AM are used synonymously, there was a distinct difference among them. RP refers to the process associated with the development of a prototype model, i.e. here the model processing is restricted till the pre-production step that could not be used as functional working objects. Thus, the progressive transformation of RP leads to the AM processes that involve the actual production of functional workpieces from prototype models. Thus, AM allows RP to evolve into rapid manufacturing (RM) with more flexibility, work freedom, and exploitation of applications in developing a layered physical object. 3D printing was found to have vast potential applications of prototyping in several industrial sectors such as pharmaceuticals, automotive, space engineering, civil constructions, art, aviation, archaeology, cosmetics, and fashion industries (Rahman et al. 2018). Nevertheless, the most attractive application of 3D printing in food manufacturing is designing foods in a customized manner that leads to the development of the food fabrication process commonly referred to as food 3D printing. 3D printing of foods has a quite huge market potential as it aids in the mass customization, personalized diets, and sustainability practices than the traditional food manufacturing technologies (Derossi et al. 2019). Thus, 3D printing of foods referred to as food layer manufacturing (FLM) involves the sequential process of fabricating three-dimensional edible constructs in a layer-by-layer manner with the capability of binding the adjacent layers through phase transitions or by chemical reactions (Nachal et al. 2019). A typical 3D printing process follows a series of well-defined steps (Figure 1.2). First, it starts with scanning of real-time objects or the creation of a 3D model using computer-aided design (CAD) software. The shape and surface characteristics are stored in a unique STL file format that is native to 3D printing technology. Later the digital representation of the stored 3D object is transformed to the sliced information using a slicing software that translates the 3D model into computer-generated codes (G and M codes). Based on which the movement arms and motors of 3D printers are controlled (Bechtold 2016). Thus, the whole printing process is controlled digitally using computers with minimal human interactions.

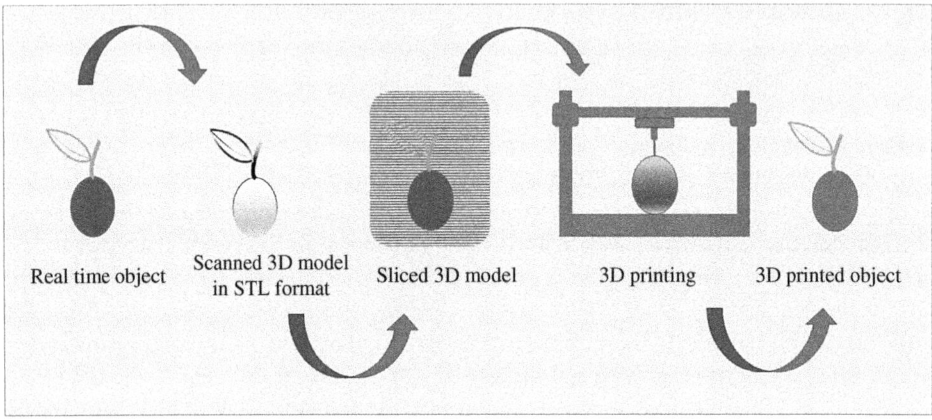

Figure 1.2 Workflow of 3D printing process.

1.3 Milestones in 3D Printing Technology

Although 3D printing received a wide attention in recent years, the technology dated back to several decades. Printing technology that uses two laser beams to fabricate the 3D objects was patented by Wyn Swainson of Denmark in the 1970s (Bechtold 2016). Later in the 1980s, another patent on 3D printing was filed by Dr. Hideo Kodama of Japan. However, this patent got rejected as the deadline for filing was passed out. After that, the next patent was awarded to Charles Chuck Hull in 1986 for his efforts in developing a stereolithography (SLA) apparatus (Beltagui et al. 2020). Hull co-founded 3D Systems, one of the leading companies in the 3D industry. Later the company introduces the first commercialized 3D printer based on SLA in 1988. Further, they have developed a new file format that was specific to 3D printing technology named STL that was understandable to 3D printers which aid in the printing of 3D objects. While in 1988 DTM Inc., developed the first 3D printer based on selective laser sintering (SLS) technique (Saptarshi and Zhou 2019). Another 3D printing technology named fused deposition modelling (FDM) which was the most commonly adopted 3D printing technique was developed by Scott and Lisa Crump in the 1980s. They received a patent on this FDM technology and co-founded Stratasys, another major player among the 3D industries (Su and Al'Aref 2018).

During the 1990s with the advancements of technology researchers of Stanford and Mellon proposed several other 3D techniques applied for micro-level casting and spraying of materials. In 1993, MIT filed a patent on inkjet technology that employs liquid-based ink for the construction of 3D objects commonly used in inkjet printers (Prasad and Smyth 2016). Later this technique was transferred and licensed to Z corporation for the development and marketing of 3D printers. Apart from materials manufacturing at the industrial level, 3D printing allows to produce consumer end products. In 2005, the RepRap project (3D printing open-source project) was started by Adrian Bowyer at the University of Bath for the development of 3D printers at a low cost that could be affordable to the consumers (Bechtold 2016).

1.4 Different Historical Eras in 3D Printing

Since the incipient of 3D printing technology, 3D printing has shaped and transformed into different forms. Various technological advancements in the development of these AM processes are summarized in the subsequent sections.

1.4.1 Ancient Age

As stated earlier, the 3D printing process was first demonstrated and documented by Kodama of MIT where he developed a method for fabrication of 3D models out of plastics through photo-hardening of photopolymers cured by ultraviolet (UV) light (Kodama 1981). Later in 1984, three researchers named Mehute, Witte, and Andre filed a patent on the STL process which was unsuccessful with the lack of business potential (Sokolov et al. 2018). After that, the STL technology was commercialized by 3D systems corporations which resulted in a viable manufacturing process for 3D printing. Meanwhile, the other 3D printing FDM technology has become popular as it paved for the production of consumer-oriented 3D printed products (Sanchez Ramirez et al. 2019). This technology involved extruding hot-melted plastics through the nozzle die thereby resulting in the deposition of layers to form 3D objects. These printers were quite large as like '1970s 5 MB hard disk' which were then gradually reduced in size with advancements in 3D printing technology.

1.4.2 Middle Age

Around the 1990s, 3D printing received a vast attention due to its advantageous features that drive researchers of different universities to start working on this emerging area. In the 1990s, EOS GmbH developed a 'stereos' system, the first commercial industrial 3D printer (Calignano et al. 2019). Then Stratasys filed a patent on FDM technology that leads to the development of domestic 3D printers. In the late 1990s, new technologies were introduced by many aspiring 3D printing companies such as dot-on-dot printing techniques that use polymer jet for the fabrication of 3D objects. One such technique is MIT's inkjet printing that uses polymer solution in a drop-on-demand (DoD) manner (Prasad and Smyth 2016). Similarly, the Fraunhofer Institute of Germany introduced selective laser melting (SLM) in 1995 which employs laser light as a curing medium. Meanwhile, the Z corporation worked in collaboration with MIT for the development and production of FDM printers on a commercial scale. Another advancement of printing technology that made its application in the biological field is in regenerative medicine that supports the growth of human organs as the Wake Forest Institute made a successful attempt in the development of tissue scaffolds (Su and Al'Aref 2018). This medieval period remains to be a golden age that promoted various advancements in 3D technologies and 3D printers.

1.4.3 Modern Age

During the start of the twenty-first century, 3D printing had moved and expanded its wing from the commercial scale and entered into the domestic level. In 2000, the workers of Object Geometries created the first inkjet 3D printer which was then commercialized by Z corporation that paves a way for the development of multi-colour 3D printer which remains one of the milestones in the evolutionary history of 3D printers (Yang et al. 2018). Later in 2001, the desktop 3D printers were becoming common and in 2002 Wake Forest Institute worked in the development of miniature 3D printed kidney that mimics the functions of the human kidney (Ledford 2015). This leads to the advancements of 3D printing in biomedicine. Around 2005, the open-source 3D printing project RepRap developed the first 3D printer capable of producing its part named RepRap Darwin (DIY 3D printers). RepRap introduces the word fused filament fabrication (FFF) that replaces the term FDM.

With the innovations of AM, the first 3D printed car was developed by Urbee in 2011, and then in 2013 3D printable gun was released (3DSourced 2021). Gradually the 3D printing moved from polymers to foods as National Aeronautics and Space Administration (NASA) experimented with 3D printing the foods for aeronauts in 2014 (Lipton et al. 2015). Meanwhile, the emergence of flexible new software enhances the mass production of 3D printers in 2017 and until to the present date. Beyond fashion jewellery and aircraft, 3D printing allows for the construction of affordable houses for the developing world. Still, many advancements are happening, and much research is going on in exploring the potential applications of 3D printing in different sectors.

1.5 Prospects of 3D Food Printing

The concept of 3D printing encompasses three key criteria: universal, practical, and efficient. With the development of information and technology, it becomes possible to print foods (one of the essential components for life) in the desired form by uploading a digital file to printers that deliver printed food (Bandyopadhyay and Heer 2018). The main purpose of applying this AM technology in food printing relies on the advantage of designing foods with newer texture, consistency, flavour, and taste with enhanced nutrition (Figure 1.3). A synergistic combination of material properties, chemical interactions, and binding mechanisms assist in achieving a stable 3D food constructs (Sun et al. 2018a). However, understanding material behaviour is cumbersome that stands to be a challenge in the food printing process since food is a complex substance with a wide variations in its physiochemical characteristics. Globally, several research works are going on in exploring the potential applications of 3D printing in the food sector. Researchers of the Netherlands Organisation for Applied Scientific Research (TNO) have explored the application of food ingredients for 3D printing and converted them to tasty, printed food products because of the health issues due to busy lifestyles and environmental concerns from depletion of resources. Similarly, NASA is exploring and developing 3D printed space foods for astronauts (Sun et al. 2015b).

The idea of customization aids in the delivery of food items as per the requirements and needs. Researchers have customized foods like pizzas and cakes with the idea of incorporating complex shapes and intricate designs using 3D printing technology. Nowadays, the consumer perception towards diet is gradually changing due to changing lifestyle, and different age grouped people require a varying degree of nutrients. Formulating a balanced diet as per individual needs and preferences is the need of the day. This technology allows for personalized meals based on age groups and helps in reduced calorie intake. Thus, the concept of 'personalized nutrition' comes into focus which makes possible the delivery of

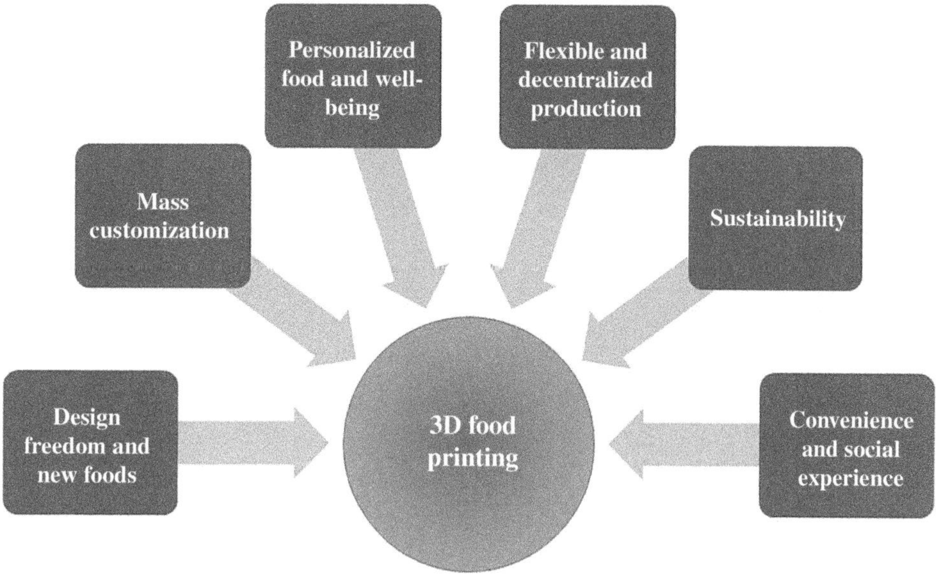

Figure 1.3 Prospects of 3D food printing.

'digitalized food'. 3D food printing reduces the multi-step process into a single step which has the potential to revolutionize the future food industry. The supply chain of 3D food printing allows consumers to have a prime role in the value addition of the end products based on their requirements which is very minimal in the case of a conventional food supply chain (Figure 1.4) (Jayaprakash et al. 2019).

1.6 Design Considerations of 3D Printer

1.6.1 Printer Configurations

Food printing is a synergistic combination of incorporating digital culinary skills with 3D printing techniques. 3D food printers are machines that have the potential capability of reproducing 3D edible constructs from a designed digitalized 3D models. In a broad sense, 3D printers are classified as cartesian, delta, polar, and selective compliance assembly robot arm (SCARA) based on the movement of printing arms (Figure 1.5) (Sun et al. 2018a).

A simple configuration of a 3D printer is a cartesian type. The movement of these printers is configured in a linear straight-line path (coordinate axes) whose movements are controlled by moving printing nozzle, printing platform, and/or both simultaneously for respective movements in X, Y, and Z direction (Horvath 2014b). Based on the motion of coordinate planes, cartesian printers are grouped as XY head printers, XZ head printers, and XYZ head printers. In the former type of XY head printer, the print head moves in the XY plane, and the print bed moves in the Z plane. Another variation in print head is 'XZ head printers' where the print head moves in XZ plane and the print bed moves in

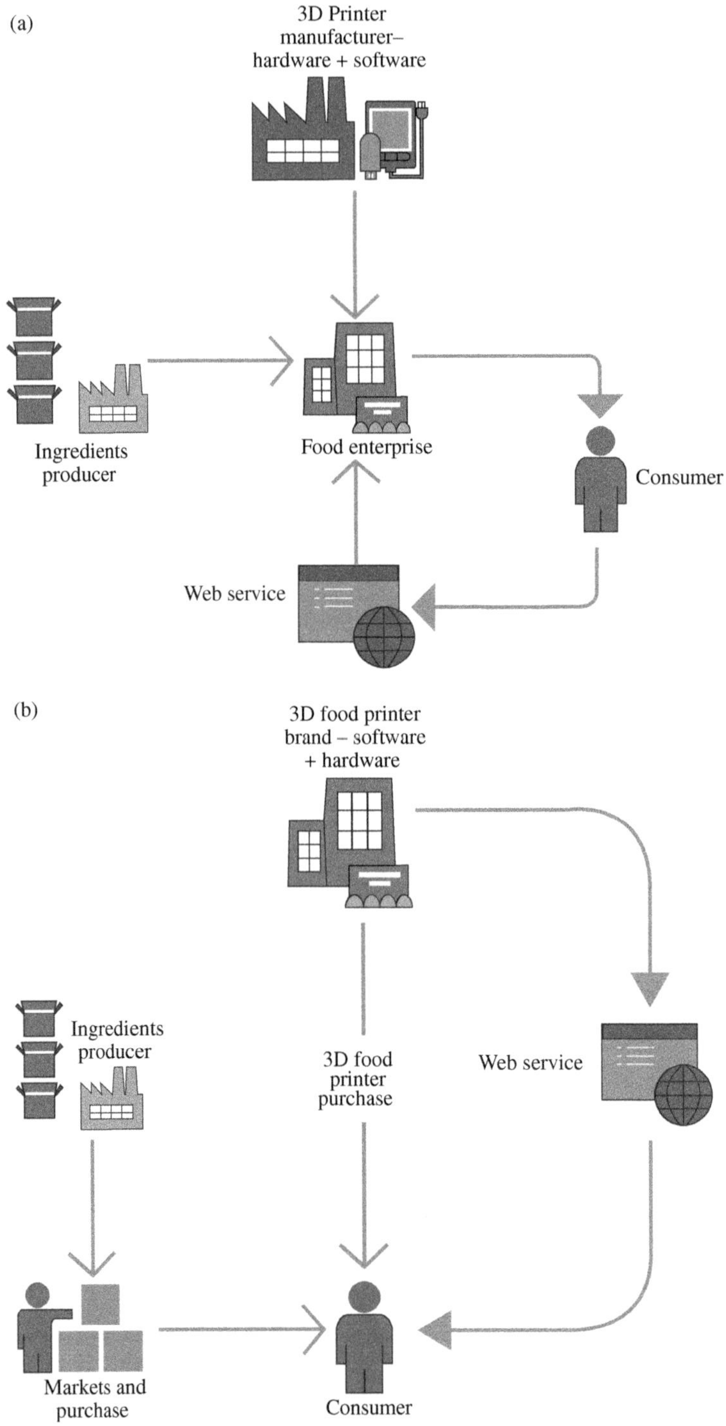

Figure 1.4 Value chain of 3D food printing. (a) consumer buying 3D printed foods; (b) consumer buying 3D food printer. *Source:* Jayaprakash et al. (2019) / With permission of Elsevier.

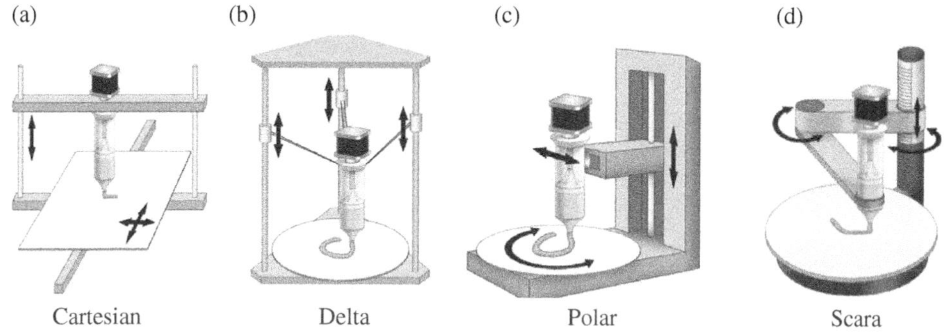

Figure 1.5 Illustration of printing movements in various printer configurations. (a) Cartesian; (b) Delta; (c) Polar; (d) SCARA. *Source:* Sun et al. (2018b) / With permission of Elsevier.

Y plane, respectively. An example of this type of printer is Choc creator, a commercial 3D food printer specifically designed for the customized fabrication of 3D constructs from chocolates. The third case is '*XYZ* print head' where the print bed remains stationary and only the print head moves in all directions as in the case of Foodini, a commercial 3D printer developed by Natural Machines (Derossi et al. 2019). Comparatively, the motion of *XY* and *XYZ* head printers remains faster than that of *XZ* head printers. However, this type of cartesian printer requires accurate and regular calibration before printing.

The second type is the delta 3D printers works on the triangular coordinate mechanism based on the Pythagoras theorem with relative movement of printing arms in three coordinate axes (*X*, *Y*, and *Z* directions) (Sun et al. 2018a). These types of printing system consist of three pairs of carriages (arms) that moves simultaneously up and down and aids in printing with a stationary print bed. Here, each pair of arms form the diagonal of a triangle and makes an angle to other planes namely *X* and *Z*. Likewise, all the three carriages move at the same time thereby aids in simultaneous printing. The major advantage of the delta type over the cartesian is its higher printing speed because of less physical loads and its ability to print bigger-sized objects especially in the *Z* direction (Horvath 2014b). However, this suffers from the limitation of low precision in printing smaller objects.

Another configuration of the 3D printer is polar which is the rarest that works based on the polar coordinate system. Here, the motion of printer arms is defined by an angle of 360° with a pre-defined centre point along radial direction while printhead moves vertically up and down thereby forming a 3D construct (Sun et al. 2018a). An example of this type of configuration is the XOCO 3D printer, a commercial chocolate printer equipped with a rotating build plate with a single supporting pillar that holds a print head and glass covering (Ontwerp 2018). Another example for polar configuration is the TNO food printer that consists of three rotating arms provided with a pair of material cartridge facilitates dual print simultaneously (Van der Linden 2015). One of the advantageous features of this particular design is its less floor space requirement and its ability to print larger-sized objects. In contrast to other configurations, polar printers can rotate and move either forward/backward and sideways. The rare availability of this type makes them expensive as it costs more than twice as that of cartesian (Derossi et al. 2019).

The last type of 3D printer is SCARA which consists of rotating elbows that move in an arc (*X* and *Y* directions) while a separate motor is provided to assist the movement in vertical (*Z*) direction. The movement along the *Z*-axis can be achieved by raising or lowering the printing platform (Sun et al. 2018a). In comparison with other configurations, SCARA is more compact and portable. An example of SCARA type is Sanna food printers equipped with eight food capsules with a temperature control unit and supported by an infrared panel for cooking printed food (Tan et al. 2018). But these systems suffer from the limitation of possessing expensive components that are not widely available as like delta and cartesian type that narrow down its use and applicability in food 3D printing.

In addition to the above context, based on the structural configuration, 3D printers are also termed as triangle structure (Prusa printer), triangle-claw structure (Rostock printer), rectangle-cassette structure (Ultimaker printer), and rectangle-pole structures (Printrobot printer) (Yang et al. 2017). With these basic configurations, 3D printers are modified and adapted for food applications. In general, food 3D printing must address the following key considerations since food material is being printed, the entire components must be food-grade; printer parts must be resistant to corrosion and should possess enough strength to wear and tear. Different printers have been studied for the printing of various food materials by several researchers. A commercial 3D printer Felix 3.0 originally designed for polymer printing was modified with a motor-driven system and adapted for extrusion-based food printing (Chen et al. 2019). Researchers used this modified system for printing soy protein gels and studied the effect of hydrocolloids on strength of 3D printed protein matrix. In another study, researchers were attempted to develop a multi extruder system that can be applied for printing 3D constructs from multi-material which has precise control over material deposition. A commercial food 3D printer, FoodBot developed by Changxing Shiyin Technology Co. Ltd. (China) was modified in this study and used for dual extrusion of composite food gel (Liu et al. 2018). 3D printed edible circuits from bread substrate were developed from a commercial desktop 3D printer, BioBot 1 (extrusion printer) (Hamilton et al. 2018). Researchers are exploring the advancements of 3D printers for food printing by modifying the structure and design of commercial 3D printers. Likewise, many studies are being conducted for the applicability and suitability of materials for 3D printing in context with printing multi-materials using multi-head printing systems.

1.6.2 Components of a Typical 3D Printer

The basic components of a food 3D printer include printing movement arms, drive unit assisted with pulley mechanism, mechanical motors and feed rollers, material dispensing unit, temperature controlling system, printing head, printing platforms, and microprocessing controller unit (Nachal et al. 2019).

1.6.2.1 Enclosure, Build Plate, and Guide Rails

Frames are the supporting structures that carry the print head and carriage arms. Supporting frames are made from metal sheets with acrylic covering provided with nuts and bolts. It supports all the other accessory components which influence the printer's stability and durability (Horvath 2014b). Based on the type of configuration of the 3D printer, the carriage arms differ in design and movement. The printing movement of the delta configuration is illustrated in Figure 1.6. These arms are facilitated with a motion mechanism achieved by linear motion guideways incorporated with GT2 belts and pulley controls. Belts assists in the smooth motion of respective arms in X, Y, and Z directions. In the case

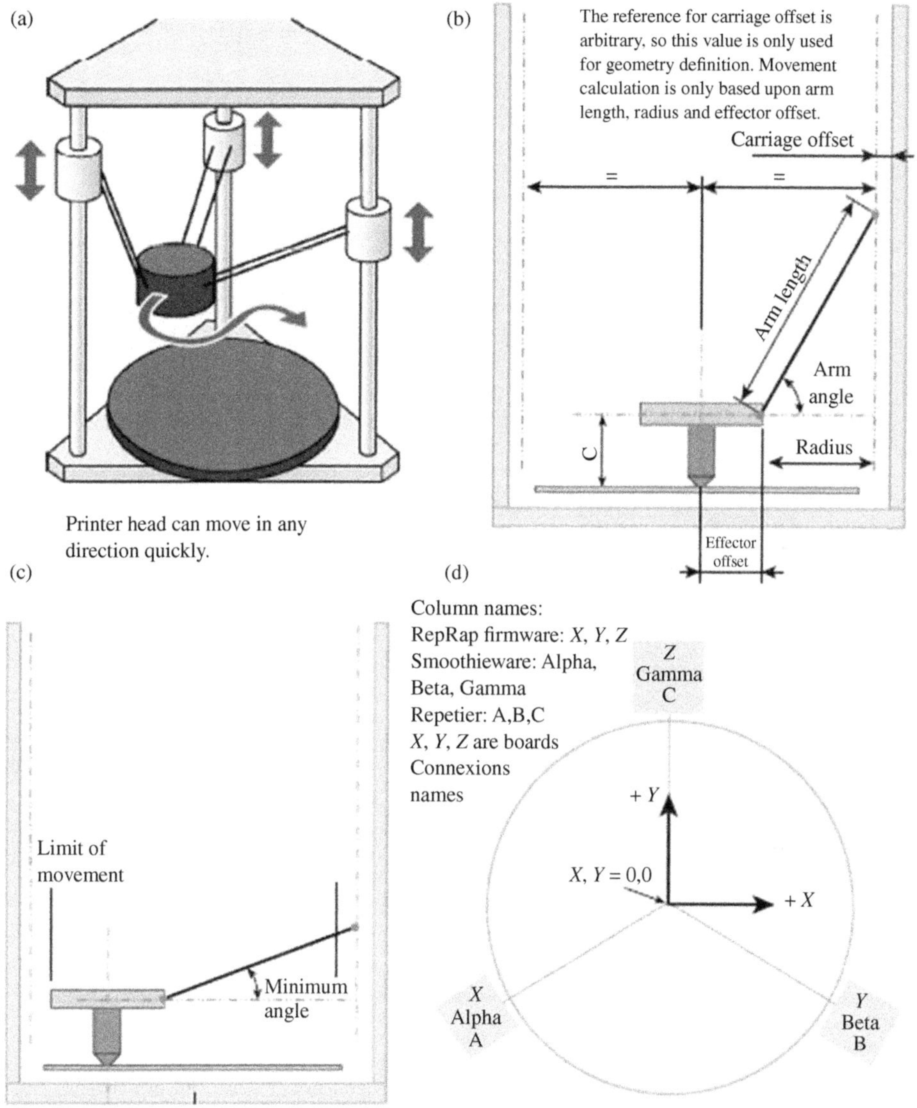

Figure 1.6 Operation of delta type 3D printer. (a) Illustration of printing movements, (b), (c), and (d) pictorial views of printing movement following Pythagorean theorem. *Source:* Derossi et al. (2019) / With permission of Elsevier.

of cartesian type printers, the assembly is equipped with threaded steel rods on which the nuts are mounted that assist movement upwards and downwards along *Z*-axis (Huang 2018). The design of the print head varies with configuration type and it consists of a feeder system mounted with printing. The method of material dispensing also varies with the printing technologies and the description of which are explained in subsequent chapters.

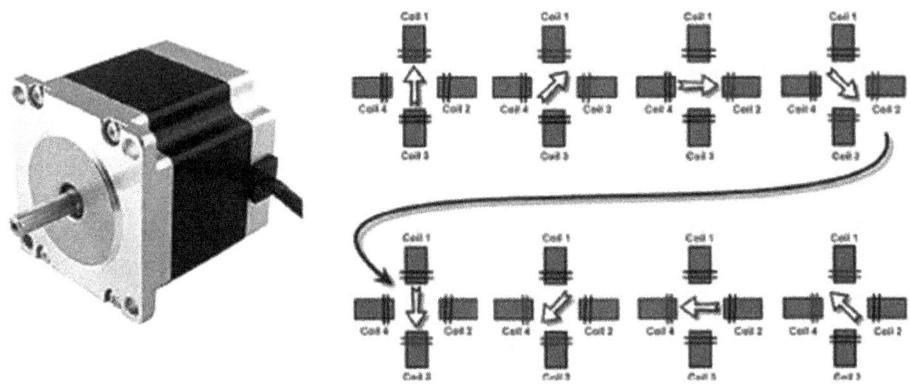

Figure 1.7 Stepper motor and its controlling mechanism. *Source:* From Derossi et al. (2019) / With permission of Elsevier.

1.6.2.2 Mechanical Drive Systems

Irrespective of the type of structural configuration, the carriage arms are connected with a mechanical drive system for assisting its motion along *X*, *Y*, and *Z* directions. The drive systems consist of stepper motor-operated in a pre-defined step of pulses (Figure 1.7). This step-up mechanism aids in the rotation of printer arms in full degrees (Horvath 2014b). In general, the step-up motors consist of a stator and rotor. Based on the structural configuration and working principle, stepper motors are of three types: permanent magnet stepper motor, variable reluctance stepper motor, and hybrid stepper motor. Among these, hybrid steppers are the most used stepper motors for 3D printing applications as it combines the advantageous features of permanent and variable stepper motors. Other characteristics that define a stepper motor includes its dimensions and rotating power (torque). The United States National Electrical Manufacturers Association (NEMA) declares a dimensional number to stepper motors named NEMA's number based on the length and associated torque (Derossi et al. 2019). Apart from guiding the movement of carriages, stepper motors are also used for controlling the amount of material deposited during printing.

1.6.2.3 Microprocessor Controlling System

The motherboard is the heart of any 3D printer, responsible for coordinating the motion of components like the extruder, print bed, motors, and sensors. End stops are provided at all the carriages that act as sensors and allow 3D printers in identifying printing locations along the three axes, preventing it from moving past its limit otherwise it could result in hardware damages (Horvath 2014b). An interface arrangement of liquid crystal display (LCD) replicating rapid prototyper (REP RAP) full graphics smart controller is provided for controlling the printing process. Firmware refers to a set of computer programmes and instructions that connects the software with hardware (Horvath and Cameron 2015). The integration of software with hardware is detailed in the subsequent sections. A 3D printer with an LCD user interface can work as a standalone machine (using an SD card), i.e., it can be controlled without connecting to the computer. Although there are different kinds of user interfaces, the most common one is a basic LCD interface operated via a knob/dial. The printer can be operated with both LCD interface as well as universal serial bus (USB) assisted interface system connected with computer. Based on the working capacity and build size, the power requirement will vary. Often the printing process was carried out by connecting the printer to a computer via a USB cable. Files can be transferred and read either through an onboard controlling unit or through a USB external interface with a computer or through both of these (Horvath 2014a). The user interface/controller allows the user to send machine-readable computer codes to the 3D printer and assists in changing some of the parameters during run time and move the print head along the $X/Y/Z$-axis. Most of the 3D printers are lab-scale and can be operated at a power supply of 110 V/230 V AC, 50/60 Hz.

1.7 Software Requirements and Hardware Integration

3D model designing involves the creation of 3D models using designing software such as AutoCAD (Autodesk), SolidWorks (Dassault Systèmes), SketchUp (Trimble), Rhinoceros 3D (Robert McNeel & Associates), etc. All this software allows the users to design their desired 3D model. After designing, the stored file can be converted to STL format to fed to a 3D printer. Model building requires professional knowledge and skilled personnel to create the desired 3D model. However, with the advancements of technology designing models have become easier. Software like 3DSlash and Tinkercad are specially designed for students and beginners (Guo et al. 2019). It also allows the users to design their model on an online platform rather than downloading the software itself. Sometimes designing a 3D model may consume time for which software like Autodesk 123D catch helps in building a 3D model from the 2D image that saves time and aids in achieving the same precision as designing from 3D model software (Parras et al. 2018).

After model designing, the stored information of the 3D model in STL format is read/sliced in a layer-by-layer manner by using appropriate slicing software. The slicing process involves the extraction of actual information of the 3D model and transforms it into G and M codes which are readable by 3D printers. G codes are the numerical language generated by CAD software that is computer readable which guides the motors and assists in the motion of printing arms to the printing region. On the other hand, M codes are the auxiliary commands that aids and assists in machine functioning (Horvath and Cameron 2015). Most commonly used slicing software includes Cura, Simplify3D, Repetier, Slic3r, Craftware, SelfCAD, SliceCrafter, and Astroprint (Table 1.1). Various process parameters that determine the final printing quality includes printing speed, retraction speed, layer height, printing temperature, nozzle size, number of outer shells, and infill percentage. Thus, slicing software allows the user to define the above printing parameters and aids in achieving good precise printing with higher resolution. All this setting information about the 3D model is fed to the printer in form of computer codes. G codes would have been written in numerical form and it slightly varies from printer to printer. The general commands commonly used are G0 (fast linear motion) and G1 (linear motion) which are referred to as the movement of the arms. For illustration, 'G0 X_{nnn} Y_{nnn} Z_{nnn} E_{nnn} F_{nnn} S_{nnn}' represents a command in which X_{nnn}, Y_{nnn}, and Z_{nnn} denote the position of X, Y, and Z-axis; E_{nnn} denotes the position of E axis, i.e. motion of print head stepper motor,

Table 1.1 Common slicing software used in 3D printing.

Software	Developer	Users	Description	Compatibility	Price
Cross platform tools					
Cura	Ultimaker	Beginners and advanced users	• Open source, fast, and easy to use • Its application is not just limited with Ultimaker 3D printer and can be used for other 3D printers as well	Windows, Mac, Linux	Free
CraftWare	CraftWare	Beginners and advanced users	• Fast and easy to use • Possess impressive G code visualizer and friendly support than other software	Windows, Mac	Free
KISSlicer	KISSlicer	Beginners and advanced users	• Easy and flexible software that results in excellent slicing	Windows, Mac, Linux, Raspberry Pi	Both free and paid
Repetier Host	Hot-World GmbH & Co. KG	Intermediate and advanced users	• It is an open-source software • Uses software Slic3r as default slicing engine	Windows, Mac, Linux	Free
Simplify3D	Simplify3D	Beginners and advanced users	• 3D printer host with premium slicer • Allows users to modify, repair, preview and slice 3D models	Windows, Mac	Paid
ReplicatorG	ReplicatorG	Beginners and advanced users	• Uses Skeinforge as slicer • Adaptable with 3D printers such as MakerBot, CupCake CNC, RepRap printers, Thing-O-Matic	Windows, Mac	Free
Slic3r	Slic3r	Advanced users and professional users	• It is an open source software • Fast and efficient that needs some tweaking initially generating good results	Windows, Mac, Linux	Free
OctoPrint	Gina Haußge	Intermediate and advanced users	• This software is based on webcam feeding of 3D models and provides constant feedback as the printing progress • Utilize powerful plug-ins that allows Ocotprint to send notifications, integrate messages, or collects print statistics	Windows, Mac, Linux, Raspberry Pi	Free

Cloud-based solutions					
3Dprinter OS	3D Control Systems Ltd.,	Advanced and professional users	• Allows users to search and upload 3D models, facilitates repair of STL errors, allows to customize settings, capable of cloud-based slicing of 3D models	Windows, Mac	Both free and paid
MatterControl	MatterHackers, Inc.	Beginners and advanced users	• Free, open-source software that allows users to queue multiple parts to be printed without pre-slicing, preview and manipulate designs and also integrates a library that enables users in organizing and searching the files	Windows, Mac, Linux	Free
Formide	Printr	Beginners	• Helps in organizing files and print jobs through its built-in cloud-based slicer Katana • It allows users to queue the print jobs, enables remote accessing, and receives instant feedback	Raspberry Pi	Free
AstroPrint	3DaGoGo, Inc.	Advanced and professional users	• Browser-based software that allows to slice, manage, and print the file only from web-enabled device	Windows, Mac, Linux	Free
SelfCAD	SelfCAD	Beginners and advanced users	• User-friendly, fully integrated online 3D modelling application • Featured with a comprehensive 3D Sketching and 3D Modelling Tools	Browser	Both free and paid
MakerBot Print	MakerBot	Beginners	• Optimizes and streamlines printing process that allows users to explore, manage, and share 3D files	Windows, Mac	Free
Tinkerine Suite	Tinkerine	Beginners	• User-friendly 3D slicing software • Minimal interface and possess an array of powerful features	Windows, Mac	Free

F_{nnn} denotes the speed of arms (mm min^{-1}) and S_{nnn} denotes the function to verify limit switch or not (S0 – do not check and S1 – check) (Guo et al. 2019). However, the entire 3D object can be printed using a complex command of G codes that requires more knowledge and skill in computer coding.

1.8 Designing, Digital Imaging, and Modelling

Nowadays, technological advancements made 3D printing easier and more accessible. 3D models can be directly downloaded in STL format from online webpages such as Thingiverse, Cults, Pinshape, Turbosquid, Youmagine, etc. (Table 1.2). These web pages allow the users to download the 3D models instantly either for free or paid. Otherwise, 3D scanners are employed for scanning of real-time objects and the scanned object in STL format was processed using slicing software.

1.8.1 Image Acquisition, Processing, and Modelling

3D scanning is used for capturing the surface characteristics of a physical object using a 3D scanner. The output of scanning is the generation of a 3D model file that can be edited using designing software or can be taken for 3D printing directly. The objects to be scanned can be a real-time object, natural scenarios (place/surroundings), or a person. The information obtained during scanning can be extracted and reproduced using a process called reverse engineering (Guo et al. 2019). Considering the affordability, portability, and versatility, reverse engineering employs two types of 3D scanning equipment namely contact and non-contact scanners (Table 1.3). Contact scanners are point-based systems that employ a probe to touch the object to be scanned. Both data collection and designing can be done simultaneously using a portable coordinate measuring machine (CMM) (Martinez-Pellitero et al. 2018). On the other hand, non-contact 3D scanners use an array of laser/ light beams to extract the surface information of the object. These types of scanners project either a laser beam or structured light on the object that aids in capturing the surface data. A web camera which on integration with software aids in transferring the reflected beam into the 3D model data for printing (Wei et al. 2019). Based on the mode of applications, 3D scanners are categorized as desktop scanners (EinScan, Solutionix 500) and handheld scanners (Sense, EinScan Pro 2X, Structure sensor). Most of the scanners are adaptable with computers while some other advanced versions employ the iPad to transfer the 3D model.

Another type of scanning technique is photogrammetry in which 3D models are obtained from 2D photographs by integrating computer vision and computational algorithms. Photogrammetry is based on analyzing several static points in the 2D photographs taken from different views so that it would detect the pixel corresponding to actual physical objects (Peña-Villasenin et al. 2020). Image processing depends on the focal length of the camera and the distortion of the lens. The major advantage of photogrammetry is its acquisition speed and model precision. However, it requires more processing time to run the algorithms that limit its applications.

Table 1.2 List of online sources for downloading 3D models.

Web page	Logo	Founder	Type	Access
Thingiverse	Thingiverse	MakerBot Industries	Repository	Free
MyminiFactory	MyMiniFactory	Romain Kidd, Sylvain Preumont	Marketplace	Both free and paid
Cults	Cults.	Hugo and Pierre	Marketplace	Both free and paid
CGTrader	cgtrader	Marius Kalytis Dalia Lasaite	Marketplace	Both free and paid
Pinshape	pinshape	Lucas Matheson, Nick Schwinghamer, and Andre Yanes	Marketplace	Both free and paid
YouMagine	YM YOUMAGINE	Erik de Bruijn	Marketplace	Both free and paid

(*Continued*)

Table 1.2 (Continued)

Web page	Logo	Founder	Type	Access
TurboSquid	TURBOSQUID	Matt Wisdom, Tom Avgikos, David Avgikos, Andy Wisdom	Repository	Both free and paid
XYZprinting 3D Gallery	XYZPRINTING	New Kinpo Group	Marketplace	Both free and paid
Threeding	threeding 3D PRINTING MARKETPLACE	Cveta-Maria Partaleva, Stan Partalev, and Tony	Marketplace	Both free and paid

Table 1.3 Different scanning technologies.

Scanning mode	Scan range	Scanning principle	Description	Commercial scanners
Non-contact	Short range	Laser triangulation-based 3D technology	Projects a laser light source onto the surface of the object and determine the amount of deformed light reflected from the surface	Sense 3D scanner
		Structured light-based 3D technology	Uses trigonometric triangulation of the structured light (white/blue). It measures the deformed light pattern from the surface of object based on digital light processing (DLP) technique	EinScan, EinScan Pro 2X
	Long range	Pulse-based laser	Based on laser beam's time of flight. The projected laser light is collected by sensor and time of travel between emission and reception provides surface details	Lidar 3D scanner
		Phase shift laser	In addition to pulse-based scanner, these systems modulate power of the laser thereby it allows comparison of the phase of the laser emitted and returned to the sensor	Lidar 3D scanner
Contact	Medium range	Coordinate measuring machine (CMM)	Involves measuring the degree of deformation of a probe by scanning several coordinate points on object's surface	Altera SL bridge CMM
		Arm-based 3D scanners	Similar to CMM with additional provisions attachable 3D scanner unit for collecting large amount of points	Hexagon's ROMER Absolute Arm scanner
		Optically tracked 3D scanners	It uses a set of cameras to capture and track the location of scan in working space	Hexagon's Leica Absolute Scanner and Creaform's Metrascan 3D scanner

1.8.2 Repairing and Post-Processing

After the acquisition of the image, the captured 3D models are processed adequately to restore better precision and higher resolution during 3D printing. Scanned 3D models are subjected to digital processing to improve the quality of the acquired image. Any 3D solid files would be stored in STL form which is described in terms of small clustered triangular facets forming the 3D object (Szilvśi-Nagy and Matyasi 2003; Wu and Cheung 2006). This format follows four basic principles such as common vertex, orientation rule, closed facets, and must have non-negative coordinate values. Some of the common STL errors (Figure 1.8) and their causes are as follows: holes – caused by missing of facets during triangulation; gaps – occurs due to dislocation of vertices forming a boundary between the pairs; dislocation – a common error that occurs during computation when the same vertex is shared by two triangular facets; reverse oriented normal – a disordered pattern occurs when triangular facets fail to follow orientation rule; overlaps – inaccuracy during computation causes merging of facets one above the other; and redundancy – independent triangular facets stick out the regular topological structure (Liu et al. 2009). Common vertex is the sharing of two edges by the adjacent triangular facets. The orientation rule implies that the normal vector of the individual facets must follow the right-hand rule and it should point towards the perimeter of the solid. The next principle is that all the triangular facets must be closed adequately without any gaps between the vertices. Finally, the model should possess a non-negative coordinate value (Tan and Chen 2016). Analysis of the STL file involves the assessment of the model whether it satisfies all the basic rules.

Two common errors that occur during processing were topological error and normal vector error. Whenever a solid model shares a common vertex, a non-manifold error would occur that results in topological errors. In such cases, a single solid model was split into

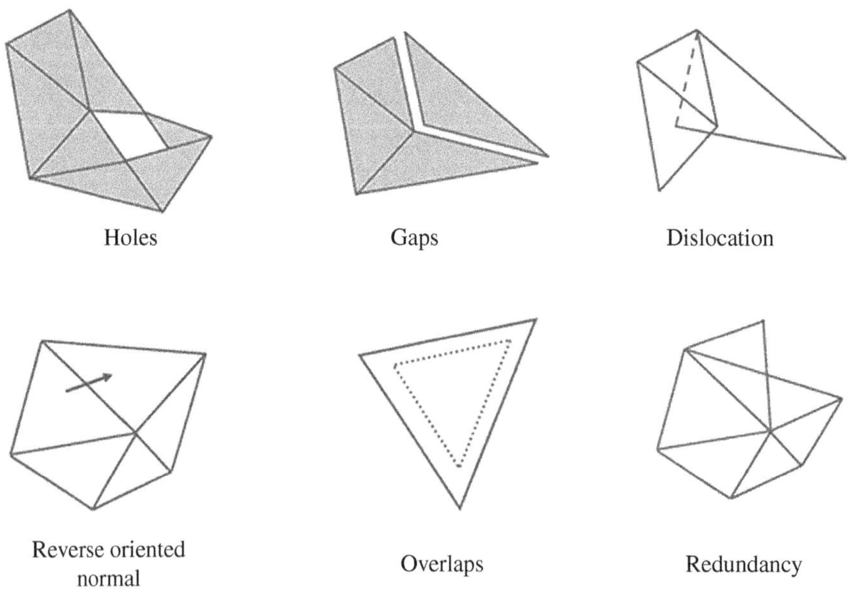

Figure 1.8 Common STL errors.

two individual models so that the model fixing and repairing would be easier. The latter type of error occurs when the model fails to follow the orientation rule. As a result, the vertices are not closed sufficiently causing normal vector errors. The repair mechanism involves the reordering of the vertex indices in an ordered pattern of triangular facets (Chao et al. 2011). Other common errors encountered during image processing are cracks, holes, voids or missing/ duplicate facets, and flipped orientations (Liu et al. 2009). All these errors fixing and repairing mechanisms are simplified and become easier with modelling software such as Fusion 360, Autodesk ReCap, Trinkle, Sculpteo, MakePrintable, MeshLab, etc. which are specially designed for repairing the STL files (Table 1.4). These software saves time and helps to improve the printing quality.

1.9 Food Printing Platforms

The food printing platform consists of a well-coordinated *XYZ* axis system, a dispensing system, and a microprocessor control unit connecting the computer with the printer. The emergence of food printing not only leads to industrial level printing but also makes the possibility of using 3D printers at domestic kitchens. Thus, food 3D printers would become a common kitchen appliance that allows the users to define their own recipes by either downloading the digital recipe files from an online source or creating a new user-defined digital recipe. Food 3D printing will be a technological cutting edge that helps the users to manipulate the number of ingredients as well as to monitor the calorie intake. Food printing recipes follow two different platforms namely universal, and user-defined platforms.

1.9.1 Universal Platform

Universal platforms are the existing open-source printing systems adapted for food applications from allied sectors such as polymer printing. The use of this type of available printing platform for food applications shortens the processing time and saves energy (Sun et al. 2015a). Most of the food printing systems available in the literature were based on universal platforms. Researchers had used these platforms for the printing of food matrices like protein pastes and cake dough and studied the effect of hydrocolloids in textural modification of different food systems (Lipton et al. 2010). A desktop 3D printer, MakerBot was adapted for food printing by modifying the existing extruder head to be suitable for food printing (Millen 2012). Although universal platforms save time and cost, these platforms were not suitable for printing a wide range of food materials since food is a complex matrix comprising of varied physical and chemical composition.

1.9.2 User-Defined Platform

These are the self-developed 3D printing platforms created for overcoming the limitations existing with the universal platforms. Based on the specific requirements of printing and the nature of food matrices, self-developed platforms are defined by users (Hao et al. 2010). These platforms allow the users to have a proportional degree of freedom that helps in

Table 1.4 Software used for repairing STL file after 3D scanning.

Web page	Features	Operating system	Access
Autodesk Meshmixer	• Helps in mesh repairing that is featured with functions such as sculpting, scaling, building, mirroring, and cutting	Windows, OS X and Linux	Free
MeshFix	• Open-source STL repair software that fixes various defects such as holes and self-intersections resulting in a water-tight triangular mesh output	Windows	Free
MeshLab	• Featured with tools for editing, managing, inspecting, texturing and converting 3D meshes	Windows, Linux, iOS, Mac OS X and Android	Free
MakePrintable	• Assess 3D design model and provides print statistics • Helps in configuring the model characteristics and yields algorithm for 3D printing	Online browser	Both free and paid
Autodesk Netfabb	• Offers tools for repairing meshes, adjusting thickness, merging, and deleting parts thereby proving powerful functioning	Windows	Paid
3D Builder	• Enables model visualization, provide editing capabilities, and helps in 3D editing and validating	Windows	Free
Blender	• Open-source 3D repairing software that supports modelling, animation, simulation, and motion tracking	Windows, OS X and Linux	Free
FreeCAD	• Open-source parametric 3D software used primarily to design and repair • Allows users to easily modify design and changing its parameters	Windows, OS X and Linux	Free
Emendo	• Allows users to automatically identify number, type and location of STL errors	Windows and OS X	Paid
Open3mod	• General-purpose 3D model viewer • Efficient tools to inspect, filtering and isolating errors	Windows	Free
Sculpteo	• Allows user to upload, repair, analyse and optimize the STL files	Online browser	Free
3D Tools	• Allows users to assess STL files based on dimensional parameters like volume and surface area prior to mesh repair	Online browser	Free

achieving flexibility in design components and working mechanisms. Further, these platforms allow for modification and up-gradation of printing systems. Researchers have developed an in-house-built customized extrusion-based 3D printer CARK (controlled AM robotic kit) specially designed for food applications (Anukiruthika et al. 2020). Here, the components of the printer were selected based on specific requirements for food printing. The developed delta-based food printer was applicable for printing a wide range of food materials such as food pastes, hydrogels, colloids, etc.

1.9.3 Applicability of User Interface Systems

To widen the applications and usage of 3D food printers, researchers are now focussing on developing a simplified user interface system that can be operated by everyone irrespective of technical skills. Most of the user-defined platforms require an adequate pre-processing of raw materials that aids in tailoring the physiochemical properties suitable for food printing (Nachal et al. 2019). Hence designing of interface system was one of the critical factors that must consider the end-users applications like domestic household level, industrial level, or lab scale. Based on this, the model design, tool path settings, and other process parameters were framed while designing the web-based user template (Evans 2012). The advantage of these systems helps in achieving personalized nutrition, customized fabrication of 3D constructs with improved texture, new flavour, and mouthfeel. With such a user-friendly interface system, consumers can directly print their food as per their individual needs and preferences.

1.10 Comparison Between Food 3D Printing and Robotic Food Manufacturing

Although food 3D printing and robotics-based food manufacturing processes were aimed at automation and reduction of workload, there exists a distinct difference between them. The former technology relies upon the consumer's needs allowing the users to manipulate ingredients while the latter involves the replacement of labour-intensive operations and automation of manual processes (Sun et al. 2018a). Baked recipes and confectionery products such as sugar cubes can be prepared either by food printing or robotics-based manufacturing. Basic ingredients remain to be same for both these processes, however, certain modifications are required for materials to be used in food printing. These modifications aid in tailoring the rheological and post-deposition requirements for achieving a stable printed food. Applications of digital gastronomy in food 3D printing help in the fabrication of foods with a different eating experience that goes beyond the taste encompassing all aspects of gastronomy. For instance, robotic-based chocolate manufacturing employs cocoa powder, butter, sugar, and full cream milk as raw materials; however, food printing uses the commercially available readymade chocolates (Sun et al. 2015b).

In terms of market value, the critical factors that distinguish food printing from conventional food processing techniques are customization, complexity, and production volume (Conner et al. 2014; Petrick and Simpson 2013). In the case of conventional processing, the unit cost would increase with the increase in complexity/ customization. Since the addition of complex designs requires more tools, energy, and labour. On the other hand, the unit cost would decrease with production volume without considering the complexity. In contrast to this scenario, the unit cost remains stable irrespective of the increase in complexity/ customization for 3D printing. This was because any changes in design could be done in the 3D model before printing which is not the case in conventional manufacturing. This feature adds value to food 3D printing to emerge as a promising technology that balances the cost as well as production volume without compromising the structural complexity of the designs (Pinkerton 2016).

1.11 Conclusion

3D printing as an AM technique is well flourished in several industrial sectors. As of food printing is concerned, FLM is at its nascent stage. Printing food is not as easy as it seems, as food is a complex substance that comprises several intrinsic and extrinsic factors. The criteria for achieving well defined 3D food structure and science behind the inherent chemical interactions and binding of the deposited layers have not yet been clarified. Researchers all around the world were working in exploring the potential applications of food printing for mass customization, personalized nutrition, intricate designing, etc. In terms of cost, food printing remains to be cheaper than conventional food processing techniques as printing combines multiple steps and eliminates human interventions. 3D food printing allows fabrication of edible 3D constructs of complex designs and helps in food customization. With the technological advancements, food printing seems to be a futuristic technology that would turn the so-called 'balanced diet' into a 'digitalized diet'.

References

3DSourced (2021). Featured stories: history of 3D printing. https://www.3dsourced.com/guides/history-of-3d-printing (accessed 12 September 2021).

Anukiruthika, T., Moses, J.A., and Anandharamakrishnan, C. (2020). 3D printing of egg yolk and white with rice flour blends. *Journal of Food Engineering* **265**: 109691. https://doi.org/10.1016/j.jfoodeng.2019.109691.

Bandyopadhyay, A. and Heer, B. (2018). Additive manufacturing of multi-material structures. *Materials Science and Engineering: R: Reports* **129**: 1–16.

Bechtold, S. (2016). 3D printing, intellectual property and innovation policy. *IIC-International Review of Intellectual Property and Competition Law* **47** (5): 517–536.

Beltagui, A., Rosli, A., and Candi, M. (2020). Exaptation in a digital innovation ecosystem: the disruptive impacts of 3D printing. *Research Policy* **49** (1): 103833.

Calignano, F., Peverini, O.A., Addamo, G. et al. (2019). High-performance microwave waveguide devices produced by laser powder bed fusion process. *Procedia CIRP* **79**: 85–88.

Chao, H., Li, Y., and Ying-ying, Z. (2011). Research on repair algorithms for hole and cracks errors of STL models. *International Conference on Information and Management Engineering*, pp. 42–47.

Chen, J., Mu, T., Goffin, D. et al. (2019). Application of soy protein isolate and hydrocolloids based mixtures as promising food material in 3D food printing. *Journal of Food Engineering* **261**: 76–86. https://doi.org/10.1016/j.jfoodeng.2019.03.016.

Conner, B.P., Manogharan, G.P., Martof, A.N. et al. (2014). Making sense of 3-D printing: creating a map of additive manufacturing products and services. *Additive Manufacturing* **1**: 64–76.

Derossi, A., Caporizzi, R., Ricci, I., and Severini, C. (2019). Critical variables in 3D food printing. In: *Fundamentals of 3D Food Printing and Applications* (eds. F.C. Godoi, B.R. Bhandari, S. Prakash and M. Zhang), 41 91. Elsevicr.

Evans, B. (ed.) (2012). 3D printer toolchain. In: *Practical 3D Printers*, pp. 27–47. Berkeley, CA: Apress https://doi.org/10.1007/978-1-4302-4393-9_2.

Guo, C., Zhang, M., and Bhandari, B. (2019). Model building and slicing in food 3D printing processes: a review. *Comprehensive Reviews in Food Science and Food Safety* **18** (4): 1052–1069. https://doi.org/10.1111/1541-4337.12443.

Hamilton, C.A., Alici, G., and Panhuis, M. (2018). 3D printing vegemite and marmite: redefining "breadboards.". *Journal of Food Engineering* **220**: 83–88.

Hao, L., Mellor, S., Seaman, O. et al. (2010). Material characterisation and process development for chocolate additive layer manufacturing. *Virtual and Physical Prototyping* **5** (2): 57–64.

Hon, K.K.B. (2007). Digital additive manufacturing: from rapid prototyping to rapid manufacturing. Proceedings of the 35th International MATADOR Conference, pp. 337–340.

Horvath, J. (ed.) (2014a). Driving your printer: G-code. In: *Mastering 3D Printing*, pp. 65–76. Berkeley, CA: Apress https://doi.org/10.1007/978-1-4842-0025-4_6.

Horvath, J. (ed.) (2014b). The desktop 3D printer. In: *Mastering 3D Printing*, pp. 11–20. Berkeley, CA: Apress https://doi.org/10.1007/978-1-4842-0025-4_2.

Horvath, J. and Cameron, R. (eds.) (2015). Controlling your 3D printer. In: *3D Printing with MatterControl*, pp. 71–83. Berkeley, CA: Apress https://doi.org/10.1007/978-1-4842-1055-0_6.

Huang, C.Y. (2018). Extrusion-based 3D Printing and Characterization of Edible Materials [University of Waterloo]. https://uwspace.uwaterloo.ca/handle/10012/12899

Jayaprakash, S., Ituarte, I.F., and Partanen, J. (2019). Prosumer-driven 3D food printing: role of digital platforms in future 3D food printing systems. In: *Fundamentals of 3D Food Printing and Applications* (eds. F.C. Godoi, B.R. Bhandari, S. Prakash and M. Zhang), pp. 331–354. Elsevier.

Jiang, J., Xu, X., and Stringer, J. (2019). Optimization of process planning for reducing material waste in extrusion based additive manufacturing. *Robotics and Computer-Integrated Manufacturing* **59**: 317–325.

Kodama, H. (1981). Automatic method for fabricating a three-dimensional plastic model with photo-hardening polymer. *Review of Scientific Instruments* **52** (11): 1770–1773.

Ledford, H. (2015). The printed organs coming to a body near you. *Nature News* **520** (7547): 273.

Ligon, S.C., Liska, R., Stampfl, J. et al. (2017). Polymers for 3D printing and customized additive manufacturing. *Chemical Reviews* **117** (15): 10212–10290.

Lipton, J., Arnold, D., Nigl, F. et al. (2010). Multi-material food printing with complex internal structure suitable for conventional post-processing. *Solid Freeform Fabrication Symposium*, pp. 809–815.

Lipton, J.I., Cutler, M., Nigl, F. et al. (2015). Additive manufacturing for the food industry. *Trends in Food Science and Technology* **43** (1): 114–123. https://doi.org/10.1016/j.tifs.2015.02.004.

Liu, F., Zhou, H., and Li, D. (2009). Repair of STL errors. *International Journal of Production Research* **47** (1): 105–118.

Liu, Z., Zhang, M., and Yang, C. (2018). Dual extrusion 3D printing of mashed potatoes/strawberry juice gel. *Lwt* **96**: 589–596.

Martinez-Pellitero, S., Cuesta, E., Giganto, S., and Barreiro, J. (2018). New procedure for qualification of structured light 3D scanners using an optical feature-based gauge. *Optics and Lasers in Engineering* **110**: 193–206.

McClements, D.J. (2017). The future of food colloids: next-generation nanoparticle delivery systems. *Current Opinion in Colloid & Interface Science* **28**: 7–14.

Millen, C.I. (2012). *The Development of Colour 3D Food Printing System: A Thesis Presented in Partial Fulfilment of the Requirements for the Degree of Master of Engineering in Mechatronics at Massey University, Palmerston North, New Zealand.* Massey University.

Nachal, N., Moses, J.A., Karthik, P., and Anandharamakrishnan, C. (2019). Applications of 3D printing in food processing. *Food Engineering Reviews* **11** (3): 123–141. https://doi.org/10.1007/s12393-019-09199-8.

Ontwerp, C.M. (2018). XOCO Chocolate printer. http://www.michielcornelissen.com/portfolio_page/xoco-chocolate-printer (accessed 19 October 2019).

Parras, D., Cavas-Martinez, F., Nieto, J. et al. (2018). Reconstruction by low cost software based on photogrammetry as a reverse engineering process. *International Conference on Virtual, Augmented and Mixed Reality*, pp. 145–154.

Peña-Villasenin, S., Gil-Docampo, M., and Ortiz-Sanz, J. (2020). Desktop vs cloud computing software for 3D measurement of building façades: the monastery of San Martín Pinario. *Measurement* **149**: 106984.

Petrick, I.J. and Simpson, T.W. (2013). 3D printing disrupts manufacturing: how economies of one create new rules of competition. *Research-Technology Management* **56** (6): 12–16.

Pinkerton, A.J. (2016). Lasers in additive manufacturing. *Optics & Laser Technology* **78**: 25–32.

Prasad, L.K. and Smyth, H. (2016). 3D Printing technologies for drug delivery: a review. *Drug Development and Industrial Pharmacy* **42** (7): 1019–1031.

Rahman, Z., Ali, S.F.B., Ozkan, T. et al. (2018). Additive manufacturing with 3D printing: progress from bench to bedside. *The AAPS Journal* **20** (6): 101.

Sanchez Ramirez, A., Islán Marcos, M.E., Blaya Haro, F. et al. (2019). Application of FDM technology to reduce aerodynamic drag. *Rapid Prototyping Journal* **25** (4): 781–791. https://doi.org/10.1108/RPJ-09-2018-0251.

Saptarshi, S.M. and Zhou, C. (2019). Basics of 3D printing: engineering aspects. In: *3D Printing in Orthopaedic Surgery* (eds. M. Dipaola and F.M. Wodajo), pp. 17–30. Elsevier.

Sokolov, P.S., Dosovitskiy, G.A., Dosovitskiy, A.E. et al. (2018). Towards new production technologies: 3D printing of scintillators. International Conference on Engineering of Scintillation Materials and Radiation Technologies, pp. 99–112.

Su, A. and Al'Aref, S.J. (2018). History of 3D printing. In: *3D Printing Applications in Cardiovascular Medicine* (eds. S.J. Al'Aref, B. Mosadegh, S. Dunham and J.K. Min), pp. 1–10. Elsevier.

Sun, J., Peng, Z., Zhou, W. et al. (2015a). A review on 3D printing for customized food fabrication. *Procedia Manufacturing* **1**: 308–319.

Sun, J., Zhou, W., Huang, D. et al. (2015b). An overview of 3D printing technologies for food fabrication. *Food and Bioprocess Technology* **8** (8): 1605–1615.

Sun, J., Zhou, W., Huang, D., and Yan, L. (2018a). 3D food printing: perspectives. In: *Polymers for Food Applications*, pp. 725–755. Springer.

Sun, J., Zhou, W., Yan, L. et al. (2018b). Extrusion-based food printing for digitalized food design and nutrition control. *Journal of Food Engineering* **220**: 1–11.

Szilvśi-Nagy, M. and Matyasi, G.Y. (2003). Analysis of STL files. *Mathematical and Computer Modelling* **38** (7–9): 945–960.

Tan, J. and Chen, J. (2016). Research and application on model repairing algorithm of 3D modelling technology. 2016 6th International Conference on Mechatronics, Computer and Education Informationization (MCEI 2016).

Tan, C., Toh, W.Y., Wong, G., and Lin, L. (2018). Extrusion-based 3D food printing - materials and machines. *International Journal of Bioprinting* **4** (2): 1–13. https://doi.org/10.18063/ijb.v4i2.143.

Tofail, S.A.M., Koumoulos, E.P., Bandyopadhyay, A. et al. (2018). Additive manufacturing: scientific and technological challenges, market uptake and opportunities. *Materials Today* **21** (1): 22–37.

Van der Linden, D. (2015). 3D food printing. TNO. https://www.tno.nl/media/2217/3d_food_printing.pdf (accessed 19 October 2019).

Wei, Y., Ding, Z., Huang, H. et al. (2019). A non-contact measurement method of ship block using image-based 3D reconstruction technology. *Ocean Engineering* **178**: 463–475.

Wu, T. and Cheung, E.H.M. (2006). Enhanced stl. *The International Journal of Advanced Manufacturing Technology* **29** (11–12): 1143–1150.

Yang, F., Zhang, M., and Bhandari, B. (2017). Recent development in 3D food printing. *Critical Reviews in Food Science and Nutrition* **57** (14): 3145–3153.

Yang, M., Lv, X., Liu, X., and Zhang, J. (2018). Research on color 3D printing based on color adherence. *Rapid Prototyping Journal* **24** (1): 37–45.

2

3D Printing Approaches

CHAPTER MENU

3D Printing of Foods, First Edition. C. Anandharamakrishnan, Jeyan A. Moses, and T. Anukiruthika.

2.1 Introduction

The advancement in technology greatly influences lifestyle and eating practices. The specific needs and dietary requirements made food scientists develop convenient processing technologies. One such approach is the 3D printing of foods that involves the layer-by-layer deposition of food material in a pre-determined dimensions and shapes using computerized software (Kuo et al. 2021). 3D printing is an additive manufacturing (AM) process that is well established in many industrial sectors such as automobile, aerospace, fashion and designing, material science, construction, and medicine (Nida et al. 2020). However, applications of 3D printing in food manufacturing are under their initial growth stage that seems to be fascinating as well as challenging. In the food sector, 3D printing is recently gained potential research interest due to its novelty, convenience, creativity, efficiency, and flexibility (Jayaprakash et al. 2020). Since food is a complex multicomponent system, it possesses unique functional properties and characteristic behaviour to external treatments such as heating, cooling, mixing, and so on. Digitalization improves industrial production by implementing a novel design process thereby increases efficiency in both the manufacturing and supply chain (Baiano 2020). Thus, 3D printing is forecasted to be a revolutionizing technology that simplifies the food supply chain. 3D printing allows fabricating the foods as per the individual's needs and requirements in a desired shape, size, colour, flavour, and nutrition. Also, 3D constructs with complex internal microstructures can be developed with greater accuracy and precision using 3D printing that are nearly impossible with conventional moulds (Mantihal et al. 2020).

The success of food printing depends on material composition, process conditions, and printing variables (Jiang et al. 2018). All these factors influence the final printability as well as edibility of the 3D constructs that makes 3D printing to be challenging when applied to foods. Hence, food printing is a multidisciplinary operation that requires knowledge of material science, software skills, mechanical as well as numerical controlling computer skills. In contrast to conventional food technologies, 3D printing allows the utilization of novel food ingredients and materials for the fabrication of foods (Derossi et al. 2020). As a characteristic feature, 3D printing uses uncommon protein sources such as fungi, algae, insects, buck wheat, and lupin seeds. As a sustainable technology, 3D printing utilizes the by-products of the food processing industrial wastes for the development of value-added sustainable foods for the future (Garcia-Oliveira et al. 2020). However, the fuller potential of this emerging technology must be studied in detail to better understand the material behaviour. In this context, food printing technologies are one of the major considerations. As not all food materials are suitable for printing that requires appropriate pre-processing that in turn depends on the specific printing technology. The present chapter provides a summary of various 3D printing techniques and their applications. As an initial attempt, the focus of this chapter is to streamline the available food printing technologies with their advantageous features and the different process variables that affect food printing. Thus, the present chapter provides a broad knowledge on the basic principle, mechanisms, and system components of food printing technologies that helps in better understanding and handling of food materials. Further, this work provides valuable insights in bridging up the gap in transformation of these lab-scale techniques to industrial level.

2.2 Additive Manufacturing

3D printing is an additive process of developing a 3D construct from a digitalized CAD 3D design with the integration of a 3D printer with a computer. Most commonly 3D printing is used for rapid prototyping that allows testing the materials before practical implementation that has advantages like the use of lesser material resources, avoids wastage, saves energy, time, and cost (Shree et al. 2020). The higher degree of freedom of this technology allows its application to a diverse range of industrial sectors. In contrast to conventional manufacturing approaches, AM can produce goods with design personalization and mass customization. Ultimately, 3D printing could greatly limit the cost of processing with efficient utilization of resources and energy (Van Wijk and Van Wijk 2015). The basic working mechanism of a 3D printer is the printing of material in a pre-determined design through appropriate conversion of digital CAD model into standard triangle language/ standard tessellation language (STL) (stereolithography [SLA]) file that initiates printing and controls the movement of coordinate axes. As discussed in the earlier chapters, the multi-axis stages used in 3D printing can be of the cartesian, delta, polar, and scara configurations (Sun et al. 2018a). Among which cartesian and delta are the most used configurations in the manufacturing sector. The concept of 3D printing is the multilayer stacking of material along X, Y, and Z axes through precise control over material deposition. As a prototyping tool, the material deposition in 3D printing occurs through appropriate phase transitions and chemical reactions (Nachal et al. 2019). Based on the nature and type of heat source and the materials employed, there are several types of the AM process. These technologies are majorly used for plastics, polymers, composites, ceramics, and cement (Pradhan et al. 2021). The American Society for Testing and Materials (ASTM) framed different standards for AM and has classified processes into seven categories as follows (Liu et al. 2018c) (Figure 2.1).

- Material extrusion: A polymeric filament is melted using an electrical heat source and the melted liquid is deposited in a layered fashion using a 3D printer through an extrusion mechanism. The deposited material gets fused together upon cooling and attains the desired 3D shape (Shahrubudin et al. 2019). A filament is delivered using a Bowden mechanism for pushing the material filament through a long flexible Teflon tube to the hot end of the printing nozzle. From the nozzle tip, the melted polymer is deposited over the printing platform. Materials such as polylactic acid (PLA) and Acrylonitrile butadiene styrene (ABS) are the most commonly used polymer filaments. A temperature controlling unit is used for varying the temperature according to the melting point of the polymers used. Commercial 3D printers that are based on extrusion technology are referred to as fused deposition modelling (FDM) and fused filament fabrication (FFF). Extrusion-based technology is comparatively cheaper than other AM processes (Vithani et al. 2019). Although the printing accuracy and precision are good, sometimes post-polishing is required to enhance the surface smoothness of the 3D printed construct.
- Powder bed fusion: In this AM process, a heat source of either hot air or a high-energy laser beam is employed to scan the 3D CAD design over the powder bed (Godoi et al. 2019). Due to which the powdered material gets melted and fused according to the shape and design of the 3D CAD model. After completion of the scanning process, the printed material is taken out from the loosened powder of the bed.

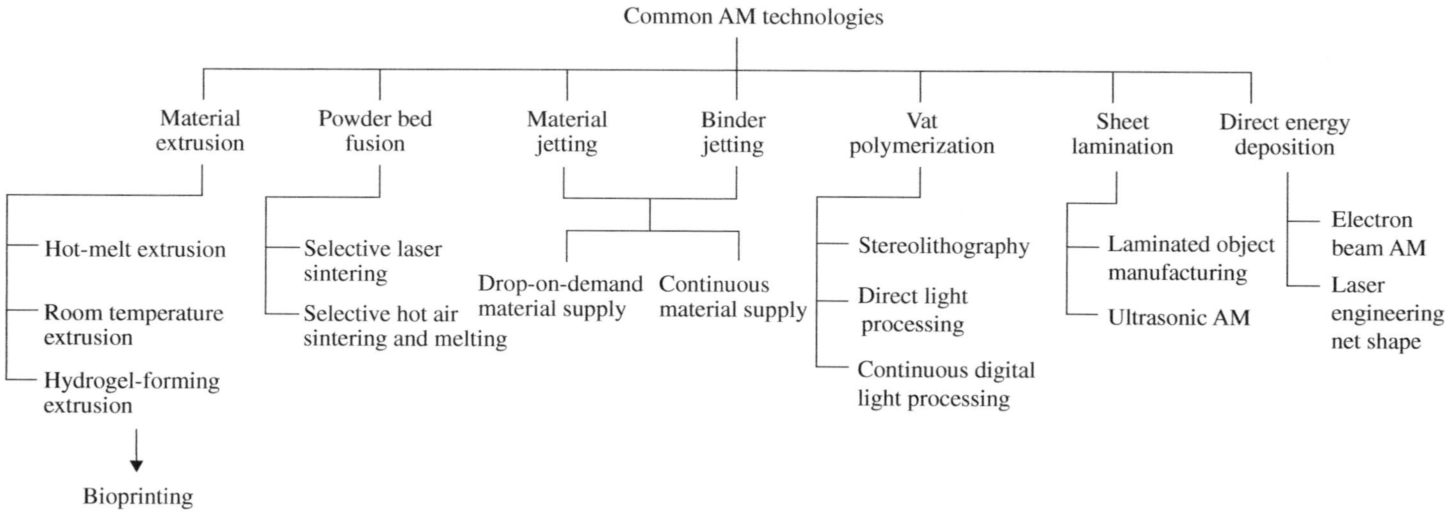

Figure 2.1 Classification of AM technologies.

- Material jetting: Materials like wax and photo-polymeric resins are the commonly used materials in this type of AM process. Here, the feed material is deposited in the form of a fine jet either in a continuous fashion or in a drop-on-demand (DoD) manner over the printing platform (Holland et al. 2019). The movement of printer arms can be operated based on the 3D model that controls the deposition of the printed layers. The melting point of the material can be precisely controlled through a temperature controller.
- Binder jetting: In the case of binder jetting, the liquid binder is sprayed over the powdered bed. Here, the binder acts as an adhesive that binds the material together. Above which the second layer of powder bed is spread over the printing platform then the liquid binder is sprayed (Sun et al. 2015). Likewise, the process is repeated continuously until the entire 3D model is printed. Commercial 3D printers such as Chef Jet and Chef Jet Pro works based on a binder jetting mechanism.
- Vat polymerization: In this process, a vat of liquid polymer-based resin mixture is used for constructing a 3D object in a layer-by-layer manner. Here, a ultraviolet (UV) light source is used for curing the polymeric resin that results in the hardening of the layer of resin (Xu et al. 2020). After the formation of each layer, the platform moves the fabricated 3D object in a downward direction. Technologies such as SLA, digital light processing (DLP), and liquid crystal printers (LCD) are based on vat polymerization techniques.
- Sheet lamination: In this AM process, a laser beam acts as a knife that cuts the materials like papers, plastics, metals, and ceramics that are suitable for thermal pressing to bind thin layers together into a 3D construct (Ngo et al. 2018). Here, the movement of the laser light is controlled by a computer that moves according to the 3D CAD design. Thus, the 3D construct can be fabricated through laser as a thermal source and the unwanted portions of the material can be easily removed after the completion of the printing process.
- Direct energy deposition/laser metal deposition (LMD): This method employs a high-energy laser source to melt the material over the processing surface. Thereafter the metal powder is injected into the melted material for the completion of the deposition process (Barro et al. 2020). The movement of the robotic arms is controlled with integration to a computer that moves according to the 2D directions that aids in the construction of a 3D object.

The above-mentioned AM process differs based on the process of formation of the 3D construct and the way of assembling the finished structure. These AM technologies can be broadly grouped based on the type of chemical transition as controlled fusion, controlled deposition, controlled cutting with lamination, and combined technologies (Wegrzyn et al. 2012). Techniques based on controlled fusion are sintering and SLA where a heat or light source is directed to fuse the powdered materials. On the other hand, extrusion and inkjet technologies are based on the controlled deposition methods where a liquid or semi-solid material is deposited in a controlled manner over the printing platform for the fabrication of a 3D object. Sheet lamination works based on the controlled cutting of thermal pressing materials followed by a lamination process to bind the layers together. Lamination-based 3D printing technologies are referred to as layered object manufacturing (LOM) that are widely applied for tissue engineering for the fabrication of 3D scaffolds in the biomedical field. Sometimes, a combination of AM processes is adapted especially in bioprinting applications for the fabrication of soft scaffolds and tissues. However, not all these AM technologies are applied to the food layer manufacturing process. The 3D printing techniques that are adapted and applied for foods are discussed in the subsequent sections of this chapter.

2.3 3D Food Printing Technologies

3D food printing is a digitalized process of fabrication of edible 3D construct in a layered manner. The subsequent layers of materials are bonded together through appropriate chemical transitions that alter the state of the material upon deposition. In contrast to the robotic manufacturing process, 3D printing of foods not only allows for automation of process but also helps in customization and personalization of diet (Sun et al. 2018b). More complex food geometries with intricate designs can be fabricated using 3D printing with accurate precision that is nearly impossible with conventional food processing techniques. Interestingly, 3D printing as a single standalone technology can integrate multiple unit operations of conventional processing into a single step (Nachal et al. 2019). Food printing follows a sequence of steps that starts with the development of 3D CAD design or scanned 3D model. The 3D model in its STL format is fed to the slicing software for retrieving the design information in machine-readable G and M codes. These computer codes are applied to the 3D food printers that assist in the printing process. A typical 3D food printer consists of the coordinate movement axes, printing platform, motor assembly, feed supply, print head, and microprocessor controller unit (Anukiruthika et al. 2020) (Figure 2.2).

In a broader sense, food printing technologies are categorized based on the nature and state of raw materials used (Feng et al. 2018). It includes 3D printing of flowable liquids, solid powders, semi-solid paste, and cell cultures. Accordingly, the AM processes that are adapted for 3D printing of liquid-based semi-solid food materials include extrusion printing, material jetting, and binder jetting. While the powder materials are 3D printed based on the sintering process. As a variant of existing 3D printing techniques, bioprinting

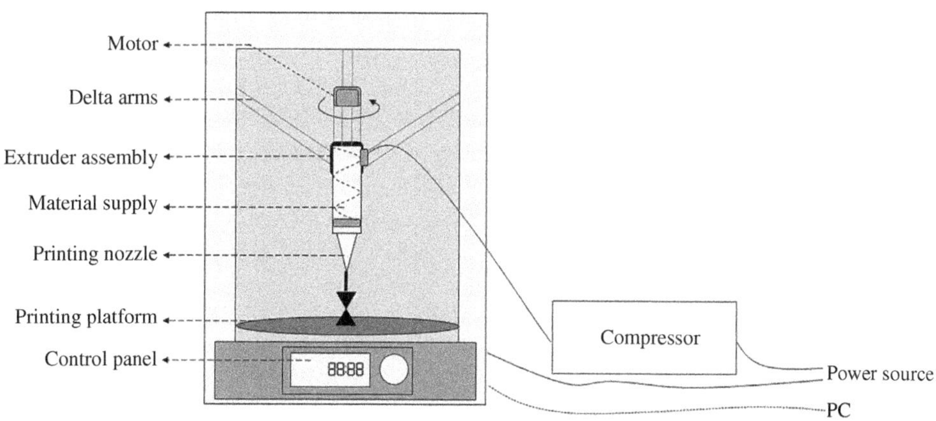

Figure 2.2 Schematic view of CARK food 3D printer. *Source:* Anukiruthika et al. (2020) / With permission of Elsevier.

involves the culturing of live cell cultures and tissue scaffolds (Godoi et al. 2016). Extrusion-based food printing is the most common technique used for a wide range of food materials such as starch (Theagarajan et al. 2020), proteins (Liu et al. 2019a), dietary fibre (Lille et al. 2018), and myofibrillar tissues (Dick et al. 2019b). This is because of its simplicity, flexibility, and ease of operations. However, these types of food materials require appropriate pre-processing for enhancing printability and post-processing for its edibility (Feng et al. 2020). Next to the extrusion, the sintering technique is widely used for the fabrication of sugar and starch-based 3D constructs. In the process of sintering, the solid particulates get compacted into a solid mass of material upon external stimuli such as either heat or pressure. Materials such as plastics, ceramics, and metals are commonly used in the sintering process. Considering food printing, sintering is used to fabricate 3D constructs from low melting powders such as sugars, fat, cocoa powder, and Nesquik without melting the material to the point of liquefaction (Sun et al. 2018a). Another powder-based technique is inkjet printing. It involves the controlled deposition and accumulation of droplets of food inks either in a continuous or DoD manner. The solid powder particles are bonded together with the printed liquid material. Commercial 3D food printers such as De Grood Innovations™ and Foodjet™ are based on inkjet printing mechanisms that are used for 2D decorations and surface filling of cakes, pastries, and biscuits (Godoi et al. 2016). Although inkjet printing can produce fine and appealing 3D constructs than extrusion-based printing, this technology is not suitable for high viscous food materials. In case of liquid binding, the printing involves the accumulation of powder layers through direct fusion with droplets of liquid binder as opposed to printing material as in case of material jetting (Holland et al. 2019). Commercial 3D printer Chef Jet™ are based on this technique and are used for printing of sugar fondants, confectioneries, and candies.

Another variant of the 3D printing technique used for printing live cell cultures is bioprinting. It involves the process of fabrication of living structures from a cell-laden bio-ink. Nowadays, scientists are more interested in the development of *in-vitro* cultured meat that has a minimal environmental impact (Post et al. 2020). 3D bioprinting is an ideal means for the construction of cell tissues, scaffolds, and organs. Various AM approaches that are used for 3D bioprinting are extrusion-based, droplet-based, and photocuring-based bioprinting. Among which extrusion-based 3D printing is widely used for the printing of turkey (Lipton et al. 2010), fish (Chen et al. 2020), beef (Dick et al. 2019b), and pork-based food products (Dick et al. 2020). Apart from extrusion-based bioprinting, other approaches like droplet-based and photocuring must be explored and studied in detail for a better understanding of the state-of-the-art of printing of *in-vitro* meat. Some of the commercial 3D printers available for bioprinting include 3D Bioplotter, NovoGen MMX Bioprinter™, and Biobot™ (Gu et al. 2019). Advancements in bioprinting for food applications would open up a new dimension of research and remains to offer a promising solution for the shortage of animal-based meat. Further, 3D printing has a less environmental impact with reduced carbon footprints and overall product life cycles.

Printability refers to the dimensional stability of the material to withstand its own weight (Kim et al. 2018a). So, the success of printability greatly depends on the material composition and printing variables. Food is a complex mixture of macro and micro constituents with varied physiochemical properties (Pérez et al. 2019). The selection of the 3D printing technique is greatly influenced by the type and state of the printing material supply. Some of the material properties are considered specifically based on the type of the printing technique. Properties such as glass transition temperature, powder characteristics, wettability, compressibility, melting and solidification behaviour, viscosity, and crystallinity highly influence the end-quality and stability of the 3D printed constructs (Nachal et al. 2019). Similar to the product characteristics, process variables such as nozzle size, nozzle height, printing speed, flow rate, motor speed, laser power, and printing temperature must be optimized for the fabrication of seamless 3D construct (Liu and Zhang 2019). Consideration of the material properties and the process variables would vary based on the 3D printing technique used. Some of the common 3D printing technologies that are widely used for food applications have been discussed in the following sections.

2.4 Extrusion-Based Printing

As stated earlier, material extrusion is based on the FDM approach that is widely used for 3D printing of thermoplastics and polymers. However, for food applications, the same principle has been adapted for layered deposition of food materials that are extruded out through the printing nozzle with precise control over the temperature (Mantihal et al. 2020). The extrusion-based 3D food printing relies on the continuous flow of material supply either in a semi-solid paste, highly viscous slurry, or in colloidal form. The solid-like elastic behaviour of the material supply is ideal for the proper extrusion (Liu et al. 2018a). Hence viscosity is one of the critical factors for the successful deposition of material through extrusion-based 3D printing. Most of the food materials could be successfully 3D printed using the extrusion technique. However, material consistency is a significant criterion since not all material supplies possess a characteristic solidification behaviour upon extrusion. Considering this issue, the material supply must be pre-processed adequately to enhance the flow behaviour and its printability (Liu et al. 2019b). The pre-processing includes the reduction of moisture from the printing material supply either by drying or by concentration. Based on the printability, the food materials are categorized as printable, non-printable, and alternative ingredients. The characteristic extrudability of non-printable and alternative ingredients can be improved through the addition of food hydrocolloids such as xanthan gum, guar gum, acacia gum, and sodium alginate as an additive into the printing material supply (Gholamipour-Shirazi et al., 2019). Material properties such as gelatinization, gelation, and emulsification are the basic mechanisms that support the stability of the printed layers in extrusion-based 3D printing.

2.4.1 Working Principle, System Components, and Process Variables

Although a range of 3D printing technologies is available for printing foods, extrusion is the commonly used technique due to its simpler design, versatility, flexibility, and scale-up for commercial applications (Beg et al. 2020). Extrusion is the process of forcing the solid/semi-solid/liquid materials through a smaller aperture thereby forming the desired shape with desired cross-sections. A typical extrusion-based 3D printer consists of a feed supply unit, extruder assembly, extrusion motor, pneumatic assembly, and a microprocessor controller unit (Figure 2.3). The 3D printing process starts with the material loading into the feed

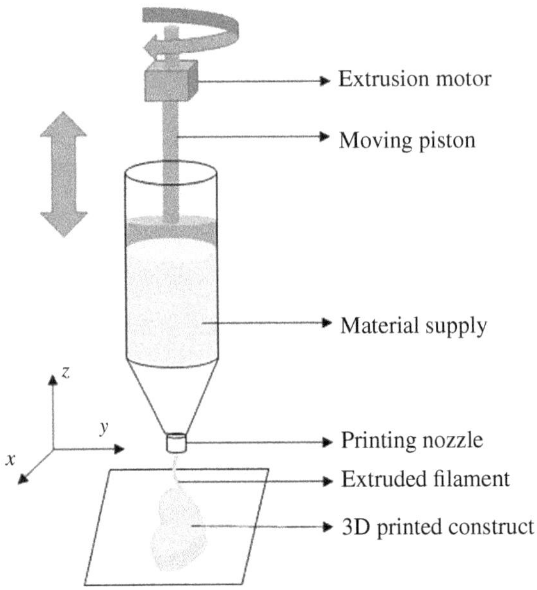

Figure 2.3 Schematic diagram of typical extrusion-based 3D printing.

supply system followed by pushing the material through the extruder assembly via printing nozzle in a controlled manner.

The movement of the printing arms and coordinate axes are controlled by a microprocessor unit in a pre-defined tool path (Guo et al. 2019b). Thus, the successively deposited layers are bonded together and get solidified forming a coherent stable solid 3D structure. Based on the extrusion mechanism, the extrusion unit can be of either a screw-based extrusion, syringe-based extrusion, pneumatic driven extrusion, or a combined mechanism (Portanguen et al. 2019) (Figure 2.4).

In the first type of screw-based feeding system, a programmed screw-driven motor assists in the continuous pushing of the material through the printing nozzle. While in a syringe-based extrusion unit, a plunger is driven by the programmed stepper motor that assists in the extrusion of material through the nozzle in a linear motion. The syringe-based extrusion system is more common for food printing applications that are easy and convenient for a small batch of samples. However, for large-scale applications, a continuous flow of material is required which can be possible with the screw-driven system (Guo et al. 2019a). As an added advantage, screw-driven extrusion results in simultaneous mixing as well as extrusion of the material. This combined unit operation ensures proper consistency and the continuous flowability of the material supply that would otherwise result in nozzle blocking as in the case of a syringe-based extrusion unit. On the other hand, the pneumatic assisted extrusion unit is almost similar to that of the syringe-based system with the replacement of the syringe with the plunger in cartridge that is pushed by the air pressure using a pneumatic compressor unit. The screw-based extrusion unit is not suitable for materials with high viscosity and mechanical strength (Liu et al. 2018a). For those samples with high viscosity, a syringe-based extrusion mechanism is used for achieving proper extrusion and good precision. While the pneumatic-based air pressure-driven extrusion is more suitable for printing low viscosity liquid samples (Sun et al. 2018b). Although extrusion-based 3D

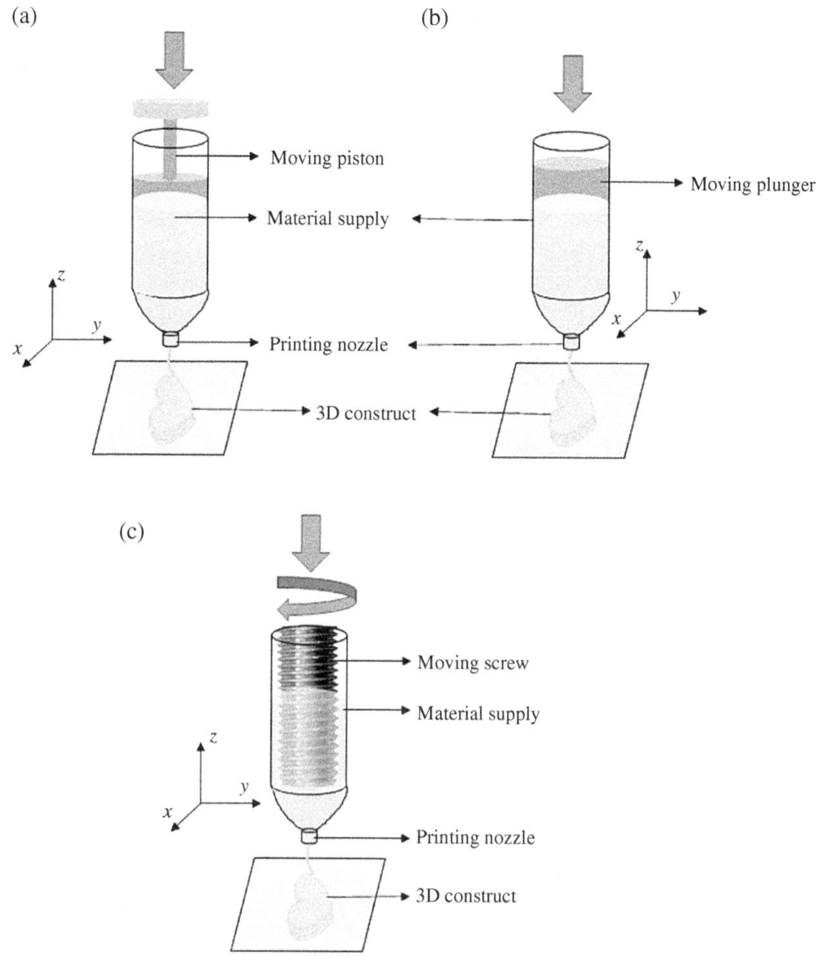

Figure 2.4 Different types of extrusion mechanism (a) piston-based extrusion system, (b) syringe-based extrusion system, and (c) screw-based extrusion system.

printing has a great scope and applicability for printing foods, the printing of delicate complex shapes is very challenging. The extrusion-based 3D constructs are more prone to structural collapse and deformation that requires additional support for stable product geometry which must be removed after 3D printing process (Liu and Zhang 2019). In case of polymer printing, the removal of additional support structures is easy, however, for food applications it greatly affects the final resolution of the printed 3D construct. Hence, more studies on material properties and post-stability behaviour are required for deeper understanding of the food printing technologies.

Material characteristics like moisture content, rheological, and thermal properties are crucial for the successful printing of 3D constructs based on extrusion printing. It was reported that the viscosity must be low enough for easy flow of the material and high enough for stable deposition and proper adhesion of the printed layers (Wilson et al. 2020). Based on the temperature used for the deposition of material, extrusion-based printing can be categorized as melting extrusion and soft material extrusion. 3D printing of chocolates and starch is known to be hot-melt extrusion (HME) that is accompanied by a phase change during printing. On the other hand, cereal-based dough and meat pastes comes under soft material extrusion that involves direct deposition without any phase change during printing (Godoi et al. 2019). As a multicomponent food system, the binding of the extruded layers is accompanied by the intermolecular interactions of macro and micro constituents. The printing precision and the resolution of the 3D constructs could be achieved through proper optimization of printing variables. The infill percentage and infill pattern are related to the complex internal microstructure and the final appearance of the 3D constructs (Severini et al. 2016). In recent days, attempts are made to determine the printability of the common daily foods includes staple food crops such as rice, millets, and wheat (Severini et al. 2018); fruits and vegetables such as lemon juice (Yang et al. 2019b), spinach (Lee et al. 2019), potato (Dankar et al. 2020), broccoli, peas, and carrot (Kim et al. 2018b; Pant et al. 2021); meat-based foods such as surimi, beef, chicken, and pork (Dick et al. 2019a; Wang et al. 2018; Wilson et al. 2020). All these studies dealt with the optimization of material concentration, process parameters and post-processing methods to deliver the customized foods. So far, the basic concept of a single extrusion system has been discussed. Focussing on the development of the novel fascinating 3D constructs, a dual extrusion system is in practice. In contrast to a single extrusion system, the dual extrusion unit consists of an additional extruder unit that works alternatively in a pre-determined interval that prints a single structure with dual material supplies. Liu et al. (2018b) reported a study on dual extrusion of mashed potato and strawberry juice gel using a commercial 3D printer (Figure 2.5). In this study, dual extrusion was performed by two different methods: the creation of the multi-part model and assigning each of them to each extruder; and creation of the single part model and assigning different roles to each extruder. The former method aids in varying the inside shapes of the model while the latter

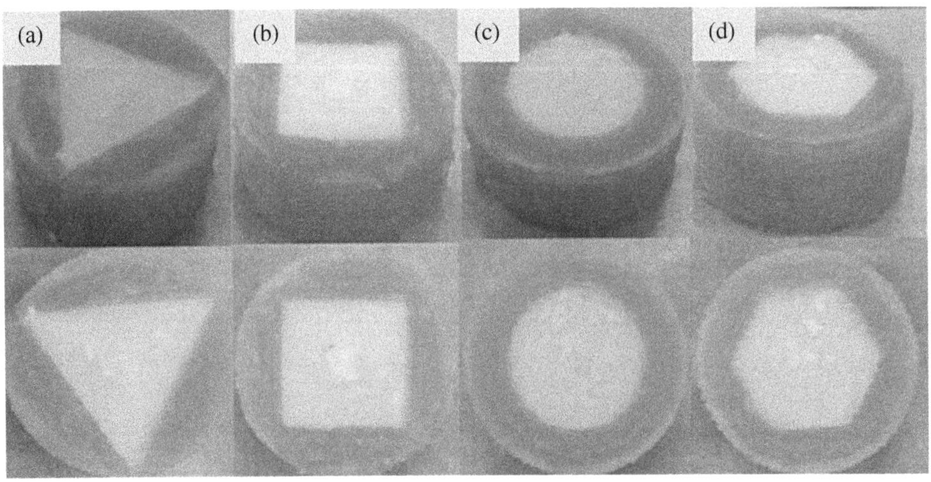

Figure 2.5 3D printed samples using dual extrusion of different internal structures (a) triangle, (b) square, (c) circle, and (d) hexagon. *Source:* From Liu et al. (2018b), Figure 02 [p. 019] / With permission of Elsevier. DOI-https://doi.org/10.1016/j.lwt.2018.06.014.

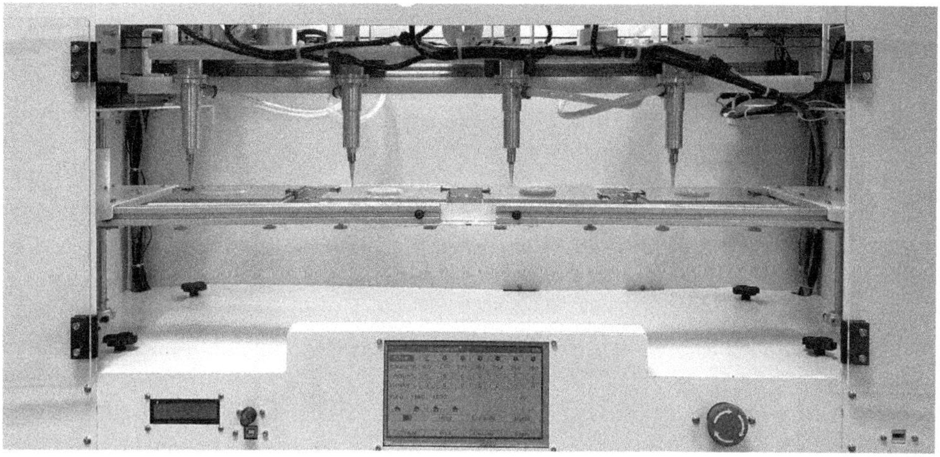

Figure 2.6 3D printing of material supply using multi-head 3D food printer MultiCARK™ (unpublished).

method helps in the creation of porous texture by varying the infill levels. In another study, edible constructs with overhanging geometries have been fabricated using a dual-printing approach (Periard et al. 2007). More recently, a dual extrusion commercial 3D printer was used to fabricate a colourful 4D ready-to-eat food using purple sweet potato puree and mashed potato (He et al. 2020). Thus, these studies proved the feasibility of the creation of multi-material constructs that has a great scope for personalized functional foods.

Another variant of the 3D printing system is the applicability of a multi-head printing unit for the simultaneous printing of food materials (Nachal et al. 2019). The printing process using an in-house designed multi-head 3D food printer is presented in Figure 2.6 (MultiCARK™ unpublished). The lower printing speed of the single head system is the major disadvantage that results in lesser production and consumes more time. The incorporation of the multiple heads eventually overcomes the above-mentioned limitations. However, the printing of multi-scale ingredients with multi-head systems adds to the complexity involved in the fabrication of 3D constructs. Hence, appropriate design modifications are required for ensuring the homogeneity of the ingredient mixtures. Conventionally, static- and agitating-mixer units are used for the mixing of food materials (Millen 2012). So, the appropriate integration of the conventional processing units with a multi-head 3D printing system would result in a homogeneous material supply. Thus, the system design of 3D printers for food applications remains a void research area that must be addressed for exploring the potential benefits of 3D food printing. Also, concerns related to operational safety and system cleanliness are the critical factors that must be considered for the safe delivery of the foods. Some of the available commercial 3D printers based on extrusion mechanism are Choc creator, Foodini, and BeeHex robot pizza printer (Sun et al. 2018a).

2.4.2 Classification of the Extrusion-Based 3D Printing System

2.4.2.1 Hot-Melt Extrusion

HME also known as FDM, was first used for the 3D fabrication of polymers and ceramics. Considering food applications, HME is used for those materials considering their melting and solidification behaviour. The process involves the deposition of melted semi-solid food from a moveable FDM print head through the hot-end nozzle tip (Mantihal et al. 2019). The deposited layers will solidify immediately after extrusion and bonded together with the previous layers upon cooling. The temperature is precisely controlled and determined based on the melting point of the food materials used. HME is mostly applied for the fabrication of 3D constructs using chocolates, starch, and protein gels (Chen et al. 2019; Hao et al. 2010; Liu et al. 2019a). Understanding the material properties is crucial for the fabrication of 3D constructs in a well-defined quality. In the case of chocolate printing, the combination of sugar with cocoa fat assists in the easy flow of the material through the extruder. The self-supported layers of chocolate rely on the thermal properties such as glass transition temperature (T_g) and melting point that are critical for the successful solidification of the material (Mantihal et al. 2017). The chocolate ink with pseudoplastic behaviour imparts conducive printability that in turn depends on the temperature used. It is essential to play around the six crystal polymorphs of the cocoa butter for achieving a stable 3D-oriented product with better texture and glossy appearance (Figure 2.7) (Lanaro et al. 2017). Some of the commercial 3D printers specific to chocolate printing are Choc Creator, ChefJet, and CocoJet. Researchers from the Massachusetts Institute of Technology (MIT) printed the melted chocolate using the direct ink writing (DIW) method using the developed 3D printer 'Digital Chocolatier' (Zoran and Coelho 2011). A similar approach has been used by 3D Food-Inks Printer for the printing of 3D-colored images on the extruded base (Golding et al. 2011). However, a post-processing step is required for the fusion of printed layers. More recently, Rando and Ramaioli (2020) studied the effect of heat transfer on the print stability of chocolate. The study investigated the correlation between the rheological and thermal properties for achieving a well-stable 3D structure. The stability criterion based on the developed yield stress during extrusion explained the stability or deformation of the printed materials.

Starch being an integral macro component of food grains that has a great scope for 3D printing. Recently, research on the fabrication of starch-based 3D constructs using HME

Figure 2.7 Hot-melt extrusion of complex 3D geometry bunny using chocolate. *Source:* From Lanaro et al. (2017), [p.30] / With permission of Elsevier. DOI-https://doi.org/10.1016/j.jfoodeng.2017.06.029.

is gaining attention. When the starch suspensions are subjected to heat treatment, the starch granules will swell with the absorption of a large amount of water and results in a thicker gel matrix through the process of gelatinization. This resulted in a starch gel that possesses characteristic viscoelastic, shear-thinning, and thixotropic behaviour that aids in the smooth continuous flow of material during extrusion and structural stability to printed layers during and after the extrusion process (Maniglia et al. 2020b). The mechanical strength and the extrudability of the starch gels explained its printability. Results showed that corn starch with 20% concentration (w/w) at 70–75 °C printing temperature and rice starch with 15–20% concentration (w/w) at 75 to 80 °C was suitable for HME (Zeng et al. 2020). A similar study on comparative analysis of printability of potato, corn, and rice starch gels has been reported (Chen et al. 2019). Nowadays, potato starch is widely used as a thickening and gelling agent in processed foods. In addition, the characteristic high degree of polymerization imparts viscosity and expansion behaviour to potato starch gels making it a suitable ingredient for the fabrication of personalized 3D printed food using HME. It was reported that potato starch of 15–25% concentration at 70 °C exhibits optimal printability. The hot-melt extruded layers of potato starch result in a uniform extrusion with a compact gel network and possess greater structural integrity (Liu et al. 2020). Likewise, detailed studies on the gelation and gelatinization of millets and pseudo-cereals would lead to the delivery of nutritious and personalized 3D printed foods that reduces the emergence of lifestyle disorders such as diabetes and obesity.

2.4.2.2 Cold Extrusion

The cold extrusion also known as room temperature extrusion (RTE) refers to the process of extrusion and deposition of materials without phase change. Most of the natively printable materials and pre-processed non-natively printable, as well as alternative ingredients, can be printed using RTE (Nachal et al. 2019). This type of extrusion is applied for the fabrication of 3D printed products that are difficult to produce by conventional food processing methods. Adoption of 3D printing would result in the production of foods with higher repeatability and smooth finish making them suitable for mass production and customization. Ready to cook (RTC) pasta products with novel 3D designs can be printed from a mixture of wheat semolina and water without the influence of temperature using RTE. Similarly, surface filling and graphical decorations of confectioneries are done using RTE (Van der Linden 2015). The semi-solid paste-like food materials are more appropriate for RTE; hence it is also referred to as soft material extrusion. The process involves the continuous extrusion of material from the moving print head that results in layered deposition of materials adhered to the preceding layers upon cooling without phase transition. The printing material supplies not only possess the adequate yield stress and elastic modulus but also shear-thinning tendency to withstand the desired shape after printing (Huang 2018). Although the temperature is less significant during RTE, the temperature must be fine-tuned that has a direct influence over the material consistency (K) and flow behaviour (n) (Hamilton et al. 2018). The printability of vegemite and marmite has been determined based on the K and n values in achieving a proper extrusion rate. Apart from the rheological properties, the particle size, crystallinity, and material composition would greatly affect the printability and quality of the 3D constructs.

Cereal and millet-based doughs, cheese, creamy peanut butter, cake frostings, jam, jelly, hummus, and Nutella are some of the common food materials that are suitable for RTE (Cohen et al. 2009; Millen 2012; Periard et al. 2007). Among the food printing technologies, extrusion-based 3D printing is widely used for a diverse range of food materials. More recently, the non-printable surimi (*Scomberomorus niphonius*) paste was printed with the addition of microbial transglutaminase as an additive (Dong et al. 2020). Results showed that the addition of transglutaminase in the range of 0.2–0.3% (w/w) enhanced the printability of surimi paste. Further, the textural properties such as hardness, cohesiveness, and resilience of the surimi gel were gradually increased with an increase in the concentration of transglutaminase up to 1.4%. The entire printing process was performed using an extrusion-based 3D printer at room temperature (20 °C). In a similar approach, the printability of the microalgae was assessed for the development of nutritious 3D printed snacks from pastry wheat flour. In this study, a dual printhead system was employed for the coaxial extrusion of microalgae-enriched snacks. Results showed that the material supplies with 3 and 4% Chlorella resulted in desired 3D construct with accurate precision (Uribe-Wandurraga et al. 2020). Other than pre-processing of the materials to enhance printability, post-processing is another significant step that ensures the safety and edibility of the 3D printed foods. Few of the commercial 3D food printers integrate the post-processing cooking step along with the printing process. However, most of the lab-scale 3D printers reported in the literature are at their initial stage that assists in the printing process alone. Hence, additional post-processing must be followed immediately after the printing process (Yang et al. 2019a). It is crucial to ensure the shape stability and structural integrity of the 3D printed layers during post-processing without any deformations. Post-processing methods such as drying, frying, baking, and freezing are common food processing methods used for cooking 3D printed samples (Krishnaraj et al. 2019). A detailed discussion on the various post-processing methods and its feasibility are presented in subsequent chapter of the present book.

2.4.2.3 Hydrogel-Forming Extrusion

Hydrogel-forming extrusion (HFE) is the process of the extrusion of hydrocolloid solutions/ dispersions into a polymeric/ hardening/ gel setting bath using a syringe-based extrusion mechanism through a moving printing nozzle (Kuo et al. 2021). Here, the solution temperature is a key criterion that determines the stability of droplets. The gel droplet's diameter ranges about 0.2–5 mm that forms a smooth distinct layer on deposition (Sun et al. 2018a). The rheological property and the gelation characteristics of the polymeric solution have significant implications on the successful printing of hydrogels. During the printing process, the polymer solution in the liquid state gets transformed into a stable gel state upon deposition. Research works on the fabrication of edible hydrogels are quite increased due to the advantage of the development of soft foods for aged people with swallowing disorders (Serizawa et al. 2014). Commercially available hydrogel printers are equipped with advanced dispensing units for the precise deposition of material. A 3D fruit printer developed by a UK firm, Dovetailed combined strawberry fruit flavour with the sodium gel for 3D printing of little spheres into a cold solution of calcium chloride bath to resemble a raspberry fruit (Molitch-Hou 2014).

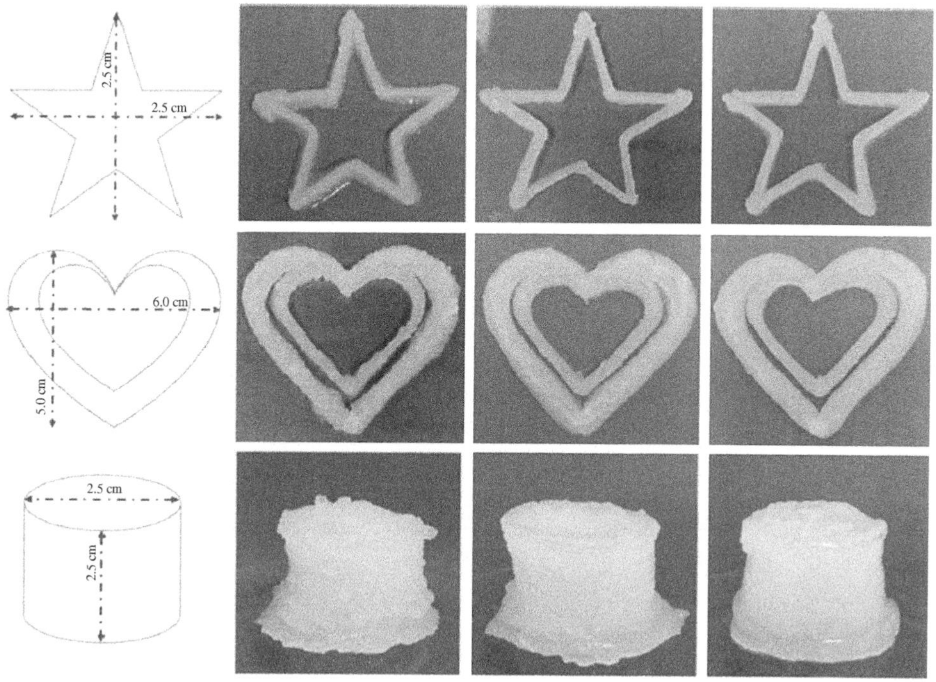

Figure 2.8 3D printed wheat starch hydrogels. *Source:* From Maniglia et al. (2020a) / With permission of Elsevier.

Recently the effect of dry heat treatment (DHT) on the 3D printing of cassava starch hydrogels has been reported (Maniglia et al. 2020b). The effect of pre-processing and the post-stability of printed cassava hydrogels were analyzed. The starch was chemically modified with prolonged exposure to DHT (4 hours at 130 °C) that resulted in higher carbonyl content and larger granule size. Thus, DHT was proved to have a significant implication on rheological properties that in turn aids in printability. Further, it was reported that the longest storage period increases the firmness of hydrogel preserves the structural integrity of printed 3D constructs. Similar results were obtained for DHT of wheat starches (Figure 2.8) (Maniglia et al. 2020a). During DHT, the molecular depolymerization was evident with a reduction in starch crystallinity. Thus, studies on starch-based hydrogels would extend the possibilities of fabrication of novel soft foods with altered textures. Hence, more research studies on the impact of pre-treatments on the chemical modification of macro components of food systems would explore the range of potential opportunities in 3D printing. In another study, the fabrication of bio-scaffold with hybrid gels of gelatin and alginate (1 : 2) at 7% resulted in a stable 3D printed matrix with the highest hardness and adhesiveness (Kuo et al. 2021). Subsequent post-processing freeze-drying of 3D printed hydrogels enhanced the mechanical properties and shelf-life of the product. Also, the microstructural analysis revealed the porous structure of the 3D printed scaffolds that has a great scope for encapsulation of bioactive compounds such as probiotics, enzymes, vitamins, minerals, enzymes, and antioxidants (Pérez-Luna and González-Reynoso 2018). However, only a few studies reported on the scope of integration of 3D printing and encapsulation. More research works are required in improving design modification of 3D printers suitable for micro- and nano-encapsulation; software integration with a high power source; material characterization and method for improving the bioavailability of micronutrients.

2.5 Selective Sintering

Selective sintering is the process of binding the powdered materials together to form a solid 3D construct using a laser or hot air as the power source. The movement of the 3D printer is controlled automatically by directing the axis movements at pre-defined points in space according to the 3D model. The power source is applied to the powdered bed that selectively fuses the powdered particles in a layer-by-layer manner for fabrication of desired 3D structure (Diaz et al. 2016). After scanning of each cross-section of structure, a new layer of powder is dropped and is sintered using the power source that leads to a 3D structure (Figure 2.9). Once the printing process is completed, the formed 3D structure is post-processed for the removal of unfused powder thereby ensuring the smooth finishing quality of the 3D prints (Diaz et al. 2017). The sintering process is mostly applied for metals and ceramics 3D printing. However, for food applications, studies are being put forth for the development of sugar-based confectionery 3D printed products (Fuh et al. 2015). The advantage of this technique when applied to foods is the use of the unfused powdered particles for the next cycle of printing and the feasibility for the fabrication of thc complex internal designs. Compare to the extrusion technology, the selective sintering process allows for the construction of overhanging 3D structures with greater resolution (Liu and Zhang 2019). However, the sintering process is limited to powdered material such as sugar, starch, and fat. This limitation can be overviewed by the design freedom, high productivity, and throughput of the sintering technology.

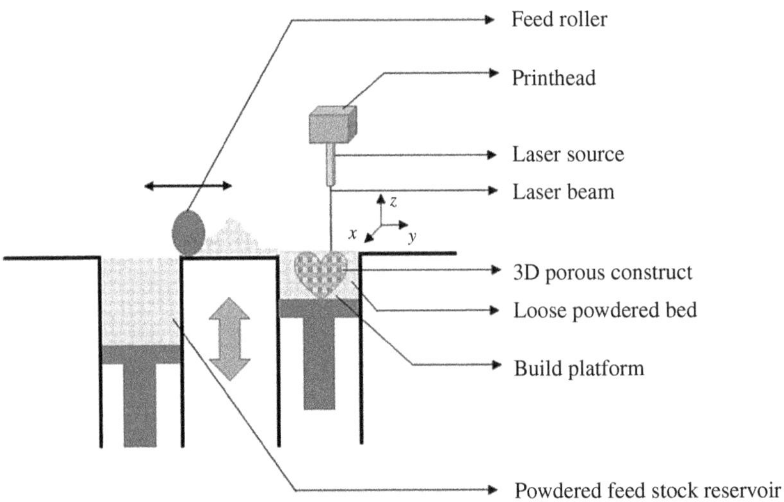

Figure 2.9 Schematic diagram of selective laser printing.

2.5.1 Working Principle, System Components, and Process Variables

As stated earlier, selective sintering is based on the powder bed fusion principle that refers to the selective consolidation of solid powdered particles in a specific area according to 3D design into a finished 3D printed construct using a thermal source. The use of a light source and the subsequent increase in printing temperature allows fabricating 3D constructs through melting and fusion of the powdered particles in a layer-by-layer manner (Vithani et al. 2019). One major advantage of this technology is the localized phase transition of powder particles that allows the reuse of the unfused feedstock materials. The basic system components of a selective sintering system include a build platform, a thermal power source, Galvano mirrors, powdered feedstock reservoir, mechanical roller, and powdered material vat (Ma et al. 2018). The printing process starts with the rising of the build platform to its uppermost position where a fresh layer of the powdered feed material is spread across the platform and flattened by the roller for uniform dispersion (Liu et al. 2017). Here the print head consists of an inbuilt thermal power source that scans across the powder and sinters it by following a pattern of 3D design. After the formation of each layer, the build platform is lowered to provide enough space for the formation of the next layer. Meanwhile, the reservoir platform ascends and spreads the next layer of material to form a new layer. Thus, the above process continues until the completion of whole 3D structure. After completion of 3D printing process, the system is allowed to cool that assists in the removal of excess un-sintered material from the fabricated 3D object. The surface finishing of the 3D printed construct can be improved by appropriate post-processing such as coating and polishing (Awad et al. 2020). These post-processing methods not only improve the appearance but also it enhances the mechanical properties like hardness and tensile strength.

Various process variables that must be considered during the sintering process are power source type (laser or thermal heaters), beam diameter, beam power, and scanning speed (Liu and Zhang 2019). The material phase change is associated with the complex interaction of powdered feed material with the light beam. Here the strength of interaction greatly depends on the energy density of the power source (Gu et al. 2012). Hence, the optimization of these process variables is crucial for attaining higher precision and resolution of 3D constructs. The energy density of the laser beam can be adjusted by varying the scan speed and laser power. A higher laser energy density can be obtained by a longer interaction time that results in the fabrication of denser 3D constructs with higher mechanical stability. On the other hand, a lesser interaction time produces a laser beam with a lower energy density that results in a porous brittle 3D construct (Amorim et al. 2014). Other process variables such as printing temperature, powdered bed thickness, space/gap between the print platform and laser head, and laser spot diameter also impact the stability of the printed structures. Apart from process variables, material properties such as particle size, bulk density, wettability, crystallinity, flowability, compressibility, glass transition temperature, melting, and solidification behaviour influence the printability of powdered materials (Yang et al. 2017). The powder bed fusion technology is useful in the fabrication of multi-material structures that requires a more detailed understanding of the chemical transitions and interactions at the molecular level of feed material and binder component. This opens an array of research opportunities in fabrication of 3D constructs using sintering technology especially in system design, material science, and processing.

2.5.2 Classification of Selective Sintering System

Based on the operation, the powder bed fusion technology has been categorized as selective laser sintering (SLS), selective laser melting (SLM), electron beam melting (EBM), and multi-jet fusion (MJF) (Awad et al. 2020). Thermoplastics, polymers, metals, and alloy powders are common feed materials employed in the sintering process. The above-mentioned printing technologies differ based on the type of feed materials used, based on the type and amount of light used for the transmission of energy during the phase change. For food printing applications, SLS is the most used sintering technique for the fabrication of 3D construct (Nachal et al. 2019). Food scientists of TNO used sugar and Nesquik powders for the fabrication of 3D structures with intricate internal designs (Gray 2010). A similar approach was used in CandyFab Project that successfully printed the complex 3D constructs from sugars (Figure 2.10) (CandyFab 2014). Another variant of SLS is the use of hot air instead of a laser and the technique is known as selective hot air sintering and melting (SHASAM) (Godoi et al. 2016). Both these technologies are well applied for foods that offer a greater degrees of freedom to fabricate complex structures in a short period without post-curing step.

2.5.2.1 Selective Laser Sintering

A laser power source is applied to a powdered bed of feed material that causes the agglomeration of the particle through the sintering process. The SLS technique works well for the construction of multi-layered 3D food constructs. The process involves the superficial melting and fusion of materials without the use of liquid binders to form a 3D construct. The researchers of the Netherlands Organization for Applied Scientific Research (TNO) used this technology for sugar-based 3D constructs with tailored nutrition and flavour (Van der Linden 2015). Works of literature were available on the development of sugar cubes and curry cubes using SLS that adds convenience to cooking (Nachal et al. 2019). SLS technique has a great control over the internal microstructure and porosity that imparts novel brittle texture. Sometimes the printed objects show a surface deformation and shrinkage

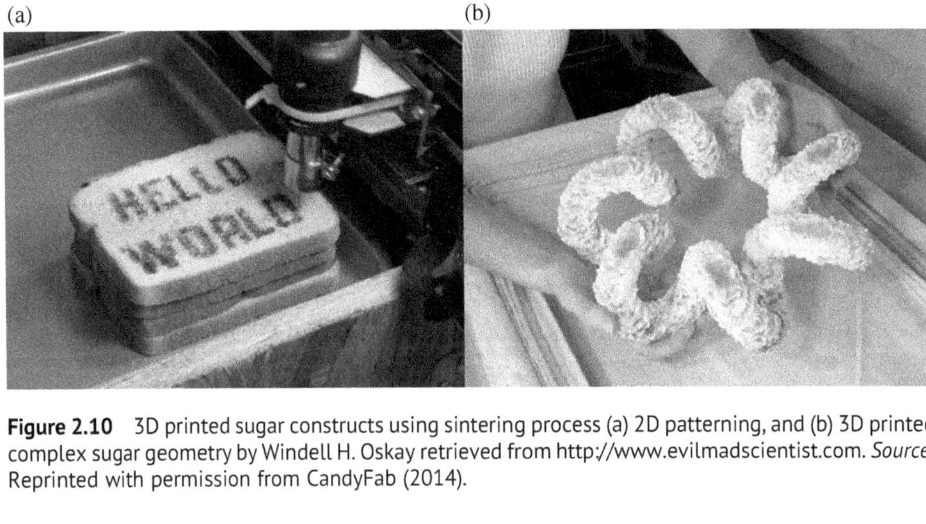

Figure 2.10 3D printed sugar constructs using sintering process (a) 2D patterning, and (b) 3D printed complex sugar geometry by Windell H. Oskay retrieved from http://www.evilmadscientist.com. *Source:* Reprinted with permission from CandyFab (2014).

due to the impact of laser irradiation of materials. Further, SLS is limited by its lower sintering speed and its suitability only for those materials that have low melting points such as sugars, lipids, and starches. The SLS technique of 3D printing is well established in the non-food sectors particularly in tissue engineering and pharmaceutical formulations (Awad et al. 2020). However, the technology must be studied in detail for broader applications of SLS in the delivery of nutraceuticals and functional bioactive compounds in 3D printed foods.

2.5.2.2 Selective Hot Air Sintering and Melting

Another variant of selective sintering is the use of hot air instead of a laser as a thermal source. The process remains the same as the hot air is used to fuse the powdered material supply. Researchers had successfully printed sugar-based 3D constructs using hot air as the heating medium for the fusion of printed layers. Powder density and compressibility are the critical factors in SHASAM that significantly affect the flowability and have an impact on the carnation of 3D design patterns (Schmid et al. 2013). In general, free-flowing powder without any solid clumps is suitable for the sintering process. Further, the powdered material supply must be adhesive enough and possess the tendency to agglomerate to get adhere to the contact points of adjacent molecules. It was reported that the layer thickness was inversely related to the mechanical strength of the 3D constructs (Amorim et al. 2014). As a result, thin layers possess higher mechanical strength with decreased porosity of the fabricated structures. Also, the particle size of the powders greatly influences the finishing quality of 3D printed samples. Diaz et al. (2018) described the method for the fabrication of multi-material 3D construct using the sintering process with a greater degree of printing precision and resolution. The material supply is comprised of structural elements and binder components. Thus, the structural element remains the base that provides bulk and scaffold function to the 3D construct and is fused by the particle-particle sintering of the binder component. It was reported that the use of a combination of two or more binders with different glass transition and melting temperatures exhibits better printing performance (Liu and Zhang 2019). In general, the glass transition and melting temperature of the binder component should range between 10 and 200 °C. So, the phase transition of the binder material can occur in less than 5 seconds while the structural element remains unaffected at temperatures below 200 °C (Diaz et al. 2018). Similar approaches have yet to be studied in detail for the optimization and characterization of structural and binder materials for improved printing performance.

2.6 Inkjet Printing

Another 3D printing technology that has been in practice for the past decade is inkjet printing. The inkjet printing technology is mainly used for surface decorations on cakes and pastries in 2D forms (Sun et al. 2015). Later, the same principle is applied for the fabrication of 3D designs over food materials. The inkjet printing involves the controlled accumulation of liquid-like slurry material supply in the form of droplets (Nachal et al. 2019). As a variant of powder-based technology, the layers of solid particles are bonded together by the printed liquid droplets ejected from the printhead. Hence the technology is also referred to as material jetting. The major advantage of this technology is the precise use of material without wastage based on a DoD manner. The print head uses two types of actuation process to spray the droplets namely thermal based actuation and piezoelectric based actuation (Vithani et al. 2019). In order to achieve the desired flowability, it is a general practice to charge the material by the addition of conductive agents into the material supply (Liu et al. 2017). Since the material supply used for inkjet printing is liquid-based, the mechanical strength to hold the 3D structures is poor (Godoi et al. 2016). Hence this technology is widely adopted for the printing of images in 2D forms. The success of printing the 2D and 3D design patterns using inkjet technology depends on the compatibility of food ink and the substrate (Holland et al. 2019). Factors such as rheological properties of food ink, printing temperature, and printing speed also significantly impact the final quality of design and printer performance.

2.6.1 Working Principle, System Components, and Process Variables

The process of ink-jetting involves the deposition of a liquid droplet on the solid material substrate using a controlled dispensing unit for the fabrication of 3D constructs (Figure 2.11). Different components of a 3D ink-jetting system include inkjet printhead, levelling roller, binder feeders, powder bed, build piston, and powder feed piston (Galeta et al. 2016). Two variations of the actuation mechanism used to initiate the flow of jet droplets are thermal and piezoelectric types. In a thermal-based inkjet system, a resistor is equipped with the printhead which generates heat resulting in the sudden creation of vapour bubbles in the feed material reservoir (Daly et al. 2015). Due to this, a small volume of a material droplet (ink) is ejected from the nozzle tip. On the other hand, the piezoelectric printhead system employs a crystal or ceramic-based piezoelectric element that transforms electrical

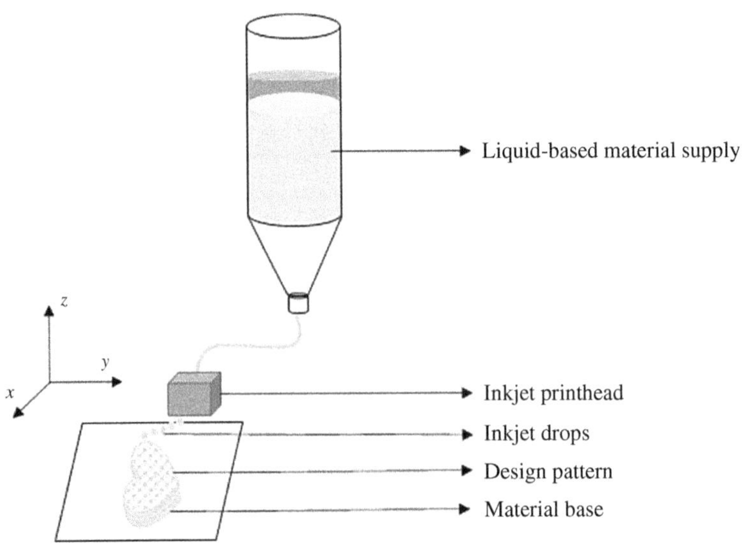

Figure 2.11 Schematic diagram of inkjet printing.

energy into mechanical energy. The associated deformation process results in the built-up of the required pressure to push the liquid in the form of droplets through the nozzle (Lee et al. 2012). The presence of the acoustic system in piezoelectric head assist in splitting the liquid into droplets. The change in the shape of liquid droplets can be evident when an electrical voltage is applied to the piezoelectric element that develops the required pressure as needed for the ejection of liquid droplets from the nozzle (Alomari et al. 2015). The less volatile liquid materials are inkjet printed at room temperature using a piezoelectric printhead. Although heat is applied to a small contact area for a short span of time, the application of thermal energy can potentially increase the local temperature of the material reservoir (Vithani et al. 2019). This sudden temperature fluctuation may affect the thermo-labile bioactive components of the food system. Hence, piezoelectric print heads are more common for pharmaceutical applications.

The most significant material properties that affect the print fidelity of the inkjet printing are the rheological and thermal behaviour of the food ink. In general, the inkjet printing system uses a low viscosity material for easy ejection of liquid through tiny orifice channels of the printhead. Material viscosity is a crucial factor that ensures the flowability of material. So, the desired material viscosity ranges between 2.8 and 6 mPas. While the viscosity above 10 mPas causes cavitation inside the printhead during the printing process and the viscosity below 2 mPas is not stable enough to form a droplet (Liu and Zhang 2019). The surface filling and decoration involve the jetting of food inks dispensing through the micro-sized channels in the range from 20 to 50 μm. About 1-pl (picolitre) of ink is dispensed during ejection of each droplets that typically range about 13 μm across (Xaar 2018). The principle of dispersion of ink involves the breaking up of a stream of droplets of the same volume with reduced surface area. The underlying process of ink-jetting technology is based on the Rayleigh–Plateau instability phenomenon (Godoi et al. 2019). Another factor that affects droplet deposition is temperature. Variations in the temperature modify the rheological properties and influence the surface energy of the food inks. Application of lower temperature reduces the surface energy and spreading tendency of food inks (Willcocks et al. 2011). The ingredient composition is another criterion that must be optimized for achieving the desired flowability for fabricating 3D designs on food substrate. Only a limited range of food materials has been analyzed for inkjet printing, in which chocolate ink is the most commonly used ingredient (Lanaro et al. 2019). However, other food materials in form of emulsion, slurry, and suspensions have a great scope for 2D and 3D inkjet printing. Future studies on the analysis of different material supplies suitable for inkjet printing are required to broaden the food applications.

2.6.2 Classification of Inkjet Printing

2.6.2.1 Drop-On-Demand Inkjet Printing

In a DoD inkjet printing method, the food ink material supply is deposited as required as discontinuous droplets in DoD at appropriate intervals. Mainly the ink is dispensed from the printhead under pressure exerted by a valve (Nachal et al. 2019). Irrespective of the flowability method, the compatibility between the ink and substrate surface is essential in order to impart the adhesiveness and binding of the printed layers. A proper flowability with appropriate viscosity must be required for seamless printing. The viscosity of the food ink is influenced by process temperature that in turn depends on the material composition which has to be optimized for achieving good printing precision and resolution. About 100–1000 heads can be used for the multi-head operation of the DoD system that works based on either using thermal or piezoelectric heads (Godoi et al. 2019). Inkjet printing is well suited for low viscosity materials that can be used for image filling and decoration. The DoD printer was used to dispense edible inks onto the food surface for the creation of appealing images and graphics (De Grood and De Grood 2013). The commercial food printer 'FoodJet' employs pneumatic membrane nozzle jets for dispersing of food ink droplets over the moving food substrate for customized designing of pizza base and biscuit filling (FoodJet 2020).

2.6.2.2 Continuous Inkjet Printing

In continuous inkjet printing, the edible ink is continuously dispensed without any pause through a piezoelectric vibrating crystal at a constant frequency. The use of a high-pressure pump creates a continuous flow that directs the liquid food ink through an orifice of smaller diameter ranges between 50 and 80 μm (Godoi et al. 2019). In some cases, the material flowability can be ensured by incorporating conductive agents into food ink to impart charges. The use of electrically charged conducting agents for improving the performance efficiency limits its application in food customization. The droplet generation rate of continuous inkjet printing is comparatively higher than the DoD inkjet printing. However, the resolution and precision of the produced images from DoD inkjet printing seem to be superior with neat dispensing than continuous inkjet printing. Willcocks et al. (2011) reported that the maximum resolution of inkjet printed images using a single head continuous ink-jetting system of about 70–90 dots per square inch (dpi).

2.7 Binder Jetting

Binder jetting is also referred to as the liquid binding where the powdered material gets deposited in a layer-by-layer manner with the selective dispensing of liquid binder based on a 3D data file. The process involves the accumulation of the powdered layers through direct fusion with the ejection of binder solution from the printhead (Le-Bail et al. 2020). Here, the binder fuses the cross-section of one layer with the succeeding layer thereby forming a 3D object (Figure 2.12). The working principle of the binder jetting technique is similar to that of inkjet printing apart from spraying the liquid binder droplets instead of food ink as in inkjet printing. Binder jetting allows us to fabricate complex 3D porous structures with precise control over the micro and macro structures of the material supply (Farzadi et al. 2014). The unfused powder materials act as a support during the fabrication of a 3D structure that can be easily removed after the completion of printing and can be recycled for further use. One of the attractive features of binder jetting technology is the production of colourful 3D structures. Variation in the liquid binder composition helps in the creation of complex, delicate, and attractive coloured 3D objects (Sun et al. 2015). Due to the spray of liquid binder over the food substrate, the fabricated 3D construct would require an additional drying process for removing the residual moisture. Further, the drying

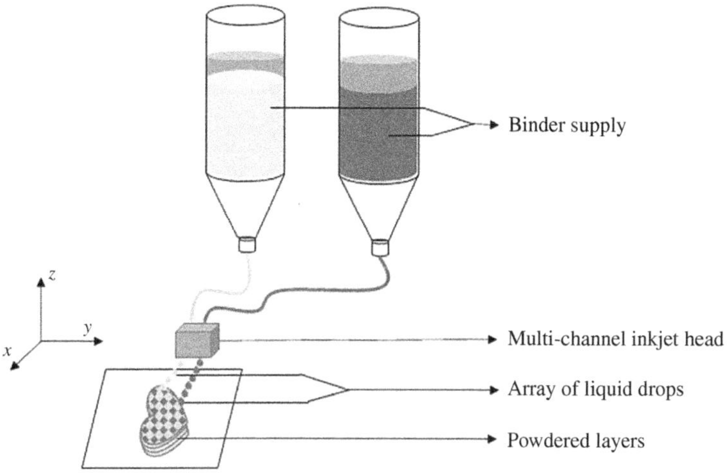

Figure 2.12 Schematic diagram of binder jetting.

treatment improves the structural stability and the mechanical integrity of the 3D construct (Vithani et al. 2019). Considering the food applications, binder jetting is limited with its applicability only to the powdered materials and the use of edible binder considering food safety.

2.7.1 Working Principle, System Components, and Process Variables

The process of binder jetting follows an even distribution of the powder layer across the printing platform and a liquid binder is sprayed onto the powdered bed to bind the two consecutive powder layers. In general, 3D printing using binder jetting technology starts by spraying a mist of water in order to stabilize the powdered material and to reduce the effects of distortion caused by spraying the liquid binder (Sun et al. 2018a). The system components of binder jetting are the same as that of a 3D ink-jetting system instead of a material reservoir, a liquid binder reservoir is used. So, the components are inkjet printhead, levelling and spreading roller, built envelop, binder feeders, powder bed, build piston, and powder feed piston (Trenfield et al. 2018). The first layer is formed by spreading powdered material of appropriate thickness onto the build plate and the subsequent ejection of the liquid binder. It is followed by lowering the powder feed piston in order to ensure the adequate gap for the formation of the next layer that gets fused with the previous layer (Ziaee and Crane 2019). The process is repeated until the pre-determined 3D object is produced based on a 3D data file. The customized 3D shapes were printed out of sugars and starch mixtures using the Z Corporation powder binder 3D printer in the edible 3D printing project (Southerland et al. 2011). Similar to inkjet printing, binder jetting offers fast fabrication at reduced operational cost. However, the 3D printing machine costs higher, and also the end quality of the fabricated 3D structure possess a rough surface. The commercial firm, 3D Systems utilizes colour jet printing technology for the fabrication of sugar-based complex 3D constructs with different flavoured binders (3D Systems 2013). This approach is well-known for the fabrication of complex sculptural cakes for special occasions and events. The developed system demonstrated a safe printing process that meets all the food safety standards.

Common process variables that influence the printing precision and accuracy are head type, droplet size and shape, droplet spacing, printing speed, velocity, and frequency of droplets (Shirazi et al. 2015). All these process variables are interrelated with the material properties. The compatibility of the liquid binder with powder bed substrate is adequate for achieving good finishing quality and resolution of the printed image. The surface tension of liquid binder and substrate material determines the interaction and fusion of printed layers onto substrate base (Liu et al. 2017). Sometimes, the compatibility of the substrate surface is improved by coating the surface with a binder or compatibility-enhancing film before the start of the printing process. Considering this aspect, binder shellac (poly 1-vinyl-2-pyrrolidone) was added into edible food ink for increasing surface compatibility (Willcocks et al. 2016). It was also reported that water-based glazing along with gums or other surfactants such as polysorbates and polyglycerol oleates were used for modifying the chocolate ink for achieving high-resolution images (Willcocks et al. 2011). Thus, the application of multilayered surfactants over the substrate surface had proved to significantly improve the compatibility yields a good precision and higher resolution of the printed images. The material's surface tension is mostly preferred to be lower than 35 dyne cm^{-1} for good compatibility (Shastry et al. 2009). Also, it was reported that a contact angle of less than 50° is desired for achieving good printability. Low polar materials such as carnauba wax were used as a coating material over the hard-panned sugar confectioneries that showed adverse effects on the accuracy and precision of printed images. Hence, hydrophilic materials are used for coating of substrate's surface in order to improve the compatibility of the water-based ink with the substrate (Liu and Zhang 2019). Thus, the material viscosity, ink density, and surface tension are significant properties that must be optimized for preventing the overflow and spreading of droplets from the nozzle. A successful 3D printed construct using binder jetting must possess adequate mechanical strength, minimal shrinkage and minimal bleeding or oozing out of liquid binder (Von et al. 2015a).

The binder concentration greatly influences the dimensional stability and precision of the printed sample. Other factors such as flowability and wettability also influence binder jetting (Godoi et al. 2016). Generally, materials with free-flowing nature are more suitable for binder jetting that possess good spread ability and packing properties. Typically, the angle of repose of material should be low enough generally smaller than 30° (Diaz et al. 2017). Adequate wettability must be preferred for binder jetting as low wettability of materials affects the powder alignment that subsequently declines printability. On the other hand, higher wettability interferes with the binding of the layers that significantly affects the end quality of printed construct. The particle size and distribution affect the binding behaviour of the water-based binder (Liu et al. 2017). As a common practice, the coarse particles are mixed with the fine powders for achieving an edible construct with compact packing. Materials with a moisture content of less than 6% are more suitable for binder jetting (Von et al. 2015b).

2.7.2 Classification of Binder Jetting

There is no distinct classification in binder jetting printing technology. Similar to inkjet printing binder jetting is used for the fabrication of complex porous internal structures.

The printhead with binder reservoir can be operated either by thermal or piezoelectric mechanism (Shirazi et al. 2015). Further, the dispersion of the binder jetting is usually performed by DoD manner since the continuous dispersion would result in moistening of powder bed that makes the post-drying to be difficult. The characteristic restructuring and amorphization of cellulose have been utilized for 3D printing applications. Holland et al. (2018) reported a study on the ball-milled amorphous cellulose powder with xanthan gum for the creation of 2D structures. In view of 3D printing, the researchers play around with the hygroscopic and thermal properties of the ink for designing a 3D crystalline network that resembles the gluten network. Results showed that heat treatment after printing aids in the recrystallization of the cellulose powder printed with liquid binder. Different kinds of food structures with novel textures can be fabricated using this approach. Further, the use of food-grade xanthan gum proved to be edible binder and its applications can be extended in printing of different food inks. The technology can only be applied for low viscous food inks that limits its applications in food printing (Nachal et al. 2019). This technology is well established for biomedical applications in the fabrication of functional drugs and pharmaceuticals. Advancements in binder jetting technology help in designing porous tablets for rapid dispersion when get contacted with a fluid medium (Prasad and Smyth 2016). However, more studies are required in exploring the binder jetting technology in the food sector. There is a great scope for binder jetting in the fabrication of soft materials. Future studies on understanding the fluid property and solvation chemistry of food inks would bridge up the void that exists with food applications of binder jetting technology.

2.8 Bio-Printing

3D bioprinting is an interdisciplinary science that is closely related to medical science, material science, and mechanical engineering. 3D bioprinting deals with a manoeuvre of cell-laden bio-inks to fabricate living structures such as 3D tissue scaffolds and organs (Dey and Ozbolat 2020). Major driving needs for the *in-vitro* culturing of the living structures are the existing huge demand for organ transplantation and tissue engineering (Matai et al. 2020). Traditional approaches of 2D culturing methods suffer from flaws during drug screening and medical examination studies. Approaches based on 3D printing helps in spatiotemporal directional manipulation of different kinds of cells and better prediction of organ models within an appropriate period (Armstrong and Stevens 2020). When applied to foods bioprinting provides a new dimension of research in the fabrication of cultured meat. Artificially cultured meat by *in-vitro* methods remains to provide a promising solution for the rising protein demand for mankind. The application of bioprinting for the fabrication of edible meat has a great scope to be the food of the future that possesses both economic as well as environmental benefits (Tomiyama et al. 2020). Most of the 3D printing studies reported in literature utilize natural meat and tested for its printability. Not many studies have been reported on the development of artificially cultured *in-vitro* meat. The rising awareness about the vegan diet explored the opportunities and feasibilities in the fabrication of meat analogues (Alexander et al. 2017). Hence, future research must be directed for a better understanding of the state-of-the-art of printability of cultured meat with special consideration in material selection, material response, safety, edibility, and acceptability of the artificial meat.

2.8.1 Working Principle, System Components, and Process Variables

3D bioprinting is an AM process that involves the fabrication of cells and biomaterials using a digital file to print organ-like structures in a layer-by-layer manner. The process of 3D bioprinting involves a sequence of steps: data acquisition, material selection, bioprinting, and functionalization (Gu et al. 2019). In data acquisition, the information about the 3D models can be obtained using the scanning techniques such as computed tomography (CT), X-ray imaging, and magnetic resonance imaging (MRI) (Ozbolat and Gudapati 2016). Then the scanned image can be reconstructed to extract the information, or the design can be modelled directly using CAD software. Thereafter the model is sliced into 2D cross-sections using appropriate slicing software (Figure 2.13). The materials for bioprinting can be selected based on the requirements of printed structures and printing approaches used. The cell cultures, hydrogels, and growth factors are the common base materials used in the bioprinting process. Based on the printing approach used, the system components of the bioprinter would vary. The printed cells are then dispersed in solutions for maturation (Koch et al. 2013). Later the cultured organ is subjected to physical and chemical simulations to determine its targeted function. A similar approach is utilized for the culturing of artificial meat.

The mixture of cell-matrix and nutrient mixture is supplied and printed from the printer cartridge in the form of layers. The selection of bio-ink is crucial for successful printing in terms of biocompatibility, printability, and mechanical integrity (Gao et al. 2019). The characteristic feature of bioprinter is the controlled and precise deposition of the biomaterials. The major advantage of bioprinting is the ability to fabricate porous structures in the desired shape with specific functionalities (Ying et al. 2018). 3D bioprinting offers rapid and customized fabrication of scaffolds with higher precision. With recent advancements in technology, the system components of the bioprinter allow for accurate control over the distribution of pore size, pore volume, and interconnectivity of the pores with network formation. Researchers of the University of Missouri used this approach for the printing of a strip of edible porcine tissue (Nachal et al. 2019). A plant-based cell-laden hydrogel was

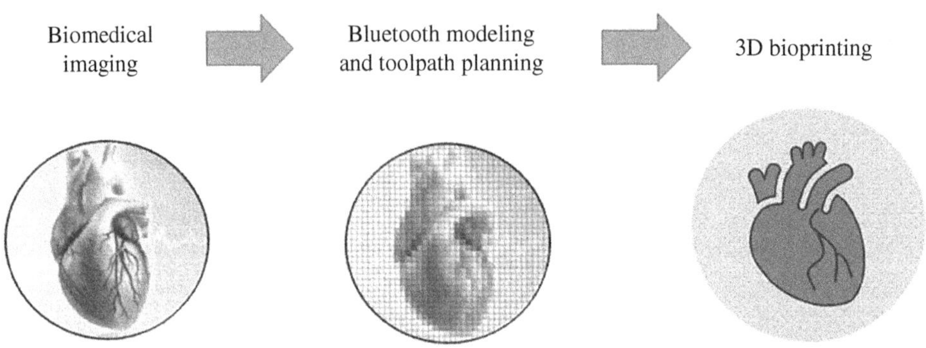

Figure 2.13 Steps involved in 3D bioprinting process.

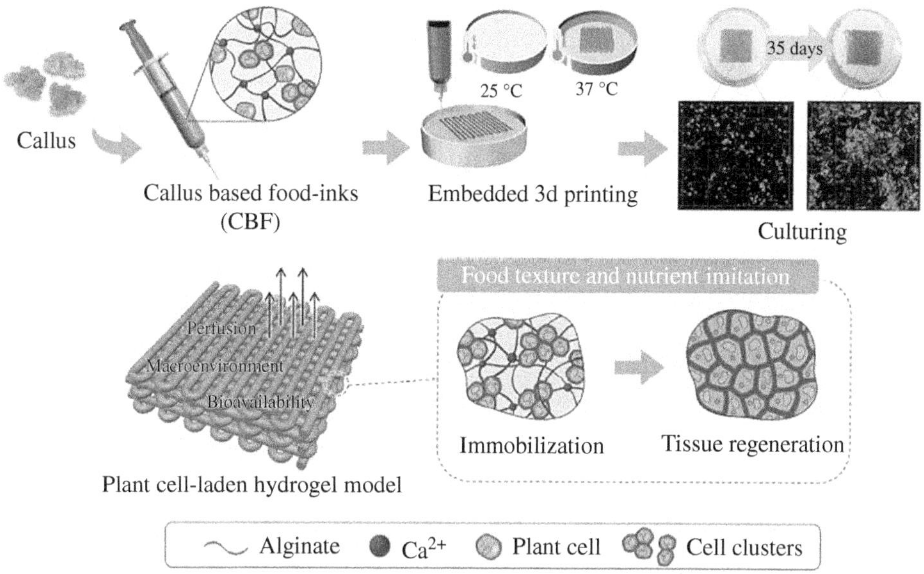

Figure 2.14 Schematic representation of integration of 3D printing with plant cell culture technologies. *Source:* From Park et al. (2020) / With permission of Elsevier.

utilized for the production of unique textured food using 3D printing (Park et al. 2020). The callus-based food ink was prepared by blending callus dispersion with 4% alginate in different ratios of 1 : 2, 1 : 1, and 2 : 1 (w/w). The process involves the curing of the callus samples using Ca^{2+} ions that aids in the formation of rigid gel suitable for 3D printing (Figure 2.14). The samples of 1 : 2 and 1 : 1 ratio showed a proper shape fidelity while the sample with excess cell concentrations i.e., 2 : 1 lowered the printing resolution and resulted in deviation from intended dimensions. Thus, the cells immobilized in hydrogel form a cluster and exhibits texture as that of plant tissues. Hence, the concept of callus-laden food ink showed a promising result in simulating the plant tissues thereby improving the textural properties of 3D printed foods. As food printing is considered, bioprinting seems to have implications on the ethical, cultural, and societal perspectives associated with animal meat consumption. The common challenges encountered during bioprinting are involved in the selection of biomaterials and the maintenance of a sterile working environment to prevent cross-contamination. The range of biomaterials available currently is very less that limits bioprinting applications in the food sector. Collagen, gelatin, fibrin, and thermoplastic polymers are a few of biomaterials used for fabrication of tissue scaffolds (Asadi et al. 2020). The above-mentioned limitations must be addressed in order to explore the research opportunities that exist with bioprinting in food applications.

2.8.2 Classification of Bioprinting

2.8.2.1 Extrusion-Based Bioprinting

Extrusion-based bioprinting also referred to as DIW is the most commonly used approach of 3D bioprinting due to its versatility and affordability (Sanz-Garcia et al. 2020). The process

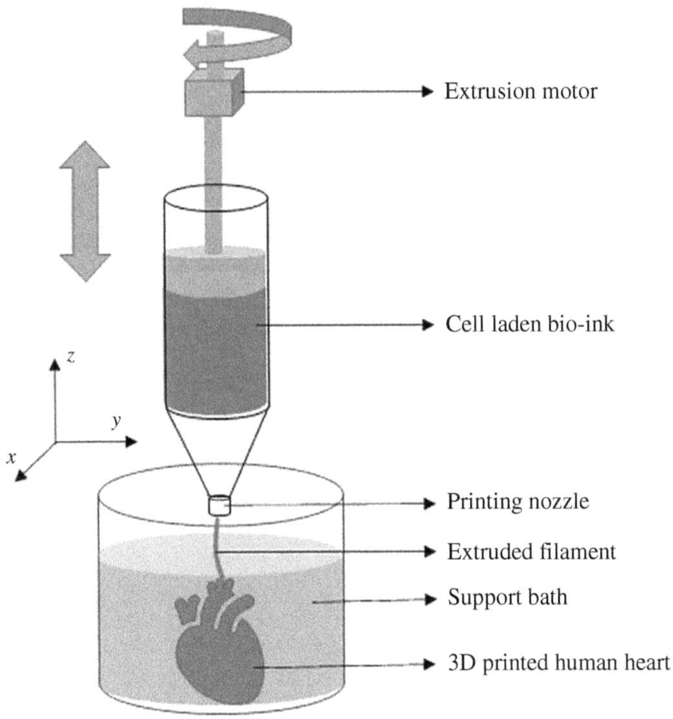

Figure 2.15 Schematic diagram of 3D bioprinting.

involves the continuous extrusion of biofilaments at a micro-scale level that is deposited over the material substrate thereby forming the desired 3D structure (Figure 2.15). The receiving substrate can be of either solid (culture dish), liquid (growth medium), or gel matrix (Ozbolat and Hospodiuk 2016). Various process parameters considered for extrusion-based 3D bioprinting includes nozzle size, extrusion rate, extrusion pressure, movement speed, printing, temperature, and path interval. Based on the method of actuation used for dispensing of biomaterial, the system can be categorized as screw-driven, piston-driven, and pneumatic-driven 3D bioprinting systems (Derakhshanfar et al. 2018). In addition, extrusion-based 3D bioprinting is used for coaxial and multi-material bioprinting based on the end use of the 3D printed scaffolds. Material properties like bio-ink viscosity and substrate properties must be considered for the successful fabrication of 3D scaffolds (Ying et al. 2018).

2.8.2.2 Droplet-Based Bioprinting

Instead of deposition of continuous biofilament, the droplet-based bioprinting involves the ejection of independent and discrete droplets onto the substrate surface thereby forming a 3D structure. Compare to extrusion technology, droplet-based bioprinting has a high throughput and produces a 3D structure with greater precision (Gudapati et al. 2016). Based on the principle of droplet formation, droplet-based bioprinting can be categorized as inkjet bioprinting, laser-assisted bioprinting, and electrohydrodynamic jetting (Gu et al. 2019).

Further inkjet bioprinting has been subdivided into DoD inkjet bioprinting and continuous inkjet bioprinting. Similarly, laser-assisted bioprinting is subdivided into laser guidance direct writing (LGDW) and laser-induced forward transfer (LIFT).

2.8.2.3 Photocuring-Based Bioprinting

Photocuring-based bioprinting is based on the photopolymerization of light-sensitive polymers for the fabrication of cell-free scaffolds (Kousaalya 2020). Compared to other approaches of bioprinting, photocuring has a higher printing speed and greater printing resolution. Based on the scanning mode, photocuring-based bioprinting can be further classified into SLA and DLP (Lim et al., 2020). In photocuring-based bioprinting, there won't be any problem of nozzle plugging and shear stress that would otherwise affect the cell viability.

2.9 Future Prospects and Challenges

3D printing is a print-and-eat technology that allows us to customize the food as per individual needs. The present chapter gives an overview of various food printing technologies. Each of them has its own advantages and disadvantages. The 3D printing of foods greatly depends on material properties and the type of binding mechanism employed. The selection of printing technology is decided by the material properties. Extrusion-based 3D printing can be adapted for most of the food materials; however, its application is limited to high viscosity foods. Sintering-based technologies best suits for the development of porous brittle 3D structures; however, they are limited with only a fewer range of powder materials. Binder jetting and ink-jet printing are well known for the carving of 3D designs for surface decorations where the compatibility of the substrate with food ink is adequate. Compared to conventional processing, 3D printing converges the multi-step processing into a single step. However, there are still many barriers in 3D printing that must be overcome for incorporating into a niche market of personalized foods. One major strength of 3D food printing is the conversion of our idea into a reality that allows us to deliver nutritious foods in desired shapes, colours, and forms. It is possible through the integration of 3D printing with digital gastronomy and culinary skills. However, for food applications, control over the process and product parameters without significant implications on end-product quality are challenging. Food is a complex matrix with varied physiochemical properties that in turn behave differently with different printing technology. It is not an easy task to print food as not all the food materials are printable that require adequate processing to make them printable. This complexity involved in the optimization of 3D printing parameters must be addressed through the development of streamlined testing methods and protocols.

In view of sustainability, 3D printing allows for usage of lesser material resources with minimal material wastage. Further, 3D printing allows for the utilization of by-products and waste streams of food industries such as fruits and vegetable peels, meat trimmings, and so on for the development of value-added 3D printed foods. 3D food printing possesses both economic and environmental benefits that reduce the carbon footprints and overall product life cycle. As the need of the hour, the rising demand for the food shortage and animal protein demand can be encountered using 3D printing by adopting bioprinting principles in developing *in-vitro* cultured artificial meats and meat analogues. Forecasting the future scale-up operations, currently, 3D printing suffers from the limitation of lower print speed and less production rate. This can be overcome by adopting multi-head printing systems. However, the extent of the feasibility of the adaption of multi-heads to different food printing technologies remains under question. The use of multiple heads gradually complicates the coordination mechanism of movement arms and the integration of 3D printers with the microprocessor controlling unit. Hence, more research works on design components, accessories, software integration, and development are essential to bridge up the gap in reducing the difficulty involved with multi-head print systems. Insights on design attributes would be useful for scaling up of process at an industrial level that results in higher operation speed and production rate. These features could gradually reduce the cost of operation as well as the price incurred with 3D printed foods.

With advancements in technology, 3D printing has a greater degrees of freedom that can be integrated with other food processing technologies. For instance, technologies such as encapsulation are well known for delivering functional foods. Similar approaches are adapted for the fabrication of functional 3D printed foods by encapsulating essential vitamins and minerals, probiotics, antioxidants, and so on. The basic coaxial-based extrusion technique available for dual material printing can be modified to integrate 3D printing and encapsulation. 3D printing could be a promising solution for addressing malnourishment. The development in biotechnology paves a way for the analysis of genetic data through the use of 'omics' technology. This concept of retrieving individual genomic data for addressing health disorders can be improved efficiently by integrating with 3D food printing (Kumar et al. 2020). Hence, 3D printing has a far way to go that has the capability to revolutionize the nutraceuticals and food industry. More research works must be carried out in these directions for exploring the potential opportunities that exist with 3D printing. Not surprisingly, 3D printers could become a part of domestic kitchen appliance in the near future that transforms dietary practices.

2.10 Conclusion

3D printing is an additive technology when applied to foods aids in the customization and personalization of diets. Various printing technologies used for food printing are extrusion-based techniques, sintering process, ink-jet printing, binder jetting, and bioprinting. The basic difference exists with the mechanism of binding of the printed layers. Various system components, working principles, and factors affecting each of the printing technologies have been discussed in the present chapter. Apart from aesthetic benefits, 3D printing of foods seems to be environmentally friendly providing several economic benefits. The printability of food material can be determined by material properties and process variables. Advancements in 3D printing can take the existing technology to a next level in the delivery of functional foods thereby reduces the incidence of malnourishment. As a smart manufacturing process, 3D printing has been considered to be a part of 4.0 industrial revolution. In this regard, 3D printing is forecasted to be adaptable in the food sector in the development of novel and sustainable foods.

References

3D Systems (2013). Technology brings edible printing to professionals and consumers alike. https://www.3dsystems.com/press-releases/3d-systems-acquires-sugar-lab (accessed 24 december 2019).

Alexander, P., Brown, C., Arneth, A. et al. (2017). Could consumption of insects, cultured meat or imitation meat reduce global agricultural land use? *Global Food Security* **15**: 22–32.

Alomari, M., Mohamed, F.H., Basit, A.W., and Gaisford, S. (2015). Personalised dosing: printing a dose of one's own medicine. *International Journal of Pharmaceutics* **494** (2): 568–577.

Amorim, F., Lohrengel, A., Neubert, V. et al. (2014). Selective laser sintering of Mo-CuNi composite to be used as EDM electrode. *Rapid Prototyping Journal* **20** (1): 59–68.

Anukiruthika, T., Moses, J.A., and Anandharamakrishnan, C. (2020). 3D printing of egg yolk and white with rice flour blends. *Journal of Food Engineering* **265**: 109691. https://doi.org/10.1016/j.jfoodeng.2019.109691.

Armstrong, J.P.K. and Stevens, M.M. (2020). Using remote fields for complex tissue engineering. *Trends in Biotechnology* **38** (3): 254–263.

Asadi, N., Del Bakhshayesh, A.R., Davaran, S., and Akbarzadeh, A. (2020). Common biocompatible polymeric materials for tissue engineering and regenerative medicine. *Materials Chemistry and Physics* **242**: 122528.

Awad, A., Fina, F., Goyanes, A. et al. (2020). 3D printing: principles and pharmaceutical applications of selective laser sintering. *International Journal of Pharmaceutics* **586**: 119594.

Baiano, A. (2020). 3D printed foods: a comprehensive review on technologies, nutritional value, safety, consumer attitude, regulatory framework, and economic and sustainability issues. *Food Reviews International*, pp. 1–31.

Barro, Ó., Arias-González, F., Lusquiños, F. et al. (2020). Effect of four manufacturing techniques (casting, laser directed energy deposition, milling and selective laser melting) on microstructural, mechanical and electrochemical properties of co-CR dental alloys, before and after PFM firing process. *Metals* **10** (10): 1291.

Beg, S., Almalki, W.H., Malik, A. et al. (2020). 3D printing for drug delivery and biomedical applications. *Drug Discovery Today* **25** (9): 1668–1681. https://doi.org/10.1016/j.drudis.2020.07.007.

CandyFab (2014). The CandyFab Project. https://candyfab.org (accessed 24 December 2019).

Chen, H., Xie, F., Chen, L., and Zheng, B. (2019). Effect of rheological properties of potato, rice and corn starches on their hot-extrusion 3D printing behaviors. *Journal of Food Engineering* **244**: 150–158. https://doi.org/10.1016/j.jfoodeng.2018.09.011.

Chen, H.-z., Zhang, M., and Yang, C. (2020). Comparative analysis of 3D printability and rheological properties of surimi gels via LF-NMR and dielectric characteristics. *Journal of Food Engineering* **292**: 110278.

Cohen, D.L., Lipton, J.I., Cutler, M. et al. (2009). Hydrocolloid printing: a novel platform for customized food production. Solid Freeform Fabrication Symposium, pp. 807–818.

Daly, R., Harrington, T.S., Martin, G.D., and Hutchings, I.M. (2015). Inkjet printing for pharmaceutics–a review of research and manufacturing. *International Journal of Pharmaceutics* **494** (2): 554–567.

Dankar, I., Haddarah, A., Sepulcre, F., and Pujolà, M. (2020). Assessing mechanical and rheological properties of potato puree: effect of different ingredient combinations and cooking methods on the feasibility of 3D printing. *Foods* **9** (1): 21.

De Grood, J.P.W. and De Grood, P.J. (2013). Method and device for dispensing a liquid. US Patent No. 8,556,392. Washington, DC: US Patent and Trademark Office.

Derakhshanfar, S., Mbeleck, R., Xu, K. et al. (2018). 3D bioprinting for biomedical devices and tissue engineering: a review of recent trends and advances. *Bioactive Materials* **3** (2): 144–156.

Derossi, A., Husain, A., Caporizzi, R., and Severini, C. (2020). Manufacturing personalized food for people uniqueness. An overview from traditional to emerging technologies. *Critical Reviews in Food Science and Nutrition* **60** (7): 1141–1159.

Dey, M. and Ozbolat, I.T. (2020). *3D Bioprinting of Cells, Tissues and Organs*. Nature Publishing Group.

Diaz, J.V., Van Bommel, K.J.C., Noort, M.W. et al. (2016). Method for the production of edible objects using sls and food products. Google Patents.

Diaz, J.V., Noort, M.W. J., and Van Bommel, K.J.C. (2017). Method for the production of an edible object by powder bed (3d) printing and food products obtainable therewith. Google Patents.

Diaz, J.V., Van Bommel, K.J.C., Noort, M.W. J. et al. (2018). Method for the production of edible objects using sls and food products. Google Patents.

Dick, A., Bhandari, B., and Prakash, S. (2019a). 3D printing of meat. *Meat Science* **153**: 35–44. https://doi.org/10.1016/j.meatsci.2019.03.005.

Dick, A., Bhandari, B., and Prakash, S. (2019b). Post-processing feasibility of composite-layer 3D printed beef. *Meat Science* **153**: 9–18. https://doi.org/10.1016/j.meatsci.2019.02.024.

Dick, A., Bhandari, B., Dong, X., and Prakash, S. (2020). Feasibility study of hydrocolloid incorporated 3D printed pork as dysphagia food. *Food Hydrocolloids* **107**: 105940.

Dong, X., Pan, Y., Zhao, W. et al. (2020). Impact of microbial transglutaminase on 3D printing quality of *Scomberomorus niphonius* surimi. *LWT* **124**: 109123.

Farzadi, A., Solati-Hashjin, M., Asadi-Eydivand, M., and Osman, N.A.A. (2014). Effect of layer thickness and printing orientation on mechanical properties and dimensional accuracy of 3D printed porous samples for bone tissue engineering. *PloS One* **9** (9): e108252.

Feng, C., Zhang, M., and Bhandari, B. (2018). Materials properties of printable edible inks and printing parameters optimization during 3D printing: a review. *Critical Reviews in Food Science and Nutrition* **8398**: 1–8. https://doi.org/10.1080/10408398.2018.1481823.

Feng, C., Zhang, M., Liu, Z. et al. (2020). Effect of drying method on post-processing stability and quality of 3D printed rose-yam paste. *Drying Technology* **39**: 1–9.

FoodJet (2020). Foodjet-Precision depositing solutions. https://www.foodjet.com (accessed 24 December 2019).

Fuh, J.Y.H., Hong, G.S., Zhou, W. et al. (2015). An overview of 3D printing technologies for food fabrication. *Food and Bioprocess Technology* **8** (8): 1605–1615. https://doi.org/10.1007/s11947-015-1528-6.

Galeta, T., Raos, P., Stojšić, J., and Pakši, I. (2016). Influence of structure on mechanical properties of 3D printed objects. *Procedia Engineering* **149**: 100–104.

Gao, Q., Niu, X., Shao, L. et al. (2019). 3D printing of complex GelMA-based scaffolds with nanoclay. *Biofabrication* **11** (3): 35006.

Garcia-Oliveira, P., Fraga-Corral, M., Pereira, A.G. et al. (2020). Solutions for the sustainability of the food production and consumption system. *Critical Reviews in Food Science and Nutrition*, pp. 1–17.

Gholamipour-Shirazi, A., Norton, I.T., and Mills, T. (2019). Designing hydrocolloid based food-ink formulations for extrusion 3D printing. *Food Hydrocolloids* **95**: 161–167. https://doi.org/10.1016/j.foodhyd.2019.04.011.

Godoi, F.C., Prakash, S., and Bhandari, B.R. (2016). 3D printing technologies applied for food design: Status and prospects. *Journal of Food Engineering* **179**: 44–54. https://doi.org/10.1016/j.jfoodeng.2016.01.025.

Godoi, F.C., Bhandari, B.R., Prakash, S., and Zhang, M. (2019). An introduction to the principles of 3D food printing. In: *Fundamentals of 3D Food Printing and Applications* (eds. F.C. Godoi, B.R. Bhandari, S. Prakash and M. Zhang), pp. 1–18. Elsevier.

Golding, M., Archer, R., Gupta, G. et al. (2011). *Design and development of a 3-D food printer. Centre of Research Excellence. Massey University.*

Gray, N. (2010). Looking to the future: creating novel foods using 3D printing. http://www.foodnavigator.com/Science-Nutrition/Looking-to-the-future-Creating-novel-foods-using-3D-printing (accessed 1 May 2020).

Gu, D.D., Meiners, W., Wissenbach, K., and Poprawe, R. (2012). Laser additive manufacturing of metallic components: materials, processes and mechanisms. *International Materials Reviews* **57** (3): 133–164.

Gu, Z., Fu, J., Lin, H., and He, Y. (2019). Development of 3D bioprinting: from printing methods to biomedical applications. *Asian Journal of Pharmaceutical Sciences* **15**: 529–557.

Gudapati, H., Dey, M., and Ozbolat, I. (2016). A comprehensive review on droplet-based bioprinting: past, present and future. *Biomaterials* **102**: 20–42.

Guo, C.-F., Zhang, M., and Bhandari, B. (2019a). A comparative study between syringe-based and screw-based 3D food printers by computational simulation. *Computers and Electronics in Agriculture* **162**: 397–404.

Guo, C., Zhang, M., and Bhandari, B. (2019b). Model building and slicing in food 3D printing processes: a review. *Comprehensive Reviews in Food Science and Food Safety* **18**: 1052–1069.

Hamilton, C.A., Alici, G., and Panhuis, M. (2018). 3D printing vegemite and marmite: redefining "breadboards.". *Journal of Food Engineering* **220**: 83–88.

Hao, L., Mellor, S., Seaman, O. et al. (2010). Material characterisation and process development for chocolate additive layer manufacturing. *Virtual and Physical Prototyping* **5**: 2759. https://doi.org/10.1080/17452751003753212.

He, C., Zhang, M., and Guo, C. (2020). 4D printing of mashed potato/purple sweet potato puree with spontaneous color change. *Innovative Food Science & Emerging Technologies* **59**: 102250.

Holland, S., Foster, T., MacNaughtan, W., and Tuck, C. (2018). Design and characterisation of food grade powders and inks for microstructure control using 3D printing. *Journal of Food Engineering* **220**: 12–19.

Holland, S., Foster, T., and Tuck, C. (2019). Creation of food structures through binder jetting. In: *Fundamentals of 3D Food Printing and Applications* (eds. F.C. Godoi, B.R. Bhandari, S. Prakash and M. Zhang), pp. 257–288. Elsevier.

Huang, C.Y. (2018). Extrusion-based 3D Printing and Characterization of Edible Materials [University of Waterloo]. https://uwspace.uwaterloo.ca/handle/10012/12899 (accessed 1 April 2020).

Jayaprakash, S., Paasi, J., Pennanen, K. et al. (2020). Techno-economic prospects and desirability of 3D food printing: perspectives of industrial experts, researchers and consumers. *Foods* **9** (12): 1725.

Jiang, H., Zheng, L., Zou, Y. et al. (2018). 3D food printing: main components selection by considering rheological properties. *Critical Reviews in Food Science and Nutrition* **59**: 1–13. https://doi.org/10.1080/10408398.2018.1514363.

Kim, H.W., Bae, H., and Park, H.J. (2018a). Classification of the printability of selected food for 3D printing: development of an assessment method using hydrocolloids as reference material. *Journal of Food Engineering* **215**: 23–32.

Kim, H.W., Lee, J.H., Park, S.M. et al. (2018b). Effect of hydrocolloids on rheological properties and printability of vegetable inks for 3D food printing. *Journal of Food Science* **83** (12): 2923–2932.

Koch, L., Gruene, M., Unger, C., and Chichkov, B. (2013). Laser assisted cell printing. *Current Pharmaceutical Biotechnology* **14** (1): 91–97.

Kousaalya, A.B. (2020). Sustainable photo-curable polymers in additive manufacturing arena: a review. In: *Sustainability & Green Polymer Chemistry Volume 1: Green Products and Processes*, ACS Symposium Series (eds. H.N. Cheng and R.A. Gross), pp. 89–98. Washington DC: ACS Publications.

Krishnaraj, P., Anukiruthika, T., Choudhary, P. et al. (2019). 3D extrusion printing and post-processing of fibre-rich snack from indigenous composite flour. *Food and Bioprocess Technology* **12** (10): 1776–1786. https://doi.org/10.1007/s11947-019-02336-5.

Kumar, S., Anukiruthika, T., Dutta, S. et al. (2020). Iron deficiency anemia: a comprehensive review on iron absorption, bioavailability and emerging food fortification approaches. *Trends in Food Science & Technology* **99**: 58. https://doi.org/10.1016/j.tifs.2020.02.021.

Kuo, C.-C., Qin, H., Cheng, Y. et al. (2021). An integrated manufacturing strategy to fabricate delivery system using gelatin/alginate hybrid hydrogels: 3D printing and freeze-drying. *Food Hydrocolloids* **111**: 106262.

Lanaro, M., Forrestal, D.P., Scheurer, S. et al. (2017). 3D printing complex chocolate objects: platform design, optimization and evaluation. *Journal of Food Engineering* **215**: 13–22. https://doi.org/10.1016/j.jfoodeng.2017.06.029.

Lanaro, M., Desselle, M.R., and Woodruff, M.A. (2019). 3D printing chocolate: properties of formulations for extrusion, sintering, binding and ink jetting. In: *Fundamentals of 3D Food Printing and Applications* (eds. F.C. Godoi, B.R. Bhandari, S. Prakash and M. Zhang), pp. 151–173. Elsevier.

Le-Bail, A., Maniglia, B.C., and Le-Bail, P. (2020). 3D printing of foods: recent developments, future perspectives and challenges. *Current Opinion in Food Science* **35**: 54–64.

Lee, B.K., Yun, Y.H., Choi, J.S. et al. (2012). Fabrication of drug-loaded polymer microparticles with arbitrary geometries using a piezoelectric inkjet printing system. *International Journal of Pharmaceutics* **427** (2): 305–310.

Lee, J.H., Won, D.J., Kim, H.W., and Park, H.J. (2019). Effect of particle size on 3D printing performance of the food-ink system with cellular food materials. *Journal of Food Engineering* **256**: 1–8. https://doi.org/10.1016/j.jfoodeng.2019.03.014.

Lille, M., Nurmela, A., Nordlund, E. et al. (2018). Applicability of protein and fiber-rich food materials in extrusion-based 3D printing. *Journal of Food Engineering* **220**: 20–27.

Lim, K.S., Galarraga, J.H., Cui, X. et al. (2020). Fundamentals and applications of photo-cross-linking in bioprinting. *Chemical Reviews* **120** (19): 10662–10694.

Lipton, J., Arnold, D., Nigl, F. et al. (2010). Multi-material food printing with complex internal structure suitable for conventional post-processing. *Solid Freeform Fabrication Symposium*, pp. 809–815.

Liu, Z. and Zhang, M. (2019). 3D food printing technologies and factors affecting printing precision. In: *Fundamentals of 3D Food Printing and Applications* (eds. F.C. Godoi, B.R. Bhandari, S. Prakash and M. Zhang), 19–40. Elsevier Inc. https://doi.org/10.1016/b978-0-12-814564-7.00002-x.

Liu, Z., Zhang, M., Bhandari, B., and Wang, Y. (2017). 3D printing: printing precision and application in food sector. *Trends in Food Science & Technology* **69**: 83–94.

Liu, Z., Zhang, M., Bhandari, B., and Yang, C. (2018a). Impact of rheological properties of mashed potatoes on 3D printing. *Journal of Food Engineering* **220**: 76–82. https://doi.org/10.1016/j.jfoodeng.2017.04.017.

Liu, Z., Zhang, M., and Yang, C. (2018b). Dual extrusion 3D printing of mashed potatoes/strawberry juice gel. *Lwt* **96**: 589–596.

Liu, C., Ho, C., and Wang, J. (2018c). The development of 3D food printer for printing fibrous meat materials. *IOP Conference Series: Materials Science and Engineering* **284** (1) https://doi.org/10.1088/1757-899X/284/1/012019.

Liu, Y., Yu, Y., Liu, C. et al. (2019a). Rheological and mechanical behavior of milk protein composite gel for extrusion-based 3D food printing. *Lwt* **102**: 338–346. https://doi.org/10.1016/j.lwt.2018.12.053.

Liu, Z., Bhandari, B., Prakash, S. et al. (2019b). Linking rheology and printability of a multicomponent gel system of carrageenan-xanthan-starch in extrusion based additive manufacturing. *Food Hydrocolloids* **87**: 413–424.

Liu, Z., Chen, H., Zheng, B. et al. (2020). Understanding the structure and rheological properties of potato starch induced by hot-extrusion 3D printing. *Food Hydrocolloids* **105**: 105812.

Ma, F., Zhang, H., Hon, K.K.B., and Gong, Q. (2018). An optimization approach of selective laser sintering considering energy consumption and material cost. *Journal of Cleaner Production* **199**: 529–537.

Maniglia, B.C., Lima, D.C., da Matta Júnior, M. et al. (2020a). Dry heating treatment: a potential tool to improve the wheat starch properties for 3D food printing application. *Food Research International* **137**: 109731.

Maniglia, B.C., Lima, D.C., Junior, M.D.M. et al. (2020b). Preparation of cassava starch hydrogels for application in 3D printing using dry heating treatment (DHT): a prospective study on the effects of DHT and gelatinization conditions. *Food Research International* **128**: 108803.

Mantihal, S., Prakash, S., Godoi, F.C., and Bhandari, B. (2017). Optimization of chocolate 3D printing by correlating thermal and flow properties with 3D structure modeling. *Innovative Food Science and Emerging Technologies* **44**: 21–29. https://doi.org/10.1016/j.ifset.2017.09.012.

Mantihal, S., Prakash, S., Godoi, F.C., and Bhandari, B. (2019). Effect of additives on thermal, rheological and tribological properties of 3D printed dark chocolate. *Food Research International* **119**: 161–169.

Mantihal, S., Kobun, R., and Lee, B.-B. (2020). 3D food printing of as the new way of preparing food: a review. *International Journal of Gastronomy and Food Science* **22**: 100260.

Matai, I., Kaur, G., Seyedsalehi, A. et al. (2020). Progress in 3D bioprinting technology for tissue/organ regenerative engineering. *Biomaterials* **226**: 119536.

Millen, C.I. (2012). *The Development of Colour 3D Food Printing System: A Thesis Presented in Partial Fulfilment of the Requirements for the Degree of Master of Engineering in Mechatronics at Massey University, Palmerston North, New Zealand.* Massey University.

Molitch-Hou, M. (2014). The 3D fruit printer and the raspberry that tasted like a strawberry. https://3dprintingindustry.com/news/3d-fruit-printer-raspberry-tasted-like-strawberry-27713 (accessed 1 September 2020).

Nachal, N., Moses, J.A., Karthik, P., and Anandharamakrishnan, C. (2019). Applications of 3D printing in food processing. *Food Engineering Reviews* **11** (3): 123–141. https://doi.org/10.1007/s12393-019-09199-8.

Ngo, T.D., Kashani, A., Imbalzano, G. et al. (2018). Additive manufacturing (3D printing): a review of materials, methods, applications and challenges. *Composites Part B: Engineering* **143**: 172–196.

Nida, S., Anukiruthika, T., Moses, J.A., and Anandharamakrishnan, C. (2020). 3D printing of grinding and milling fractions of rice husk. *Waste and Biomass Valorization*, pp. 1–10. https://doi.org/10.1007/s12649-020-01000-w.

Ozbolat, I. and Gudapati, H. (2016). A review on design for bioprinting. *Bioprinting* **3**: 1–14.

Ozbolat, I.T. and Hospodiuk, M. (2016). Current advances and future perspectives in extrusion-based bioprinting. *Biomaterials* **76**: 321–343.

Pant, A., Lee, A.Y., Karyappa, R. et al. (2021). 3D food printing of fresh vegetables using food hydrocolloids for dysphagic patients. *Food Hydrocolloids* **114**: 106546.

Park, S.M., Kim, H.W., and Park, H.J. (2020). Callus-based 3D printing for food exemplified with carrot tissues and its potential for innovative food production. *Journal of Food Engineering* **271**: 109781.

Pérez, B., Nykvist, H., Brøgger, A.F. et al. (2019). Impact of macronutrients printability and 3D-printer parameters on 3D-food printing: a review. *Food Chemistry* **287** (October 2018): 249–257. https://doi.org/10.1016/j.foodchem.2019.02.090.

Pérez-Luna, V.H. and González-Reynoso, O. (2018). Encapsulation of biological agents in hydrogels for therapeutic applications. *Gels* **4** (3): 61.

Periard, D., Schaal, N., Schaal, M. et al. (2007). Printing food. Proceedings of the 18th Solid Freeform Fabrication Symposium, pp. 564–574.

Portanguen, S., Tournayre, P., Sicard, J. et al. (2019). Toward the design of functional foods and biobased products by 3D printing: a review. *Trends in Food Science & Technology* **86**: 188–198.

Post, M.J., Levenberg, S., Kaplan, D.L. et al. (2020). Scientific, sustainability and regulatory challenges of cultured meat. *Nature Food* **1** (7): 403–415.

Pradhan, R.A., Rahman, S.S., Qureshi, A., and Ullah, A. (2021). Biopolymers: opportunities and challenges for 3D printing. *Biopolymers and Their Industrial Applications*, pp. 281–303.

Prasad, L.K. and Smyth, H. (2016). 3D Printing technologies for drug delivery: a review. *Drug Development and Industrial Pharmacy* **42** (7): 1019–1031.

Rando, P. and Ramaioli, M. (2020). Food 3D printing: effect of heat transfer on print stability of chocolate. *Journal of Food Engineering* **294**: 110415.

Sanz-Garcia, A., Sodupe-Ortega, E., Shimizu, T. et al. (2020). A versatile open-source printhead for low-cost 3D microextrusion-based bioprinting. *Polymers* **12** (10): 2346.

Schmid, M., Amado, F., Levy, G. et al. (2013). Flowability of powders for selective laser sintering (SLS) investigated by round robin test. High Value Manufacturing: Advanced Research in Virtual and Rapid Prototyping: Proceedings of the 6th International Conference on Advanced Research in Virtual and Rapid Prototyping, p. 95.

Serizawa, R., Shitara, M., Gong, J. et al. (2014). 3D jet printer of edible gels for food creation. *Behavior and Mechanics of Multifunctional Materials and Composites* **2014** (9058): 90580A.

Severini, C., Derossi, A., and Azzollini, D. (2016). Variables affecting the printability of foods: preliminary tests on cereal-based products. *Innovative Food Science and Emerging Technologies* **38**: 281–291. https://doi.org/10.1016/j.ifset.2016.10.001.

Severini, C., Azzollini, D., Albenzio, M., and Derossi, A. (2018). On printability, quality and nutritional properties of 3D printed cereal based snacks enriched with edible insects. *Food Research International* **106** (November 2017): 666–676. https://doi.org/10.1016/j.foodres.2018.01.034.

Shahrubudin, N., Lee, T.C., and Ramlan, R. (2019). An overview on 3D printing technology: technological, materials, and applications. *Procedia Manufacturing* **35**: 1286–1296. https://doi.org/10.1016/j.promfg.2019.06.089.

Shastry, A.V., Collins, T.M., Suttle, J.M. et al. (2009). Edible inks for ink-jet printing on edible substrates. Google Patents.

Shirazi, S.F.S., Gharehkhani, S., Mehrali, M. et al. (2015). A review on powder-based additive manufacturing for tissue engineering: selective laser sintering and inkjet 3D printing. *Science and Technology of Advanced Materials* **16** (3).

Shree, M.V., Dhinakaran, V., Rajkumar, V. et al. (2020). Effect of 3D printing on supply chain management. *Materials Today: Proceedings* **21**: 958–963.

Southerland, D., Walters, P., and Huson, D. (2011). Edible 3D printing. *NIP & Digital Fabrication Conference* **2011** (2): 819–822.

Sun, J., Peng, Z., Zhou, W. et al. (2015). A review on 3D printing for customized food fabrication. *Procedia Manufacturing* **1**: 308–319.

Sun, J., Zhou, W., Huang, D., and Yan, L. (2018a). 3D food printing: perspectives. In: *Polymers for Food Applications* (ed. T.J. Gutierrez), pp. 725–755. Springer.

Sun, J., Zhou, W., Yan, L. et al. (2018b). Extrusion-based food printing for digitalized food design and nutrition control. *Journal of Food Engineering* **220**: 1–11. https://doi.org/10.1016/j.jfoodeng.2017.02.028.

Theagarajan, R., Moses, J.A., and Anandharamakrishnan, C. (2020). 3D extrusion printability of rice starch and optimization of process variables. *Food and Bioprocess Technology* **13**: 1048. https://doi.org/10.1007/s11947-020-02453-6.

Tomiyama, A.J., Kawecki, N.S., Rosenfeld, D.L. et al. (2020). Bridging the gap between the science of cultured meat and public perceptions. *Trends in Food Science & Technology* **104**: 144–152.

Trenfield, S.J., Madla, C.M., Basit, A.W., and Gaisford, S. (2018). Binder jet printing in pharmaceutical manufacturing. In: *3D Printing of Pharmaceuticals*, AAPS Advances in the Pharmaceutical Sciences Series, vol. **31** (eds. A.W. Basit and S. Gaisford), 41–54. Springer https://doi.org/10.1007/978-3-319-90755-0_3.

Uribe-Wandurraga, Z.N., Zhang, L., Noort, M.W.J. et al. (2020). Printability and physicochemical properties of microalgae-enriched 3D-printed snacks. *Food and Bioprocess Technology* **13** (11): 2029–2042.

Van der Linden, D. (2015). 3D food printing. TNO. https://www.tno.nl/media/2217/3d_food_printing.pdf (accessed 19 October 2019).

Van Wijk, A.J.M. and Van Wijk, I. (2015). *3D Printing with Biomaterials: Towards a Sustainable and Circular Economy*. IOS press.

Vithani, K., Goyanes, A., Jannin, V. et al. (2019). An overview of 3D printing technologies for soft materials and potential opportunities for lipid-based drug delivery systems. *Pharmaceutical Research* **36** (1): 4.

Von, H.K.W., Von, H.E.M., Williams, D.X., et al. (2015a). Method for producing a three-dimensional food product. Google Patents.

Von, H.K.W., Von, H.E. M., Williams, D.X., et al. (2015b). Procede de production d'un produit alimentaire en trois dimensions. Google Patents.

Wang, L., Zhang, M., Bhandari, B., and Yang, C. (2018). Investigation on fish surimi gel as promising food material for 3D printing. *Journal of Food Engineering* **220**: 101–108. https://doi.org/10.1016/j.jfoodeng.2017.02.029.

Wegrzyn, T.F., Golding, M., and Archer, R.H. (2012). Food layered manufacture: a new process for constructing solid foods. *Trends in Food Science & Technology* **27** (2): 66–72.

Willcocks, N.A., Shastry, A., Collins, T.M. et al. (2011). High resolution ink-jet printing on edibles and products made. Google Patents.

Willcocks, N.A., Shastry, A.V., Collins, T. M. et al. (2016). Edibles comprising a high resolution image. Google Patents.

Wilson, A., Anukiruthika, T., Moses, J.A., and Anandharamakrishnan, C. (2020). Customized shapes for chicken meat–based products: feasibility study on 3d-printed nuggets. *Food and Bioprocess Technology* **13**: 1–16.

Xaar (2018). A guide to industrial inkjet. https://www.xaar.com/media/1830/xaar-inkjet-guide-eng.pdf (accessed 23 December 2020).

Xu, X., Awad, A., Martinez, P.R. et al. (2020). Vat photopolymerization 3D printing for advanced drug delivery and medical device applications. *Journal of Controlled Release* **329**: 743–757. https://doi.org/10.1016/j.jconrel.2020.10.008.

Yang, F., Zhang, M., and Bhandari, B. (2017). Recent development in 3D food printing. *Critical Reviews in Food Science and Nutrition* **57** (14): 3145–3153.

Yang, F., Zhang, M., Fang, Z., and Liu, Y. (2019a). Impact of processing parameters and post-treatment on the shape accuracy of 3D-printed baking dough. *International Journal of Food Science and Technology* **54** (1): 68–74. https://doi.org/10.1111/ijfs.13904.

Yang, F., Guo, C., Zhang, M. et al. (2019b). Improving 3D printing process of lemon juice gel based on fluid flow numerical simulation. *LWT* **102**: 89–99.

Ying, G.-L., Jiang, N., Maharjan, S. et al. (2018). Aqueous two-phase emulsion bioink-enabled 3D bioprinting of porous hydrogels. *Advanced Materials* **30** (50): 1805460.

Zeng, X., Chen, H., Chen, L., and Zheng, B. (2020). Insights into the relationship between structure and rheological properties of starch gels in hot-extrusion 3D printing. *Food Chemistry* **342**: 128362.

Ziaee, M. and Crane, N.B. (2019). Binder jetting: a review of process, materials, and methods. *Additive Manufacturing* **28**: 781–801.

Zoran, A. and Coelho, M. (2011). Cornucopia: the concept of digital gastronomy. *Leonardo* **44** (5): 425–431.

3

Food Components and Their Role in Printability

CHAPTER MENU

3.1 Recipes in 'Print and Eat Technology'

Food is an essential component of life comprised of various macro and micronutrients. The complex nature of food makes digestion to be pretty complicated. The composition of a diet plays a major role in fulfilling the daily needs and energy requirements. Macronutrients include carbohydrates, proteins, fat, and dietary fibre. On the other hand, vitamins and minerals are grouped as micronutrients. Nowadays the dietary pattern of food is kept on changing with busy and hectic lifestyle (Phan and Stodolska 2019). Formulation of a convenient diet without compromising nutrition is the need of the hour. As a printandeat technique, 3D food printing promises to be a futuristic technology that addresses the nutritional needs and the dietary requirements of the modern world. The science of printing is quite interesting as food printing allows consumers to design their own recipe. Food printing is an iteration of novelty and creativity that mingle the digitalized intricate designing with traditional cuisines (Anukiruthika et al. 2020). The success of a food recipe depends on consumer perception, characteristics of material supply and acceptability of end-product. Printing of food allows for flexibility and more degrees of freedom to test inexperienced food material supplies like alternative/ uncommon food ingredients (Nachal et al. 2019). Researchers have classified printing recipes as element and traditional recipes. Element-based recipes were based on a pre-set of standardized formulation and material

3D Printing of Foods, First Edition. C. Anandharamakrishnan, Jeyan A. Moses, and T. Anukiruthika.

dispensing that determines the sensory attributes of food. Experiments were conducted by Massachusetts Institute of Technology (MIT) researchers on element-based printing recipes using Virtuoso Mixer™, Robotic Chef™, and Digital Fabricator™ printers. On the other hand, traditional recipes aimed at developing a customized food by modifying the existing printing recipes. Examples of this category include the fabrication of customized chocolates and sugar sculptures (Sun et al. 2015). Mainly traditional printing helps in the formulation and design of new recipe based on consumers preference. A combination of element and traditional based recipes would open up for exemplar dishes out of 3D printing.

3.2 Role of Food Constituents

Consumer's preference towards new recipes is decided by five key factors that include price, convenience, palatability, experience, and nutrition. An American survey on food habits revealed that 47% of consumers considered themselves to be health-conscious while the other 35% of people described themselves as ingredient sensitive (Sun et al. 2018). Healthy dietary concepts are the trending topics of the day that turned the attention towards personal foods and health care products. A healthy diet consists of a proportional amount of macro and micro food constituents that are essential for the normal metabolism of the body. Each of the food constituents is equally important and has its functionality (Sammugam and Pasupuleti 2019). Most of the energy is obtained from dense food constituents such as carbohydrates and lipids. Carbohydrates form a major group of energy-giving macronutrients that comprises sugars, starches and cellulose. Cereal grains, starchy vegetables, roots and tubers, and fruits are rich in carbohydrates. Another dense food constituent is lipids that comprise of fatty acids and triglycerides which are present in nuts, oilseeds, dairy products, and flesh foods. Proteins are a group of amino acids that forms a vital part of regular diet obtained from pulses and legumes, red meat, and flesh products. Other than macro elements, fruits, and vegetables also possess an ample quantity of essential vitamins and minerals that are required for revitalizing the body. Any discrepancy in these constituents will lead to an imbalance of nutrients causing a nutritional deficiency. This scenario increases the incidence of physiological disorders like obesity, cardio diseases, and diabetes (Yahia et al. 2019). A properly balanced diet comprises of all the food constituents in an adequate amounts. With changing climatic conditions and environment, the requirement of nutrients varies from person to person which could not be met by regular food pattern and diet. Food printing is a novel approach for addressing the varying dietary patterns and needs, it aids in mass customization and personalization of foods. Apart from nutritional benefits, each of these constituents significantly contributes to food structures and textures that impart palatability and mouthfeel. Thus, 3D food printing remains to be a single solution for multiple constraints.

3.3 Panorama of Food Printing

As a digitally controlled robotic process, food 3D printing dealt with the sustainable approaches in food manufacturing that would transform the traditional food supply chain.

It allows for customized designing of foods, development of prototype models for educating the pre-school children, personalization of diets by imparting new textures, and taste suitable for people of different age groups (Godoi et al. 2019; Perez et al. 2019). Conventional food production involves the development of limited products with internal designs that restricts the cost. While 3D printing of foods helps in the mass production of designed foods that integrates the artistic skills of culinary professionals without a hike in the cost of production. Being a rapid prototyping tool, food printing paves a way for delivering nutrients in fascinating shapes. In contrast to conventional processing techniques, food printing requires a considerable amount of pre-processing in order to tailor its physicochemical properties thereby making it suitable for layer-by-layer fabrication. Studies were reported on the delivery of fruits and vegetables in attractive shapes thus making the kids consume nutritious food (Severini et al. 2018). It aids in changing the consumer's mindset that transforms their dislikeness towards a product into likeness. Further food printing helps in cutting down the calorie intake with less fat, less sugar, and less salt and leads to the fabrication of customized patterned food in layers (Prakash et al. 2019). Microstructural designing of food is also possible with the development of printing technology as 3D printing could be used for modifying the fill density of a product that gives a distinct mouthfeel. This micro-scale printing also helps to incorporate food bioactive compounds into the 3D printed products. Further 3D printing helps to deliver texture modified foods for people with swallowing disorders. All these potential advantages of food printing would lead to the development of a convenient food product that holds a huge market value. Post-processing of printed foods will add upon taste and contributes to structural integrity to retain its shape. Thus, it is essential to understand the role of each constituent in printability and the response of 3D structures when subjecting to post-processing treatments. The structural chemistry of food constituents towards 3D printing is discussed in the subsequent sections.

3.4 Insights on the Printability of Different Food Constituents

3.4.1 Carbohydrates and Starch

Carbohydrates are chemical organic compounds comprised of simple sugars and polysaccharides (starch and non-starch). The latter group of carbohydrates takes more time to get digested than simple sugars and are referred to as dietary carbohydrates. The concept of well-being and healthy food provokes the fascination towards fibre-rich diet possessing a low glycaemic response (Amicucci et al. 2019). The digestible polysaccharide, starch comprises of linear amylose and non-linear amylopectin fractions. These are homopolymers made of glucosidic bonds forming the backbone of starch molecules. Non-starch polysaccharides are grouped as cellulosic and non-cellulosic (pectin and hemicellulose) that forms the major part of the plant cells. The polymeric network of polysaccharides and their physicochemical characteristics makes them more suitable for food printing. Most of the 3D printing works reported in the literature were carried out with starch-based food materials in assessing its behaviour towards printability. On comparing with the other food constituents, carbohydrate-based printing material supplies possess more tendency of gelation that

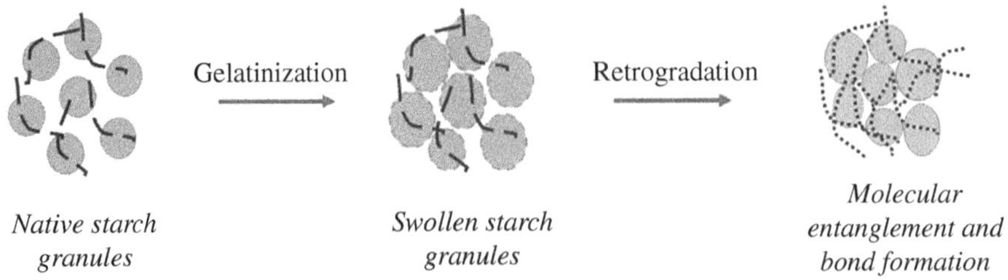

Figure 3.1 Schematic representation of gelation mechanism of starch granules.

tailors the consistency and material flowability and hence the layered deposition of 3D constructs (Figure 3.1).

Researchers are quite interested in exploring the printability of common daily foods such as cereals and millets, legumes and pulses, fruits and vegetables, dairy products and meat products. Rice a staple food of South Asia is widely consumed and forms a major part of the regular diet (Ramadoss et al. 2019). Huang et al. (2019) have reported a study on the assessment of printability of brown rice and evaluated its effect on end-product quality. In this study, ready-to-cook (RTC) brown rice flour was prepared and used for printing trials. Material supply was pre-gelatinised in order to enhance the chemical integrity of starch molecules. Granule size, amylose, and amylopectin content of flour determined its pasting behaviour (Huang et al. 2019). This could be correlated with the swelling power and gelling behaviour of starch molecules. The formation of hydrogen bonds and corresponding molecular entanglement had a significant impact on textural properties such as the hardness and gumminess of the 3D printed sample.

In another study, mashed potato along with potato starch (0, 1, 2, and 4%, w/w) was used for 3D printing. Raw potatoes were steam cooked and ground to form smooth paste making them suitable for extrusion-based printing. Application of heat causes the molecules to swell due to which the amorphous nature of the starch molecules was increased with a decrease in their crystallinity. Upon subsequent cooling, the starch mixture would reverse the process to regain its crystalline nature that leads to the formation of a gel. The strong association of hydrogen bonds between water and starch forms a dense network imparting a smooth texture for material supply making it easy for extrudability and hence 3D printing of starch-based food mixtures (Liu et al. 2018).

Southerland et al. (2011) reported a study on printing a mixture of starch, sugar, and mashed potato as dry fractions using a Z Corporation 3D printer to fabricate teeth like prototypes based on sugar crystals. These type of 3D printers employs a heat source for binding and fusion of materials based on sintering process thereby forming a 3D edible construct. Sintering is a solid-state process involving the melting of solid particles to their melting point that leads to the formation of grain boundaries between grain-void interfaces (Figure 3.2). In the case of sugar crystals, the molecular voids open out forming the fusion of particles together with the application of heat and slight pressure. 3D printing of sugars greatly depends on properties like crystallinity, melting point, source and extent of heating, glass transition temperature (T_g), compressibility, solubility, and material density. These parameters greatly influence the porosity of fabricated structures and hence the texture

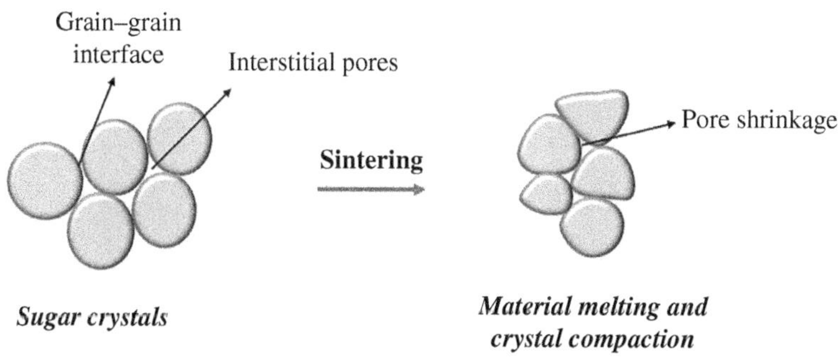

Figure 3.2 Schematic representation of sintering process.

of 3D printed candies could be controlled. Similarly, 3D systems had fabricated sugar cubes from milk chocolates, sweet, and sour candies (Sun et al. 2015).

3.4.2 Proteins and Amino Acids

Proteins are a complex group of biopolymers that consists of long chains of linear and branched amino acids linked together by peptide bonds. Chemical break down of proteins could result in a shorter chain amino acid that forms a precursor to nucleic acid, antigen, antibodies, hormones, and co-enzymes. Proteins play a vital role in imparting structural integrity to cell walls and stiffness to tissues. Other functional role includes regulation of physiological activities, maintenance of normal pH, acts as chemical messenger, and storage pool (Hoffman 2019). Legumes and pulses are a good source of proteins from a plant source. Proteins from the animal source include eggs, red meat, fish, and dairy products such as cheese and whey powders. Based on the amino acid profile, proteins from animal sources are considered as complete proteins while plant proteins are considered incomplete due to lack of one or more essential amino acids. Various factors that affect the functionality and availability of proteins includes heat, pH (acid/ base), enzymatic hydrolysis that leads to denaturation, aggregation, and cross-linking of proteins (Lassé et al. 2015). These properties in turn affect the printability of native proteins. Proteins being an important macronutrient for bodybuilding, several kinds of research were being conducted focussing on the effect of physicochemical properties on the printability of protein-based food systems. When subjected to external stress such as heat, acidic/ basic conditions, and mechanical agitation proteins undergo denaturation. The production process of various food products such as cheese, tofu, minced meat, etc., involves simultaneous denaturation and coagulation of protein molecules that enhance the textural properties of the final product (Xiong 2018). Protein denaturation refers to the breakdown of secondary and tertiary structures that has a significant effect on modifying the texture of food products and hence mouth feel.

Modification of textures of food being one of the potential applications of food 3D printing, the printability of the processed cheese was examined by Le Tohic et al. (2018) in which the effect of the printing process on structural properties of hot-melt cheese was reported (Figure 3.3). Different cheese samples (untreated cheese, melted cheese, and printed cheese) were analysed for their hardness. Results showed that printed cheese

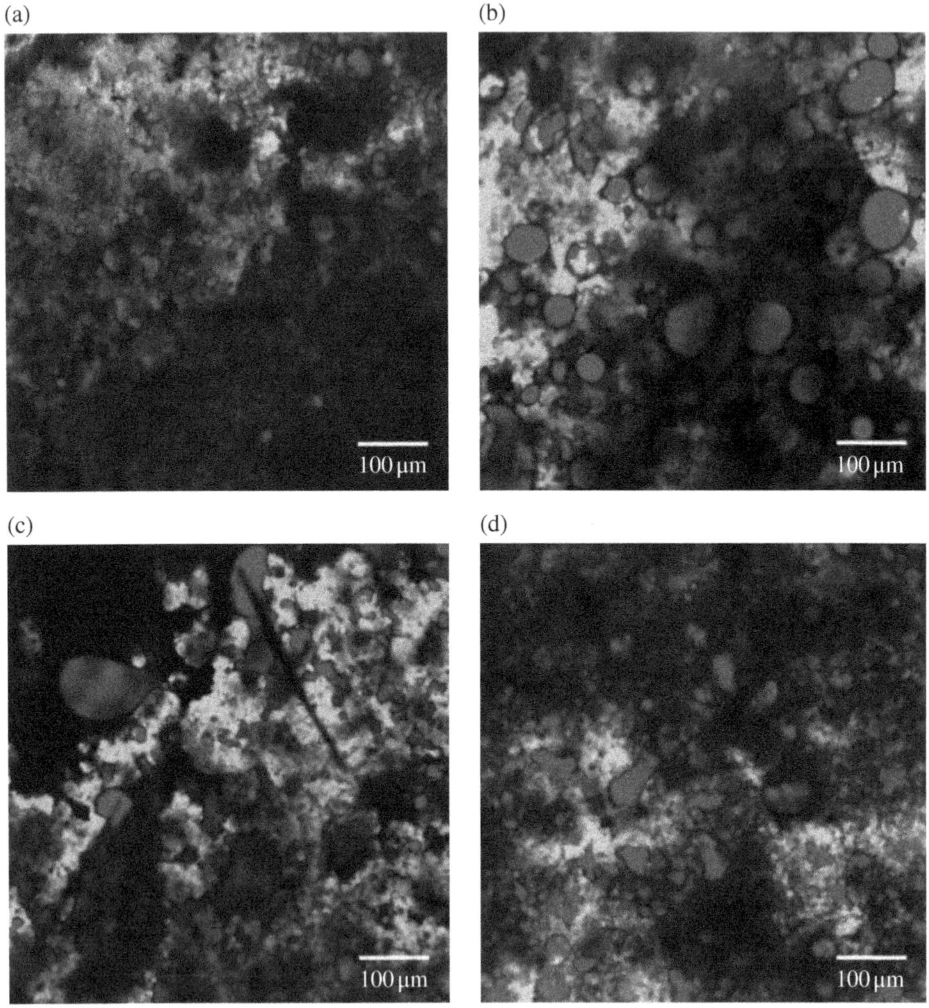

Figure 3.3 Confocal micrographs of cheese sample showing distribution of fat globules (pigmented spots) in a protein matrix (a) untreated cheese, (b) melted cheese, (c) low speed printed cheese, (d) high speed printed cheese. *Source:* From Le Tohic et al. (2018), Figure 05 [p. 61] / With permission of Elsevier. DOI-https://doi.org/10.1016/j.jfoodeng.2017.02.003.

was found to be less hard than other samples. The lesser hardness of the printed cheese was attributed to the material's meltability, and the associated shear stress exhibited during extrudability. The weaker bonds of casein molecules of the printed cheese were responsible for its soft texture and ease of flowability. Combined effects of material shearing during printing and solidification of which during the after-printing process resulted in the coalescence of fat globules. The resulted food system with disrupted fat globules embedded in the protein matrix imparts the tendency of melt-in-mouth texture to the 3D printed cheese. Proteins alone are not conducive for 3D printing, proteins along with other food constituents either with lipids or starch aids in printability as they act as a plasticizing agent. This could be due to the synergistic effects of proteins in combination with other macronutrients that increase crosslinking and hence improve the strength of the gel matrix. These characteristics greatly influence materials viscosity and stability of printed structures.

In another work, Dong et al. (2019) reported a study on the printability of surimi gel with sweet potato starch as a potential structural enhancer for 3D printing. Conventional processing of multi-step surimi preparation deteriorates the gel quality that may not possess enough strength for 3D constructs. Hence, sweet potato starch (0, 2, 6, 8, and 10%, w/w) was used for enhancing the rheological properties of surimi gel. Results showed that surimi gel with 8% sweet potato starch possesses good gel strength with a softer texture. This was due to the cross-linking of myofibrillar fish proteins with sweet potato starch that forms a uniform aggregate structure making it suitable for 3D printing.

Another protein-rich formula was made from soy protein isolate (SPI) by utilizing the functional properties of co-blending of gelatin with sodium alginate that imparts a stable structure to the 3D constructs made from SPI (Chen et al. 2019). The physical interactions of the peptide bonds of gelatin along with sodium salts of alginic acid at its melting temperature could form a mesh-like network that provides strength to the SPI printing mixture. However, results showed that the addition of these co-blended systems to SPI does not cause chemical cross-linking among the protein subunits, but it had a significant effect on improving the textural properties of 3D printed geometries. An increase in the concentration of gelatin had a notable effect on the material's flowability that in turn resulted in improved hardness and chewiness of the SPI gels. Thus, SPI with 2, 6, and 10% (w/v) of gelatin was found to be printable with better resolution (Chen et al. 2019). This study reveals the potential utilization of hydrogels in protein-based food systems for 3D printing applications.

3.4.3 Lipids and Fatty Acids

Another major food constituent that regulates and assist in printability is lipids. Lipids are organic compounds either in form of fat or oil that are essential for storing energy in the human body. It consists of fatty acids, triglycerides (an ester derived glycerol with three fatty acids), and phospholipids. Based on the presence of double bonds in the carbon chain, fatty acids are classified as saturated and unsaturated fatty acids. Dietary fat from plant-based sources includes oil seeds, nuts, and fruits (olive, palm, and avocado). While animal-based sources such as meat, fish (salmon and mackerel), eggs, and dairy products like butter and margarine also possess a considerable amount of fat. Various factors that affect the functionality of dietary fat include solidification/ meltability, crystal structure/ polymorphism, globule size, esterification, level of hydrogenation, fatty acid composition, and its distribution within triacylglycerol (Devi and Khatkar 2016). Any changes in these properties have a significant effect on the printability of material supply.

An illustration for characterizing the printability of lipid-based material is chocolates which consist of a mixture of cocoa solids, milk fat possessing several triglycerides. Based on thermostability, chocolates are categorized into six different forms. The highly polymorphic nature of the fat in the chocolate was responsible for its varied physiochemical properties. Among all the six forms of triglycerides, the Vth form seems to be stable possessing β_2 crystal with a melting point of around 35 °C. These forms are commonly used in food industries and are referred to as tempered chocolates (Lanaro et al. 2019). They possess high-quality colour and gloss with desired hardness and melting point suitable for conching and tempering. The crystallinity of fat plays an important role in flowability and stability which can be correlated with its printability. In a study conducted by Mantihal et al. (2017), the extrudability of the chocolate was linked with the melting temperature of the fat crystals (29–32 °C). For the printing of chocolates, magnesium stearate was added as an emulsifier in order to maintain stable nucleation of β crystals as 3D printing eliminates tempering which is an essential step to achieve the desired rheology of chocolates. Stearic acid, a saturated fat act as a lubricant that imparts anti-sticking characteristics and aids in regulating the flow of chocolate mixture through the printing nozzle. Results showed that the addition of magnesium stearate has no impact on the thermal properties but in turn, it delays the crystallization thereby enhancing post-deposition of the chocolates after printing, which is an essential criterion in determining the end quality (solidification, mechanical strength, and appearance) of printed 3D constructs from chocolates (Mantihal et al. 2017).

Researchers were working on tailoring the nutritional profile of foods using 3D food printing. In context with this, a study was conducted in analysing the effect of fat from varied animal sources and analyzed the integration of which for the printing of fibrous beef (Dick et al. 2019). Natively presented undesirable saturated fat in beef was removed and a desired fatty acid profile composed of polyunsaturated fatty acid from the pig (lard) possessing 39% of saturated non-hydrogenated fat was incorporated as a separate layer into the beef using 3D printing. This study evaluates the effect and amount of lard fat as interlayers on the printability of beef (Figure 3.4). Results showed that lipids were responsible for the smooth flow of the fibrous meat paste through the printing nozzle. However, the fatty layers sandwiched between the meat layers resulted in coalescence due to the instability of fat globules when subjected to heat treatment. Further, textural properties of the printed beef were greatly influenced by the presence of lard fat at interlayers, as the fat content proportionally decreases the hardness and chewiness of the cooked samples. This study describes the significance of fat in assisting the printing of meat fibrils and its effect on end-product quality (Dick et al. 2019; Wood et al. 2008).

In another study, the effect of the addition of oil on flexibility and flowability of material supply was investigated (Liu et al. 2019). A composite dough was prepared from wheat flour, mango powder, olive oil along with water and the change in viscosity of dough was studied in correlation with total solids content. The addition of 2% (w/w) of olive oil into the material supply enhanced the 3D printing performance of dough that resulted in a better surface finish of the 3D printed samples. This was due to the presence of the higher amount of unsaturated fatty acids of olive oil that imparts lubrication and plasticizing behaviour of the material supply. In general, unsaturated fats have more spreading power than saturated ones, thus the addition of olive oil had a significant impact on the final baking quality of 3D printed samples such as tenderness, mouthfeel, appearance, and colour of the product (Liu et al. 2019). Fats and oils are considered to be the most prominent ingredients in tuning the physiochemical attributes of the material supply towards 3D printing. Along with other food constituents, lipids impart a creamy texture to the printing mixture thereby enhances dough formation. This was due to the presence of β′ crystal of the fatty acids that remains to be more stable than α and β forms. Although the portion of lipids was taken in smaller proportions in the material supply, these are the key component in tailoring the printability of the whole printing mixture.

Figure 3.4 Multi-material 3D printed meat (a) Raw and cooked samples, (b) 3D printed meat paste and lard interlayers. *Source:* From Dick et al. (2019), Figure 03 [p. 13] / With permission of Elsevier. DOI-https://doi.org/10.1016/j.meatsci.2019.02.024.

3.4.4 Dietary Fibre

Dietary fibres are a group of indigestible roughage portion of a meal comprised of carbohydrate components that are not hydrolysed by the intestinal enzymes. This class of macromolecules possess a vital functionality in regulating the normal physiology of the human body (Zinina et al. 2019). A considerable portion of dietary fibres in a regular diet helps in maintaining gut health with reduced incidence of diseases related to sedentary workstyle such as cardiovascular disease (CVD), diabetes, and obesity. Based on solubility, these portion of indigestible carbohydrate polymers is categorised as soluble and insoluble. The former class of soluble dietary fibre (SDF) includes pectin, gum, and mucilage while the latter group of insoluble dietary fibre (IDF) includes cellulose, hemicellulose, and lignin (Dai and Chau 2017). With increasing awareness, the dietary concepts are changing from refined food towards whole food for accomplishing the potential benefits of dietary fibres. Sources of fibre-rich foods include cereal grains (wheat, oats, barley, psyllium husk), pulses (soy, peas and beans) and legumes, fruits, and vegetables (Mehta et al. 2019). In this regard, printing foods that are rich in fibres are quite interesting as other than providing the nutritional benefits it also possesses organoleptic value as it imparts texture and flavour to the foods and remains to be an active ingredient in personalizing the diet. In this regard, a study on the printability of fibre-rich foods is quite adequate as it provides a way for the customization of foods and hence personalized nutrition.

Only a few studies were reported on revealing the complexity involved in the 3D printing of fibres. Lille et al. (2018) conducted a study on correlating the material supply for the development of novel 3D structures with a food formula rich in fibre and protein with reduced-fat and sugars. The ingredients of the printing mixture composed of starch, nanocellulose fibres and milk powder. These ingredients were tested individually as well as in combination for obtaining stable 3D structures after printing. Considering the printability of fibres, the gel matrix was reinforced with cellulose nanofibers (CNF). Printing of fibres showed that there was a discontinuous material flow of the CNF gel system that seems to block the printing nozzle due to its larger fibre particles, causing phase separation and material flocculation (Lille et al. 2018). This study provides a comparative assessment of ingredients and showed that 3D printing of fibres was quite difficult than that of starch and protein-based food systems. Further, the study highlights a key challenge associated with the printing of fibre-rich food systems providing an overall idea of the applicability of diverse food materials for 3D printing.

In another study, a fibrous cellular material freeze-dried spinach was characterized for microstructural composition and the suitability of which as food ink was assessed for 3D printing (Lee et al. 2019). The effect of particle size on printing performance was investigated and the addition of 10% (w/w) xanthan gum showed a promising result in retarding the swelling effects of spinach powder by increasing its water holding capacity. Here, printability was demonstrated by the shear modulus and the extrudability. Results showed that the mechanical stability of the printed structures was improved with the higher particle size of spinach powder (Figure 3.5). Further, this study emphasized particle size as an important variable in assessing the printability of the food ink. However, the study was not focused on revealing the science behind the printing of fibrous material. In another study, researchers have made successful attempts in the printing of fibre and protein-rich indigenous composite flour (Krishnaraj et al. 2019). Here, natively printable composite flour composed of barnyard millet, green gram, fried gram, and ajwain seeds were subjected for 3D printing to develop a nutritious snack for personalized nutrition. The formulated composite flour contains a concoction of macronutrients with 11% and 18% of fibre and protein, respectively. The physicochemical composition of the indigenous native flour was responsible for its printability. The presence of a small fraction of lipid in the millets was adequate for aiding the smooth flow of printing material supply as it results in the plasticizing effect that eases the material flowability through the printing nozzle (Krishnaraj et al. 2019). Further, the ingredient composition of the formulated material supply well assists in overcoming other undesirable effects of agglomeration and flocculation of fibre during printing.

Except for handful of works, there is only less information available on printing characteristics of fibre-rich material supply. With the available literature, it was evident that the printability of the dense nature of fibrous material could be correlated with physical attributes such as particle size, surface corrugation, and material density (Nachal et al. 2019).

Figure 3.5 Effect of particle size on 3D printing of fibrous spinach powder (a) 307 μm, (b) 259 μm, (c) 172 μm, (d) 50 μm. *Source:* From Lee et al. (2019), Figure 03 [p. 05] / With permission of Elsevier. DOI-https://doi.org/10.1016/j.jfoodeng.2019.03.014.

Certain pre-processing was adequate for fibre-rich material in order to make it suitable for 3D printing applications. This includes size reduction and particle breakdown with the regulation of total solids content that in turn determines the printing quality of fibre-rich material supply. Still, the biophysics underlying the printability of fibres has not yet been well explored. More studies must be taken forward for exploring the inherent potential benefits of 3D printing of fibre-rich material supplies.

3.4.5 Other Additives

Another diverse class of polysaccharides that are widely employed in the food industry are the hydrocolloids which form a group of long-chain hydrophilic polymers (Li and Nie 2016). The presence of one or more hydroxyl groups rendered them water-loving thereby forming

Table 3.1 Classification of food hydrocolloids.

Origin	Examples of hydrocolloids
Plant	Gum arabic, basil seed gum, gum karaya, konjac, locust bean gum, flaxseed gum, guar gum, and starches
Animal	Chitin, chitosan, and gelatin
Seaweeds	Agar, alginate, xylan, carrageenan, and ulvan
Microbial	Gellan gum, tara gun, xanthan gum, dextran, curdlan, and pullulan
Synthetic	Carboxymethyl cellulose, methylcellulose, hydroxypropyl methylcellulose, and other cellulose derivatives

a colloidal suspension when added into the food mixture. Hydrocolloids are derived either from plants/ seaweeds (xanthan gum, carrageenan, gum arabic, locust bean gum, guar gum, gellan gum, glucomannan, and pectin) or animal sources (gelatin, chitin, and chitosan) (Mehta et al. 2019) (Table 3.1). They exhibit a varied range of functional property as they can act as thickeners, stabilizers, foaming agents, emulsifiers, and gelling agents that help in tailoring the rheology of food systems. Concerning 3D printing, hydrocolloids are commonly used as food additives for making the material supply attain a conducive consistency suitable for 3D printing by altering its flow and viscosity. Further many studies were reported on food hydrocolloids as reference material for 3D food printing. Kim et al. (2018) conducted a study on a comparative assessment of various hydrocolloids such as guar gum, gellan gum, xanthan gum, locust bean gum, hydroxypropyl methylcellulose (HPMC), methylcellulose (MC), and gelatin for their applicability in food printing. The study was performed for framing a classification system based on printability assessment. Among all the hydrocolloids considered for the study, MC was found to be capable of simulating the handling properties and deformation behaviour of foods and possess good mechanical stability which was comparable to most of the food systems with diverse viscosity. Thus, a classification system was made based on dimensional and storage stability as grade A to D with grade A more suitable for 3D printing that requires less extrudable force as with cheese. On the other hand, the latter category of grade D includes materials like cookie dough which requires more force to extrude out of the printing nozzle (Kim et al. 2018).

In another study, Vancauwenberghe et al. (2017) developed a pectin-based food stimulant for 3D printing applications. The study correlates the effect of the degree of methoxylation of pectin with a concentration of Ca^{2+} ions as a cross-linking agent that aids in printability. Results showed that the formulation of food ink was greatly influenced by the density of the polymeric network and its degree of cross-linking with Ca^{2+} ions. The addition of sugar to the food stimulant affects the viscoelastic behaviour and hence could be positively correlated with printability. As the incorporation of sugar dehydrates the polymeric matrix thereby enhancing the gelation mechanism with the formation of hydrogen bonds that would result in higher young's modulus with firmer pectin-based gel (Vancauwenberghe et al. 2017). Further, the study demonstrates the production of porous aerated 3D structure with the use of bovine serum albumin (BSA) and remains to be a base for future works for the development of food stimulants with added flavours and textures for food 3D printing.

Basic mechanisms responsible for the gelation of hydrogels are ionotropic cross-linking, chemical cross-linking and coacervate formation. Among which ionotropic cross-linking is the most commonly employed technique for hydrogel formation. The physiochemical characteristics of the hydrocolloids were greatly affected by several factors such as temperature, melting and solidification, ionic strength, gelation behaviour, and cross-linking (Godoi et al. 2016). Hydrocolloids in the form of hydrogels are gaining attention in recent days as it possesses the centric value as that of dietary fibres. A study was reported on the satiety effects of hydrocolloids, as it could result in slowing down of enzymatic actions or could delay gastric emptying (Li and Nie 2016). As an important application, hydrocolloids are utilized for increasing the fibre content of the diet. Along with other major food constituents, hydrocolloids play a significant role in modifying the sensorial aspects of food and hence found to possess a key role in regulating the dietary aspects of food for providing a healthy 3D printed customized diet.

3.5 Classification of Foods Based on Their Printability

The nature of the material supply is an essential criterion in determining the final quality of the 3D printed edible constructs. Printability refers to the inherent ability of the material to withstand its own weight upon deposition over the printing platform (Kim et al. 2018). A material to possess good printability must have enough dimensional and mechanical stability to retain its shape after printing. Based on the nature and printability of the material supply, the ingredients for 3D printing can be categorized as natively printable, non-printable traditional materials and alternative ingredients (Sun et al. 2018). The first class of natively printable materials include chocolates, cheese, cake frostings, hydrogels, etc., which are innately printable on their own without the addition of additives. This class of material supply requires less processing with ease of handling during printing. A second class of traditional materials includes staple cereals, fruits and vegetables, eggs and flesh foods that are common in our day-to-day life but are non-printable on its own. This kind of material requires certain pre-processing such as the addition of food additives or mixing a proportionate amount of natively printable materials for converting them into a printable form. The most common application is the use of food hydrocolloids as additives for making non-printable materials printable. The third category includes alternate ingredients like insects, algae, fungi, and lupin seeds which are quite a good source of nutrients but are uncommon foods (Sun et al. 2015). They are widely used as a novel approach for the development of sustainable food to tackle the ever-growing food demand. A detailed description of the classification of materials based on printability was discussed in the subsequent chapters.

3.6 Conclusion

Food 3D printing implies a novel way of customizing foods based on individual preferences. The final acceptability of any food recipe was based on the ingredient's choice that makes a wholesome food. The role of various food constituents in terms of printing was well discussed in the present chapter. The chemical and structural constituents responsible for the printability of macronutrients provides insights for the development of 3D printed functional foods based on age group and dietary pattern. The study of the individual food constituent and its response towards printability helps to combine the ingredients that pave for the exploration of new dishes by utilizing the synergistic benefits of each constituent. Thus, apart from the nutritional benefits, each of the macro elements of food has its role in imparting printability and acts as a key in modifying the sensorial attributes of food.

References

Amicucci, M.J., Nandita, E., and Lebrilla, C.B. (2019). Function without structures: the need for in-depth analysis of dietary carbohydrates. *Journal of Agricultural and Food Chemistry* **67** (16): 4418–4424.

Anukiruthika, T., Moses, J.A., and Anandharamakrishnan, C. (2020). 3D printing of egg yolk and white with rice flour blends. *Journal of Food Engineering* **265**: 109691. https://doi.org/10.1016/j.jfoodeng.2019.109691.

Chen, J., Mu, T., Goffin, D. et al. (2019). Application of soy protein isolate and hydrocolloids based mixtures as promising food material in 3D food printing. *Journal of Food Engineering* **261**: 76–86. https://doi.org/10.1016/j.jfoodeng.2019.03.016.

Dai, F.-J. and Chau, C.-F. (2017). Classification and regulatory perspectives of dietary fiber. *Journal of Food and Drug Analysis* **25** (1): 37–42.

Devi, A. and Khatkar, B.S. (2016). Physicochemical, rheological and functional properties of fats and oils in relation to cookie quality: a review. *Journal of Food Science and Technology* **53** (10): 3633–3641.

Dick, A., Bhandari, B., and Prakash, S. (2019). Post-processing feasibility of composite-layer 3D printed beef. *Meat Science* **153**: 9–18. https://doi.org/10.1016/j.meatsci.2019.02.024.

Dong, X., Huang, Y., Pan, Y. et al. (2019). Investigation of sweet potato starch as structural enhancer for 3D printing of Scomberomorus niphonius surimi. *Journal of Texture Studies* **50**: 316–324. https://doi.org/10.1111/jtxs.12398.

Godoi, F.C., Prakash, S., and Bhandari, B.R. (2016). 3d printing technologies applied for food design: status and prospects. *Journal of Food Engineering* **179**: 44–54.

Godoi, F.C., Bhandari, B.R., Prakash, S., and Zhang, M. (2019). An introduction to the principles of 3D food printing. In: *Fundamentals of 3D Food Printing and Applications* (eds. F.C. Godoi, B.R. Bhandari, S. Prakash and M. Zhang), pp. 1–18. Elsevier.

Hoffman, J.R. (2019). *Dietary Supplementation in Sport and Exercise: Evidence, Safety and Ergogenic Benefits*. Routledge.

Huang, M., Zhang, M., and Bhandari, B. (2019). Assessing the 3D printing precision and texture properties of brown rice induced by infill levels and printing variables. *Food and Bioprocess Technology* **12**: 1–12. https://doi.org/10.1007/s11947-019-02287-x.

Kim, H.W., Bae, H., and Park, H.J. (2018). Classification of the printability of selected food for 3D printing: development of an assessment method using hydrocolloids as reference material. *Journal of Food Engineering* **215**: 23–32.

Krishnaraj, P., Anukiruthika, T., Choudhary, P. et al. (2019). 3D extrusion printing and post-processing of fibre-rich snack from indigenous composite flour. *Food and Bioprocess Technology* **12** (10): 1776–1786. https://doi.org/10.1007/s11947-019-02336-5.

Lanaro, M., Desselle, M.R., and Woodruff, M.A. (2019). 3D printing chocolate: properties of formulations for extrusion, sintering, binding and ink jetting. In: *Fundamentals of 3D Food Printing and Applications* (eds. F.C. Godoi, B.R. Bhandari, S. Prakash and M. Zhang), pp. 151–173. United Kingdom: Academic Press.

Lassé, M., Deb-Choudhury, S., Haines, S. et al. (2015). The impact of pH, salt concentration and heat on digestibility and amino acid modification in egg white protein. *Journal of Food Composition and Analysis* **38**: 42–48.

Le Tohic, C., O'Sullivan, J.J., Drapala, K.P. et al. (2018). Effect of 3D printing on the structure and textural properties of processed cheese. *Journal of Food Engineering* **220**: 56–64.

Lee, J.H., Won, D.J., Kim, H.W., and Park, H.J. (2019). Effect of particle size on 3D printing performance of the food-ink system with cellular food materials. *Journal of Food Engineering* **256**: 1–8. https://doi.org/10.1016/j.jfoodeng.2019.03.014.

Li, J.-M. and Nie, S.-P. (2016). The functional and nutritional aspects of hydrocolloids in foods. *Food Hydrocolloids* **53**: 46–61.

Lille, M., Nurmela, A., Nordlund, E. et al. (2018). Applicability of protein and fiber-rich food materials in extrusion-based 3D printing. *Journal of Food Engineering* **220**: 20–27.

Liu, Z., Zhang, M., Bhandari, B., and Yang, C. (2018). Impact of rheological properties of mashed potatoes on 3D printing. *Journal of Food Engineering* **220**: 76–82. https://doi.org/10.1016/j.jfoodeng.2017.04.017.

Liu, Y., Liang, X., Saeed, A. et al. (2019). Properties of 3D printed dough and optimization of printing parameters. *Innovative Food Science and Emerging Technologies* **54**: 9–18. https://doi.org/10.1016/J.IFSET.2019.03.008.

Mantihal, S., Prakash, S., Godoi, F.C., and Bhandari, B. (2017). Optimization of chocolate 3D printing by correlating thermal and flow properties with 3D structure modeling. *Innovative Food Science and Emerging Technologies* **44**: 21–29. https://doi.org/10.1016/j.ifset.2017.09.012.

Mehta, N., Chatli, M.K., Kumar, P. et al. (2019). Development of dietary fiber-rich meat products: technological advancements and functional significance. In: *Bioactive Molecules in Food*, Reference Series in Phytochemistry (eds. J.M. Mérillon and K. Ramawat). Cham: Springer https://doi.org/10.1007/978-3-319-78030-6_9.

Nachal, N., Moses, J.A., Karthik, P., and Anandharamakrishnan, C. (2019). Applications of 3D printing in food processing. *Food Engineering Reviews* **11** (3): 123–141. https://doi.org/10.1007/s12393-019-09199-8.

Perez, B., Nykvist, H., Brogger, A.F. et al. (2019). Impact of macronutrients printability and 3D-printer parameters on 3D-food printing: a review. *Food Chemistry* **287** (October 2018): 249–257. https://doi.org/10.1016/j.foodchem.2019.02.090.

Phan, M.M. and Stodolska, M. (2019). Food practices, dietary patterns, and leisure among Mexican Americans in the Midwestern US. *Annals of Leisure Research* **23**: 1–18.

Prakash, S., Bhandari, B.R., Godoi, F.C., and Zhang, M. (2019). Future outlook of 3D food printing. In: *Fundamentals of 3D Food Printing and Applications* (eds. F.C. Godoi, B.R. Bhandari, S. Prakash and M. Zhang), pp. 373–381. Elsevier.

Ramadoss, B.R., Gangola, M.P., Agasimani, S. et al. (2019). Starch granule size and amylopectin chain length influence starch in vitro enzymatic digestibility in selected rice mutants with similar amylose concentration. *Journal of Food Science and Technology* **56** (1): 391–400.

Sammugam, L. and Pasupuleti, V.R. (2019). Balanced diets in food systems: emerging trends and challenges for human health. *Critical Reviews in Food Science and Nutrition* **59** (17): 2746–2759.

Severini, C., Derossi, A., Ricci, I. et al. (2018). Printing a blend of fruit and vegetables. New advances on critical variables and shelf life of 3D edible objects. *Journal of Food Engineering* **220**: 89–100. https://doi.org/10.1016/j.jfoodeng.2017.08.025.

Southerland, D., Walters, P., and Huson, D. (2011). Edible 3D printing. *NIP & Digital Fabrication Conference* **2011** (2): 819–822.

Sun, J., Zhou, W., Huang, D. et al. (2015). An overview of 3D printing technologies for food fabrication. *Food and Bioprocess Technology* **8** (8): 1605–1615.

Sun, J., Zhou, W., Huang, D., and Yan, L. (2018). 3D food printing: perspectives. In: *Polymers for Food Applications* (ed. T.J. Gutierrez), pp. 725–755. Springer.

Vancauwenberghe, V., Katalagarianakis, L., Wang, Z. et al. (2017). Pectin based food-ink formulations for 3-D printing of customizable porous food simulants. *Innovative Food Science and Emerging Technologies* **42** (June): 138–150. https://doi.org/10.1016/j.ifset.2017.06.011.

Wood, J.D., Enser, M., Fisher, A.V. et al. (2008). Fat deposition, fatty acid composition and meat quality: a review. *Meat Science* **78** (4): 343–358.

Xiong, Y.L. (2018). Muscle proteins. In: *Proteins in Food Processing* (ed. R.Y. Yada), pp. 127–148. Elsevier.

Yahia, E.M., Garcia-Solis, P., and Celis, M.E.M. (2019). Contribution of fruits and vegetables to human nutrition and health. In: *Postharvest Physiology and Biochemistry of Fruits and Vegetables* (ed. E.M. Yahia), pp. 19–45. Elsevier.

Zinina, O., Merenkova, S., Tazeddinova, D. et al. (2019). Enrichment of meat products with dietary fibers: a review. *Agronomy Research* **17** (4): 1808–1822. https://doi.org/10.15159/ar.19.163.

4

Factors Affecting the Printability of Foods

CHAPTER MENU

3D Printing of Foods, First Edition. C. Anandharamakrishnan, Jeyan A. Moses, and T. Anukiruthika.

4.1 Introduction

3D printing is the next-generation modern technology used for the customized production of manufacturing goods. As an additive manufacturing process, 3D printing found its applications in a diverse range of industrial sectors (Nida et al. 2020). In the food industry, 3D printing is gaining attention in recent days due to its enhanced features of customization and personalization of foods. The major reasons for gaining research interest in 3D printing and its tremendous growth in the food sector include the simplification of the supply chain, convenience of production, extended shelf-life, customized food designs, and personalized nutrition (Liu et al. 2017). Common food printing technologies in practice are extrusion technology, selective sintering, powder bed technology, and inkjet printing. Each of these technologies has its advantages in the creation of unique textured foods depending on material characteristics and printing variables. Since foods are biologically active ingredients, the 3D printing of foods remains a challenging task in creating a complex stable 3D construct. The feed material used for 3D printing has been referred to as 'material supply' in upcoming chapters to maintain consistency and to avoid confusion. Based on printability, the material supply has been grouped as natively printable materials, non-printable materials, and alternative food ingredients (Sun et al. 2015). Among the existing 3D printing technologies, extrusion technology is the most utilized technique for the printing of foods. The success of 3D printing of material depends on print quality, printing accuracy, printing precision, and resolution. Hence, most of the research works focused on improving the print fidelity of the 3D construct. It was reported that the print quality depends on the nature of raw material, machine configurations, and printing variables (Sun et al. 2018a). The selection of food material and the adjustment of the formulation of material supply to enhance printability plays a crucial role in achieving a stable 3D construct.

Broadly speaking, the factors affecting the printability of the material can be categorized as intrinsic and extrinsic factors. Here, the intrinsic factors refer to those properties related to material characteristics influencing the printability of the material. The intrinsic factors include material flowability, apparent viscosity, complex viscosity, shear stress, shear rate, viscoelastic moduli, crystallinity, agglomeration, material hardness, adhesiveness, cohesiveness, springiness, resilience, wettability, bulk density, particle size, the surface to volume ratio, surface tension, glass transition temperature, melting point, and solidification behaviour (Nachal et al. 2019). On the other hand, extrinsic factors refer to those external variables that affect the 3D printing process. This includes printing speed, nozzle height, nozzle size, extrusion rate, extrusion pressure, and extrusion motor speed for extrusion-based 3D printing; scanning speed, laser source, energy density, laser spot diameter for powder-based technology; flow rate, droplet size, printhead size, printing velocity, and resonance frequency of the head for inkjet printing and binder jetting (Liu and Zhang 2019). The main focus of the present chapter is to provide a broad knowledge of how the material composition affects the printing variables. The state-of-the-art of printing of foods and optimization of the 3D printing process in context with different intrinsic and extrinsic factors has been detailed. Thus, the present chapter on factors affecting printability would believe to provide an overview on a better understanding of food printing and delivers valuable insights on the optimization strategies for future 3D printing studies.

4.2 Factors That Affect Extrusion 3D Printing

Printability refers to the balance among the material properties with appropriate consistency for smooth flowability through a printing nozzle and adequate mechanical strength supporting the printed layers to hold its shape and own weight without deformation (Caporizzi et al. 2019; Kim et al. 2019). The material properties and process variables are the two significant criteria that have a significant role in determining the success of the process of 3D printing. The present section deals with the detailed discussion on these two criteria that are being referred to as intrinsic and extrinsic properties. The effect of macro and micro constituents of material supply on printability has been described as intrinsic properties. The intrinsic properties can be grouped as physical, rheological, textural, frictional, and thermodynamic properties. The extrinsic properties are those that deal with the optimization of printing variables related to design parameters. It includes nozzle diameter, nozzle height, flow rate, extrusion pressure, extrusion motor speed, nozzle height, and nozzle movement speed. Each of the 3D printing technologies has its own scope and applicability based on the intrinsic properties of material supplies. In this regard, the associated printing variables specific to each technology also varied. Thus, the optimization of the 3D printing process mainly focuses on the following three things: material characterization, material optimization, and optimization of process variables. The first half of the present chapter discusses different attributes on the characterization of material supply with printability. The second half of this chapter deals with the optimization of selected material supply and optimization of printing variables. Since extrusion is the commonly used technique for printing foods, the present chapter mainly describes the process attributes that are influencing extrusion-based food printing. Also, a short brief discussion on attributes influencing the printability of the other printing technologies is presented as well. Further, the success of the developed 3D printed product greatly depends on consumer preference and acceptability. Thus, a highly acceptable 3D printed product must possess high printing precision, accuracy, distinct shape, definite layers, and mechanical stability as well as integrity.

4.3 Intrinsic Properties

4.3.1 Physical Properties

The effect of different food components and their role on printability has been discussed in the previous chapter. The tendency of these macro and micro constituents of the food towards printability would be greatly affected by the pre-processing methods such as mixing, size reduction, homogenization, concentration, drying, and so on. All these unit operations have a significant impact on the physical properties such as the size of the dry powdered particles, bulk density, porosity, and flowability (Nachal et al. 2019). Hence the determination of these material characteristics would provide a broad knowledge on formulation and modification of the material supply in tailoring the printability. Therefore, analysis of these physical properties is adequate before testing the printability of the material supply. These physical properties have a significant implication on process parameters such as feed rate, material spreading, filament/ thread thickness, and layer density (Holland et al. 2019). Considering the physical properties, a study was reported on the effect of particle size (259 and 307 μm) on the 3D printing of spinach. In this study, the spinach powder (20%, w/w) was 3D printed by incorporating it with the xanthan gum (10%, w/w). Results inferred that the increased particle size significantly affected the material composition and bulk density of the spinach powder (Lee et al. 2019). Further, the increased particle size increased the storage and loss moduli of the material supply. On the other hand, a reduction in the particle size had increased the water and oil holding capacity of the material supply. Thus, the study draws a significant inference on determining the prerequisite parameters such as milling time and percent incorporation of raw material that is required for achieving the desired consistency of the material supply.

The bulk density of the powdered sample can be determined by filling a known amount of sample in a known graduated measuring cylinder. Then the cylinder is tapped lightly until no evident decrease in the sample volume. Later, the bulk density can be calculated by dividing the sample mass by the measuring cylinder volume (Eq. (4.1)) (Theagarajan et al. 2020).

$$\text{Bulk density}\left(\text{g ml}^{-1}\right) = \frac{\text{Mass of the powdered sample}}{\text{Volume of the measuring cylinder}} \tag{4.1}$$

On the other hand, true density refers to the ratio of the sample mass to the true volume of the powdered sample. While the true volume of the biomaterials can be determined using the toluene displacement method from which the true density can be calculated using Eq. (4.2).

$$\text{True density}\left(\text{g ml}^{-1}\right) = \frac{\text{Mass of the powdered sample}}{\text{True volume occupied by the sample}} \tag{4.2}$$

The porosity, i.e. void fraction refers to the ratio of the intergranular void gap to the total amount of space occupied by the sample which can be calculated using Eq. (4.3). The knowledge on the porosity of the raw materials would help in understanding the extrusion pattern of the dough system.

$$\text{Void fraction} = \frac{\text{True density} - \text{bulk density}}{\text{True density}} \times 100 \tag{4.3}$$

The particle size distribution of the material is another criterion that significantly affects the precision of printing and accuracy of 3D printed structures (Godoi et al. 2016). Conventionally the average size of the particle can be determined using the standard ASABE method. However, for convenient rapid analysis sophisticated instruments such as laser-based light scattering particle size analyzer is being commonly used nowadays for determining the particle size (Nida et al. 2020). In a study on 3D printing of rice starch, the significance of these physical properties in the optimization of the printing process has been discussed. From the particle size distribution curve, the average particle size of rice starch used for 3D printing was found to be 0.31 mm. Results confirmed the significance of the particle size in tailoring the pasting properties of the semi-solid dough that in turn influences the printability (Theagarajan et al. 2020). From the gastronomic viewpoint, the physiochemical properties of raw material (powder samples) and the changes in the surface-to-volume ratio influences the bioavailability of the nutrients. The effect of macronutrients on printability and the formulation of material composition in the enhancement of 3D printing has been explained in the previous chapter. Further, these physical attributes of the raw material would have significant control over the flowability of the material supply through the nozzle otherwise it could result in nozzle clogging.

The physiochemical analysis of the spinach powder showed that the bulk density was increased with increased particle size. This significant increase in the bulk density of the finest particle was associated with lower porosity (Lee et al. 2019). The densification of particles with smaller spaces indicates a relatively finer distribution of particles. Thus, the reduction in the powder volume fraction with increased bulk density significantly affected the viscoelastic behaviour of the material supply. Another physical attribute that is influenced by the particle size is the water holding capacity (WHC) that represents the ability of the powder to retain water without the application of external force except for atmospheric pressure and gravity. The WHC of the raw material can be determined by the procedure reported by Chen et al. (2015). Initially, the empty weight of the centrifuge tubes was noted and then about 1.5 g of powdered sample is transferred into it. After the addition of 30 ml of deionized water, the centrifuge tubes were placed in the water bath for about 1 hour at 60 °C followed by placing in the cold water for 30 minutes. Later the samples were centrifuges at 10 000 rpm for 20 minutes. The resulting supernatant was removed and the centrifuge tubes containing the residue were weighed. The WHC can be calculated using Eq. (4.4).

$$\text{WHC}\left(\text{g}\,\text{g}^{-1}\right) = \frac{\text{Weight of the residue} - \text{weight of the sample}}{\text{Weight of the sample}} \tag{4.4}$$

The increase in the particle size of spinach powder increased the WHC mainly because of the differences in the powder characteristics such as surface area, porosity, bulk density, the impact of grinding, and sieving. Likewise, the oil holding capacity (OHC) of the raw material can be determined by centrifuging the sediment left over from the supernatant. The OHC represents the material's ability to sustain its own oil under the application of external force. First, the tubes with the sediment were placed in the water bath for about 1 hour at 25 °C. Later it was centrifuged at 3000 rpm for 20 minutes. The OHC can be calculated using Eq. (4.5) (Lee et al. 2019).

$$\text{OHC}\left(\text{g}\,\text{g}^{-1}\right) = \frac{\text{Weight of the sediment from supernatant} - \text{weight of the sample}}{\text{Weight of the sample}} \tag{4.5}$$

Results showed that the OHC of spinach powder was proportional to the particle size. The increase in bulk density of the particles as a result of sieving significantly decreased the porosity and the capillary action of fibre that leads to reduced entrapment of oil. Another powder attribute is the water solubility index (WSI) that refers to the leakage of the soluble component due to the binding of moisture present in the sample by the hydrophilic groups. WSI of the raw material can be determined by incubating the sample (0.7 g) with 50 ml of added deionized water in a water bath for about 1 hour at 80 °C. Later the tubes were centrifuged at 12 000 rpm for about 10 minutes. After centrifugation, the supernatant was transferred to a beaker and evaporated at 105 °C in the oven. The final weight of the beaker after the evaporation of the liquid supernatant was noted for determining the WSI. Thus, the WSI can be calculated using Eq. (4.6) (Lee et al. 2019).

$$\text{WSI}\ (\%) = \frac{\text{Weight of beaker after evaporation} - \text{weight of the supernatant}}{\text{Weight of the sample}} \times 100 \tag{4.6}$$

Results showed that there was no systematic trend found with WSI due to the sieving process with the lowest values were observed with the finest sample.

So far, the physical properties of the raw material supplies are discussed. Other than these aspects, the quality of the 3D printing is evaluated through the first look of the 3D printed layers. The acceptability of 3D-printed foods mainly relies on the visual appearance and the colour of the product. The print quality of the 3D printed rice starch using different sized nozzles was assessed visually by evaluating the following attributes: dimensional stability, layer definition, texture, shape, finishing quality, thread quality, binding ability, appearance, and overall acceptability (Figure 4.1) (Theagarajan et al. 2020). Likewise, the colour of the raw material, 3D printed samples, and post-processed samples were analyzed through a conventional approach based on L^*, a^*, b^* scale values (Ghazal et al. 2019; He et al. 2020b). Based on the visual sensory evaluation of the printed constructs, the quality of surface finishing can be assessed that remains to be one of the significant aspects in determining the success of the developed 3D printed construct.

4.3.2 Rheological Properties

Rheology is the science that deals with the flow and deformation of matter. Understanding the rheological behaviour of food is significant in determining the process design and flow conditions (Nijdam et al. 2021b). Especially in 3D printing, the rheological properties of the material are one the most important factors that decide its printability. Design parameters such as syringe pump sizing, feeder capacity, printing time, layer thickness, and stabilization are based on the rheological attributes such as viscosity, shear rate, yield stress, and viscoelastic moduli (Dankar et al. 2018). These complex material properties of the fluid and soft solid food materials are characterized using steady shear measurements. These viscometric functions are analyzed using a special system configuration using parallel plate geometry or cone plate geometry. The rheological analysis can be grouped into steady shear measurements and oscillatory dynamic measurements. The dynamic oscillatory analysis is comparatively easier to perform and is more reliable than steady shear measurements (Katsaros et al. 2020). A comparison of the material properties would draw a significant conclusion in determining the material response and its behaviour that are conducive for 3D printing. Based on the consistency and state of the material supply, it has been reported that high viscous materials are suitable for extrusion-based printing while low viscous materials can be printed using inkjet printing (Nachal et al. 2019). Hence, the rheology of material supply is most important not only in the selection of suitable 3D printing technology but also in achieving post-deposition stability.

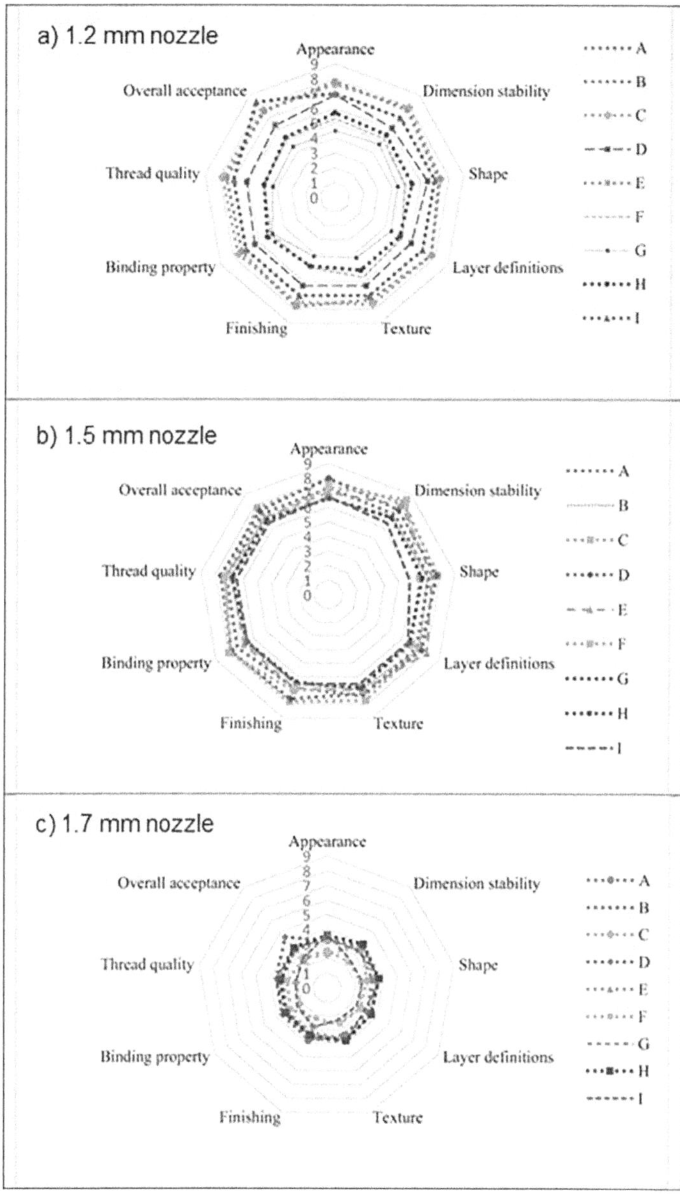

Code	Printing speed (mm/min)	Motor speed (rpm)
A	2200	240
B	2200	180
C	2200	120
D	1500	240
E	1500	180
F	1500	120
G	800	240
H	800	180
I	800	120

Figure 4.1 Sensory evaluation of 3D printed rice starch of varied printing conditions. *Source:* Theagarajan et al. (2020).

4.3.2.1 Steady Shear Rheology

For a smooth extrusion of soft solids through a small orifice printing nozzle, a certain amount of mechanical force is required. The magnitude of this mechanical force depends on the rheological attributes as well as the geometry of the material dispensing unit (Guo et al. 2019). In this context, many studies have been reported correlating the rheological behaviour and mechanical stability with the structural integrity of 3D constructs. The material's viscosity must be low as possible to aid in a continuous flow through a small orifice printing nozzle and must be high as possible to hold the successively deposited 3D printed threads with adequate adhesiveness (Kim et al. 2019). Most of the 3D printing studies stated that the shear-thinning behaviour of the printing material attributed to non-Newtonian flow is best suited for 3D printing by extrusion principle. The flow curves demonstrate the dependence of shear stress and apparent viscosity on the applied shear rates (Liu et al. 2017). In non-Newtonian behaviour, the increase in the material's apparent viscosity with proportional increment in shear rate is described as a shear thickening and vice versa is known to be shear thinning. Most of the material supplies used for extrusion-based 3D printing exhibit a shear-thinning characteristic synonymously referred to as pseudoplastic behaviour with exception of some materials like cheese that exhibits shear thickening behaviour (Kern et al. 2018). The structured fluids or semi-solid food materials such as pastes, sauces, mayonnaise, spreads, and ice cream exhibits shear-thinning characteristics (Martínez-Monzó et al. 2019). Sometimes these materials also tend to be highly thixotropic that results in irreproducible rheological measurements due to disruptions of microstructure that occurred during the sample loading and analysis. The viscoelastic liquid state of the material supply at low shear rates attribute to shear-thinning while the change from viscoelastic liquid to a solid with fluctuations in shear stress and a shear rate of the sample denoting shear thickening behaviour (Bähler et al. 2015). The measured steady shear values are fitted to the different rheological models to determine the flow behaviour index (n) and material consistency (k). Among various rheological models, the most often reported models in 3D printing studies are the Power law model and Herschel–Bulkley model. The flow curve behaviour of egg yolk and egg white fractions using Power law model showed that all the formulations of material supplies possess shear thinning pseudoplastic behaviour with $n < 1$ (Figure 4.2) (Anukiruthika et al. 2020). This resulted in easy extrusion of the material through the printing nozzle that assists in 3D printing process.

Both these attributes play an important role in assessment of the printability of the material supply by measuring the desired extrusion rates with a small value of k seems to be beneficial in maintaining the continuity of flowability of the material supply (Liu et al. 2020a). The effect of printing temperature on 3D printing of commercial potato puree in context with rheological and mechanical properties was reported. The viscosity–temperature profiles, apparent flow curves, and dynamic oscillatory measurements were analyzed; and results inferred that the printability was directly proportional with consistency index and inversely related with the flow behaviour of potato puree (Martínez-Monzó et al. 2019). The reversible gelation behaviour of the sodium caseinate dispersions was assessed for 3D printing applications. The dispersing characteristics of the printing material supply were described by a modified Poiseuille equation used for power-law fluids and to set the extrusion pressure and line speed. The k indicates the viscosity at a shear rate of s^{-1}; if $n < 1$ indicates the shear-thinning nature of the material. The volumetric flow rate (Q) in a circular tube can be described by the modified Poiseuille equation (Eq. (4.7)) (Schutyser et al. 2018).

$$Q = \frac{n\pi R^3}{1 + 3n}\left(\frac{R\,\Delta\,P}{2\,mL}\right)^{1/n} \tag{4.7}$$

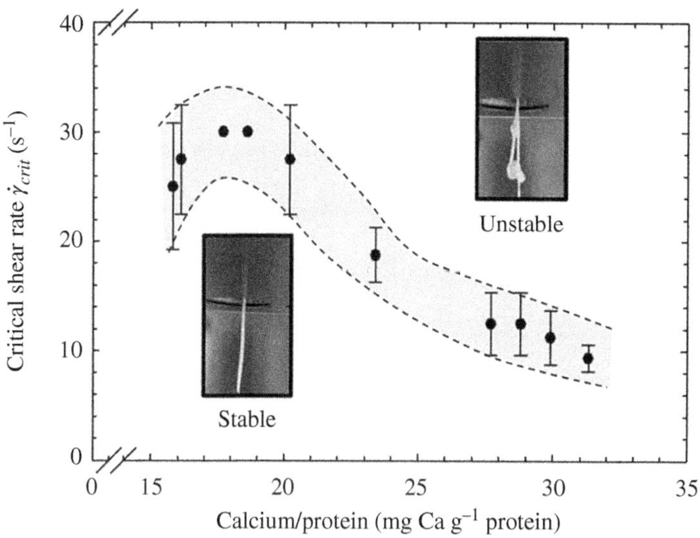

Figure 4.2 Effect of calcium content on critical shear rate of the model cheese. *Source:* From Kern et al. (2018) / With permission of Elsevier.

where Q is the volumetric flow rate ($m^3\ s^{-1}$), ΔP is the pressure difference that exists with air pressure and gravity (Pa), L is the length of the tube (m), and R is the radius of the tube (m). The study suggested that the spatial distribution of the oil droplets could provide greater freedom in designing the customized nutritious foods. In another study on the printability of alternative ingredient mushroom, the shear stress and shear rate of the material during its transit through the nozzle were described in terms of pressure drop and volumetric flow rate (Eqs. (4.8) and (4.9)) (Keerthana et al. 2020).

$$\tau = \frac{\Delta P r}{2L} \tag{4.8}$$

$$\gamma = \frac{4Q}{\pi r^3} \tag{4.9}$$

where τ is the shear stress (Pa), r is the radius of the printing nozzle (mm), and γ is the shear rate of the material (s^{-1}). Results showed that the material formulation with 20% mushroom powder exhibits desired extrusion using a 1.28 mm nozzle with corresponding shear stress and shear rate in the range of 3.52 to 1.23 kPa and 11.33 to 48.59 s^{-1}, respectively. The calcium and protein content of cheese was found to have a significant impact on printing temperature and flowability (Kern et al. 2018). The critical shear rates of model cheese samples were plotted based on pressure profile and visual observation of extruded material (Figure 4.3). When the shear rate exceeds critical shear rate, there occurs melt fracture and phase separation indicating the unstable region where oscillating pressure profile showed flow instabilities. The equilibrated pressure profile with the homogeneous appearance of material represents the stable region. While the grey region marked by dotted lines in the plot represents the transition zone of the material supply that was not capable of 3D printing showing surface distortion of the extruded filaments referred to

(a)

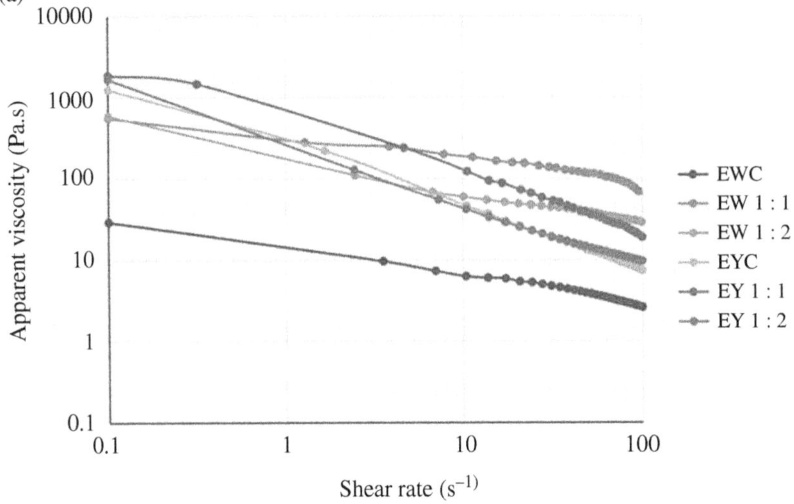

(b)

Material supply	Power law fitting		
	n	k (Pa.s^n)	R^2
EWC	0.65 ± 0.005^c	14.42 ± 0.15^a	0.994
EW1 : 1	0.75 ± 0.01^d	293.38 ± 2.15^e	0.985
EW1 : 2	0.65 ± 0.02^c	145.26 ± 1.29^d	0.998
EYC	0.39 ± 0.004^a	142.43 ± 0.13^c	0.958
EY1 : 1	0.41 ± 0.004^{ab}	362.4 ± 0.31^f	0.962
EY1 : 2	0.43 ± 0.001^b	136.62 ± 0.22^b	0.987

Figure 4.3 Steady shear rheology of egg yolk and egg white material supplies. (a) Apparent viscosity curve and (b) fitting of Power-law model. *Source:* Anukiruthika et al. (2020) / With permission of Elsevier.

as 'shark-skinning'. The critical shear rate was increased with an increase in calcium/protein content that in turn affected by elastic modulus and loss tangent. Thus, proper optimization of material supply is adequate for better printing performance.

4.3.2.2 Dynamic Shear Rheology

The shear modulus of the hydrocolloid methyl cellulose was measured by a time sweep of about 100 seconds with strain value ranging from 0.001 to 10% at 25 °C based on the shear rate controlled rotational test. The linear elastic region of the stress/strain plot gives the shear modulus (Eq. (4.10)) (Kim et al. 2018a).

$$G = \frac{F/A}{\Delta x/h} \tag{4.10}$$

where G is the shear modulus (Pa), F is the tangential force acting on the material (N), A is the cross-sectional area of material that is being sheared (m^2), Δx is the horizontal distance moved by the sample due to shear (m), and h is the height of the sample (m). Results showed that the increase in the concentration of hydrocolloid proportionally increased the shear modulus. The obtained values of the shear modulus of tested samples were consistent with the results of dimensional stability. Materials with a higher value of shear modulus hold their shape from deformation and associated impact during fabrication of complex geometry with concentrated load (Zhang et al. 2018). Thus, the deformation of the 3D printed construct can be evaluated by determining the shear modulus that in turn assists in the evaluation of the printability.

The linear viscoelastic (LVE) range of the material supply can be determined through the stress sweep test (Zhang et al. 2018). The LVE region can be derived from the proportionality of preset and measured parameters of the shear stress-strain curve. From the LVE range, the small amplitude oscillatory frequency sweeps at constant stress determines the storage modulus, i.e. elastic modulus (G') and loss modulus, i.e. viscous modulus (G''). The G' represents the material's resistance towards elastic deformation represents the elastic solid-like behaviour indicating the mechanical stability and integrity of the material supply. On the other hand, G'' represents the material's viscous response that can be defined as the ratio of material's stress to strain under experimental process conditions (Yang et al. 2018b).

Another attribute is the loss factor also referred to as the damping factor ($\tan \delta = G''/G'$) that reveals the relationship between the G' and G'' is being used to illustrate the viscoelastic behaviour. The higher value of loss factor ($\tan \delta > 1$) at lower frequencies of <1 Hz indicates the higher predominance of $G'' > G'$ showing that the material is viscous with a more fluid-like behaviour. On the other way around, when the $G' > G''$, the material is considered as elasticity dominant (gel-like structure) with $\tan \delta < 1$ indicates the solid-like behaviour of the material with poor fluidity (Liu et al. 2018c). Among these conditions, the predominance of G'' of the material supply favors the stability of the deposited layers. The frequency dependence of the G' and G'' provides information about the structural state of the material (Liu et al. 2019b). The material supplies that are independent of frequency over a large period indicate a solid-like behaviour as in the case of a true gel system. In contrast, the materials that are strongly dependent on frequency indicate the material structure with molecular entanglements. This type of frequency-dependent nature of the material supplies behaves more like solid as well as like liquid at higher and lower frequencies, respectively. The viscoelasticity of the dough with different proportions of ground chicken meat was analyzed through steady and oscillatory dynamic shear measurements (Figure 4.4) (Wilson et al. 2020). Results showed that chicken-based material supply was elasticity dominant with more solid-like behaviour that indicates the transition in material fluidity that is conducive for 3D printing.

Similarly, the small amplitude oscillatory strain measurements with temperature sweep help in determining the thermo-rheological properties of the materials like hydrogel, chocolate, and cheese (Chen et al. 2019; Kern et al. 2018; Maniglia et al. 2020). The temperature sweep test was performed in a controlled strain rheometer with serrated plate geometry. The temperature where $G' = G''$ is referred as gel–sol transition temperature ($\vartheta_{gel-sol}$). Results showed that use of citric acid instead of lactic acid for acidification of model cheese

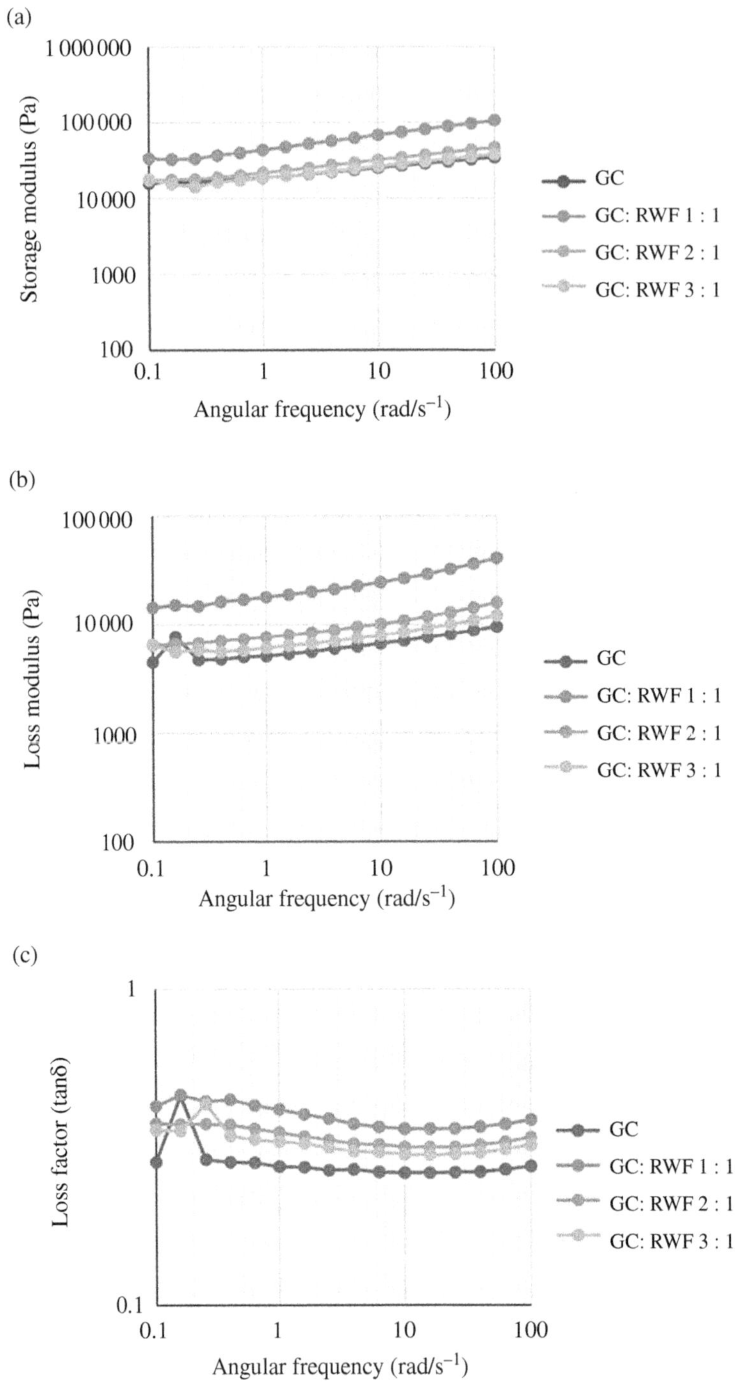

Figure 4.4 Dynamic oscillatory rheology of chicken-based material supply. (a) Storage modulus, (b) loss modulus, and (c) loss tangent. *Source:* Wilson et al. (2020).

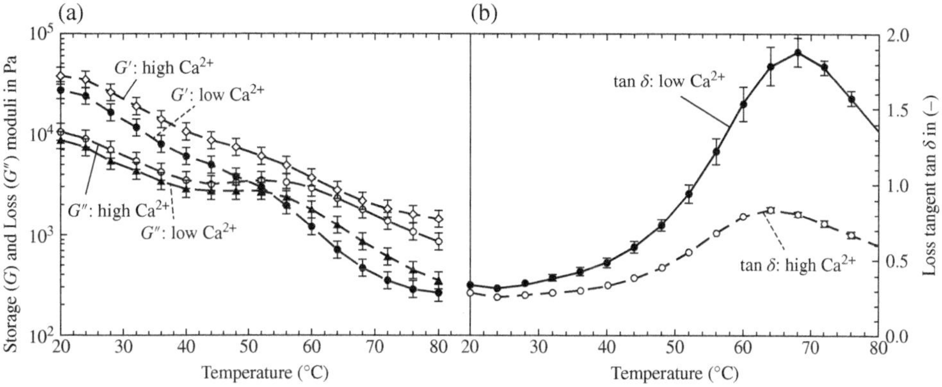

Figure 4.5 Temperature sweep plot of model cheese. (a) Storage modulus and loss modulus and (b) loss tangent of low calcium model cheese (solid symbols – 18 ± 2 mg Ca g^{-1} protein; pH 5.5 LA) and high calcium model cheese (hollow symbols – 31 ± 2 mg Ca g^{-1} protein; pH 6.3 LA). *Source:* Kern et al. (2018) / With permission of Elsevier.

was responsible for significant decrease in the G'. With lowering of pH, the $\vartheta_{gel-sol}$ of cheese was significantly decreased (Figure 4.5). From the observations, it was confirmed that the G' at 20.5 and 60 °C; and $\vartheta_{gel-sol}$ was positively correlated with pH, calcium content, and calcium/ protein interactions while these characteristics were negatively correlation with tanδ at 20.5 and 60 °C (Kern et al. 2018).

Another factor is the complex modulus that determines the mechanical rigidity or stiffness of the material. The complex shear modulus (G^*) can be related to the storage modulus and loss modulus (Eq. (4.11)) (Nijdam et al. 2021a).

$$G^* = \sqrt{(G')^2 + (G'')^2} \tag{4.11}$$

The G^* is useful in comparative analysis of the apparent viscosity of the different material supplies. Those material supplies that formed a dimensionally stable triple steps structure covered a wide range of complex moduli, yield stress, yield strain, and loss tangent. For instance, the peanut butter 3D printed at 20 °C possessed a G^* of 260 000 Pa and yield stress of 42 Pa while for 15% gelatin the G^* and yield stress values were found to be 5100 and 3900 Pa, respectively (Nijdam et al. 2021a). All the tested material supplies were combined by correlating the rheological attributes by calculating the ratio of the weight of the 3D structure (ρgh) to the G' that falls within the window of dimensional stability (Figure 4.6). Thus, this work presented the characterization of the material supply in a practical timeframe that assists in a reasonable prediction of dimensional stability covering a wide range of diverse foods of different consistency from gels to pastes.

4.3.2.3 Yield Stress

Yield stress is another important parameter in the assessment of extrusion behaviour that indicates the minimum force required for initiating the material flow (Wilson et al. 2020). During the printing process, the extrusion does not remain constant that varies based on the input process parameters. The material supplies with high yield stress require greater

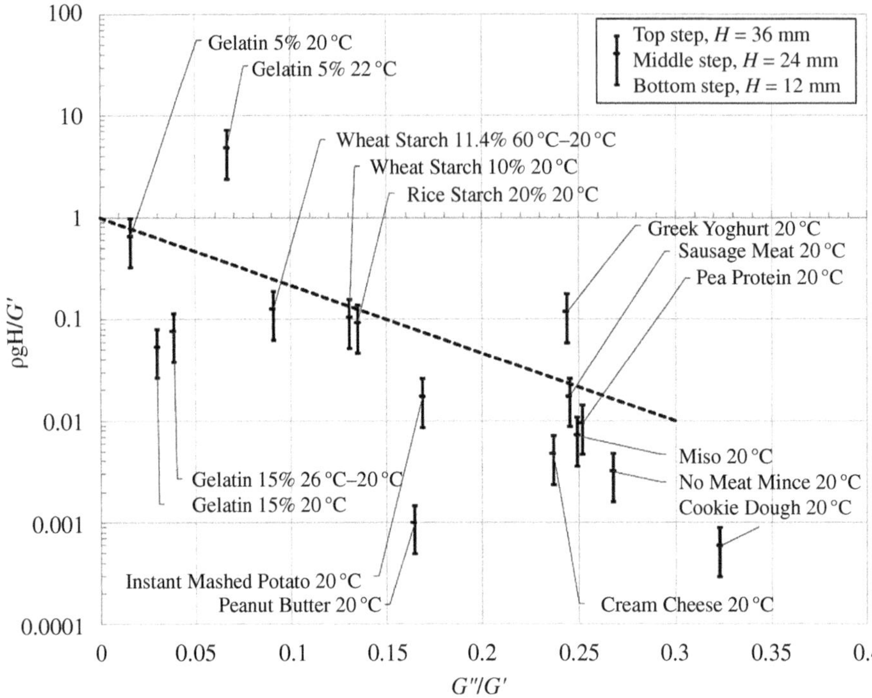

Figure 4.6 Window of dimensional stability of different material supplies based on a tolerance of 5% deformation in an hour. *Source:* Nijdam et al. (2021a) / With permission of Elsevier.

pressure to overcome the material's yield stress. Thus, insufficient extrusion pressure could significantly affect the material flow rate that may result in under extrusion with broken filaments. The relation between the applied pressure and yield stress of the material supply can be described by Eq. (4.12).

$$P_{\min} = \left(\frac{4L}{D}\right)\tau_{\text{yield}} \tag{4.12}$$

where $P_{\min}$ is the minimum force (N), L is the nozzle length (m), D is the nozzle diameter (m) and τ_{yield} is the yield stress (Pa). The yield stress reflects the post-print stability of the material supply to withstand the structure of the deposited layers. Thus, a highly desirable material supply for 3D printing applications must possess minimum yield stress to initiate the flow and strong enough to resist any deformation after 3D printing (Liu et al. 2020a). The lowest yield stress along with lowest G' indicates the poor stability of the material supply to withstand the printed 3D structures.

The point where the G' and G'' start to leave the plateau represents the yield stress which can be determined from stress sweep analysis (Lille et al. 2018). For the cereal-based probiotics incorporated dough, the yield stress was increased with a decrease in dough's water content and increase in calcium caseinate in the formulation (Zhang et al. 2018). Thus, the yield stress was related to the resistance of 3D printed structures against material disintegration due to successively deposited layers and gravity effects. In a study performed by Lille

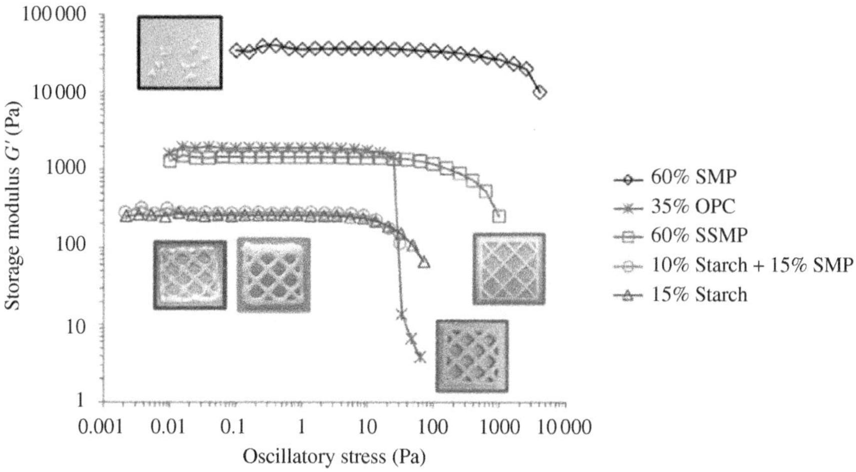

Figure 4.7 Oscillatory stress sweep measurements for different material supplies and the corresponding 3D printed samples. *Source:* From Lille et al. (2018) / With permission of Elsevier.

et al. (2018), the rheological parameters (G', G'', phase angle, and yield stress were determined from the stress sweep measurements that confirmed the correlation existing between these attributes and extrusion pressure (Figure 4.7). These results clearly indicated the existence of the larger deformations with the extrusion of material supply during stress sweep measurements that often requires working under LVE range.

For milk protein food simulant, the linearity limit (LVE) was turned out to be under 1% where the applied strain was small enough to affect the microstructure of protein matrix and the G' was found to be constantly higher than the G'' (Liu et al. 2018a). Beyond this LVE range, the deformation of a structural network of protein gel was evident with a monotonic decrease in G' and G''. The intersection of G' and G'' moved from 30% strain to 55% with gradual rise in the proportion of whey protein isolate (WPI). This indicates that the milk protein concentrate (MPC) was more delicate and vulnerable to the applied external force but WPI could assist in easy flowability of printing mixture during extrusion process. The G'' becomes dominant when the mass ratio of MPC and WPI changed from 4 to 3 that represents the transition of material to a liquid-like behaviour. These findings were consistent with the results of Vancauwenberghe et al. (2017) where the overshoot of G'' of pectin gel was mainly because of additional friction resulting from the preliminary damage of gel just before its collapse under large deformation.

4.3.2.4 Complex Viscosity

The resistance of the material flowability in its structured state, i.e. the viscous or elastic flow resistance to the oscillating movement is referred to as complex viscosity (Wilson et al. 2020). The composition of the sample influences viscoelastic properties thereby determines the printability. The amount of water present in the material supply reflects changes in dynamic oscillatory measurements. It was reported that the reducing level of the water content of the cereal-based dough increased the G' and G'' along with substantial increase in

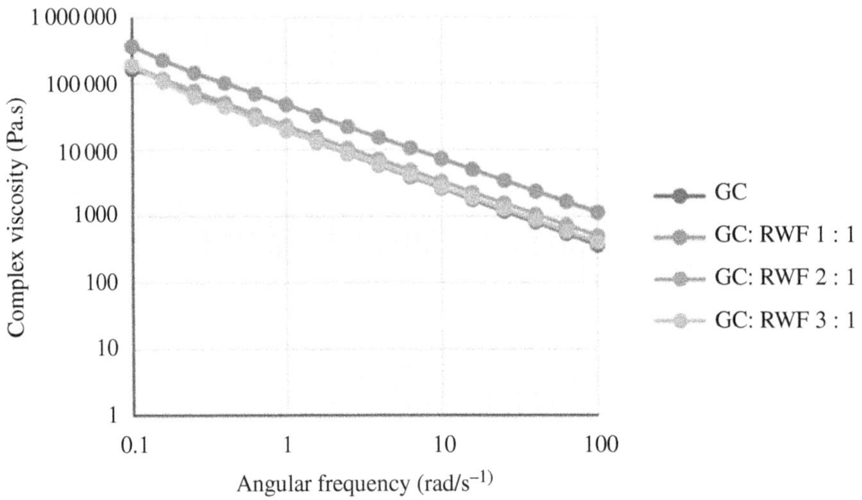

Figure 4.8 Complex viscosity of chicken-based material supply. *Source:* Wilson et al. (2020).

complex viscosity (η^*) at 1 Hz frequency that favoured printability of probiotics incorporated material supply (Zhang et al. 2018). Here, the water acts as an inert filler in the cereal dough that proportionally changes the viscoelastic properties and imparts lubrication thereby enhancing the relaxation of components (Mastromatteo et al. 2013). The effect of the influence of food ingredients on η^* of hydrocolloid mixture was studied. The η^* of the vegetable ink with different hydrocolloids was inversely related with the angular frequency ω that translates the pseudoplastic nature of the material supply (Kim et al. 2018b). Similar results were observed with the η^* measurements of brown rice paste and chicken meat-based material supply (Huang et al. 2019; Wilson et al. 2020). The complex viscosity measurement of chicken-based material supply was presented in Figure 4.8.

The complex interactions of these components with the food system influence its rheological attributes. Results showed that the tested hydrocolloids were stable to sugars, organic acids, and ionic electrolytes. The food constituents along with polysaccharide hydrocolloids would compete for water molecule thereby results in reduced η^* as in the case of hydroxypropyl methylcellulose (HPMC) solution with added acetic acid and NaCl. The hydrodynamic volume of the material can be directly related to the material's viscosity that depends on variation in the ionic strength and pH resulted in the loss of η^*. In contrast, the addition of sucrose led to a uniform bending of the methoxy group that increases hydrophobic interactions in contact surfaces of HPMC due to which the viscosity was increased (Kim et al. 2018b). This study inferred that the rheological properties greatly depend on the nature of hydrocolloids as well as the composition of the vegetable powder.

4.3.2.5 Thixotropy and Creep Recovery

The material supplies with appropriate thixotropic behaviour are most beneficial for 3D printing by extrusion principle. Thixotropy reflects the characteristic behaviour of the sample whose viscosity decreased with shear rate upon extrusion; however, the same

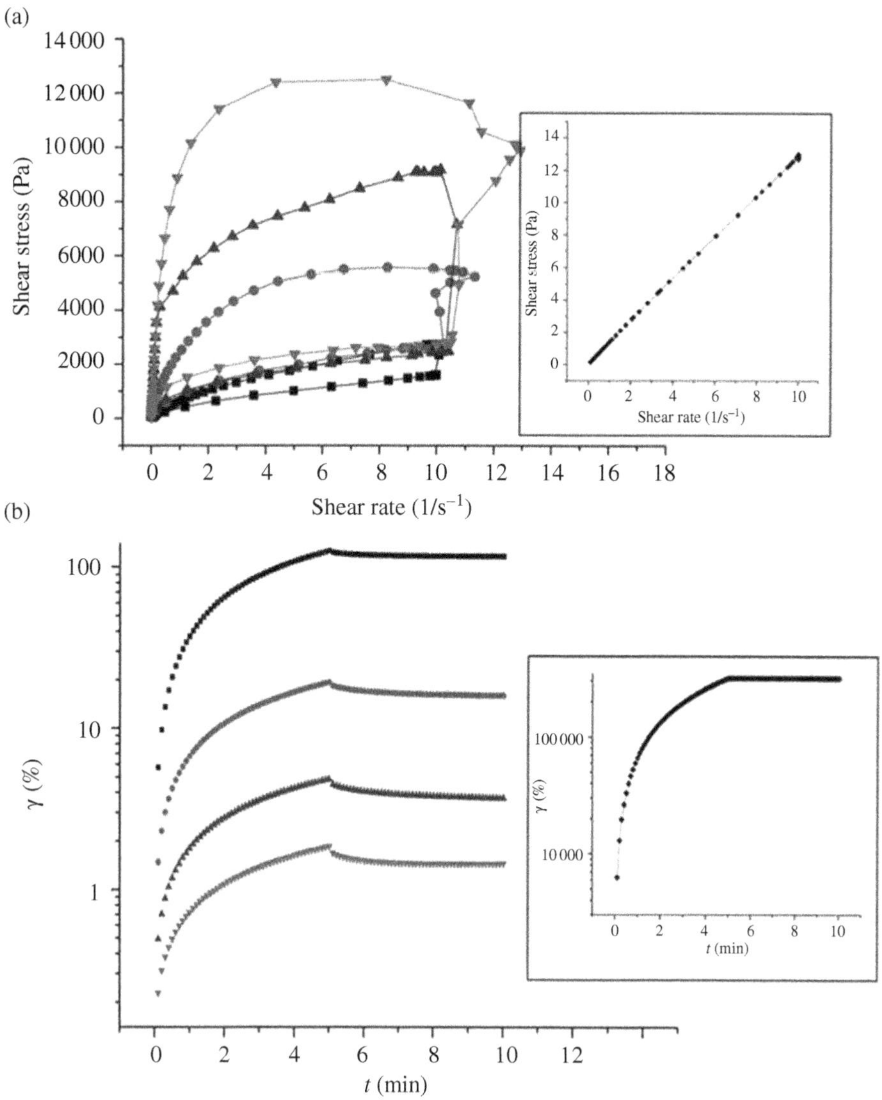

Figure 4.9 Rheological analysis of milk protein composite gel. (a) Thixotropic hysteresis loop stress ramps and (b) creep recovery curves (□, 350 g l^{-1}, ○, 400g l^{-1}; ▽, 450 g l^{-1}; △, 500 g l^{-1}; ◇, pure NaCN dispersion). *Source:* Liu et al. (2019b) / With permission of Elsevier.

material could be able to regain its viscosity and recover to its gel structure on the removal of applied shear stress (Keerthana et al. 2020). Thixotropy can be determined by the hysteresis loop test. Materials with a larger hysteresis area indicate stronger thixotropic nature. The milk protein composite gel of 350 g l^{-1} protein content showed a poor shear tolerance with a smaller loop area (Figure 4.9) (Liu et al. 2019b). In another study, the effect of processing methods of potato puree on 3D printing was studied. Results showed that potato puree treated with microwave heating possessed higher thixotropic areas than that of the boiling.

The larger loop area is directly related to the proportional amount of energy required to break up the material's internal structure (Dankar et al. 2020). This represents the stronger inner tendency of the material for achieving stable 3D constructs. The impact of microfluidization on the WPI emulsion was examined. Results showed that thixotropic behaviour was increased with increased oil fractions of emulsion. The sample with the highest oil fraction (0.6) showed stronger thixotropy that represents its higher resistance to time-dependent flow (Liu et al. 2019c). Thus, the tendency of the material showing lower viscosity (pseudoplastic behaviour) that resolidify upon resting was because of thixotropic property and strong internal elasticity with higher yield stress values (Dankar et al. 2020). These characteristics make the material supply more suitable for smooth desired extrusion.

The change in strain (γ) over time is referred to as creep which can be determined through creep-recovery analysis. The percentage of γ as a function of time represents the extent of deformation with constant stress. The recovery of the samples after high shear can be expressed as the recovery index (Eq. (4.13)). The creep-recovery test of sodium caseinate (NaCN) dispersions showed poor resistance with the irreversible recovery of structure upon removal of stress (Liu et al. 2019b). Results showed that the recovery rate was increased with an increasing proportion of MPC. Thus, material supply with higher protein content ($500\,\mathrm{g\,l^{-1}}$) showed the highest resistance, and lower protein content ($350\,\mathrm{g\,l^{-1}}$) showed poor resistance to stress.

$$\text{Recovery index} = \frac{\eta_{\infty} - \eta_{\text{shear}}}{\eta_0 - \eta_{\text{shear}}} \times 100 \tag{4.13}$$

where η_0 is the initial apparent viscosity in its structured state at low oscillation, η_{∞} is the apparent viscosity at the end in its restructured state during equilibrium, and η_{shear} is the apparent viscosity at the end of high shear. The plot of recovery index versus phase angle provides clear knowledge of the printability of material supply. Although the 4% guar gum and 4% locust bean gum resulted in a recovery percentage of 80 and 100% respectively, they were non-printable. Whereas 2% gellan gum with a low recovery index showed higher printability (Gholamipour-Shirazi et al. 2019). Thus, analysis of thixotropic and creep behaviour provides significant information about the nature of the printability of the material.

4.3.2.6 Qualitative and Quantitative Assessment of Rheology

The printability of different food constituents such as starch, protein, and fibre-rich material supplies has been studied (Lille et al. 2018). From the observations, an optimum printing formula was developed with semi-skim milk powder and the results dictate the relation among the visual appearance and shape stability of the 3D constructs with the yield stress. The cereal-based probiotics incorporated 3D structures have been 3D printed wherein the rheological measurements stated that the dynamic oscillatory measurements were related to the printability in terms of qualitative manner but not quantitatively (Zhang et al. 2018). Thus, most of the reported studies on 3D printing has qualitatively assessed the printability in terms of the extrusion performance through visual observation for nozzle clogging and discontinuous flow rate (Chen et al. 2019; Liu et al. 2018c; Wang et al. 2018). Considering this void an attempt was made to quantitatively assess the printability of material supply. In a more recent study, the tomato paste was used as a model system for investigating the correlation between print stability, rheological properties, and dispensability (Zhu et al. 2019).

Results showed a linear correlation among the flow stress, shear viscosity, and print stability. With the increase in the flow stress, the extrusion pressure was also increased, and the obtained results were found to be consistent with aqueous-based food formulations. However, the fat-based formulations could exhibit different printing behaviour. More studies are yet needed in this direction on quantitative assessment of the rheological properties which could provide more insights on a deeper understanding of material behaviour towards 3D printing.

4.3.3 Mechanical Properties

Along with the rheological property, the mechanical properties such as extrusion force and textural parameters play a significant role in the manufacturing, storage, and handling of food products (Yang et al. 2018b).

4.3.3.1 Extrusion Assay

In extrusion-based 3D printing, the material supplies generally experience relatively high mechanical shear forces and pressure during the extrusion. The extrudability of the material supply was evaluated using a forward extrusion assay (Garcia-Segovia et al. 2020). This involves the measurement of applied compression force required to extrude the material out through the nozzle (Martínez-Monzó et al. 2019). The area under the curve plateau of the force and time plot represents the extent of force required that is indirectly related to the material consistency suitable for 3D printing. The mean extrusion force observed for 3D printing of commercial potato puree sample was found to be in the range of 6.1 68 N. With the increase in the content of dehydrated potato in puree, the force required for extrusion got decreased with decreasing extrusion temperature. Thus, printing temperature interlinks the material consistency with the mechanical behaviour thereby determining the suitability of the material for 3D printing. The deformation behaviour of the designed 3D model can be assessed by determining Young's modulus. Here, Young's modulus is proportional to the relative density of the 3D model (Eq. (4.14)) (Vancauwenberghe et al. 2018).

$$\frac{E_c}{E^*} = \frac{3}{2}\left(\frac{\rho}{\rho^*}\right)^3 \tag{4.14}$$

where E_c is the compressive Young's modulus of cellular plant material and E^* is the compressive Young's modulus of the material. The relative density $\left(\frac{\rho}{\rho^*}\right)$ of the 3D structure can be expressed in terms of porosity (φ) (Eq. (4.15)).

$$\frac{\rho}{\rho^*} = 1 - \varphi \tag{4.15}$$

The equivalent Young's modulus of the designed 3D structure can be measured from the force exerted on the displaced surface of the material and the associated deformation (Eq. (4.16)).

$$E = \frac{\sigma}{\varepsilon} = \frac{F/A}{dL/L} \tag{4.16}$$

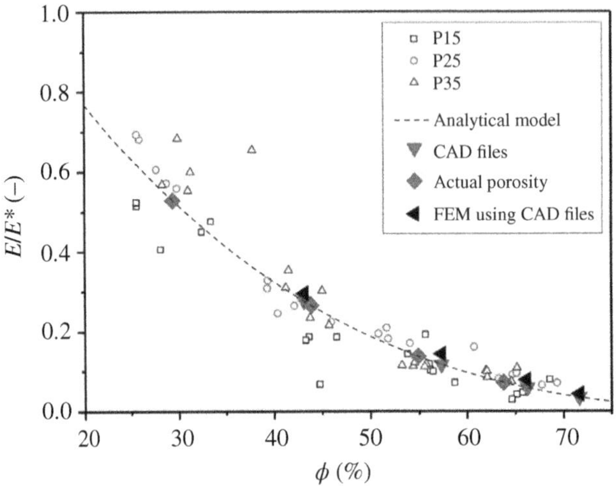

Figure 4.10 Comparison of the relative Young's modulus of the 3D printed pectin construct using analytical model and FEM irrespective of the formulation of food-ink. *Source:* Vancauwenberghe et al. (2018) / With permission of Elsevier.

where E is the engineering Young's modulus (Pa), σ is the stress experienced by the material (Pa), ε is the strain experienced by the material, F is the reaction force (N), A is the projected area of the 3D design model perpendicular to the applied force (m^2), L is the initial length of the 3D model (m), and dL is change in length after deformation (m). The relative Young's modulus of the honeycomb infill structure used for 3D printing of pectin followed a decreasing trend with increasing cell size that was proportional to the material's porosity (Vancauwenberghe et al. 2018). The study inferred that the honeycomb infill pattern of pectin-based 3D constructs was subjected to bending deformation across the cell walls of the material. Further, the spreading and shrinkage of the printed structures can be characterized by modified design through modelling of solidification of gel using a multiphase model that relates the pore size, volume, deviations of wall thickness, and porosity during 3D printing with ink concentration (Figure 4.10). The Young's modulus values better fitted the experimental results indicating the significance of porosity which was influenced by the spreading and the shrinkage of 3D printed structures. Hence, the porosity of the material supplies must be taken into account for predicting Young's modulus. Another mechanical attribute is the rupture stress that represents the network strength of the gel matrix. It was reported that the concentrations of sugar and bovine serum albumen had a negative effect on the rupture test of the printed 3D construct (Vancauwenberghe et al. 2017). Similarly, the stirring and $CaCl_2$ also influence the rupture stress of the material supply to some extent.

4.3.3.2 Textural Profile Analysis

The textural properties of the material supplies and the 3D printed foods are adequate to assess the mastication behaviour for determining the people's acceptability. Among the textural attributes, hardness is one of the significant attributes that represents the amount of force required by the sample to attain a certain degree of deformation (Anukiruthika et al.

2020). The hardness refers to the force applied to push the material out through the printing nozzle. Gumminess measures the amount of energy required for the breakdown of semi-solid materials into a stable state to consistency in a swallowable form. The effect of printing variables such as nozzle size, infill density, and perimeter on the hardness and gumminess of the brown rice paste was examined (Huang et al. 2019). Both the hardness and gumminess showed a decreasing trend with decreasing nozzle size, infill density, and perimeter. This indicates that the texture has decreased sharply during the first compression and then changed slowly with increased void percentage. Interestingly, the number of slices indirectly affected the textural characteristics. A smaller-sized nozzle could create an object with a softer texture and vice versa. Further, the way of deposition of the 3D printing technique could influence the hardness of the 3D constructs. On comparing the moulded product and 3D printed product, the printed sample was less hard (Liu et al. 2018b). The associated shear effects induced during extrusion of material supply were mainly responsible for the reduced hardness of the 3D printed sample (Le Tohic et al. 2018). Among the tested printing variables, infill density has the strongest linear effect on gumminess than nozzle size and perimeter. For materials like chocolate, a snap test is being performed to determine its mechanical properties. A good snap is an indicator of the material's ability to break apart easily (Mantihal et al. 2017, 2019a). The snap-ability of the chocolate is greatly influenced by infill density and its pattern (Figure 4.11). The differences in the amount of force needed to snap the specimen depending on the applied intensity and stability of infill structures. The presence of crisscross infill structure of the star and Hilbert curve (HC) imparts to its mechanical strength than honeycomb (HNY) pattern. This was mainly due to the greater number of bonding zones that exist with the star and Hilbert curve infill pattern.

Apart from printing variables, the composition of material supply also significantly affects the textural attributes. The hardness of the material supply strongly depends on protein denaturation and starch gelatinization (Theagarajan et al. 2020; Xu et al. 2020). Springiness refers to the potential characteristic of the material supplies to get back to their original structure after deformation on the removal of the applied force while cohesiveness represents the adhesive forces acting within the sample (Anukiruthika et al. 2020). The hardness of the sample was increased initially and then decreased with increasing egg white protein (EWP) content. However, springiness and cohesiveness were not increased with increased EWP. Regarding springiness, the addition of EWP significantly affected the microstructure of the printing mixture (Liu et al. 2019a). Further, the addition of EWP increased the chewiness and resilience of the material supply. Considering the printing temperature (45 °C), an increased proportion of EWP preventing denaturation thereby improves solidification of EWP gel that assists in supporting its own weight. In another study, the mechanical strength (hardness, adhesiveness, cohesiveness, springiness, chewiness, and resilience) of dough was studied with the increased content of ground chicken (Wilson et al. 2020). The least adhesiveness with higher hardness was observed for material supply with least amount of ground chicken. Higher cohesiveness of 2 : 1 ratio of ground chicken to refined wheat flour was attributed to the higher printability. Regardless of increased proportion of ground chicken, there was no significant difference observed with springiness and resilience of the material supply. Thus, textural properties are more closely correlated with the rheological properties that decides the printability of the testing material.

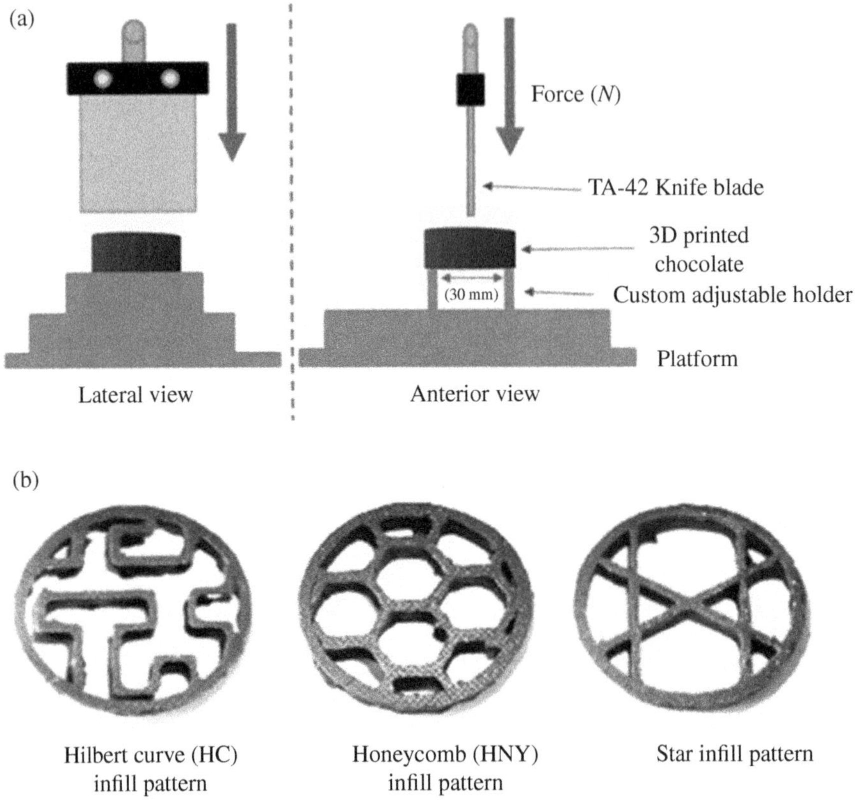

Figure 4.11 Testing of snap-ability of 3D printed chocolates. (a) Illustration of positioning of samples for snap test. *Source:* Mantihal et al. (2019a) / With permission of Elsevier and (b) 3D printed chocolate samples with different infill patterns. *Source:* From Mantihal et al. (2019a), Figure 09 [p. 33] / With permission of Elsevier. DOI-https://doi.org/10.1016/j.foodres.2018.12.034.

The effect of post-processing on the textural attributes of 3D printed snacks was studied. With the increase in the percentage of insect enrichment, the hardness of the baked snacks was increased (Severini et al. 2018a). In contrast, the behaviour of hardness was not linear with the baking temperature that showed a decreased hardness at intermediate temperatures. So, the higher hardness at lower temperature was mainly because of limited expansion that led to greater rigidity while higher hardness at higher temperature was due to hardening of external crust (Figure 4.12). The insect enrichment improved the water evaporation that aids in the creation of porous structure, the resulted microstructure had more rigidity.

It was also stated that the hardness was also dependent on pasting properties of starch and coagulation of protein as well as chemical interactions between insect proteins and lipids which has been increased with insect enrichment (Alam et al. 2019). In another study on the development of 3D printed snacks from indigenous composite flour where the effects of three different post-processing methods on hardness were examined. Results showed that deep-fried samples had a relatively higher hardness due to crust dehydration followed by

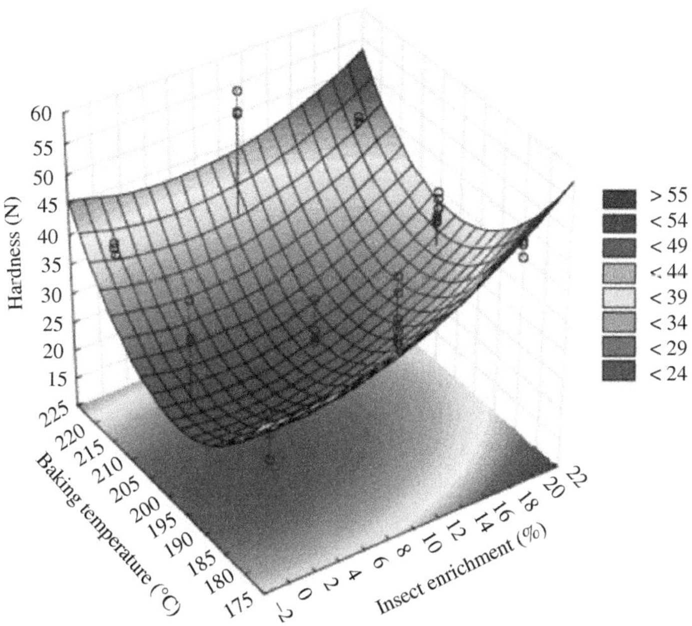

Figure 4.12 Three-dimensional plot showing the effects of baking temperature and levels of insect enrichment on hardness of 3D printed baked samples. *Source:* Severini et al. (2018a) / With permission of Elsevier.

microwave dried and hot air dried-deep fried samples (Krishnaraj et al. 2019). On the other hand, crispiness represents the fracturability that associates with the breaking and disintegration of food. Fracturability of the post-processed samples was positively correlated with the hardness showing the highest values for microwave dried samples followed by deep-fried and hot air dried-deep fried. Thus, textural properties are influenced by material composition, consistency, and printing variables that determine the mechanical strength and stability of the 3D printed product.

4.3.4 Frictional Properties

Frictional properties deal with the introduction of sliding forces over a hard surface that provides knowledge on the design of equipment and material conveyance. The rheological properties are well correlated with the sensory attributes of food. Similarly, tribology is the branch of science that deals with friction, lubrication, and wear (Sethupathy et al. 2020). Tribological measurements help in the quantification of the frictional force acting on the object. Recently, tribology has gained research interest in exploring the frictional behaviour of foods, especially in food texture studies. The common engineering properties of foods using texture analyzer, rheometers, and viscometers are based on bulk disintegration and shear deformation (Nguyen et al. 2016). The measurements based on these instruments are directly linked with the bulk phase deformation that is related only to the textural characteristics. While the specific sensory attributes of foods such as creaminess, smoothness, and slipperiness are based on the frictional behaviour of foods during oral processing (Prakash et al. 2013). Thus, tribological studies provide a relationship between the frictional coefficient, viscosity, and sliding speed that is represented in the Stibeck curves. An ideal Stibeck curve of a Newtonian fluid is comprised of three distinct regimes: boundary, mixed, and hydrodynamic. When the sliding speed is relatively low, the frictional behaviour falls in the boundary regime; if the sliding speed changes, the frictional coefficient remains constant dominated by interactions between the sliding surfaces. In contrast, higher sliding speeds are observed in the hydrodynamic regime where the lubricant get completely separated from the sliding surface and the frictional response is dominated by the viscous drag. On the other hand, the transition between the boundary and hydrodynamic regimes is referred to as a mixed regime where the frictional coefficient decreases with the separation of sliding surfaces by the testing sample (Chen and Stokes 2012).

The coefficient of friction of EWP was determined as a function of sliding speed for assessing the lubrication properties of the printing system. Due to the complex composition, the EWP material supply used for 3D printing did not fit the traditional Stribeck curve. The frictional curves of five different concentrations of EWP exhibit a 'stick and slide' pattern (Liu et al. 2019a). When the sliding speed was in the range of 100–500 $\mu m\ s^{-1}$, the particles pass through the gap between the sliding surfaces that result in less lubrication with increase in coefficient of friction. On the other hand, the coefficient of friction of EWP was decreased with an increase in sliding speed when it reaches 500 $\mu m\ s^{-1}$ (Scaraggi and Persson 2014). However, with an increase in the sliding speed of >100 000 $\mu m\ s^{-1}$ the coefficient of friction was increased as the curve entered the hydrodynamic regime. It was reported that highly viscous samples exert a low coefficient of friction as in the case of dairy products (Nguyen et al. 2016). With the increase in the EWP, the coefficient of friction was significantly increased from 0.29 to 0.37 under 50 000 $\mu m\ s^{-1}$ due to a substantial increase in the particle size of the printing mixture. As a function of printing temperature, the coefficient of friction was increased with the rise in the temperature from 20 to 60 °C; later it got decreased when the temperature reached 60 °C. This showed the transition of a liquid matrix of EWP to a gel system at 70 °C where the coefficient of friction was decreased due to squeezing out of the sample between the sliding surface. On the other hand, the coefficient of friction remained stable during the cooling stage at a temperature from 40 to 20 °C and then increased when the temperature drops below 20 °C (Liu et al. 2019a). From the observation, the highest coefficient of friction was observed at 10 °C. The associated changes in the frictional behaviour were mainly because of the gelation characteristics of gelatin in the EWP printing mixture.

In another study, the lubrication properties of the chocolate mixture were examined for 3D printing. The obtained frictional curve of 3D printing chocolate mixture with and without additives did not resemble the Stribeck curve (Figure 4.13) (Mantihal et al. 2019b). This deviation was mainly because of the existence of the particulate structure of chocolate and also the addition of plant sterol and magnesium stearate as additives that enhanced the particulate structure of the chocolate matrix. As the entrainment speed was increased, the coefficient of friction was low at the boundary regime. The coefficient of friction was increased on the transition to a mixed regime where the sliding was increased from 0.1 to 100 $mm\ s^{-1}$. This was due to the slow change in the slope of the frictional curve along with a steady increase in the frictional coefficient with an increase in the sliding speed (Carvalho-da-Silva

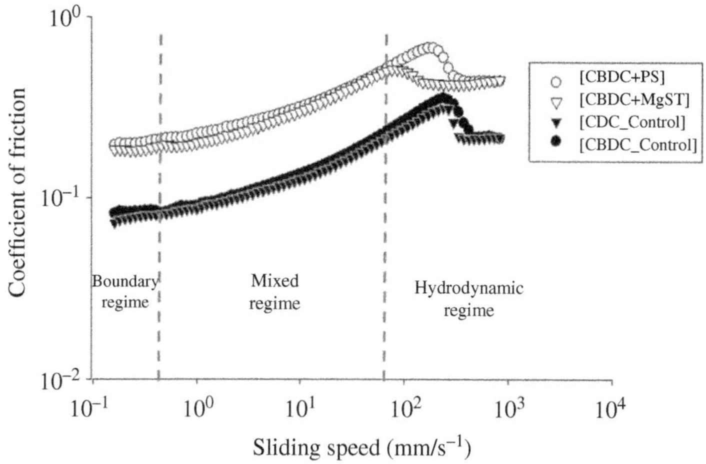

Figure 4.13 Coefficient of friction curve of different samples of 3D printed chocolate at constant temperature 35 °C. *Source:* Mantihal et al. (2019b) / With permission of Elsevier.

et al. 2013). Further, the increased coefficient of friction was because of the pure elastic response of the chocolate mixture below the critical frictional force. On comparing the effects of additive, the chocolate mixture with added plant sterol showed the highest coefficient of friction than magnesium stearate due to its larger particle size. In contrast, the control sample without any additives had a lower coefficient of friction indicating greater lubrication than samples added with additives. Thus, the incorporation of additives contributes to the lower tendency of slippage during the material flow by extrusion (Muredzi et al. 2013). From the sensory perception, the chocolate mixture with plant sterol had higher friction that contributes to grainy textural mouthfeel to the 3D printed chocolates. However, the reduction in the size of the particulates of the 3D printing mixture greatly helped in overcoming the graininess of the chocolate. The knowledge about the frictional behaviour of material supplies used for 3D printing helps in a deeper understanding of the rheological properties in turn assists in the optimization of material formulation for 3D printing. In addition, the tribological studies provide insights about sensory attributes of the food matrix that aids in understanding of the taste of the 3D printed food which is crucial for consumer's acceptance and marketability.

4.3.5 Thermal Properties

The thermal properties of the biomaterials are measured as a function of temperature (−150 to 1600 °C) and time. Some of the common thermal analysis techniques used for characterization of polymer composites include differential scanning calorimetry (DSC) based on heat difference, thermogravimetric analysis (TGA) based on mass/ weight, thermomechanical analysis (TMA) based on dimensional deformation, dynamic mechanical analysis (DMA) based on mechanical stiffness and damping, differential thermal analysis (DTA) based on temperature difference and dielectric thermal analysis (DEA) based on dielectric permittivity and loss factor (Saba et al. 2017). Among which DSC and TGA are the most commonly used techniques for the characterization of material supply as reported in 3D printing studies. The temperature of printing material supply is interrelated with the thermal properties such as melting point, gelatinization entropy, crystallinity, thermal conductivity, thermal diffusivity, and heat resistance; rheological properties such as flowability, viscosity, and consistency; and textural properties such as hardness and mouthfeel (Nachal et al. 2019). Thus, thermal properties have significant implications on the finishing quality of the 3D printed constructs especially for those materials printed using hot-melt extrusion. Chocolates are one such material that requires optimal printing temperature for successive deposition of layers. In a more recent study on the 3D printing of chocolate, the thermal flow curves were analyzed for correlating the chocolate crystalline forms with printability (Rando and Ramaioli 2020). With an increase in the temperature above 29 °C, the heat flows were gradually decreased showing the endothermic transition due to the melting of cocoa butter crystals in V form. Consequently, during the cooling stage, the exothermic transition occurred between 0 and 24 °C with a temperature peak found at 14.35 °C. From the DSC thermograms, the enthalpy of crystallization and the specific value of the enthalpy of crystallization was found to be 45.78 and 121.11 $J\,g^{-1}$, respectively. Among the crystalline type, form V β is the most favourable crystal in cocoa butter that provides required properties such as snap, glossiness, and smooth mouthfeel to the printed 3D constructs (Lanaro et al. 2017).

In another study, the peak melting point (T_p) corresponding to this crystalline form lies in the range of 32–34 °C (Mantihal et al. 2019b). Results showed that T_p found around 33.2 °C and the addition of additives (plant sterol and magnesium stearate) in its powder form had no influence over the melting behaviour of cocoa fat. Also, the thermal properties were not affected by the printing process. Although there was a significant difference between the melting temperature of the sample in the feed syringe and 3D printed chocolate samples, the difference was comparable. Considering the chocolate types, the melting point of the control samples of Callebaut dark chocolate (CBDC) was slightly lower than the control samples of Cadbury dark chocolate (CDC). This was because of the differences in the crystal size and polymorphs of the individual types (Tan and Kerr 2018). On the other hand, the melting enthalpy of the chocolate showed a significant difference with the addition of additives. The higher melting enthalpy value of the samples with additives was mainly due to a rise in the particulate materials. There was no significant difference observed in the melting enthalpies between the additives (plant sterol and magnesium stearate) which indicates that these additives did not influence the thermal properties. Considering the effect of extrusion, the enthalpy of the pre-crystalized samples was not affected during the 3D printing process (Mantihal et al. 2019b). Thus, the study concluded that both plant sterol and magnesium stearate could be used as flow enhancers in 3D printing of chocolate and the printing conditions must be set below the melting point of the fat that retained the seed for achieving a stable crystal growth and nucleation.

In a more recent study, TGA was used for characterizing the thermal properties of a 3D printing mixture of complex coacervates from gelatin-gum Arabic oil (Guo et al. 2021). The event of thermal mass loss (about five events) experienced by the microcapsules of the printing mixture was evident from the TGA curve (Figure 4.14). The first mass loss event was observed to be 8% between 25 and 120 °C due to the evaporation of the adsorbed water from the surface of the microcapsules. The second mass loss event was found to be 1% that was recorded in the temperature range of 120–175 °C mainly because of loss of unencapsulated

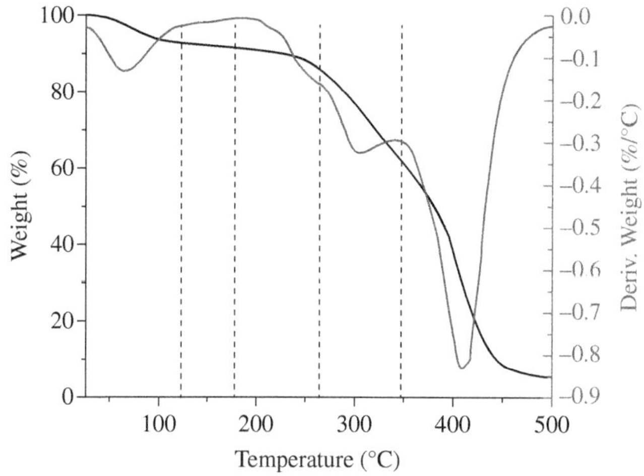

Figure 4.14 TGA thermograms of microcapsules of buckwheat based complex coacervate. *Source:* Guo et al. (2021) / With permission of Elsevier.

aroma compounds. Subsequently, the third, fourth, and fifth event occurred in the temperature range of 175 and 500 °C resulting in 86% mass loss. This higher mass loss was mainly due to the series of events: destruction and depolymerization of the functional groups of hydrocolloids (Singh et al. 2020); the dehydration and depolymerization of CHO rings of gum Arabic (de Barros Fernandes et al. 2016); and thermal degradation of a polymeric network of gelatin (Inal and Mulazimouglu 2019). In addition to the above mass loss, another 5% mass loss was encountered during the temperature range of 175–260 °C mainly attributed to the release of embedded aroma and volatile flavour compounds from essential oils used (Guo et al. 2021). Thus, the heat triggered the release of the pigment and essential oil from complex coacervates of gelatin-gum Arabic oil was characterized using TGA. In a similar study, the thermal stability of Bovine skin gelatin (BSG)/ zinc oxide (ZnO)/ clove essential oil (CEO) nanocomposite film was assessed from TGA curves showed 2–3 stages of thermal decomposition (Ahmed et al. 2020). The weight loss of the first decomposition was related to the thermal degradation of bound water, glycerol and proteins with low molecular weight (Hoque et al. 2011). In the second stage, the decomposition was associated with the degradation of larger-sized protein fractions (Tongnuanchan et al. 2015). It was inferred that the reinforcement of ZnO enhanced the thermal stability of the gelatin films. The impregnation of ZnO nanoparticles on the gelatin matrix acts as a thermal insulator to the volatile substances that were produced during decomposition by heat. Thus, the thermal attributes such as entropy, enthalpy, melting, and crystallization points are significant in determining the chemical state and molecular stability that in turn influences the post-deposition characteristics of the 3D printing material supply.

4.3.6 Dielectric Properties

The dielectric properties of the material provide knowledge about the heating characteristics of the material under the microwave field. The common dielectric properties are dielectric constant (ε'), dielectric loss (ε''), and dielectric loss tangent ($\tan \varepsilon = \varepsilon''/\varepsilon'$) (Kumar and Karim 2019). The ε' represents the ability of the material to absorb electrical energy shows the polarizable nature of the molecules. While ε'' represents the amount of energy loss by the material that describes the ability of molecules to convert the electromagnetic energy into heat energy (Munoz et al. 2018). On the other side, $\tan \varepsilon$ represents the energy dissipation of material where the electromagnetic energy is transformed into heat energy at specified power and temperature. The higher $\tan\varepsilon$ of the material indicates effectiveness of absorption of microwave energy that leads to faster temperature rise. The amount of water content of the material supply greatly influences the dielectric properties.

The effect of microwave-salt synergistic pre-processing on the printability of soy protein isolate (SPI) and the strawberry mixture was assessed for printing performance (Fan et al. 2020). Since microwave pre-processing was done, the dielectric properties of the material supply have been examined. The addition of salt along with the increased microwave power level was inversely related with the ε' and directly proportional to the ε''. This observed trend of the dielectric properties was due to the effects of conductivity of the salt solution under the electromagnetic field. In addition, the microwaves promoted molecular motion thereby enhancing the combination of brine with the other molecular components (Chandrasekaran et al. 2013). This resulted in the immobilization of the water molecules that had significantly reduced their response towards electric field polarization. Further, the effect of microwaves in evaporating the water also led to decreased ε' (Pei et al. 2015). The dielectric characteristics of the sweet potato puree were studied with the addition of NaCl (He et al. 2020a). With increased content of salt from 0 to 4%, the ε' decreased by 6.7% while ε'' increased by 166.3%. This was mainly because of the dipole rotation and ionic conductivity of water molecules present in the material supply (Wray and Ramaswamy 2015). When a portion of water content was replaced with fructose syrup, the ε' was decreased while the ε'' has less effect. The less effects with added fructose content than that of salt content was because of the lower ionic conductivity of fructose (Wang et al. 2011).

Similarly, the dielectric properties of the effect of two different pretreatments such as microwave heating (MH) and ultrasound–microwave combined heating (UMCH) of wheat starch–papaya mixture were studied (Xu et al. 2019). The effective absorption of the microwave was increased with ultrasound treatment of material supply. This leads to a sudden rise in temperature that resulted in the distortion of the crystalline region of starch granules before hydration of the amorphous phase disintegrating the granular structure (Munoz et al. 2018). The phase transition of the particles from ordered to the disordered state during the heating of water leads to thermal gelatinization. With the increase in microwave power, the temperature of the system increased rapidly. On comparing, the temperature of UMCH treated samples was higher than that of MH samples. Thus, the dielectric properties of the material supply were greatly influenced by the structural state of the material. In another study, the effect of microwave vacuum drying on the dielectric properties of mango juice gel was reported (Yang et al. 2018a). Water showed the strongest absorption ability towards microwave energy. With the increase in drying time, both the ε' and ε'' of the mango juice gel was decreased. This indicates the weaker absorption ability of the mango juice gel which was mainly because of the composition of gel with added potato starch as an additive. The microwave vacuum post-treatment of material supply resulted in minor shrinkage and optimization process showed that 4-minute microwave vacuum drying was more suitable for 3D printed mango juice gel.

4.4 Extrinsic Properties

As stated earlier, the extrinsic properties are those that deal with the optimization of printing variables related to design parameters. Each of the 3D printing technologies has its own scope and applicability based on the intrinsic properties of material supplies. In this regard, the associated printing variables specific to each technology also varied. Thus, the optimization of the 3D printing process mainly focuses on the following three things: material characterization, material optimization, and optimization of process variables. In earlier sections of the present chapter different attributes on the characterization of material supply with printability were discussed. The present section deals with the optimization of selected material supply and optimization of process variables. The success of the developed 3D printed construct greatly depends on consumer preference and acceptability. A highly acceptable 3D printed product must possess high printing precision, accuracy, distinct shape, definite layers, and mechanical stability as well as integrity.

4.4.1 Optimization of Material Supply

Material selection is a significant aspect that dealt with specific 3D printing technologies. Based on the printability, the material supplies used for 3D printing are grouped as printable, non-printable, and alternative ingredients. A detailed discussion on the science of the printability of each of these material supplies has been presented in the subsequent chapters. Regardless of the category of material supply, the composition of material supply is formulated in such a way that supports 3D printing. The printing material supply is a complex mixture of macro food constituents such as carbohydrates, proteins, fibres and lipids (Godoi et al. 2016). The carbon chains of these food constituents remain the backbone of the printing system. Carbohydrates are biopolymers that range from simple sugars to complex polysaccharides. Similarly, the protein-based material supply comprises a sequence of peptide chains and amino acids (Lille et al. 2018). Although both these components exhibit positive effects on printability, chemically the mixture lacks the binding ability and adhesive property. These properties are most adequate for achieving a stable structure that can be incorporated through the addition of food additives (Kim et al. 2018b). Starch-based materials and food hydrocolloids are most often used as additives in the printing mixture where the associated gelation mechanism imparts adequate printability (Chen et al. 2019). Although lipids form a small fraction of material supply, the presence of lipids must be essential in tailoring the rheological properties through its plasticization behaviour (Anvari and Joyner 2019). In a more recent study, the rice starch from indigenous var. *Ponni* was examined for its printability without any additives. The associated pasting behaviour of the rice starch imparts printability that favours extrusion-based 3D printing (Theagarajan et al. 2020). Interestingly in a recent study, brown rice gels were used for 3D printing with the hydrocolloids like xanthan gum (XG), guar gum (GG), XG–GG mixture, and sodium carboxymethyl cellulose (NCMC) (Huang et al. 2020). The pre-processing of the brown rice mixture improves flowability and provides post-printing stability. Among the tested samples, the mixture printed with XG–GG showed better printing with appropriate hardness and smooth texture. Further, the microstructural analysis revealed the tightly connected network of smooth starch molecules without obvious deformations. Although rice starch is printable, the incorporation of hydrocolloids aids in achieving a stable post-deposited 3D construct.

The importance of pre-processing on the 3D printing of egg fractions was examined by Anukiruthika et al. (2020). In this study, the printability of non-printable egg fractions was studied with rice flour in different proportions. On comparing the egg fractions, the egg yolk showed greater printability than egg white due to its viscoelastic nature. Further, the lipid fractions of egg yolk act as a lubricant and imparts plasticization behaviour to egg yolk which is completely absent in egg white (Anvari and Joyner 2019). The addition of rice flour had a linear positive effect on printability with an increased proportion of rice flour irrespective of egg white and egg yolk material supply. The combination of egg yolk proteins (plasma and granules) and rice starch had a synergistic effect in enhancing printability through its binding properties. In another study, the heat-induced egg yolk paste without any additives was 3D printed (Xu et al. 2020). The associated gel stability of egg yolk was examined at different temperatures (72, 76, 80, and 84 °C). Results showed that gelation characteristics, rheological attributes, and mechanical stability of the printing mixture varied with heat treatments. The heat-induced denaturation of proteins altered its distribution and changed the mobility of egg fractions resulting in the reduction of the intramolecular β-sheet structures. Among which the egg yolk treated at 76 °C for 8 minutes showed the best printing performance with higher shape accuracy. At this condition, the sample exhibits shear-thinning characteristics and possesses a solid-like behaviour that aids in the extrudability of the material supply. Likewise, studies were reported on the optimization of printing formula for baking of cookies and the development of 3D printed snacks from vegetables (Derossi et al. 2018; Severini et al. 2018b). Also, studies were reported on 3D printing of myofibrillar tissues with suitable hydrocolloids and adequate pre-processing of the raw material prior to 3D printing (Dick et al. 2020; Lipton et al. 2010).

In a more recent study, the printability of hydrocolloid mixture (xanthan/ konjac gums) was demonstrated by correlating the material properties with visual contour 2D plots (Garcia-Segovia et al. 2020). In 2D contour plots, those points that had the same response were connected to form contour lines that correspond to the shape parameters, i.e. height and area of the figure. From the contour plot, the lesser observation corresponds to the height and area of the figure after a 1-hour period. Results showed that after 1 hour the samples maintained a height of 1.7 cm with an area < 12.5 cm^2 and also the sample possess a height of 1.8 cm with a range of area from 11.5 to 12.5 cm^2 after 5 minutes. Similar contour plots were drawn for the rheological and textural parameters (Figure 4.15). Both rheological parameters (loss tangent, viscosity, and storage modulus) and textural parameters (maximum force and slope gradient) provided a good fit concerning the shape parameters. Thus, the study inferred that the correlation of the rheological properties of the gel system with its key parameters helps in ensuring the printability of material supply under controlled process conditions.

In another study on 3D printing of mashed potato, the effect of variation in material formulation on printing behaviour was presented in terms of printing percentage (Liu et al.

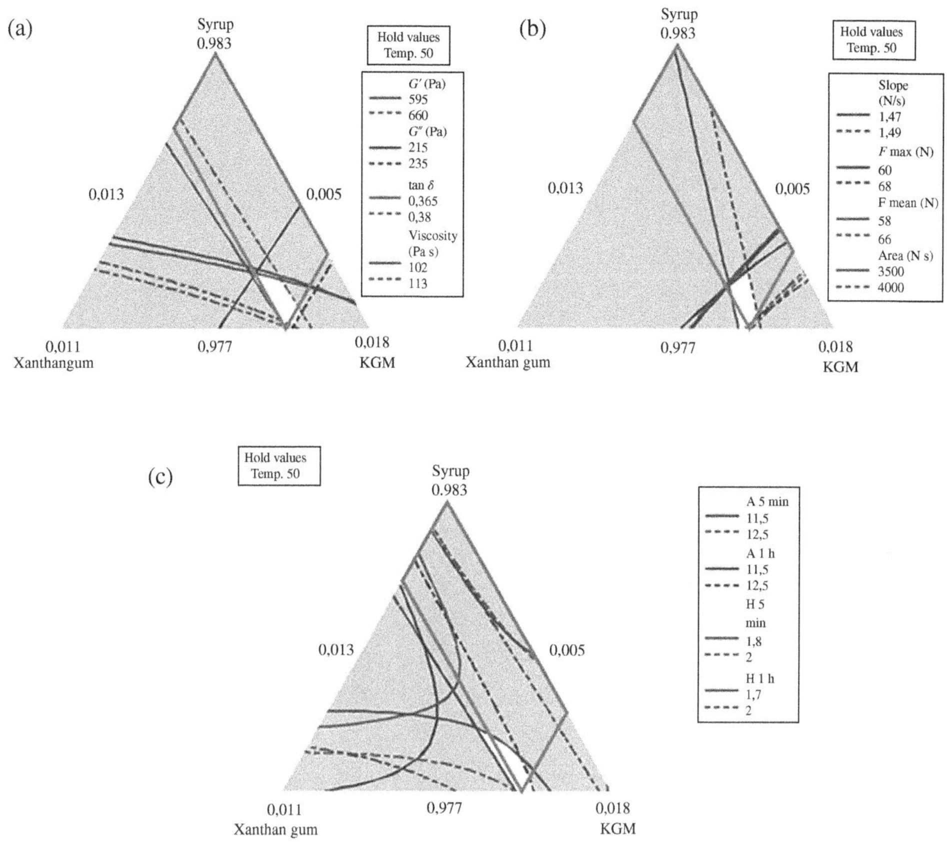

Figure 4.15 Two-dimensional contour plots of 3D printed xanthan/konjac gum mixture. (a) Rheological parameters, (b) textural parameters, and (c) shape parameters. *Source:* Garcia-Segovia et al. (2020) / With permission of Elsevier.

2020b). Different ratios of water to potato flakes were evaluated by 3D printing of a hollow cylinder (30 mm diameter and 50 mm height) based on the printing performance of each composition (Figure 4.16). Thus, the viscosity, material consistency, and flow behaviour of each formulation were correlated with the printability of the material supply.

4.4.2 Optimization of 3D Printing Process Variables

4.4.2.1 Nozzle Size and Nozzle Height

It is essential to ensure the continuous flowability of material through the small orifice of the printing nozzle. The printability of the material supply can also be determined by the rheological attributes of the material composition. It was stated that the sample viscosity must be low as possible to assist in material flow through the printing nozzle and high as possible to impart adhesiveness to the printed layers without deformation (Kim et al. 2019). The size of the printing nozzle used during printing must be proper enough for ensuring continuous flow without nozzle clogging. Further, the print fidelity of the 3D printed

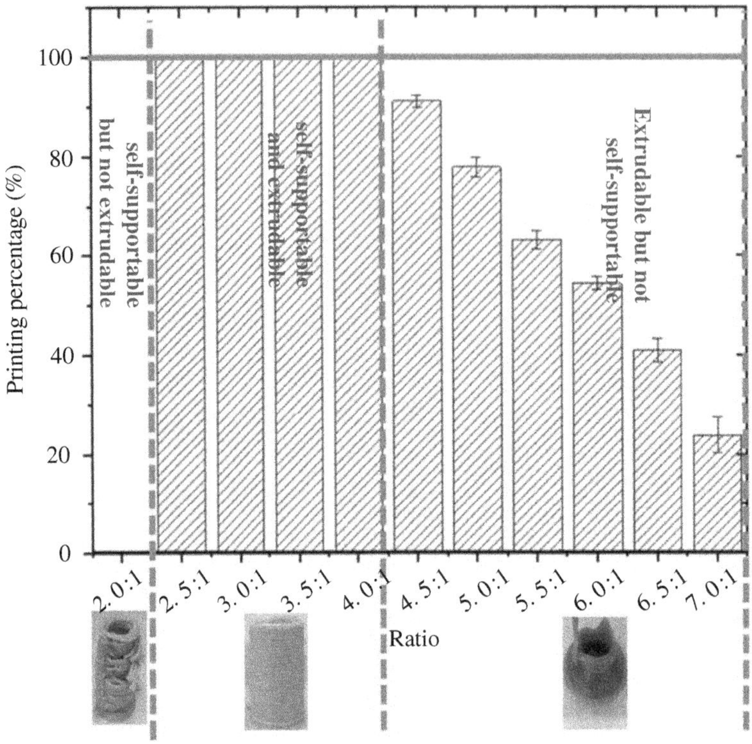

Figure 4.16 Different categories of 3D printing behaviour of mashed potato based on its printing percentage. *Source:* From Liu et al. (2020b) / With permission of Elsevier.

construct depends on the resolution and printing accuracy that in turn influenced by the nozzle size (diameter). As a well-known fact, a smaller nozzle could produce fine smooth thread with high layer definition through the desired deposition of material while a larger sized nozzle could deposit more material producing larger threads that deteriorate layer definition (Liu aand Zhang 2019). However, it depends on the material composition as well as the particle size of the material. Considering extrusion-based 3D printing, nozzle height is the measurement of the space from the printing nozzle tip to the top of the deposited layer. An optimal nozzle height is adequate for achieving proper 3D printed thread. The inadequate nozzle height led to the piercing of the 3D printed layers while maximum space results in inaccurate threads that fall in the space rather than deposition over the printed layer (Krishnaraj et al. 2019). Hence in a 3D printing process, before optimization of the nozzle size, it is essential to optimize the nozzle height as it directly influences the shape of 3D printed geometry. An equation was proposed for estimating the critical nozzle height by Eq. (4.17) (Wang and Shaw 2005).

$$h_c = \frac{V_d}{v_n D_n} \tag{4.17}$$

where h_c is the critical nozzle height (mm), V_d is the volumetric extrusion rate (mm^3 s^{-1}), v_n is the nozzle movement speed, i.e. printing speed (mm s^{-1}), and D_n is the nozzle diameter (mm). From this equation, it was clear that the critical nozzle height is directly related to the extrusion rate and inversely related to printing speed and nozzle size. On average, the critical nozzle height was found to be 0.4 mm for egg white and egg yolk material supplies of different formulations with rice flour blends (Anukiruthika et al. 2020). For desired printing, the nozzle height should not be greater than the critical nozzle height as it is like a cut-off value. It was stated that the thread thickness should be the same as that of nozzle diameter. However, in a practical situation there occurs discrepancies due to material swelling and shrinkage with the use of improper nozzle size. This was mainly because of the differences exist with the density of the material formulations.

In another study, the layer height (nozzle height) was varied as 25, 50, 75, and 100% as that of the nozzle diameter (0.33, 0.84, and 1.28 mm) (Krishnaraj et al. 2019). Results showed that nozzle height of 25 and 50% resulted in the piercing of the pre-printed layers while 100% nozzle height, i.e., nozzle height equal to that of the nozzle diameter leads to the incongruence of the printed structure. In contrast, nozzle height of 75% yielded in good 3D structure without significant layer disruptions. Among the tested nozzles, smaller nozzles of 0.84 mm produced fine smooth layers with higher printing resolution. However, the nozzle with a bigger diameter (1.28 mm) showed good print fidelity than the smaller nozzle (0.82 mm) for the mushroom-based material supply (Keerthana et al. 2020). This was mainly because of the microstructural characteristics of the material supply that composed of a considerable amount of fibrous content. Due to which the smaller nozzle resulted in under extrusion of the material supply. Further, the larger-sized nozzle withstands a higher shear rate that could conveniently overcome the material stickiness and particle aggregation. These results were found to be consistent with the 3D printing of the chicken where the material supply comprised of myofibrillar tissues (Wilson et al. 2020). Thus, nozzle height and size of the printing nozzle are the most significant factors in the material optimization for 3D printing by extrusion principle.

4.4.2.2 Printing Speed

The dynamic working range of the 3D printer was assessed by considering the effect of the speed of printing over the 3D printing process (Wang et al. 2018). The nozzle movement speed must be optimal enough for the ideal deposition of material that directly influences the extrusion rate. The higher printing speed results in dragging and pulling of layers lead to a discontinuous thread. While lower printing speed produces a wavy thread with over deposition of material. Both these printing conditions clearly indicate the mismatch exists between the speed of printing and the extrusion rate of the material supply. The critical nozzle movement (v_N, mm s^{-1}) can be determined from material flow rate (Q, cm^3 s^{-1}) and nozzle diameter (D_N, mm) (Eq. (4.18)) (Liu and Zhang 2019). Further, the range of printing speed varies with the nature and composition of the material supply. The composite millet-based flour with the particle size of $\leq$0.2 mm showed desired printability at the highest printing speed of 2400 mm minute^{-1} through optimizing the motor speed of the extruder to a maximum of 300 rpm (Krishnaraj et al. 2019). This significantly reduced the printing time that gives great scope for the commercial large-scale production of 3D printed foods. However, this varies with the type of material supply used for 3D printing. In another study, the optimal printing speed was evident at 600 and 800 mm minute^{-1} for the protein-based egg material supply where the maximum printing speed used was

1200 mm minute^{-1} (Anukiruthika et al. 2020). The higher speed of printing of 1200 mm minute^{-1} produced a discontinuous layer for the egg-based material supply, hence further printing speeds were not considered. These results were consistent with fibrous material supplies as well (Keerthana et al. 2020; Lille et al. 2018).

$$v_N = \frac{4Q}{\pi D_N^2} \tag{4.18}$$

In the optimization of 3D printing of fish surimi gel, the printing speed was ranged from 1 to 40 mm s^{-1} which in turn limited by the extrusion motor speed (Wang et al. 2018). When the printing speed was kept higher, the nozzle height was found to be much closer to the actual nozzle height of 5 mm. This represents the consistent flow of the material without any forced flow. However, with respect to printing speed, the flow instabilities were evident when the speed of printing was low (20 mm s^{-1}) resulting in the formation of coiled threads. On the other hand, higher printing speed (32 mm s^{-1}) causes the dragging of the extruded threads resulting in the breaking of printed threads. An ideal printing condition was optimized at 28 mm s^{-1} where the flow was very stable showing higher resolution and printing accuracy with fewer broken threads. The printing speed of the EWP mixture was ranged from 0 to 100 mm s^{-1} that was significantly affected by critical nozzle height and the rest of the variables were kept as constant (Liu et al. 2019a). Results showed that there was no obvious forced flow of the material supply when the printing speed was higher. This reflects that the determined nozzle height was consistent with the actual nozzle height as used in the experiments (3 mm). With a comparatively lower printing speed of 60 mm s^{-1}, the printed layers could not maintain the shape due to inconsistent flow with unsolidified threads. On the other side, higher printing speeds of 80 and 90 mm s^{-1} showed dragging of extruded threads lead to broken filaments. The average printing speed in the tested range of printing speed, i.e. 70 mm s^{-1} resulted in higher printing resolution and accuracy with lesser deviation from the design model. Thus, the printing speed must be optimized in relation to other printing parameters like nozzle size, layer height, and extrusion rate.

4.4.2.3 Extrusion Rate

In extrusion-based 3D printing, extrusion rate is one of the significant factors that reflects the amount of deposited material. The extrusion rate can be optimized by varying the extrusion motor speed and extrusion pressure that is directly proportional to the extrusion rate. Theoretically, the extrusion rate can be calculated from the nozzle height and nozzle diameter (Eq. (4.19)) (Yang et al. 2018b).

$$V_d = \frac{\pi}{4} v_n D_n^2 = \frac{\pi}{4} v_n h_c^2 \tag{4.19}$$

Based on the above equation, the extrusion rate of lemon juice gel was optimized based on printing speed for different nozzle diameters (Figure 4.17). Results showed a linear increase in extrusion rate with both nozzle diameter and printing speed that indicates higher deposition of material. As expected, when the printing speed was increased, the printing efficiency was higher. But the predicted extrusion rate was lesser than the predicted value. Concerning the print quality, there was no significant difference in the printing quality of the fabricated 3D construct up to the printing of 35 mm s^{-1}. Thus, the optimized

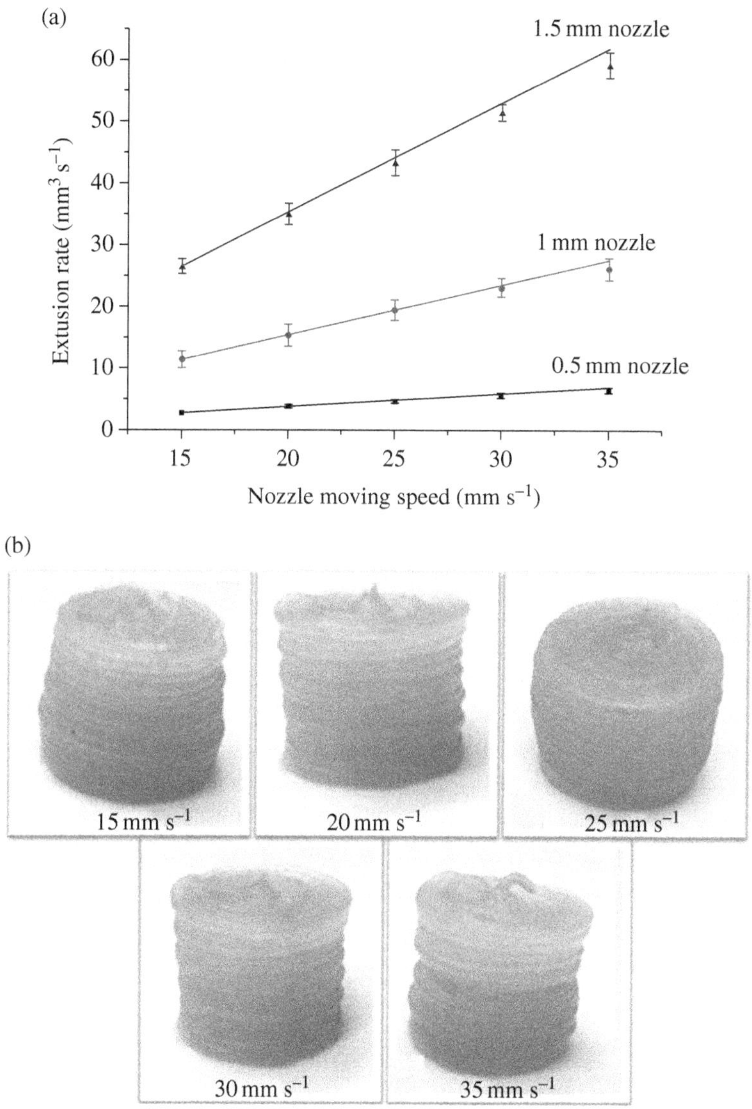

Figure 4.17 3D printed lemon juice gel. (a) Relationship between extrusion rate and printing speed. *Source:* Yang et al. (2018b) / With permission of Elsevier and (b) 3D printed constructs of different printing speeds. *Source:* From Yang et al. (2018b), Figure 04 [p. 32] / With permission of Elsevier. DOI-https://doi.org/10.1016/j.lwt.2017.08.054.

extrusion rate was found to be 24 $\text{mm}^3\,\text{s}^{-1}$ for the printing speed of 30 mm s^{-1} and 1 mm nozzle diameter.

The extrusion rate (deposition rate) of the cylindrical filaments of the blend of fruit and vegetable mixture was determined using Eq. (4.20) (Derossi et al. 2018).

$$\text{Deposition rate}\left(\text{mm}^3\,\text{s}^{-1}\right) = \pi\left(\frac{N}{2}\right)^2 \times S \times \text{flow} \tag{4.20}$$

where N is the nozzle diameter (mm), S is the printing speed (mm s^{-1}), and flow represents the material flow (%). The optimal extrusion rate was practically determined using a line test where a linear relation was derived for the extrusion rate and thickness of the printed thread of the EWP mixture (Liu et al. 2019a). The diameter of the printed thread was larger than that of the used nozzle diameter (2.5 mm) at higher extrusion rates of 0.006 and 0.007 $cm^3\ s^{-1}$. This was due to the result of increased nozzle size and extrusion pressure that leads to the spreading of the printing mixture. On the other hand, there was no extrusion of the material when the extrusion rate was lowest. This was attributed to the decreased mass of material resulting in reduced thread diameter than the actual nozzle size. Thus, for the EWP mixture, the optimized extrusion rate was found to be 0.004 $cm^3\ s^{-1}$. A similar trend was observed for the egg white and egg yolk fractions where the thread thickness was increased with extrusion rate irrespective of proportions of rice flour blends (Figure 4.18) (Anukiruthika et al. 2020). For the optimized material supply of chicken meat-based printing mixture, the rate of extrusion was found in the range of 3.52–14.08 $mm^3\ s^{-1}$ for 0.82 mm nozzle and 7.79–31.17 $mm^3\ s^{-1}$ for 1.22 mm nozzle (Wilson et al. 2020). The printability of low-fat chicken material supply was attributed to its collagen content that acts as a plasticizing agent (Choi et al. 2016). The hydrophilic-hydrophobic balance of the fibrous meat constituents resulted in the formation of a stable emulsion that improves the extrudability of the material supply.

4.4.2.4 Printing Rate

The printing rate also referred to as flow rate can be expressed as the amount of material printed in a certain period of time. The printing rate is the mass flow rate of the material determined by Eq. (4.21) (Mantihal et al. 2017). Similar to the volumetric extrusion rate, the printing rate is also influenced by the speed of printing, extrusion motor speed, and extrusion pressure (Liu and Zhang 2019). The flow rate of different formulations of sodium caseinate dispersions showed a nonlinear increase with the applied pressure that indicates the shear-thinning nature of the material supplies (Figure 4.19; Schutyser et al. 2018). When a smaller-sized nozzle is used, the 3D model can be sliced into a greater number of layers that correspond to greater printing time. Further, the flow rate of the material is varied with the nozzle size as well as nozzle height.

The maximum printing rate was observed for highest the motor speed of 240 rpm and the lowest printing speed of 800 mm $minute^{-1}$ from the larger sized printing nozzle of 1.7 mm diameter during the optimization of 3D printing of rice starch (Figure 4.20) (Theagarajan et al. 2020). Thus, the printing rate of the sample greatly affects the shape accuracy and resolution of the 3D printed construct.

The printing rate for the mushroom-based printing mixture was found to be in the range of 0.2–0.6 g $minute^{-1}$ for a 1.28 mm nozzle and 0.037–0.21 g $minute^{-1}$ 0.82 mm nozzle (Keerthana et al. 2020). Results showed that the material supply with 15 and 20% incorporated mushroom powder with 1.28 mm nozzle at 800 mm $minute^{-1}$ printing speed had good print fidelity with a flow rate of 0.29–0.44 g $minute^{-1}$. The increased proportion of mushroom powder with wheat flour significantly affected the gluten network that helped to overcome stickiness and contributes to flowability. Among the tested printing conditions, the material supply with 20% mushroom powder was found to be optimal with an average flow rate of 0.043 g $minute^{-1}$ that exhibits better printing characteristics. In another study on 3D

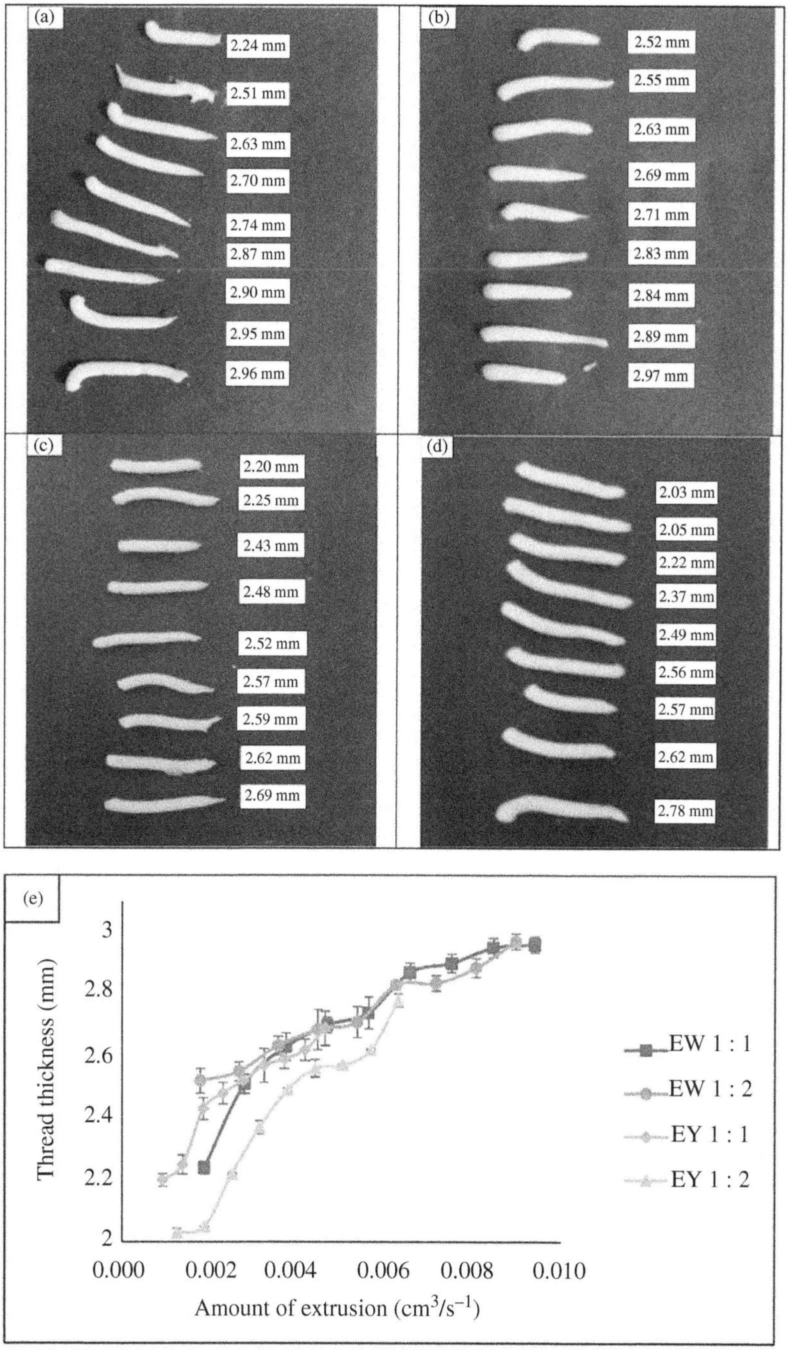

Figure 4.18 Optimization of extrusion rate of egg white (EW) and egg yolk (EY) fractions. (a) EW 1 : 1 *Source:* From Anukiruthika et al. (2020), Figure 03 [p. 009] / With permission of Elsevier. DOI-https://doi.org/10.1016/j.jfoodeng.2019.109691, (b) EW 1 : 2, (c) EY 1 : 1, and (d) EY 1 : 2 (images are not to the scale); (e) Relationship between extrusion rate and thread thickness. *Source:* Anukiruthika et al. (2020) / With permission of Elsevier.

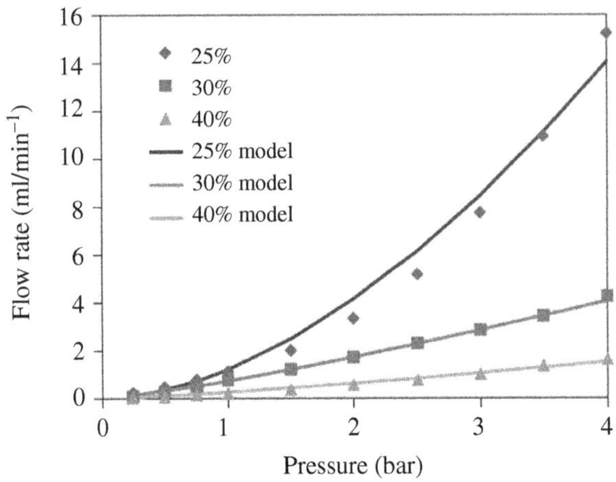

Figure 4.19 Flow rate as a function of dispenser pressure of sodium caseinate solution of different concentrations. *Source:* Schutyser et al. (2018) / With permission of Elsevier.

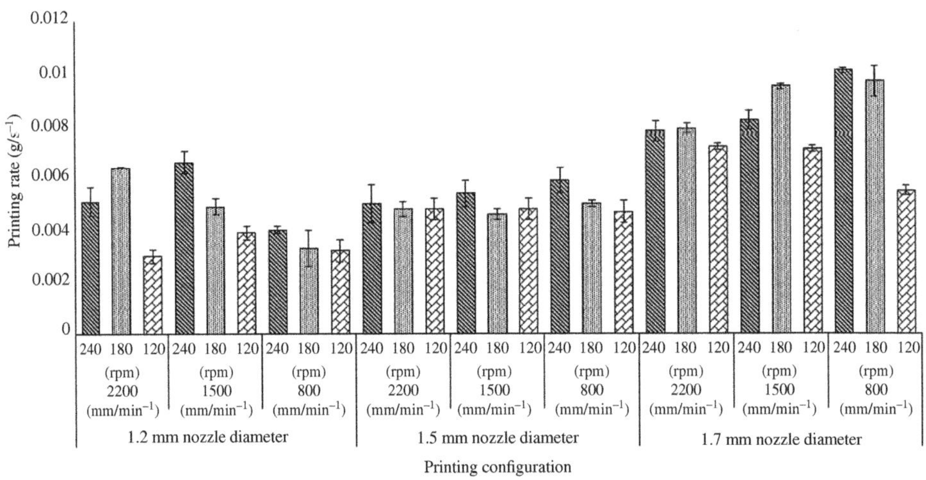

Result as observed for three print speeds (800,1500, and 2200 mm/min^{-1}), three motor speeds (120,1801 and 240 rpm), and three nozzle diameters (1.2, 1.5, and 1.7 mm).

Figure 4.20 Effect of printing variables on the printing speeds of 3D printing of rice starch. *Source:* Theagarajan et al. (2020).

printing of chicken meat, the printing rate was decreased with proportional increase in the chicken meat into the refined wheat flour (Wilson et al. 2020). However, the flow rate had a positive correlation with printing speed.

$$\text{Printing rate}\left(\text{g min}^{-1}\right) = \frac{\text{Weight of the 3D printed sample}}{\text{Time taken for 3D printing of sample}} \tag{4.21}$$

4.4.2.5 Infill Percentage and Infill Pattern

Infill percentage or infill density is a common factor for all types of 3D printing technologies that refers to the amount of material that constitutes the internal structure (Nachal et al. 2019). Infill percentage is a significant factor that determines the stability and mechanical strength of the internal structure by varying the void fraction and mass of the inner part of the 3D model. It was reported that the proportional increase in the percentage of infill significantly improved the mechanical stability of the 3D printed mashed potato (Liu et al. 2018c). Similarly, the textural properties such as hardness and gumminess of 3D printed brown rice were closely related to infill density. On the other hand, the infill pattern denotes the design pattern of the internal structure that mainly provides support to the printed constructs. Three different infill patterns (star, honeycomb, and Hilbert curve) with varied infill percentages (5, 30, 60, and 100%) were used for 3D printing of chocolate (Mantihal et al. 2019a). Results showed that star and honeycomb infill internal designs were the most stable and possess tough structures even at a reduced infill density of 60% (Figure 4.21). Further, the snap test indicated that 3D printed chocolate even at 100% infill density was less hard than the casted samples. Thus, the variation in the infill percentage aids in the creation of complex intricate designs and internal geometries with modified textures.

In another study, the effect of internal support structures on the printing performance of chocolate was reported (Mantihal et al. 2017). The support structures have a significant role in preventing the disintegration of the 3D printed construct (Figure 4.22). Results indicated that samples with cross-support possessed higher snap-ability with a firmer texture than parallel support while the sample devoid of support required relatively lesser force to break.

The viability of the probiotics was analyzed in two different cylindrical 3D constructs of 60% concentric infill and 30% honeycomb (Zhang et al. 2018). The viability of cells was affected by the surface-to-volume ratio of the printed 3D constructs that significantly accelerated the baking process (Sun et al. 2018b). The viable counts of the honeycomb infill structure after the baking process were found to be 2-log higher than the concentric infill structure. In a study on the development of 3D printed snacks from composite flour, the infill percentage is kept at 100% for all the printing trials and the infill pattern was selected as honeycomb (Krishnaraj et al. 2019). The printing behaviour of brown rice gel with different hydrocolloids was 3D printed into a hollow cylinder with a 25% infill percentage (Huang et al. 2020). In this study, the viscosity of the material supply was related to the stability of the printed lines. The 3D printed cylinder with a mixture of xanthan gum and guar gum produced continuous filaments and tightly connected microstructure showing good internal stability. This infill pattern was mainly because it provides cross-support to the 3D construct. The effect of three different infill densities (50, 75, and 100%) with the rectilinear infill pattern was analyzed in 3D printing of beef (Dick et al. 2019). In this study, the meat paste and lard were 3D printed to form a multi-material 3D construct. There was no significant difference observed with the interaction of fat percentage and infill density on the dimensional deviation of the printed sample. However, significant deviations were observed with the increasing infill density from 50 to 100% due to a proportional increase in the weight of the printed construct that attributes to a rise in the structure's density.

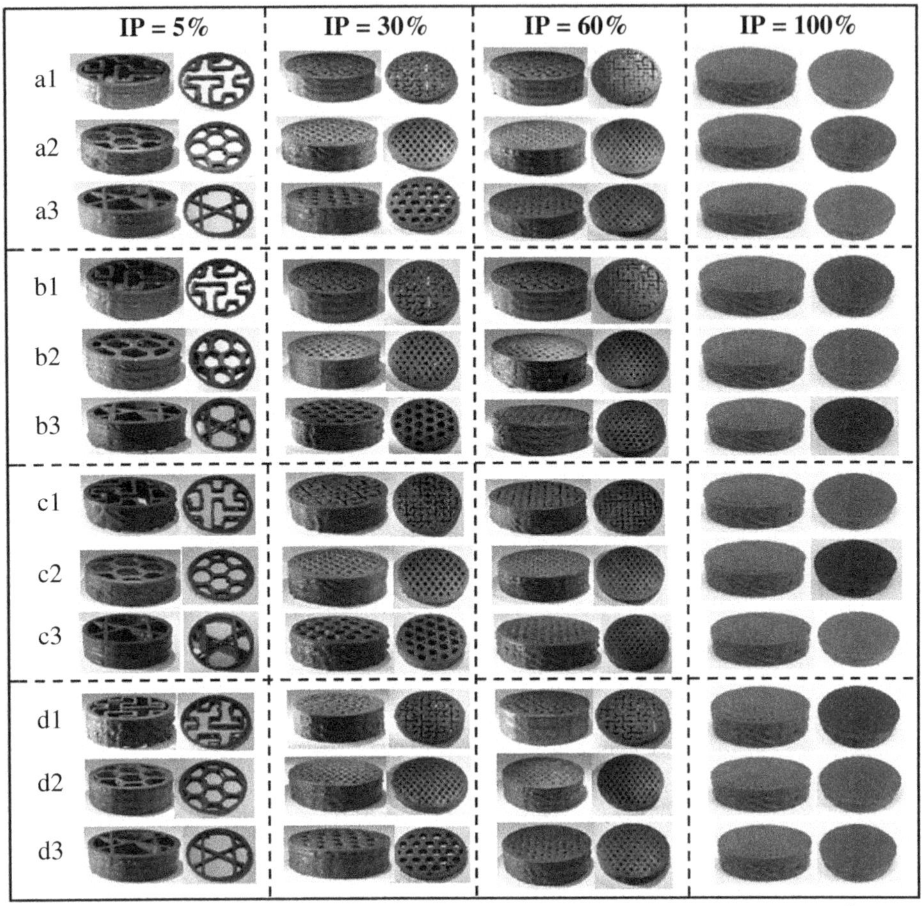

Figure 4.21 3D printed constructs of different chocolate formulations (a)–(d) with varied infill percentage (IP) and infill patterns (1–3). (a) Choc-1_control, (b) Choc-2_Control, (c) Choc-2 + MgST, and (d) Choc-2 + PS; 1, Hilbert curve; 2, honeycomb; and 3, star. *Source:* From Mantihal et al. (2019a), Figure 05 [p. 24] / With permission of Elsevier. DOI-https://doi.org/10.1016/j.foodres.2018.12.034.

Figure 4.22 3D printed chocolates of different model designs. (a) Hexagonal geometry with cross-support, (b) hexagonal geometry with parallel support, and (c) hexagonal geometry devoid of support. *Source:* From Mantihal et al. (2017), Figure 07 [p. 020] / With permission of Elsevier. DOI-https://doi.org/10.1016/j.ifset.2017.09.012.

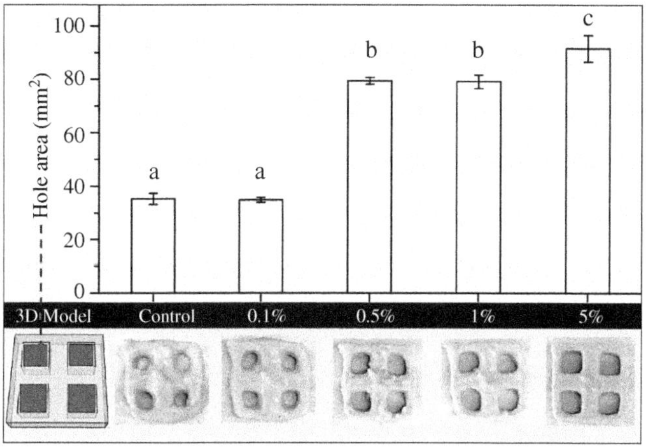

Figure 4.23 Printing performance of yellow flesh peach incorporated buckwheat dough with different levels of microcapsules and their measured hole areas. *Source:* From Guo et al. (2021) / With permission of Elsevier.

In an interesting study, the 3D printing performance of buckwheat flour was analyzed in combination with yellow-fleshed peach powder and sodium alginate (Guo et al. 2021). A porous 3D construct was printed with varying levels of microcapsules (complex coacervates with pigments and flavoured oil). The printability of the material supply was assessed by determining hole area of the internal structure. During the printing of a porous structure, the print path follows a sequential movement of the piston resulting in a back pumping-extrusion cycle. At these conditions, low viscous material supplies could result in the collapse of the internal structure showing poor self-stability with a small hole area. Results showed that increase in the proportion of microcapsules, the outer layers of the 3D construct became smoother with distinct sharp corners and edges (Figure 4.23). This clearly indicates the enhancement of mechanical properties of the wheat dough with added microcapsules that improved the self-stability.

It was observed that an increased proportion of microcapsules significantly increased the hole area, especially at 0.5% that in turn is associated with rheological properties showing higher printing performance (Guo et al. 2021). In another study on the development of programmable texture modified 3D construct, a parallelepiped shape with internal cube-shaped pores with varied fractions of porosity (18–54%) was analyzed (Derossi et al. 2020b). The overall volume of the 3D printed snack was relatively smaller than the virtual model with a fractional deviation of porosity fraction in the range of 10 and 140% indicating a greater increase in porosity. The effect of baking was analyzed through the relationship between the weight and porosity fraction of raw and cooked samples (Figure 4.24). The weight of the raw samples was linearly decreased with number of pores while the porosity fraction significantly increased between 35 and 60% for the samples after cooking.

4.4.2.6 Extruder Offset and Retraction Length

Extruder offset is one of the design variables that are more particular to multi-extruder 3D printers. During simultaneous printing of multiple material supplies, the levelling of the

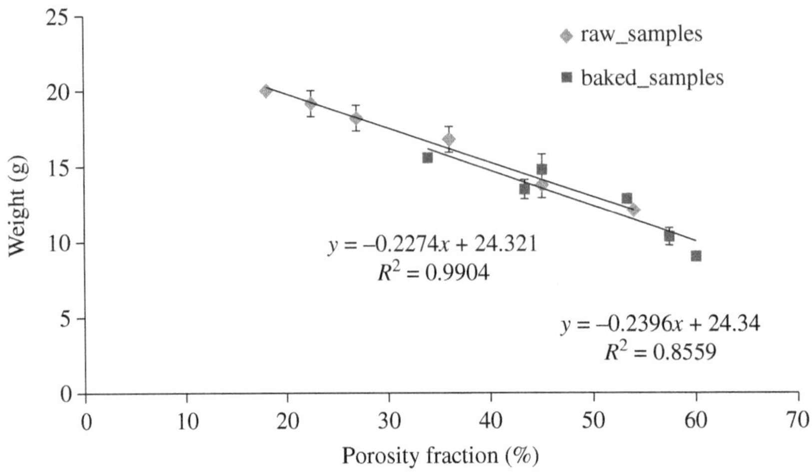

Figure 4.24 Relationship between weight and porosity fraction of raw and cooked 3D printed cereal dough-based material supply. *Source:* Derossi et al. (2020b) / With permission of Elsevier.

extruders is essential for the successful deposition of the material (Ghazal et al. 2019). Thus, the extruder heads must be parallel to the deposition stage or printing platform with an equal distance. In general, the extruder offset of the first print head is usually set as (0,0) while the extruder offset of the successive print heads is optimized based on the previous print head. So, the optimization of the extruder offset of the print heads is adequate as the printer firmware could not handle the displacement of each of the additional nozzle by itself (Tan et al. 2018). Improper extruder offset settings may lead to large deviations of the 3D construct from the desired one. The deformation results mainly due to improper material deposition either because of overlapping or material gap occur in the junction of 3D structures. The extruder offset of the second extruder was found to be at $x \sim 66.5$ mm and $y \sim 0.8$ mm with respect to (0,0) for dual extrusion of mashed potato along with strawberry juice gel (Liu et al. 2018d).

Another design attribute related to material deposition is the retraction length that must be optimized to prevent the oozing effects (Dankar et al. 2018). As in the case of dual extrusion, the second print head must be dormant while the first print head is working and vice-versa. This indicates that the extruder must have precise control to stop the flow of material (oozing) during the dormant stage. Thus, the extruder must control the flowability of the material during the dormant as well as in the active stage. This aspect is more closely correlated with the rheological properties of the material supply (Severini et al. 2018b). During testing materials like blueberry jam showed very low viscosity that resulted in oozing out of the material under gravity. Even the mashed potato and strawberry juice gel had oozing effects during the initial stage of printing. Later the material supplies become stable with time due to its viscoelasticity (Sweeney et al. 2017). Thus, it is critical to retract the material from oozing by optimizing the flow properties and composition of the printing mixture. The retraction length was optimized by setting the length to a positive value in the slicing software that automatically directs the extruder to be reversed by the given distance before its travel and during the dormant stage. Automatically, the retraction would be compensated again before the next extrusion cycle. Results indicated that strawberry juice gel had more oozing effects than mashed potato, hence a larger retraction length value of 2 mm was set for strawberry juice gel while for mashed potato was 1.5 mm (Liu et al. 2018d). This was mainly because of the more elastic nature of strawberry juice gel than that of mashed potato characterized with lower loss tangent.

The effect of retraction distance, speed of retraction, and travel speed on the morphology of wheat-based dough samples were analyzed (Derossi et al. 2020a). Results showed that the print quality was negatively affected by retraction length while the retraction speed had positive effects. There was an excess deposition of material due to effects of retraction movements; however, a gradual increase in retraction speed reduced the over deposition of material. Thus, the increase in retraction distance negatively affected print quality while the subsequent increase in retraction speed helped to overcome the negative effects of retraction distance that resulted in a 3D construct with minor discrepancies when retraction distance was 1 mm. The desirability profile of the perimeter of 3D constructs was presented in Figure 4.25. Results indicated that maximum desirability of 0.833 was obtained for the following conditions with a 1 mm retraction distance, 200 mm s^{-1} travel speed, and 105 mm s^{-1} retraction speed.

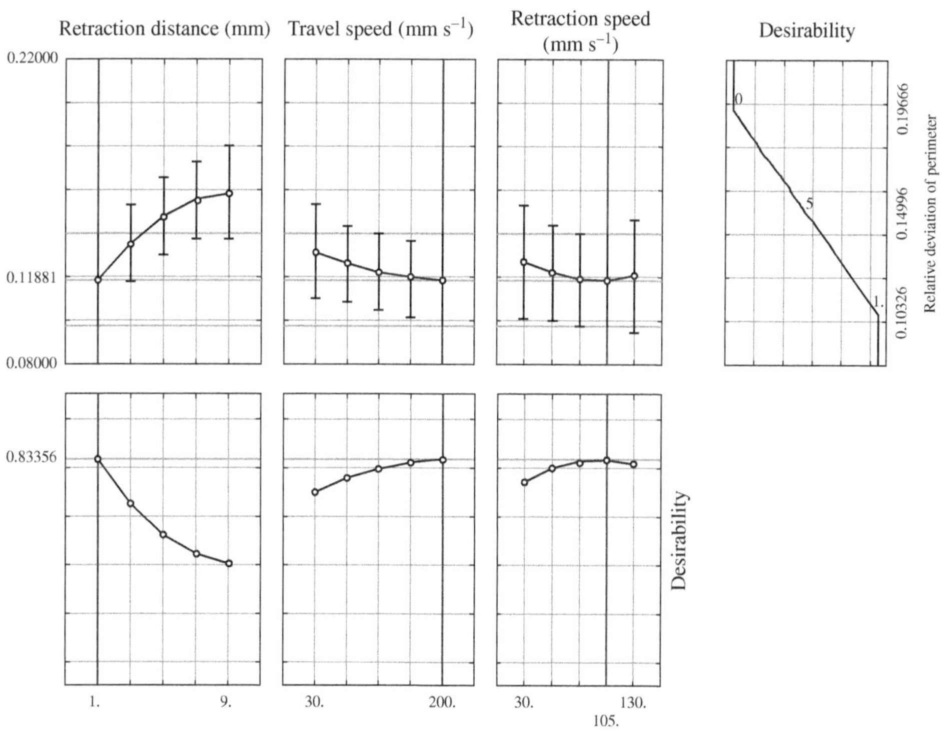

Figure 4.25 Desirability profile for relative deviation of perimeter of 3D construct as function of retraction distance, travel speed, and retraction speed. *Source:* Derossi et al. (2020a) / With permission of Elsevier.

4.5 Factors Affecting Other 3D Printing Technologies

Although extrusion technology is most widely used, it is essential to know the other critical process variables of other food printing technologies like selective laser sintering (SLS), binder jetting, and inkjet printing. From the literature review, it was clear that only a handful of studies were available on other food 3D printing technologies. So, the present section on critical variables that are affecting other food 3D printing technologies would be helpful in process optimization for future 3D printing studies on selective sintering, inkjet printing, and binder jetting.

4.5.1 Selective Laser Sintering

Sintering is the process of selective consolidation of powdered materials in a specific pattern based on the applied thermal source (hot air or laser) (Diaz et al. 2017). The critical process variables in SLS are laser power (P), laser energy density (E_{v}), laser scanning speed (V_{s}), laser spot size (S_{d}), printing temperature, and glass transition temperature (T_{g}) of feed powder. Other factors that affect the print accuracy of SLS 3D printed construct are STL file conversion, slicing process, machine resolution, the thickness of the layer, shrinkage of material, beam offset, scanning pattern, length to width ratio, scanning angle, size of the laser beam spot, the temperature of the bed and hatch distance (Ali et al. 2019). Further, the characteristics of the laser beam are another criterion that governs the effectiveness of interaction among the materials. As the feed powder material absorbs light at a specific wavelength based on its optical properties, the laser beam significantly influences the powder agglomeration. The laser wavelength required for sintering depends on the diameter of the laser beam as well as the material components in the formulation. The surface quality of the SLS printed construct is influenced by the pulse durations and frequencies. The energy density of the laser beam can be defined as input energy per unit area (Eq. (4.22)) (Charoo et al. 2020).

$$\text{Energy density} = \frac{\text{Laser power}}{\text{Scan speed} \times \text{hatch distance}} \tag{4.22}$$

The above equation can be used when the hatch distance (mm) is smaller than the effective diameter of the laser beam, i.e. laser spot size. In case if the hatch distance is equal to or greater than that of the effective diameter of the laser, Eq. (4.23) is used for determining the laser energy density.

$$\text{Energy density} = \frac{\text{Laser power}}{\text{Scan speed} \times \text{effective laser diameter}} \tag{4.23}$$

As stated, the laser scanning speed has a greater impact on the consolidation of powder surface by laser energy density. The relationship among these factors is given by Eq. (4.24) (Awad et al. 2020).

$$E_{\mathrm{v}} = \frac{P}{S_{\mathrm{d}} \times V_{\mathrm{s}}} \tag{4.24}$$

Other than these factors, the dimensional accuracy of the 3D printed SLS construct requires higher process accuracy that in turn affected by material shrinkage. The critical process variables such as laser power and length of scan impact shrinkage along X-direction; laser power and beam speed impacts shrinkage in Y-direction; and speed of laser beam, bed temperature, and hatch spacing influence shrinkage in the Z-direction. The temperature of the bed is usually set just above the T_g for amorphous materials since the polymers are highly viscous at this temperature that helps in material consolidation. On the other hand, the bed temperature is set a few degrees lower than that of the T_m for the crystalline polymers. In the case of semi-crystalline materials, the optimal bed temperature is set close to the T_g which can be calculated using a simple Fox equation (Eq. (4.25)) (Awad et al. 2020).

$$\frac{1}{T_g} = \frac{W_{f_1}}{T'_g} + \frac{W_{f_2}}{T''_g} \tag{4.25}$$

where W_{f_1} and W_{f_2} are the weight fractions of each polymer, and T'_g and T''_g are the T_g values of each polymer, respectively. A different range of the powdered food materials was tested for SLS printing. A hollow cube was 3D printed using SLS with 50% laser power and 50 layers of 0.3 mm from a mixture of 50% semolina and 50% binder (9 : 1 of polydextrose: palm oil powder) (Diaz et al. 2018). While the same material supply was used for printing of TNO logo with varied process variables such as 60% laser power, 5 × square layers with a layer thickness of 0.3 mm. Similarly, the fruit powders and legume powders were tested for SLS printing with the view of developing an instant fruit beverage and legume soup, respectively. The high solubility of the SLS printed construct makes it feasible to be used in preparing instant beverages when mixed with water.

4.5.2 Inkjet Printing and Binder Jetting

Both ink-jetting printing and binder jetting are similar technologies with the exception that inkjet printing dispenses the printing material while binder jetting ejects the drops of binder along with feed supply (Trenfield et al. 2018). The binder jetting is based on the adhesion of the powered bed using a spray binder (Ziaee and Crane 2019). The key factors that determine the process accuracy of inkjet printing and binder jetting are feed material characteristics such as particle size, shape, distribution, and additives; binder characteristics such as the method of binding, surface tension, mechanism of the droplet formation, and droplet infiltration kinetics; printing parameters including impact velocity, solid binder, layer thickness, saturation, and orientation of printing; post-processing parameters such as depowdering and densification process (Lv et al. 2019). All these factors significantly influence the density, porosity, mechanical strength, surface roughness, accuracy, and printing resolution of the 3D construct. An edible substrate was surface decorated using high-resolution inkjet printing (Shastry et al. 2009). The material supply used was a wax-based edible ink that contains a colourant, fat/wax dispersible carrier, and a wax material base. The piezoelectric printhead was employed to produce confectionery products like wax polished sugar shell surface with a printing resolution as high as 300 dpi.

Three major parameters that are employed for predicting droplet jetting behaviour of inkjet material supply are R_e/W_e, droplet spreading, and droplet splashing. The ratio of the Reynolds number (R_e) to the Weber number (W_e) represents the trends of viscous force, inertial force, and surface tension acting on fluid flow (Eq. (4.26)). (Godoi et al. 2019).

$$\frac{R_e}{W_e} = \frac{\sqrt{\gamma \rho l_c}}{\eta} \tag{4.26}$$

where γ is the fluid's surface tension, ρ is the fluid density, l_c is the characteristic length (typically, nozzle or droplet diameter), and η is the fluid's viscosity. The droplet spreading of fluid in absence of solidification (ε) can be derived using Eq. (4.27).

$$\varepsilon = \frac{r_{max}}{r} = \left[\frac{We^2 + 12}{3(1 - \cos\theta) + 4\frac{We^2}{R_e^{1/2}}}\right]^{1/2} \tag{4.27}$$

where r_{max} is the maximum splat radius, r is the radius of the initial droplet, and θ is the equilibrium angle of contact of the droplet with the substrate. The splashing of liquid droplets upon jetting would result in dimensional instability and the absence of uniformity. In general, splashing occurs when the splashing of liquid droplets (K) exceeds its critical value (K_c) (Eq. (4.28)) (Godoi et al. 2019).

$$K = W_e R_e^{1/4} \tag{4.28}$$

The formation and behaviour of the liquid jet and droplets were highly influenced by surface tension, inertia, and viscosity that can be quantified using dimensionless Ohnesorge number (Oh) (Eq. (4.29)) (Daly et al. 2015). The Oh represents the behaviour of the liquid jet emerging from printing nozzle that in turn based on physical properties of the liquid and droplet size (Derby 2010) and is independent of driving conditions, i.e. the velocity of liquid (Figure 4.26) (McKinley and Renardy 2011).

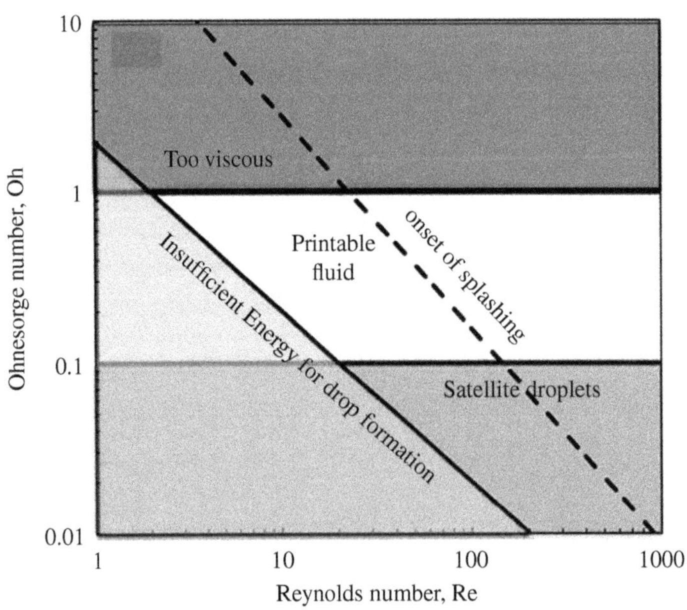

Figure 4.26 Schematic representation of operating regime of inkjet printing for stable operation. *Source:* McKinley and Renardy (2011).

$$\mathrm{Oh} = \frac{\eta}{\sqrt{\gamma \rho l_c}} \tag{4.29}$$

If Oh is too high (Oh > ~1) then the viscous force prevents the separation of liquid droplets while Oh is too low (Oh < ~0.1) then the jet would form a large number of satellite droplets. The jettability (Z) of the material is the inverse of Oh as given in Eq. (4.30). A Z value in the range from 1 to 10 is an indication of the material's ability to produce stable droplets. Thus, the success of jetting depends on molecular weight and the polymer concentration (Holland et al. 2018).

$$Z = \frac{1}{\mathrm{Oh}} = \frac{\mathrm{R_e}}{\sqrt{\mathrm{W_e}}} = \frac{\sqrt{\gamma \rho l_c}}{\eta} \tag{4.30}$$

Considering the powder binding in the jetting process, the powder behaviour is equally important as that of liquid droplets. Techniques such as Schulze annular ring tester, Jenike's technique are useful in determining the mechanical properties of the material. The flowability index (ff_c) value is a dimensionless number that gives the relationship between the consolidation stress (σ_1) of bulk powder to the unconfined yield stress (σ_c) (Eq. (4.31)) (Holland et al. 2019).

$$ff_c = \frac{\sigma_1}{\sigma_c} \tag{4.31}$$

The powder material will not flow when $ff_c < 1$ and when $ff_c > 10$, it represents the free-flowability. While the index value in the range between 4 and 10 indicates 'good flow' that shows the applicability of powder material for binder jetting and the value is <4 indicating the 'cohesive' nature that is detrimental for the binder jetting process especially in disturbing the stability of lower layers during spreading. In addition, a small angle of repose value was found to yield a good flowability (Holland et al. 2019). Other than these factors, the sorption and desorption isotherms provide useful information about the post-processing of solvent removal after the binder jetting process. The Hershey's shell topping was 3D printed using inkjet printing based on the piezoelectric mechanism (Lanaro et al. 2019). Also, a commercial 3D printer FoodJet was used for the fabrication of chocolate-based 3D constructs that resembles the chocolate flavour same as that of traditional processing.

4.6 Conclusion

The consumer's need for innovative food products drives the food industry to switch over to 3D printing. As a fascinating technology 3D printing allows to fabricate the foods based on individual likeness that transforms the imaginary idea into reality. However, in practical situations there occurs discrepancies in the reproduction of the 3D construct as that of the imaginary idea. The accuracy of the production process involves the optimization of process and product variables. As food is a complex multicomponent system, the formulation of which is one of the key criteria in deciding the success of printability. Further, the selected 3D printing technologies are another determinant in determining the printability of the material supply. Hence, understanding the material properties and its response towards 3D printing is vital. So, the present chapter describes the various factors that affect the 3D printing process. In brief, the different range of factors affecting the printing process is broadly grouped as intrinsic and extrinsic properties. The material characterization that reflects the properties related to the material supply used for 3D printing is considered intrinsic whereas the material optimization and optimization of process variables with design parameters are considered extrinsic. Thus, the various properties that have been studied for assessing the printability of the material supplies are discussed in detail. Certainly, the present chapter provides broad knowledge on the basic process of optimization of the 3D printing process and remains a valuable base for future 3D printing studies in the creation of novel complex innovative functional 3D food constructs.

References

Ahmed, J., Mulla, M., Joseph, A. et al. (2020). Zinc oxide/clove essential oil incorporated type B gelatin nanocomposite formulations: a proof-of-concept study for 3D printing applications. *Food Hydrocolloids* **98**: 105256.

Alam, M.R., Scampicchio, M., Angeli, S., and Ferrentino, G. (2019). Effect of hot melt extrusion on physical and functional properties of insect based extruded products. *Journal of Food Engineering* **259**: 44–51.

Ali, S.F.B., Mohamed, E.M., Ozkan, T. et al. (2019). Understanding the effects of formulation and process variables on the printlets quality manufactured by selective laser sintering 3D printing. *International Journal of Pharmaceutics* **570**: 118651.

Anukiruthika, T., Moses, J.A., and Anandharamakrishnan, C. (2020). 3D printing of egg yolk and white with rice flour blends. *Journal of Food Engineering* **265**: 109691. https://doi.org/10.1016/j.jfoodeng.2019.109691.

Anvari, M. and Joyner, H.S. (2019). Concentrated emulsions as novel fat replacers in reduced-fat and low-fat Cheddar cheeses. Part 2. Large amplitude oscillatory shear behavior. *International Dairy Journal* **91**: 137–146. https://doi.org/10.1016/j.idairyj.2018.08.018.

Awad, A., Fina, F., Goyanes, A. et al. (2020). 3D printing: principles and pharmaceutical applications of selective laser sintering. *International Journal of Pharmaceutics* **586**: 119594.

Bähler, B., Back, R., and Hinrichs, J. (2015). Evaluation of oscillatory and shear strain behaviour for thermo-rheological plasticisation of non-ripened cheese curd: effect of water, protein, and fat. *International Dairy Journal* **46**: 63–70.

de Barros Fernandes, R.V., Borges, S.V., Silva, E.K. et al. (2016). Study of ultrasound-assisted emulsions on microencapsulation of ginger essential oil by spray drying. *Industrial Crops and Products* **94**: 413–423.

Caporizzi, R., Derossi, A., and Severini, C. (2019). Cereal-based and insect-enriched printable food: from formulation to postprocessing treatments. Status and perspectives. In: *Fundamentals of 3D Food Printing and Applications* (eds. F.C. Godoi, B.R. Bhandari, S. Prakash and M. Zhang). Elsevier Inc. https://doi.org/10.1016/B978-0-12-814564-7.00004-3.

Carvalho-da-Silva, A.M., Van Damme, I., Taylor, W. et al. (2013). Oral processing of two milk chocolate samples. *Food & Function* **4** (3): 461–469.

Chandrasekaran, S., Ramanathan, S., and Basak, T. (2013). Microwave food processing—a review. *Food Research International* **52** (1): 243–261.

Charoo, N.A., Barakh Ali, S.F., Mohamed, E.M. et al. (2020). Selective laser sintering 3D printing–an overview of the technology and pharmaceutical applications. *Drug Development and Industrial Pharmacy* **46** (6): 869–877.

Chen, J. and Stokes, J.R. (2012). Rheology and tribology: two distinctive regimes of food texture sensation. *Trends in Food Science & Technology* **25** (1): 4–12.

Chen, Y., Zhang, B.-C., Sun, Y.-H. et al. (2015). Physicochemical properties and adsorption of cholesterol by okra (*Abelmoschus esculentus*) powder. *Food & Function* **6** (12): 3728–3736.

Chen, H., Xie, F., Chen, L., and Zheng, B. (2019). Effect of rheological properties of potato, rice and corn starches on their hot-extrusion 3D printing behaviors. *Journal of Food Engineering* **244**: 150–158. https://doi.org/10.1016/j.jfoodeng.2018.09.011.

Choi, Y.-S., Han, D.-J., Choi, J.-H. et al. (2016). Effect of chicken skin on the quality characteristics of semi-dried restructured jerky. *Poultry Science* **95** (5): 1198–1204.

Daly, R., Harrington, T.S., Martin, G.D., and Hutchings, I.M. (2015). Inkjet printing for pharmaceutics - a review of research and manufacturing. *International Journal of Pharmaceutics* **494** (2): 554–567.

Dankar, I., Haddarah, A., Omar, F.E.L. et al. (2018). 3D printing technology: the new era for food customization and elaboration. *Trends in Food Science and Technology* **75**: 231–242. https://doi.org/10.1016/j.tifs.2018.03.018.

Dankar, I., Haddarah, A., Sepulcre, F., and Pujolà, M. (2020). Assessing mechanical and rheological properties of potato puree: effect of different ingredient combinations and cooking methods on the feasibility of 3D printing. *Foods* **9** (1): 21.

Derby, B. (2010). Inkjet printing of functional and structural materials: fluid property requirements, feature stability, and resolution. *Annual Review of Materials Research* **40**: 395–414.

Derossi, A., Caporizzi, R., Azzollini, D., and Severini, C. (2018). Application of 3D printing for customized food. A case on the development of a fruit-based snack for children. *Journal of Food Engineering* **220**: 65–75.

Derossi, A., Paolillo, M., Caporizzi, R., and Severini, C. (2020a). Extending the 3D food printing tests at high speed. Material deposition and effect of non-printing movements on the final quality of printed structures. *Journal of Food Engineering* **275**: 109865.

Derossi, A., Caporizzi, R., Paolillo, M., and Severini, C. (2020b). Programmable texture properties of cereal-based snack mediated by 3D printing technology. *Journal of Food Engineering* **289**: 110160.

Diaz, J.V., Noort, M.W.-J., and Van Bommel, K.J.C. (2017). Method for the production of an edible object by powder bed (3d) printing and food products obtainable therewith. Google Patents.

Diaz, J.V., Van Bommel, K.J.C., Noort, M.W.-J. et al. (2018). Method for the production of edible objects using sls and food products. Google Patents.

Dick, A., Bhandari, B., and Prakash, S. (2019). Post-processing feasibility of composite-layer 3D printed beef. *Meat Science* **153**: 9–18. https://doi.org/10.1016/j.meatsci.2019.02.024.

Dick, A., Bhandari, B., Dong, X., and Prakash, S. (2020). Feasibility study of hydrocolloid incorporated 3D printed pork as dysphagia food. *Food Hydrocolloids* **107**: 105940.

Fan, H., Zhang, M., Liu, Z., and Ye, Y. (2020). Effect of microwave-salt synergetic pre-treatment on the 3D printing performance of SPI-strawberry ink system. *LWT* **122**: 109004.

Garcia-Segovia, P., Garcia-Alcaraz, V., Balasch-Parisi, S., and Martinez-Monzó, J. (2020). 3D printing of gels based on xanthan/konjac gums. *Innovative Food Science & Emerging Technologies* **64**: 102343.

Ghazal, A.F., Zhang, M., and Liu, Z. (2019). Spontaneous color change of 3D printed healthy food product over time after printing as a novel application for 4D food printing. *Food and Bioprocess Technology* **12** (10): 1627–1645.

Gholamipour-Shirazi, A., Norton, I.T., and Mills, T. (2019). Designing hydrocolloid based food-ink formulations for extrusion 3D printing. *Food Hydrocolloids* **95**: 161–167. https://doi.org/10.1016/j.foodhyd.2019.04.011.

Godoi, F.C., Prakash, S., and Bhandari, B.R. (2016). 3D printing technologies applied for food design: status and prospects. *Journal of Food Engineering* **179**: 44–54.

Godoi, F.C., Bhandari, B.R., Prakash, S., and Zhang, M. (2019). An introduction to the principles of 3D food printing. In: *Fundamentals of 3D Food Printing and Applications* (eds. F.C. Godoi, B.R. Bhandari, S. Prakash and M. Zhang), 1–18. Elsevier.

Guo, C.-F., Zhang, M., and Bhandari, B. (2019). A comparative study between syringe-based and screw-based 3D food printers by computational simulation. *Computers and Electronics in Agriculture* **162**: 397–404.

Guo, C., Zhang, M., and Devahastin, S. (2021). Color/aroma changes of 3D-printed buckwheat dough with yellow flesh peach as triggered by microwave heating of gelatin-gum Arabic complex coacervates. *Food Hydrocolloids* **112**: 106358.

He, C., Zhang, M., and Devahastin, S. (2020a). Investigation on spontaneous shape change of 4D printed starch-based purees from purple sweet potatoes as induced by microwave dehydration. *ACS Applied Materials & Interfaces* **12** (34): 37896–37905.

He, C., Zhang, M., and Guo, C. (2020b). 4D printing of mashed potato/purple sweet potato puree with spontaneous color change. *Innovative Food Science & Emerging Technologies* **59**: 102250.

Holland, S., Foster, T., MacNaughtan, W., and Tuck, C. (2018). Design and characterisation of food grade powders and inks for microstructure control using 3D printing. *Journal of Food Engineering* **220**: 12–19.

Holland, S., Foster, T., and Tuck, C. (2019). Creation of food structures through binder jetting. In: *Fundamentals of 3D Food Printing and Applications* (eds. F.C. Godoi, B.R. Bhandari, S. Prakash and M. Zhang), pp. 257–288. Elsevier.

Hoque, M.S., Benjakul, S., Prodpran, T., and Songtipya, P. (2011). Properties of blend film based on cuttlefish (*Sepia pharaonis*) skin gelatin and mungbean protein isolate. *International Journal of Biological Macromolecules* **49** (4): 663–673.

Huang, M., Zhang, M., and Bhandari, B. (2019). Assessing the 3D printing precision and texture properties of brown rice induced by infill levels and printing variables. *Food and Bioprocess Technology* **12**: 1–12. https://doi.org/10.1007/s11947-019-02287-x.

Huang, M., Zhang, M., and Guo, C. (2020). 3D printability of brown rice gel modified by some food hydrocolloids. *Journal of Food Processing and Preservation* **44**: e14502.

Inal, M. and Mulazimouglu, G. (2019). Production and characterization of bactericidal wound dressing material based on gelatin nanofiber. *International Journal of Biological Macromolecules* **137**: 392–404.

Katsaros, G., Tsoukala, M., Giannoglou, M., and Taoukis, P. (2020). Effect of storage on the rheological and viscoelastic properties of mayonnaise emulsions of different oil droplet size. *Heliyon* **6** (12): e05788.

Keerthana, K., Anukiruthika, T., Moses, J.A., and Anandharamakrishnan, C. (2020). Development of fiber-enriched 3D printed snacks from alternative foods: a study on button mushroom. *Journal of Food Engineering*. https://doi.org/10.1016/j.jfoodeng.2020.110116.

Kern, C., Weiss, J., and Hinrichs, J. (2018). Additive layer manufacturing of semi-hard model cheese: effect of calcium levels on thermo-rheological properties and shear behavior. *Journal of Food Engineering* **235**: 89–97. https://doi.org/10.1016/j.jfoodeng.2018.04.029.

Kim, H.W., Bae, H., and Park, H.J. (2018a). Classification of the printability of selected food for 3D printing: development of an assessment method using hydrocolloids as reference material. *Journal of Food Engineering* **215**: 23–32.

Kim, H.W., Lee, J.H., Park, S.M. et al. (2018b). Effect of hydrocolloids on rheological properties and printability of vegetable inks for 3D food printing. *Journal of Food Science* **83** (12): 2923–2932.

Kim, H.W., Lee, I.J., Park, S.M. et al. (2019). Effect of hydrocolloid addition on dimensional stability in post-processing of 3D printable cookie dough. *LWT* **101**: 69–75.

Krishnaraj, P., Anukiruthika, T., Choudhary, P. et al. (2019). 3D extrusion printing and post-processing of fibre-rich snack from indigenous composite flour. *Food and Bioprocess Technology* **12** (10): 1776–1786. https://doi.org/10.1007/s11947-019-02336-5.

Kumar, C. and Karim, M.A. (2019). Microwave-convective drying of food materials: a critical review. *Critical Reviews in Food Science and Nutrition* **59** (3): 379–394.

Lanaro, M., Forrestal, D.P., Scheurer, S. et al. (2017). 3D printing complex chocolate objects: platform design, optimization and evaluation. *Journal of Food Engineering* **215**: 13–22. https://doi.org/10.1016/j.jfoodeng.2017.06.029.

Lanaro, M., Desselle, M.R., and Woodruff, M.A. (2019). 3D printing chocolate: properties of formulations for extrusion, sintering, binding and ink jetting. In: *Fundamentals of 3D Food Printing and Applications* (eds. F.C. Godoi, B.R. Bhandari, S. Prakash and M. Zhang), pp. 151–173. Elsevier.

Le Tohic, C., O'Sullivan, J.J., Drapala, K.P. et al. (2018). Effect of 3D printing on the structure and textural properties of processed cheese. *Journal of Food Engineering* **220**: 56–64.

Lee, J.H., Won, D.J., Kim, H.W., and Park, H.J. (2019). Effect of particle size on 3D printing performance of the food-ink system with cellular food materials. *Journal of Food Engineering* **256**: 1–8. https://doi.org/10.1016/j.jfoodeng.2019.03.014.

Lille, M., Nurmela, A., Nordlund, E. et al. (2018). Applicability of protein and fiber-rich food materials in extrusion-based 3D printing. *Journal of Food Engineering* **220**: 20–27.

Lipton, J., Arnold, D., Nigl, F. et al. (2010). Multi-material food printing with complex internal structure suitable for conventional post-processing. Solid Freeform Fabrication Symposium, pp. 809–815.

Liu, Z. and Zhang, M. (2019). 3D food printing technologies and factors affecting printing precision. In: *Fundamentals of 3D Food Printing and Applications* (eds. F.C. Godoi, B.R. Bhandari, S. Prakash and M. Zhang). Elsevier Inc. https://doi.org/10.1016/b978-0-12-814564-7.00002-x.

Liu, Z., Zhang, M., Bhandari, B., and Wang, Y. (2017). 3D printing: Printing precision and application in food sector. *Trends in Food Science & Technology* **69**: 83–94.

Liu, Y., Liu, D., Wei, G. et al. (2018a). 3D printed milk protein food simulant: improving the printing performance of milk protein concentration by incorporating whey protein isolate. *Innovative Food Science and Emerging Technologies* **49**: 116–126. https://doi.org/10.1016/j.ifset.2018.07.018.

Liu, Z., Bhandari, B., Prakash, S., and Zhang, M. (2018b). Creation of internal structure of mashed potato construct by 3D printing and its textural properties. *Food Research International* **111**: 534–543.

Liu, Z., Zhang, M., Bhandari, B., and Yang, C. (2018c). Impact of rheological properties of mashed potatoes on 3D printing. *Journal of Food Engineering* **220**: 76–82. https://doi.org/10.1016/j.jfoodeng.2017.04.017.

Liu, Z., Zhang, M., and Yang, C. (2018d). Dual extrusion 3D printing of mashed potatoes/strawberry juice gel. *LWT* **96**: 589–596.

Liu, L., Meng, Y., Bhandari, B. et al. (2019a). Correction to 3D printing complex egg white protein objects: properties and optimization. *Food and Bioprocess Technology* **12** (2): 280.

Liu, Y., Yu, Y., Liu, C. et al. (2019b). Rheological and mechanical behavior of milk protein composite gel for extrusion-based 3D food printing. *LWT* **102**: 338–346. https://doi.org/10.1016/j.lwt.2018.12.053.

Liu, Y., Zhang, W., Wang, K. et al. (2019c). Fabrication of gel-like emulsions with whey protein isolate using microfluidization: rheological properties and 3D printing performance. *Food and Bioprocess Technology* **12** (12): 1967–1979.

Liu, Z., Bhandari, B., and Zhang, M. (2020a). Incorporation of probiotics (*Bifidobacterium animalis* subsp. *Lactis*) into 3D printed mashed potatoes: effects of variables on the viability. *Food Research International* **128**: 108795.

Liu, Z., Zhang, M., and Ye, Y. (2020b). Indirect prediction of 3D printability of mashed potatoes based on LF-NMR measurements. *Journal of Food Engineering* **287**: 110137.

Lv, X., Ye, F., Cheng, L. et al. (2019). Binder jetting of ceramics: powders, binders, printing parameters, equipment, and post-treatment. *Ceramics International* **45** (10): 12609–12624.

Maniglia, B.C., Lima, D.C., da Matta Júnior, M. et al. (2020). Dry heating treatment: a potential tool to improve the wheat starch properties for 3D food printing application. *Food Research International* **137**: 109731.

Mantihal, S., Prakash, S., Godoi, F.C., and Bhandari, B. (2017). Optimization of chocolate 3D printing by correlating thermal and flow properties with 3D structure modeling. *Innovative Food Science and Emerging Technologies* **44**: 21–29. https://doi.org/10.1016/j.ifset.2017.09.012.

Mantihal, S., Prakash, S., and Bhandari, B. (2019a). Textural modification of 3D printed dark chocolate by varying internal infill structure. *Food Research International* **121**: 648–657.

Mantihal, S., Prakash, S., Godoi, F.C., and Bhandari, B. (2019b). Effect of additives on thermal, rheological and tribological properties of 3D printed dark chocolate. *Food Research International* **119**: 161–169.

Martínez-Monzó, J., Cárdenas, J., and García-Segovia, P. (2019). Effect of temperature on 3D printing of commercial potato puree. *Food Biophysics*: 1–10. https://doi.org/10.1007/s11483-019-09576-0.

Mastromatteo, M., Guida, M., Danza, A. et al. (2013). Rheological, microstructural and sensorial properties of durum wheat bread as affected by dough water content. *Food Research International* **51** (2): 458–466.

McKinley, G.H. and Renardy, M. (2011). Wolfgang von ohnesorge. *Physics of Fluids* **23** (12): 127101.

Munoz, I., Gou, P., Picouet, P.A. et al. (2018). Dielectric properties of milk during ultra-heat treatment. *Journal of Food Engineering* **219**: 137–146.

Muredzi, P., Nyahada, M., and Mashswa, B. (2013). Effects of soya bean meal feed properties on extrusion failures and implementing a solution. Case study Monmouth Path's Investments (Pvt) Ltd, Harare, Zimbabwe. *International Journal of Nutrition and Food Sciences* **2** (2): 60–69. https://doi.org/10.11648/j.ijnfs.20130202.16.

Nachal, N., Moses, J.A., Karthik, P., and Anandharamakrishnan, C. (2019). Applications of 3D printing in food processing. *Food Engineering Reviews* **11** (3): 123–141. https://doi.org/10.1007/s12393-019-09199-8.

Nguyen, P.T.M., Bhandari, B., and Prakash, S. (2016). Tribological method to measure lubricating properties of dairy products. *Journal of Food Engineering* **168**: 27–34.

Nida, S., Anukiruthika, T., Moses, J.A., and Anandharamakrishnan, C. (2020). 3D printing of grinding and milling fractions of rice husk. *Waste and Biomass Valorization*, pp. 1–10. https://doi.org/10.1007/s12649-020-01000-w.

Nijdam, J.J., Agarwal, D., and Schon, B.S. (2021a). Assessment of a novel window of dimensional stability for screening food inks for 3D printing. *Journal of Food Engineering* **292**: 110349.

Nijdam, J.J., LeCorre-Bordes, D., Delvart, A., and Schon, B.S. (2021b). A rheological test to assess the ability of food inks to form dimensionally stable 3D food structures. *Journal of Food Engineering* **291**: 110235.

Pei, S., Zhang, J., Gao, M. et al. (2015). A facile hydrothermal approach towards photoluminescent carbon dots from amino acids. *Journal of Colloid and Interface Science* **439**: 129–133.

Prakash, S., Tan, D.D.Y., and Chen, J. (2013). Applications of tribology in studying food oral processing and texture perception. *Food Research International* **54** (2): 1627–1635.

Rando, P. and Ramaioli, M. (2020). Food 3D printing: effect of heat transfer on print stability of chocolate. *Journal of Food Engineering* 110415.

Saba, N., Jawaid, M., and Sultan, M.T.H. (2017). Thermal properties of oil palm biomass based composites. In: *Lignocellulosic Fibre and Biomass-Based Composite Materials* (eds. M. Jawaid, P.M. Tahir and N. Saba), pp. 95–122. Elsevier.

Scaraggi, M. and Persson, B.N.J. (2014). Theory of viscoelastic lubrication. *Tribology International* **72**: 118–130.

Schutyser, M.A.I., Houlder, S., de Wit, M. et al. (2018). Fused deposition modelling of sodium caseinate dispersions. *Journal of Food Engineering* **220**: 49–55.

Sethupathy, P., Moses, J.A., and Anandharamakrishnan, C. (2020). Food oral processing and tribology: instrumental approaches and emerging applications. *Food Reviews International* **37**: 1–34.

Severini, C., Azzollini, D., Albenzio, M., and Derossi, A. (2018a). On printability, quality and nutritional properties of 3D printed cereal based snacks enriched with edible insects. *Food Research International* **106** (November 2017): 666–676. https://doi.org/10.1016/j.foodres.2018.01.034.

Severini, C., Derossi, A., Ricci, I. et al. (2018b). Printing a blend of fruit and vegetables. New advances on critical variables and shelf life of 3D edible objects. *Journal of Food Engineering* **220**: 89–100. https://doi.org/10.1016/j.jfoodeng.2017.08.025.

Shastry, A.V, Collins, T.M., Suttle, J.M., et al. (2009). Edible inks for ink-jet printing on edible substrates. Google Patents.

Singh, J., Vennapusa, J.R., and Chattopadhyay, S. (2020). Protein-polysaccharide based microencapsulated phase change material composites for thermal energy storage. *Carbohydrate Polymers* **229**: 115531.

Sun, J., Zhou, W., Huang, D. et al. (2015). An overview of 3D printing technologies for food fabrication. *Food and Bioprocess Technology* **8** (8): 1605–1615.

Sun, J., Zhou, W., Huang, D., and Yan, L. (2018a). 3D food printing: perspectives. In: *Polymers for Food Applications* (ed. T.J. Gutierrez), pp. 725–755. Springer.

Sun, J., Zhou, W., Yan, L. et al. (2018b). Extrusion-based food printing for digitalized food design and nutrition control. *Journal of Food Engineering* **220**: 1–11.

Sweeney, M., Campbell, L.L., Hanson, J. et al. (2017). Characterizing the feasibility of processing wet granular materials to improve rheology for 3D printing. *Journal of Materials Science* **52** (22): 13040–13053.

Tan, J. and Kerr, W.L. (2018). Determination of chocolate melting properties by capacitance based thermal analysis (CTA). *Journal of Food Measurement and Characterization* **12** (1): 641–649.

Tan, C., Toh, W.Y., Wong, G., and Lin, L. (2018). Extrusion-based 3D food printing – materials and machines. *International Journal of Bioprinting* **4** (2): 1–13. https://doi.org/10.18063/ijb.v4i2.143.

Theagarajan, R., Moses, J.A., and Anandharamakrishnan, C. (2020). 3D extrusion printability of rice starch and optimization of process variables. *Food and Bioprocess Technology.* https://doi.org/10.1007/s11947-020-02453-6.

Tongnuanchan, P., Benjakul, S., Prodpran, T., and Nilsuwan, K. (2015). Emulsion film based on fish skin gelatin and palm oil: physical, structural and thermal properties. *Food Hydrocolloids* **48**: 248–259.

Trenfield, S.J., Madla, C.M., Basit, A.W., and Gaisford, S. (2018). Binder jet printing in pharmaceutical manufacturing. In: *3D Printing of Pharmaceuticals* (eds. A.W. Basit and S. Gaisford), pp. 41–54. Springer.

Vancauwenberghe, V., Katalagarianakis, L., Wang, Z. et al. (2017). Pectin based food-ink formulations for 3-D printing of customizable porous food simulants. *Innovative Food Science and Emerging Technologies* **42** (June): 138–150. https://doi.org/10.1016/j.ifset.2017.06.011.

Vancauwenberghe, V., Delele, M.A., Vanbiervliet, J. et al. (2018). Model-based design and validation of food texture of 3D printed pectin-based food simulants. *Journal of Food Engineering* **231**: 72–82. https://doi.org/10.1016/j.jfoodeng.2018.03.010.

Wang, J. and Shaw, L.L. (2005). Rheological and extrusion behavior of dental porcelain slurries for rapid prototyping applications. *Materials Science and Engineering: A* **397** (1–2): 314–321. https://doi.org/10.1016/j.msea.2005.02.045.

Wang, R., Zhang, M., Mujumdar, A.S., and Jiang, H. (2011). Effect of salt and sucrose content on dielectric properties and microwave freeze drying behavior of re-structured potato slices. *Journal of Food Engineering* **106** (4): 290–297.

Wang, L., Zhang, M., Bhandari, B., and Yang, C. (2018). Investigation on fish surimi gel as promising food material for 3D printing. *Journal of Food Engineering* **220**: 101–108. https://doi.org/10.1016/j.jfoodeng.2017.02.029.

Wilson, A., Anukiruthika, T., Moses, J.A., and Anandharamakrishnan, C. (2020). Customized shapes for chicken meat–based products: feasibility study on 3D-printed nuggets. *Food and Bioprocess Technology* **13**: 1–16.

Wray, D. and Ramaswamy, H.S. (2015). Novel concepts in microwave drying of foods. *Drying Technology* **33** (7): 769–783.

Xu, K., Zhang, M., and Bhandari, B. (2019). Effect of novel ultrasonic-microwave combined pretreatment on the quality of 3D printed wheat starch-papaya system. *Food Biophysics* **15**: 1–12.

Xu, L., Gu, L., Su, Y. et al. (2020). Impact of thermal treatment on the rheological, microstructural, protein structures and extrusion 3D printing characteristics of egg yolk. *Food Hydrocolloids* **100**: 105399.

Yang, F., Zhang, M., and Liu, Y. (2018a). Effect of post-treatment microwave vacuum drying on the quality of 3D-printed mango juice gel. *Drying Technology* **37**: 1–9. https://doi.org/10.1080/07373937.2018.1536884.

Yang, F., Zhang, M., Bhandari, B., and Liu, Y. (2018b). Investigation on lemon juice gel as food material for 3D printing and optimization of printing parameters. *LWT - Food Science and Technology* **87**: 67–76. https://doi.org/10.1016/j.lwt.2017.08.054.

Zhang, L., Lou, Y., and Schutyser, M.A.I. (2018). 3D printing of cereal-based food structures containing probiotics. *Food Structure* **18**: 14–22.

Zhu, S., Stieger, M.A., van der Goot, A.J., and Schutyser, M.A.I. (2019). Extrusion-based 3D printing of food pastes: correlating rheological properties with printing behaviour. *Innovative Food Science & Emerging Technologies* **58**: 102214.

Ziaee, M. and Crane, N.B. (2019). Binder jetting: a review of process, materials, and methods. *Additive Manufacturing* **28**: 781–801.

5

Printability and Techniques

CHAPTER MENU

3D Printing of Foods, First Edition. C. Anandharamakrishnan, Jeyan A. Moses, and T. Anukiruthika.

5.1 Introduction

Three-dimensional (3D) printing has gained an immense scope and research attention in the food sector for manufacturing foods in a desired customized form. As a solid-free form of fabrication, 3D printing allows us to control the shape and texture of the developed food as per the individual's nutritional requirement. Due to this advantage, 3D printing has been considered as a paradigm of the industrial revolution 4.0 (Shahrubudin et al. 2019). The growing innovations aids in the creation of impossible food structures and designs of the conventional technology into possible using 3D printing. Thus, 3D printing as a print-and-eat technology has precise control over the portion size, composition, texture, flavour, colour, and shape that possess both economic as well as environmental benefits. 3D printing is not only used for manufacturing foods but also for designing food packages (Nida et al. 2020). The 3D printing technique has greater versatility and design freedom over the other manufacturing processes. All these features are realizable when considering the feasibility of materials used for 3D printing. It is a well-known fact that food is a multicomponent system that behaves differently making food printing to be challenging. However, the success of 3D printing of any material supplies relies on the composition, temperature, and additives. Hence the optimization of process and product variables are crucial for the success of the 3D printing process (Pérez et al. 2019). As an initial attempt, most of the research works focus on analyzing the printability of the selected material and the evaluation of printing variables. In this regard, studies were reported on the optimization of natively printable materials such as starch and cereal-based dough (Theagarajan et al. 2020); fibre and protein-rich material supply (Krishnaraj et al. 2019; Lille et al. 2018); non-printable egg yolk and egg white fractions, beef, chicken, fish, and pork (Dick et al. 2020; Wilson et al. 2020; Xu et al. 2020); and alternative food ingredients like insects (Caporizzi et al. 2019). However, there is no protocol procedures or standards available for the qualitative and quantitative assessment of printability. This void misleads the analytical methods employed for assessing printability. Each of the studies reported in the literature claimed to follow varied methods for determining printability. Thus, it is essential to streamline the existing procedures used for analyzing printability to explore the vast research opportunities in the 3D printing of foods.

The most widely used food printing technique is extrusion-based 3D printing that has been evolved from the fused deposition modeling (FDM) approach. Other than extrusion technology, selective sintering, binder jetting technology, and inkjet printing are also applied for food 3D printing applications. Technically each of the food printing approaches possesses different printing variables that in turn varied based on the material supply and 3D printing technique used. In a broader sense, printability represents the ability of the materials to retain the intact deposited 3D layers and its own weight without any deformation (Kim et al. 2018a). The present chapter focuses on the rationale of the 3D printing procedures adopted for determining the printability of materials. Based on the 3D printing studies reported in past years, the present chapter aimed at the development of a set of standards as a base for testing printability. Undoubtedly, the present chapter on printability and techniques would provide valuable insights on the material design and optimization for the development of novel 3D printed food constructs in near future.

5.2 Printability and Material Characteristics

The quantification of printability remains to be challenging as it is greatly influenced by several process variables such as nozzle size, nozzle height, clearance gap, printing speed, rate of scanning, and flow rate on the printing precision, accuracy, and dimensional stability (Godoi et al. 2016). As stated earlier, the optimization of the printing process is varied based on the type of printing technology used. Since extrusion printing is the most applied food printing technique, the present chapter focuses majorly on the testing procedures and protocols with consideration only to extrusion technique. Food additives play a significant role in imparting printability and enhancing the mechanical strength and structural integrity of 3D constructs (Gholamipour-Shirazi et al. 2019). Hydrocolloids are hydrophilic polymers that possess several functional properties due to their viscoelastic nature, emulsification, dispersion, gelation, and thickening to name a few (Zhang et al. 2020). Food hydrocolloids could mimic a different range of food materials of varied consistency and texture. Hence, hydrocolloids are used as a simplified food model system for systematic research. Considering this, Kim et al. (2018a) reported a study on the hydrocolloids as reference material to manifest a classification system and explain the underlying techniques for assessment of the printability. This was the only study to report a detailed comparison of material properties with the dimensional stability and handling of the material supplies for extrusion-based 3D printing. About 12 food samples with different rheological properties were used for examining the hydrocolloid reference material to simulate those food samples. As a result, a classification system of methylcellulose as a reference material was established. Based on the capability of the material's concentration to hold the shape, the material (concentration >12%) suitable for printing of 80 mm concentric cylinder can be used for fabrication of complex 3D constructs marked as grade A; while material (concentration 11–12%) suitable for printing of 40 mm concentric cylinder can be used for fabricating simple 3D models that were marked as grade B; 3D constructs with materials of concentration range from 9 to 10% were used for printing of 20 mm concentric cylinder can be used for designing of custom cookies marked as grade C; and materials (concentration <9%) that were impossible to construct 3D structures had marked as grade D and used for surface decorations. A similar classification system was established for printability assessment by considering the handling properties. The grading follows a material concentration of <6% can be easily extruded (grade A); 6–10% can be extruded (grade B); 11–15% difficult to extrude (grade C); and >15% extremely difficult to extrude (grade D). Food materials such as cream cheese and chocolate jam were categorized as grade A; grade B were bean paste and cheddar cheese; grade C was cookie dough; and grade D were sugar paste and marzipan.

Similar approaches were used for evaluating the print quality of the 3D constructs. As a simpler method, print fidelity was assessed qualitatively based on visual examinations. Different range of arbitrary designs had been reported in works of literature including the fabrication of miniature Australia map using egg white foam (Kouzani et al. 2016), fish-shaped 3D design by a mixture of pumpkin and fish puree (Kouzani et al. 2017), coral reef from tofu noodles (Mizrahi et al. 2016), honeycomb fondants using insect powder (Soares et al. 2014), sugar-based Chinese calligraphy characters (Leung 2017), flower vase using rice paste (Tanaka et al. 2015), apple logo using mashed potato (Liu et al. 2018), and pyramid from vegemite and marmite (Hamilton et al. 2018). Various significant attributes that are considered for quality assessment during the printing process include deposition uniformity, material backup, nozzle clogging, lag time, structural deformation, and post-print stability (Vancauwenberghe et al. 2017). Later, another approach of printing simple geometric design was employed to amalgamate the assessment of printability of material supply. This includes the one-dimensional (1D), two-dimensional (2D), and 3D stability of the printed constructs (Huang 2018). As the next level of quantification, the texture of the printed 3D construct was assessed to determine the print quality. The process variables such as infill percentage, infill pattern, layer height, porosity, powder characteristics, and bulk density are some of the significant factors that influence the texture of the 3D printed construct (Nachal et al. 2019). The printability of any material supply is well correlated with its rheological properties. It was reported that the viscosity of the sample to be printed must be as low as possible to enhance the flowability and should be as high as possible to sustain the successively deposited threads. Hence, the deformation behaviour of the testing material supply for 3D printing can be depicted by measuring the shear modulus by quantifying the printability of the material (Chen et al. 2020). Likewise, some of the characterization imaging techniques such as scanning electron microscopy (SEM), nuclear magnetic resonance imaging, X-ray diffraction (XRD), and confocal laser scanning microscopy (CLSM) are commonly used for evaluating the chemical nature and structure of the raw materials used for 3D printing (Anukiruthika et al. 2020b; Sanchez et al. 2020). A detailed discussion on characterization techniques in context with printability has been presented in the following sections.

5.3 Material Characterization Techniques

The print quality of the fabricated 3D constructs is related to their shape, colour, size, and texture. These features are interrelated with the organization of food components that can be studied at the microscopic, macroscopic, and molecular levels (Jiang et al. 2019). The interactions among the food constituents such as starch, protein, and lipids during the processing, i.e., printing tailors the texture of the resulted 3D construct. The material response of the printing material supply under external stress during the extrusion or sintering or jetting is one of the significant criteria in determining the final print quality. The intermolecular bonding, chemical reactions, and phase transitions associated with the deposition and fusion of material dictate the mechanical stability of the 3D printed layers. So, the extent of the chemical interactions in determining the material suitability for 3D printing can be assessed based on the particle size and distribution, porosity, bulk density, wettability, crystallinity, textural properties such as graininess, adhesiveness, chewiness, gumminess and hardness, optical characteristics, rheological properties such as apparent viscosity, complex viscosity, material consistency, flow behaviour, shear stress, yield stress, and shear modulus (Dankar et al. 2018; Mantihal et al. 2017). Various factors that affect the 3D printing process have been discussed in the earlier chapter. Before the optimization of these process variables, material characterization and its feasibility for 3D printing must be studied. In this regard, various kinds of microscopic and imaging techniques are in practice for the analysis of the food structures at the microscopic level. It includes SEM, transmission electron microscopy, confocal laser microscopy, and stereomicroscopy. Other characterization techniques such as XRD, nuclear magnetic resonance, differential scanning calorimetry (DSC), and Fourier transform infrared spectroscopy (FTIR) are used to study the crystal morphology, chemical structure, and material state at the molecular levels (Ahmed et al. 2020; Chuanxing et al. 2018; Liu et al. 2020c). Different characterization techniques that are commonly used in 3D printing studies have been briefed in the following sections.

5.3.1 Structural Imaging

5.3.1.1 Scanning Electron Microscopy

SEM has been used to detect the molecular interactions and the associated formation of a 3D network during processing (Lee et al. 2019). SEM allows visualizing the surface morphology of the modified food components especially those materials that have undergone pre-treatments such as size reduction, comminution, and shearing, drying, and concentration to name a few. Further, the printability of the material supplies was improved with the addition of hydrocolloids as additives. The incorporation of hydrocolloids would have greater implications on the material properties (Kim et al. 2019). In this regard, the SEM micrographs are used for examining the structural characteristics of the formed gels. The effect of 3D printing on the granular structure of starch molecules in a pea protein matrix was analyzed using SEM micrographs (Chuanxing et al. 2018). The granular structure of the native and the 3D printed potato starch revealed the surface morphology. Results showed that native starch molecules had an ellipsoid structure, smooth surface with varied sizes. While the number of starch granules with smooth and complete surfaces had gradually declined in 3D printed samples and resulted in agglomeration. The breakdown of the particles and the surface roughness was mainly because of the external force exerted during the printing process. The degree of cross-linking between the starch molecules was increased first and then subsequently decreased with increasing concentrations of pea protein from 0 to 8%. The addition of the pea protein resulted in a continuous protein matrix where the starch granules are uniformly distributed thereby promoting the cross-linking between the starch molecules. At higher pea protein concentrations of >2%, the protein molecules compete with the starch for absorption of moisture. This scarcity of moisture hinders the process of complete gelatinization of starch resulted in the weakening of the crosslinks among the starch granules. The effects of oil phase fraction on the 3D printability of emulsion prepared from whey protein (10%, w/v) were reported (Liu et al. 2019b). Interestingly, the printability of the emulsion was correlated with microstructural and rheological properties. The SEM morphologies of gel-like whey protein emulsions with oil fractions ranging from 0.4 to 0.6 possess solid-like characteristics with increasing oil content. This confirms the material suitability for 3D printing by extrusion possessing sufficient viscosity well enough to flow through the nozzle and to get adhere with the successively printed layers. A transformation in the material consistency from soft and smooth to a semi-solid behaviour was observed with increasing oil content. In addition, the oil droplets became more aggregated and closely packed with increasing oil content from 0.4 to 0.5. As a result, a rough and wrinkled microstructure was evident mainly because of the high impact shear forces during the microfluidization of the emulsions.

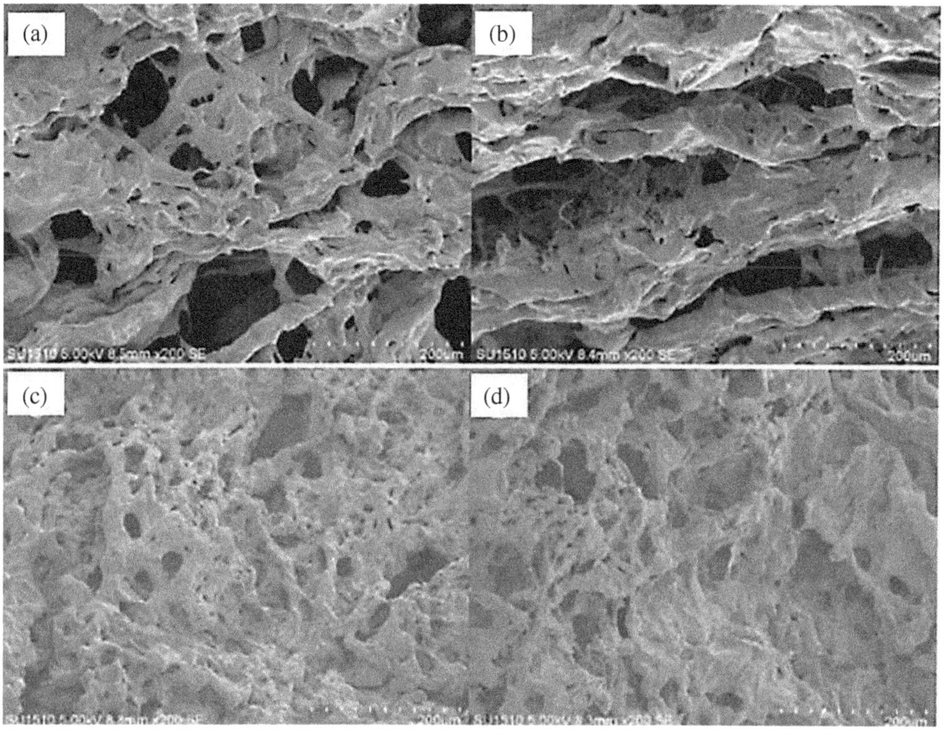

Figure 5.1 Morphology of 3D printed surimi gel with different levels of NaCl. (a) Control, (b) 0.5 NaCl/100 g, (c) 1 g NaCl/100 g, and (d) 1.5 g NaCl/100 g. *Source:* From Wang et al. (2018), Figure 04 [p. 105] / With permission of Elsevier. DOI-https://doi.org/10.1016/j.jfoodeng.2017.02.029.

The characterization of the 3D network of the gel system is an important attribute that determines its functional properties. The addition of NaCl into fish surimi gels significantly improved the firmness of the matrix. The control sample possesses uneven shapes with a compact structure while the NaCl treated samples revealed an aggregated structure that clearly indicates the formation of crosslinks among the myofibrillar proteins. The associated chemical bonding of free amino acids to the protein network reduced the empty spaces that transformed the aggregated matrix into a gel network with fine strands which assists in the printability of fish surimi gel (Figure 5.1). The improved gel structure was well correlated with the gel strength and the water holding capacity of the surimi gel. Thus, the microstructure of the material supply contributes significantly to the shape stabilization and retention of 3D structures (Wang et al. 2018).

The morphology of the 3D printed potato starch samples showed a spongy structural network with different extent of cell density (Liu et al. 2020b). This indicates the structural transformation of starch molecules due to gelatinization at hot-melt printing conditions. The associated interactions among the diffused starch chains led to the cross-linked structural network (Huang et al. 2017). The morphology of the printed starch gel was greatly influenced by increasing concentrations (10–30%) of potato starch at constant printing temperature (70 °C). Larger cells with thin cell walls were evident at the low starch concentration (10%). With increasing starch concentrations, the cells became smaller with thicker cell walls. These results showed that samples with higher cell density could impart higher mechanical strength to the 3D printed gel systems (Guo et al. 2016). At constant starch concentrations, the increase in printing temperatures could significantly affect printability. At 15% potato starch samples, the starch molecules were found to be embedded in the gelatinized matrix at 60 °C. At these conditions, insufficient gelatinization limits the diffusion of the starch chains that hinders the entanglement of molecular chains and hydrogen bonding. With higher printing temperatures greater than 65 °C, there were no remains of the starch molecules while smaller cell sizes and higher cell density were evident at even greater temperatures up to 75 °C. A less uniform dispersion was observed at temperatures higher than 80 °C that attributes to the flow instability and the lesser molecular interactions (Figure 5.2). Thus, the morphological results of starch gel system indicate that both the starch concentrations and printing temperatures are crucial for achieving uniform gel matrix providing acceptable printability of the material supply.

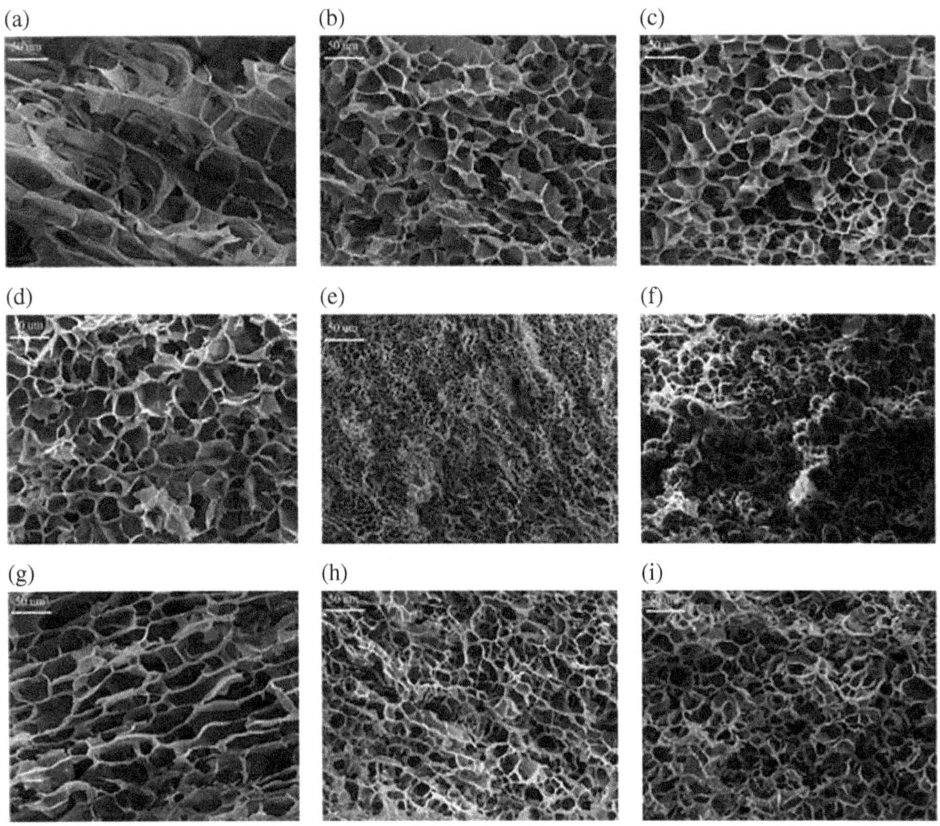

Figure 5.2 Morphology of 3D printed potato starch sample at different conditions. (a) 10%/70 °C, (b) 15%/70 °C, (c) 20%/70 °C, (d) 25%/70 °C, (e) 30%/70 °C, (f) 15%/60 °C, (g) 15%/65 °C, (h) 15%/75 °C, and (i) 15%/80 °C. *Source:* From Liu et al. (2020b), Figure 01 [p. 003] / With permission of Elsevier. DOI-https://doi.org/10.1016/j.foodhyd.2020.105812.

5.3.1.2 X-ray Microtomography

Nuclear magnetic resonance (NMR) imaging is based on proton density and relaxation dynamics of the protons of food components. This technique differs from X-ray imaging in the aspect that the images produced using X-rays are based on the absorption of the X-ray energy (Asadchikov et al. 2019). Tomography is a technique referred to as the quantitative description of a thin layer or slice of the sample. It involves the image acquisition of several radiographs of the same sample under different orientations followed by the reconstruction of the individual radiographs (du Plessis et al. 2020). Thus, the resultant 3D mapping provides information about the structural density of the object. In 3D printing studies, the microstructure characterization using 3D tomographic images aids in assessing the tomography scanned object resolution and the virtual model resolution on the accuracy of the fabricated 3D printed construct. Further, the stereological analysis of the food components is considered as a large-scale finite element analysis of the reconstructed 3D image that predicts the mechanical behaviour of the material suitable for printing applications (Falcone et al. 2006). The macro and microstructure of the 3D printed pectin-based food simulant were characterized using X-ray microtomography. The reconstructed image was used to measure the volume, pore size distribution, porosity, and thickness distribution. The average thickness of the object (δ_{av}) was described by parallel plate model as given in Eq. (5.1) (Vancauwenberghe et al. 2018)

$$\delta_{av} = \frac{2}{S_{obj}/V_{obj}} \tag{5.1}$$

where S_{obj}/V_{obj} is the surface to volume ratio of the solid 3D printed object. The micro-computed tomography (CT) images of the pectin simulant showed a small visual deviation from the design model. This deviation in the shape was mainly because of the spreading of food inks that significantly affect the resolution of the 3D construct (Kim et al. 2018a). The composition of the formulated food ink has a significant impact on the thickness distribution of the structure. The variation in the distribution of structure thickness was because of the spreading tendency of the sample that in turn related with the viscosity (Bakarich et al. 2013). The 3D construct with an open structure and large cell size resulted in lesser thickness distribution leads to less overspreading of the food ink due to its highly porous structure influencing the material deposition (Figure 5.3). Results of the average object thickness showed that the average thickness was significantly higher with structures having 2 mm cell size while for the rest of the structures there was a slight decrease in the average thickness of the wall with the cell size (Vancauwenberghe et al. 2018). These results were consistent with the findings of the thickness distribution of the structure. The 2D analysis of the honeycomb (infill pattern) cells of the fabricated 3D structures aids in the determination of the distribution of pore size. The average pore size (diameter) of the 3D printed constructs was linearly increased with the cell size. From the micro-CT imaging, the study inferred that the spreading of the printing mixture during the printing process was related to the shrinkage of the gel induced after post-treatment. Due to which the average wall thickness, porosity, pore size, and volume of the fabricated 3D construct had a slight deviation from the predicted structural design.

The process steps of image processing for the determination of pore volume were presented in Figure 5.4 (Vancauwenberghe et al. 2017). Similarly, the pore size distribution

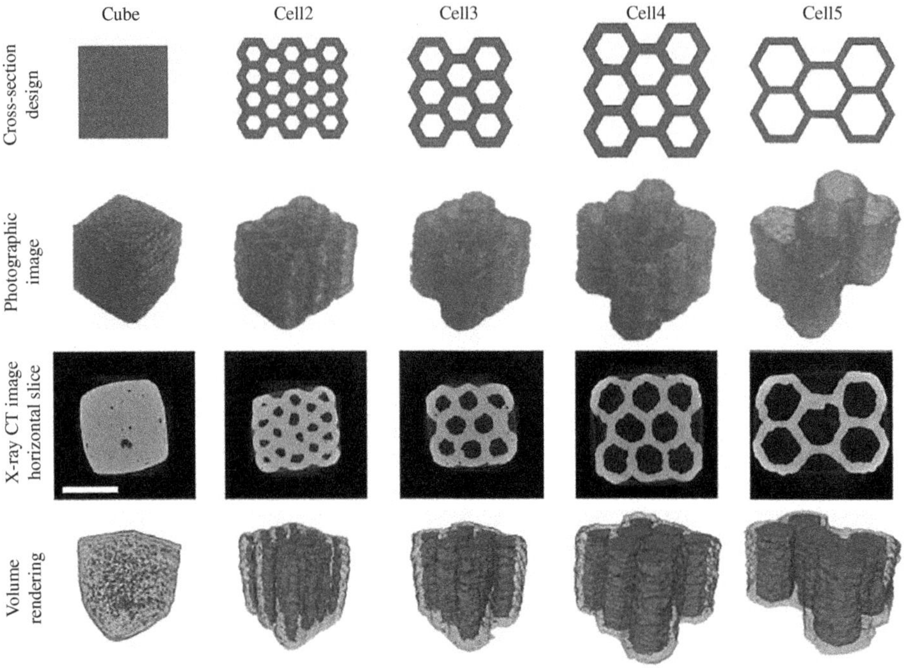

Figure 5.3 Pectin-based model 3D constructs of cube and honeycomb structures of concentration 25 g l^{-1} pectin. *Source:* From Vancauwenberghe et al. (2018), [p. 35] / With permission of Elsevier. DOI-https://doi.org/10.1016/j.jfoodeng.2018.03.010.

of the printing mixture was determined by taking into account the equivalent diameter (d_{eq}). The composition of the printing mixture on the macro and microstructure of the 3D printed pectin gels was studied (Figure 5.5). The addition of sugar had both positive and negative effects on the build quality of the food ink. Discrepancies such as irregular extrusion, automatic stalling of material supply, and partial spreading of food ink were evident for material supplies with 15 g l^{-1} pectin and 55 g l^{-1} pectin. This was due to the higher viscosity of the material that significantly affects the final shape of the 3D construct. In another study, the dimensional measurements of the virtual model and the 3D construct were analyzed using X-ray tomography (Severini et al. 2018b). Results showed that the overall dimensions of the virtual model and the printed sample were in agreement with each other. A slight deviation in the diameter of the bottom layer was mainly because of the system design of the 3D printer. Since the commercial 3D printer used for printing of thermoplastic materials such as polylactic acid (PLA) and acrylonitrile butadiene styrene (ABS) was adapted for 3D printing of blend of fruits and vegetable mixture, the system has a limitation to be used with food system (Severini et al. 2016). The toolpath generation during the slicing process was mainly based on the properties of thermoplastics that overviewed the physiochemical properties of the material supply such as ink viscosity, wettability, density, and stickiness. The material deposition was based on the polymeric deposition that resulted in a discrepancy of the material deposition of the formulated food ink. Hence studies

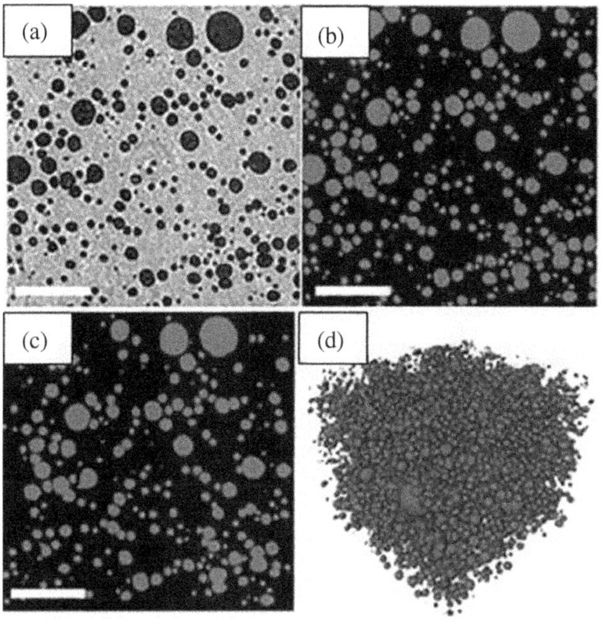

Figure 5.4 3D pore analysis of 3D printed pectin gel. (a) Cross-sectional image after filtering, (b) global thresholding, (c) de-speckle, separation, and border kill treatments, and (d) volume rendering of pore distribution. *Source:* From Vancauwenberghe et al. (2017), Figure 03 [p. 142] / With permission of Elsevier. DOI-https://doi.org/10.1016/j.ifset.2017.06.011.

focussing on system design and improvement of the commercial 3D printers are adequate for food applications.

Severini et al. (2018a) evaluated the printability of insect protein-enriched dough samples through dimensional and microstructural analysis using X-ray microtomography. The 2D analysis was performed on five different slices from the bottom to the top of the 3D construct chosen at every 150 images. Similarly, for 3D analysis, about 300 cross-sectional images were transformed into binary images using Otsu's method. From the acquired image, the size distribution of the solid phase of the 3D constructs was obtained employing imaging software with the function structure thickness. While the ratio of the volume of the solids to the total volume of the solids gave the volumetric fraction of the 3D constructs (Figure 5.6). Results showed that the height and overall diameter of the 3D constructs were well matched with the design of the virtual model. However, some discrepancies were observed within the levels of the insect enrichment.

There was an enlargement in the diameter of the first layer of the printed snack while the height got a significant reduction from the actual measurement of 20 to 18.9 mm. Using the obtained results, the overall volume of the 10 and 20% insect enriched 3D construct was found to be in the range of 4.56 and 5.18 dm^3 respectively. The calculated value of the overall volume was close to the volume of the virtual design model of about 4.99 dm^3. Results inferred that the reduction in the consistency of the material supplies with insect enrichment levels enhanced the amount of material deposition. As a result, the solid phase of

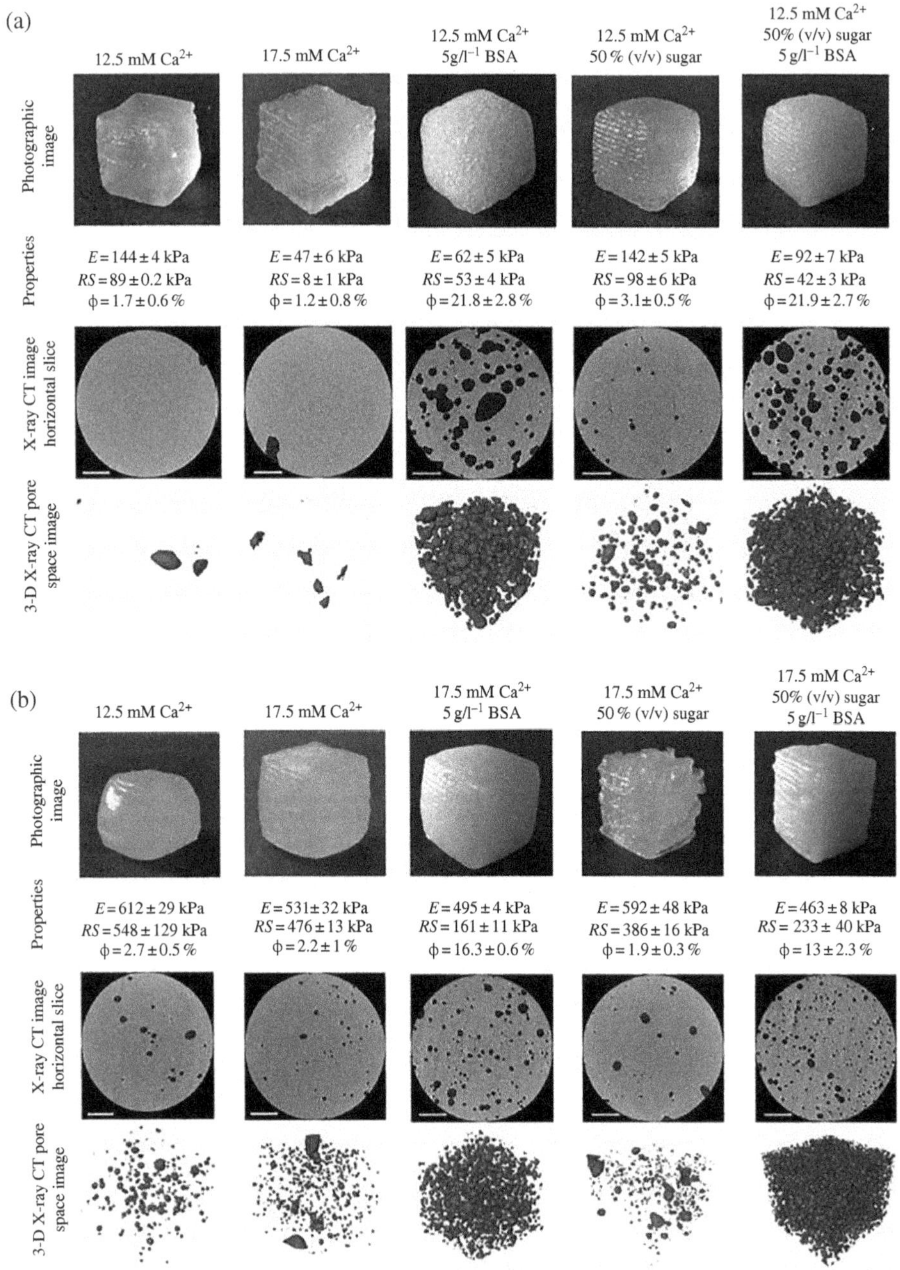

Figure 5.5 Macro and microstructural analysis of 3D printed pectin gel. (a) 15 g pectin l^{-1} and (b) 55 g pectin l^{-1}. *Source:* From Vancauwenberghe et al. (2017), Figure 06&07 [p. 144] / With permission of Elsevier. DOI-https://doi.org/10.1016/j.ifset.2017.06.011.

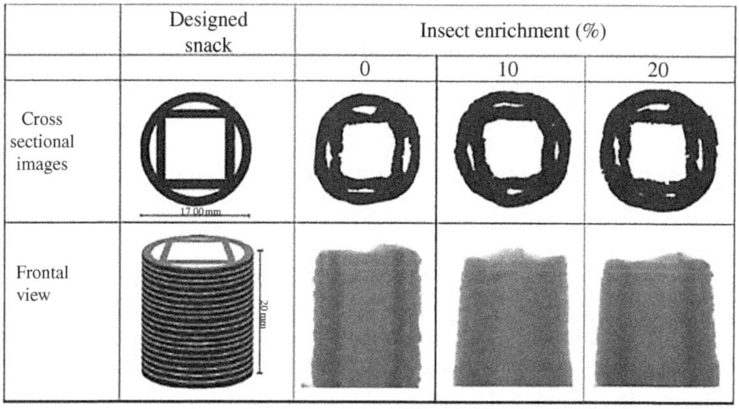

Figure 5.6 Micro CT images of insect enriched 3D printed snacks. *Source:* From Severini et al. (2018a) / With permission of Elsevier.

the 3D printed snacks occupied a higher volumetric fraction of 68.6% than that of the virtual model of 36%. This mismatch was mainly attributed to the material deposition behaviour governed by 3D printer movement which was optimized for the printing of thermoplastic materials (Severini et al. 2016). The filament thickness distribution showed a double bell curve for all the formulations which was well characterized with the virtual design model. Results showed that the thickness of the single filament was much larger than the size of the printing nozzle. This was due to the material deposition behaviour and design features of the slicing software used that estimates the printing variables by considering the properties of the thermoplastic materials. The differences with a structural thickness of the 3D constructs of varying levels of insect enrichment were clearly visualized (Figure 5.7). Variation in colour represent the range of the varying thickness of the 3D construct. Results showed that the conjunction zones were relatively thicker than the outer shell and inner parts.

5.3.1.3 Confocal Laser Scanning Microscopy

CLSM is an advanced form of fluorescence microscopy. The CLSM allows the analysis of thin optical sections of the thick intact specimens that use a laser spot to scan the object across the field of view (Sharif et al. 2020). The advantage of CLSM to use fluorescent labels for the detection of specific food components made it to be a useful characterization technique in 3D printing studies. The CLSM has been used for visualizing the microstructure of the cheese samples: untreated cheese (UC) sample, melted cheese (MC) sample, and cheese extruded at two different extrusion rates (4 and 12 ml minute^{-1}) (Le Tohic et al. 2018). A mixture of fast green FCF and Nile red dye has been used to strain the fat and protein phases of samples. The microstructural analysis revealed a homogeneous distribution of circular fat globules (in red colour) on a continuous phase of protein matrix (in green colour). Results showed that the fat globules of MC were significantly larger than that of UC. This was due to the coalescence of the fat globules upon thermal treatment. The low-speed printed cheese (LSPC) showed an altered microstructure with discontinuous protein phase with interstitial fat globules with larger non-spherical morphologies. Similar morphology

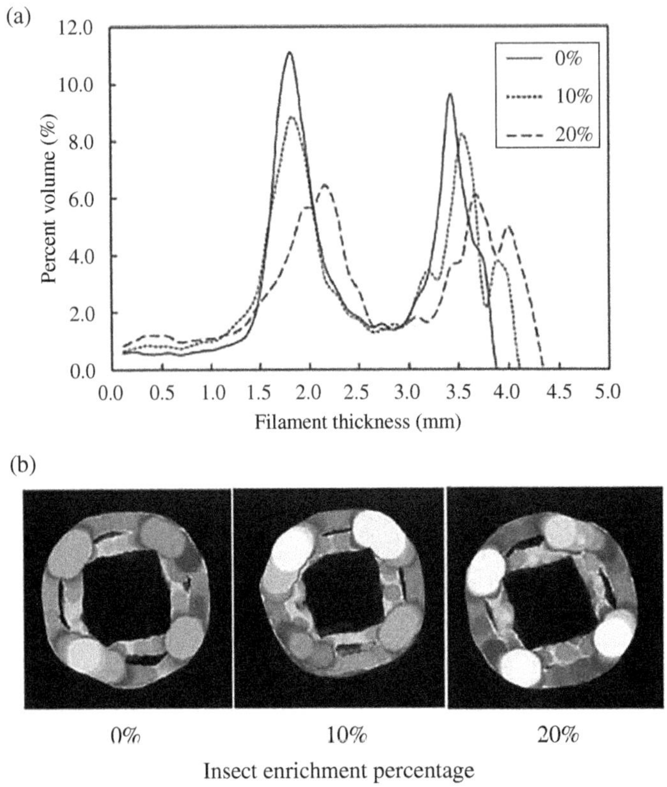

Figure 5.7 Structural properties of insect enriched 3D printed snacks. (a) Distribution function of filament thickness vs percent volume. *Source:* Severini et al. (2018a) / With permission of Elsevier and (b) images showing distribution of filament thickness. *Source:* From Severini et al. (2018a), Figure 02 [p. 671] / With permission of Elsevier. DOI-https://doi.org/10.1016/j.foodres.2018.01.034.

was observed with high-speed printed cheese (HSPC); however, the fat globules were relatively smaller and possess uniform distribution than LSPC. This distinct size difference was due to the shear intensity during the extrusion of the cheese. The lower shear forces facilitate the coalescence of the fat globules while higher shear forces significantly reduce the fat globule size due to the breakdown of molecules. The altered microstructure with the disrupted protein phase, fat globules size and its morphology impart a soft texture to the cheese. The higher adhesiveness of the printed cheese with less stickiness was because of the release of surface fat during shearing. The homogenous dispersion of fat in a continuous protein matrix provides a harder texture to the cheese. The colour differences among the cheese samples were related to the droplet sizes. A larger-sized droplet led to the dark colour of samples (MC, LSPC, and HSPC) than that of UC; similarly, the HSPC was bluer than that of UC mainly because of the high interstitial fat globules. Thus, CLSM can be conveniently used to reveal the microstructural properties of the material supply that in turn correlated with the modification of texture. The textural alteration and associated flowability are significant aspects that aids in determining the printability of the cheese samples.

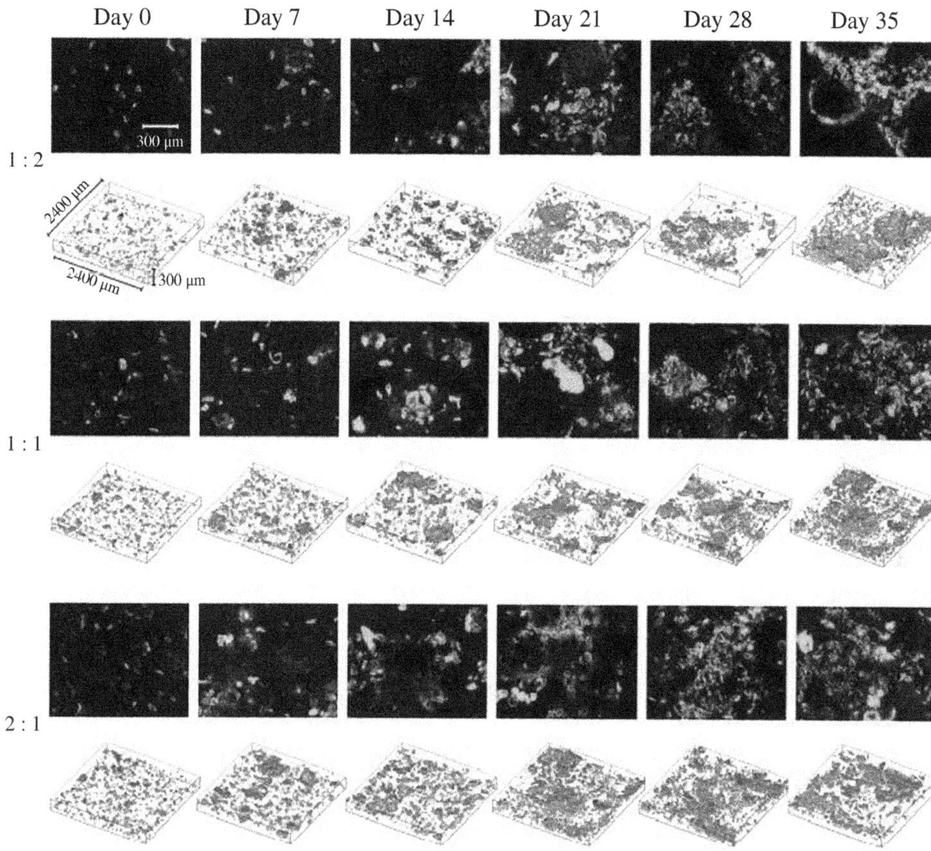

Figure 5.8 Confocal images showing distribution of carrot callus in callus-based food ink of different alginate to cell contents. *Source:* From Park et al. (2020), Figure 07 [p. 32] / With permission of Elsevier. DOI-https://doi.org/10.1016/j.jfoodeng.2019.109781.

Recently, the CLSM was used for assessing the cell proliferation within the cell-loaded callus-based food (CBF) inks (Park et al. 2020). The CBF gels were stained using fluorescein diacetate (FDA) and propidium iodide (PI) to distinguish the live cells (green) and dead cells (red), respectively. The CBF samples were monitored for 35 days in a 1-week interval. Initially, the CBF gel was viewed as a homogeneous dispersion of cells with distinct smaller cell clusters that are clearly distinct from callus. The CLSM results showed that that the cell proliferation was activated at specific sites instead of uniform occurrence over the gel matrix (Figure 5.8). This was due to the immobilization of each cell and microclusters on the alginate gel system that hinders the intercellular interactions. Hence, the cell matrix has no access to endogenous substances that were significant for the morphogenesis and proliferation of cells. Comparing the CBF prepared at different alginate to cell content (1 : 2, 1 : 1, and 2 : 1, w/w), the initial cell concentration only affects the cell growth rate but not the final cell volume. Results showed that 1 : 2 CBF gel had a very low initial concentration of cells that grew slowly and characterized with loosely arranged cells on day 35. On the other hand, CBF samples of 1 : 1 and 2 : 1 showed enough cell proliferation by day

28 and were characterized with a uniform distribution at day 35 as stated earlier. This clearly showed the dissociation of cells immobilized in hydrogel when cell density was exceeded. However, the cell clusters remain free without adhering to each other which fails to form a rigid tissue (Park et al. 2020). These results put forth the need for further investigations on the mechanism of binding among the immobilized cell clusters in hydrogel to promote cell adhesion and to prevent dissociation of cells for simulating the plant tissues.

In another study, the distribution of proteins and fat in dough samples was determined using CLSM through staining with fluorescein isothiocyanate isomer I (FITC) (protein) and Nile red (fat particles). From the CSLM images, the higher intensity and uniformity indicate a finer network with higher mechanical strength and adhesiveness (Figure 5.9). Hence, the samples with higher intensity and uniformity were most conducive for successful 3D printing. Both protein and fat distribution were characterized with different intensities (Yang et al. 2018a). The microstructure of the printing material supply containing a dispersion of vegetable particles in a hydrocolloid matrix was visualized using CLSM (broccoli powder, carrot powder, and spinach powder) (Kim et al. 2018b). The dye Rhodamine B was used that works by binding to the hydrophobic sites of the protein molecules in vegetable powder.

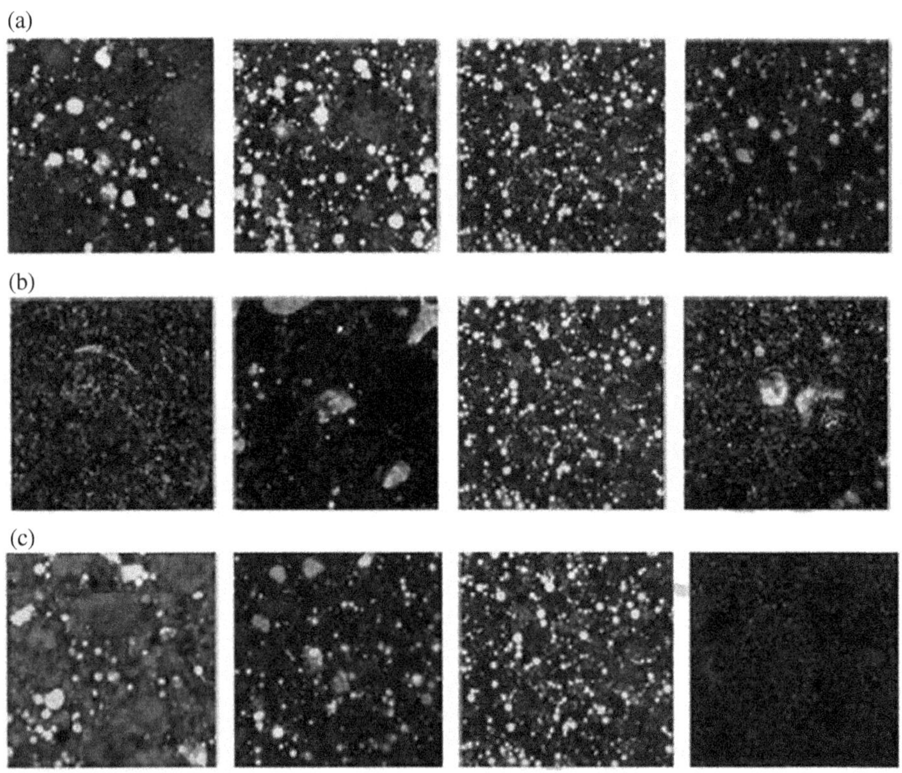

Figure 5.9 Protein and fat distribution of different formulation of baking dough. (a) Sucrose group: 3.3, 5, 6.6, 8.2 g/100 g of formulation, (b) butter group: 0, 3, 6, 9 g/100 g, and (c) low gluten flour group: 36, 42, 488, 54 g/100 g. *Source:* From Yang et al. (2018a), Figure 04 [p. 25] / With permission of Elsevier. DOI-https://doi.org/10.1016/j.ifset.2018.01.001.

Results of CLSM showed that the vegetable powder samples with 10% concentration (w/w) in xanthan gum (XG) matrix had intact matrix while in hydroxypropyl methylcellulose (HPMC), the vegetable particles were completely swollen. With increasing the concentration of vegetable powders from 10 to 30%, the XG dispersions with three different powder mixtures significantly suppressed the powder swelling at the higher concentrations. In contrast, the vegetable powders remained swollen even at higher concentrations in HPMC dispersions. This was mainly because of the difference in the properties of hydrocolloids confirming the fact that hydrocolloids with higher water hydration capacity would prevent swelling of starch particles in a gel-like matrix containing dispersed solid particles (Kim et al. 2018b).

5.3.2 Crystal Morphology

5.3.2.1 X-ray Diffraction

XRD is a non-destructive analytical technique commonly used for characterizing crystalline materials. Material characterization is a significant aspect of the selection of raw materials for 3D printing. XRD patterns provide information about crystallinity, chemical structures, phases, average grain size, crystal orientations, and crystal defects (Bunaciu et al. 2015). The XRD analysis is more useful in the sintering process and hot-melt extrusion where the printing material supply undergoes phase transition with subsequent melting and solidification upon deposition. The crystal morphology of the 3D printing mixture of potato starch and pea protein was analyzed (Chuanxing et al. 2018). The XRD patterns indicate a direct relationship between the size of the crystallization and the peak width. Results of XRD analysis showed that pea protein had no crystalline structure, when added with starch it weakened the diffraction peak at 17°. The complex interaction among protein and starch molecules during gelatinization strengthens the diffraction peak at 20° with no new peaks appeared with the 3D printed sample mixture. This clearly shows that there was no significant interactions between the starch and pea protein during gelatinization. Thus, there was no formation of new substance with crystalline structure and the amorphous structure of the starch molecules remains unchanged.

The XRD patterns of the potato starch samples under different hot-melt extrusion conditions have been reported (Liu et al. 2020b). Results confirmed the vanishing of the diffraction peaks of potato starch of 10–30% concentration under the hot-melt extrusion process that destroys the original crystalline structure. At a higher printing temperature of 60 °C, the potato starch samples still exhibit the B-type crystalline pattern as that of the original starch (Zhang et al. 2019). With the increase in the printing temperature from 60 to 80 °C, the diffraction peak intensity gradually decreased and finally disappears. This clearly indicates that the printing temperature has control over the crystallinity of the printing material supply (potato starch). Hence, the higher temperature could completely gelatinize the starch-based material supply that shows the significance of temperature over the printability. There was a reduction in the gelatinization enthalpy (ΔH) with the dry heat treatment of the wheat starch (Maniglia et al. 2020). The gelatinization enthalpy is an indirect prediction of the extent of the range of crystalline regions of starch molecules that are broken during heating. Hence, the reduction in the ΔH shows less crystallinity indicating the formation of weaker bonds. These results were consistent with the XRD patterns of the maize starch (Lei

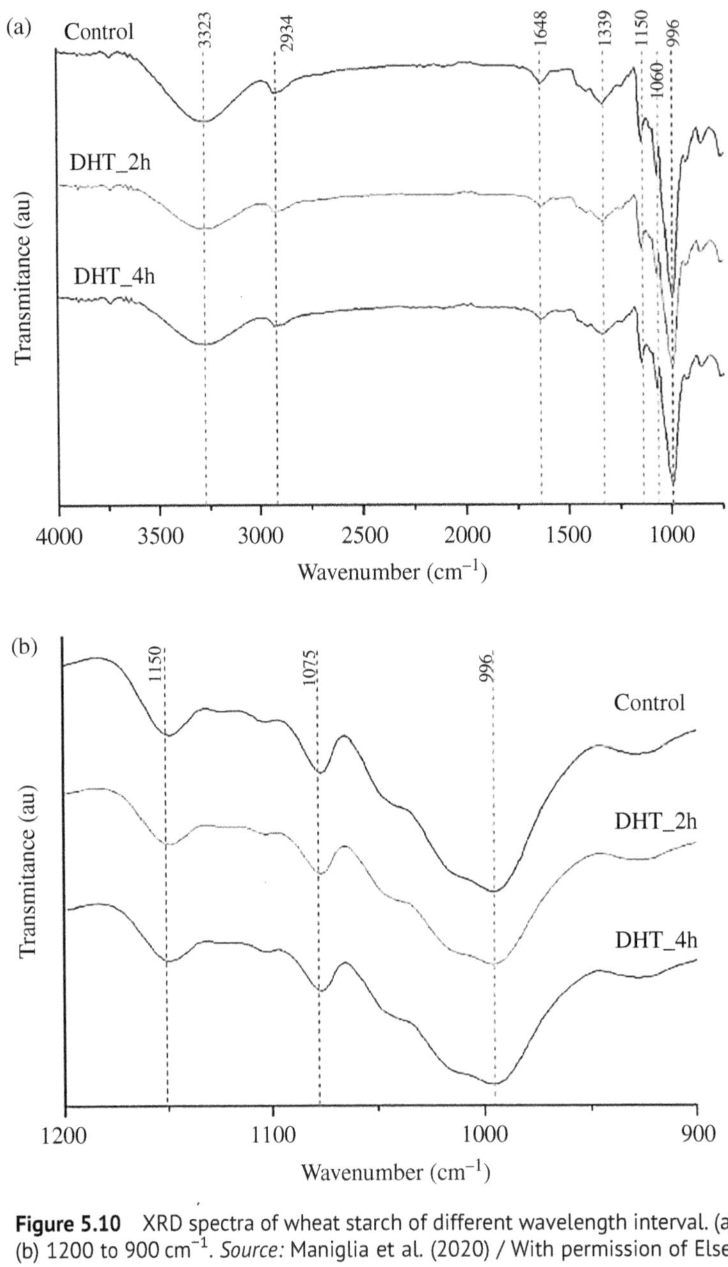

Figure 5.10 XRD spectra of wheat starch of different wavelength interval. (a) 4000 to 900 cm^{-1} and (b) 1200 to 900 cm^{-1}. *Source:* Maniglia et al. (2020) / With permission of Elsevier.

et al. 2020). A typical 'A' type diffraction pattern of wheat starch samples with singlet peaks, doublet peak, and two weak peaks were observed at 15° and 22.8°, 16.8°, and 17.8°, 5.6°, and 20°, respectively (Figure 5.10). The relative crystallinity of the dry heat-treated samples for 4 h was relatively lower than the control and the dry heat-treated samples for 2 h. These results confirm the reduction in the crystalline zones of wheat starch due to dry heat treatment of the starch granules because of the reduction in Maltese cross brightness in modified starch (Maniglia et al. 2020).

The diffractogram of different varieties of rice paste (waxy rice, indica rice, and japonica rice) material supply showed a typical 'A' type diffraction pattern (Liu et al. 2020c). Results depicted that the highest amount of amylopectin in waxy rice was responsible for greater crystallinity than the other two rice varieties (Lawal et al. 2011). Irrespective of the rice varieties, the relative intensity of the XRD diffraction peaks was increased with concentrations of sodium alginate. This was due to the formation of hydrogen bonds among the sodium alginate and rice starch that had a positive effect on the crystallinity of the rice paste sample (Figure 5.11). However, no obvious variations were observed in the crystallinity of the rice samples indicating the addition of sodium alginate had not affected the inner crystalline portions of the starch molecules. In a recent study, the bioactive nanocomposite film based on mixture of bovine skin gelatine (BSG)/zinc oxide (ZnO)/clove essential oil (CEO) was explored for the bioprinting applications (Ahmed et al. 2020). The XRD patterns of the nanocomposite film showed a characteristic peak at 2θ of 10.4, 12.7, 22.8. 32.2, 34.8, 36.8, and

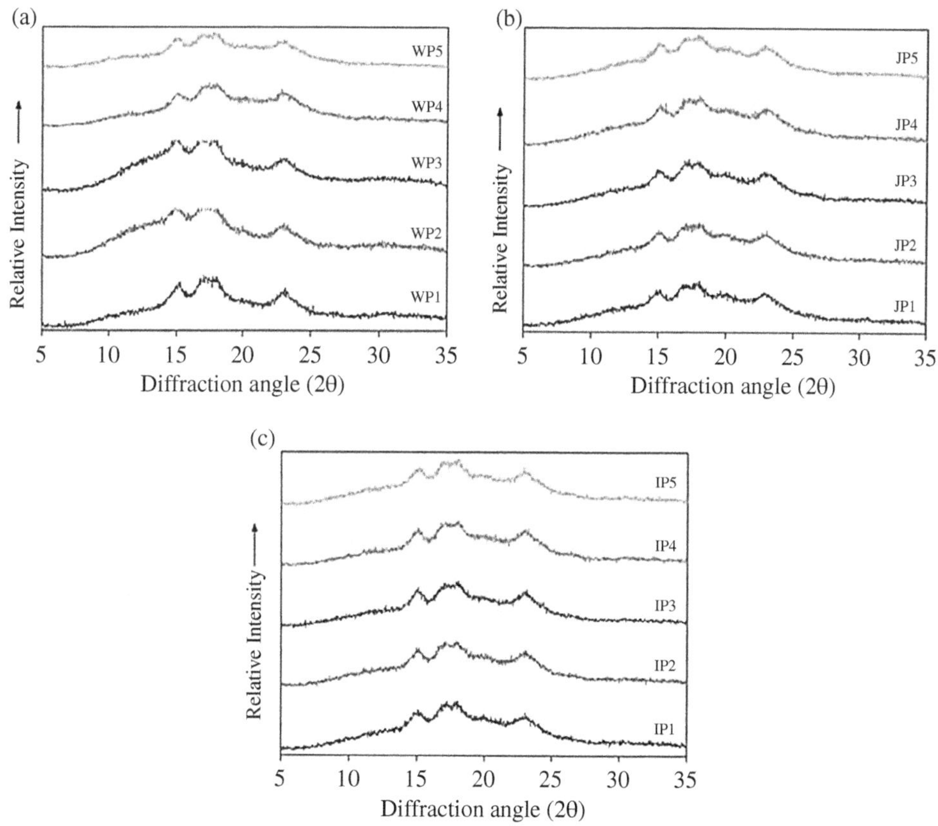

Figure 5.11 XRD spectra of varieties of rice paste with different levels of sodium alginate. (a) Waxy rice paste, (b) japonica rice paste, and (c) indica rice paste. *Source:* Liu et al. (2020c) / With permission of Elsevier.

56.5 that corresponds to various planes of ZnO nanocomposites, respectively (Sangeetha et al. 2011). These diffraction peaks clearly indicate the presence of the crystalline nature of ZnO in nanocomposite films. Thus, the crystalline nature of the material supplies is determined using X-ray diffractograms that aids in assessing the chemical characteristics of the samples for 3D printing.

5.3.2.2 Small-Angle X-ray Scattering

The small-angle X-ray scattering (SAXS) is a quantification technique employed to measure the density differences of the sample at the nano-level (Pauw 2013). The SAXS helps in the analysis of the periodic structure and particle diameter at the nanoscale, mesoscopic pore size, shape, and its distribution (Chi et al. 2017). The submicroscopic structure of the 3D printed sample was assessed based on the electron density using SAXS. The analysis involves the transformation of SAXS curves into the Kratky plots based on the Lorentz equation. From the Kratky plots, the aggregation nature of the particles of the material supplies used for 3D printing was revealed. A peak at the finite q range of the Kratky plot shows the presence of the heterogenous aggregated structure of the gel system (Chi et al. 2018). The nano-aggregated structure of the potato starch samples at different hot-melt printing conditions was assessed using SAXS curves (Liu et al. 2020b). The Kratky plots of the potato starch samples represent the uniform gel structure of the 3D constructs with no detectable peak observed in the lower q region of the samples of 10–30% potato starch. The crystalline ordered structure of the starch molecules was disintegrated at the hot-melt printing conditions that resulted in the formation of a uniform network of samples during 3D printing. However, the characteristic peaks were still observed at $0.7\,\text{nm}^{-1}$ at 60 °C depicting the employed printing temperature was not sufficient to disintegrate the lamellar structure of potato starch molecules. This clearly shows that printing temperatures higher than 60 °C, i.e. 65–80 °C could lead to the disappearance of the characteristic peak with the destruction of the original lamellar structure thereby formed a uniform and stable gel matrix suitable for 3D printing.

The SAXS curve of the potato starch mixture is fitted to the Ornstein–Zernike (OZ) equation (Eq. (5.2)) for measuring the correlation length (ξ) that reflects the size of pores of the material supplies formed by the chain entanglement (Shibayama 2012). It was reported that the higher degree of the entanglement of the gel indicates lower pore size which in turn is directly related to the ξ indicating a lower correlation length (Shibayama 2011). It was observed that the ξ was declined with an increase in potato starch concentration from 10 to 30% indicating a dense gel matrix with a smaller pore size as a result of a greater degree of the chain entanglement. On the other hand, the ξ was initially increased and then gradually decreased with an increase in temperature of printing from 65 to 80 °C at constant 15% potato starch (Figure 5.12). From this observation, it was clear that temperature as high as 70 °C is significant for gelatinization; however, only temperatures greater than 70 °C could allow for the availability of more free starch molecules leads to the development of a homogenous gel system conducive for 3D printing.

$$I(q) = \frac{I_{OZ}(0)}{1 + \xi^2 q^2} \qquad (5.2)$$

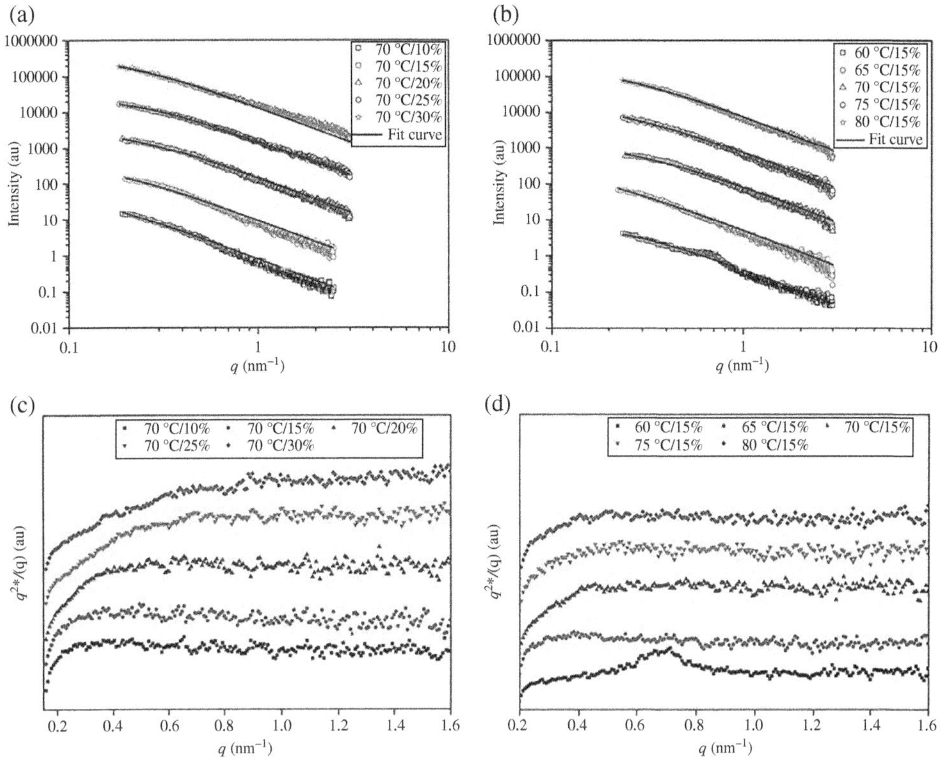

Figure 5.12 SAXS analysis of 3D printed potato starch samples of different hot-melt extrusion conditions. (a) and (b) SAXS patterns; (c) and (d) Kratky plots. *Source:* Liu et al. (2020b) / With permission of Elsevier.

where $I(q)$ is the intensity of scattering, q is the scattering vector, and $I_{OZ}(0)$ is the zero-scattering vector associated with the osmotic bulk modulus K_{OZ}.

The inhomogeneity of the polymeric gel matrix (phase-separated systems) attributes to a surge in the lower q region of the scattering curves that is referred to as the Picot–Benoit effect. In such samples, the OZ equation is not relevant which requires the use of other fitting models like the Debye–Bueche (DB) equation (Eq. (5.3)).

$$I(q) = \frac{I_{DB}(0)}{\left[1 + \xi^2 q^2\right]^2} + Aq^{-\delta} \tag{5.3}$$

where $\frac{I_{DB}(0)}{1 + \left(\xi^2 q^2\right)^2}$ is the Debye–Bueche model function, ξ is the characteristic length indicating the degrees of inhomogeneity and $Aq^{-\delta}$ is the Power-law function with A and δ are prefactor and exponent (Crompton et al. 2009). Both DB and OZ equations are used to analyze quantitatively the submicroscopic structure of the rice starch and corn starch gels (Zeng et al. 2020). Results showed that the corn starch possesses no distinct peak at the lower q region indicating the gel homogeneity at concentrations ranging from 10 to 30%. Decreasing correlation length with increased concentration of starch was related to the smaller pore size with the whole network forms a dense network. This was due to the enhancement of molecular entanglement and the hydrogen bond interactions with increased concentrations of corn starch. On the other hand, rice gel had a weak scattering peak at 0.4 nm^{-1} that indicates the structural inhomogeneity (Karino et al. 2006). A less uniform distribution of electron clouds was observed for rice starch as a result of strong hydrogen bond interactions among the molecules showing a V-type crystallinity embedded in the rice starch gel matrix. With the increase in rice starch concentration, the characteristic length was decreased indicating the enhancement of the homogeneity of the gel system.

There was no scattering peak observed with the printing temperature in the range of 65–85 °C for corn starch at low q and at 0.7 nm^{-1}. This clearly indicates the homogeneous gel system with the disappearance of the lamellar structure of starch with increasing temperature. The correlation length of the corn starch was increased initially and then declined with increasing temperature due to less influence over the molecular entanglement at low temperatures that hindered cross-linking of the molecules (Zeng et al. 2020). On the other hand, the temperature rise disrupts the corn starch leads to the complete stretching of molecular chains forming a homogenous compact gel system. An infirm scattering peak was observed at 0.7 nm^{-1} and printing temperature of 65 and 70 °C showing incomplete gelatinization of the rice starch molecules. While the scattering peak was disappeared at the same q value at the printing temperature of 75–85 °C (Figures 5.13 and 5.14). However, the peak was more obvious at lower q of 0.4 nm^{-1} showing the presence of a new ordered molecular structure with the disruption of lamellar structure. Thus, the SAXS

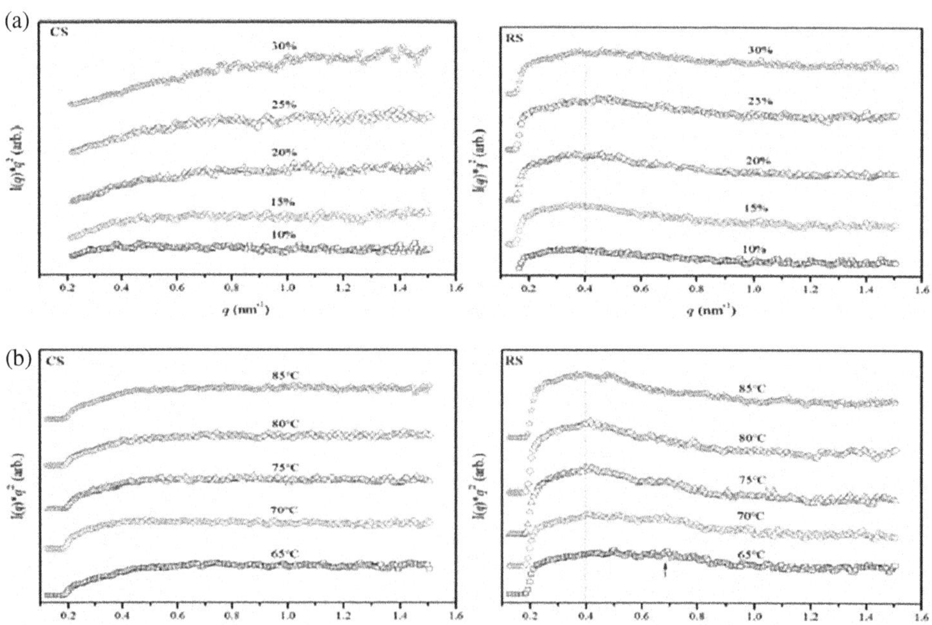

Figure 5.13 Kratky plots of 3D printed corn starch and rice starch samples. (a) Different starch concentrations and (b) different printing temperatures. *Source:* Zeng et al. (2020) / With permission of Elsevier.

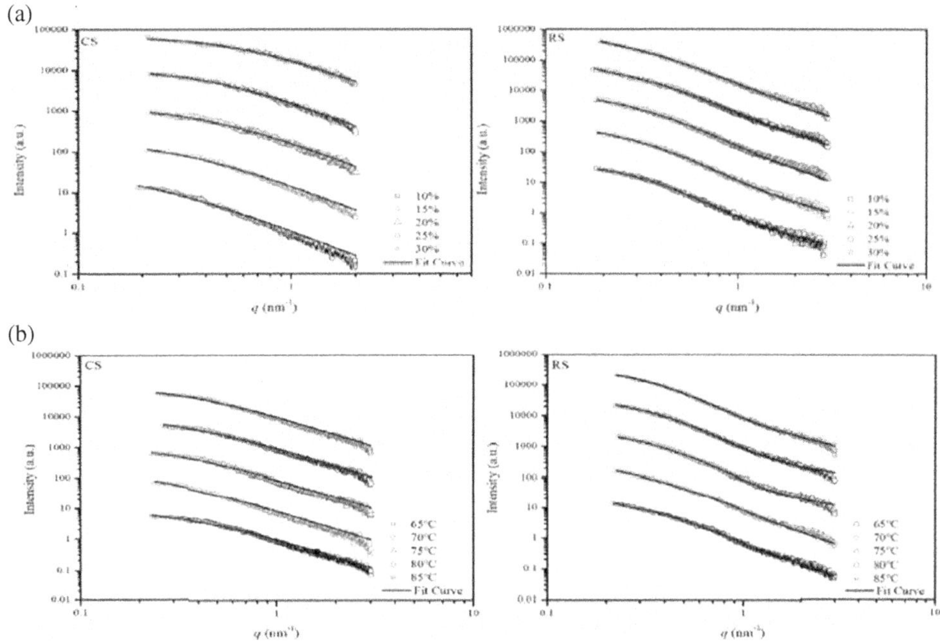

Figure 5.14 SAXS of 3D printed corn starch and rice starch samples. (a) Different starch concentrations and (b) different printing temperatures. *Source:* Zeng et al. (2020) / With permission of Elsevier.

measurements and analytical plots of material supplies would help in the understanding of the chemical state of the material supplies at the submicroscopic levels which could be correlated with the printability.

5.3.3 Molecular and Chemical Analysis

5.3.3.1 Nuclear Magnetic Resonance Imaging

The NMR analysis is commonly reported in 3D printing studies for understanding the status of the moisture content of the material supply. The investigations on the variations in the distribution and the movement of the protons of food constituents such as lipids, proteins, structural, and bulk water were highly correlated with the textural parameters (Sun et al. 2020). The resulting NMR relaxation plot helps in the prediction of the firmness and the elasticity of the material supply that in turn useful in determining the suitability of the material applied for 3D printing. Flowability is another crucial factor that influences the printability of the samples. The chemical status of the water and its distribution have a close relationship with the microstructure and rheology. The transverse relaxation time (T_2) peaks were determined for the potato starch incorporated lemon juice gel (Yang et al. 2018b). Each of the peaks possesses three distinct relaxation populations centred at 0.3–20 ms (T_{21}), 20–200 ms (T_{22}), and 200–1000 ms (T_{23}) that indicates the bound water, partially immobilized water, and free water, respectively. NMR relaxation data showed that the relative signal intensity of the free water was close to 0 with no significant differences

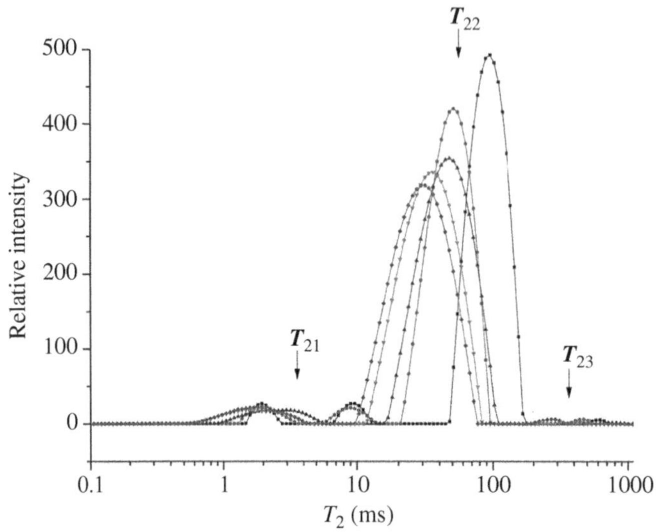

Figure 5.15 NMR spectra of 3D printed lemon juice gel with different concentrations of potato starch (□, 10 g/100 g; ○, 12.5 g/100 g; ▽, 15 g/100 g; △, 17.5 g/100 g; ◇, 20 g/100 g). *Source:* Yang et al. (2018b) / With permission of Elsevier.

observed with an increase in the proportion of potato starch. On the other hand, the signal intensity of the partially immobilized water significantly reduced with the increase in the concentration of potato starch. This was due to the presence of abundant chances of the interactions among the starch and water molecules. The existence of the adequate quantity of water for the gelation process would maximizes the gel strength of the material supply. The NMR signal of the tightly bounded water became smaller with the time and with increase in the starch concentration, the two peaks had been changed into a single peak. From NMR spectra, it was clear that T_2 decreased with increased content of starch indicating the change of partially immobilized water into its bound forms (Figure 5.15). This clearly shows the possibility of interactions between the water and starch molecules that leads to the formation of a dense network with a proportional rise of starch. This in turn improves the apparent viscosity, storage modulus, hardness, and springiness of the lemon juice gel thereby enhancing its printability.

A small T_2 of the NMR signal implies that the degree of freedom of moisture content is low and the moisture is tightly bound with non-aqueous food components. On the other hand, a large T_2 peak showed that moisture is not well-bounded with non-aqueous food components with higher degrees of freedom (Yang et al. 2019). Compared to wheat starch, the samples with added potato starch possess relatively higher fluidity with a lesser extent of bound or immobilized water. These samples with bound or immobilized water possess better formability that is desired for extrusion printing. In contrast, the relative intensity of the free water with added sweet potato starch was close to zero that indicates the presence of internal water in the bound and partially immobilized forms. This results in solid-like viscoelastic behaviour of the material supply that is comparatively difficult to extrude. The addition of NaCl increased the T_{22} relaxation components showed the alteration of water–protein interactions in the fish surimi gel (Wang et al. 2018). In contrast, the relaxation component T_{23} gradually declined with NaCl addition mainly because of the swelling behaviour of myofibrillar proteins. This swelling phenomenon induces the shift of free water to immobilize water thereby decreasing the availability of water. Thus, the moisture status of the gel is related to textural properties that correspond to the gel strength of the sample used for printing.

The distribution of water in food materials is closely associated with its physical properties. It was reported that in a strong gel matrix, the mobility of water was restricted within the gel structure (Yang et al. 2018b). The components with different relaxation times represent the proton populations with different rotational mobilities. The water distribution in milk protein gel used for 3D printing was analyzed using NMR (Liu et al. 2019a). The milk composite gels had different water distribution at 0–10 ms (T_{21}) and 10–100 ms (T_{22}) that corresponds to immobilized and partially immobilized fractions of water, respectively (Figure 5.16). Results showed that T_{21} was gradually increased and T_{22} was decreased and ultimately reached 0 ms with an increase in protein content of gel. This relative intensity of the peaks indicates the formation of the denser gel with an increased protein content of the material supplies for 3D printing. The dispersion of various types of water in different levels of orange concentrate–wheat starch (OC–WS) mixture was analyzed (Azam et al. 2018b).

The water dynamics could be assessed from the peak and the signal amplitude area represents the relative content of water (Zhu 2017). Results showed that all the samples of OC–WS mixture possessed the intensity of the free water close to 0. The samples with the higher starch proportioned mixture in the peak T_{21} (0.8–1.54 ms) reflect the tightly

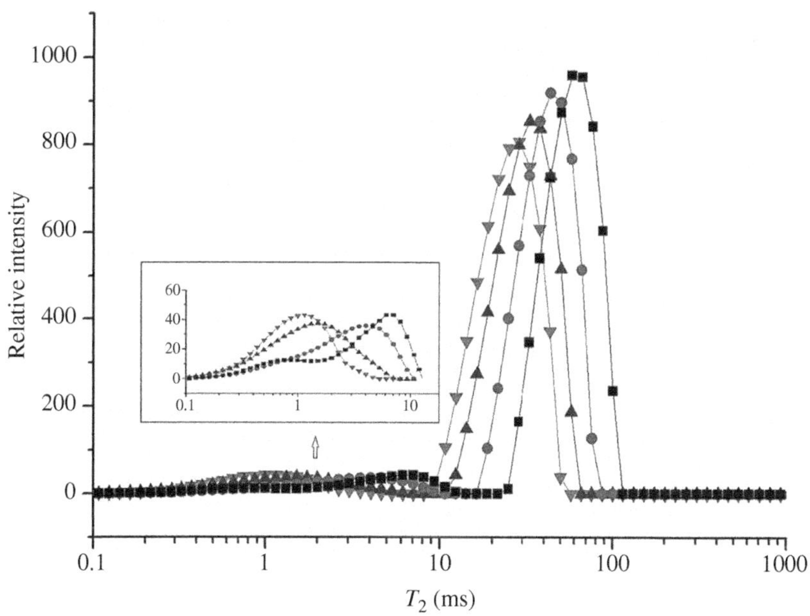

Figure 5.16 NMR spectra of milk gels with different total protein contents (□, 350 g l^{-1}; ○, 400 g l^{-1}; ▽, 450 g l^{-1}; △, 500 g l^{-1}). *Source:* Liu et al. (2019a) / With permission of Elsevier.

bound immobilized water. The mechanical strength of the printing material supply was imparted by the conversion of maximum of free and immobilized water into its bound forms. As expected, 30% WS bound mixture had maximum conversion followed by 25, 20, and 15%. From T_{23} peak, it was observed that with increase in the WS, the T_2 value was decreased. This represents the conversion of partially bound water into its bound forms. During the agitation and thermal treatment, the available moisture from OC diffuses into the wheat starch. Thus the diffused water molecules interact with the hydroxyl groups of amylose and amylopectin of wheat starch thereby forming a gel matrix with closer network. The amount of water available for gelation process could be related with the yield stress of the mixture in turn reflects the printability. When enough water is available for gelation process, a maximum yield stress can be attained; hence the mixture with higher WS (30%) does not possess enough water for maximum gelation that resulted in poor extrudability of the sample. Results showed that the overall conversion of partially immobilized water into its bound forms enhances the formation of a closer and denser water–starch network (Azam et al. 2018b). This eventually increased the yield stress, apparent viscosity, hardness, adhesiveness, and chewiness of the material supply thereby improve its printability.

5.3.3.2 Fourier Transform Infrared Spectroscopy

FTIR is an analytical non-destructive technique that provides useful information about the molecular and chemical microstructure of heterogeneous food materials (Falcone et al. 2006). In 3D printing studies, FTIR spectra give information about the material interactions and resultant products of the selected raw materials for 3D printing. The FTIR spectra of the lyophilized rice paste of three different varieties were found to be similar to each other (Liu et al. 2020c). This was due to the absence of the generation of new functional groups with the incorporation of sodium alginate. The absorption spectra ranged from 1640 to 1660 cm^{-1} that corresponds to dipole-dipole vibrations of H bond correlated with the water absorption by the amorphous portion of starch compounds. Further, the lesser wavenumbers represent the existence of strong interactions among the ingredients (Shalviri et al. 2010). On the other hand, a broader peak was observed in the range of 3200–3600 cm^{-1} corresponds to the vibrational stretching of the hydroxyl groups on the starch. The addition of sodium alginate leads to the formation of a stronger hydrogen bond with a shifting of the absorption peak to a lower wavenumber. With the increase in the concentration of sodium alginate, the hydroxyl stretching peak of all the varieties of rice samples became tapered. From these results, it can be concluded that added sodium alginate had interacted with the starch of rice varieties thereby forming a hydrogen bond that contributed to a shift in absorption spectrum (Koh et al. 2009). In another study, similar rice varieties (waxy rice, indica rice, and japonica rice) were used for 3D printing along with potato starch as an additive (Liu et al. 2020d). As stated earlier, a broad peak corresponds to vibration of O—H stretching that falls in the range of 3200–3600 cm^{-1}. The absorption peaks at 2928 and 1655 cm^{-1} correspond to the C—H bond and hydrogen bond vibrations, respectively. While the absorption peaks at 1080 cm^{-1} correspond to the coupling of the C—O, C—C, and O—H bond bending and stretching, and 1154 cm^{-1} attributed to the bending and asymmetrical stretching of glycosidic C—O—C bond (Wang et al. 2014). The lower wavenumbers of 859 and 927 cm^{-1} belong to the characteristic α configurations and α-1,4-

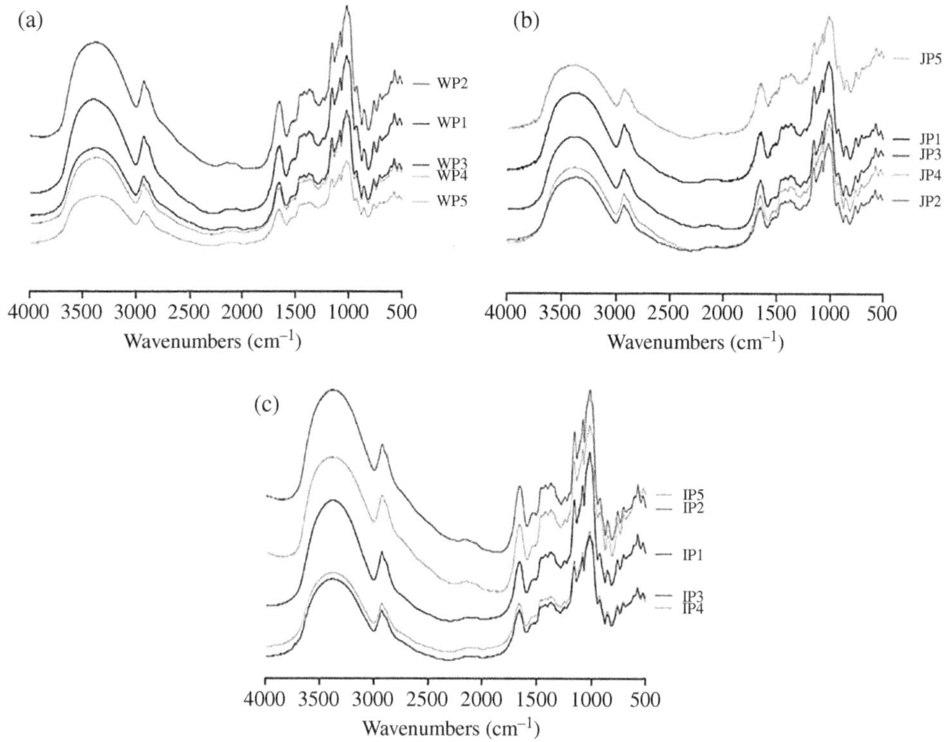

Figure 5.17 FTIR spectra of varieties of rice paste with different levels of sodium alginate. (a) Waxy rice paste, (b) japonica rice paste, and (c) indica rice paste. *Source:* Liu et al. (2020c) / With permission of Elsevier.

glycosidic bond. Results inferred that raw printed samples differed with the absorbance ratio at 1045 and 1022 cm^{-1} indicating the crystalline and short-range order of the starch molecules (Figure 5.17). On comparing all the varieties, waxy rice showed the highest absorbance ratio at 0.976 followed by indica at 0.960 and japonica at 0.956. These spectral results indicate that the degree of short-range order and crystallinity of the starch molecules of waxy rice greater than that of indica and japonica varieties. Hence the 3D constructs from waxy variety had the most ordered compact structure which was in agreement with the morphological analysis.

In another study, the IR absorbance bands of freeze-dried potato starch samples were assessed for the short-range ordered pattern. The absorption peaks at 1045 and 1022 cm^{-1} were associated with the ordered amorphous nature of the potato starch, respectively. Hence the intensity ratio $\left(R_{\frac{1045}{1022}}\right)$ was used for measuring the short-range order of the starch molecules (Liu et al. 2019d; Zhang et al. 2013). Comparing the native and 3D printed potato starch samples, the $R_{\frac{1045}{1022}}$ of 3D printed samples were lower than native potato starch samples that indicates the destruction of the native short-range order of the starch. However, there was no significant difference exist with intensity ratio among the 3D printed samples of different concentrations of potato starch (10–30%). On the other hand, at the same starch concentration (15%), the degree of short-range order was decreased with increase in the printing temperature from 60 to 65 °C. However, no change was observed in short-range order with further increase in temperature up to 80 °C. Thus, the FTIR spectral data reveals that the short-range order of the potato starch samples was greatly influenced by the printing temperature (Liu et al. 2020b). Hence, the successful 3D printing of the material supply relies on the printing temperature of the hot-melt extrusion. The change in the chemical structure of potato starch with the proportional addition of pea protein was analyzed using FTIR (Chuanxing et al. 2018). Similar absorption peaks were observed with native potato starch and 3D printed pea-protein incorporated samples. A strong and broad absorption band was observed at 3700–3000 cm^{-1} due to vibrational stretching of O—H of potato starch. Similarly, the absorption IR spectral peak at 2900 cm^{-1} belongs to the vibrational stretching of CH_2 (Zhang et al. 2012). The intensity of the absorption peak was inversely related to the transmittance (%) of the samples. Compared to native potato starch, the absorption peak intensity of 3D printed samples decreased with increasing pea protein content. In addition to the absorption bands of native potato starch, new peaks were observed at 2855 and 1745 cm^{-1} for 3D printed samples that attribute to the vibrational stretching of C—H and C=O. Further, the addition of pea protein changed the intensity of newly observed peaks of 3D printed samples that indicates the characteristic peaks of the lipids (Dzhatdoeva et al. 2016). The spectral results clearly showed that there were not even a few formations of new substances with the 3D printed materials.

5.3.4 Thermal Analysis

5.3.4.1 Differential Scanning Calorimetry

DSC gives knowledge about the phase change of materials that helps in food product development. In 3D printing, the thermodynamic factors such as melting temperature (T_m), enthalpy (H), glass transition temperature (T_g), the temperature of crystallization (T_c), and heat capacity (C_p) influence the material state and flowability of the printing material supply (Leyva-Porras et al. 2020). DSC allows the detection of qualitative and quantitative chemical transitions in a wide range of temperatures from −90 to 550 °C. The thermodynamic properties of various soy protein hydrocolloid mixtures were analyzed using DSC thermograms (Chen et al. 2019). The results of the DSC profile showed that gelatin had an endothermic heat flow in the range of 0–120 °C (Figure 5.18). On varying the concentrations of hydrocolloid gelatin, the intensity of the endothermic peak and the temperature got varied. Rheological analysis of the soy protein mixture showed that mechanical properties and the viscosity of the material supply decreased at this temperature range. Thus, the gelatin began to dissolve at temperature above 30 °C that soften the material supply. The temperature corresponding to the endothermic peak was considered as the melting temperature of gelatin. Under this range of melting temperature, the sodium alginate and the peptide chains of gelatin forms a network structure that allows a greater number of soy protein molecules got winded into the mesh structure (He et al. 2016). Lower concentrations of gelatin (2%) tailors the rheology of the soy protein isolate that helps in the smooth extrusion of the material supply at 25 °C. On the other hand, higher concentrations of gelatin above 6% significantly affect both the rheological and thermal properties. The material supply was non-melting at temperatures below melting point while the sample becomes softer when

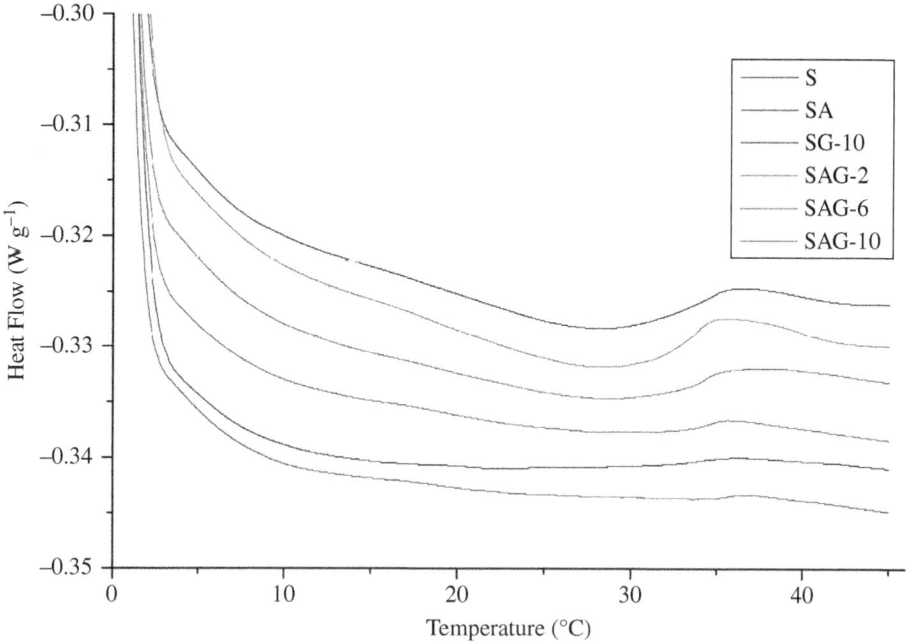

Figure 5.18 DSC thermograms of 3D printing material supply of SPI and SPI mixtures. *Source:* Chen et al. (2019) / With permission of Elsevier.

the temperature transcends the melting point (Liu et al. 2017). Although the rise in temperature affects the consistency, the material supply still retains its mechanical property. The associated transition in the material state helps in the easy flowability of the material supply through the printing nozzle.

In another study, the DSC thermograms of different varieties of rice paste have been studied (Liu et al. 2020c). The presence of a singular endothermic peak depicted in the DSC profile of each variety of rice paste is ascribed to the melting of starch crystals and double helices (Wang et al. 2019). With the addition of the same amount of sodium alginate to the rice paste, the transition temperature was varied. Thus, the rice pastes of indica variety had greater endothermic transition temperature (T_g) than the japonica and waxy varieties. This was mainly due to the higher content of partially immobilized water in the indica variety. The glass transition temperature was increased, and the transition enthalpy was decreased for both waxy and indica varieties when the sodium alginate concentrations was lower than 0.75%. In contrast, the increase in the level of sodium alginate from 0 to 0.5% for japonica variety, the endotherms trend towards decreasing temperature and the enthalpy of transition was increased. Thus, the DSC thermograms confirm the disturbance that exists during the coupling of the crystalline and amorphous matrix with the inclusion of hydrocolloid sodium alginate. The thermal property T_g is one of the significant parameters that have a control over the self-stability of the material supply to hold up its own mass after post-deposition (Liu et al. 2017). The moisture content difference of the rice flour attributes to the different endothermic transition temperatures. From the DSC profile, the T_g for waxy, japonica, and indica rice flour were found at 102, 101, and

95 °C, respectively. It was reported that carbohydrates are printable with sufficient mechanical integrity with the modification of material supply with the addition of gelling agents or water (Godoi et al. 2016). The plasticization behaviour of water was responsible for depressing the T_g of food polymers. The thermogravimetric analysis (TGA) of the rice flour showed that the curves were completely overlapped with one another before its degradation with no significant difference with the degradation and thermal stability. The thermal degradation was found to be at 309.91, 308.05, and 308.06 °C, respectively. On comparing, the waxy-variety had greater thermal stability than the japonica and indica variety at higher temperatures of 309–500 °C that significantly reduced the mass loss. From DSC results, it was inferred that 3D constructs from waxy rice would remain more stable than other two varieties in withstanding the structural stability during post-processing treatments (Liu et al. 2020d).

5.4 Assessment of Printability

5.4.1 Line Test

Line test is one of the simple basic tests used for determining the optimal extrusion rate of material supply in accordance with nozzle movement and printing speed. A well-optimized condition would result in lines with a diameter equal to that of the nozzle diameter (size) (Hao et al. 2010; Wang et al. 2018). As a modification of the line test, a report was available on the printing of dotted lines for determining the accuracy of the extruder (Ghazanfari et al. 2017). In general, the extruder requires a considerable amount of time to build up enough force to conquer the yield stress of the material supply. Li et al. (2015) reported a study on the comparison of working of different extruder designs (needle value, auger valve, and piston). Results concluded that the auger valve offered the best extrusion with the desired start and stop accuracy along with a consistent flow rate. The printing performance of the line test would result in three different extrusion conditions (Anukiruthika et al. 2020a). First is the over extrusion of the material supply where the printed line thickness would be much larger than the nozzle diameter. The resulted 3D printed object remains swollen that deteriorates the fine details of the printed layers. The second one is the under extrusion just opposite of the over extrusion where the material supply results in a discontinuous line. The printed line is broken as a result of a mismatch between the flow rate and the nozzle movement speed that led to insufficient extrusion of the material supply (Krishnaraj et al. 2019). In some cases, minor under extrusion is evident where the printed lines are stretched and dragged resulting in the rounding of the sharp angle and edges of the design model (Huang 2018). The third result is the situation where the printed thread thickness is equal or slightly larger than the nozzle diameter referred as desired extrusion (Keerthana et al. 2020). A slightly larger line thickness than the actual model would help in the proper adhesion of the printed layers.

The nozzle size and the material's extrusion rate are the most critical parameters in controlling the desired diameter of the printed line. The line diameter can be predicted as a function of volumetric extrusion rate and the cross-sectional area of the printing nozzle using Eq. (5.4) (Huang 2018).

$$\frac{Q}{v} = A = \frac{\pi r h}{4} \tag{5.4}$$

where Q is the steady-state volumetric flow rate of the material supply ($mm^3\ s^{-1}$), v is the nozzle movement or printing speed ($mm\ s^{-1}$), and r and h are the width and the height of the elliptical cross-section (mm), respectively. Rearranging Eq. (5.4), the width of the cross-section is given by Eq. (5.5).

$$r = \sqrt{\frac{4}{\pi a}\frac{Q}{v}} \tag{5.5}$$

where a is the aspect ratio, i.e. $a = h/r$. When $a = 1$ the cross-section is a circle and there exists horizontal spreading of the line when $a < 1$. Thus, the aspect ratio and line thickness would influence the printability and shape fidelity of the material supply. Different conditions of line test and cylinder test at varied extrusion rate for the optimization of the lemon juice gel are presented in Figure 5.19 (Yang et al. 2018b).

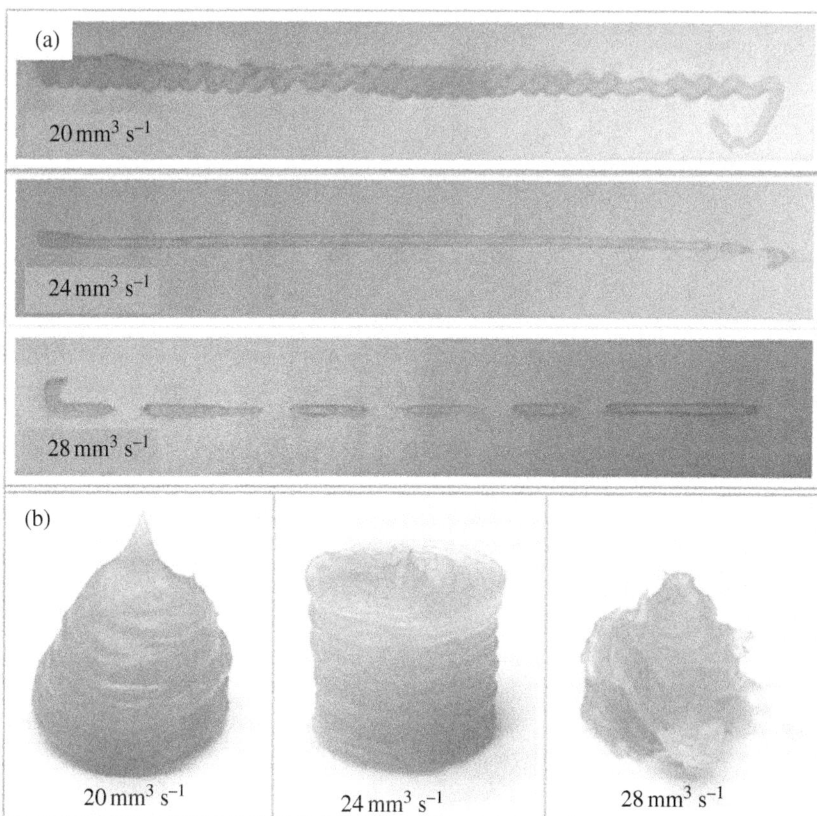

Figure 5.19 Printability test used for testing lemon juice gel. (a) Line test and (b) cylinder test (extrusion rate 20 $mm^3\ s^{-1}$ – over extrusion; 24 $mm^3\ s^{-1}$ – desired extrusion; and 28 $mm^3\ s^{-1}$ – under extrusion). *Source:* From Yang et al. (2018b), Figure 07 [p. 35] / With permission of Elsevier. DOI-https://doi.org/10.1016/j.lwt.2017.08.054.

5.4.2 Lattice Test

Lattices are simple 2D designs that are used to evaluate print quality. In the lattice test, the width and the line spacing of the printed 2D structure are measured and compared with the design model to assess the printability of the material supply (He et al. 2016; Zhao et al. 2015). The line spacing is based on the input printing variables including infill percentage and infill pattern. As stated earlier, the line thickness can be estimated as a function of volumetric extrusion rate and nozzle speed (O'Bryan et al. 2017). While the line spacing can be assessed quantitatively through image analysis. The printability of the material supply can be evaluated through image analysis of the pore shape in the lattice (Ouyang et al. 2016). Thus, printability can be calculated as the ratio of circularity of the perfect square over the actual circularity of the pore shape. The circularity (C) of an enclosed area can be given by Eq. (5.6) (Huang 2018)

$$C = \frac{4\pi A}{L^2} \tag{5.6}$$

where A is the cross-sectional area of the printed construct and L is the perimeter of the printed construct. Circles have the highest circularity with $C = 1$ and when closed the C value to 1, more similar to the shape to the circle. Thus, printability based on the pore shape is given by Eq. (5.7).

$$\text{Printability} = \frac{\pi}{4}\frac{1}{C} = \frac{L^2}{16A} \tag{5.7}$$

For achieving perfect print fidelity, the interconnected channels or pore shape must be sharp enough to demonstrate the fine details. In the case of lattice, the pore shape must be a perfect square that yields a quantitative value of printability as 1. A larger printability value represents a greater degree of gelation of the food ink and is vice versa. The optical images of the printed 3D constructs can be assessed using image processing software such as Image pro plus and ImageJ. The 2D lattice test used for printing of semi-solid dough sample is presented in Figure 5.20 (Derossi et al. 2020).

5.4.3 Cylinder Test

Basic 3D shapes such as cube, cylinder, and cone are generally printed for determining the mechanical strength and suitability of material consistency to withstand the stacked layers (Liu et al. 2019c). In general, the cylinder was 3D printed to assess the effects of material viscosity and infill levels on maximum build height. The infill percentage refers to the fraction of the interior to be filled with the material during 3D printing (Mantihal et al. 2019). Thus, the stability of the stacked layers one above the other is assessed based on the built height of the 3D construct. The 3D printing of cylindrical-shaped dough samples with different infill percentages showing its stability is presented in Figure 5.21 (Liu et al. 2019c).

5.4.4 Extrusion Test

The extrusion test is generally used to measure the steady-state extrusion rate of the material supply and the time required to reach steady-state conditions (Nida et al. 2020). The

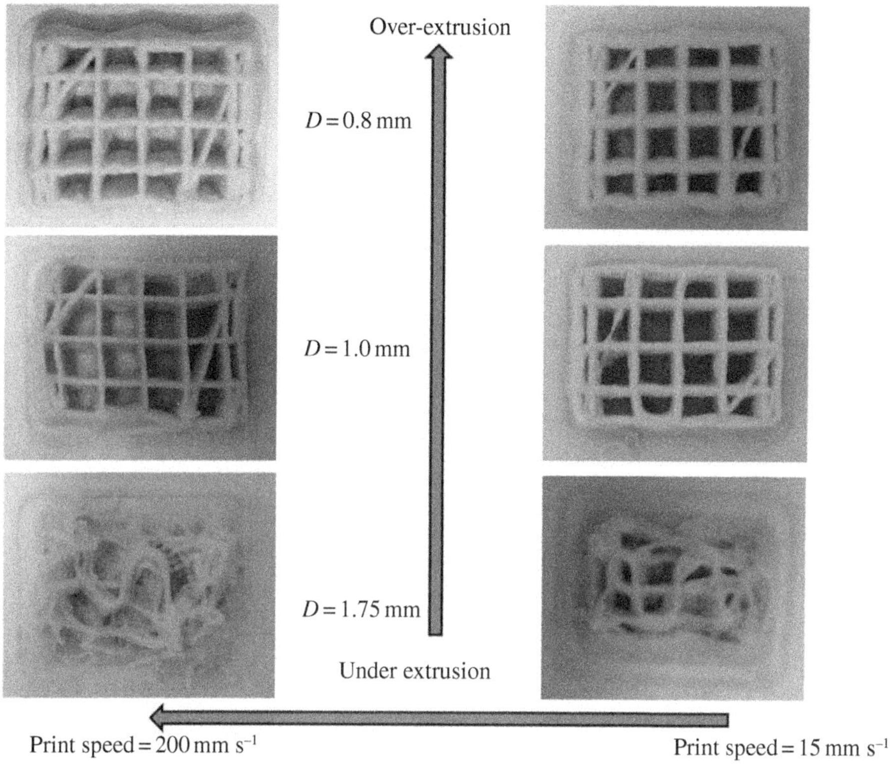

Figure 5.20 Lattice tests of 3D printed dough obtained by modulating the filament diameter and printing speed. *Source:* From Derossi et al. (2020), Figure 05 [p. 34] / With permission of Elsevier. DOI-https://doi.org/10.1016/j.jfoodeng.2019.109865.

viscoelastic shear-thinning behaviour of the sample is considered to be best suited for 3D printing by extrusion. A sufficient amount of time is required for the pressure to build up in order to push the viscoelastic material out of the printing nozzle. In the extrusion test, the mass of the material that exit through the printing nozzle was recorded over a period of 10 minutes at different extrusion motor speeds (Huang 2018). This estimates the extrusion rate of the material supply over time. The changes in the height of the stacked 3D layers of the fruit-based snacks were modelled using Eq. (5.8) (Derossi et al. 2018)

$$H(t) = H_{\infty}\left[1 - e^{-kt^{n}}\right] \tag{5.8}$$

where $H(t)$ is the height of the 3D printed snack after some time t and H_{∞} is the height of the 3D printed snack at the end of the printing process. The constant k (s^{-1}) and n (dimensionless) represents the location factor and shape factor, respectively. The extrusion time lag (t_{lag}) can be calculated as the time required for reaching 90% of M_{∞} as presented in Eq. (5.9) (Peleg and Normand 2013).

$$t_{lag} = \left(\frac{\ln 0.1}{-k}\right)^{1/n} \tag{5.9}$$

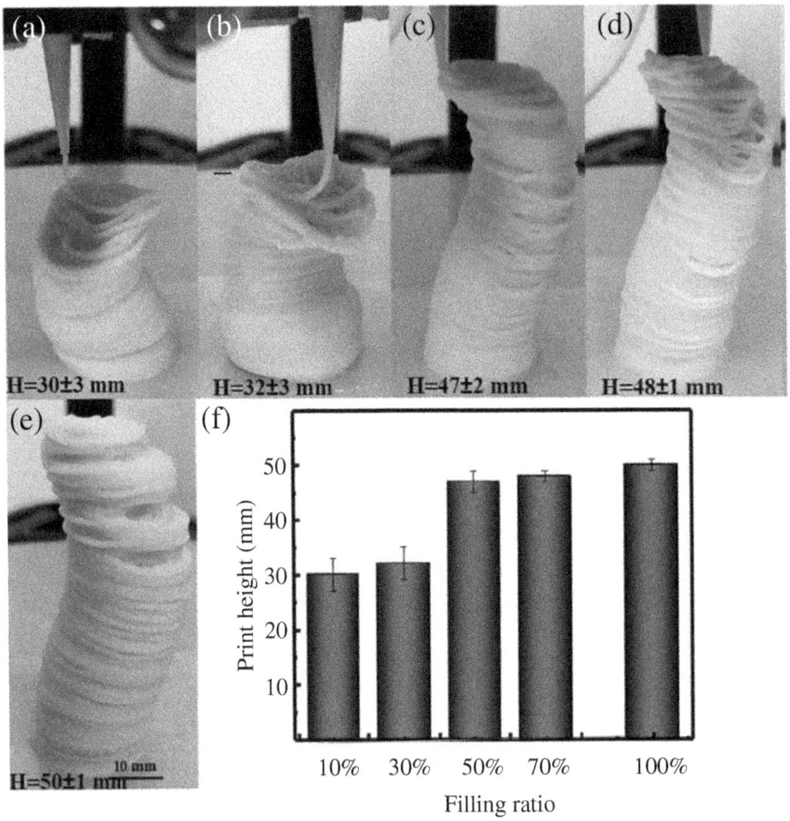

Figure 5.21 3D printing of cylindrical shaped dough samples with different infill percentage. (a) 10%, (b) 30%, (c) 50%, (d) 70%, and (e) 100%, and (f) print height Vs infill percentage at 0.58 mm nozzle size, 600 kPa printing pressure, and 6 mm/s printing speed. *Source:* From Liu et al. (2019c) / With permission of Elsevier.

Similarly, the variation in the weight of the sample as a function of time can be fitted using Eq. (5.10) (Derossi et al. 2018).

$$W(t) = k \times t \tag{5.10}$$

where $W(t)$ is the weight function at time t (s), and k is the rate constant (g s^{-1}).

5.4.5 Assessment of the Dimensional Stability

The dimensional stability of the material supply was determined by measuring the deformation behaviour of the 3D construct. The process involves the 3D printing of a cylindrical structure using a 3D printer. A lightweight plate of mass of 0.061 g was placed into the piston of the material supply unit to protect the top surface of the printed cylindrical structure from deformation (Figure 5.22). After the 3D printing process, the fabricated structures were left undisturbed for about 1 hour at 25 °C. Later the degree of deformation of the cylindrical structure was assessed using data obtained from 3D scanning of the structures (Kim et al. 2018a). When the rate of deformation was lesser than 10%, the height of the cylindrical

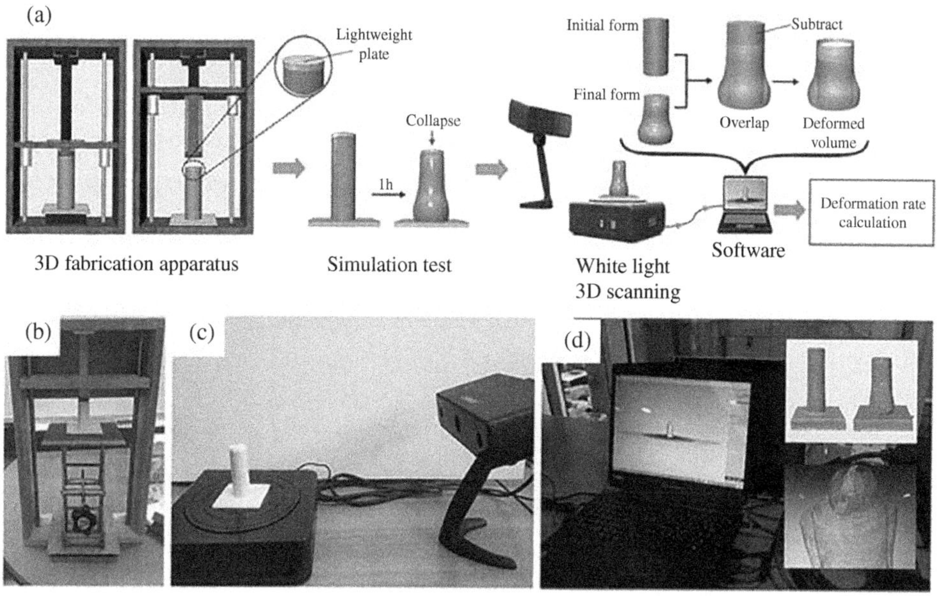

Figure 5.22 Schematic diagram of experimental setup used for analyzing the dimensional stability of 3D constructs. (a) and (b) 3D fabrication apparatus, (c) white light-based 3D scanning system, and (d) image processing for dimensional stability assessment. 2018a *Source:* From Kim et al. (2018a), Figure 01 [p. 29] / With permission of Elsevier. DOI-https://doi.org/10.1016/j.jfoodeng.2017.07.017.

structure was gradually raised from 20 to 80 mm for further observation to determine the deformation difference. Thus, the deformation rate was calculated using Eq. (5.11).

$$\text{Deformation rate (\%)} = \frac{V_D}{V} \times 100 \tag{5.11}$$

where V_D is the volume of the deformed shape from the initial form and V is the volume of the 3D original fabricated shape.

Various concentrations of methylcellulose (5–14%) were tested for dimensional stability with tested food materials (Figure 5.23). The deformation behaviour of the formulated methylcellulose formulations well simulates the deformation behaviour of the experimental foods. However, the fabrication of structures with low height imparts difficulty in classification based on printability. Results showed that degree of deformation significantly different for peanut butter and cheddar cheese with 20 mm height. On the other hand, with increased fabrication height to 40 mm, the cheddar cheese exhibited greater structural stability than peanut butter. Methylcellulose-based reference solutions of varying concentrations were used to fabricate cylindrical structures of different build heights. The cylindrical structures with different heights of 20, 40, and 80 mm showed deformation under 10% were found to be 9, 11, and 13%, respectively. Besides lower built height (20 mm), a significant collapse occurred with chocolate jam (46.93%), ketchup (27.36%), and Greek yoghurt (39.26%). These results indicate that these experimental foods must require further

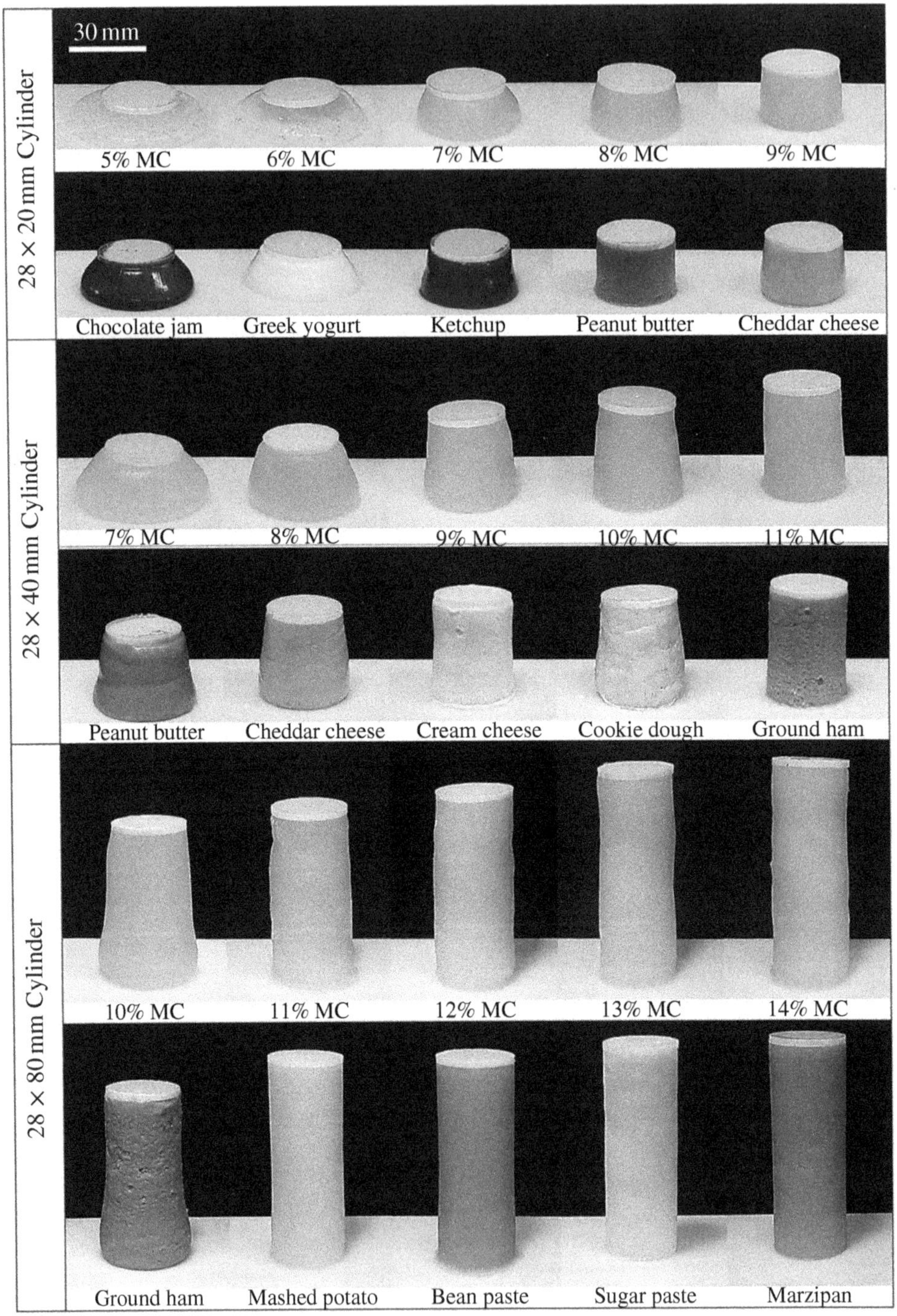

Figure 5.23 Deformation behaviour of different levels of methylcellulose and experimental foods. *Source:* From Kim et al. (2018a), Figure 04 [p. 32] / With permission of Elsevier. DOI-https://doi.org/10.1016/j.jfoodeng.2017.07.017.

processing to enhance printability for 3D applications. However, these materials could be applied for surface decorations and 2D design patterning.

Similar structures fabricated from peanut butter, cream cheese, cheddar cheese, and cookie dough with 20 mm height hold their shape with <10% deformation rate. However, these experimental foods were deformed with an increase in the built height to 40 mm. Considering these results, cheddar cheese and the 9% methylcellulose constitute a single group. On the other hand, the cookie dough and cream cheese were grouped along with the 10% methylcellulose. The change in deformation rate was too large for the classification of peanut butter with increasing methylcellulose concentration from 10 to 11%. When the fabrication height was 80 mm the deformation rate was significantly low for most of the experimental foods except ground ham. Among which the sugar paste and marzipan showed the lowest deformation rate of 0.43 and 0.26% respectively. Hence, they were grouped with the 14% methylcellulose with a deformation rate of 0.83% (Kim et al. 2018a). Thus, the proposed method for the classification of food inks based on printability requires a broad knowledge of the rheological characteristics of the material supply.

In another study, a customized cylindrical rig was used for testing the dimensional stability of wheat starch gel, HPMC, and gelatin (Figure 5.24) (Nijdam et al. 2021b).

Among the tested samples, wheat starch gel showed a significant initial instantaneous deformation upon dispensing of material with a small subsequent time-dependent deformation showing the dominance of initial elastic deformation (Figure 5.25). In contrast, the HPMC gel showed a time-dependent deformation with less initial instantaneous deformation indicating the dominance of viscoelastic deformation. On the other hand, the gelatin showed both modes of deformation that were obvious at lower concentrations. Thus, the

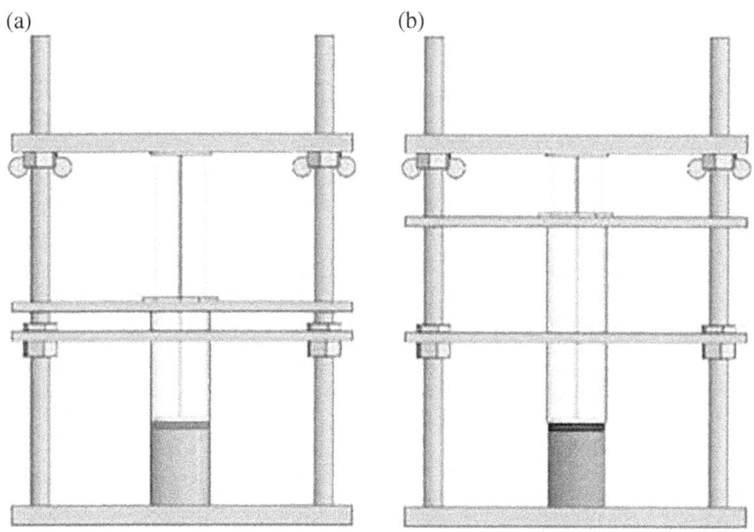

Figure 5.24 Experimental setup of a cylindrical rig used for assessing dimensional stability of food ink (a) before dispensing and (b) after dispensing of material. *Source:* Nijdam et al. (2021b) / With permission of Elsevier.

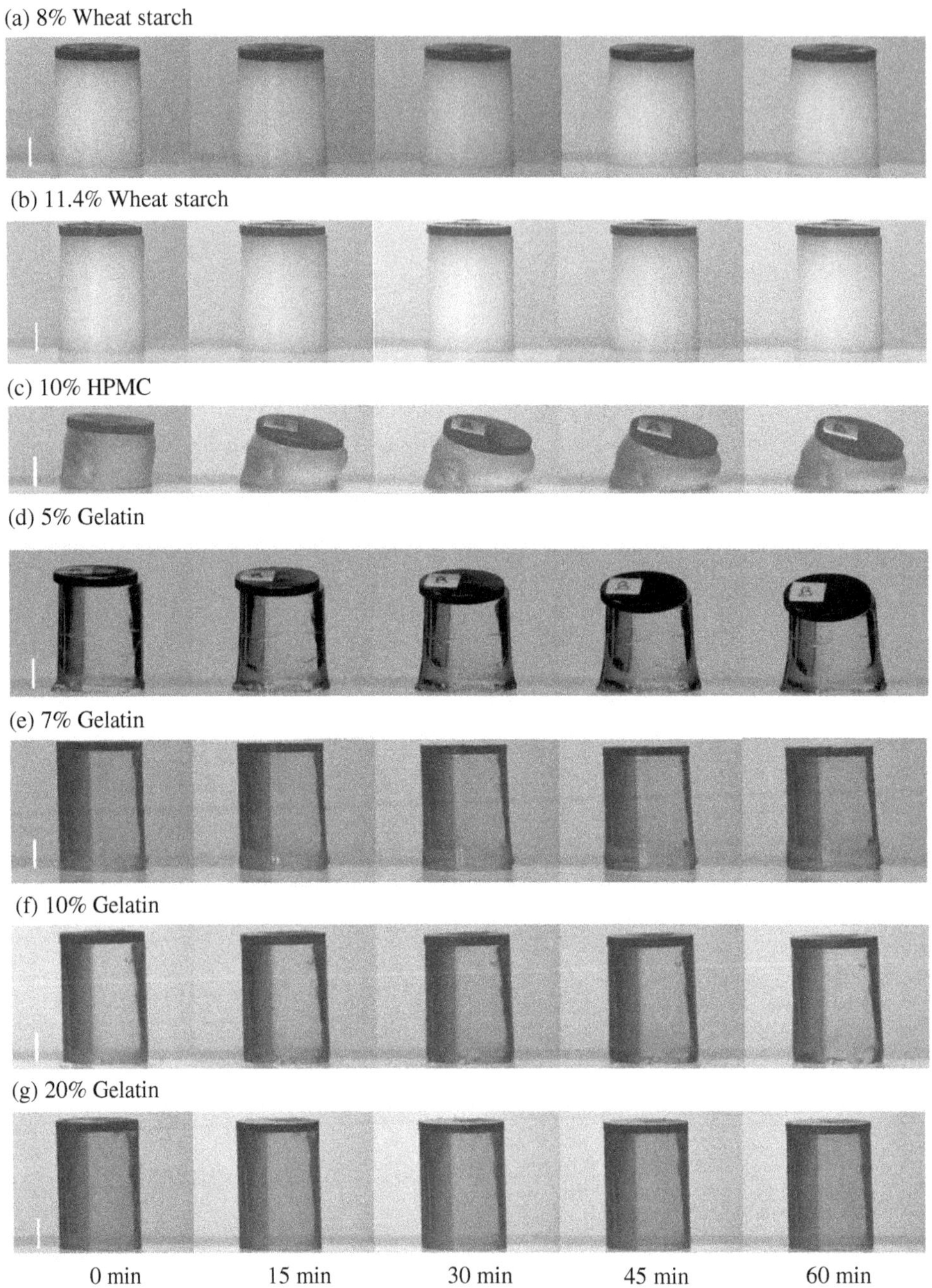

Figure 5.25 Images representing the deformation of 3D printed cylinders of different material supplies over different time periods. a) 8% wheat starch, b) 11.4% wheat starch, c) 10% HPMC, d) 5% gelatin, e) 7% gelatin, f) 10% gelatin, and g) 20% gelatin. *Source:* Nijdam, LeCorre-Bordes, et al., 2021.

difference in the deformation behaviour of the material supplies was related to its printability and post-print stability.

In a more recent study, the material supply was assessed for printability through a novel approach of 'window of dimensional stability'. A triple-step structure (3 ×3 grid of 12 mm squares) was used for qualitative assessment of the printability of a wide range of food materials (Figure 5.26) (Nijdam et al. 2021a). Results showed that Greek yoghurt and 5% gelatin

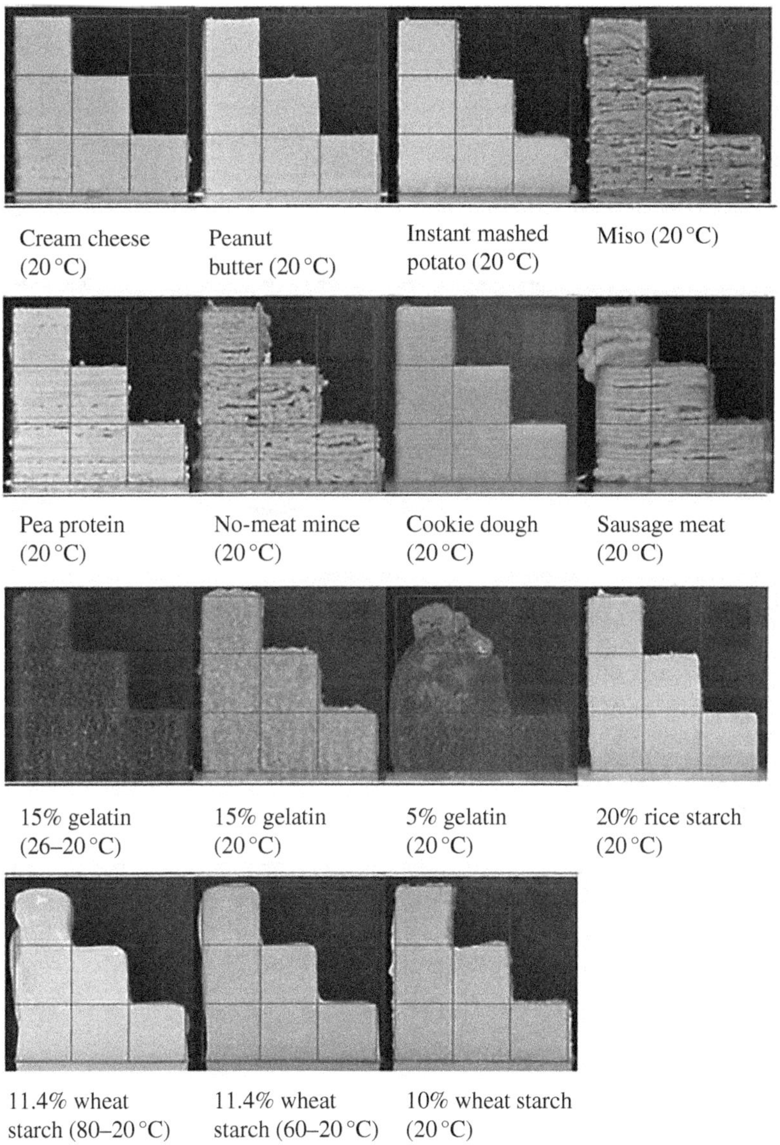

Figure 5.26 Images showing the elevation view of 3D printed triple-steps of different printed samples. *Source:* From Nijdam et al. (2021a), Figure 01 [p. 4] / With permission of Elsevier. Doi-https://doi.org/10.1016/j.jfoodeng.2020.110349.

had very poor dimensional stability while other materials such as cream cheese, instant mashed potato, and cookie dough possessed good dimensional stability with a stable 3D printed triple steps. The dimensional stability was explained by rheological properties such as complex moduli, yield stress, yield strain, and loss tangent.

5.4.6 Assessment of the Handling Properties

The handling properties of the material supply were assessed by performing a forward extrusion assay that measures the extrusion force needed by the 3D printer to push the material out of the nozzle. For this assay, an extrusion test cell with two different reservoir diameters (RD) of 25 and 30 mm and a nozzle aperture size of 3 and 4 mm was used. About 30 ml of the material supply was extruded from the test cell using a texture analyzer (Kim et al. 2018a). Thus, the extrusion hardness is measured as the maximum load marked on the graph, i.e. the force recorded during compression of the material supply. Results showed that the force required to extrude the material supply was increased with decreasing nozzle aperture size (NAS) and increasing RD at the same extrusion speed. The extrusion limit of the 3D printer was greatly influenced by the extrusion rate that in turn depends on printing speed, reservoir capacity, and nozzle type. Thus, the manufacturing capacity in terms of the rate of printing, resolution, and efficiency of a 3D printer depends on the handling properties of the material supply. The simulation results of the methylcellulose and experimental foods showed that the values of extrusion force for the tested methylcellulose mixture were in agreement with the results of the dimension stability test. However, these values were different for the experimental foods, i.e. the deformation rate of the mashed potato (5.85%) was close to that of the bean paste (5.15%) but the extrusion force values were different for these two samples (Kim et al. 2018a). These differences were mainly due to the variation in the constituents of the food system. The reference material methylcellulose was made of a single component whereas the experimental foods contain multiple constituents such as sugars, proteins, and lipids that impart unique characteristics and complex interactions within the food system.

5.5 Printability Evaluation of 3D Printed Constructs

The quality of the fabricated 3D objects is influenced by a combination of process and product variables. The effect of pre- and post-processing on the characteristics of the material supply is a significant criterion that assesses print fidelity (Wilson et al. 2020).

5.5.1 Shape Resemblance

The accuracy refers to how close the measured value with the true value. In 3D printing, the true value refers to the actual dimensions of the computer-aided design (CAD) model used. Hence, printing accuracy refers to how close the dimension of the fabricated object is close to that of the designed model. On the other hand, precision implies the repeatability of the measurement. Hence printing precision refers to the consistency in the measurements that translates the reliability. The printing precision of the food-based material supplies can be calculated using Eq. (5.12) (Liu et al. 2019a).

$$\text{Printing precision (\%)} = \frac{D_{\mathrm{r}}}{D_{\mathrm{n}}} \times 100 \tag{5.12}$$

where D_{r} is the diameter of the printed thread after the equilibration period and D_{n} is the diameter of the printing nozzle. The printing precision was reported for extrusion printing of milk protein composite gel and mushroom-based dough. Results inferred that for the milk protein gel, the printing performance, and printing precision were increased with a proportional rise in the total protein content from 350 to 450 $\mathrm{g\,l^{-1}}$ but decreased when the total protein content of material supply exceeds 500 $\mathrm{g\,l^{-1}}$ (Liu et al. 2019a). These results were consistent with the study performed on assessing the printability of mushroom-based dough (Keerthana et al. 2020). The complex interactions of fibrous content of incorporated mushroom powder with the glutenous wheat network were responsible for imparting mechanical stability and printing precision of 3D constructs.

5.5.2 Printing Percentage

The printing percentage was used as an index to assess the printing performance of different formulations of mashed potatoes (Eq. (5.13)) (Liu et al. 2020a).

$$\text{Printing percentage} = \frac{T_{\text{total}} - T_{\text{left}}}{T_{\text{total}}} \times 100\% \tag{5.13}$$

where T_{total} is the total time required for printing of a 3D construct and T_{left} is the left-over time when the product deposition does not correspond to the bottom layer. A printing percentage of 100% depicts that the 3D construct was totally built while a small value depicts that printed 3D construct could collapse easily.

5.5.3 Dimensional Deviation and Appearance

The effects of processing on the final quality of the 3D printed constructs are represented by shape fidelity and percentual variation as given in Eqs. (5.14) and (5.15) (Vieira et al. 2020). Here, the processed and the unprocessed dimensions represent the post-treatments such as cooking.

$$\text{Shape fidelity (\%)} = \frac{\text{Measured dimension of printed construct}}{\text{Theoritical dimension of the 3D model}} \times 100 \tag{5.14}$$

$$\text{Variation (\%)} = \frac{\text{Processed dimension} - \text{unprocessed dimension}}{\text{Unprocessed dimension}} \times 100 \tag{5.15}$$

A similar equation as that of Eq. (5.14) was used for assessing the shape fidelity of the phase-separated food inks prepared from whey protein isolate and guar gum (Oliveira et al. 2020). The methodology followed for evaluating the printability of the ink in this study was presented in (Figure 5.27). In another study, the dimensional stability of the cookie dough has been evaluated by assessing all the three dimensions of processed and unprocessed 3D constructs. The deformation was calculated using Eq. (5.16) (Pulatsu et al. 2020).

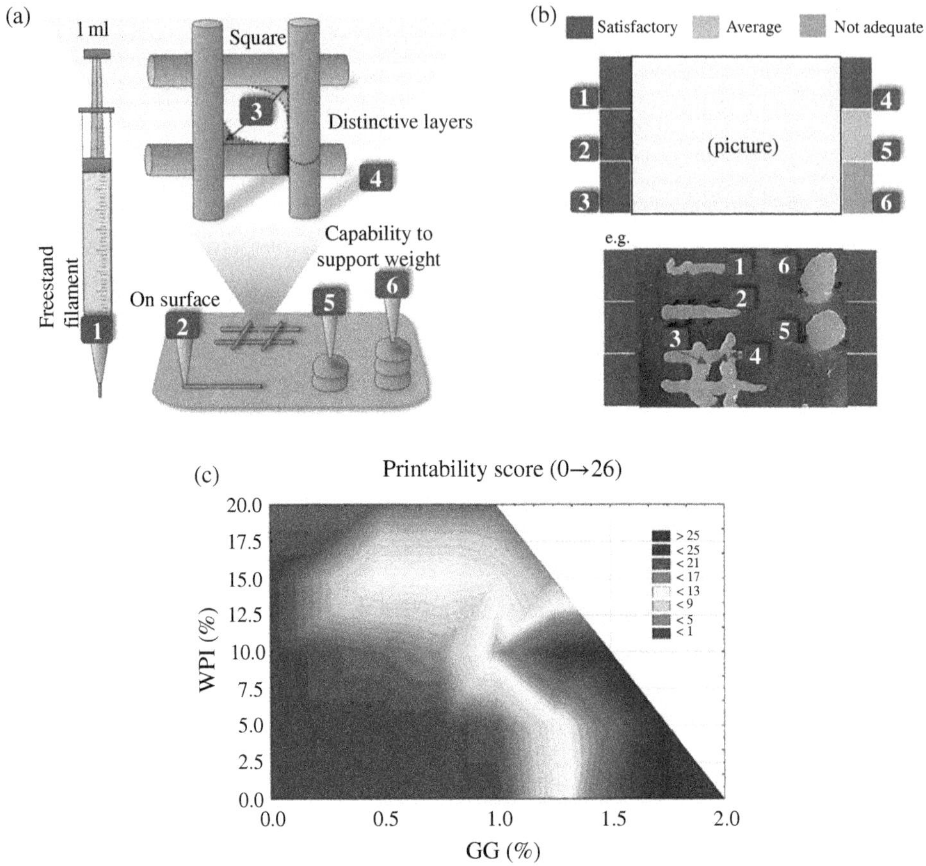

Figure 5.27 Methodology used for evaluating printability of WPI/GG inks. (a) Printing process, (b) representation of qualitative results, and (c) printability scores. *Source:* From Oliveira et al. (2020) / With permission of Elsevier.

$$\Delta X\,(\%) = \frac{X_{\text{unprocessed}} - X_{\text{processed}}}{X_{\text{unprocessed}}} \times 100 \tag{5.16}$$

where $X_{\text{unprocessed}}$ and $X_{\text{processed}}$ represent the X dimension of 3D constructs (mm). The $\Delta X > 0$ indicates shrinkage of the sample while <0 represents the collapse or spreading of the samples during the post-processing treatment. A similar equation can be applied for determining the deformation of the 3D construct in Y and Z directions. In another study, the deformation of the 3D construct was expressed in terms of height as the ratio of the height of the printed object after 1 hour to the height of the printed object immediately after the printing process. Similarly, the printability was determined based on the built height of the 3D structure (Eq. (5.17)) (Azam et al. 2018a).

$$\text{Printability}\,(\%)\,(\text{in terms of built height}) = \frac{\text{Achieved height of the printed object}}{\text{Target height of the object}} \times 100 \tag{5.17}$$

5.5.4 Dimensional Stability

The printability of the material supply is influenced by the printing technology employed. The dimensional stability of the 3D constructs varied based on the characteristics of the material supply. Soft materials are those that possess self-supporting ability without any phase change upon deposition (Gross et al. 2014; Khan et al. 2015). Hence, the dimensional stability of these types of samples directly affects the printing resolution before 3D printing. In contrast, for those materials that are suitable for hot-melt extrusion, the accuracy of the process is influenced by a change in phase of material processes such as solidification upon cooling after melting process, cross-linking, gelling ability, and photo-polymerization. Hence, the dimensional stability of fully gelled food ink affects the accuracy of the 3D construct. The study on classification of printability based on methylcellulose as reference material inferred that the dimensional stability of food ink was affected by compressive stress and bending stress that in turn interlinked with the rheology of printing mixture (Kim et al. 2018a). Therefore, the assessment of printability of material supply involves the careful optimization of design variables and material properties.

5.5.5 Ternary Representation of Printability

In a more recent study, the effect of material composition on the printability of food paste was represented through a ternary diagram (Liu and Ciftci 2020). The water phase consists of water and corn starch in the ratio of 8.5 : 10, w/w whereas the oil phase of the material supply consists of canola oil with whey protein isolate (WPI) as an emulsifier. From the printability tests, the following observations were drawn: the sample which cannot be mixed as a whole part is placed in region A; the sample that can flow through the orifice of the nozzle that is extremely delicate to hold its shape is placed in region B; the sample that can be 3D printed to the desired shape without storage is placed in region C and with 24-hour storage at 4 °C is placed in region D; the sample that is very hard to flow through the nozzle is placed in region E. It was observed that among the five different regions of the ternary diagram, only the printing conditions at regions B and C resulted in better printability through adjusting the printing speed (80–120) and extrusion rate (60–150) (Figure 5.28). Results inferred that the higher content of whey protein isolates negatively influenced the process of printing. It was found that when the ratio of whey protein isolate reduced from 25 to 20, the surface finishing and smoothness of filaments have improved that result in the holding of 3D printed shape without spreading.

5.5.6 Correlation of Printability and Rheology

The effect of material composition and its rheological properties have been correlated with printability (Álvarez-Castillo et al. 2020). In this study, the printability of different biopolymer/glycerol dough was assessed at 20 °C. Here, the printability of the material supply was described in terms of precision and shape stability into five levels (Figure 5.29). From the printability results, two different printability maps were developed by correlating the material composition with rheological properties. The first map defines the viscoelastic properties where the material supplies could result in the desired extrusion with retention of the 3D shape. It was inferred that the material supplies with viscoelastic moduli lesser than

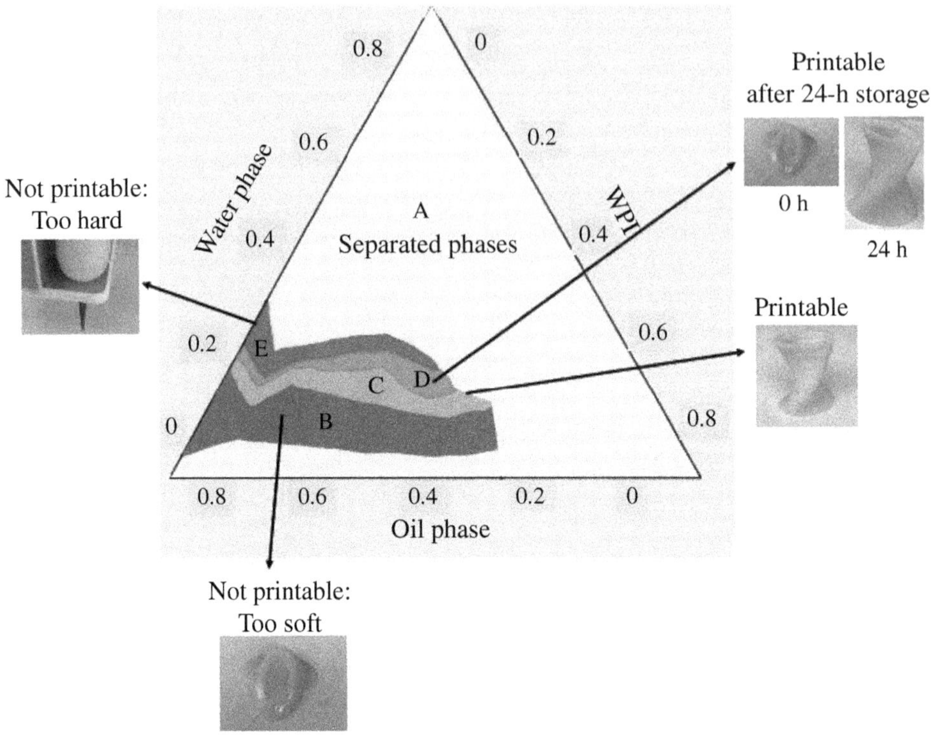

Figure 5.28 Ternary diagram showing the printability of the high oil composition material supply. *Source:* From Liu and Ciftci (2020) / With permission of Elsevier.

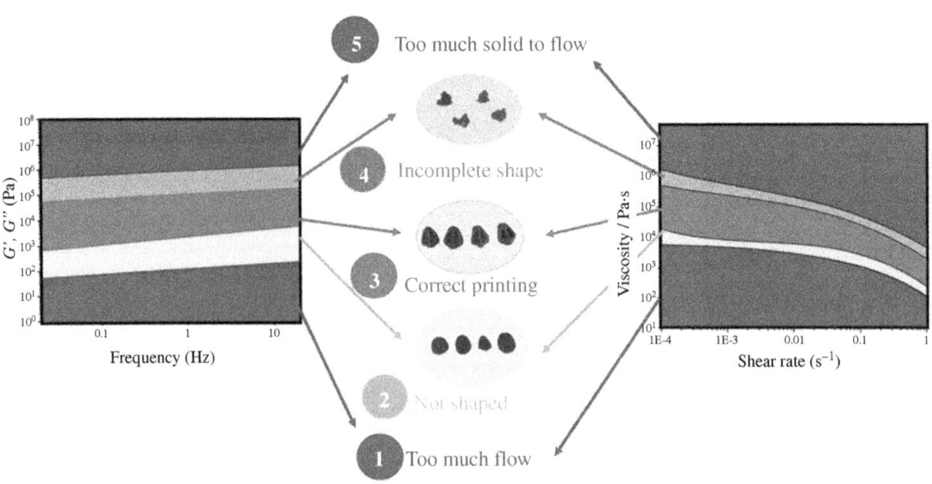

Figure 5.29 Schematic representation of different levels of printability of protein-based dough in relation with frequency moduli and viscosity. *Source:* From Álvarez-Castillo et al. (2020) / With permission of Elsevier.

10^3 Pa or higher than 10^3 Pa were considered to be too weak to hold the shape or too hard to extrude through the nozzle, respectively. On the other hand, the second map was obtained from the flowability test based on the shear rate. It was found that all the printability frontiers tend to reduce viscosity with an increased shear rate. Thus, a greater slope was determined at higher shear rates with a reduction in the printability from level 3.

5.5.7 Rational Approach for Printability

Conventionally, the 3D food printing process is optimized through a trial-and-error approach based on the material's extrudability. However, in case if material passes extrusion but later suffers from instability of the 3D printed layers, then the whole process must be reformulated. This consumes more time as well as the energy that leads to material wastage as well. By considering this, a rational approach for optimization of the 3D printing process has been proposed (Zhu et al. 2019). This study attempted quantitative measurement of the rheological properties of material supplies for correlating with printability. The study inferred that flow stress could be used as a quantification factor for testing extrudability as well as post-deposition stability of the aqueous-based food pastes (Figure 5.30). The proposed approach helps in determining the paramount flow stress of the material supply as well as the maximum-built height of the 3D construct which could help in evaluating the suitability of the material for 3D printing. However, the proposed approach was useful only with room temperature extrusion where no phase transition occurs. Hence, this

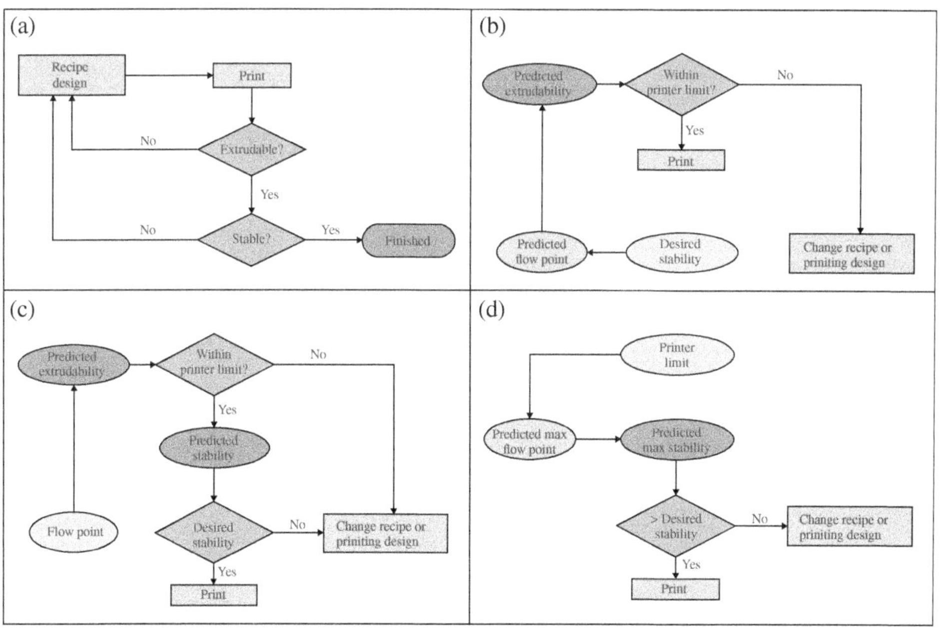

Figure 5.30 Scheme of flow chart used for decision making of 3D food printing design and selection of material for aqueous-based paste. (a) Conventional trail-and-error approach, (b)–(d) new guideline based on rational approach. *Source:* Zhu et al. (2019) / With permission of Elsevier.

approach must further be investigated to frame a protocol that aids in the decision-making of material selection for 3D printing.

As a basic criterion, the suitability of the material for 3D printing can be determined by adopting the aforementioned printability tests as presented in this chapter followed by an evaluation of the end-quality of the 3D printed construct.

5.6 Conclusion

3D printing of foods involves a complex interaction of macro and micro constituents. So, the proper selection of printing technology based on the nature of material supply and end-product requirements must be adequate. The printability reflects the material properties that are well enough to withstand the successive layers with adequate dimensional stability and handling properties. The diverse material properties such as microstructural features, crystallinity, melting and solidification behaviour, water mobility, wettability, and particle size distribution influences the printability. These properties of the raw materials used for 3D printing can be efficiently determined using different characterizing techniques as discussed. The protocol for the optimization of material supply for 3D printing involves the assessment of the material nature before printing, material response during 3D printing, and post-deposition behaviour of the material supply. The successful deposition of material supply relies on the combined rheological and textural properties. Hence, assessment of printability requires material characterization and analysis of viscoclastic and mechanical properties of the material supply besides optimization of printing parameters such as nozzle height, printing speed, nozzle diameter, material extrusion rate, and extrusion pressure. Thus, the present chapter summarizes various material properties and their impact on printability in context with characterization techniques employed in 3D printing studies as reported in the literature. In addition, the basic printability tests used for assessing printability and evaluation of the print quality of the 3D constructs have been presented. Undoubtedly, the present chapter would provide valuable insights on improving the current optimization procedures of material supply for 3D printing. This would open a new dimensional array of testing protocols of printability that aids in the establishment of standard procedures for the 3D printing of foods.

References

Ahmed, J., Mulla, M., Joseph, A. et al. (2020). Zinc oxide/clove essential oil incorporated type B gelatin nanocomposite formulations: a proof-of-concept study for 3D printing applications. *Food Hydrocolloids* **98**: 105256.

Álvarez-Castillo, E., Oliveira, S., Bengoechea, C. et al. (2020). A rheological approach to 3D printing of plasma protein based doughs. *Journal of Food Engineering*: 110255.

Anukiruthika, T., Moses, J.A., and Anandharamakrishnan, C. (2020a). 3D printing of egg yolk and white with rice flour blends. *Journal of Food Engineering* **265**: 109691. https://doi.org/10.1016/j.jfoodeng.2019.109691.

Anukiruthika, T., Priyanka, S., Moses, J.A., and Anandharamakrishnan, C. (2020b). Characterisation of green nanomaterials. In: *Green Nanomaterials* (eds. K. Knoerzer and K. Muthukumarappan), pp. 43–79. Springer.

Asadchikov, V.E., Buzmakov, A.V., Dyachkova, I.G. et al. (2019). Study of properties of materials by absorption and diffraction X-ray microtomography. *Inorganic Materials* **55** (15): 1458–1464.

Azam, R.S.M., Zhang, M., Bhandari, B., and Yang, C. (2018a). Effect of different gums on features of 3D printed object based on vitamin-D enriched orange concentrate. *Food Biophysics* **13** (3): 250–262.

Azam, S.M.R., Zhang, M., Mujumdar, A.S., and Yang, C. (2018b). Study on 3D printing of orange concentrate and material characteristics. *Journal of Food Process Engineering* **41** (5): 1–10. https://doi.org/10.1111/jfpe.12689.

Bakarich, S.E., Beirne, S., Wallace, G.G. et al. (2013). Extrusion printing of ionic--covalent entanglement hydrogels with high toughness. *Journal of Materials Chemistry B* **1** (38): 4939–4946.

Bunaciu, A.A., Udristioiu, E.G., and Aboul-Enein, H.Y. (2015). X-ray diffraction: instrumentation and applications. *Critical Reviews in Analytical Chemistry* **45** (4): 289–299.

Caporizzi, R., Derossi, A., and Severini, C. (2019). Cereal-based and insect-enriched printable food: from formulation to postprocessing treatments. Status and Perspectives. In: *Fundamentals of 3D Food Printing and Applications*. Elsevier Inc. https://doi.org/10.1016/B978-0-12-814564-7.00004-3.

Chen, J., Mu, T., Goffin, D. et al. (2019). Application of soy protein isolate and hydrocolloids based mixtures as promising food material in 3D food printing. *Journal of Food Engineering*. https://doi.org/10.1016/j.jfoodeng.2019.03.016.

Chen, H., Zhang, M., and Yang, C. (2020). Comparative analysis of 3D printability and rheological properties of surimi gels via LF-NMR and dielectric characteristics. *Journal of Food Engineering*, p. 110278.

Chi, C., Li, X., Zhang, Y. et al. (2017). Digestibility and supramolecular structural changes of maize starch by non-covalent interactions with gallic acid. *Food & Function* **8** (2): 720–730.

Chi, C., Li, X., Zhang, Y. et al. (2018). Understanding the mechanism of starch digestion mitigation by rice protein and its enzymatic hydrolysates. *Food Hydrocolloids* **84**: 473–480.

Chuanxing, F., Qi, W., Hui, L. et al. (2018). Effects of pea protein on the properties of potato starch-based 3D printing materials. *International Journal of Food Engineering* **14** (3).

Crompton, K.E., Forsythe, J.S., Horne, M.K. et al. (2009). Molecular level and microstructural characterisation of thermally sensitive chitosan hydrogels. *Soft Matter* **5** (23): 4704–4711.

Dankar, I., Haddarah, A., Omar, F.E.L. et al. (2018). 3D printing technology: the new era for food customization and elaboration. *Trends in Food Science and Technology* **75**: 231–242. https://doi.org/10.1016/j.tifs.2018.03.018.

Derossi, A., Caporizzi, R., Azzollini, D., and Severini, C. (2018). Application of 3D printing for customized food. A case on the development of a fruit-based snack for children. *Journal of Food Engineering* **220**: 65–75.

Derossi, A., Paolillo, M., Caporizzi, R., and Severini, C. (2020). Extending the 3D food printing tests at high speed. Material deposition and effect of non-printing movements on the final quality of printed structures. *Journal of Food Engineering* **275**: 109865.

Dick, A., Bhandari, B., Dong, X., and Prakash, S. (2020). Feasibility study of hydrocolloid incorporated 3D printed pork as dysphagia food. *Food Hydrocolloids* **107**: 105940.

Dzhatdoeva, A.A., Polimova, A.M., Proskurnina, E.V. et al. (2016). Determination of lipids and their oxidation products by IR spectrometry. *Journal of Analytical Chemistry* **71** (6): 542–548.

Falcone, P.M., Baiano, A., Conte, A. et al. (2006). Imaging techniques for the study of food microstructure: a review. *Advances in Food and Nutrition Research* **51**: 205–263.

Ghazanfari, A., Li, W., Leu, M.C., and Hilmas, G.E. (2017). A novel freeform extrusion fabrication process for producing solid ceramic components with uniform layered radiation drying. *Additive Manufacturing* **15**: 102–112.

Gholamipour-Shirazi, A., Norton, I.T., and Mills, T. (2019). Designing hydrocolloid based food-ink formulations for extrusion 3D printing. *Food Hydrocolloids* **95**: 161–167. https://doi.org/10.1016/j.foodhyd.2019.04.011.

Godoi, F.C., Prakash, S., and Bhandari, B.R. (2016). 3d printing technologies applied for food design: status and prospects. *Journal of Food Engineering* **179**: 44–54. https://doi.org/10.1016/j.jfoodeng.2016.01.025.

Gross, B.C., Erkal, J.L., Lockwood, S.Y. et al. (2014). *Evaluation of 3D Printing and its Potential Impact on Biotechnology and the Chemical Sciences*. ACS Publications.

Guo, L., Hu, J., Zhang, J., and Du, X. (2016). The role of entanglement concentration on the hydrodynamic properties of potato and sweet potato starches. *International Journal of Biological Macromolecules* **93**: 1–8.

Hamilton, C.A., Alici, G., and Panhuis, M. (2018). 3D printing vegemite and marmite: redefining "breadboards.". *Journal of Food Engineering* **220**: 83–88.

Hao, L., Mellor, S., Seaman, O. et al. (2010). Material characterisation and process development for chocolate additive layer manufacturing. *Virtual and Physical Prototyping* **5** (2): 57–64.

He, Y., Yang, F., Zhao, H. et al. (2016). Research on the printability of hydrogels in 3D bioprinting. *Scientific Reports* **6**: 29977.

Huang, C.Y. (2018). Extrusion-based 3D printing and characterization of edible materials [University of Waterloo]. https://uwspace.uwaterloo.ca/handle/10012/12899

Huang, J., Wei, M., Ren, R. et al. (2017). Morphological changes of blocklets during the gelatinization process of tapioca starch. *Carbohydrate Polymers* **163**: 324–329.

Jiang, H., Zheng, L., Zou, Y. et al. (2019). 3D food printing : main components selection by considering rheological properties. *Critical Reviews in Food Science and Nutrition* **59**: 1–13. https://doi.org/10.1080/10408398.2018.1514363.

Karino, T., Shibayama, M., Okumura, Y., and Ito, K. (2006). SANS study on pulley effect of slide-ring gel. *Physica B: Condensed Matter* **385**: 807–809.

Keerthana, K., Anukiruthika, T., Moses, J.A., and Anandharamakrishnan, C. (2020). Development of fiber-enriched 3D printed snacks from alternative foods: a study on button mushroom. *Journal of Food Engineering*. https://doi.org/10.1016/j.jfoodeng.2020.110116.

Khan, F., Tanaka, M., and Ahmad, S.R. (2015). Fabrication of polymeric biomaterials: a strategy for tissue engineering and medical devices. *Journal of Materials Chemistry B* **3** (42): 8224–8249.

Kim, H.W., Bae, H., and Park, H.J. (2018a). Classification of the printability of selected food for 3D printing: development of an assessment method using hydrocolloids as reference material. *Journal of Food Engineering* **215**: 23–32.

Kim, H.W., Lee, J.H., Park, S.M. et al. (2018b). Effect of hydrocolloids on rheological properties and printability of vegetable inks for 3D food Printing. *Journal of Food Science* **83** (12): 2923–2932.

Kim, H.W., Lee, I.J., Park, S.M. et al. (2019). Effect of hydrocolloid addition on dimensional stability in post-processing of 3D printable cookie dough. *LWT* **101**: 69–75.

Koh, L.W., Kasapis, S., Lim, K.M., and Foo, C.W. (2009). Structural enhancement leading to retardation of in vitro digestion of rice dough in the presence of alginate. *Food Hydrocolloids* **23** (6): 1458–1464.

Kouzani, A.Z., Adams, S., Oliver, R. et al. (2016). 3D printing of a pavlova. 2016 IEEE Region 10 Conference (TENCON), pp. 2281–2285.

Kouzani, A.Z., Adams, S., Whyte, D.J. et al. (2017). 3D printing of food for people with swallowing difficulties. DesTech 2016: Proceedings of the International Conference on Design and Technology, pp. 23–29.

Krishnaraj, P., Anukiruthika, T., Choudhary, P. et al. (2019). 3D extrusion printing and post-processing of fibre-rich snack from indigenous composite flour. *Food and Bioprocess Technology* **12** (10): 1776–1786. https://doi.org/10.1007/s11947-019-02336-5.

Lawal, O.S., Lapasin, R., Bellich, B. et al. (2011). Rheology and functional properties of starches isolated from five improved rice varieties from West Africa. *Food Hydrocolloids* **25** (7): 1785–1792.

Le Tohic, C., O'Sullivan, J.J., Drapala, K.P. et al. (2018). Effect of 3D printing on the structure and textural properties of processed cheese. *Journal of Food Engineering* **220**: 56–64.

Lee, J.H., Won, D.J., Kim, H.W., and Park, H.J. (2019). Effect of particle size on 3D printing performance of the food-ink system with cellular food materials. *Journal of Food Engineering* **256**: 1–8. https://doi.org/10.1016/j.jfoodeng.2019.03.014.

Lei, N., Chai, S., Xu, M. et al. (2020). Effect of dry heating treatment on multi-levels of structure and physicochemical properties of maize starch: a thermodynamic study. *International Journal of Biological Macromolecules* **147**: 109–116.

Leung, P.Y.V. (2017). Sugar 3D printing: additive manufacturing with molten sugar for investigating molten material fed printing. *3D. Printing and Additive Manufacturing* **4** (1): 13–18.

Leyva-Porras, C., Cruz-Alcantar, P., Espinosa-Solis, V. et al. (2020). Application of differential scanning calorimetry (DSC) and modulated differential scanning calorimetry (MDSC) in food and drug industries. *Polymers* **12** (1): 5.

Li, W., Ghazanfari, A., Leu, M.C., and Landers, R.G. (2015). Methods of extrusion on demand for high solids loading ceramic paste in freeform extrusion fabrication. Solid Freeform Fabrication Symposium, Austin, TX.

Lille, M., Nurmela, A., Nordlund, E. et al. (2018). Applicability of protein and fiber-rich food materials in extrusion-based 3D printing. *Journal of Food Engineering* **220**: 20–27.

Liu, L. and Ciftci, O.N. (2020). Effects of high oil compositions and printing parameters on food paste properties and printability in a 3D printing food processing model. *Journal of Food Engineering*, p. 110135.

Liu, Z., Zhang, M., Bhandari, B., and Wang, Y. (2017). 3D printing: Printing precision and application in food sector. *Trends in Food Science & Technology* **69**: 83–94.

Liu, Z., Zhang, M., Bhandari, B., and Yang, C. (2018). Impact of rheological properties of mashed potatoes on 3D printing. *Journal of Food Engineering* **220**: 76–82. https://doi.org/10.1016/j.jfoodeng.2017.04.017.

Liu, Y., Yu, Y., Liu, C. et al. (2019a). Rheological and mechanical behavior of milk protein composite gel for extrusion-based 3D food printing. *LWT* **102**: 338–346. https://doi.org/10.1016/j.lwt.2018.12.053.

Liu, Y., Zhang, W., Wang, K. et al. (2019b). Fabrication of gel-like emulsions with whey protein isolate using microfluidization: rheological properties and 3D printing performance. *Food and Bioprocess Technology* **12** (12): 1967–1979.

Liu, Y., Liang, X., Saeed, A. et al. (2019c). Properties of 3D printed dough and optimization of printing parameters. *Innovative Food Science & Emerging Technologies* https://doi.org/10.1016/J.IFSET.2019.03.008.

Liu, Y., Chen, L., Xu, H. et al. (2019d). Understanding the digestibility of rice starch-gallic acid complexes formed by high pressure homogenization. *International Journal of Biological Macromolecules* **134**: 856–863.

Liu, Z., Zhang, M., and Ye, Y. (2020a). Indirect prediction of 3D printability of mashed potatoes based on LF-NMR measurements. *Journal of Food Engineering*; 110137.

Liu, Z., Chen, H., Zheng, B. et al. (2020b). Understanding the structure and rheological properties of potato starch induced by hot-extrusion 3D printing. *Food Hydrocolloids*, pp. 105812.

Liu, Y., Tang, T., Duan, S. et al. (2020c). Effects of sodium alginate and rice variety on the physicochemical characteristics and 3D printing feasibility of rice paste. *LWT* **127**: 109360.

Liu, Y., Tang, T., Duan, S. et al. (2020d). Applicability of rice doughs as promising food materials in extrusion-based 3D printing. *Food and Bioprocess Technology* **13** (3): 548–563.

Maniglia, B.C., Lima, D.C., da Matta Júnior, M. et al. (2020). Dry heating treatment: a potential tool to improve the wheat starch properties for 3D food printing application. *Food Research International* **137**: 109731.

Mantihal, S., Prakash, S., Godoi, F.C., and Bhandari, B. (2017). Optimization of chocolate 3D printing by correlating thermal and flow properties with 3D structure modeling. *Innovative Food Sciencc and Emerging Technologies* **44**: 21–29. https://doi.org/10.1016/j.ifset.2017.09.012.

Mantihal, S., Prakash, S., and Bhandari, B. (2019). Textural modification of 3D printed dark chocolate by varying internal infill structure. *Food Research International* **121**: 648–657.

Mizrahi, M., Golan, A., Mizrahi, A.B. et al. (2016). Digital gastronomy: methods & recipes for hybrid cooking. Proceedings of the 29th Annual Symposium on User Interface Software and Technology, pp. 541–552.

Nachal, N., Moses, J.A., Karthik, P., and Anandharamakrishnan, C. (2019). Applications of 3D printing in food processing. *Food Engineering Reviews* **11** (3): 123–141. https://doi.org/10.1007/s12393-019-09199-8.

Nida, S., Anukiruthika, T., Moses, J.A., and Anandharamakrishnan, C. (2020). 3D printing of grinding and milling fractions of rice husk. *Waste and Biomass Valorization*: 1–10. https://doi.org/10.1007/s12649-020-01000-w.

Nijdam, J.J., Agarwal, D., and Schon, B.S. (2021a). Assessment of a novel window of dimensional stability for screening food inks for 3D printing. *Journal of Food Engineering* **292**: 110349.

Nijdam, J.J., LeCorre-Bordes, D., Delvart, A., and Schon, B.S. (2021b). A rheological test to assess the ability of food inks to form dimensionally stable 3D food structures. *Journal of Food Engineering* **291**: 110235.

O'Bryan, C.S., Bhattacharjee, T., Hart, S. et al. (2017). Self-assembled micro-organogels for 3D printing silicone structures. *Science Advances* **3** (5): e1602800.

Oliveira, S.M., Fasolin, L.H., Vicente, A.A. et al. (2020). Printability, microstructure, and flow dynamics of phase-separated edible 3D inks. *Food Hydrocolloids* **109**: 106120.

Ouyang, L., Yao, R., Zhao, Y., and Sun, W. (2016). Effect of bioink properties on printability and cell viability for 3D bioplotting of embryonic stem cells. *Biofabrication* **8** (3): 35020.

Park, S.M., Kim, H.W., and Park, H.J. (2020). Callus-based 3D printing for food exemplified with carrot tissues and its potential for innovative food production. *Journal of Food Engineering* **271**: 109781.

Pauw, B.R. (2013). Everything SAXS: small-angle scattering pattern collection and correction. *Journal of Physics: Condensed Matter* **25** (38): 383201.

Peleg, M. and Normand, M.D. (2013). Modeling of fungal and bacterial spore germination under static and dynamic conditions. *Applied and Environmental Microbiology* **79** (21): 6765–6775. https://doi.org/10.1128/AEM.02521-13.

Pérez, B., Nykvist, H., Brøgger, A.F. et al. (2019). Impact of macronutrients printability and 3D-printer parameters on 3D-food printing: a review. *Food Chemistry* **287** (October 2018): 249–257. https://doi.org/10.1016/j.foodchem.2019.02.090.

du Plessis, A., Tshibalanganda, M., and le Roux, S.G. (2020). Not all scans are equal: X-ray tomography image quality evaluation. *Materials Today Communications* **22**: 100792.

Pulatsu, E., Su, J.-W., Lin, J., and Lin, M. (2020). Factors affecting 3D printing and post-processing capacity of cookie dough. *Innovative Food Science & Emerging Technologies* **61**: 102316.

Sanchez, P.D.C., Hashim, N., Shamsudin, R., and Nor, M.Z.M. (2020). Applications of imaging and spectroscopy techniques for non-destructive quality evaluation of potatoes and sweet potatoes: a review. *Trends in Food Science & Technology* **96**: 208–221.

Sangeetha, G., Rajeshwari, S., and Venckatesh, R. (2011). Green synthesis of zinc oxide nanoparticles by aloe barbadensis miller leaf extract: structure and optical properties. *Materials Research Bulletin* **46** (12): 2560–2566.

Severini, C., Derossi, A., and Azzollini, D. (2016). Variables affecting the printability of foods: preliminary tests on cereal-based products. *Innovative Food Science and Emerging Technologies* **38**: 281–291. https://doi.org/10.1016/j.ifset.2016.10.001.

Severini, C., Azzollini, D., Albenzio, M., and Derossi, A. (2018a). On printability, quality and nutritional properties of 3D printed cereal based snacks enriched with edible insects. *Food Research International* **106** (November 2017): 666–676. https://doi.org/10.1016/j.foodres.2018.01.034.

Severini, C., Derossi, A., Ricci, I. et al. (2018b). Printing a blend of fruit and vegetables. New advances on critical variables and shelf life of 3D edible objects. *Journal of Food Engineering* **220**: 89–100. https://doi.org/10.1016/j.jfoodeng.2017.08.025.

Shahrubudin, N., Lee, T.C., and Ramlan, R. (2019). An overview on 3D printing technology: technological, materials, and applications. *Procedia Manufacturing* **35**: 1286–1296. https://doi.org/10.1016/j.promfg.2019.06.089.

Shalviri, A., Liu, Q., Abdekhodaie, M.J., and Wu, X.Y. (2010). Novel modified starch-xanthan gum hydrogels for controlled drug delivery: synthesis and characterization. *Carbohydrate Polymers* **79** (4): 898–907.

Sharif, N., Khoshnoudi-Nia, S., and Jafari, S.M. (2020). Confocal laser scanning microscopy (CLSM) of nanoencapsulated food ingredients. In: *Characterization of Nanoencapsulated Food Ingredients* (ed. S.M. Jafari), pp. 131–158. Elsevier.

Shibayama, M. (2011). Small-angle neutron scattering on polymer gels: phase behavior, inhomogeneities and deformation mechanisms. *Polymer Journal* **43** (1): 18–34.

Shibayama, M. (2012). Structure-mechanical property relationship of tough hydrogels. *Soft Matter* **8** (31): 8030–8038.

Soares, S. and Forkes, A. (2014). Insects Au gratin-an investigation into the experiences of developing a 3D printer that uses insect protein based flour as a building medium for the production of sustainable food. DS 78: Proceedings of the 16th International Conference on Engineering and Product Design Education (E&PDE14), Design Education and Human Technology Relations, University of Twente, The Netherlands, 04-05.09. 2014, pp. 426–431.

Sun, Y., Zhang, M., and Chen, H. (2020). LF-NMR intelligent evaluation of rheology and printability for 3D printing of cookie dough pretreated by microwave. *LWT* **132**: 109752.

Tanaka, H., Asano, Y., Watanabe, M., and Masumori, A. (2015). Food printing technologies out of white rice. *NIP & Digital Fabrication Conference* **2015** (1): 289–292.

Theagarajan, R., Moses, J.A., and Anandharamakrishnan, C. (2020). 3D extrusion printability of rice starch and optimization of process variables. *Food and Bioprocess Technology*. https://doi.org/10.1007/s11947-020-02453-6.

Vancauwenberghe, V., Katalagarianakis, L., Wang, Z. et al. (2017). Pectin based food-ink formulations for 3-D printing of customizable porous food simulants. *Innovative Food Science and Emerging Technologies* **42** (June): 138–150. https://doi.org/10.1016/j.ifset.2017.06.011.

Vancauwenberghe, V., Delele, M.A., Vanbiervliet, J. et al. (2018). Model-based design and validation of food texture of 3D printed pectin-based food simulants. *Journal of Food Engineering* **231**: 72–82. https://doi.org/10.1016/j.jfoodeng.2018.03.010.

Vieira, M.V., Oliveira, S.M., Amado, I.R. et al. (2020). 3D printed functional cookies fortified with *Arthrospira platensis*: evaluation of its antioxidant potential and physical-chemical characterization. *Food Hydrocolloids* **107**: 105893.

Wang, S., Luo, H., Zhang, J. et al. (2014). Alkali-induced changes in functional properties and in vitro digestibility of wheat starch: the role of surface proteins and lipids. *Journal of Agricultural and Food Chemistry* **62** (16): 3636–3643.

Wang, L., Zhang, M., Bhandari, B., and Yang, C. (2018). Investigation on fish surimi gel as promising food material for 3D printing. *Journal of Food Engineering* **220**: 101–108. https://doi.org/10.1016/j.jfoodeng.2017.02.029.

Wang, H., Xiao, N., Wang, X. et al. (2019). Effect of pregelatinized starch on the characteristics, microstructures, and quality attributes of glutinous rice flour and dumplings. *Food Chemistry* **283**: 248–256.

Wilson, A., Anukiruthika, T., Moses, J.A., and Anandharamakrishnan, C. (2020). Customized shapes for chicken meat–based products: feasibility study on 3D-printed nuggets. *Food and Bioprocess Technology*, pp. 1–16.

Xu, L., Gu, L., Su, Y. et al. (2020). Impact of thermal treatment on the rheological, microstructural, protein structures and extrusion 3D printing characteristics of egg yolk. *Food Hydrocolloids* **100**: 105399.

Yang, F., Zhang, M., Prakash, S., and Liu, Y. (2018a). Physical properties of 3D printed baking dough as affected by different compositions. *Innovative Food Science and Emerging Technologies* **49** (2017): 202–210. https://doi.org/10.1016/j.ifset.2018.01.001.

Yang, F., Zhang, M., Bhandari, B., and Liu, Y. (2018b). Investigation on lemon juice gel as food material for 3D printing and optimization of printing parameters. *LWT - Food Science and Technology* **87**: 67–76. https://doi.org/10.1016/j.lwt.2017.08.054.

Yang, F., Guo, C., Zhang, M. et al. (2019). Improving 3D printing process of lemon juice gel based on fluid flow numerical simulation. *LWT* **102**: 89–99.

Zeng, X., Chen, H., Chen, L., and Zheng, B. (2020). Insights into the relationship between structure and rheological properties of starch gels in hot-extrusion 3D printing. *Food Chemistry* **342**: 128362.

Zhang, B., Ni, B., Lu, S. et al. (2012). Synthesis and characterization of a novel potato starch derivative with cationic acetylcholine groups. *International Journal of Biological Macromolecules* **50** (3): 701–706.

Zhang, B., Li, X., Liu, J. et al. (2013). Supramolecular structure of A-and B-type granules of wheat starch. *Food Hydrocolloids* **31** (1): 68–73.

Zhang, B., Zhou, W., Qiao, D. et al. (2019). Changes in nanoscale chain assembly in sweet potato starch lamellae by downregulation of biosynthesis enzymes. *Journal of Agricultural and Food Chemistry* **67** (22): 6302–6312.

Zhang, H., Zhang, F., and Yuan, R. (2020). Applications of natural polymer-based hydrogels in the food industry. In: *Hydrogels Based on Natural Polymers*, pp. 357–410. Elsevier.

Zhao, Y., Li, Y., Mao, S. et al. (2015). The influence of printing parameters on cell survival rate and printability in microextrusion-based 3D cell printing technology. *Biofabrication* **7** (4): 45002.

Zhu, F. (2017). NMR spectroscopy of starch systems. *Food Hydrocolloids* **63**: 611–624.

Zhu, S., Stieger, M.A., van der Goot, A.J., and Schutyser, M.A.I. (2019). Extrusion-based 3D printing of food pastes: correlating rheological properties with printing behaviour. *Innovative Food Science & Emerging Technologies* **58**: 102214.

6

Natively Printable Foods

CHAPTER MENU

6.1 Introduction

Food fabrication is an art that deals with the science of the material response for constructing a three-dimensional object in a layered fashion one above the other with the enhanced levels of customization (Kailash et al. 2020). This approach to the food manufacturing process promises to revolutionize the food industry of the modern world by transforming virtual imagination into reality through innovations in culinary skills with digital gastronomy approaches (Gouton et al. 2020). As the name implies, 3D printing is a newly emerging field that receives great interest in recent days because of its advantages of providing foods based on consumer's preferences and requirements. However, the science of printing foods or printing on foods in 2D forms dates back to several decades. This form of 2D printing includes the decoration of foods with creamy toppings like cakes, pizza bases, cookies, and pastries products (Pallottino et al. 2016). The advanced level of 2D printing is the 3D forms that require huge efforts and immense attention for achieving the desired printing precision and stability which in turn depends on the process and product attributes (Pérez et al. 2019).

The choice of ingredients selected for the material formulation is one of the important aspects that decide printability (Nachal et al. 2019). Material supplies for 3D food printing are categorized as natively printable, traditionally non-printable, and alternative ingredients. Natively printable materials are food systems with an inherent material property conducive for 3D printing. Some of the natively printable materials are chocolates, cheese, cake

3D Printing of Foods, First Edition. C. Anandharamakrishnan, Jeyan A. Moses, and T. Anukiruthika.

frostings, hydrogels, sugar confectioneries, and starch-based material supplies (Sun et al. 2015). However, a certain amount of pre-processing is required for tailoring the desired rheological properties and post-deposition stability of the 3D constructs out of natively printable materials (Huang et al. 2019). Different printing technologies that are commonly used for printing foods are extrusion-based printing, selective laser sintering, binder jetting, and ink-jet printing. Various critical process parameters like printing temperature, pressure, print speed, nozzle size, printer configurations, heat source, laser power, binder type, and printing time determine the final print fidelity (Severini et al. 2016). Some of the material properties that are interrelated with the printing variables are material density, particle size, porosity, meltability, crystallinity, thermal conductivity and diffusivity, mass to volume ratio, surface area, absorption, permeability, viscosity, compressibility, adhesion, wettability, and solidification behaviour (Lanaro et al. 2017; Mantihal et al. 2017). Specific to the type of 3D printing technique used, the printability of the same material would behave differently for each technique. For instance, chocolate could be 3D printed using all the above-mentioned printing techniques each would result in a varied texture which cannot be achieved with the conventional processing of chocolates (Lanaro et al. 2019). Thus, natively printable material supplies have a great degrees of freedom that forms the foundation of food formulations used for 3D printing. The present chapter focuses on the feasibility of the natively printable materials with insights on the applicability of such material supplies in 3D printing through a concise overview of the science of printability as reported by different studies in the literature.

6.2 Natively Printable Materials as Basic Food 3D Printing Formulations

As stated earlier, natively printable materials possess a desirable viscoelastic property suitable for printing on their own without the addition of binders and adhesives. This category includes hydrogels, cake frostings, cheese, hummus, ice creams, powdered sugars, and starch-based material supplies (Sun et al. 2018b). Although these materials exhibit printability, they do not represent whole food. This arises a need for the optimal formulation of such materials to form a wholesome product. In this regard, a combination of butter, egg yolk, and sugar was used for developing a natively printable cookie dough. Results showed that the addition of yolk improves the dimensional stability of the printed cookies while butter and sugar aids in tailoring the flowability of the material for extrusion (Lipton et al. 2010). This kind of batter formulations requires an additional post-processing step of cooking immediately after printing thereby maintaining the post-deposition of material until consumption. Therefore, it is important to play around the material composition, consistency, and solidification for achieving greater printing precision and accuracy. In another study by Southerland et al. (2011) a mixture of sugar crystals, mashed potato, and starch in dry forms was tested for printing using Z corporation 3D printer through binding technology. A semi-solid flowable material supply used for 3D printing is commonly referred to as food inks. Voon et al. (2019) broadly categorized the food inks based on the nature and source of raw materials as confectionery, dairy, hydrogels, plant, and meat-based formulations considering the mechanical and rheological properties. Among which confectionery, dairy, and hydrogels are known to be natively extrudable whereas the latter food inks of plant and animal-based formulations are referred as non-natively extrudable. In this study, materials are mainly categorized focussing on its applicability suitable for extrusion-based printing, a commonly used printing technology.

Another class of natively printable materials is hydrogels which are known to be soft foods. Hydrogels are the 3D networks of one or more hydrophilic polymeric fractions that possess a characteristic swelling behaviour and a good water holding capacity (Gulrez et al. 2011). The complex interaction of hydrogen ions would result in the formation of a 3D network through physical or chemical cross-linking of the polymers. This group of polymers includes hydrocolloids such as starch, xanthan gum, guar gum, gum arabic, locust bean gum, gum karaya, and other cellulose derivatives which are traditionally used as food thickeners whereas another group of hydrocolloids such as gelatin, alginate, carrageenan, pectin, gellan, and agar is commonly used as gelling agents in the food industry (Zhang et al. 2020). This diverse range of applications of food hydrocolloids and the resulting hydrogels increased the uprising research on this group of materials for 3D printing. The first edible matrix out of hydrocolloids was demonstrated using a combination of xanthan gum and gelatin with flavour concentrates resulted in a modified texture ranging from weak to firm matrix (Cohen et al. 2009). This shows the flexibility of the hydrogels towards 3D printing based on which a wide range of products with desired texture attributes can be prepared. Recently, many researches are going on in exploring the potentiality of the applications of hydrocolloids in the preparation of customized foods with altered texture (Le Tohic et al. 2018; Mantihal et al. 2019a). Thus, hydrocolloids are used as additives in enhancing the printability of non-natively printable food ingredients such as plant and meat-based food inks (Dick et al. 2019; Park et al. 2020). The success of which greatly depends on the optimization of printing parameters, simulation and optimization of food inks with different concentrations and types of hydrocolloids, investigation on the suitability of the printing technique for hydrogels.

6.3 Printability: Concepts and Underlying Mechanisms

Food printing is a complex process associated with macro and micro components of the material supplies. The response of various chemical components that aid in 3D printing has been discussed in the earlier chapter. The present section on printability helps in understanding the mechanism associated with the printability of natively printable materials. In this regard, the innate tendency of the natively printable material is associated with the formation of complex 3D networks of the biopolymers upon external stresses like heating, cooling, swelling, acid–base interactions, UV radiations, catalytic cross-linkers, etc. (Kern et al. 2018; Zheng et al. 2019). Based on the nature of the material supply, the mechanism of formation of the 3D network and its strength would greatly differ. Gelation, gelatinization, cross-linking (chemical-induced, enzyme induced, and ionotropic), acid induction, and complex coacervation are the common mechanisms that translate the printable nature of biopolymers which are being responsible for the printability of natively printable materials (Voon et al. 2019). An in-depth visualization at the micro-level reveals the association of a carbon chain as the backbone that supports the printability of the natively printable materials. This group of polymers comprises carbohydrates as well as proteins as base components. The carbohydrate polymers consist of linear and branched chains of amylose and amylopectin. The molecular arrangement of these components determines the functional properties like gelatinization and gelation behaviour (Hu et al. 2020; Zhuang et al. 2020). Various critical factors that affect gelation include temperature, viscosity, flowability, mechanical stability, crystallinity, meltability, solidification, and retrogradation behaviour (Liu et al. 2017). Most of these materials exhibit a common characteristic of hydrophilicity and a good water holding capacity that makes it a suitable candidate for 3D printing that enhances the printability when mixed with other materials (Theagarajan et al. 2020). These are salient features that demonstrate the material properties of extrusion-based food printing (room temperature and hot-melt type), a form of fused deposition modelling (FDM) approach.

Another class of food components that forms the carbon chain backbone are polypeptides. These protein polymers exhibit viscoelasticity forming an emulsion and are known to form a liquid-like continuous protein matrix that enables printability (Phuhongsung et al. 2020). This group of natively printable materials includes dairy and dairy products such as casein, whey protein, cheese, butter, etc. Considering the polypeptide chain, the amount of protein content is a significant factor affecting the shape and structure of 3D constructs (Chen et al. 2019a). It was found that the lesser the protein content higher the fluidity resulting in a weaker gel and chain density. The precision and mechanical stability during printing can be achieved by increasing the protein content of the gel matrix (Chen et al. 2019b). It was reported that the addition of whey protein concentrate reduces the viscosity of milk protein paste which in turn enhances the printability of protein mixtures for extrusion-based printing of milk protein concentrates.

Modification of the structural behaviour and properties of polymeric fractions can be achieved by the so-called process of cross-linking through fusion with the formation of a covalent bond resulting in a complex 3D polymeric network. Based on the process of induction, cross-linking can be either chemical or enzymatic or ionic induced which is further broadly categorized as *ex-situ* and *in-situ* based on the means of initialization of gelation process possessing a wide range of applications in the food systems (Lee and Mooney 2012). The process of cross-linking through a complex interaction of oppositely charged ions is known as ionotropic gelation, a common gelation technique employed for polysaccharide and polypeptide fractions (Bocchetta 2020). Intending to develop a customizable porous food stimulant, Vancauwenberghe et al. (2018) used a mixture of pectin, calcium chloride, bovine serum albumin, and sugar syrup concentrates as printing material supply. Results showed that the galacturonic acid component of hetero-polysaccharide pectin was responsible for the firmness and strength of the gel matrix whereas pectin along with sugar crystals improves the viscosity suitable for printing applications. The use of bovine serum albumin helps in promoting aeration through the entrapment of the air bubbles that ultimately resulted in porous water-based food stimulants. In this study, hydrocolloids were used for developing a food stimulant and its applicability for food printing applications was analyzed. The basic mechanism underlying is the gelation which occurs due to the cross linking of the calcium ions with the carbonyl group of the low methoxylated pectin molecules. This form of gelation mechanism is being referred to as the 'egg-box model' that mainly depends on the ionic strength, pH, calcium ion concentration, degree of methoxylation, and source of pectin (Cao et al. 2020). Thus, this study provides an in-depth analysis of the printability of hydrocolloids that in turn provides insights on future applications of printing soft food matrices. Thus, 3D printing could be used in the fabrication of low caloric density food with modified textures providing satiety without compromising the flavours and nutritional components.

6.4 Types of Natively Printable Materials

Based on the diverse nature of food materials, it is cumbersome to provide a single umbrella of a classification system for natively printable materials. Hence a broad division of natively printable materials was presented based on the source of origin and type of the basic component of the food system (Figure 6.1). The present section does not intend in merely listing the natively printable materials reported on literature but to provide insights on the

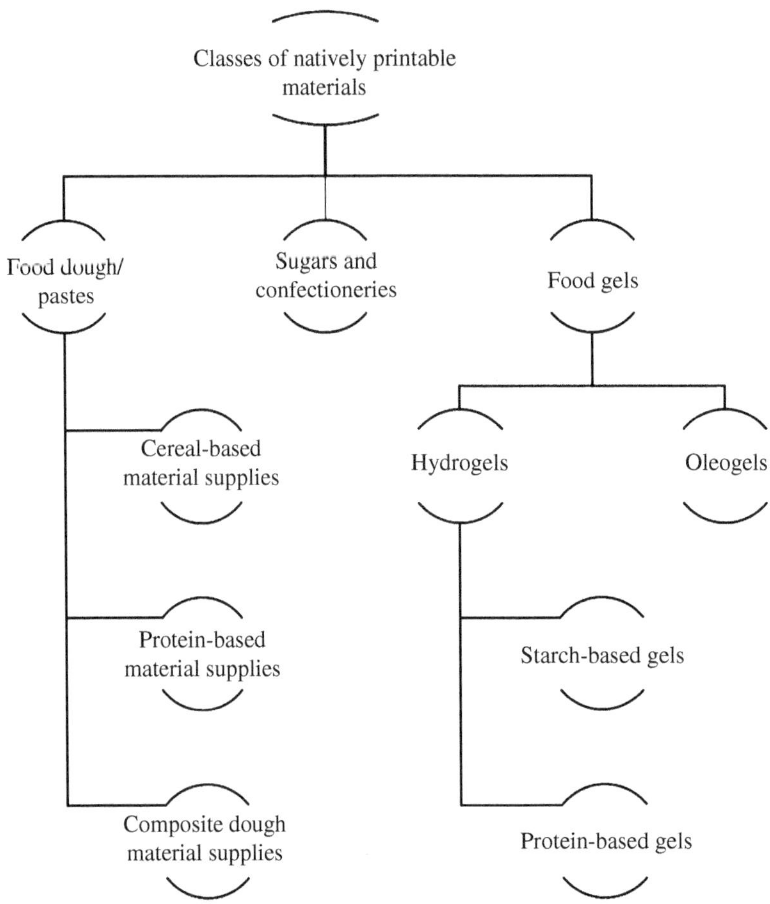

Figure 6.1 Broad classification of natively printable materials.

feasibility of these material supplies focussing on the futuristic applications in the delivery of nutritious digitalized foods. Microstructural modifications of the food systems correlating the printability and post-deposition stability are discussed in the following sections. A summary of the mechanism of printability of natively printable materials is presented in Table 6.1.

6.4.1 Cereal-Based Material Supplies

As of food printing is concerned most of the works were reported on 3D printing of chocolates, sugar-based materials, and gel matrices. Since these are food items that are known to be conveniently natively printable with less pre-processing. On the other hand, cereal-based material supplies are semi-solid/ paste that requires adequate mixing and appropriate pre-processing for achieving uniform consistency and better flowability. It is well known that food grains form a major portion of our daily diet. To develop a whole food formulation, nowadays research works are being focussed in exploring the printability of starch-based foods as the base for the next-generation digitalized recipes using 3D printing technology. The major component of food grains is the amylum, a polymeric carbohydrate comprised of numerous glucose units joined together by glycosidic bonds (Huang and Perdon 2020). These fractions of cereal grains form an integral component that assists in the printability of starch-based natively printable materials. Considering the nature of material supplies, a semi-solid paste is being formed out of dry fractions of cereal grains by mixing the powder with water as solvent through appropriate pre-processing. The formed dough must be viscous enough to be flowable through a smaller orifice printing nozzle and high enough to withstand the weight of the stacked layers without buckling and sagging to yield a 3D construct of good resolution which could be achieved by correlating printability with the rheological properties (Chen et al. 2019b). Studies were reported that starch-based material supplies must be of elasticity dominant with loss tangent ($\tan \delta$) < 1 possessing higher yield stress to be resulted in a continuous thread/ filament without breaking during printing (Huang et al. 2019; Liu et al. 2020).

Most of the starch-based material supplies were suitable for extrusion-based printing. Material supplies with fractions of whole cereal grains (dough) are subjected to room temperature extrusion, i.e. the deposition of material through the printing nozzle without the influence of temperature (Theagarajan et al. 2020), whereas the material supplies with specific components for instance the chocolates are more suitable for hot-melt extrusion (Lanaro et al. 2019). Thus, temperature plays a crucial role in estimating the end quality of the product and its application. In this regard, studies were reported on the comparison of different starch systems such as rice, corn, and potato starches (Chen et al. 2019b). Here, the rheological characteristics of cereal-based starch matrices were evaluated along with the plant-based starch matrix concerning the printability using the hot-extrusion technique. This study provides significant remarks in developing the individualized starch-based foods and process conditions were optimized for respective starch matrices. Results showed that printability was greatly influenced by thermo-rheological properties of starch systems that differ based on the source of starch. Hence, printability with good resolution and shape stability was obtained at 80 °C for rice starch of 12–25% (w/w); 75 °C for corn starch of 20–25% (w/w); and 70 °C for potato starch of 15–20% (w/w) (Figure 6.2). During processing when

Table 6.1 Summary of the mechanism of printability of natively printable materials.

Category	Natively printable material	Extrusion temperature employed	Material composition	Pre-processing	Stability of printed structures	Results	References
Cereal-based material supplies	Cereal-based ink	Room temperature extrusion (21 ± 2 °C)	Wheat flour, wheat starch, and egg white protein	Kneading	Protein hydration and starch swelling provides structural integrity to printed structures	Dimensional attributes and quality behaviours of 3D printed cereal structures are quantitatively studied	Fahmy et al. (2020)
	Individual gels made out of selected cereal grains	Room temperature extrusion (25 °C)	Black rice Job's tear seeds Mung bean Brown rice Buckwheat	Steaming of size reduced cereal grains	Starch gelation	Among the tested materials, mung bean gel yields the strongest gel while job's tear seeds possessed weak structure	Guo et al. (2020)
	Starch gel	Hot-melt extrusion (45 °C)	Potato starch Wheat starch Corn starch	Thermal treatment	Gelatinization	Among the tested starches, wheat starch possessed a regular network of microstructure, lower viscosity with better extrudability	Zheng et al. (2019)
	Indigenous composite flour	Room temperature extrusion (25 °C)	Green gram, fried gram, barnyard millet, and ajwain seeds	Size reduction	Hydration of macro components of starch and dietary fibre imparts stability and printability	Fibre and protein-rich indigenous composite flour was printed without any additives	Krishnaraj et al. (2019)
	Rice dough			–			

	Room temperature extrusion (25 °C)	Rice varieties used in the study: japonica rice, indica rice, and waxy rice		Water holding capacity of rice dough was correlated with the material's viscosity that determines the stability of printed structures	Effect of material formulation on the quality of the 3D printed rice-based foods was examined	Liu et al. (2020)
Brown rice paste	Room temperature extrusion (25 °C)	Brown rice powder, xanthan gum-guar gum mix	Cooling (4 °C for 24 h)	Sol–gel transition and gelation imparts mechanical strength to 3D structures	Results showed that the textural properties like hardness and gumminess were strongly correlated with the printing variables such as infill density, perimeters, and nozzle size	Huang et al. (2019)
Rice starch	Room temperature extrusion (25 °C)	Indigenous rice variety *ponni*	–	Hydrophilicity of rice starch exhibits a strong intermolecular interaction forming a gel network that aids in printability	This study provides a scientific base for future 3D printing of foods using rice starch	Theagarajan et al. (2020)
Potato, rice, and corn starches	Hot-melt extrusion (70–85 °C)	–	Thermal treatment	Gelatinization of starch molecules provides strength to	Starch suspensions of concentration 15–25% (w/w) heated to 70–85 °C	Chen et al. (2019b)

(*Continued*)

Table 6.1 (Continued)

Category	Natively printable material	Extrusion temperature employed	Material composition	Pre-processing	Stability of printed structures	Results	References
					3D printed structures	possessed excellent extrudability and mechanical integrity	
Protein-based material supplies	Cheese	Hot-melt extrusion (60 °C)	Cheese with different calcium levels of 15.8 ± 0.9 to 31 ± 2 mg of Ca/g of protein	Melting	Crosslinking of calcium ions stabilizes the casein gels enhancing the printability	Among the tested samples, the gel–sol transition ranged from 52 to 60 °C that significantly decreased with decreasing calcium levels	Kern et al. (2018)
	Cheese	Hot-melt extrusion (75 °C)	Untreated cheese, melted cheese, 3D printed cheese at different extrusion rates	Melting	Hydrophobic interactions of the casein molecules provide stability to 3D printed structures upon cooling	This study showed that the textural properties like hardness could be tailored through 3D printing	Le Tohic et al. (2018)
Sugar and confectioneries	Chocolate	Hot-melt extrusion (32 °C) with heated printer bed system (15–16 °C)	Cadbury dark chocolate and Callebaut dark chocolate	Melting	Solidification of the melted 3D printed chocolate imparts the mechanical strength	Sensory and textural properties of 3D printed chocolates were significantly altered by varying the infill percentage	Mantihal et al. (2019a)
	Chocolate	Hot-melt extrusion (32–40 °C)	Cadbury milk chocolate	Melting	Changes in the crystalline structure during melting and solidification imparts the mechanical stability	The successful deposition of chocolate depends upon the material characteristics and printing variables	Hao et al. (2010)

Food paste and puree	Mashed potato	Room temperature extrusion (25 °C)	Potato (steam-cooked), 15% trehalose (w/w), and potato starch (0, 1, 2, and 4%)	Cooking and refrigeration	Addition of potato starch leads to the formation of denser network that aids in stability of 3D printed mashed potato	This study establishes a significant correlation between the material properties and printability with increased proportion of potato starch	Liu et al. (2018a)
	Commercial potato puree	Hot extrusion (10, 20, and 30 °C)	Potato puree and whole milk	Gelation	Increase in the solid content had a significant impact on the rheology of formulations that imparts structural stability	Results showed that printability increased with consistency index and decreased with the behaviour index	Martínez-Monzó et al. (2019)
	Potato puree	Room temperature extrusion (20 °C)	Cooked potato, carrot puree, olive oil, butter, alginate, and agar	Gelatinization	Starch gelatinization improves the consistency of mixture that aids in printability	The effect of different cooking methods (microwave heating and boiling) on microstructural changes of starch for 3D printing was studied	Dankar et al. (2020)
Composite dough	Cookie dough	Room temperature extrusion	Rice flour, wheat flour, tapioca flour, butter and shortening, sugar,	Mixing	Interactions among starch and protein matrix aids in structural	Functionality of the starch molecules greatly affected physiochemical	Pulatsu et al. (2020)

(Continued)

Table 6.1 (Continued)

Category	Natively printable material	Extrusion temperature employed	Material composition	Pre-processing	Stability of printed structures	Results	References
			milk powder, and water		formation of 3D networks	property of printing mixture	
	Baking dough	Room temperature extrusion (25 °C)	Icing sugar, butter, egg, low gluten flour, and water	Mixing of ingredients and post-process fast cooling (after printing)	Hydration of macro components imparts printability	Post-process fast cooling of the 3D printed structures (−65 °C for approximately 10 min) was beneficial in structure and shape stability during baking	Yang et al. (2019a,b)
	Composite dough	Room temperature extrusion (25 °C)	Wheat flour, freeze-dried mango powder, olive oil, and water	Mixing	Addition of oil imparts flexibility to the glutenous flour that aids in printability	Effect of the material composition on the 3D printing characteristics of dough was analyzed	Liu et al. (2019c)
	Baking dough	Room temperature extrusion (25 °C)	Low gluten flour, icing sugar, butter, egg, and water	Mixing	Enhancement of binding capacity due to molecular interactions of hydroxyl groups aids in printability	A pseudoplastic gel with higher gel strength, extrudability, elasticity, and a lower ductility were desirable for achieving higher print fidelity	Yang et al. (2019a)

	Cookie dough	Room temperature extrusion (25 °C)	Wheat starch, low gluten flour, sugar, butter, and egg	Microwave treatment	The thermal effects of microwave treatment impart partial gelatinization enhancing the viscosity forming a gel matrix	Microwave pre-treatment significantly increased the degree of gelatinization and improved viscosity, hardness, and elasticity of dough	Sun et al. (2020)
	Plasma protein-based dough	Room temperature extrusion (20 °C)	Porcine plasma protein, pea protein concentrates, and soy protein isolate	Mixing	Protein-protein interactions had been greatly hindered by glycerol imparts mobility to the polymeric chains	Dough with porcine plasma protein between 42.5 and 47.5% (w/w) showed higher printing performance. Also, mixed dough with pea protein concentrate and soy protein isolate of 10–15% (w/w) possessed better printability	Álvarez-Castillo et al. (2020)
Food gels	Milk protein composite gel	Room temperature extrusion (25 °C)	Milk protein concentrate and sodium caseinate	Gelation	Milk protein concentrate strengthen the entanglement of the sodium caseinate network forming 3D structures	Results showed that milk protein gel containing 400–450 g^{-1} of total protein possessed best print fidelity	Liu et al. (2019a)
	Milk protein food simulant	Room temperature extrusion (25 °C)	Whey protein isolate, milk protein concentrate,	Gelation	Addition of whey protein isolate altered the inner structure of protein	Milk protein concentrate: whey protein isolate of 5 : 2 proportion	Liu et al. (2018c)

(*Continued*)

Table 6.1 (Continued)

Category	Natively printable material	Extrusion temperature employed	Material composition	Pre-processing	Stability of printed structures	Results	References
			glycerol, and xanthan gum		mixture resulted in liquid-like protein matrix enhancing printability	showed better printing performance	
	Pea protein gel	Room temperature extrusion	Pea protein, xylose, protease, glycerol, and potato starch	Gelatinization	Maillard reaction had combined the functionalities of peptides and sugars resulting in a novel protein-saccharide 3D conjugate	Xylose Maillard reaction products had the highest degree of cross-linking showed a strongest adhesion and cohesion with better plasticity of the material	Zhou et al. (2020)
	Pea protein gel	Room temperature extrusion (25 °C)	Pea protein, alginate powder, and calcium chloride	Gelation	Physical cross-linking of Ca^{2+} with protein network resulted in increased covalent bond formation that enhances the printability	Alginate to pea protein proportion of 80 : 20 showed better printing performance	Oyinloye and Yoon (2020)
	Hydrogel	Syringe-based pump system was used for pre-heated solution to deposit on the platform at 0 degrees	Agar and gelatin (used individually)	Gelation	The sol–gel transition resulted in 3D cross-linked network imparts the printability of gel systems	Tested formulation of gel systems had hardness ranged from 8 to 45 kPa possessing a soft jelly-like texture	Serizawa et al. (2014)

Carrot moulds	Room temperature extrusion	Carrot puree, Guar gum, xanthan gum, and gelatin	Gelation	Formation of cross-linking among the hydroxyl compounds resulted in a gel network	Results showed that the samples with gelatin and xanthan gum possessed higher hardness and densest texture	Strother et al. (2020)
Hydrogel	Extrusion-based printing (25 and 50 °C)	Xanthan and Konjac gum	Gelation	Sol–gel transition of hydrocolloids imparts strength to 3D structures	Increasing syrup content resulted in a weak gel. Material composition had significant impact on textural and rheological properties	Garcia-Segovia et al. (2020)
Soy protein isolate gel	Room temperature extrusion (25 °C)	Soy protein isolate, xanthan gum, and sodium chloride	Microwave treatment	Ionic bonding and hydrophobic interactions of salt with protein and polysaccharides resulted in viscous gel matrix	Soy protein isolate gel with xanthan gum and NaCl solution at 1 g/100 ml showed better printability	Phuhongsung et al. (2020)
Hydrogel	Room temperature extrusion (20 °C)	Cassava starch	Dry heat treatment	Dry heat treatment resulted in intermolecular associations among starch granules replacing the hydroxyl groups by carbonyl groups thereby enhancing the pasting properties	Dry heat treatment 3D printed samples exhibit better printability than native starch samples	Maniglia et al. (2020)

(Continued)

Table 6.1 (Continued)

Category	Natively printable material	Extrusion temperature employed	Material composition	Pre-processing	Stability of printed structures	Results	References
	Hydrogel	Room temperature extrusion (25 °C)	Xanthan gum, locust bean gum, carrageenan, pectin, agar, sodium alginate, guar gum, gum Arabic, maltodextrin, iota-carrageenan, and gelatin (used individually)	Gelation	Hydration of the compounds resulted in cross-linking of polymeric strands forming a gel network	Results showed that the phase angle and relaxation exponent could be used for understanding of viscoelastic behaviour of the material supplies	Gholamipour-Shirazi et al. (2019)
	Hydrogels	Room temperature extrusion (20 °C)	Cassava starch	Ozonation	With the increase in concentration of ozonated cassava, starch-starch interactions become more dominant with greater amount of water immobilization resulted in improved viscosity	This study highlights the effects of ozone treatment in starch modification that can be used for production of 3D printed starch-based food with different industrial applications	Maniglia et al. (2019)

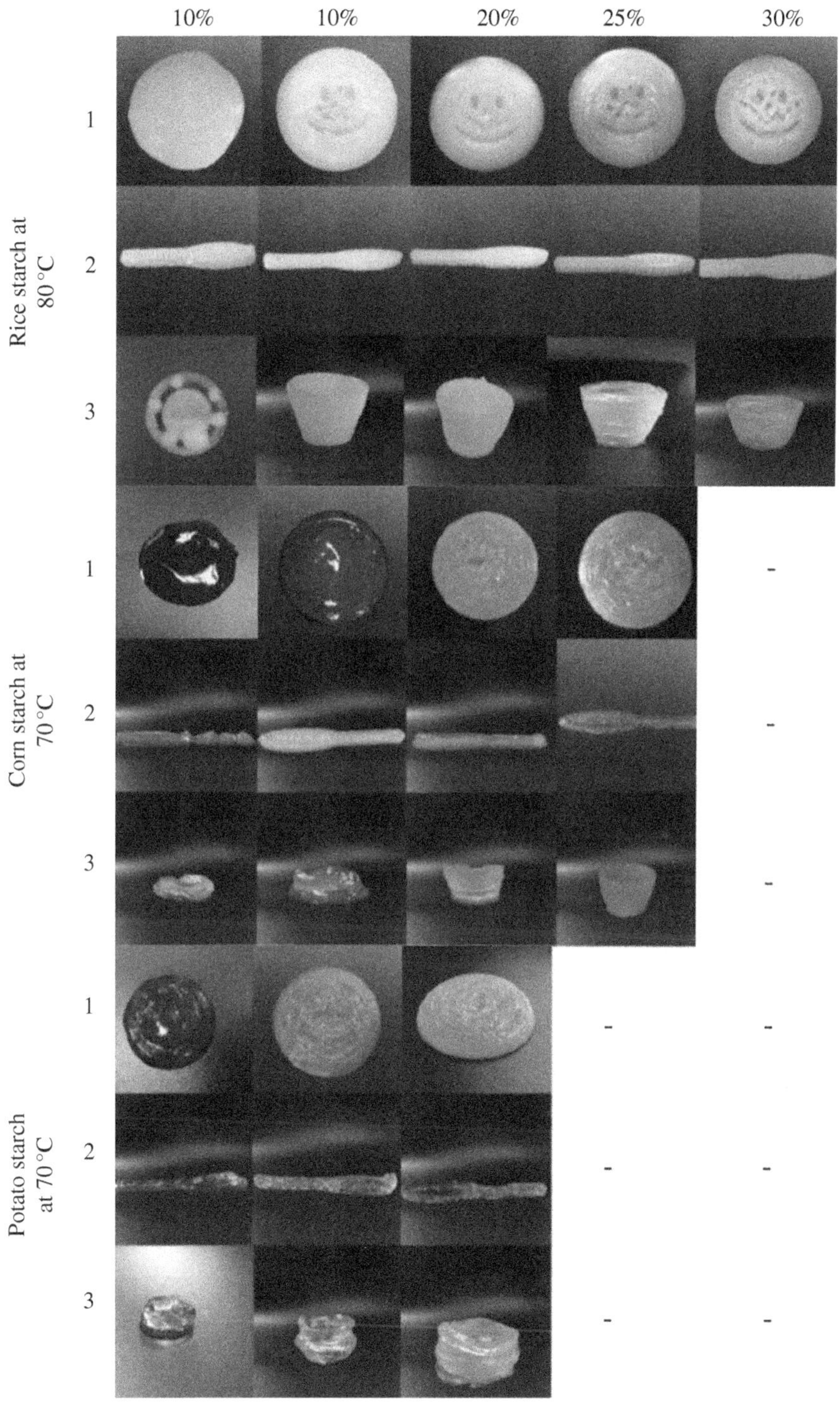

Figure 6.2 Different concentrations of 3D printed starch samples using hot-melt extrusion. *Source:* From Chen et al. (2019b), Figure 06 [p. 29] / With permission of Elsevier. DOI-https://doi.org/10.1016/j.jfoodeng.2018.09.011.

the material was subjected to thermal treatment, starch undergoes gelatinization resulting in a disrupted crystalline structure due to the diffusion of amylose from the swollen starch granules (Wang et al. 2018). Thus, the molecular disruption and rearrangement of the amylose and amylopectin fractions of the starch results in the formation of a complex 3D matrix forming a gel matrix upon cooling that aids in printing and post-deposition stability.

Most recently, a study on the effect of the rice starch from indigenous south Indian variety (*Ponni*) on the extrusion printability was reported. The scientific rationale of the tendency of the rice starch towards printing was examined with varying levels of printing variables such as extrusion motor speed, printing speed, and nozzle size. The study confirmed that amylose and amylopectin fractions; and their molecular leaching and swelling behaviour illustrates the printability of the rice dough (Theagarajan et al. 2020). Further, this study demonstrates the commercial scope for the development of formulations for complementary foods using starch-based foods using 3D printing. In another study, different process variables affecting the printability of brown rice were examined using brown rice paste obtained from the ready-to-eat brown rice flour with xanthan gum and guar gum as additives (Huang et al. 2019). This study showed the importance of pre-processing in tailoring the rheological and printing attributes of the starch-based material supply from brown rice. Further, this study provides useful information for the development of the 3D constructs with complex internal structures by customizing the void space and infill density. Another study on the effect of varietal differences of rice on 3D printing was studied by Liu et al. (2020). The material formulation and the printability were optimized for waxy rice, japonica rice, and indica rice. Results showed that waxy rice exhibited better printability with the highest precision followed by indica rice and japonica rice whereas after the post-processing of steaming, the printed 3D constructs from waxy rice were deformed indicating that the materials printed out of indica and japonica rice were the promising materials for 3D printing that withstands the post-processing treatments. These studies on starch-based material supplies provide valuable insights for the production of wholesome starchy foods using 3D printing and explore the potential benefits of natively printable materials in improving the printability of non-printable materials. In this context, the effect of material formulations on the printability of baking dough was reported (Yang et al. 2018). Results showed that variations in the proportions of raw materials of baking dough had a significant effect on the printing quality of 3D printed constructs (Figure 6.3).

6.4.2 Sugar-Based Material Supplies

Sugar-based material supplies are known for their flexibility and a higher degrees of freedom of operation towards 3D printing. Materials of this category include confectionery items that can be 3D printed with high levels of customization (Piyush et al. 2020). For instance, chocolates are viscoelastic materials comprised of cocoa, milk solids, and lipid components that could be 3D printed using the extrusion-based technique, powder-based sintering technique, and ink-jet printing which demonstrates its versatility and applicability. Many of the commercial 3D printers were specifically designed to fabricate chocolates (Lanaro et al. 2019). Due to which this class of materials is getting more attention in the fabrication of customized foods with tailored nutrition using 3D printing that would govern

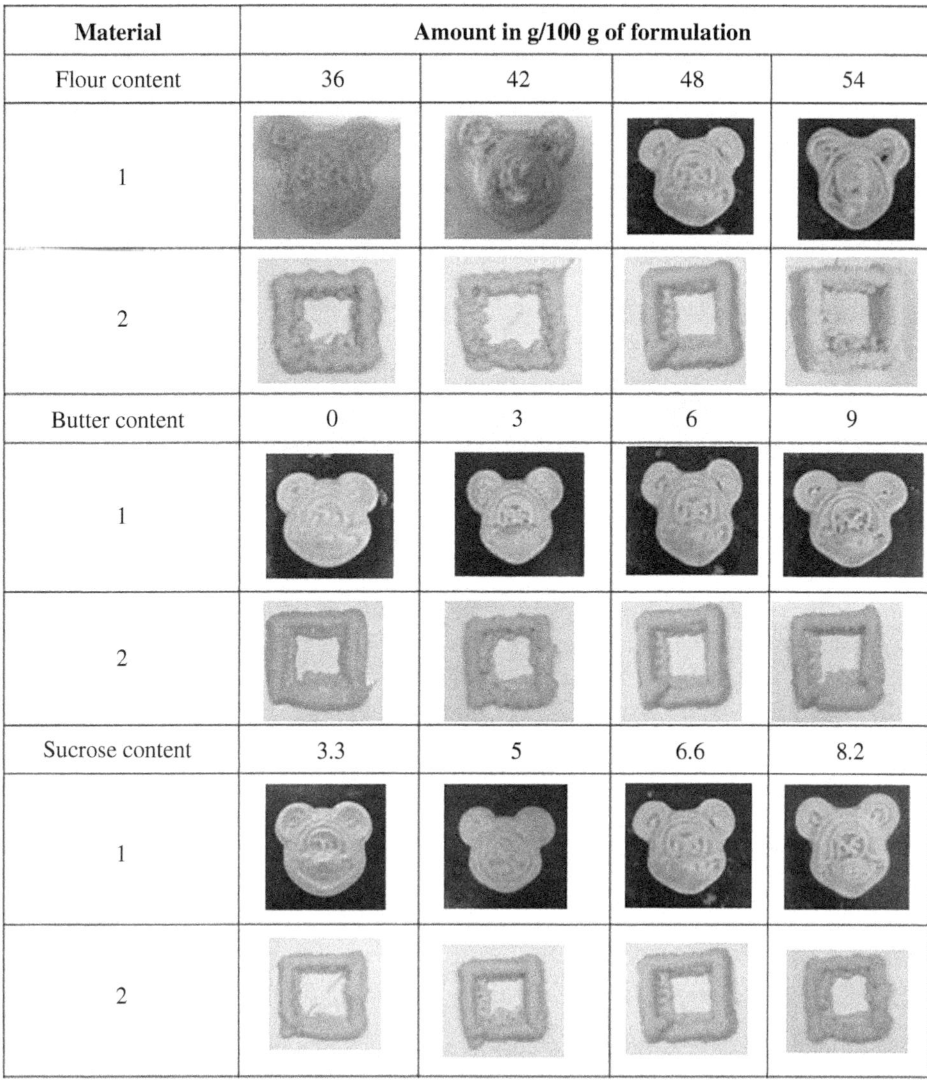

Figure 6.3 Effect of material formulations on 3D printing of baking dough. *Source:* From Yang et al. (2018), Figure 05 & 07 & 08 [p. 26 & 28 & 29] / With permission of Elsevier. DOI-https://doi.org/10.1016/j.ifset.2018.01.001.

the confectionery industry in near future. The success of post-deposition of 3D printed chocolates depends on the precise control over temperature; particle size optimization; the effect of emulsifiers to enhance the flowability and to maintain consistency. A commercial 3D chocolate printer PORIMY prints chocolate with the incorporation of magnesium stearate up to 5% (w/w) as a lubricant to enhance the material's flowability and controls the stickiness (Mantihal et al. 2019a). 3D printing allows for the construction of complex structures whose fabrication is not possible using conventional moulds. With this advantage, more research works were reported on the 3D printing of chocolates focussing on improving the stability of 3D constructs with varied support structures. Mantihal et al. (2019b) analyzed the rheological and thermal profile of chocolate inks with magnesium stearate as an additive. Fabrication of hexagonal constructs with the cross-support was more stable than that of the parallel-support in terms of snap quality. Recently, a new approach on the printing of chocolate was proposed by Karyappa and Hashimoto (2019) where the chocolate was directly printed in its liquid phase at room temperature using direct ink writing (DIW) 3D printer, a complementary method for hot-melt extrusion printing. In this cold extrusion method, chocolate ink was prepared from readily available products such as chocolate syrups and pastes mixed with the varied proportions of cocoa powder to appropriate consistency for printing. Thus, this study was proved to be a simple alternate method for printing chocolates with modified textures.

Another printing technology commonly used for printing chocolates in dry powder forms is sintering and binding techniques (Fuh et al. 2015). The former involves the application of heat source either hot air or laser in a layered fashion forming 3D constructs whereas the latter applies a liquid binder for imparting required adhesion of layers to bind together. The effectiveness of these methods was greatly influenced by the distance between the print head and the print bed. However, additional post-processing of removal of the excess powders from the constructed 3D printed object is adequate. The sintering process involves the agglomeration of powder materials such as sugar crystals and cocoa powder through agglomeration due to the short range of physical and chemical forces resulting in the phase change and structural modification yielding a 3D object (Lanaro et al. 2019). Oskay and Edman (2009) reported the first 3D printing on sugar constructs using sintering technology. Another study was reported on printing the commercial chocolate product Nesquik using the selective laser sintering technique (Sol et al. 2015). In general, sintering occurs below the material's melting point that eventually causes the agglomeration due to reduction of total surface energy resulting in a porous 3D printed matrix whereas liquid binding occurs due to the molecular enlargement during wetting of granules that makes the particles to conglomerate thereby forming a 3D object (Nachal et al. 2019).

Inkjet printing is another 3D printing technique where the material supply in the droplet form at the micro-level is allowed to deposit on the printing platform in a layer-by-layer manner. This technique is an age-old one used for decoration of the foods in 2D forms; however, for 3D printing, the same principle was followed that allows for higher printing speed and greater printing precision than extrusion-based 3D printing. Based on the method of initialization of liquid droplets, inkjet printing can be of the thermal and piezo static methods. Takagishi et al. (2018) reported a study on inkjet printing of chocolates based on the electrostatic principle using Hershey's shell topping. FoodJet, a commercial 3D printer is mainly designed for toppings and decorations on foods using chocolates. In contrast to the powder-based techniques, inkjet printing could able to produce a much similar flavour and taste as that of the conventionally processed foods due to the common nature of the feed supply. Sugars and sugar-based products are the all-time favourites for people of all age groups. By the implementation of 3D printing technology, more complex products could be positively fabricated using a variety of the printing techniques that imparts different texture, flavour, and tastes.

6.4.3 Gel-Based Food Systems

Food gels are the polymeric 3D network formed by polysaccharides and proteins possessing several functionalities such as modulation of the textural characteristics and sensorial properties, maintenance of the metastable food structures, increasing the shelf life, tailoring physiochemical properties through the enhancement of satiety, replacement of fat and improves the bioavailability (Cao and Mezzenga 2020). In recent years, food gels have become the topic of interest due to their versatile applications in food systems. Hence, more researches are being focused on the fabrication of gel systems from a natural source that eliminates the effects of the synthetic counterparts. Based on the nature of the medium like the incorporation of air, water, and oil, gel systems are referred to as aerogels, hydrogels, and oleogels, respectively (Rebaka et al. 2020). With the advent of 3D food printing, the soft matrix of the gel systems can be easily structured and designed to fabricate complex 3D structures with tailored biological properties. The changing consumer attitude and increasing awareness about the personal health urged the food industries to deliver foods focussing on special demands of varied range of consumers like children, adolescents, adults, athletes, pregnant women, and elderly people (Zhang et al. 2020). Thus, 3D food printing aids in the delivery of customized foods with precise nutrition in a diversified form suitable for all age groups based on each of the individual requirements. One such application of 3D printing is the design and fabrication of a gel system with modified textures for elderly people suffering from dysphagia (Strother et al. 2020). The feasibility of food hydrocolloids in the 3D printing of pork as dysphagia food was studied recently for people with swallowing disorders (Dick et al. 2020). Many other foods like potato puree (Martínez-Monzó et al. 2019), fish surimi gel (Chen et al. 2020), brown rice (Huang et al. 2019), lemon juice gel (Yang et al. 2019b), mashed potato/ strawberry juice gel (Liu et al. 2018b), whey protein isolate (Liu et al. 2019b), soy protein gel (Phuhongsung et al. 2020), and milk protein composite gels (Liu et al. 2019a) were also employed for 3D printing with the addition of hydrocolloids for gel formation and to enhance the printability.

The effect of different hydrocolloids such as xanthan gum, guar gum, agar, sodium carboxymethyl cellulose, and a mixture of xanthan gum and guar gum on the printability of brown rice gel was reported (Huang et al. 2020). Results confirmed that the addition of hydrocolloids had a significant implication on the rheological properties that enable the printability of brown rice gel. Thus, the present study demonstrates the applicability of brown rice for the preparation of ready-to-eat 3D printed food thereby opening a new dimension of futuristic foods from staple food crops. This translates the significance of hydrocolloids in the food system. In this context, applications of hydrocolloids are being explored revealing an array of functional properties in food systems. Kim et al. (2018) had used hydrocolloids as reference material for the assessment of printability. This study establishes a classification system for printability based on the dimensional stability and degree of handling of methylcellulose. Results of this assessment help to compare the rheological properties of different food systems such as ketchup, yoghurt, cream cheese, cheddar cheese, mashed potato, chocolate jam, peanut butter, ground ham, marzipan, sugar paste, bean paste, and cookie dough, and its end applications based on its printability. Similarly, the compatibility of peptides for 3D printing had been investigated for the formulation of protein-based model systems from dairy ingredients (Liu et al. 2018c, 2019a).

Hydrogels are characterized as an intermediate substance with an aqueous phase between solid and liquid possessing elastic and viscous behaviour (Banerjee and Bhattacharya 2012). Based on the physical structure of the biopolymeric network hydrogels are categorized as weak, pseudo, and strong gels (Nazir et al. 2017). The printing technique employed for the 3D printing of hydrogels under a supporting bath is termed hydrogel-forming extrusion (Sun et al. 2018a). The technique of hydrogel-forming extrusion is described in the earlier chapter on 3D printing approaches. A simple schematic representation of methods of fabrication of hydrogel is presented in Figure 6.4. In this regard, several studies were reported on the 3D printing of hydrocolloids that imply the potential applications of its incorporation in food systems.

The printability of gels based on the sugar syrup, xanthan, and konjac gum have been studied by Garcia-Segovia et al. (2020). This study draws out the interrelation among the printing variables concerning the printing of gel systems thereby providing insights for the formulation of personalized foods. In a recent study, pre-treatments like ozonation and dry heat had been employed for modifying the starch-producing hydrogels from cassava. Results showed that pre-treatments had a significant effect in improving the pasting properties of starch hydrogels resulting in improved printing performance. The effect of the pre-treatments of ozonation (Figure 6.5) and dry heat treatment (Figure 6.6) on starch modification of cassava starch and its applicability for 3D printing of cassava hydrogels were reported (Maniglia et al. 2019, 2020).

In another study, the effect of microwave pretreatment of wheat flour for starch modification was reported to increase the printing accuracy with increased power levels (Figure 6.7) (Sun et al. 2020). Similarly, the suitability of casein whey protein suspensions for the development of individualized foods using 3D printing was illustrated (Daffner et al. 2020). Results showed that protein aggregation is one of the influential factors that decides the physical stability of casein-whey protein gels which in turn helps in assessing its printability.

Liu et al. (2019d) developed a gel model system composed of carrageenan, xanthan, and starch. The formulation of the developed gel system was studied for its mechanical strength. Results showed that the thermo-responsive behaviour of the carrageenan resulted in a firm brittle gel structure that aids in printability. The combination of xanthan gum and starch increased the gelation temperature, yield stress, and viscosity with a characteristic shear thinning behaviour that aids in the flowability. Thus, the higher storage modulus of the multicomponent gel system helps in the shape retention of the printed 3D structures (Figure 6.8). All these studies reveal the functional properties of the hydrocolloids that can be added as additives for improving printing behaviour in food formulations. Thus, hydrocolloids are promissory components for 3D food printing.

Another interesting class of gel systems is oleogels which receives attention in recent days. When the liquid component of the hydrogels is replaced with organic solvents like oil, the gel system is termed as oleogels (Co and Marangoni 2018). They have an extensive range of applications in food industries as edible gels and in pharmaceutical industries as drug delivery systems. Nowadays, more works are focused on the applications of 3D printing for the delivery of bioactive components through biocompatible oleogels derived from vegetable oil. These are a semisolid system where the organic liquid phase is immobilized in a 3D network comprised of a self-assembled interlinked network of polymers that exhibit viscoelasticity, a characteristic tendency of material that is conducive for 3D printing. Two

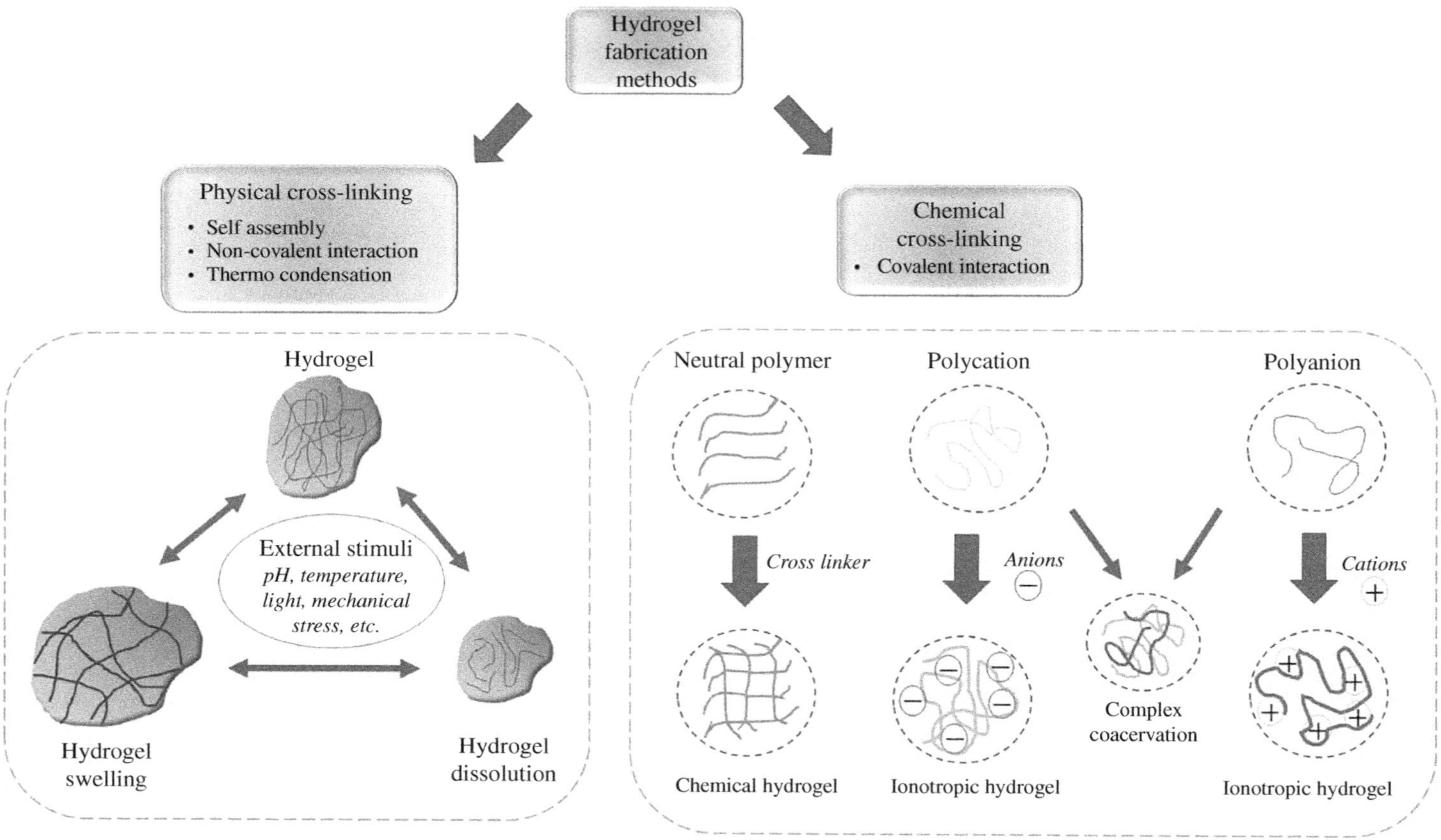

Figure 6.4 Methods employed for fabrication of hydrogels.

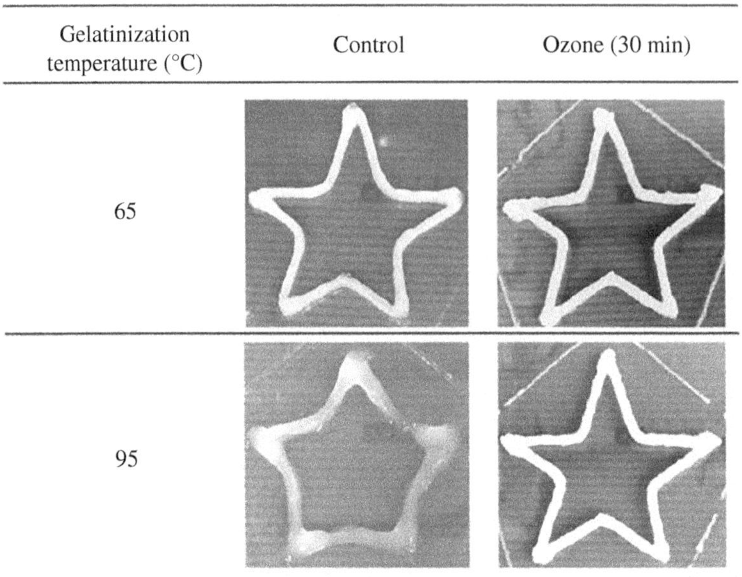

Figure 6.5 Effect of ozonation pre-treatment on 3D printability cassava starch hydrogel. *Source:* From Maniglia et al. (2019), Figure 10 [p. 1096] / With permission of Elsevier. DOI-https://doi.org/10.1016/j.ijbiomac.2019.07.124.

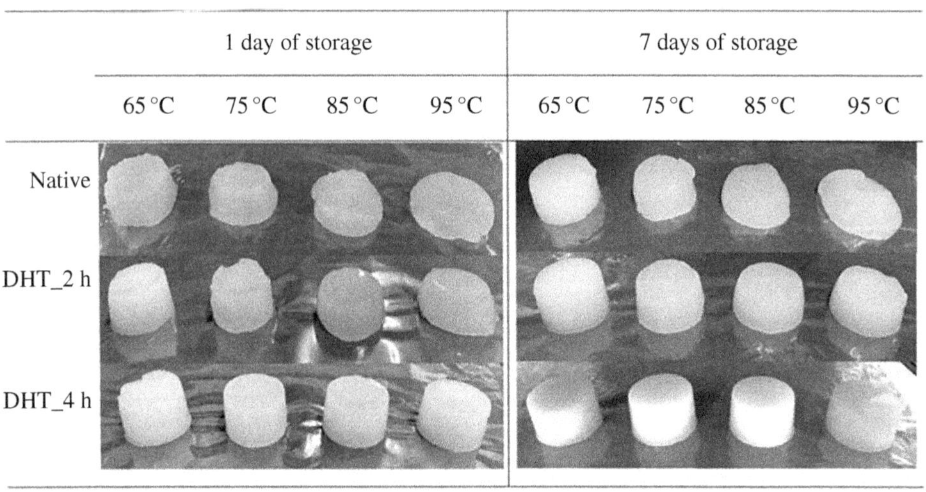

Figure 6.6 Images of 3D printed starch-modified hydrogel of different periods of dry heat treatment (DHT). *Source:* From Maniglia et al. (2020), Figure 07 [p. 33] / With permission of Elsevier. DOI-https://doi.org/10.1016/j.foodres.2019.108803.

classes of oleogelators are low-molecular-weight oleogelators and polymeric oleogelators (Rebaka et al. 2020). The former class of material includes alkanes, waxes, fatty acids, and fatty alcohols, phytosterols, ceramides, esters, lecithin, surfactants, triterpenoid, monoglyceride, and triacylglycerol derivatives. While the latter group of polymeric oleogelators

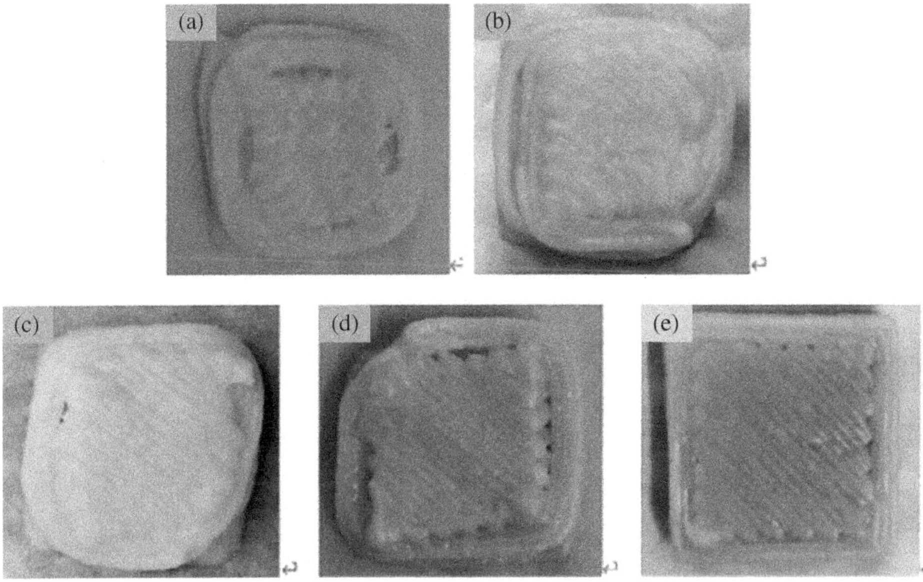

Figure 6.7 Effect of microwave power levels on the quality of 3D printed cookie dough. (a) Control, (b) 40 W, (c) 50 W, (d) 60 W, (e) 70 W. *Source:* From Sun et al. (2020), Figure 01 [p. 04] / With permission of Elsevier. DOI-https://doi.org/10.1016/j.lwt.2020.109752.

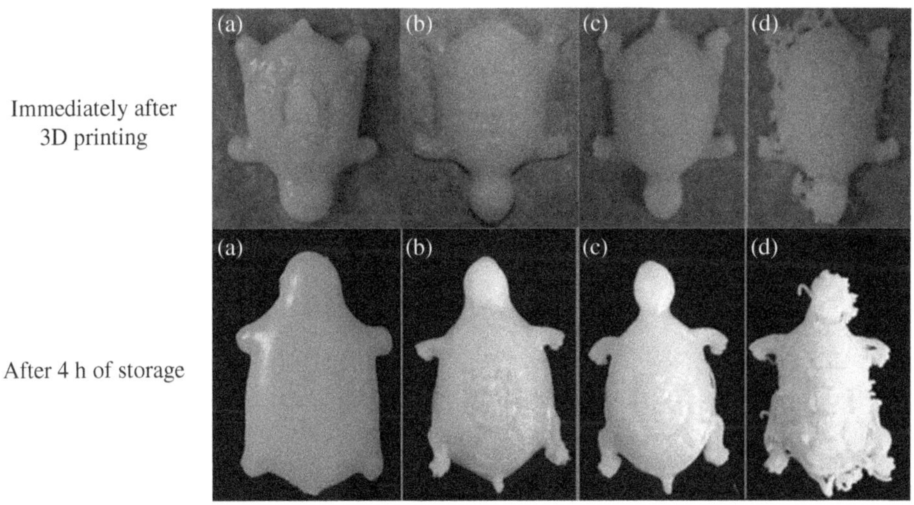

Figure 6.8 3D printed milk protein gels of different protein contents. (a) 350 g l^{-1}, (b) 400 g l^{-1} (c) 450 g l^{-1}, (d) 500 g l^{-1}. *Source:* From Liu et al. (2019a), [p. 31] / With permission of Elsevier. DOI-https://doi.org/10.1016/j.lwt.2018.12.053.

comprises cellulose derivatives, water-soluble polysaccharides, and proteins. Thus, oleogels are identified as an efficient drug delivery vehicle with a significant place in food and pharmaceutical industries for oil structuring. More recently, Cotabarren et al. (2019) explored the potential applicability of oleogels incorporated with liposoluble active ingredients for extrusion-based 3D printing. The study involved the assessment of printability of oleogel formulation from a mixture of monoglycerides and phytosterols. Results showed that the mixtures containing 30 and 40% (w/w) phytosterol/oleogel formulated with 10 and 20% (w/w) monoglycerides exhibit better printability corresponding to the lowest gelation temperature. Thus, gels are a group of polymers with several functionalities and flexibility suitable for 3D food printing. Upon realizing the potential benefits of gel systems more studies are required for the efficient delivery of bioactive ingredients through oral formulations using 3D printing technology (Zhang et al. 2020).

6.5 Insights and Scope for Commercialization

Natively printable materials are those food and food components that possess inherent characteristics of printability that can be printable without the addition of food additives. Due to this attribute, natively printable materials can be used as additives for enhancing the printability of non-natively printable traditional food ingredients (Dick et al. 2019; Anukiruthika et al. 2020). Materials under this class of printability have a great scope of applications in the future food industry. One of the main components under this category is starch. This fraction of component can be obtained from rice, cassava, and potatoes that forms the integral portion of our regular diet. Thus, the milling fractions of starch from food processing units can be efficiently utilized for the production of wholesome foods using 3D printing (Feng et al. 2020). This form of value addition improves the edibility of materials at a relatively low processing cost. Henceforth more studies on starch-based material supplies are being reported in the literature that finds applications in infant diet formula and the formulation of health mix. In this regard, there rises the research needs to explore the starch from unconventional food sources such as potato peels, millets, and whole wheat fractions possessing dietary fibres that adds upon the value to the final product. Also, the utilization of these waste fractions from conventional processing using 3D printing greatly reduces the food stream waste that paves a way for the sustainable production and eating practices.

Other than starch, proteins and polypeptides from dairy sources like cheese also comes under the natively printable category. Hot-melt extrusion printing of cheese helps to develop a model cheese with a modified texture that gives a different mouthfeel (Kern et al. 2018). Similarly, dairy ingredients like casein and whey protein concentrate, proteins from plant sources like soy protein concentrates were studied for their applicability for 3D printing (Chen et al. 2019a; Liu et al. 2018c, 2019b). These studies provide insights into the development of novel protein-based 3D printed foods that are wholesome, convenient, and nutritious. This form of proteins possesses greater value in health drinks for athletes, infant diet formula, and protein mix. More studies are required in examining the printability of protein-based food materials that has an integral part in personalized food. In complementary to the conventional food processing units, implementation of 3D printing technology in processing line forecast to yield multiple benefits such as less usage of raw materials, value addition of food, efficient utilization of by-products, less consumption of available resources, automation, and precise control of processes.

6.6 Concluding Remarks

The present chapter provides an in-depth discussion on the printability of natively printable materials and their scope in the future food industry. Studies on the natively printable materials can draw a significant result in the transformation of our regular diet into digitalized recipes using 3D printing. Further, the utilization of staple food crops such as cereal, root tubers in 3D printing helps in the formulation of novel foods with enhanced nutrition and customized delivery of bioactive ingredients. In the long-term run, 3D printing is forecasted to rejuvenate the food manufacturing practices through the integration of multiple steps by a single unit thereby simplifies the food supply chain. More research studies on the printability of indigenous food materials native to each geographic location have to be undertaken in order to accomplish the full benefits of 3D printing in the food sector. Through the scientific rationale of printability of natively printable materials, the complexity associated with the printability of non-printable materials can be overcome. As a standalone food technology, 3D printing can reform the food industry through the design, fabrication, innovation, and customization of foods using natively printable materials as a base ingredient in the food formulations used for 3D printing.

References

Álvarez-Castillo, E., Oliveira, S., Bengoechea, C. et al. (2020). A rheological approach to 3D printing of plasma protein based doughs. *Journal of Food Engineering* **288**: 110255.

Anukiruthika, T., Moses, J.A., and Anandharamakrishnan, C. (2020). 3D printing of egg yolk and white with rice flour blends. *Journal of Food Engineering* **265**: 109691. https://doi.org/10.1016/j.jfoodeng.2019.109691.

Banerjee, S. and Bhattacharya, S. (2012). Food gels: gelling process and new applications. *Critical Reviews in Food Science and Nutrition* **52** (4): 334–346.

Bocchetta, P. (2020). Ionotropic gelation of chitosan for next-generation composite proton conducting flat structures. *Molecules* **25** (7): 1632.

Cao, Y. and Mezzenga, R. (2020). Design principles of food gels. *Nature Food* **1** (2): 106–118.

Cao, L., Lu, W., Mata, A. et al. (2020). Egg-box model-based gelation of alginate and pectin: a review. *Carbohydrate Polymers* **1**: 116389.

Chen, J., Mu, T., Goffin, D. et al. (2019a). Application of soy protein isolate and hydrocolloids based mixtures as promising food material in 3D food printing. *Journal of Food Engineering* https://doi.org/10.1016/j.jfoodeng.2019.03.016.

Chen, H., Xie, F., Chen, L., and Zheng, B. (2019b). Effect of rheological properties of potato, rice and corn starches on their hot-extrusion 3D printing behaviors. *Journal of Food Engineering* **244**: 150–158. https://doi.org/10.1016/j.jfoodeng.2018.09.011.

Chen, H.-z., Zhang, M., and Yang, C. (2020). Comparative analysis of 3D printability and rheological properties of surimi gels via LF-NMR and dielectric characteristics. *Journal of Food Engineering* **292**: 110278.

Co, E.D. and Marangoni, A.G. (2018). Oleogels: an introduction. In: *Edible Oleogels* (eds. A.G. Marangoni and N. Garti), pp. 1–29. Elsevier.

Cohen, D.L., Lipton, J.I., Cutler, M. et al. (2009). Hydrocolloid printing: a novel platform for customized food production. Solid Freeform Fabrication Symposium, pp. 807–818.

Cotabarren, I.M., Cruces, S., and Palla, C.A. (2019). Extrusion 3D printing of nutraceutical oral dosage forms formulated with monoglycerides oleogels and phytosterols mixtures. *Food Research International* **126**: 108676.

Daffner, K., Vadodaria, S., Ong, L. et al. (2020). Design and characterization of casein-whey protein suspensions via the pH-temperature-route for application in extrusion-based 3D-Printing. *Food Hydrocolloids* **112**: 105850.

Dankar, I., Haddarah, A., Sepulcre, F., and Pujolà, M. (2020). Assessing mechanical and rheological properties of potato puree: effect of different ingredient combinations and cooking methods on the feasibility of 3D printing. *Foods* **9** (1): 21.

Dick, A., Bhandari, B., and Prakash, S. (2019). Post-processing feasibility of composite-layer 3D printed beef. *Meat Science* **153**: 9–18. https://doi.org/10.1016/j.meatsci.2019.02.024.

Dick, A., Bhandari, B., Dong, X., and Prakash, S. (2020). Feasibility study of hydrocolloid incorporated 3D printed pork as dysphagia food. *Food Hydrocolloids* **107**: 105940.

Fahmy, A.R., Becker, T., and Jekle, M. (2020). 3D printing and additive manufacturing of cereal-based materials: quality analysis of starch-based systems using a camera-based morphological approach. *Innovative Food Science & Emerging Technologies* **63**: 102384.

Feng, C., Zhang, M., Bhandari, B., and Ye, Y. (2020). Use of potato processing by-product: effects on the 3D printing characteristics of the yam and the texture of air-fried yam snacks. *LWT* **125**: 109265.

Fuh, J.Y.H., Hong, G.S., Zhou, W. et al. (2015). An overview of 3D printing technologies for food fabrication. *Food and Bioprocess Technology* **8** (8): 1605–1615. https://doi.org/10.1007/s11947-015-1528-6.

Garcia-Segovia, P., Garcia-Alcaraz, V., Balasch-Parisi, S., and Martinez-Monzó, J. (2020). 3D printing of gels based on xanthan/konjac gums. *Innovative Food Science & Emerging Technologies* **64**: 102343.

Gholamipour-Shirazi, A., Norton, I.T., and Mills, T. (2019). Designing hydrocolloid based food-ink formulations for extrusion 3D printing. *Food Hydrocolloids* **95**: 161–167. https://doi.org/10.1016/j.foodhyd.2019.04.011.

Gouton, M.-A., Dacremont, C., Trystram, G., and Blumenthal, D. (2020). Validation of food visual attribute perception in virtual reality. *Food Quality and Preference* **87**: 104016.

Gulrez, S.K.H., Al-Assaf, S., and Phillips, G.O. (2011). Hydrogels: methods of preparation, characterisation and applications. In: *Progress in Molecular and Environmental Bioengineering-from Analysis and Modeling to Technology Applications* (ed. A. Carpi), pp. 117–150. Croatia: InTech.

Guo, C., Zhang, M., and Devahastin, S. (2020). 3D extrusion-based printability evaluation of selected cereal grains by computational fluid dynamic simulation. *Journal of Food Engineering* **286**: 110113.

Hao, L., Mellor, S., Seaman, O. et al. (2010). Material characterisation and process development for chocolate additive layer manufacturing. *Virtual and Physical Prototyping* **5** (2): 57–64.

Hu, W.-X., Chen, J., Zhao, J.-W. et al. (2020). Effect of the addition of modified starch on gelatinization and gelation properties of rice flour. *International Journal of Biological Macromolecules* **153**: 26–35.

Huang, V.T. and Perdon, A.A. (2020). Major changes in cereal biopolymers during ready-to-eat cereal processing. In: *Breakfast Cereals and How They Are Made* (eds. A.A. Perdon, S.L. Schonauer and K.S. Poutanen), pp. 109–140. Elsevier.

Huang, M., Zhang, M., and Bhandari, B. (2019). Assessing the 3D printing precision and texture properties of brown rice induced by infill levels and printing variables. *Food and Bioprocess Technology* **12**: 1–12. https://doi.org/10.1007/s11947-019-02287-x.

Huang, M., Zhang, M., and Guo, C. (2020). 3D printability of brown rice gel modified by some food hydrocolloids. *Journal of Food Processing and Preservation* **44**: e14502.

Kailash, M., Kavitha, L., Madan, S.V., and Ravichandran, C. (2020). 3D food printing: a technology for fabricating customized artistic food. In: *Emerging Technologies in Food Science* (eds. M. Thakur and V.K. Modi), pp. 273–287. Springer.

Karyappa, R. and Hashimoto, M. (2019). Chocolate-based ink three-dimensional Printing (Ci3DP). *Scientific Reports* **9** (1): 1–11.

Kern, C., Weiss, J., and Hinrichs, J. (2018). Additive layer manufacturing of semi-hard model cheese: effect of calcium levels on thermo-rheological properties and shear behavior. *Journal of Food Engineering* **235**: 89–97. https://doi.org/10.1016/j.jfoodeng.2018.04.029.

Kim, H.W., Bae, H., and Park, H.J. (2018). Classification of the printability of selected food for 3D printing: development of an assessment method using hydrocolloids as reference material. *Journal of Food Engineering* **215**: 23–32.

Krishnaraj, P., Anukiruthika, T., Choudhary, P. et al. (2019). 3D extrusion printing and post-processing of fibre-rich snack from indigenous composite flour. *Food and Bioprocess Technology* **12** (10): 1776–1786. https://doi.org/10.1007/s11947-019-02336-5.

Lanaro, M., Forrestal, D.P., Scheurer, S. et al. (2017). 3D printing complex chocolate objects: platform design, optimization and evaluation. *Journal of Food Engineering* **215**: 13–22. https://doi.org/10.1016/j.jfoodeng.2017.06.029.

Lanaro, M., Desselle, M.R., and Woodruff, M.A. (2019). 3D printing chocolate: Properties of formulations for extrusion, sintering, binding and ink jetting. In: *Fundamentals of 3D Food Printing and Applications* (eds. F.C. Godoi, B.R. Bhandari, S. Prakash and M. Zhang), pp. 151–173. Elsevier.

Le Tohic, C., O'Sullivan, J.J., Drapala, K.P. et al. (2018). Effect of 3D printing on the structure and textural properties of processed cheese. *Journal of Food Engineering* **220**: 56–64.

Lee, K.Y. and Mooney, D.J. (2012). Alginate: properties and biomedical applications. *Progress in Polymer Science* **37** (1): 106–126.

Lipton, J., Arnold, D., Nigl, F. et al. (2010). Multi-material food printing with complex internal structure suitable for conventional post-processing. Solid Freeform Fabrication Symposium, pp. 809–815.

Liu, Z., Zhang, M., Bhandari, B., and Wang, Y. (2017). 3D printing: printing precision and application in food sector. *Trends in Food Science & Technology* **69**: 83–94.

Liu, Z., Zhang, M., Bhandari, B., and Yang, C. (2018a). Impact of rheological properties of mashed potatoes on 3D printing. *Journal of Food Engineering* **220**: 76–82. https://doi.org/10.1016/j.jfoodeng.2017.04.017.

Liu, Z., Zhang, M., and Yang, C. (2018b). Dual extrusion 3D printing of mashed potatoes/strawberry juice gel. *LWT* **96**: 589–596.

Liu, Y., Liu, D., Wei, G. et al. (2018c). 3D printed milk protein food simulant: improving the printing performance of milk protein concentration by incorporating whey protein isolate.

Innovative Food Science and Emerging Technologies **49**: 116–126. https://doi.org/10.1016/j.ifset.2018.07.018.
Liu, Y., Yu, Y., Liu, C. et al. (2019a). Rheological and mechanical behavior of milk protein composite gel for extrusion-based 3D food printing. *LWT* **102**: 338–346. https://doi.org/10.1016/j.lwt.2018.12.053.
Liu, Y., Zhang, W., Wang, K. et al. (2019b). Fabrication of gel-like emulsions with whey protein isolate using microfluidization: rheological properties and 3D printing performance. *Food and Bioprocess Technology* **12** (12): 1967–1979.
Liu, Y., Liang, X., Saeed, A. et al. (2019c). Properties of 3D printed dough and optimization of printing parameters. *Innovative Food Science & Emerging Technologies* https://doi.org/10.1016/J.IFSET.2019.03.008.
Liu, Z., Bhandari, B., Prakash, S. et al. (2019d). Linking rheology and printability of a multicomponent gel system of carrageenan-xanthan-starch in extrusion based additive manufacturing. *Food Hydrocolloids* **87**: 413–424. https://doi.org/10.1016/j.foodhyd.2018.08.026.
Liu, Y., Tang, T., Duan, S. et al. (2020). Applicability of rice doughs as promising food materials in extrusion-based 3D printing. *Food and Bioprocess Technology* **13** (3): 548–563.
Maniglia, B.C., Lima, D.C., Junior, M.D.M. et al. (2019). Hydrogels based on ozonated cassava starch: effect of ozone processing and gelatinization conditions on enhancing 3D-printing applications. *International Journal of Biological Macromolecules* **138**: 1087–1097.
Maniglia, B.C., Lima, D.C., Junior, M.D.M. et al. (2020). Preparation of cassava starch hydrogels for application in 3D printing using dry heating treatment (DHT): a prospective study on the effects of DHT and gelatinization conditions. *Food Research International* **128**: 108803.
Mantihal, S., Prakash, S., Godoi, F.C., and Bhandari, B. (2017). Optimization of chocolate 3D printing by correlating thermal and flow properties with 3D structure modeling. *Innovative Food Science and Emerging Technologies* **44**: 21–29. https://doi.org/10.1016/j.ifset.2017.09.012.
Mantihal, S., Prakash, S., and Bhandari, B. (2019a). Textural modification of 3D printed dark chocolate by varying internal infill structure. *Food Research International* **121**: 648–657.
Mantihal, S., Prakash, S., Godoi, F.C., and Bhandari, B. (2019b). Effect of additives on thermal, rheological and tribological properties of 3D printed dark chocolate. *Food Research International* **119**: 161–169.
Martínez-Monzó, J., Cárdenas, J., and García-Segovia, P. (2019). Effect of temperature on 3D printing of commercial potato puree. *Food Biophysics*: 1–10. https://doi.org/10.1007/s11483-019-09576-0.
Nachal, N., Moses, J.A., Karthik, P., and Anandharamakrishnan, C. (2019). Applications of 3D printing in food processing. *Food Engineering Reviews* **11** (3): 123–141. https://doi.org/10.1007/s12393-019-09199-8.
Nazir, A., Asghar, A., and Maan, A.A. (2017). Food gels: gelling process and new applications. In: *Advances in Food Rheology and Its Applications* (ed. J. Ahmed), 335–353. Elsevier.
Oskay, W. and Edman, L. (2009). *The CandyFab Project.* Evil Scientist Laboratories.
Oyinloye, T.M. and Yoon, W.B. (2020). Stability of 3D printing using a mixture of pea protein and alginate: precision and application of additive layer manufacturing simulation approach for stress distribution. *Journal of Food Engineering* **288**: 110127.
Pallottino, F., Hakola, L., Costa, C. et al. (2016). Printing on food or food printing: a review. *Food and Bioprocess Technology* **9** (5): 725–733.

Park, S.M., Kim, H.W., and Park, H.J. (2020). Callus-based 3D printing for food exemplified with carrot tissues and its potential for innovative food production. *Journal of Food Engineering* **271**: 109781.

Pérez, B., Nykvist, H., Brøgger, A.F. et al. (2019). Impact of macronutrients printability and 3D-printer parameters on 3D-food printing: A review. *Food Chemistry* **287** (October 2018): 249–257. https://doi.org/10.1016/j.foodchem.2019.02.090.

Phuhongsung, P., Zhang, M., and Devahastin, S. (2020). Investigation on 3D printing ability of soybean protein isolate gels and correlations with their rheological and textural properties via LF-NMR spectroscopic characteristics. *LWT* **122**: 109019.

Piyush, C., Kumar, R., and Kumar, R. (2020). 3D printing of food materials: a state of art review and future applications. *Materials Today: Proceedings* **33**: 1463–1467.

Pulatsu, E., Su, J.-W., Lin, J., and Lin, M. (2020). Factors affecting 3D printing and post-processing capacity of cookie dough. *Innovative Food Science & Emerging Technologies* **61**: 102316.

Rebaka, V.P., Rachamalla, A.K., Batra, S., and Subbiah, N. (2020). State of the art and new perspectives in oleogels and applications. In: *Sustainable Green Chemical Processes and their Allied Applications*, pp. 151–182. Springer.

Serizawa, R., Shitara, M., Gong, J. et al. (2014). 3D jet printer of edible gels for food creation. *Behavior and Mechanics of Multifunctional Materials and Composites* **2014** (9058): 90580A.

Severini, C., Derossi, A., and Azzollini, D. (2016). Variables affecting the printability of foods: preliminary tests on cereal-based products. *Innovative Food Science and Emerging Technologies* **38**: 281–291. https://doi.org/10.1016/j.ifset.2016.10.001.

Sol, I.E.J., der Linden, D., & Van Bommel, K.J.C. (2015). 3D food printing: the barilla collaboration. Feb-2015.

Southerland, D., Walters, P., and Huson, D. (2011). Edible 3D printing. *NIP & Digital Fabrication Conference* **2011** (2): 819–822.

Strother, H., Moss, R., and McSweeney, M.B. (2020). Comparison of 3D printed and molded carrots produced with gelatin, guar gum and xanthan gum. *Journal of Texture Studies* **51**: 852–860.

Sun, J., Peng, Z., Yan, L. et al. (2015). 3D food printing—an innovative way of mass customization in food fabrication. *International Journal of Bioprinting* **1** (1): 27–38.

Sun, J., Zhou, W., Huang, D., and Yan, L. (2018a). 3D food printing: perspectives. In: *Polymers for Food Applications* (ed. T.J. Gutierrez), pp. 725–755. Springer.

Sun, J., Zhou, W., Yan, L. et al. (2018b). Extrusion-based food printing for digitalized food design and nutrition control. *Journal of Food Engineering* **220**: 1–11. https://doi.org/10.1016/j.jfoodeng.2017.02.028.

Sun, Y., Zhang, M., and Chen, H. (2020). LF-NMR intelligent evaluation of rheology and printability for 3D printing of cookie dough pretreated by microwave. *LWT* **132**: 109752.

Takagishi, K., Suzuki, Y., and Umezu, S. (2018). The high precision drawing method of chocolate utilizing electrostatic ink-jet printer. *Journal of Food Engineering* **216**: 138–143.

Theagarajan, R., Moses, J.A., and Anandharamakrishnan, C. (2020). 3D extrusion printability of rice starch and optimization of process variables. *Food and Bioprocess Technology* https://doi.org/10.1007/s11947-020-02453-6.

Vancauwenberghe, V., Delele, M.A., Vanbiervliet, J. et al. (2018). Model-based design and validation of food texture of 3D printed pectin-based food simulants. *Journal of Food Engineering* **231**: 72–82. https://doi.org/10.1016/j.jfoodeng.2018.03.010.

Voon, S.L., An, J., Wong, G. et al. (2019). 3D food printing: a categorised review of inks and their development. *Virtual and Physical Prototyping* **14** (3): 203–218.

Wang, S., Chao, C., Xiang, F. et al. (2018). New insights into gelatinization mechanisms of cereal endosperm starches. *Scientific Reports* **8** (1): 1–8.

Yang, F., Zhang, M., Prakash, S., and Liu, Y. (2018). Physical properties of 3D printed baking dough as affected by different compositions. In: *Innovative Food Science and Emerging Technologies*, vol. **49**, Issue 2017. Elsevier Ltd. https://doi.org/10.1016/j.ifset.2018.01.001.

Yang, F., Zhang, M., Fang, Z., and Liu, Y. (2019a). Impact of processing parameters and post-treatment on the shape accuracy of 3D-printed baking dough. *International Journal of Food Science and Technology* **54** (1): 68–74. https://doi.org/10.1111/ijfs.13904.

Yang, F., Guo, C., Zhang, M. et al. (2019b). Improving 3D printing process of lemon juice gel based on fluid flow numerical simulation. *LWT* **102**: 89–99.

Zhang, H., Zhang, F., and Yuan, R. (2020). Applications of natural polymer-based hydrogels in the food industry. In: *Hydrogels Based on Natural Polymers* (ed. Y. Chen), 357–410. Elsevier.

Zheng, L., Yu, Y., Tong, Z. et al. (2019). The characteristics of starch gels molded by 3D printing. *Journal of Food Processing and Preservation* **43** (7): e13993.

Zhou, Q., Wang, M., Li, H. et al. (2020). Application of Maillard reaction product of xylose--pea protein enzymatic hydrolysate in 3D printing. *Journal of the Science of Food and Agriculture* **100** (7): 2982–2990.

Zhuang, X., Jiang, X., Zhou, H. et al. (2020). Insight into the mechanism of physicochemical influence by three polysaccharides on myofibrillar protein gelation. *Carbohydrate Polymers* **229**: 115449.

7

Pre-Processing of Non-Printable Foods

CHAPTER MENU

7.1 Introduction

3D food printing is an emerging food processing technology that has an immense scope for the fabrication of novel customized and personalized foods. The technology of printing foods is not native to the food industry in turn it is an applied technology inherited from other allied industrial sectors such as bioengineering, automobiles, aerospace, civil engineering, fashion and design, textiles, military, and defence (Nida et al. 2021). Although the printing of foods is not as easy as it sounds, the major criteria for 3D food printing are printability, applicability, and post-printing/ post-processing characteristics of the material supplies (Nachal et al. 2019). The post-stability of the printed material depends on the biochemical compounds of food material, method of printing, printing conditions, and printing variables (Godoi et al. 2016). As discussed in the previous chapters, food is a complex material composed of various macro and microelements with several functionalities making the food to behave differently. Most of the traditional food cuisines and recipes as such are not applicable for 3D printing. This class of materials is referred to as traditionally non-printable food materials. However, these materials can be 3D printed with appropriate pre-processing that alters the material properties by imparting necessary phase changes and chemical transitions thereby enhancing the printability (Wegrzyn et al. 2012). Materials of this category include staple food ingredients of our regular diet such as whole grains, pulses, legumes, fruits and vegetables, meat, and seafoods (Sun et al. 2018b).

3D Printing of Foods, First Edition. C. Anandharamakrishnan, Jeyan A. Moses, and T. Anukiruthika.

The process of food printing follows a series of steps starting from the selection/ designing of 3D model, slicing, loading of feed for the initiation of 3D printing (Baiano 2020). A superficial view over the material characteristics depicts that most of the foods belonging to this category are highly perishable materials which possess a large amount of moisture. This clearly shows the necessity of appropriate pre-processing of the material that alters its physicochemical properties making them suitable for 3D printing (Voon et al. 2019). Various material attributes that must be considered for tailoring the desirable properties for 3D printing include viscosity, meltability, solidification, gelation, crystallinity, etc. In this regard, several studies were attempted in the printing of fibre-rich fruits and vegetables, pulses, and meat products. As an initial step, the preparation of raw material involves either drying of ingredients followed by hydration or grinding into a paste/ puree to which a considerable amount of food additives is added for achieving a desirable consistency and flowability (Dick et al. 2019a; Oliveira et al. 2020). Considering the printing technologies, extrusion-based 3D printing is one among the several food printing technologies that best suits the materials of this category. This is because these materials are not composed of pure single component/ fractions like sugar, starch, or proteins. It is a complex mixture of carbohydrates, proteins, and fat with a considerable portion of essential vitamins and minerals. Hence, 3D printing of natively non-printable materials is a multiplex process that requires careful pre-processing of raw ingredients for the successful deposition of layers (Derossi et al. 2019). Further, these types of materials must be subjected to suitable post-processing like cooking for its edibility. One of the attractive features of 3D printing specifically to this class of materials is the modification of sensorial attributes through precise control over nutrition. As a result, it enhances the delivery of nutritional foods in an attractive form to children, provision of texture modified foods for elderly people, and fabrication of healthy foods through cut-off of fat/ cholesterol in a regular diet (Prakash et al. 2019). All these applications will lead to a new dimension of novel 3D printed foods with greater flexibility and convenience. Although studies were reported on the feasibility of printing of traditionally non-printable materials, it is essential to have more studies in explaining the ingredient formulation, its interactions, and underlying biochemical transitions that enhance the printability. The present chapter deals with the state-of-the-art of printing non-printable food materials, scientific rationale involved in the pre-processing of material supplies, and its impact on printability followed by a summary of a diverse range of non-printable food materials that have been successfully 3D printed as reported in literatures. With great efforts, we had tried to put forth a guideline on the underlying biophysics of non-printable materials and its post-stability against the natively printable ingredients providing valuable insights for diversification of cuisines of 3D printed foods from non-printable materials.

7.2 Natively Non-Printable Materials

Foods such as cereals, pulses, legumes, fruits and vegetables, seafood, and meat products are largely consumed by people all over the world. With the technological advancements, 3D printing upon integration with digital gastronomy aids in the utilization of these materials for the fabrication of novel foods in attractive shapes, structures, tastes, and mouthfeel (Sun et al. 2018a). Nowadays, researches are being focused on the development

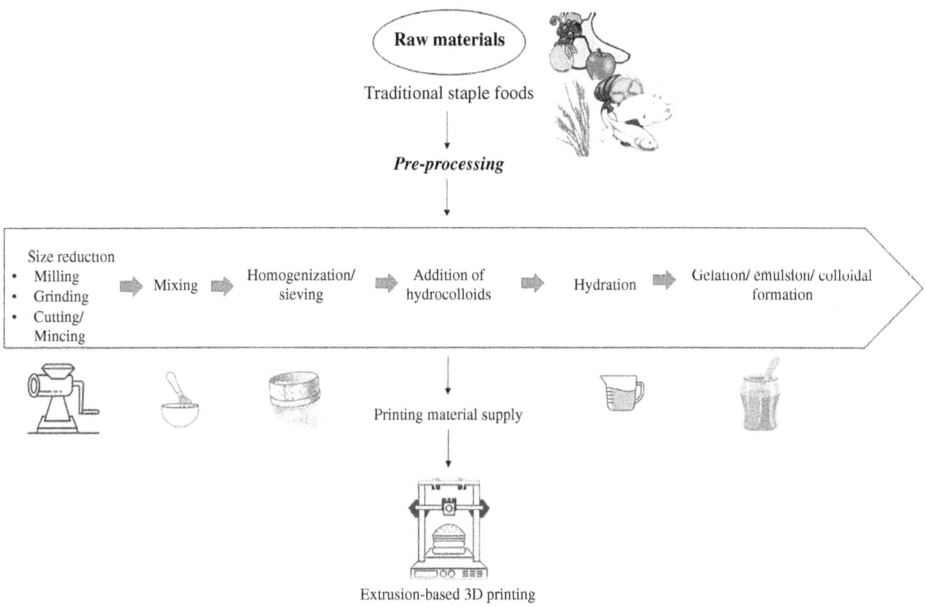

Figure 7.1 Schematic flow process of 3D printing of natively non-printable materials.

of traditional cuisines through 3D food printing. The present section deals with the nature and characteristic attributes of the natively non-printable materials. The schematic flow process of 3D printing of natively non-printable materials is shown in Figure 7.1. Further, the key facts associated with the formulation to impart printability have been discussed.

7.2.1 Traditional Foods: What Makes Them 'Non-Printable'?

Traditional food ingredients of our daily diet include vegetables, fruits, meat, and seafood. This class of materials is a rich source of fibres (vegetables and fruits), myofibrillar proteins (meat products), and moisture content that seems to be less desirable for 3D printing. Hence, they are brought under the category of non-printable food materials. This is because they contain relatively less functional ingredients such as carbs, proteins, and fat which are responsible for the gelation, hydration, lubrication, agglomeration, and solidification mechanisms that impart printability of material supplies (Anukiruthika et al. 2020). Nevertheless, these materials are a good source of antioxidants with ample amount of vitamins and minerals which are essential for revitalizing the body. It is a well-known fact that fibre fractions of foods are difficult to extrude since they are a portion of roughage that does not possess any key functionalities to enhance the flowability and printability (Lille et al. 2018). However, the portion of fibre fractions is printable in combination with other polymeric fractions such as starch and hydrocolloids. Even though the fibrous fractions are non-printable they possess several health benefits such as it helps to maintain bowel health, lowers cholesterol, controls blood sugar levels, reduces the risk of obesity, diabetes, and cardiovascular diseases (Yegin et al. 2020). Due to these functional attributes, people are becoming more health-conscious realizing the potential benefits and significance of consuming dietary fibres. 3D printing as a paradigm shift in food processing helps to provide personalized food that enables the addition of functional ingredients or elimination of undesirable component from the food formulation (Kailash et al. 2020). As a result, studies were attempted in 3D printing of fibres for imparting the feeling of satiety that helps to reduce food intake and cut off the calories. In a recent study, fibre enriched nutritious snack was 3D printed from indigenous composite millet flour (Krishnaraj et al. 2019). The successful printing of this material supply relies on the synergism of gelation nature of millets and characteristic protein hydration of pulses that supports and enhances the printability of the fibrous fractions without the addition of additives.

Because of the fibrous nature of vegetables and meat products, it is adequate to fine-tune the physicochemical properties to enhance its printability. To achieve the desired consistency and flowability for applicability to extrusion-based 3D printing, it is crucial to calibrate the portions of the solid content of the material formulation (Ricci et al. 2019). The total solids present in the material supplies is a significant criterion that influences the post-stability of the 3D constructs. With great efforts, studies are being put forward in analysing the printability of the plant and animal emulsions focussing on the production of texture-modified foods. With the incorporation of hydrocolloids, it is possible to produce plant and animal-based gel systems possessing the inherent ability to promote 3D printing. Likewise, studies were reported on 3D printing of lemon juice gel (Yang et al. 2019), mashed potato/strawberry juice gel (Liu et al. 2018b), fish surimi (Dong et al. 2019; Wang et al. 2018), vegemite and marmite (Hamilton et al. 2018), meat products such as beef (Dick et al. 2019b), pork (Dick et al. 2020), chicken meat (Wilson et al. 2020, 2021), scallop, and turkey (Lipton et al. 2010). All these studies reveal the feasibility of 3D printing in the production of wholesome traditional foods with altered textural, sensorial, and enhanced nutritional properties through the application of food hydrocolloids.

7.2.2 Role of Food Hydrocolloids in Improving Printability

The non-printable materials can be 3D printed through tailoring its rheological attributes such as viscosity, consistency, and flowability by the incorporation of hydrocolloids (Burey et al. 2008). As the name implies, hydrocolloids are hydrophilic polymers containing hydroxyl groups. This group of compounds comprises polysaccharides and proteins that are derived from either plants, animals, or microbial sources (Table 7.1). The role of food hydrocolloids in food assembly and structuring seems to be remarkable with a wide range of applications such as increased water binding capacity, act as a gelling and thickening agent, stabilization of emulsions, helps in adhesion, flocculation, and foam stabilization (Dickinson 2009; Williams and Phillips 2009). Hence food hydrocolloids are considered as a skeleton of food structures (Lu et al. 2020). Due to its enormous benefits hydrocolloids are becoming a part of nearly every processed food for imparting organoleptic properties. In 3D food printing, hydrocolloids are the key ingredients that enhance the printability of non-printable food materials. The present section explains the mechanism of interactions of the hydrocolloids with food components with the enhancement of the print stability. Based on the nature of interactions, Gao et al. (2017) outlined three categories of interactions of hydrocolloids with food components: hydrocolloid-hydrocolloid interaction, hydrocolloid-small

Table 7.1 Summary of the role of commercially important hydrocolloids in foods.

Source	Part from which derived	Hydrocolloid	Food category	Function
Botanical	Plants	Starch	Salad dressing, sauces, soups, jam, jellies, ice-cremes, cakes, candies, muscle foods, and culture dairy products	Thickening and gelling agent
		Pectin	Bakery fillings, frozen dairy desserts, jam, jellies, and marmalades	Stabilizer, thickening, and gelling agent
		Cellulose	Bakery products, whipped toppings, edible coatings, confectioneries, and reformed food products	Stabilizer, thickening agent, gelling agent, and whipping aid
	Tree gum exudates	Gum Arabic/ gum acacia	Soups, desserts, beverages, dressings, spreads, candies, puddings, and fillings	Stabilizer, binder, emulsifier, thickening and gelling agent
		Gum karaya	Desserts, sauces, egg-based liqueurs, nut coatings, and snacks	Texturizer, adhesive, binding agent, suspension agent, and thickening agent
		Gum tragacanth	Ice-creams, frozen desserts, candies, fondant, and cakes	Emulsifying and thickening agent
		Gum ghatti	Candies, chewing gum, pastilles, bakery products, and confectioneries	Emulsifier, stabilizer, thickener, coating, and gelling agent
	Seeds	Guar gum	Bakery products and fillings, culture dairy products, frozen dairy desserts, and salad dressing	Thickener, emulsifier, gelling, thickening, and preserving agent
		Locust bean gum	Culture dairy products, salad dressing, frozen dairy desserts, and bakery fillings	Stabilizer, gelling, and thickening agent
		Tara gum	Baked goods, dairy products, frozen desserts, jellies, processed meat products, sauces, and dressings	Thickener and stabilizer

(Continued)

Table 7.1 (Continued)

Source	Part from which derived	Hydrocolloid	Food category	Function
		Tamarind gum xyloglucan	Jam, jelly, marmalade, candy, chewing gum, beverages, and bakery products	Thickening, gelling, stabilizing, adhesive, and binding agent
		Soy protein	Infant food formulas, dairy products, beverages, and health mix	Texturizer, emulsifier, whipping agent, binding agent, gelling, and thickening agent
	Tubers	Konjac glucomannan	Bakery products and fillings	Protective and edible coatings, stabilizer, emulsifier, and gelling agent
Animal	Collagen	Gelatin	Candies, desserts, processed meat products, ice cream, and dairy products	Clarifying agent, protective coating, stabilizer, and thickening agent
	Chitin	Chitosan	Cakes and pastries, dairy products	Protective coatings, emulsifier, texturizer, humectant, surfactants, stabilizer, gelling, and thickening agents
	Connective tissues	Hyaluronan (hyaluronic acid)	Jelly, dairy products, beverages, and health drinks	Thickener and stabilizer
	Egg	Egg white protein	Sauces, custards, cakes, and confectioneries	Thickening, gelling and foaming agent, emulsifier, coagulant, binding agent, and texture enhancer
	Milk	Caseinate	Dairy products, jellies, candies, and confectioneries	Stabilizer, thickening, and gelling agents
		Whey protein	Bakery products, beverages, and soft drinks	Texture modifier, thickening agent, encapsulating agent, surface active agent, and gelling agent
Algal	Red sea weeds	Agar (agarose)	Jam, jelly, soft candies, fruit juices, ice-creams, icings, puddings, custard, and dairy products	Coagulator, thickening agent, emulsifier, stabilizer, and gelling agent
		Carrageenan	Milk products, beverages, dressings, meat and fish products, bakery, and confectioneries	Thickening, gelling, stabilizing, and suspending agent

	Brown seaweeds	Alginate	Restructured foods, meat products, syrups, salad dressing, muscle foods, bakery fillings, and frozen dairy desserts	Stabilizer, emulsifier, chelating agent, encapsulation agent, suspension agent, thickening, and gelling agent
Microbial	*Xanthomonas campestris*	Xanthan gum	Salad dressing, salad, gravies, desserts, bakery products, frozen dairy desserts, and culture dairy products	Thickener, suspension agent, emulsifier, thickener, stabilizer, and binding agent
	Alcaligenes faecalis	Curdlan	Meat products, fish paste, sausages, hamburgers, and noodles	Stabilizer, texture modifier, fat mimetics, stabilizer, and gelling agent
	Sphingomonas elodea	Gellan gum	Bakery fillings, jam, jelly, beverages, dairy products, desserts, and sauces	Thickener, stabilizer, binding agent, and gelling agent
	Acetobacter and *Agrobacterium*	Cellulose	Bakery products, confectioneries, dressings, and edible coatings	Stabilizer, thickening agent, gelling agent, and whipping aid
	Bacillus subtilis	Hyaluronan	Jelly, dairy products, beverages, and health drinks	Thickener and stabilizer
	Lactobacillus, Leuconostoc mesenteroides, Streptococcus mutans	Dextran	Gums, jam, jellies, candies, and ice-cream	Thickener, emulsifier, texturizer, and gelling agent

molecules interaction, and hydrocolloids-ions interaction. All these categories have specific applications in tuning the quality and functionality of the food systems.

The hydrocolloid-hydrocolloid interaction refers to the non-covalent interactions among the biopolymers like carbohydrates, starches, and proteins. The underlying concept of this type of interaction relies on the synergistic benefits of combined hydrocolloids in providing the desired consistency of material formulations rather than employing individual hydrocolloids (Gao et al. 2017). This synergism involves lowering entropy upon mixing that would result in the phase separation. Hence, the thermodynamic compatibility of the chemicals is the most important factor when employing a mixture of hydrocolloids for enhancing structural integrity. The phase separation can be of two types: segregative (dominant homotypic interaction) and associative (dominant heterotypic interaction) (Cao et al. 2015). The segregative type of phase separation occurs between similarly charged hydrocolloids such as two neutral hydrocolloids or one charged and the other one is neutral. An illustration of this category is the interaction among the two negatively charged hydrocolloids like gum Arabic and pectin. On the other hand, associative phase separation occurs between the oppositely charged hydrocolloids like ionic polysaccharides and protein molecules. This process is also termed complex coacervation where electrostatic interaction is the primary reaction that determines the phase change (Jiang et al. 2019). Example of this type of interactions includes gelatin-pectin hydrogels used for the development of low-calorie soft foods (Wu et al. 2014). Recently, more researches are carried out in the non-covalent interactions of non-mammalian gelatin with the polysaccharides for improving the gel performance. It was reported that the addition of divalent ions like Ca^{2+} results in the aggregation of the pectin-gelatin hydrocolloid system through van der Waals forces and ionic crosslinking. Hence the addition of an appropriate amount of Ca^{2+} regulates the gelation mechanism of pectin and gelatin by increasing the viscosity, melting temperature, and hardness of gelatin (Huang et al. 2020). This study provides insights into the development of a modified gel system from non-mammalian source of gelatin which finds potential applications in the field of low-sugar foods.

The second category of hydrocolloid food component interaction is hydrocolloid-small molecules, where the small molecules include fatty acids, phytochemicals, food flavours, colourants, vitamins, and biosurfactants. These components of smaller food molecules possess either a hydrophilic or hydrophobic nature results in hydrogen bonding or/ and hydrophobic interaction when interacts with hydrocolloids (Patel et al. 2013). This leads to the alteration of the nutritional, organoleptic, and functional properties of the colloidal food system (Zhang et al. 2020). This type of hydrocolloid interactions is used as a potential food vehicle for bioactive compounds. Potential applications of this type of interactions are the use of polysaccharides as encapsulating agents for flavours, antioxidants, fatty acids, probiotics, and several other food ingredients (Burnside 2014). Researches on these applications pave a way for the integration of 3D food printing for the delivery of nutraceuticals.

The third type is the hydrocolloid-ion interaction where the complex forms of mineral elements present in the foods interact with the hydrocolloids. Majorly the mineral elements exist as compounds, complexes, and/ or free ions in foods (Damodaran et al. 2007). These forms of trace elements at very low concentrations exhibit functional roles such as food structuring and assembly. Based on the valency, the functional properties of trace elements will vary. For example, the monovalent ions such as H^+, K^+, Na^+, and I^- can have either specific or non-specific interactions when encounters hydrocolloids. As a result, these ions would have implications on the solubility and gelation of polysaccharides and proteins through hydration effects or by electrostatic interactions (Cao et al. 2016). The process involves the protonation of carboxylic groups by H^+ ions followed by the hydrogen bonding or hydrophobic interactions forming a 3D network results in the induction of weak gels as in the case of alginate, pectin, and carboxymethyl cellulose (Ström et al. 2014). On the other hand, the multivalent ions like Ca^{2+}, Mg^{2+}, Fe^{2+}, Cu^{2+}, and Zn^{2+} form a complex and acts as a crosslinker for biopolymers. The combination of Ca^{2+} ions with alginates or pectin results in the formation of dimers that resembles an egg box (Cao et al. 2020). This process of gelation involves the following steps of mono-complexation, dimerization followed by lateral aggregation. It was stated that the addition of Ca^{2+} to pectin solution results in phase transition from sol to gel state which imparts mechanical stability to food inks for 3D printing. Based on this principle, a customizable water-based food stimulant was fabricated from a mixture of low methoxylated pectin, bovine serum albumin, $CaCl_2$, and sugar syrup (Vancauwenberghe et al. 2017). Results of this study are useful in the designing of internal patterns with defined microstructures and customized porous texture using 3D printing. Thus, the role of hydrocolloids in food structuring is found to be remarkable in the development of customized soft foods. The common molecular interactions among macro and micro components that enhance the structural stability and quality of the end-products of food systems are hydrogen bonding, covalent bonding, hydrophilic and hydrophobic interactions, electrostatic interaction, coordination forces, and $\pi - \pi$ stacking (Patel 2013). Considering the 3D printing, factors such as temperature, viscosity, concentrations are the most important that decide the end applications of the hydrocolloids.

7.2.3 Role of Other Additives

Other than food hydrocolloids and pure food components such as starch, food fractions like a portion of rice flour, wheat flour, refined wheat flour, composite millet flour, and similar other cereal-based materials imparts synergistic benefits in enhancing the printability of the non-printable foods (Nachal et al. 2019). Interestingly, these group of food fractions although seems to be printable on its own they are quite unstable yielding weak 3D constructs (Theagarajan et al. 2020). When these class of materials are considered as the main printing ingredient, food hydrocolloids are adequate for improving the structural integrity and mechanical stability of printing mixtures (Liu et al. 2020a). On the other hand, when these group of food fractions is not used as the main printing ingredient (for instance printing of non-printable foods such as fruits and vegetables, meat products), they can be incorporated into printing material supply for tailoring the rheological properties thereby enhances the extrudability of non-printable materials (Keerthana et al. 2020). A study was reported on the printing of egg yolk and egg white fractions through the incorporation of raw rice flour without any processing (Anukiruthika et al. 2020). The combination of rice starch along with the lipoprotein fractions of egg yolk and white fractions forms a dough with water enhancing the pseudoplastic behaviour of the material supply imparting printability. In another study, the ground chicken meat was pre-processed along with refined wheat flour with an adequate quantity of water forming a semi-solid paste (Wilson et al. 2020). The formed emulsion improves flowability and enhances the printability of the chicken meat. A similar approach was adopted for the fibre enrichment of the 3D printed meat through the addition of the composite flour (millet and pulse-based flour) into the ground chicken (Wilson et al. 2021). This approach has significantly enhanced the dietary fibre content of the 3D chicken construct. In another study, the non-printable potato peel fines were 3D printed with whole wheat flour (Madhumitha et al. 2021). This study unwinds the applicability of waste fractions of the food processing line into value-added functional 3D printed product. As a sustainable approach, milling fractions of broken wheat were 3D printed along with incorporated grape pomace powder for developing functional 3D printed cookies (Jagadiswaran et al. 2021). Hence, the food fractions of the cereals and pulses form a separate class of materials that could be promisingly used as food additives in 3D printing. The use of these materials does not affect the sensorial attributes of the regular cusines that can be conveniently processed using 3D printing without employing hydrocolloids. Despite, the method and intensity of processing have also significantly influenced the printability when these class of materials are used for 3D printing. Further, detailed studies on the sensorial attributes correlating printability and rheology of the materials are required for 3D food printing applications.

7.3 Pre-Processing and Formulations for 3D Printing

The success of 3D food printing over other food processing technologies relies on the implementation of imagination into reality through digitalized cooking and gastronomy practices. Most of the foods of our regular diet come under the category of non-printable food materials. Research works are being carried out in exploring the potential applications of 3D printing of these staple traditional food materials. The present section explains the nature of different classes of traditionally non-printable food materials. Various approaches of pre-processing employed for achieving the material consistency that translates the printability are briefed (Table 7.2).

7.3.1 Plant-Based Cellular Foods

Fruits and vegetables deserve an incredible place in our regular diet that is furnished with sugars, essential vitamins, minerals, bioactive compounds, and antioxidants (Singh et al. 2016). However, these materials are non-printable due to their fibrous content and higher moisture content. Challenges associated with the 3D printing of fruits and vegetables are the low material viscosity and greater biological variability (Ricci et al. 2019). In this cohort, it is essential to unwind the potential characteristics and suitability of fruits and vegetables for this print-and-eat technology. Studies were focussing on revolutionizing the balanced diet through 3D printing of fruits and vegetables without deterioration of its nutritional and sensorial properties thereby ensuring a healthy diet fascinatingly. In this regard, material properties are the basic factor that decides the final printability (Perez et al. 2019). The proportion of total solids and water content is the key criteria that greatly influence the rheological properties of material supplies possessing a significant impact on printability. To date, there is no protocol framed out on the state-of-the-art of printing of non-printable materials. This is because of changing proximate composition of materials, geographic

Table 7.2 Printability of the traditional food materials.

Material supply	Core component	Food additive	Pre-processing	Mechanism that enhances printability	Post-processing	Highlights	References
Fruits and vegetables							
Lemon juice	Citric acid	Potato starch	Homogenization, steam cooking, and cooling	Gelation	—	Printing of low viscous food ink	Yang et al. (2018b)
Orange concentrate	Citric acid	Wheat starch	Mixing, steam cooking, and cooling	Gelation	—	Development of 3D printed orange leather	Azam et al. (2018b)
Orange concentrate	Citric acid	15% Wheat starch and 1% gum (gum Arabic/ guar gum/ k-carrageenan/ xanthan gum)	Mixing, steam cooking, and cooling	Gelation	Addition of 1 ml of liquid vitamin D	Fabrication of fortified 3D printed orange leather	Azam et al. (2018a)
Potato flakes-strawberry juice	Starch-vitamin C	Potato starch, xanthan gum, and k-carrageenan	Mixing and boiling (mashed potato-gum); homogenization and steam cooking (strawberry juice gel)	Gelatinization	—	Dual extrusion of mashed potato/ strawberry juice gel	Liu et al. (2018b)
Vegetable ink (broccoli/ spinach/ carrot)	Polyphenols and fibre	Hydroxypropyl methylcellulose, locust bean gum, xanthan gum, guar gum, NaCl, and acetic acid	Sieving of vegetable powders, hydration of gums followed by mixing with vegetable powders	Hydration and gelation	—	Application of powder-based texturization technology in 3D printing	Kim et al. (2018)

(*Continued*)

Table 7.2 (Continued)

Material supply	Core component	Food additive	Pre-processing	Mechanism that enhances printability	Post-processing	Highlights	References
Fruits and vegetable blend (banana, mushroom, white beans, lemon juice, and non-fat milk)	Vitamins, minerals, and polyphenols	Pectin	Blending, sieving, mixing, and cooling	Gelation	—	Development of fruit-based snacks for children	Derossi et al. (2018)
Spinach powder	Fibre	Xanthan gum	Dry sieving of spinach powder followed by addition of xanthan gum solution	Hydration and gelation	—	3D printing of cellular fibrous materials	Lee et al. (2019)
Meat and meat products							
Fish surimi	Proteins and amino acids	NaCl	Mincing, centrifugation, mixing with the additive, and refrigeration	Colloidal formation	—	Surimi gel could be a promising material for 3D printing	Wang et al. (2018)
Fish surimi	Proteins and amino acids	Sweet potato starch and NaCl	Mincing, centrifugation, mixing with the additive and refrigeration, incubation, and cooking	Colloidal formation	—	Addition of starch imparts mechanical post-print stability to surimi	Dong et al. (2019)

Beef-lard	Proteins, saturated and unsaturated fat, minerals, and vitamins	NaCl and guar gum	Freezing and refrigeration followed by mincing then mixed with additives	Emulsion	Sous-vide cooking	Development of composite layer 3D printed beef	Dick et al. (2019b)
Scallop and turkey with bacon fat	Proteins, saturated and unsaturated fat, minerals, and vitamins	Transglutaminase	Additives are added to minced meat (puree)	Emulsion	Baking/ slow cooking/ frying	Development of multi-material 3D printed food with complex internal structure	Lipton et al. (2010)
Pork	Proteins, saturated and unsaturated fat, minerals, and vitamins	Xanthan gum, guar gum, and NaCl	Trimming, cutting, mincing, and freezing followed by the addition of hydrocolloids and refrigeration	Emulsion	Thermal cooking	Development of texture modified 3D printed pork as dysphagia food	Dick et al. (2020)
Egg yolk and egg white powders	Ovalbumin, ovotransferrin, ovomucoid, ovoglobulin, and phospholipids	Rice flour	Refractance window drying of egg powders, homogenization, and mixing with rice flour	Hydration and gelation	—	Comparative study on printability of egg yolk and egg white	Anukiruthika et al. (2020)
Egg yolk paste	Lipoproteins and glycoproteins	—	Steam cooking	Gelation	—	3D printing of yolk without any additives	Xu et al. (2020)

growing conditions, and seasonal variability of fruits and vegetables. Innovations in the structuring of foods through 3D printing using fruits and vegetables can be accomplished through appropriate pre-processing such as drying, gelation, size reduction, and mixing which ensures the required consistency of the material supplies (He et al. 2019). Modern food processing is aimed at the preservation of perishable commodities through the value addition of raw materials (Saha 2020). Different processed items based on fruits and vegetables include minimally processed products, canned products, dried and frozen products, sauce and ketchup, syrup and concentrates, soup mixes, squash, and ready-to-drink juice products. With the increasing health consciousness, industries focus on the manufacturing of a blend of fruit and vegetables with the added bioactive compounds (Majerska et al. 2019). Products of this category include pulps, purees, and smoothies composed of a higher amount of total solids content with abundant nutrients in their concentrated forms. This kind of materials with the highest solids contents suits best for extrusion-based 3D printing that opens a novel approach of 3D printing of fruits and vegetables based soft foods.

Considering 3D food printing, formulation of material supplies is the most important step that tailors the desired rheological properties. 3D printing of fruit and vegetables in their processed forms such as purees, paste, and smoothies can be done through the addition of hydrocolloids as additives. In the process of gelation of hydrocolloids, temperature plays a crucial role in the phase transition and the formation of 3D networks (Godoi et al. 2019). Thus, hydrocolloids are helpful in the powder-based texturization technology that aids in the optimization of material formulation for 3D printing. In this regard, the printability of the lemon juice gel was analysed with different concentrations of potato starch (Yang et al. 2018b). Potato starch was added as an additive into the lemon juice for imparting a translucent flexible chewy structure. Various printing parameters for the direct ink printing were examined and the results showed that the lemon juice gel with 15 g/100 g of potato starch exhibits a desired rheological and mechanical stability to the 3D printed constructs. The feasibility of 3D printing of fruits and vegetable blend was reported in relevance with the industrial applications in consideration with microbial safety (Severini et al. 2018). The printing formulation consisted of carrot, avocado, pear, kiwi, and broccoli raab leaves. As a pre-processing step, the study employed a juice processor for the separation of solid-phase from liquid-phase followed by homogenization at 6400 rpm. Around 1% of fish collagen was added to the obtained solid mixture for enhancing the viscosity and flowability. This resultant mixture was stored at 4 °C for about 2 hours for the cool setting of smoothie that is desired for 3D printing. It was inferred that 3D printing does not significantly affect the sensorial properties, antioxidant capacity, and the total phenolic content of the fruits and vegetable blend. This study provides valuable insights for the fabrication of soft foods using 3D printing with special concern over the sanitation and safety issues. Recently, more researches are carried out in printing of cellular fibrous materials (Park et al. 2020). Lee et al. (2019) successfully 3D printed spinach (20%, w/w) with 10% (w/w) xanthan gum as an additive. In this study, the printability was correlated with the particle size, where size reduction is employed as a pre-processing step used for enhancing the rheological properties. An increase in the particle size of the material supplies significantly improved the post-stability of the 3D printed cellular food materials. The proportion of the solids content of vegetable powder (spinach, carrot, and broccoli) on the printability, rheology, and hydration properties with different hydrocolloid matrices was examined (Kim et al. 2018). Among the tested hydrocolloids, the food ink with xanthan gum exhibits low hardness during extrusion resulting in smooth flowability with higher printing resolution. All these studies reveal that fruits and vegetables are also promising candidates suitable for 3D printing with appropriate pre-processing. In this context, more studies are required in exploring the material characteristics for the development of a customized vegan 3D printed food product with an in-depth analysis on the formulation of printing material supplies by considering food safety and allergies.

7.3.2 Animal-Based Cellular Foods

Flesh-based animal products are a secondary class of food materials being a part of our regular diet. Foods of this category include meat and meat-based products, seafood, and eggs. These materials are mainly comprised of essential amino acids, peptides, and fatty acids (Tahergorabi and Hosseini 2017). Nowadays, studies were attempted in the characterization of flesh foods and their applicability for 3D printing on realizing their health and nutritional benefits. As a fascinating technology, 3D food printing aids in the fabrication of convenient RTC 3D printed meat products with enhanced levels of customization and nutrition. This technology allows designing a novel food product with varied infill structures. As an initial attempt, Dick et al. (2019a) analysed the feasibility of the potential of 3D food printing from multi-materials based on the individual requirements. Fleshy foods are non-printable because of their fibrous nature that can easily clog the extruder nozzle. Printability of these ingredients mandatorily undergoes a pre-processing step of grinding/ mincing to achieve a homogenous mixture with smooth flowable consistency for desired extrusion (Liu et al. 2018a). Studies were reported on 3D printing of fish surimi with NaCl and sweet potato starch as structural enhancer tailoring the rheological properties of the mined fish (Dong et al. 2019; Wang et al. 2018). These additives increase the viscosity of the printing mixture which is essential to hold the 3D printed shape after deposition (Figure 7.2). It was observed that the addition of starch increases the water holding capacity and helps in preventing cooking loss. Microstructural analysis reveals the formation of the 3D cross-links through uniform dispersion of starch molecules among the fish proteins (Fan et al. 2017). This exhibits a tight association of polymeric network that enhances the gelation of fish proteins. These studies provide a guidance for the 3D printing of fish in gel forms with higher flexibility that satisfies the special needs of all age groups. Recently, the

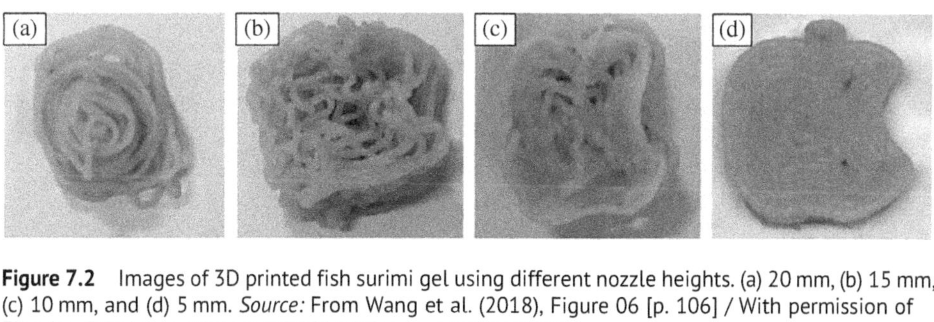

Figure 7.2 Images of 3D printed fish surimi gel using different nozzle heights. (a) 20 mm, (b) 15 mm, (c) 10 mm, and (d) 5 mm. *Source:* From Wang et al. (2018), Figure 06 [p. 106] / With permission of Elsevier. DOI-https://doi.org/10.1016/j.jfoodeng.2017.02.029.

feasibility of the incorporation of hydrocolloids for the 3D printing of pork was examined (Dick et al. 2020). This study analysed the effect of hydrocolloids (xanthan gum and guar gum) individually as well as in combination on the printability of pork for developing a dysphagia food. Results confirmed the modification of texture with the addition of a blend of hydrocolloid mixture to the pork paste that imparts a lower hardness during extrusion. Hence the cooked 3D printed pork was categorized as a transitional food within the International Dysphagia Diet Standardization Initiative (IDDSI) suitable for elderly people with swallowing difficulties.

Eggs are another class of food items that are also non-printable. Hardly few studies were reported on the applicability of eggs as a potential ingredient for 3D printing. It was reported that the addition of eggs imparts mechanical strength to printing dough (Liu et al. 2019; Yang et al. 2018a). More recently, the applicability of egg fractions for 3D food printing was studied (Anukiruthika et al. 2020). The researchers correlated the printability with the macro and micro components of egg fractions. This is the first study to report on the comparative analysis of egg yolk and egg white fractions with rice flour as an additive. Eggs as such on their own are not printable because of the lack of carbohydrates which remains as a backbone to enhance the printability. From the nutritional point of view, eggs are a first-class protein with higher bioavailability than any other animal proteins. Egg in its powder form was subjected to 3D printing through appropriate reconstitution with water and incorporation of rice flour. The starch present in the rice flour tends to improve the dough forming characteristics of egg proteins (Saleh 2018). This study draws significant inference over the nature of printability of egg proteins through drying as a pre-processing step for reduction of moisture content as well as to increase the total solids content. However, the drying temperature plays a significant role in deriving the egg powders without deterioration in their functional properties otherwise it undergoes denaturation that affects the printability (Lai et al. 2010). The dispersion of crystalline egg proteins on the starch molecules of rice flour yields desirable viscoelasticity to the printing material supply for desired extrusion (Dalbhagat et al. 2019). On comparing the egg yolk and white fractions, 3D constructs from egg yolk possess higher mechanical stability than egg white in combination with rice flour. This study highlights the importance of fat globules of yolk providing a plasticizing effect which significantly improves the printability of egg yolk than the egg white.

Taking the advantage of the gelation behaviour of egg yolk, Xu et al. (2020) investigated the effects of thermal-induced gelation of egg yolk on its printability. Results showed that with increasing treatment time, the rheological characteristics of the egg yolk were improved exhibiting a solid-like behaviour suitable for 3D printing. Among the thermal treatments with different time and temperatures, the yolk subjected to 76 °C for 8 and 10 minutes and the one subjected to 80 °C for 6 minutes exhibits good print fidelity (Figure 7.3). All the above studies on eggs provide insights for the development of 3D printed customized eggs and the potential ability of the egg to be added as a promising ingredient in food formulation for 3D printing. Also, 3D printing of eggs helps in reducing the processing loss due to breakage during transportation and to deliver the best proteinaceous food in attractive forms to children and adolescents (Anukiruthika et al. 2020). These studies put forth the need for more researches in the fabrication of customized designer eggs fortified with functional ingredients such as omega 3 fatty acids and vitamin D upon

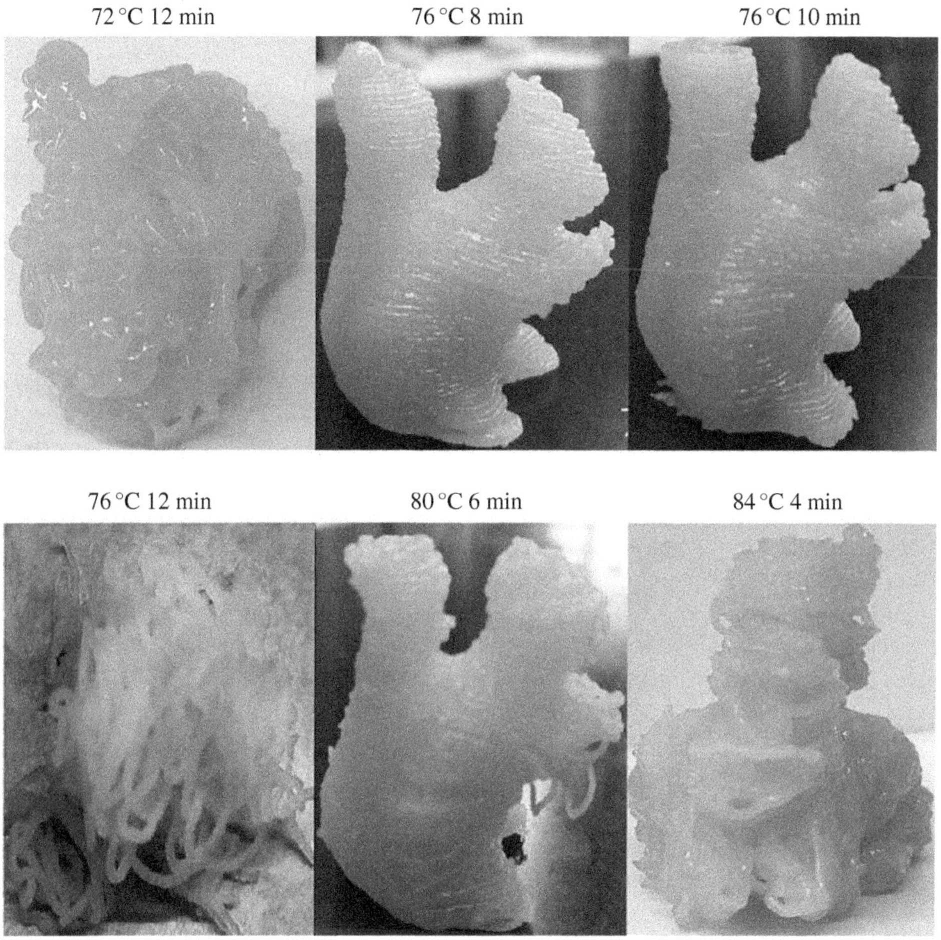

Figure 7.3 Images of 3D printed objects from egg yolk paste at different heat treatments. *Source:* From Xu et al. (2020), Figure 01 [p. 03] / With permission of Elsevier. DOI- https://doi.org/10.1016/j.foodhyd.2019.105399.

integration of 3D printing with microencapsulation techniques. Thus, animal-derived foods are having great scope for 3D printing with wide opportunities in microstructural designing of foods possessing several nutritional and health benefits.

7.4 Post-Printing Stability of the Printed 3D Constructs

Apart from chocolates and other processed food products like smoothies, puree, and gels, all the other materials that are 3D printed requires cooking for their edibility. The most important criterion of the material supplies for 3D printing are its printability, applicability, and post-processing (Nachal et al. 2019). The latter terminology 'post-processing' refers to the cooking processes such as boiling, baking, frying, and drying. The selection of the post-processing method depends on the extent of pre-processing and the nature of the raw ingredients in the material supply. Irrespective of the printability, dough-based printable, as well as non-printable materials, require post-processing. Hence the distinction of materials about printability refers to the inherent ability of the raw materials to exhibit desirable viscoelastic properties for 3D printing. Whereas post-processing is meant to retain the post-printing stability of the 3D constructs. In this regard, both pre- and post-processing are the critical factors that determine the final quality of 3D printed foods for consumption (He et al. 2019).

As stated earlier, hydrocolloids are the structuring ingredients that play a key role in food printing. As a single ingredient hydrocolloid have a dual role that aids in tailoring the flowability of raw material supply for 3D printing and improves mechanical stability to hold the printed structures during the thermal treatment after 3D printing. In an attempt to prevent the deformation of the printed constructs, Kim et al. (2019) studied the effects of hydrocolloids (methylcellulose and xanthan gum) on the dimensional stability of the cookie dough during baking. Pre-printing characterization of material supplies showed that the dough with xanthan gum exhibits lower extrudability than the methylcellulose which was reflected from the poor printing performance of dough with xanthan gum. This is due to the higher shear modulus of the xanthan gum incorporated dough that possesses higher structural retention than methylcellulose. However, the post-processing studies revealed that dough samples with xanthan gum possessed good mechanical strength even at its low concentration of 0.5 g/100 g. This study highlights the importance of pre- and post-processing in retaining the edibility of 3D printed food. In agreement with this context, Lipton et al. (2010) enhanced the printability of meat-based material supply using transglutaminase enzyme. The addition of transglutaminase to individual printing mixture (turkey, scallop, and celery) imparts flowability by increasing the viscosity of food inks. The multi-material printed structure immediately after printing was chilled in a freezer for ensuring the geometrical stability followed by baking at 250 °C. While slow cooking and deep frying were performed immediately after printing. This shows the significance of post-processing to attain structural integrity to the 3D printed structures.

Achieving post-stability of 3D printed foods is a tedious task that depends on several factors such as selection of ingredients, material composition, printing parameters, processing conditions (pre- and post-processing), mechanical handling, and storability. The drawing of definite procedures for 3D printing is merely strenuous because of the complex behaviour of macro and micro components of foods (Perez et al. 2019). Considering these aspects, 3D printing is at its nascent stage yet has a long way to go for implementation at a large scale (Dankar et al. 2018). Major barriers to the large-scale implementation of 3D food printing are the consumption of more time for printing. Efforts are being taken by the scientific community in utilizing the fuller potential of 3D printing through introducing more than one printing head. Multi-head 3D printing systems seem to be advantageous in terms of reduced processing time with greater productivity (Nachal et al. 2019). Thus, the natively printable and non-printable materials are useful in the fabrication of novel 3D printed structures based on individual likeness. With advancements in technology, 3D printers can become a domestic kitchen appliance at an affordable price (Rubio and Hurtado 2019). In near future, 3D food printing becomes the leading food processing technology with interdisciplinary approaches of food technology and robotic engineering. 3D printing could revolutionize the food industries through the delivery of nutraceuticals via foods that reflects the old saying of Hippocrates "Let food be thy medicine, and let medicine be thy food".

7.5 Scope of Non-Printable Materials for 3D Printing Applications

3D printing of the staple foods of our day-to-day life is possible with the incorporation of additives like hydrocolloids or by combining with natively printable starch sources such as rice flour, wheat flour, and millet flour (Huang 2018). Various methods of processing employed for the enhancement of printability of fruits and vegetables, seafood, eggs, meat, and its products have been discussed. This clearly shows the hidden opportunities in traditional foods for the fabrication of customized foods. The present section deals with the potential applications of non-printable materials and their scope towards 3D printing. As the foremost application, 3D printing allows us to deliver a customized food in terms of taste, texture, and visual attributes. It is possible to have a healthy diet without compromising the individual likeness of all age groups (Derossi et al. 2020). In this regard, Derossi et al. (2018) reported a case study on the development of 3D printed fruit-based snacks for children. The formulation includes fresh bananas, dried mushrooms, lemon juice, canned white beans, and dried non-fat milk that had been designed to provide around 5–10% of energy, iron, calcium, and vitamin D for infants and pre-school children (Figure 7.4).

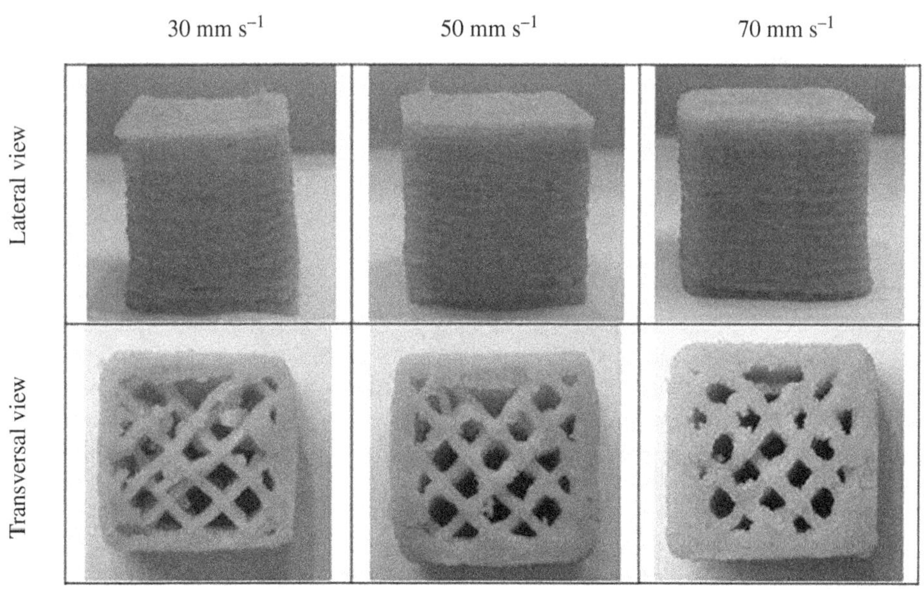

Figure 7.4 3D printed snacks from fruits and vegetable blend at different printing speeds. *Source:* From Derossi et al. (2018) From Derossi, A et al (2018), Figure 10 [p. 40] / With permission of Elsevier. DOI-https://doi.org/10.1016/j.jfoodeng.2017.05.015.

Similarly, the internal fat composition of meat was altered with the filling of lard using 3D printing (Dick et al. 2019b). A lean beef-lard composite layer 3D printed meat product was produced using 3D printing that shows the opportunities to customize the infill material providing a healthy option to the diet conscious people. In this way, the texture of the meat was modified using hydrocolloids for the designing of a dysphagia diet for people with swallowing difficulties.

Another perspective of food 3D printing from an industrial point of view is its resource utilization and sustainability. As an additive manufacturing technology, 3D printing allows the efficient use of the available material resources and energy (Baiano 2020). Recently with the increasing apprehension with the reduction of renewable resources, sustainable approaches are being implemented in industries. In this point of view, food industries are not the least that are taking serious actions to decrease the processing loss and to reduce the food waste. With the concept of 'zero waste circular economy' 3D printing aids in the effective utilization of the food processing wastes such as milling fractions of rice, bran from wheat, peels, pomace, skins, seeds of fruits and vegetables, and trimmings of flesh foods (Van Wijk and Van Wijk 2015). The generated secondary waste streams of food materials remain as a source of antioxidants, bioactive compounds, exhibits antimicrobial activity, and can sometimes be used as flavouring agents. Thus, the by-products of the main processing line are conveniently processed and transformed to a self-stable material in a dried powder forms and can be added to the material supply of 3D printing. More recently, air-fried snack was developed from potato flakes with wheat flour as structural enhancer (Liu et al. 2020b). This study provides insights into the development of texture-modified foods from unconventional food sources using 3D printing. Various process parameters employed for the extrusion-based 3D printing of the chicken meat were successfully optimized (Wilson et al. 2020). This study used refined wheat flour as an additive with different levels of ground chicken meat. Results showed that the 3D printed chicken construct with optimized process conditions was post-processed using hot-air drying followed by deep frying to develop a 3D printed chicken nugget with customized shapes. Later, the same group of researchers successfully developed fibre enriched chicken meat constructs by incorporating the ground chicken meat into the natively printable composite flour as formulated by Krishnaraj et al. (2019) (Wilson et al. 2021). Results showed that the developed fibre enriched 3D printed chicken possesses 5.36% of dietary fibre that contributes to more than 10% of daily dietary intake. Thus, these studies provide novel insights on the reutilization of meat trimmings from meat processing into a value-added 3D printed food with enriched dietary fibre. Thus, food waste containing essential micronutrients can be efficiently utilized in regular foods for direct consumption rather than using for pharmaceutical applications through 3D printing technology. Production of novel 3D printed foods from food wastes has a great scope and marketing potential contributing to a greener economy.

7.6 Conclusion

The printability of traditionally non-printable food materials is feasible with appropriate pre-processing of the material supply. Irrespective of the printability, the post-stability of the printed 3D constructs is crucial in achieving 3D printed foods. Based on individual likeness and health status, staple foods of our regular diet have the potential to exhibit printability through the incorporation of food hydrocolloids. The present chapter summarizes the characteristic non-printable nature of foods and the biophysics involved in enhancing their printability. Further, a comparison of the natively printable and non-printable material is presented explaining the synergistic benefits of enhancement of printability of traditional foods using natively printable ingredients. Applicability of the non-printable foods for 3D printing adds to a newer taste and variety to the traditional food cuisines. Nevertheless, 3D food printing helps in the efficient utilization of valuable food wastes for consumption demonstrating the enhanced levels of customization without loss of nutrients and sensorial properties. Thus, 3D printing as a print-and-eat technology restructures the organization of food processing distribution and supply chain leading to a sustainable circular economy.

References

Anukiruthika, T., Moses, J.A., and Anandharamakrishnan, C. (2020). 3D printing of egg yolk and white with rice flour blends. *Journal of Food Engineering* **265**: 109691. https://doi.org/10.1016/j.jfoodeng.2019.109691.

Azam, R.S.M., Zhang, M., Bhandari, B., and Yang, C. (2018a). Effect of different gums on features of 3D printed object based on vitamin-D enriched orange concentrate. *Food Biophysics* **13** (3): 250–262.

Azam, S.M.R., Zhang, M., Mujumdar, A.S., and Yang, C. (2018b). Study on 3D printing of orange concentrate and material characteristics. *Journal of Food Process Engineering* **41** (5): 1–10. https://doi.org/10.1111/jfpe.12689.

Baiano, A. (2020). 3D printed foods: a comprehensive review on technologies, nutritional value, safety, consumer attitude, regulatory framework, and economic and sustainability issues. *Food Reviews International*, pp. 1–31.

Burey, P., Bhandari, B.R., Howes, T., and Gidley, M.J. (2008). Hydrocolloid gel particles: formation, characterization, and application. *Critical Reviews in Food Science and Nutrition* **48** (5): 361–377.

Burnside, E. (2014). Hydrocolloids and gums as encapsulating agents. In: *Microencapsulation in the Food Industry* (eds. A.G. Gaonkar, N. Vasisht, A.R. Khare and R. Sobel), 241–252. Elsevier.

Cao, L., Lu, W., Mata, A. et al. (2020). Egg-box model-based gelation of alginate and pectin: a review. *Carbohydrate Polymers* **242**: 116389.

Cao, Y., Fang, Y., Nishinari, K., and Phillips, G.O. (2016). Effects of conformational ordering on protein/polyelectrolyte electrostatic complexation: ionic binding and chain stiffening. *Scientific Reports* **6**: 23739.

Cao, Y., Wang, L., Zhang, K. et al. (2015). Mapping the complex phase behaviors of aqueous mixtures of κ-carrageenan and type B gelatin. *The Journal of Physical Chemistry B* **119** (30): 9982–9992.

Dalbhagat, C.G., Mahato, D.K., and Mishra, H.N. (2019). Effect of extrusion processing on physicochemical, functional and nutritional characteristics of rice and rice-based products: a review. *Trends in Food Science and Technology* **85**: 226–240. https://doi.org/10.1016/j.tifs.2019.01.001.

Damodaran, S., Parkin, K.L., and Fennema, O.R. (2007). *Fennema's Food Chemistry*. CRC press.

Dankar, I., Haddarah, A., Omar, F.E.L. et al. (2018). 3D printing technology: the new era for food customization and elaboration. *Trends in Food Science and Technology* **75**: 231–242. https://doi.org/10.1016/j.tifs.2018.03.018.

Derossi, A., Caporizzi, R., Azzollini, D., and Severini, C. (2018). Application of 3D printing for customized food. A case on the development of a fruit-based snack for children. *Journal of Food Engineering* **220**: 65–75.

Derossi, A., Caporizzi, R., Ricci, I., and Severini, C. (2019). Critical variables in 3D food printing. In: *Fundamentals of 3D Food Printing and Applications* (eds. F.C. Godoi, B.R. Bhandari, S. Prakash and M. Zhang), pp. 41–91. Elsevier.

Derossi, A., Husain, A., Caporizzi, R., and Severini, C. (2020). Manufacturing personalized food for people uniqueness. An overview from traditional to emerging technologies. *Critical Reviews in Food Science and Nutrition* **60** (7): 1141–1159.

Dick, A., Bhandari, B., Dong, X., and Prakash, S. (2020). Feasibility study of hydrocolloid incorporated 3D printed pork as dysphagia food. *Food Hydrocolloids* **107**: 105940.

Dick, A., Bhandari, B., and Prakash, S. (2019a). 3D printing of meat. *Meat Science* **153**: 35–44. https://doi.org/10.1016/j.meatsci.2019.03.005.

Dick, A., Bhandari, B., and Prakash, S. (2019b). Post-processing feasibility of composite-layer 3D printed beef. *Meat Science* **153**: 9–18. https://doi.org/10.1016/j.meatsci.2019.02.024.

Dickinson, E. (2009). Hydrocolloids as emulsifiers and emulsion stabilizers. *Food Hydrocolloids* **23** (6): 1473–1482.

Dong, X., Huang, Y., Pan, Y. et al. (2019). Investigation of sweet potato starch as structural enhancer for 3D printing of *Scomberomorus niphonius* surimi. *Journal of Texture Studies* **50**: 316–324. https://doi.org/10.1111/jtxs.12398.

Fan, M., Hu, T., Zhao, S. et al. (2017). Gel characteristics and microstructure of fish myofibrillar protein/cassava starch composites. *Food Chemistry* **218**: 221–230.

Gao, Z., Fang, Y., Cao, Y. et al. (2017). Hydrocolloid-food component interactions. *Food Hydrocolloids* **68**: 149–156.

Godoi, F.C., Bhandari, B.R., Prakash, S., and Zhang, M. (2019). An introduction to the principles of 3D food printing. In: *Fundamentals of 3D Food Printing and Applications* (eds. F.C. Godoi, B.R. Bhandari, S. Prakash and M. Zhang), pp. 1–18. Elsevier.

Godoi, F.C., Prakash, S., and Bhandari, B.R. (2016). 3D printing technologies applied for food design: status and prospects. *Journal of Food Engineering* **179**: 44–54.

Hamilton, C.A., Alici, G., and Panhuis, M. (2018). 3D printing vegemite and marmite: redefining "breadboards.". *Journal of Food Engineering* **220**: 83–88.

He, C., Zhang, M., and Fang, Z. (2019). 3D printing of food: pretreatment and post-treatment of materials. *Critical Reviews in Food Science and Nutrition* **60**: 1–14.

Huang, C.Y. (2018). Extrusion-based 3D printing and characterization of edible materials [University of Waterloo]. https://uwspace.uwaterloo.ca/handle/10012/12899 (accessed 23 August 2020).

Huang, S., Tu, Z., Sha, X. et al. (2020). Gelling properties and interaction analysis of fish gelatin—low-methoxyl pectin system with different concentrations of Ca2+. *LWT* **132**: 109826.

Jagadiswaran, B., Alagarasan, V., Palanivelu, P. et al. (2021). Valorization of food industry waste and by-products using 3D printing: a study on the development of value-added functional cookies. *Future Foods* **4**: 100036.

Jiang, H., Zheng, L., Zou, Y. et al. (2019). 3D food printing: main components selection by considering rheological properties. *Critical Reviews in Food Science and Nutrition* **59** (14): 2335–2347.

Kailash, M., Kavitha, L., Madan, S.V., and Ravichandran, C. (2020). 3D food printing: a technology for fabricating customized artistic food. In: *Emerging Technologies in Food Science* (eds. M. Thakur and V. Modi). Singapore: Springer https://doi.org/10.1007/978-981-15-2556-8_25.

Keerthana, K., Anukiruthika, T., Moses, J.A., and Anandharamakrishnan, C. (2020). Development of fiber-enriched 3D printed snacks from alternative foods: a study on button mushroom. *Journal of Food Engineering* **287**: 110116. https://doi.org/10.1016/j.jfoodeng.2020.110116.

Kim, H.W., Lee, I.J., Park, S.M. et al. (2019). Effect of hydrocolloid addition on dimensional stability in post-processing of 3D printable cookie dough. *LWT* **101**: 69–75.

Kim, H.W., Lee, J.H., Park, S.M. et al. (2018). Effect of hydrocolloids on rheological properties and printability of vegetable inks for 3D food printing. *Journal of Food Science* **83** (12): 2923–2932.

Krishnaraj, P., Anukiruthika, T., Choudhary, P. et al. (2019). 3D extrusion printing and post-processing of fibre-rich snack from indigenous composite flour. *Food and Bioprocess Technology* **12** (10): 1776–1786. https://doi.org/10.1007/s11947-019-02336-5.

Lai, K.M., Chuang, Y.S., Chou, Y.C. et al. (2010). Changes in physicochemical properties of egg white and yolk proteins from duck shell eggs due to hydrostatic pressure treatment. *Poultry Science* **89** (4): 729–737.

Lee, J.H., Won, D.J., Kim, H.W., and Park, H.J. (2019). Effect of particle size on 3D printing performance of the food-ink system with cellular food materials. *Journal of Food Engineering* **256**: 1–8. https://doi.org/10.1016/j.jfoodeng.2019.03.014.

Lille, M., Nurmela, A., Nordlund, E. et al. (2018). Applicability of protein and fiber-rich food materials in extrusion-based 3D printing. *Journal of Food Engineering* **220**: 20–27.

Lipton, J., Arnold, D., Nigl, F. et al. (2010). Multi-material food printing with complex internal structure suitable for conventional post-processing. *Solid Freeform Fabrication Symposium* **9**: 809–815.

Liu, C., Ho, C., and Wang, J. (2018a). The development of 3D food printer for printing fibrous meat materials. *IOP Conference Series: Materials Science and Engineering* **284** (1): 012019. https://doi.org/10.1088/1757-899X/284/1/012019.

Liu, Y., Liang, X., Saeed, A. et al. (2019). Properties of 3D printed dough and optimization of printing parameters. *Innovative Food Science & Emerging Technologies* **54**: 9–18. https://doi.org/10.1016/J.IFSET.2019.03.008.

Liu, Y., Tang, T., Duan, S. et al. (2020a). Effects of sodium alginate and rice variety on the physicochemical characteristics and 3D printing feasibility of rice paste. *LWT* **127**: 109360.

Liu, Z., Dick, A., Prakash, S. et al. (2020b). Texture modification of 3D printed air-fried potato snack by varying its internal structure with the potential to reduce oil content. *Food and Bioprocess Technology* **13** (3): 564–576.

Liu, Z., Zhang, M., and Yang, C. (2018b). Dual extrusion 3D printing of mashed potatoes/strawberry juice gel. *LWT* **96**: 589–596.

Lu, W., Nishinari, K., Matsukawa, S., and Fang, Y. (2020). The future trends of food hydrocolloids. *Food Hydrocolloids* **103**: 105713.

Madhumitha, M., Abinash, V., Rahul, T. et al. (2021). Valorization of food industry waste streams using 3D printing: a study on noodles prepared from potato peel waste. *Food and Bioprocess Technology* https://doi.org/10.1007/s11947-021-02675-2.

Majerska, J., Michalska, A., and Figiel, A. (2019). A review of new directions in managing fruit and vegetable processing by-products. *Trends in Food Science & Technology* **88**: 207–219.

Nachal, N., Moses, J.A., Karthik, P., and Anandharamakrishnan, C. (2019). Applications of 3D printing in food processing. *Food Engineering Reviews* **11** (3): 123–141. https://doi.org/10.1007/s12393-019-09199-8.

Nida, S., Anukiruthika, T., Moses, J.A., and Anandharamakrishnan, C. (2021). 3D printing of grinding and milling fractions of rice husk. *Waste Biomass Valorization* **12**: 81–90. https://doi.org/10.1007/s12649-020-01000-w.

Oliveira, S.M., Fasolin, L.H., Vicente, A.A. et al. (2020). Printability, microstructure, and flow dynamics of phase-separated edible 3D inks. *Food Hydrocolloids* **109**: 106120.

Park, S.M., Kim, H.W., and Park, H.J. (2020). Callus-based 3D printing for food exemplified with carrot tissues and its potential for innovative food production. *Journal of Food Engineering* **271**: 109781.

Patel, A. (2013). Novel colloidal structures from food-grade materials for applications in functional foods. In: *InsideFood symposium : book of proceedings. Presented at the COST FA1001 workshop: InsideFood symposium* (eds. B. Nicolai and L. Piazza). Leuven, Belgium: KU Leuven.

Patel, A.R., Heussen, P.C.M., Dorst, E. et al. (2013). Colloidal approach to prepare colour blends from colourants with different solubility profiles. *Food Chemistry* **141** (2): 1466–1471.

Perez, B., Nykvist, H., Brogger, A.F. et al. (2019). Impact of macronutrients printability and 3D-printer parameters on 3D-food printing: a review. *Food Chemistry* **287** (October 2018): 249–257. https://doi.org/10.1016/j.foodchem.2019.02.090.

Prakash, S., Bhandari, B.R., Godoi, F.C., and Zhang, M. (2019). Future outlook of 3D food printing. In: *Fundamentals of 3D Food Printing and Applications* (eds. F.C. Godoi, B.R. Bhandari, S. Prakash and M. Zhang), pp. 373–381. Elsevier.

Ricci, I., Derossi, A., and Severini, C. (2019). 3D printed food from fruits and vegetables. In: *Fundamentals of 3D Food Printing and Applications*, vol. **153** (eds. F.C. Godoi, B.R. Bhandari, S. Prakash and M. Zhang), pp. 117–149. Elsevier Ltd https://doi.org/10.1016/b978-0-12-814564-7.00005-5.

Rubio, E. and Hurtado, S. (2019). 3D food printing technology at home, domestic application. In: *Fundamentals of 3D Food Printing and Applications* (eds. F.C. Godoi, B.R. Bhandari, S. Prakash and M. Zhang), 289–329. Elsevier.

Saha, D. (2020). Food processing: understanding common threads. In: *Economics of the Food Processing Industry. Themes in Economics (Theory, Empirics, and Policy)*. Singapore: Springer https://doi.org/10.1007/978-981-13-8554-4_2.

Saleh, M. (2018). Wheat batter physical properties as influenced by starch/flour types and egg contents. *Journal of Food Measurement and Characterization* **12** (2): 800–807. https://doi.org/10.1007/s11694-017-9694-z.

Severini, C., Derossi, A., Ricci, I. et al. (2018). Printing a blend of fruit and vegetables. New advances on critical variables and shelf life of 3D edible objects. *Journal of Food Engineering* **220**: 89–100. https://doi.org/10.1016/j.jfoodeng.2017.08.025.

Singh, J.P., Kaur, A., Shevkani, K., and Singh, N. (2016). Composition, bioactive compounds and antioxidant activity of common Indian fruits and vegetables. *Journal of Food Science and Technology* **53** (11): 4056–4066.

Ström, A., Schuster, E., and Goh, S.M. (2014). Rheological characterization of acid pectin samples in the absence and presence of monovalent ions. *Carbohydrate Polymers* **113**: 336–343.

Sun, J., Zhou, W., Huang, D., and Yan, L. (2018a). 3D food printing: perspectives. In: *Polymers for Food Applications* (ed. T.J. Gutierrez), pp. 725–755. Springer.

Sun, J., Zhou, W., Yan, L. et al. (2018b). Extrusion-based food printing for digitalized food design and nutrition control. *Journal of Food Engineering* **220**: 1–11.

Tahergorabi, R. and Hosseini, S.V. (2017). Proteins, peptides, and amino acids. In: *Nutraceutical and Functional Food Components* (ed. C.M. Galanakis), pp. 15–38. Elsevier.

Theagarajan, R., Moses, J.A., and Anandharamakrishnan, C. (2020). 3D extrusion printability of rice starch and optimization of process variables. *Food and Bioprocess Technology* **13**: 1048–1062. https://doi.org/10.1007/s11947-020-02453-6.

Van Wijk, A.J.M. and Van Wijk, I. (2015). *3D Printing with Biomaterials: Towards a Sustainable and Circular Economy*. IOS press.

Vancauwenberghe, V., Katalagarianakis, L., Wang, Z. et al. (2017). Pectin based food-ink formulations for 3-D printing of customizable porous food simulants. *Innovative Food Science and Emerging Technologies* **42** (June): 138–150. https://doi.org/10.1016/j.ifset.2017.06.011.

Voon, S.L., An, J., Wong, G. et al. (2019). 3D food printing: a categorised review of inks and their development. *Virtual and Physical Prototyping* **14** (3): 203–218.

Wang, L., Zhang, M., Bhandari, B., and Yang, C. (2018). Investigation on fish surimi gel as promising food material for 3D printing. *Journal of Food Engineering* **220**: 101–108. https://doi.org/10.1016/j.jfoodeng.2017.02.029.

Wegrzyn, T.F., Golding, M., and Archer, R.H. (2012). Food layered manufacture: a new process for constructing solid foods. *Trends in Food Science & Technology* **27** (2): 66–72.

Williams, P.A. and Phillips, G.O. (2009). Introduction to food hydrocolloids. In: *Handbook of Hydrocolloids* (eds. G.O. Philips and P.A. Williams), pp. 1–22. Elsevier.

Wilson, A., Anukiruthika, T., Moses, J.A., and Anandharamakrishnan, C. (2020). Customized shapes for chicken meat-based products: feasibility study on 3D-printed nuggets. *Food and Bioprocess Technology* **13**: 1–16.

Wilson, A., Anukiruthika, T., Moses, J.A., and Anandharamakrishnan, C. (2021). Preparation of fiber-enriched chicken meat constructs using 3D printing. *Journal of Culinary Science & Technology*, pp. 1–12.

Wu, B., Degner, B., and McClements, D.J. (2014). Soft matter strategies for controlling food texture: formation of hydrogel particles by biopolymer complex coacervation. *Journal of Physics: Condensed Matter* **26** (46): 464104.

Xu, L., Gu, L., Su, Y. et al. (2020). Impact of thermal treatment on the rheological, microstructural, protein structures and extrusion 3D printing characteristics of egg yolk. *Food Hydrocolloids* **100**: 105399.

Yang, F., Zhang, M., Prakash, S., and Liu, Y. (2018a). Physical properties of 3D printed baking dough as affected by different compositions. *Innovative Food Science & Emerging Technologies* **49**: 202–210. https://doi.org/10.1016/j.ifset.2018.01.001.

Yang, F., Guo, C., Zhang, M. et al. (2019). Improving 3D printing process of lemon juice gel based on fluid flow numerical simulation. *LWT* **102**: 89–99.

Yang, F., Zhang, M., Bhandari, B., and Liu, Y. (2018b). Investigation on lemon juice gel as food material for 3D printing and optimization of printing parameters. *LWT- Food Science and Technology* **87**: 67–76. https://doi.org/10.1016/j.lwt.2017.08.054.

Yegin, S., Kopec, A., Kitts, D.D., and Zawistowski, J. (2020). Dietary fiber: a functional food ingredient with physiological benefits. In: *Dietary Sugar, Salt and Fat in Human Health* (eds. H.G. Preuss and D. Bagchi), pp, 531–555. Elsevier.

Zhang, H., Zhang, F., and Yuan, R. (2020). Applications of natural polymer-based hydrogels in the food industry. In: *Hydrogels Based on Natural Polymers* (ed. Y. Chen), pp, 357–410. Elsevier.

8

Alternative Ingredients Used in Food Printing

CHAPTER MENU

8.1 Introduction

The present decade is witnessing the decline of renewable resources due to human interventions and urbanization. This poses a serious threat to the food supply chain with significant implications on food production due to unpredictable weather conditions (D'Amour et al. 2020). Recently, the department of economic and social affairs of the United Nations forecasted that the world population could reach 9.8 billion in the year 2050 and 11.2 billion in the year 2100 (United Nations 2017). This trend will result in the scarcity of food sources that rises the need for sustainable food processing technologies. Considering these aspects, 3D printing is a novel technology that is gaining research interest in the food sector in recent days. As a sustainable approach, 3D food printing seems to provide a promising solution for multiple concerns like climate change, population growth, food demand, food wastage, and shortage of natural resources as a single standalone technology (Jayaprakash et al. 2019). The nutritional and food security can be conserved by exploring the novel food sources, with reduction of food wastes and efficient utilization of available material resources. 3D printing

3D Printing of Foods, First Edition. C. Anandharamakrishnan, Jeyan A. Moses, and T. Anukiruthika.

as an additive technology claims to overcome the above-mentioned issues possessing several socio-economic benefits (Herrero et al. 2020). In a broader sense, material supplies used for 3D food printing can be categorized based on printability as natively printable, traditionally non-printable, and alternative ingredients. The latter class of the materials includes the use of uncommon, unexplored, and under-utilized food sources as a base material for 3D printing that paves a way for the novel fabrication of sustainable foods (Nachal et al. 2019).

The greater usage of land and dwindling pastures in rearing animals leads to the exploitation of the environment through rising carbon footprints (Spring 2019). Also, the lack of proteins in vegan and vegetarian diets remains a concern that results in serious health implications. Hence, proteins from other alternative sources must be explored to address these issues. The alternative sources of proteins are insects, lupines, seaweeds, algae, and fungi which remain as a good source of proteins and fibre (Sun et al. 2018b). With greater degrees of freedom, 3D printing can be efficiently utilized for the printing of these uncommon novel food sources to design a nutrient-dense 3D printed food. Also, the waste streams from food processing lines such as fruits and vegetable peels, meat trimmings, and fish fillets can be efficiently utilized as a material supply for 3D printing (Ricci et al. 2019). These materials can be processed into a dried powder forms that could be added to the food formulation for printing using an extrusion-based approach. The material supply can be in a semisolid form either as puree or dough that can be extruded through the printing nozzle in the desired shape. The printing movements can be guided by integrating the 3D printer with computerized slicing software that aids in the deposition of material in a layered fashion (Theagarajan et al. 2020). Alternative food ingredients have a great scope for marketability providing customized foods in a preferred shape, taste, texture, and colour with precise nutrition control. Thus, the growing awareness of environmental sustainability urges the food industry to seek novel alternative food ingredients (Prakash et al. 2019). The present chapter discusses the opportunities that exist in the 3D printing of alternative food ingredients. Also, the need and driving force for the alternative food sources are briefed with the illustrations of state-of-the art of its printability. Further, the challenges associated with the implementation of new food ingredients in the diet have been stated. The present chapter is aimed to provide insights on the feasibilities of the different novel material sources to be used as material supply for 3D printing of foods.

8.2 Alternative Food Sources and the Sustainability Perspective

A recent survey by Food and Agriculture Organization (FAO) stated that about 820 million people in the world are still hungry without adequate required daily food (FAO 2019). This present scenario poses a great challenge to attain the 'Zero Hunger Target' by 2030. In addition to hunger, the prevalence of malnourishment in developing countries still worsens the existing condition that rises the need for sustainable alternative production and eating practices. In this viewpoint, it is evident that the current food habits are not sustainable which has been the outcome of lethargic human activities. Some of the factors that drive the need

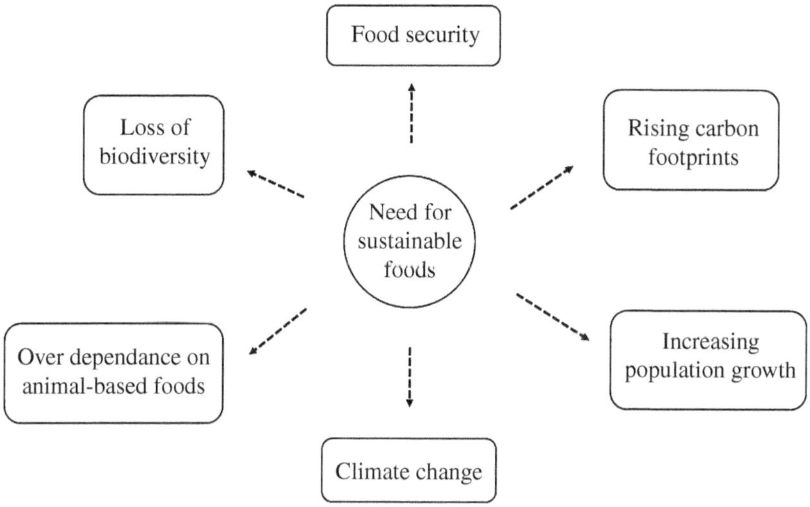

Figure 8.1 Driving forces for the emergence of novel sustainable foods.

for changing the existing dietary practices and the need for alternative foods are presented in Figure 8.1. The aforementioned factors are having a great impact on the environment and economy. The prevalence of food security depends on food availability, food accessibility, food stability, and food utilization (Padmavathy and Poyyamoli 2011). All these terms cover the needs from the individual level to the population group that is essential to reach sustainability in the diet.

More awareness on cultivational, productional, and processing of foods have to be enforced to establish sustainability and restoration of dwindling natural resources. In this regard, the solution to ensure food security is the exploration of novel alternative ingredients as a sustainable food of the future (McClements 2020). The global alternative proteins market is expected to reach about $17.9 billion by 2025 during the year 2019–2025 with a compound annual growth rate (CAGR) of 9.5%. In terms of volume, the market has been expected to reach 9787.4 kilo metric ton (KMT) by 2025 during the year 2019–2025 with a CAGR of 5.2% (Meticulous Research 2019). The global opportunity analysis and industry forecast categorized the alternative proteins based on their market growth as emerging alternative protein market (insect protein), adolescent alternative protein market (plant-based proteins like rice, corn, potato, and pea), and matured alternative protein market (plant-based proteins like soy, wheat, and canola; mycoprotein like mushroom and algae). The segment-wise market growth rate of alternative food ingredients is presented in Table 8.1. As a holistic approach, it has been put forth that the reduction in the overconsumption of calories, proteins, and animal-based foods would have a positive impact on the environment (Ranganathan et al. 2016). Thus, a shift or transition in the existing dietary practices is adequate for conserving biodiversity. Diversification of diets is one of the simpler approaches that can be effectively implemented for reducing the overdependency on animal-based foods. Thus, the emergence of novel protein sources such as insect proteins, lupin seeds, seaweeds, and algae would remain as promissory ingredients with the potential to replace animal-based foods (van Huis and Oonincx 2017; Sun et al. 2018a). It has been

Table 8.1 Recent trends on the market growth rate of alternative proteins.

Alternative proteins	Market value		CAGR (%)	Highlights	References
	From	To			
Plant-based proteins	US$8.96 billion in 2019	US$14.32 billion by 2025	8.1	Largest share of global alternative protein market	Bombe (2020)
Insect proteins	US$0.152 billion in 2018	US$1.4 billion by 2026	38	Rising demand to be used in sports nutrition industry	Watson (2019)
Aquatic plant proteins (seaweeds)	US$5.9 billion in 2019	US$11.9 billion by 2027	9.1	Rising demand for consumption of seaweed-based snacks	Market Analysis Report (2020)
Microbial proteins (single cell protein)	US$5.3 billion in 2017	US$8.69 billion by 2023	8.6	Demand for cost-effective alternative foods	GlobeNewswire (2019)
Cultured meat market	US$0.0163 billion in 2022	US$0.0198 billion by 2027	4	Demand for vegan meat like products	BCC Research (2019)

forecasted that the market value of meat substitutes would reach US$30.92 billion by 2026 from the US$10.1 billion in 2018 (Statista 2019). From the sensorial aspects, researchers are now focusing on developing meat analogues from vegan sources that mimic meat products. Plant-derived proteins such as soy protein, pea protein, wheat gluten, casein, and caseinates can be processed efficiently using spinning and extrusion techniques for producing imitation meat that resembles the fibrous and chewy texture of meat (Boukid 2020). Thus, alternative food ingredients assure sustainability with less carbon foot prints and possess a great scope to be used as the future foods.

8.3 Rationale of Alternative Material Supplies

The ever-growing human population has put forth greater land pressure with larger demand for foods. Our daily food choices have a great impact on the environment and climate change. With changing lifestyles, the dietary practices and preferences are also changing with a direct or indirect effect on food production (van Huis and Oonincx 2017). Considering the average footprints of the current dietary practices, it has been estimated that animal products account for about 43–87% of the environmental burden from an individual person (Davis et al. 2016). Various negative outcomes on the environment are greenhouse gas emissions, rising carbon footprints, climate change, shortage of water, and depletion of fossil fuels. This gives an alarm for the restoration of the natural resources to humankind that receives the attention of food scientists and researchers to focus on the implementation of sustainable practices (Hartmann and Siegrist 2017). Efforts are made through the innovations in food technologies to meet consumer demands without a serious threat to biodiversity. An in-depth view on food sustainability showed that the rising demand for animal meat-based products is the major cause for the resultant negative outcomes on the environment (Adesogan et al. 2020). From a nutritional point of view, proteins are one of the essential nutrients for growth and metabolism that are mostly acquired from animal-based foods. Hence, there is no point in a cut off of the protein intake in turn some other alternative proteins can be utilized. A transition from animal-based proteins to other sources is adequate for the conservation of biodiversity (Siegrist and Hartmann 2019).

The common existing source of proteins is meat and meat products, eggs, and dairy products. Plant and marine-based proteins are not fully explored for the production of foods. It has been expected that the protein demand would increase by more than 75% from the current levels in 2050. This scenario results in the per capita increase from 28 kg in 2007 to 42 kg in 2050 in developing countries than the developed countries which would be 80–91 kg (Herrero et al. 2015). It was reported that around 80% of the agricultural land is used for animal grazing of which 3.400 million ha as green pastures and 500 million ha as crop fields (Herrero et al. 2016). From this statistic, it was clear that rearing animals would significantly decrease the food supply for human consumption as most of the crops and seeds were taken as animal feed (Mottet et al. 2017). It was reported that the production of 1 kg of meat requires about 5–40 times more water than that for producing 1 kg of cereals (González et al. 2020). Ultimately, this would result in increasing the production cost of meat. Climate change is another significant factor that must be considered in conserving biodiversity. Intending to set up a global action plan for mitigation of climate change, around 195 countries had agreed at the Paris climate conference held in December 2015. It was enforced to frame out rules for limiting climate change to <2 °C above the pre-industrial levels thereby increasing the limit to 1.5 °C. This would result in significant reductions in the negative impacts on the environment (Wollenberg et al. 2016). The rearing of livestock for bioconversion to meat is one of the major contributors to CO_2. Globally, around 20 billion domesticated animals can produce about 5.6–7.5 Gt CO_2 equivalents/year, in which about 64–78% of emissions are contributed by cattle. Thus, the greenhouse emissions are 100% higher than the growing of vegetables. It has been reported that the production of 1 kg of beef requires about 50 times more land than vegetables. Reduction in the consumption of meat greatly reduces the environmental pressure (van Huis and Oonincx 2017). Another measure for the mitigation of the climate change is the use of feed having a low impact on environment for the aquaculture, rearing of pigs, and poultry thereby improves biodiversity.

The percentage of greenhouse gas emissions varies with the type of livestock. Considering the sustainability, the efficiency of the production of meat-based food decreases in the following order: chicken, pork, and beef (Wegier et al. 2018). Thus, the production of chicken is most efficient than the other two. In addition, the method of the productional system significantly affects the CO_2 contribution within the same animal type for instance rising/rearing of chicken. Westhoek et al. (2014) reported that a 50% reduction in the production of meat, eggs, and dairy products would result in a 25–40% reduction of CO_2 emissions in the European Union. The ever-growing global demand for meat-based products exploits the agricultural land that leads to seeking novel alternative food sources. Some of the alternative food ingredients are seaweeds, duckweed, rapeseed, micro and macroalgae, fungi, and insects (van Huis and Oonincx 2017). Another interesting approach is the development of cultured meat, i.e., meat grown in a laboratory (*in-vitro*) that simulates the texture of the real meat. These cultured meats have a less environmental impact than conventionally produced meat (Post et al. 2020). With the technological advancements, studies were attempted in the development of meat analogues of vegetarian origin that mimics the texture and flavour of the meat (He et al. 2020). All these approaches will lead to sustainable production and consumption practices. Thus, the environmental burden caused by the production of meat-based products can be significantly reduced by replacing meat with alternative food ingredients. Therefore, efforts are taken by the scientific fraternity in the development of alternative protein-based foods without having much impact on nutrition behaviour. From the environmental point of view, the alternative ingredients have a flourishing scope for the development of novel foods in near future (Lonnie and Johnstone 2020). However, consumer food behaviour in terms of willingness to reduce meat consumption, the acceptance of meat substitutes and foods from alternative food ingredients, taste preferences, customs and social norms remains as metal barriers for the transition towards sustainable food consumption (Possidónio et al. 2020). Along with the scientific reports and studies, the government bodies must undertake appropriate measures for implementation of sustainable practices in considering the animal welfare and environmental benefits.

8.4 Innovative Food Sources

Growing concerns on environmental impact and climate change due to industrialization and urbanization urged the need for the exploration of the applicability of alternative ingredients for food production (Fasolin et al. 2019). As the name implies, alternative food materials are derived from sources that are uncommon, unexplored, and under-utilized. Although alternative ingredients have not yet gained industrial importance, these ingredients remain as a valuable source of macro and micronutrients. It is equally important to unwind the potential benefits of utilizing alternative food materials for preserving food security as well as to conserve biodiversity (Joyce et al. 2012). The present section on innovative food sources is aimed at providing an overview of the available material sources that can be positively utilized for food production.

8.4.1 Uncommon Food Sources

Nowadays, people are becoming more cautious about the intake of nutrients and the proportions of calories from food intake. Increasing awareness on the health status leads to misconceptions on the protein quality of animal and plant-based sources. It is well-known that plants possess a relatively lower content of proteins than animals. Hence, plant proteins are referred to as 'incomplete' as they lack in one or more essential amino acids (Wu 2013). Nevertheless, plant-based sources like cereals, pulses, legumes, and oilseeds are nutritionally equivalent to animal-based proteins. A proper combination of protein from two or more plant sources remains complementary to each other providing the amino acid that is limiting in another source (Wu et al. 2014). The applicability of these plant-derived proteins would have a positive impact in reducing the environmental issues and problems related to ethical concerns. Some of the plant-derived proteins that have the potential to be used as meat substitutes are rice, wheat, barley, sorghum, millets, oats, soy, pea, chia, peanut, hemp seed, flaxseed, canola, rapeseed, legumes, lentils, peanuts, quinoa, and amaranth seeds (Nadathur et al. 2016). Recently, these materials are gaining potential research interest for the manufacturing of processed products such as protein concentrates, isolates, hydrolysates, and textured vegetable proteins (Gençdaug et al. 2020). Alternative use of products of plant-based origin helps in adding variety to vegetarian diets. Also, people who suffer from lactose intolerance consider these products to be their source of protein in their diet.

Amaranth seeds are pseudocereals composed of rich profiles of amino acids with predominant lysine and methionine (Caselato-Sousa and Amaya-Farfán 2012). Also, it remains to be a good source of insoluble fibre, lipids and fatty acids, minerals, and vitamins. In recent years, amaranth seeds are greatly utilized in regular diets due to their antioxidant, anticancer, and anti-allergic properties. Therefore, amaranth seeds and sprouts are used as key functional ingredients in the production of cookies and noodles (Tripathi and Mogra 2020). Leafy vegetables like moringa, spinach, lettuce, cabbage, kales, microgreens, and watercress are considered as a potential source of essential vitamins and minerals in addition to the ample amount of proteins. However, they remain hidden which does not deserve a major portion of our diet. Therefore, these uncommon food materials must be explored and can be used as a promising alternative ingredient in food preparations and formulations. Among which moringa leaves constitute a balanced amount of proteins with high fibre content (Gopalakrishnan et al. 2016). In its powder form, moringa leaves are used to fortify the bread, biscuits, cookies, yoghurt, cheese, and soup mix that remains as a potential ingredient in personalized food preparations (Oyeyinka and Oyeyinka 2018). Lupins are an interesting source of proteins that constitute about 35–40% of dry weight have a great scope to be added in smoothies and spreads (Derossi et al. 2020). Albumins and globulins are the major proteins predominant in lupin seeds of plant origin. Research studies are being carried out on the prebiotic activity of the yellow lupin polysaccharides (Thambiraj et al. 2018). Due to its unique amino acid profile, lupins are being used in formulations of yoghurt (Hickisch et al. 2016), bakery products (Villarino et al. 2015), and meat patties (Danowska-Oziewicz and Kurp 2017).

Another alternative source of proteins for replacing meat is insects. The consumption of insects as food is common in Eastern and African countries. Around 2000 species of insects are considered edible possessing high amounts of proteins, fat, essential vitamins, and minerals (Fasolin et al. 2019). Some of the common insects that are consumed as food are cricket (*Gryllodes sigillatus*), grasshopper (*Schistocerca gregaria*), mealworm (*Tenebrio molitor*), and honeybee brood (*Apis mellifera*) to name a few. The protein and fat composition of insects vary from 7 to 91% and 5 to 40% of dry weight, respectively (van Huis 2016; Roos 2018). Interestingly, some species of insects possess zinc and iron that are nearly similar to that of beef and far higher than pork and chicken (Mwangi et al. 2018). Studies were reported on the utilization of insects as supplementary foods in improving gut health (Stull et al. 2018). Because of sustainability, rearing insects are environmentally advantageous than animal husbandry and agriculture. As stated, the growth of insects has less impact in reducing soil fertility and deforestation with low carbon footprints (Oonincx 2017). Recently, many studies are carried out in unwinding the functional properties like solubility, gelling, and emulsifying properties of insects for food applications (Mishyna et al. 2019).

However, this class of proteins from arthropods has controversial issues with cultural habits and customs. Consumer's attitude and sensorial attributes are also important criteria in using insects as supplements. Efforts were made in delivering insects in their processed forms as powders that could marginally improve the consumer's acceptance (Hartmann and Siegrist 2016; Piha et al. 2018). Moreover, insects have a high feed conversion efficiency, short life cycles, and high reproduction rate that upgrades insects to be an eco-friendly and economically viable source of proteins for human consumption.

8.4.2 Unexplored Food Sources

Another class of alternative proteins that remains unexplored are marine and land-based sources including seaweeds, algae, fungi, yeast, and bacteria. These materials are receiving attention in recent days because of their nutritional content as well as environmental benefits. Seaweeds and algae are renewable resources that have a great scope in the food industry due to their sustainability. Various forms of microalgae include *Chlorella*, *Spirulina*, *Haematococcus*, *Schizochytrium*, *Dunaliella*, *Porphyridium*, *Scenedesmus*, *Odontella*, and *Aphanizomenon* (Mobin and Alam 2017). Among which *Chlorella* and *Spirulina* are the most prevalent microalgae that are gaining commercial interest to be used as supplements in dried powder forms. Spirulina powder is added to formulations of customized foods especially in flavoured shake-type powder (Santos et al. 2016), smoothies (Castillejo et al. 2018), and bread (De Marco et al. 2014). The next category of the seaweeds are macroalgae. Major groups of macro algae that are widely used in food preparations all over the world with commercial interests are Nori (*Puphyra* sp.), Kombu (*Saccharina japonica*), and Wakame (*Undaria pinnatifida*) (MacArtain et al. 2007). Seaweeds are a rich source of polysaccharides (76% dry weight), most of which are grouped as hydrocolloids. Also, they contain proteins in the range of 9.71–24.8% for brown seaweeds, 47% for red seaweeds, and 26% for green seaweeds on a dry weight basis (Paiva et al. 2014). The proteins of seaweeds exhibit a similar amino acid profile that is close to legumes and eggs, hence seaweeds are greatly valued to be used in food formulations. Due to their functional properties, macroalgae are used in the production of seaweed crisps, biscuits and cookies, different kinds of pasta, instant mashed potatoes, fruit puddings, and salads (Derossi et al. 2020).

Mushrooms are the macroscopic fruiting body of fungus which are considered to be a low-cost alternative protein. Among the identified 2000 species, about 25 species are edible that are cultivated commercially (Keerthana et al. 2020). The most common mushrooms that are consumed as food are *Agaricus bisporus*, *Pleurotus* sp., and *Lentinus edodes*. Nutritionally mushrooms are far superior to common cereals and pulses that possess about 19–37% (dry weight) of proteins with a complete amino acid profile and a significant proportion of dietary fibres (Bach et al. 2017). Apart from nutritional and sensorial properties, mushrooms are known for their medicinal value and bioactive compounds. Mushrooms can be easily cultivated at low cost in layers of organic solid materials generally food waste that does not require any soil preparation. Hence, mushrooms are a potential alternative ingredient that not only reduces the damage to the environment but also helps in the utilization of agricultural biomass and agro-industrial waste as substrate for their growth (González et al. 2020). Due to their economic viability mushrooms are used in the production of meat analogues (Stephan et al. 2018). Commercially, mushroom powders are incorporated as a potential functional ingredient in food preparations like soup mix, cookies, biscuits, crisps, vegan nuggets, patties, and burgers (Salehi 2019). Although this class of alternative proteins from algae and fungi seems to have greater nutritional value, mushrooms are known for consuming as whole food instead of the value-added products which has not yet been explored well (Reis et al. 2017). Hence, more studies must be directed in exploring the protein value, bioavailability, and concerns on the safety of consuming algae and fungi.

8.4.3 Under-Utilized Food Sources

This class includes a special class of food materials that are almost common, however, less utilized or known. Some of the plant-based foods like millets, tubers, and legumes are considered as the neglected and under-utilized crop species (NUCS) (Hunter et al. 2019). These materials possess a significant amount of vital minerals, essential vitamins, amino acids, and fatty acids that are nearly equivalent to flesh foods. It is adequate to explore the feasibility of these under-utilized foods which have the potential to transform the monotonous diet of staple foods (Muthamilarasan and Prasad 2020). Further, utilization of this class of foods would significantly reduce the burden of environmental pressure and climate change that helps in conserving biodiversity and food security. Under-utilized foods that belong to tubers are sweet potato (*Ipomoea batatas*), cassava (*Manihot esculenta*), taro (*Colocasia esculenta*), Indian shot (*Canna indica*), arrowroot (*Maranta arundinacea*), elephant foot (*Amorphophallus paeoniifolius*), gonala (*Dioscorea spicata*), and air potato (*Dioscorea bulbifera*) (Muthamilarasan et al., 2019). Some of the under-utilized millets are finger millet (*Eleusine coracana*), foxtail millet (*Setaria italica*), kodo millet (*Paspalum scrobiculatum*), and barnyard millet (*Echinochloa esculenta*) (Katoch, 2020). In the case of legumes, cowpea (*Vigna unguiculata*), Bambara groundnut (*Vigna subterranean*), adzuki beans (*Vigna angularis*), rice bean (*Vigna umbellata*), jack bean (*Canavalia ensiformis*), winged bean (*Psophocarpus tetragonolobus*), kidney bean (*Phaseolus vulgaris*), lima bean (*Phaseolus lunatus*), velvet bean (*Mucuna pruriens*), faba bean (*Vicia faba*), and horsegram (*Macrotyloma uniflorum*) remain under-utilized (Chivenge et al., 2015; Gulzar and Minnaar, 2017).

Another group of interesting materials is the by-products and co-products from the food processing industry. During the processing of foods, many streams of by-products are generated during the production of mainstream products (Cecilia et al., 2019). For instance, peel, rind, seeds, pomace are the by-products of fruits and vegetables processing while husk, bran, broken kernels are the milling fractions of the cereals, pulses, and millets. Similarly, the meat coproducts include visceral organs, tissues, skins and trimmings, blood, and fish fillets (Anzani et al. 2020). These by-products remain to be a potential source of dietary fibres and contain polyphenols, vitamins, and mineral elements possessing several functional properties like antioxidant and antimicrobial properties with a greater industrial value (Plazzotta et al. 2017). However, these by-products are less valued and utilized as animal feed. Due to their economic significance, these materials are used as substrates for fermentation, biorefineries, fertilizers, and manures (Ahuja et al. 2020; Sadh et al. 2018; Teigiserova et al. 2019). All these applications mask the applicability of the by-products to be used as food for humans. On realizing the nutritional benefits of these by-products, studies were focused on the valorization of the agro-industrial waste for producing value-added products paving a way for sustainable eating practices (Mirabella et al. 2014). In this regard, the potato peel powder and watermelon rind powder are used for the development of functional cookies and other bakery products (Jeddou et al. 2017; Naknaen et al. 2016). Also, the pomace of grapes and citrus fruits are utilized in the extraction of functional compounds such as phenolics and flavours (Meini et al. 2019). Due to its gelling properties, apple pomace is used as thickening agent in processed food products like stirred-type yoghurt (Wang et al. 2020). Similarly, the by-products from meat processing (tissues, organs, trimmings) can be efficiently utilized as source material for the preparations of patties and nuggets rather than to dump into the environment (Anzani et al. 2020).

8.5 3D Printing of Alternative Ingredients

3D printing is an emerging breakthrough technology in the food sector that helps in the deployment of artistic culinary skills in the design and fabrication of foods. Printing of foods provides a greater degrees of freedom to customize the foods in terms of geometry, shape, size, colour, texture, taste, flavour, material composition, and nutrient proportion (Feng et al. 2018). An overview of opportunities existing with alternative food ingredients using 3D printing is presented in Figure 8.2. Compare to conventional food processing technologies, 3D printing is far advantageous possessing nutritional, environmental, social, and economic benefits. However, 3D printing is a developing technology; it has a long way to go in the fabrication of sustainable foods from alternative ingredients. Various needs and the significance of utilization of alternative foods have been discussed in the previous sections. Thus, alternative food sources are gaining scope for marketability and sustainability. Considering 3D printing, the printable characteristics of these materials are a significant factor that determines the success of sustainable foods from 3D printing (Perez et al. 2019). In recent days, studies were put forth in the incorporation of these alternative materials in the conventional preparation of foods. Even though the addition of alternative ingredients in food formulations seems to be easy, the process of 3D printing is varied from that of conventional processing. 3D printing of foods depends on several intrinsic and extrinsic factors. The behavioural response of material towards 3D printing is an important criterion for successful deposition of food layers for which material composition is a key determining factor (Voon et al. 2019). Only a handful of studies have been reported on 3D printing from

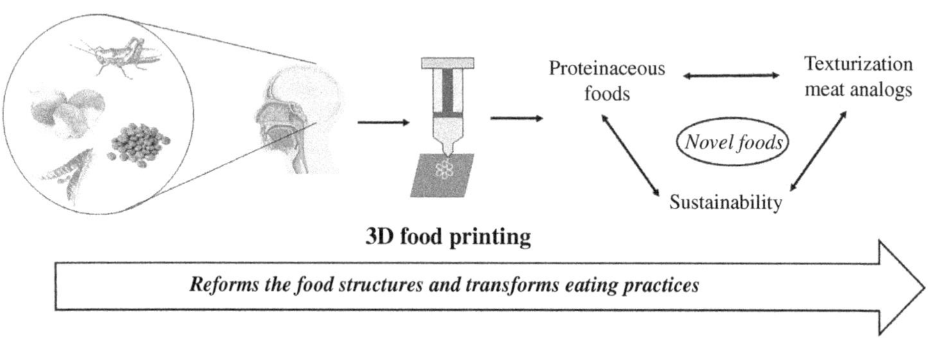

Figure 8.2 Overview of opportunities existing with alternative food ingredients using 3D printing.

alternative food materials. The focus of these studies was centred on the enrichment of nutrients and providing sustainable foods using alternative ingredients. In this regard, studies were reported on the enrichment of cookies with insect proteins (Severini et al. 2018), the development of fibre and protein-enriched snack from mushroom (Keerthana et al. 2020), and the fortification of cookies with microbial protein (Vieira et al. 2020). Also, the agro-industrial by-products are utilized for the development of fried snacks using 3D printing (Liu et al. 2020).

8.5.1 Insects as Food

With growing concern on the supply of proteins to feed the global population, researches are carried out in the fabrication of foods from the alternative ingredient. Among the alternative sources, insect proteins are uncommon materials that are being focused on in recent days to combat protein deficiency (Fasolin et al. 2019). Insects possess a good amount of protein along with other bioactive compounds. Although insects are nutritionally superior food, the percentage of acceptance of insects as food remains low among consumers. Attempts are being made to deliver insect proteins in the processed powder forms for improving the consumer's acceptability (Hartmann and Siegrist 2017). 3D printing is a novel fascinating technology that uses raw material with appropriate pre-processing that has the potential to mask the ingredients (insect) and tailors the nutrition. To explore the nutritive and eco-friendly nature of entomophagy, the 'Insects Au Gratin' project was proposed using 3D printing (Soares et al. 2011). The proposal put forth the sustainability of insects that can convert the vegetation into edible protein furnished with micronutrients. Nutritionally, a number of four crickets provides calcium equivalent to that of one glass of milk, similarly, dung beetles contain more iron than that beef in weight basis. Further, the rearing of insects can generate $1/10^{th}$ of methane produced by animal husbandry in rising 1 kg of meat under less water consumption. The advancement of the print-and-eat technology provides opportunities in designing a cereal-based snack enriched with insect proteins (Soares et al. 2011). Insects are added to the printing material supply along with common baking ingredients like icing sugar, butter, cheese, water, and additives to fabricate 3D edible ready-to-cook (RTC) and ready-to-eat (RTE) constructs.

3D printing allows the utilization of insects in their modified forms which seems to be the best way to gain acceptance of the consumers who were not accustomed in eating the whole insects. The wheat flour enriched with ground larvae of yellow mealworms (*T. molitor*) is used for developing a 3D printed snack through baking (Severini et al. 2018). The addition of 20 g of insect powder/100 g of flour (d.b.) resulted in a softer dough that aids the ease of the extrusion process. Results showed a proportional increase of the amino acid content from 32.5% (control) to 38.2 and 41.3 g/100 g protein for dough with 10 and 20% incorporated insect powder. This study provides insights into the feasibility of the development of sustainable insect-based foods using 3D printing. Another approach in enriching the foods with insects is the fractionization of the nutrients instead of ground whole insect powder. Characterization of the individual components of insect fractions helps to develop a multilayered personalized food through protein cross-linking and sol–gel transition mechanism (Derossi et al. 2018). Insects are not only a rich source of proteins but also contain fat. Some of the insects like a wasp, caterpillar, and palm weevil larvae contain about 60–70% of fat with a high energy value. Insects contain a higher amount of polyunsaturated fatty acids than chicken and red meat (Dobermann et al. 2017). In addition, insects possess a notable amount of fibre among which chitin is the most common. Chitin is *N*-acetyl glucosamine that forms the major component of exoskeleton that are indigestible to human. However, chitin can help in improving the immune response against parasitic infections. Although insects are an excellent source of proteins, an extensive research is required on the solubility, extraction, and digestibility of insects. Studies were reported on the consumer perception towards consumption of insects with the incorporation of insects invisibly in its modified forms greatly improves its acceptability (Le Goff and Delarue 2017). Thus, 3D printing provides a promising way for enrichment of dietary proteins from insects that can be delivered in an attractive fascinating manner.

8.5.2 Microorganisms as Food

Another interesting source of alternative foods is microbial origin composed of valuable proteins and polysaccharides. Microorganisms are well known for their beneficial aspects in food fermentation. In this regard, researchers are focused on the extraction of macro components from microbes. These fractions are valued for their functional properties that can be used as hydrocolloids in food systems (Razavi 2019). Foods of microbial origin that can be used as food are fungi, yeast, bacteria, and algae-based foods/ food products. Considering 3D printing, these sources can be efficiently utilized in their processed forms usually dry powder by adding into the material supplies. Hardly one or two reports were available in the applicability of microbial-based material supplies. Mushrooms are popularly consumed as food all over the world for their nutritional, organoleptic, and therapeutic properties. Some of the edible mushrooms that are cultivated commercially are button mushroom, Portobella, Oyster, Crimini, and Enoki. Also, the mushroom is cultivated largely for its market value that requires less land than agricultural crops leading to sustainability. Commercially, mushrooms in their dried forms are added to curry mix, smoothie, chips, and bakery products. More recently, researchers were attempted to print mushroom-based material supply for developing a fibre and protein-rich snack (Keerthana et al. 2020). This was the first study to report on the 3D printing of alternative ingredient mushrooms. Although the mushroom was pre-processed (dried and sieved), the fibrous portion of the mushroom showed less printability. It was stated that mushroom is non-printable on its own; however, it can be efficiently incorporated into cereal dough for improving printability (Figure 8.3). Results confirmed that the snack developed from mushroom can provide energy of about 342.82 kcal/100 g. Further, the consumer acceptance of the developed product in two variants (sweet and spiced products) was analyzed. Among the variants, the spiced snack was sensorially more acceptable than sweet snack. Thus, this study provides valuable insights in the development of 3D printed mushroom-based snacks as a food of the future.

Another promising alternative ingredient is the microalgae that are furnished with valuable bioactive compounds. This includes high-value proteins, polyunsaturated fatty acids, vitamins, minerals, polyphenols, and carotenoids. Commonly, algal biomass is incorporated into regular food preparations such as bread, breakfast cereals, pasta, cookies, jellies, desserts, and beverages. On realizing its potential application more recently, Vieira et al. (2020) explored the feasibility of incorporation of microalgae *Arthrospira platensis* in the

Printing speed (mm min^{-1})	Material formulation with different proportions of mushroom powder (%) using 1.28 mm nozzle				
	5	10	15	20	25
200					
Observation	All the material formulations resulted in over extrusion with poor resolution of the printed samples				
400					
Observation	All the material formulations possess higher stickiness and resulted in over extrusion				
600	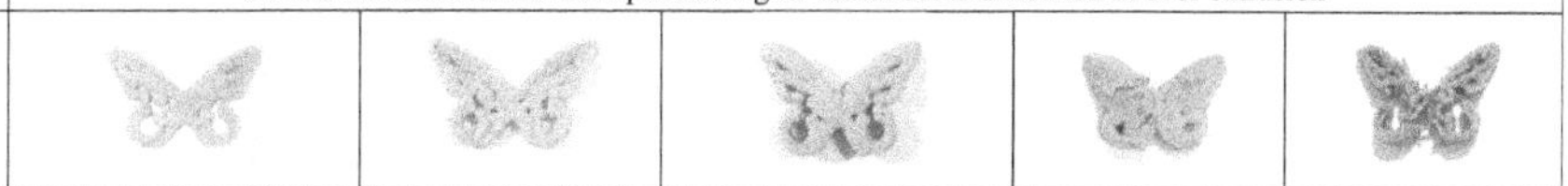				
Observation	Over extrusion				Under extrusion
800	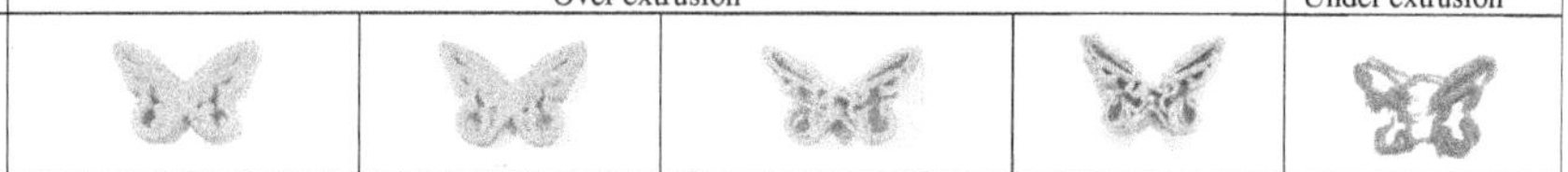				
Observation	Over extrusion		Ceaseless flow of material but the % of mushroom was less in amount	Desired extrusion	Under extrusion
1000	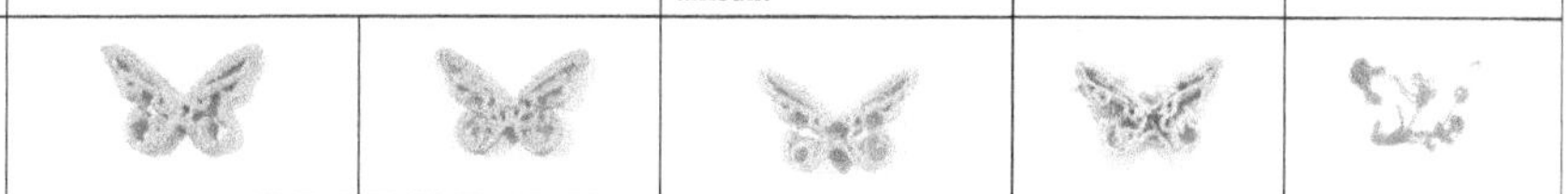				
Observation	Over extrusion	Extrusion was good but the % of mushroom was less in amount	Layers tend to break	Under extrusion	No extrusion

Figure 8.3 Optimization of mushroom based 3D printed material supply of different formulations and printing variables. *Source:* From Keerthana et al. (2020), [p.06] / With permission of ELSEVIER. DOI-https://doi.org/10.1016/j.jfoodeng.2020.110116

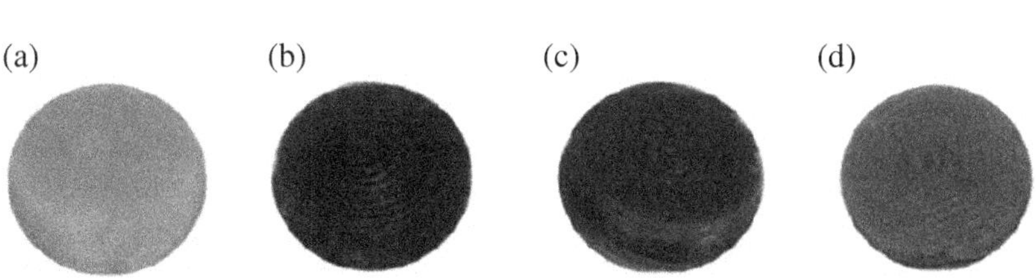

Figure 8.4 3D printed functional cookie fortified with *Arthrospira platensis* of different forms (a) control, (b) biomass, (c) free extract, and (d) encapsulated extract. *Source:* From Vieira et al. (2020), Figure 04 [p. 08] / With permission of Elsevier. DOI-https://doi.org/10.1016/j.foodhyd.2020.105893.

development of functional cookies using 3D printing. This study analyzed three different approaches of incorporation of microalgae into the material supply: dried biomass, freeze-dried biomass extract, and encapsulated biomass extract (Figure 8.4). All the forms resulted in a dough formulation with good print fidelity confirming that *A. platensis* can be used as a promising ingredient for 3D printing. In addition, the cookie prepared from the encapsulated biomass showed an improved antioxidant activity and colour stability during the storage period than dried and freeze-dried extracts. Thus, this study put forth an innovative approach of integration of 3D printing and encapsulation technologies for the development of novel functional food with antioxidant properties. Thus, macro and microalgal biomass remains as a good source of functional compounds possessing food and nutraceutical applications (Sudhakar et al. 2019). However, an in-depth analysis on the extraction, purification, digestibility, storability, and safety are required when microorganisms are used as food. Likewise, more studies on 3D printing are needed in unwinding the potential applicability and opportunities existing in the use of microorganisms as alternative ingredients for 3D food printing.

8.5.3 By-products of Fruits and Vegetables Processing

The processing of food involves the transformation of raw feed into finished foods from 'farm to fork'. In the process of the production of foods, many streams of waste are generated that are dumped into land posing a serious threat to the environment (Lee et al. 2020). This enforces the need for sustainable production and eating practices. Due to which food industries are focusing on the implementation of 'circular economy' because of reducing food waste and greenhouse gas emissions with maximal resource recovery (Maina et al. 2017). Annually the amount of food residues generated from fruits and vegetable processing; meat processing is equivalent to millions of tonnes. To combat this situation, studies were focused on the utilization of these by-products and co-products as a new ingredient for food preparations. Nevertheless, these residues possess high nutritive value containing dietary fibres, polyphenols, vitamins, and minerals (Masood et al. 2020). Despite these ingredients are less valued and used as animal feed. Nowadays, industries are focusing on sustainable approaches with the motto of 'waste to wealth' to explore the economic value of these by-products. Various forms of food residues from fruits and vegetable processing includes peels, kernels, seeds, rind, pips, and pomace. While the meat processing co-products are organs, blood, bone, hide, skin, and shells. These by-products have a great scope for extraction of polysaccharides and proteins possessing several functional properties such as gelling, thickening, binding, and emulsifying agents (Premkumar and Vasudevan 2018).

Studies were reported on the incorporation of these ingredients in the replacement of cereal flour for baked goods and cereal bars (Bouazizi et al. 2020). Due to their functional attributes, these residues were also used in RTC gravies, chutney, soup, and sauces as a thickening agent. A fortified functional cookie was prepared using apple pomace flour (Zlatanović et al. 2019). In the process of production of apple juices and cider, a large quantity of pomace is generated which contains dietary fibre, flavonols, and phenolic compounds. Results showed that conventional processing of cookies with 50% of apple pomace possesses 21 g/100 g of dietary fibre with improved sensorial properties and antioxidant activities. In another study, the skins and seeds of tomatoes were studied for their physicochemical properties to be used as fortifying agent (Aksoylu Özbek et al. 2020). It was inferred that the pomace and oil obtained from dried pomace have a valuable component like flavonoids, carotenoids, and polyphenols which has a great scope for food fortification. In a similar approach, grape pomace was employed for the replacement of flour in the development of functional cookies (Theagarajan et al. 2019). Results showed that cookies with 6% (w/w) of grape pomace possess higher antioxidant capacity and storability.

Although these studies were done through conventional processing methods, the same principle of waste utilization can be adapted using 3D printing.

In this regard, 3D printing seems to be a sustainable processing technique that allows for the efficient utilization of raw materials resulting in reduced waste generation. In addition, this print-and-eat technology allows us to use the food industry residue for developing a novel food out of waste. More recently, Liu et al. (2020) developed a 3D printed air-fried snack from potato industry processing waste. Potatoes are widely used for the production of chips and French fries that leaves the fines of peel and tubers as waste. Hence, the recovery of these fractions leads to a greener valorization approach for the fabrication of innovative foods. Results showed that 3D printing is a promising eco-friendly technology that utilizes by-products of potato processing in the development of novel texture-modified snacks with newer tastes. In another recent report, the effect of size fractions of potato peel was efficiently used along with whole wheat flour in the production of 3D printed noodles (Madhumitha et al. 2021). Results showed that fine fractions of potato peel powder of particles less than 0.125 mm exhibited better printability than coarse fractions of size ranging from 0.325 to 0.125 mm. This was because of the higher fibre content of the coarse potato peel powder (4.13%) than that of fine fractions (1.2%) (Figure 8.5). The fibre portion of the food exhibits poor plasticity and structural integrity and hence a relatively lower fibre content of fine fractions showed enhanced printability of potato peel powder using 0.82 mm nozzle at 600 mm minute^{-1} printing speed, 60% flow, and 300 rpm extrusion motor speed. Further, the study developed instant RTC 3D printed noodles with good cooking quality and textural attributes that were comparable to the conventionally prepared commercial noodles (Madhumitha et al. 2021). Interestingly, functional 3D printed cookies were developed by utilizing the waste grape pomace and broken wheat fractions (Jagadiswaran et al. 2021). The optimal printing was obtained for material supplies printed at 400 mm minute^{-1}, 600 rpm motor speed, and 1.28 mm nozzle. Among the tested formulations, the material supply with 6% grape pomace scored higher in sensory evaluation. Thus, the developed 3D printed pomace cookies were rich in dietary fibre, proteins, and bioactive compounds. Accordingly, similar approaches must be adapted in the valorization of agro-industrial waste using 3D printing. The materials of this category must undergo a pre-processing step for enhancement of printable characteristics. However, attention must be paid while recovery of waste in regard with safety and stability of the 3D printed constructs.

8.5.4 Others

Alternative ingredients that are derived from fungi, algae, macro, and micro seaweeds, hemp, flax, and lupine seeds are quite a rich source of dietary fibre and proteins that can be potentially used in food 3D printing applications. However, these classes of materials are not commonly used in regular diets. On realizing the scope of processing, nutritional benefits, sustainability, and efficiency, these ingredients are useful in developing customized and personalized food formulations that remain as a promising solution in combating depletion of renewable resources and to satisfy global hunger. More research needs are required in exploring the applicability of these ingredients in printing mixture, physicochemical properties of materials, post-processing characteristics, sensorial attributes, and acceptability of the 3D printed foods from alternative ingredients.

Nozzle size (mm)	Flow (%)	Printing speed ($mm\ min^{-1}$)			
		300	600	900	1200
1.28	60	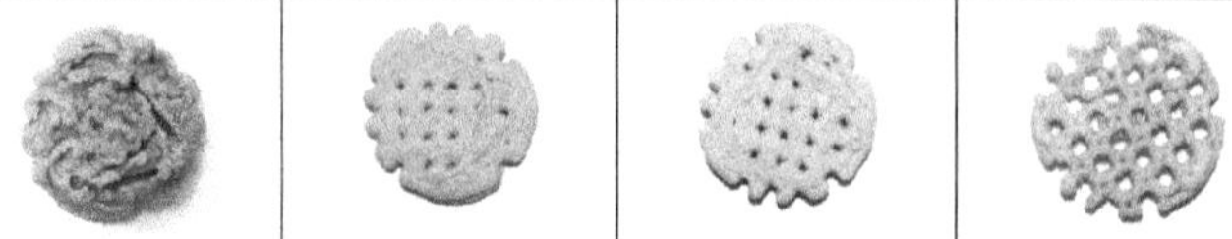			
Deposition rate ($mm^3\ s^{-1}$)		3.858	7.716	11.574	15.432
Observation		Over extrusion with poor resolution and less stable structures and therefore 80% flow would also result in over extrusion			
0.82	60	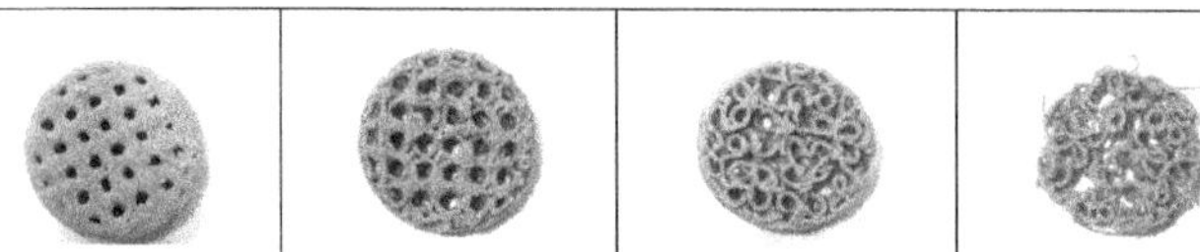			
Deposition rate ($mm^3\ s^{-1}$)		1.582	3.168	4.752	6.332
	80	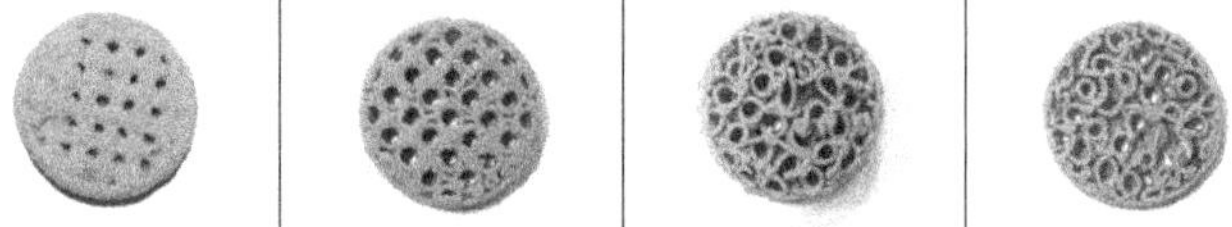			
Deposition rate ($mm^3\ s^{-1}$)		2.11	4.224	6.336	8.448
Observation		Increased flow % resulted in a higher deposition of material. With an increase in the printing speed, there was a discrepancy in the shape and structure of the printed object. The desired extrusion was obtained at 600 $mm\ min^{-1}$ printing speed with 60% material flow			
0.5	60	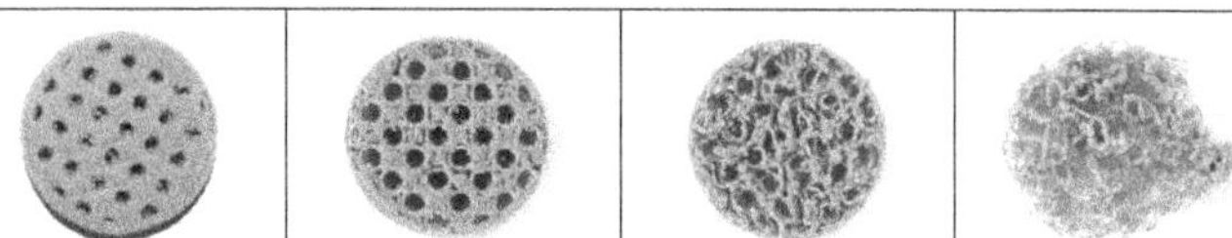			
Deposition rate ($mm^3\ s^{-1}$)		0.589	1.178	1.767	2.356
	80	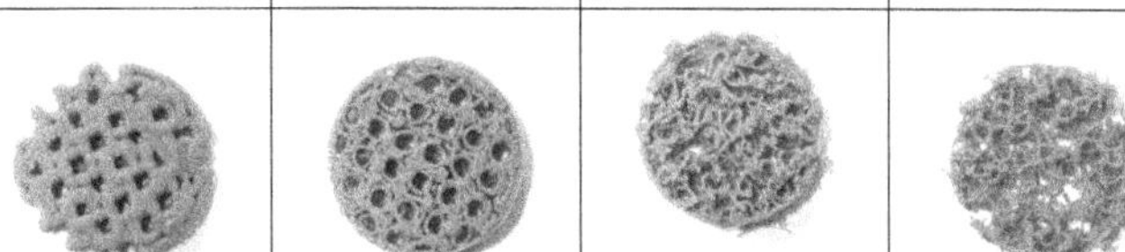			
Deposition rate ($mm^3\ s^{-1}$)		0.785	1.57	2.356	3.141
Observation		Lower printing speed resulted in over extrusion of material while higher printing speed resulted in under extrusion. The optimized desired extrusion was obtained at 80% material flow with the smaller sized nozzle of 0.5 mm			

Figure 8.5 Process optimization of printing variables of size fractions of potato peel powder. *Source:* From Madhumitha et al. (2021) with permission of Springer nature.

8.6 Future Trends and Perspectives

3D food printing forecasted to be a disruptive revolutionary technology that allows us to produce foods sustainably. Rather than manufacturing foods from common food sources, 3D food printing provides us opportunities to utilize innovative alternative ingredients. As a greener approach, the use of alternative food sources preserves food security and biodiversity (Caporizzi et al. 2019). Emergent proteins of this category include plant-based sources, seaweeds, microalgae, fungi, and insects that are rich sources of proteins. Possessing a high feed conversion rate, these novel alternative ingredients have a less environmental impact with reduced CO_2 emissions and climate change (Westhoek et al. 2014). Moreover, alternative food ingredients remain a good source of macro and micronutrients with high profiles of amino acids. Due to the unique texture of alternative ingredients and superior nutritional quality, conventionally these ingredients are used for the development of meat analogues as a substitute for animal proteins (He et al. 2020). Utilization of these ingredients for 3D printing opens up a new dimension of food constructs with high potential to mimic the textural and sensorial attributes of the desired animal proteins. This remains to give a promissory solution to a rising protein demand and food shortage. Further, the land pressure and water consumption are relatively less with these alternative food ingredients further integration with 3D printing assists in a sustainable productional approach.

Alternative food ingredients also contain a range of micronutrients possessing several nutritional and functional properties. In this regard, 3D food printing allows us to restructure the food formulation with the incorporation of the essential bioactive components for the enrichment of the 3D printed constructs with additional functionalities and nutritional benefits (Godoi et al. 2018). Integration of food printing with advanced biotechnological approaches like computational genomics aids in the modification of food constituents at the molecular level tailors the personalization of the foods based on the health status of the individuals (Kumar et al. 2020). The transformation of the balanced diet to a digitalized diet could positively address the nutritional deficiencies both at the individual level and population group. Hence the potential functional properties of these alternative food sources such as gelling properties, hydration, swelling mechanism, pH response, crystallinity, and thermal stability yet need to be explored for 3D printing applications that help in the development of future sustainable foods (Nachal et al. 2019). In complementation with the conventional food processing techniques, 3D printing could unravel the potential feasibilities of alternative food ingredients for improved food production.

8.7 Challenges and Limitations

Food processing is a complex interdisciplinary field where the success of the developed food product depends on consumer perception and acceptance. With the rising need and demand for food customization and personalization in terms of sensory and nutritional aspects, 3D printing is gaining the attention of food technologists in recent days. Alternative food sources are those food ingredients that are uncommon, unexplored, and underutilized with potential nutraceutical benefits. Although 3D printing delivers nutritious alternative food ingredients in fascinating shapes and forms with newer tastes and mouthfeels, consumer acceptance is a key criterion that determines the sustainability of the product. The challenge associated with the utilization of alternative ingredients for 3D printing is food neophobia which is the unwillingness of individuals to try new foods (Melgar-Lalanne et al. 2019). Thus, consumer aversion remains a strong metal barrier for the evolution of the novel 3D printed foods from alternative food ingredients such as insects, lupine seeds, fungi, and algae. Increasing awareness about the environmental impact of meat production, climatic changes, and depletion of fossil fuels are mandatory to explore the motivational food choices existing with alternative food ingredients.

The next challenge that exists with the exploration of the alternative ingredients for 3D printing is safety and edibility. Although these novel proteins are a good candidate for the nutritional delivery of bioactive compounds, the risk associated with the absorption and bioavailability is not well understood (Derossi et al. 2020). Further, the toxicity and allergenicity involved in the consumption of alternative food ingredients are yet to be studied. Hence, more studies have to be directed in analyzing the fate of the emergent novel proteins in the human gastrointestinal tract for unwinding the potential feasibilities of 3D printing for the development of sustainable futuristic foods. Another limitation with the usage of the alternative ingredients is their unique flavour and taste. This adds an unusual smell to the food formulations when incorporated with alternative ingredients that increase the dislikeness of the product. This issue can be overcome by the synergistic approach of other food processing technologies like encapsulation for masking the unpleasant flavour and odour (Nachal et al. 2019). Instead of using whole forms of alternative ingredients, the modified forms greatly reduce the aversion of the products. Thus, materials of this category of alternative food sources require appropriate pre- as well as post-processing in addition to the optimization of printing variables for the successful printability and development of the novel 3D printed foods.

8.8 Conclusion

Food layer manufacturing is a sustainable approach utilizing novel innovative food sources for the production of foods that has significantly lowers CO_2 footprints. The fabrication of 3D constructs from alternative ingredients remains to offer a promising solution in satisfying global hunger, especially the malnourishment and nutritional deficiency among infants and adolescents. Different food sources that are remained to be uncommon, unexplored, and underutilized have been briefed in the present chapter along with potential opportunities that exist for 3D printing with nutritional functionalities. The long-term applicability of these alternative food sources relies on the investigation of toxicity and allergenicity. Considering the advantages of alternative ingredients, only a handful of studies were reported in the literature explaining the scope of utilizing alternative foods for 3D printing. However, more studies have to be undertaken for exploring the applications of sustainable foods as a healthier choice with the potential to replace conventional food practices using 3D printing. Further, food printing allows for the valorization of food waste in transforming the waste into wealth leading to a greener circular economy.

References

Adesogan, A.T., Havelaar, A.H., McKune, S.L. et al. (2020). Animal source foods: sustainability problem or malnutrition and sustainability solution? Perspective matters. *Global Food Security* **25**: 100325.

Ahuja, I., Dauksas, E., Remme, J.F. et al. (2020). Fish and fish waste-based fertilizers in organic farming–with status in Norway: a review. *Waste Management* **115**: 95–112.

Aksoylu Özbek, Z., Çelik, K., Günç Ergönül, P., and Hepçimen, A.Z. (2020). A promising food waste for food fortification: characterization of dried tomato pomace and its cold pressed oil. *Journal of Food Chemistry & Nanotechnology* **6** (1): 9–17.

Anzani, C., Boukid, F., Drummond, L. et al. (2020). Optimising the use of proteins from rich meat co-products and non-meat alternatives: nutritional, technological and allergenicity challenges. *Food Research International* **137**: 109575.

Bach, F., Helm, C.V., Bellettini, M.B. et al. (2017). Edible mushrooms: a potential source of essential amino acids, glucans and minerals. *International Journal of Food Science & Technology* **52** (11): 2382–2392.

BCC Research (2019). *Synthetic (Cultured) Meat: Technologies and Global Markets.* BCC Publishing. https://www.bccresearch.com/market-research/biotechnology/synthetic-cultured-meat.html#:~:text=The global synthetic (cultured) meat,for the period 2022-2027 (accessed 21 October 2020).

Bombe, K. (2020). *Alternative Proteins: The Market at the Cusp of Multi-Billion Dollar Growth.* Meticulous Market Research Pvt. Ltd. https://www.globenewswire.com/news-release/2020/02/26/1991190/0/en/Alternative-Proteins-The-Market-at-the-Cusp-of-Multi-billion-Dollar-Growth.html (accessed 21 October 2020).

Bouazizi, S., Montevecchi, G., Antonelli, A., and Hamdi, M. (2020). Effects of prickly pear (*Opuntia ficus-indica* L.) peel flour as an innovative ingredient in biscuits formulation. *LWT* **124**: 109155.

Boukid, F. (2020). Plant-based meat analogues: from niche to mainstream. *European Food Research and Technology* **247**: 1–12.

Caporizzi, R., Derossi, A., and Severini, C. (2019). Cereal-based and insect-enriched printable food: from formulation to postprocessing treatments. status and perspectives. In: *Fundamentals of 3D Food Printing and Applications* (eds. F.C. Godoi, B.R. Bhandari, S. Prakash and M. Zhang). Elsevier Inc. https://doi.org/10.1016/B978-0-12-814564-7.00004-3.

Caselato-Sousa, V.M. and Amaya-Farfán, J. (2012). State of knowledge on amaranth grain: a comprehensive review. *Journal of Food Science* **77** (4): R93–R104.

Castillejo, N., Martinez-Hernández, G.B., Goffi, V. et al. (2018). Natural vitamin B12 and fucose supplementation of green smoothies with edible algae and related quality changes during their shelf life. *Journal of the Science of Food and Agriculture* **98** (6): 2411–2421.

Cecilia, J.A., Garcia-Sancho, C., Maireles-Torres, P.J., and Luque, R. (2019). Industrial food waste valorization: a general overview. In: *Biorefinery* (eds. J.-R. Bastidas-Oyanedel and J.E. Schmidt), pp. 253–277. Springer.

Chivenge, P., Mabhaudhi, T., Modi, A.T., and Mafongoya, P. (2015). The potential role of neglected and underutilised crop species as future crops under water scarce conditions in Sub-Saharan Africa. *International Journal of Environmental Research and Public Health* **12** (6): 5685–5711.

D'Amour, C.B., Pandey, B., Reba, M. et al. (2020). Urbanization, processed foods, and eating out in India. *Global Food Security* **25**: 100361.

Danowska-Oziewicz, M. and Kurp, L. (2017). Physicochemical properties, lipid oxidation and sensory attributes of pork patties with lupin protein concentrate stored in vacuum, modified atmosphere and frozen state. *Meat Science* **131**: 158–165.

Davis, K.F., Gephart, J.A., Emery, K.A. et al. (2016). Meeting future food demand with current agricultural resources. *Global Environmental Change* **39**: 125–132.

De Marco, E.R., Steffolani, M.E., Martinez, C.S., and León, A.E. (2014). Effects of spirulina biomass on the technological and nutritional quality of bread wheat pasta. *LWT-Food Science and Technology* **58** (1): 102–108.

Derossi, A., Caporizzi, R., Azzollini, D., and Severini, C. (2018). Application of 3D printing for customized food. A case on the development of a fruit-based snack for children. *Journal of Food Engineering* **220**: 65–75.

Derossi, A., Husain, A., Caporizzi, R., and Severini, C. (2020). Manufacturing personalized food for people uniqueness. An overview from traditional to emerging technologies. *Critical Reviews in Food Science and Nutrition* **60** (7): 1141–1159.

Dobermann, D., Swift, J.A., and Field, L.M. (2017). Opportunities and hurdles of edible insects for food and feed. *Nutrition Bulletin* **42** (4): 293–308.

FAO (2019). The state of food security and nutrition in the world. Safeguarding against economic slowdowns and downturns. http://www.fao.org/state-of-food-security-nutrition#:~:text=Prevalence of undernourishment (PoU)&text=This means that today%2C a,nine people in the world (accessed 21 October 2020).

Fasolin, L.H., Pereira, R.N., Pinheiro, A.C. et al. (2019). Emergent food proteins–towards sustainability, health and innovation. *Food Research International* **125**: 108586.

Feng, C., Zhang, M., and Bhandari, B. (2018). Materials properties of printable edible inks and printing parameters optimization during 3D printing: a review. *Critical Reviews in Food Science and Nutrition* **8398**: 1–8. https://doi.org/10.1080/10408398.2018.1481823.

Gençdaug, E., Görgüç, A., and Yilmaz, F.M. (2020). Recent advances in the recovery techniques of plant-based proteins from agro-industrial by-products. *Food Reviews International* **37**: 1–22.

GlobeNewswire (2019). Protein extracts from single cell protein sources market is expected to advance at a cagr of 8.6% in coming years: P&S intelligence. Prescient & Strategic Intelligence. https://www.globenewswire.com/news-release/2019/05/08/1819107/0/en/Protein-Extracts-from-Single-Cell-Protein-Sources-Market-is-Expected-to-Advance-at-a-CAGR-of-8-6-in-Coming-Years-P-S-Intelligence.html (accessed 21 October 2020).

Godoi, F.C., Bhandari, B.R., Prakash, S., and Zhang, M. (2018). *Fundamentals of 3D Food Printing and Applications*. Academic Press.

González, A., Cruz, M., Losoya, C. et al. (2020). Edible mushrooms as a novel protein source for functional foods. *Food & Function* **11**: 7400–7414.

Gopalakrishnan, L., Doriya, K., and Kumar, D.S. (2016). Moringa oleifera: a review on nutritive importance and its medicinal application. *Food Science and Human Wellness* **5** (2): 49–56.

Gulzar, M. and Minnaar, A. (2017). Underutilized protein resources from African legumes. In: *Sustainable Protein Sources* (eds. S.R. Nadathur and L. Scanlin), pp. 197–208. Elsevier.

Hartmann, C. and Siegrist, M. (2016). Becoming an insectivore: results of an experiment. *Food Quality and Preference* **51**: 118–122.

Hartmann, C. and Siegrist, M. (2017). Consumer perception and behaviour regarding sustainable protein consumption: a systematic review. *Trends in Food Science & Technology* **61**: 11–25.

He, J., Evans, N.M., Liu, H., and Shao, S. (2020). A review of research on plant-based meat alternatives: driving forces, history, manufacturing, and consumer attitudes. *Comprehensive Reviews in Food Science and Food Safety* **19** (5): 2639–2656.

Herrero, M., Wirsenius, S., Henderson, B. et al. (2015). Livestock and the environment: what have we learned in the past decade? *Annual Review of Environment and Resources* **40**: 177–202.

Herrero, M., Henderson, B., Havlik, P. et al. (2016). Greenhouse gas mitigation potentials in the livestock sector. *Nature Climate Change* **6** (5): 452–461.

Herrero, M., Thornton, P.K., Mason-D'Croz, D. et al. (2020). Innovation can accelerate the transition towards a sustainable food system. *Nature Food* **1** (5): 266–272.

Hickisch, A., Beer, R., Vogel, R.F., and Toelstede, S. (2016). Influence of lupin-based milk alternative heat treatment and exopolysaccharide-producing lactic acid bacteria on the physical characteristics of lupin-based yogurt alternatives. *Food Research International* **84**: 180–188.

van Huis, A. (2016). Edible insects are the future? *Proceedings of the Nutrition Society* **75** (3): 294–305.

van Huis, A. and Oonincx, D.G.A.B. (2017). The environmental sustainability of insects as food and feed: a review. *Agronomy for Sustainable Development* **37** (5): 43.

Hunter, D., Borelli, T., Beltrame, D.M.O. et al. (2019). The potential of neglected and underutilized species for improving diets and nutrition. *Planta*, pp. 1–21.

Jagadiswaran, B., Alagarasan, V., Palanivelu, P. et al. (2021). Valorization of food industry waste and by-products using 3D printing: a study on the development of value-added functional cookies. *Future Foods* **4**: 100036.

Jayaprakash, S., Ituarte, I.F., and Partanen, J. (2019). Prosumer-driven 3D food printing: role of digital platforms in future 3D food printing systems. In: *Fundamentals of 3D Food Printing and Applications* (eds. F.C. Godoi, B.R. Bhandari, S. Prakash and M. Zhang), pp. 331–354. Elsevier.

Jeddou, K.B., Bouaziz, F., Zouari-Ellouzi, S. et al. (2017). Improvement of texture and sensory properties of cakes by addition of potato peel powder with high level of dietary fiber and protein. *Food Chemistry* **217**: 668–677.

Joyce, A., Dixon, S., Comfort, J., and Hallett, J. (2012). Reducing the environmental impact of dietary choice: perspectives from a behavioural and social change approach. *Journal of Environmental and Public Health* **2012** *7 pages* https://doi.org/10.1155/2012/978672.

Katoch, R. (2020). Underutilized rops: an overview. In: *Ricebean*. Singapore.: Springer https://doi.org/10.1007/978-981-15-5293-9_1.

Keerthana, K., Anukiruthika, T., Moses, J.A., and Anandharamakrishnan, C. (2020). Development of fiber-enriched 3D printed snacks from alternative foods: a study on button mushroom. *Journal of Food Engineering* https://doi.org/10.1016/j.jfoodeng.2020.110116.

Kumar, S., Anukiruthika, T., Dutta, S. et al. (2020). Iron deficiency anemia: a comprehensive review on iron absorption, bioavailability and emerging food fortification approaches. *Trends in Food Science & Technology* https://doi.org/10.1016/j.tifs.2020.02.021.

Le Goff, G. and Delarue, J. (2017). Non-verbal evaluation of acceptance of insect-based products using a simple and holistic analysis of facial expressions. *Food Quality and Preference* **56**: 285–293.

Lee, J.-K., Patel, S.K.S., Sung, B.H., and Kalia, V.C. (2020). Biomolecules from municipal and food industry wastes: an overview. *Bioresource Technology* **298**: 122346.

Liu, Z., Dick, A., Prakash, S. et al. (2020). Texture modification of 3D printed air-fried potato snack by varying its internal structure with the potential to reduce oil content. *Food and Bioprocess Technology* **13** (3): 564–576.

Lonnie, M. and Johnstone, A.M. (2020). The public health rationale for promoting plant protein as an important part of a sustainable and healthy diet. *Nutrition Bulletin* **45** (3): 281–293.

MacArtain, P., Gill, C.I.R., Brooks, M. et al. (2007). Nutritional value of edible seaweeds. *Nutrition Reviews* **65** (12): 535–543.

Madhumitha, M., Abinash, V., Rahul, T. et al. (2021). Valorization of food industry waste streams using 3D printing: a study on noodles prepared from potato peel waste. *Food and Bioprocess Technology*.

Maina, S., Kachrimanidou, V., and Koutinas, A. (2017). A roadmap towards a circular and sustainable bioeconomy through waste valorization. *Current Opinion in Green and Sustainable Chemistry* **8**: 18–23.

Market Analysis Report (2020). Commercial Seaweeds Market Size, Share & Trends Analysis Report By Product (Brown Seaweeds, Red Seaweeds, Green Seaweeds), By Form (Liquid, Powdered, Flakes), By Application, By Region, And Segment Forecasts, pp. 2020–2027. https://www.grandviewresearch.com/industry-analysis/commercial-seaweed-market#:~:text=The global commercial seaweeds market,9.1%25 over the forecast period.&text=The industry in the U.S.,USD 311.4 million in 2019 (21 October 2020).

Masood, F., Haque, A., Ahmad, S., and Malik, A. (2020). Potential of food processing by-products as dietary fibers. In: *Functional Food Products and Sustainable Health* (eds. S. Ahmad and N.A. Al-Shabib), 51–67. Singapore: Springer, Springer https://doi.org/10.1007/978-981-15-4716-4_5.

McClements, D.J. (2020). Future foods: a manifesto for research priorities in structural design of foods. *Food & Function* **11** (3): 1933–1945.

Meini, M.-R., Cabezudo, I., Boschetti, C.E., and Romanini, D. (2019). Recovery of phenolic antioxidants from Syrah grape pomace through the optimization of an enzymatic extraction process. *Food Chemistry* **283**: 257–264.

Melgar-Lalanne, G., Hernández-Álvarez, A.-J., and Salinas-Castro, A. (2019). Edible insects processing: traditional and innovative technologies. *Comprehensive Reviews in Food Science and Food Safety* **18** (4): 1166–1191.

Meticulous Research (2019). *Alternative Protein Market - Global Opportunity Analysis and Industry Forecast (2019-2025)*. Meticulous Market Research Pvt. Ltd. https://www.meticulousresearch.com/product/alternative-protein-market-4985/?utm_source=globnewswire.com&utm_medium=pressrelease2&utm_campaign=paid (accessed 21 October 2020).

Mirabella, N., Castellani, V., and Sala, S. (2014). Current options for the valorization of food manufacturing waste: a review. *Journal of Cleaner Production* **65**: 28–41.

Mishyna, M., Martinez, J.-J.I., Chen, J., and Benjamin, O. (2019). Extraction, characterization and functional properties of soluble proteins from edible grasshopper (*Schistocerca gregaria*) and honey bee (*Apis mellifera*). *Food Research International* **116**: 697–706.

Mobin, S. and Alam, F. (2017). Some promising microalgal species for commercial applications: a review. *Energy Procedia* **110**: 510–517.

Mottet, A., de Haan, C., Falcucci, A. et al. (2017). Livestock:on our plates or eating at our table? A new analysis of the feed/food debate. *Global Food Security* **14**: 1–8.

Muthamilarasan, M. and Prasad, M. (2020). Small millets for enduring food security amidst pandemics. *Trends in Plant Science* **26**: 33–40.

Muthamilarasan, M., Singh, N.K., and Prasad, M. (2019). Multi-omics approaches for strategic improvement of stress tolerance in underutilized crop species: a climate change perspective. In: *Advances in Genetics*, vol. **103** (ed. D. Kumar), pp. 1–38. Elsevier.

Mwangi, M.N., Oonincx, D.G.A.B., Stouten, T. et al. (2018). Insects as sources of iron and zinc in human nutrition. *Nutrition Research Reviews* **31** (2): 248–255.

Nachal, N., Moses, J., Karthik, K., and Anandharamakrishnan, C. (2019). Applications of 3D printing in food industry: a review. *Food Engineering Reviews* **11**: 123–141.

Nadathur, S., Wanasundara, J.P.D., and Scanlin, L. (2016). *Sustainable Protein Sources.* Academic Press.

Naknaen, P., Itthisoponkul, T., Sondee, A., and Angsombat, N. (2016). Utilization of watermelon rind waste as a potential source of dietary fiber to improve health promoting properties and reduce glycemic index for cookie making. *Food Science and Biotechnology* **25** (2): 415–424.

Oonincx, D.G.A.B. (2017). Environmental impact of insect production. In: *Insects as Food and Feed: From Production to Consumption* (eds. A. Van Huis and J.K. Tomberlin), pp. 79–93. Wageningen Academic Publishers.

Oyeyinka, A.T. and Oyeyinka, S.A. (2018). *Moringa oleifera* as a food fortificant: recent trends and prospects. *Journal of the Saudi Society of Agricultural Sciences* **17** (2): 127–136.

Padmavathy, K. and Poyyamoli, G. (2011). Alternative farming techniques for sustainable food production. In: *Genetics, Biofuels and Local Farming Systems. Sustainable Agriculture Reviews*, vol. **7** (ed. E. Litchtfouse), 367–424. Dordrecht: Springer https://doi.org/10.1007/978-94-007-1521-9_13.

Paiva, L., Lima, E., Patarra, R.F. et al. (2014). Edible Azorean macroalgae as source of rich nutrients with impact on human health. *Food Chemistry* **164**: 128–135.

Perez, B., Nykvist, H., Brogger, A.F. et al. (2019). Impact of macronutrients printability and 3D-printer parameters on 3D-food printing: a review. *Food Chemistry* **287** (October 2018): 249–257. https://doi.org/10.1016/j.foodchem.2019.02.090.

Piha, S., Pohjanheimo, T., Lahteenmaki-Uutela, A. et al. (2018). The effects of consumer knowledge on the willingness to buy insect food: an exploratory cross-regional study in Northern and Central Europe. *Food Quality and Preference* **70**: 1–10.

Plazzotta, S., Manzocco, L., and Nicoli, M.C. (2017). Fruit and vegetable waste management and the challenge of fresh-cut salad. *Trends in Food Science & Technology* **63**: 51–59.

Possidónio, C., Prada, M., Graça, J., and Piazza, J. (2020). Consumer perceptions of conventional and alternative protein sources: a mixed-methods approach with meal and product framing. *Appetite* **156**: 104860.

Post, M.J., Levenberg, S., Kaplan, D.L. et al. (2020). Scientific, sustainability and regulatory challenges of cultured meat. *Nature Food* **1** (7): 403–415.

Prakash, S., Bhandari, B.R., Godoi, F.C., and Zhang, M. (2019). Future outlook of 3D food printing. In: *Fundamentals of 3D Food Printing and Applications* (eds. F.C. Godoi, B.R. Bhandari, S. Prakash and M. Zhang), pp. 373–381. Elsevier.

Premkumar, J. and Vasudevan, R.T. (2018). Bioingredients: functional properties and health impacts. *Current Opinion in Food Science* **19**: 120–128.

Ranganathan, J., Vennard, D., Waite, R. et al. (2016). *Shifting Diets for a Sustainable Food Future*, 831878447–1548317259. World Resources Institute.

Razavi, S.M.A. (2019). Introduction to emerging natural hydrocolloids. In: *Emerging Natural Hydrocolloids: Rheology and Functions* (ed. S.M.A. Razavi), 1–52. USA: John Wiley & Sons, Inc.

Reis, F.S., Martins, A., Vasconcelos, M.H. et al. (2017). Functional foods based on extracts or compounds derived from mushrooms. *Trends in Food Science & Technology* **66**: 48–62.

Ricci, I., Derossi, A., and Severini, C. (2019). 3D printed food from fruits and vegetables. In: *Fundamentals of 3D Food Printing and Applications*, vol. **153**, Issue 1 (eds. F.C. Godoi, B.R. Bhandari, S. Prakash and M. Zhang), 117–149. Elsevier Ltd. https://doi.org/10.1016/b978-0-12-814564-7.00005-5.

Roos, N. (2018). Insects and human nutrition. In: *Edible Insects in Sustainable Food Systems* (eds. A. Halloran, R. Flore, P. Vantomme and N. Ross), 83–91. Cham: Springer https://doi.org/10.1007/978-3-319-74011-9_5.

Sadh, P.K., Duhan, S., and Duhan, J.S. (2018). Agro-industrial wastes and their utilization using solid state fermentation: a review. *Bioresources and Bioprocessing* **5** (1): 1.

Salehi, F. (2019). Characterization of different mushrooms powder and its application in bakery products: a review. *International Journal of Food Properties* **22** (1): 1375–1385.

Santos, T.D., de Freitas, B.C.B., Moreira, J.B. et al. (2016). Development of powdered food with the addition of Spirulina for food supplementation of the elderly population. *Innovative Food Science & Emerging Technologies* **37**: 216–220.

Severini, C., Azzollini, D., Albenzio, M., and Derossi, A. (2018). On printability, quality and nutritional properties of 3D printed cereal based snacks enriched with edible insects. *Food Research International* **106** (November 2017): 666–676. https://doi.org/10.1016/j.foodres.2018.01.034.

Siegrist, M. and Hartmann, C. (2019). Impact of sustainability perception on consumption of organic meat and meat substitutes. *Appetite* **132**: 196–202.

Soares, S., Kupfer, P., Nicholls, B. et al. (2011). Insects Au Gratin. 2012.

Spring, Ú.O. (2019). Climate-smart agriculture and a sustainable food system for a sustainable-engendered Peace. In: *Climate Change, Disasters, Sustainability Transition and Peace in the Anthropocene. The Anthropocene: Politik—Economics—Society—Science*, vol. **25** (eds. H.G. Brauch, U.O. Spring, A.E. Collins and S.E.S. Oswald), 95–123. Cham: Springer https://doi.org/10.1007/978-3-319-97562-7_5.

Statista (2019). Food and nutrition: forecasted market value of plant-based meat worldwide from 2018 to 2026. https://www.statista.com/statistics/877369/global-meat-substitutes-market-value (accessed 21 October 2020)

Stephan, A., Ahlborn, J., Zajul, M., and Zorn, H. (2018). Edible mushroom mycelia of *Pleurotus sapidus* as novel protein sources in a vegan boiled sausage analog system: functionality and sensory tests in comparison to commercial proteins and meat sausages. *European Food Research and Technology* **244** (5): 913–924.

Stull, V.J., Finer, E., Bergmans, R.S. et al. (2018). Impact of edible cricket consumption on gut microbiota in healthy adults, a double-blind, randomized crossover trial. *Scientific Reports* **8** (1): 1–13.

Sudhakar, M.P., Kumar, B.R., Mathimani, T., and Arunkumar, K. (2019). A review on bioenergy and bioactive compounds from microalgae and macroalgae-sustainable energy perspective. *Journal of Cleaner Production* **228**: 1320–1333.

Sun, J., Zhou, W., Huang, D., and Yan, L. (2018a). 3D food printing: perspectives. In: *Polymers for Food Applications* (ed. T.J. Gutierrez), 725–755. Springer.

Sun, J., Zhou, W., Yan, L. et al. (2018b). Extrusion-based food printing for digitalized food design and nutrition control. *Journal of Food Engineering* **220**: 1–11.

Teigiserova, D.A., Hamelin, L., and Thomsen, M. (2019). Review of high-value food waste and food residues biorefineries with focus on unavoidable wastes from processing. *Resources, Conservation and Recycling* **149**: 413–426.

Thambiraj, S.R., Phillips, M., Koyyalamudi, S.R., and Reddy, N. (2018). Yellow lupin (*Lupinus luteus* L.) polysaccharides: antioxidant, immunomodulatory and prebiotic activities and their structural characterisation. *Food Chemistry* **267**: 319–328.

Theagarajan, R., Malur Narayanaswamy, L., Dutta, S. et al. (2019). Valorisation of grape pomace (cv. Muscat) for development of functional cookies. *International Journal of Food Science & Technology* **54**: 1299–1305.

Theagarajan, R., Moses, J.A., and Anandharamakrishnan, C. (2020). 3D extrusion printability of rice starch and optimization of process variables. *Food and Bioprocess Technology* https://doi.org/10.1007/s11947-020-02453-6.

Tripathi, P. and Mogra, R. (2020). Organoleptic and nutritional evaluation of value added noodles from amaranth seed flour. *International Research Journal of Pure and Applied Chemistry* **21** (13): 39–47. https://doi.org/10.9734/irjpac/2020/v21i133023.

United Nations (2017). World population projected to reach 9.8 billion in 2050 and 11.2 billion in 2100. Department of Economic and Social Affairs, United Nations. https://www.un.org/development/desa/en/news/population/world-population-prospects-2017.html (6 October 2020).

Vieira, M.V., Oliveira, S.M., Amado, I.R. et al. (2020). 3D printed functional cookies fortified with *Arthrospira platensis*: evaluation of its antioxidant potential and physical-chemical characterization. *Food Hydrocolloids* **107**: 105893.

Villarino, C.B.J., Jayasena, V., Coorey, R. et al. (2015). The effects of lupin (*Lupinus angustifolius*) addition to wheat bread on its nutritional, phytochemical and bioactive composition and protein quality. *Food Research International* **76**: 58–65.

Voon, S.L., An, J., Wong, G. et al. (2019). 3D food printing: a categorised review of inks and their development. *Virtual and Physical Prototyping* **14** (3): 203–218.

Wang, X., Kristo, E., and LaPointe, G. (2020). Adding apple pomace as a functional ingredient in stirred-type yogurt and yogurt drinks. *Food Hydrocolloids* **100**: 105453.

Watson, J. (2019). insect protein market is expected to reach USD 1.4 billion by 2026. Reports and Data. https://www.globenewswire.com/news-release/2019/10/07/1925879/0/en/Insect-Protein-Market-is-expected-To-Reach-USD-1-4-Billion-By-2026-Reports-And-Data.html (6 October 2020).

Wegier, A., Alavez, V., Pérez-López, J. et al. (2018). Beef or grasshopper hamburgers: the ecological implications of choosing one over the other. *Basic and Applied Ecology* **26**: 89–100.

Westhoek, H., Lesschen, J.P., Rood, T. et al. (2014). Food choices, health and environment: effects of cutting Europe's meat and dairy intake. *Global Environmental Change* **26**: 196–205.

Wollenberg, E., Richards, M., Smith, P. et al. (2016). Reducing emissions from agriculture to meet the 2 C target. *Global Change Biology* **22** (12): 3859–3864.

Wu, G. (2013). *Amino Acids: Biochemistry and Nutrition*. CRC Press.
Wu, G., Fanzo, J., Miller, D.D. et al. (2014). Production and supply of high-quality food protein for human consumption: sustainability, challenges, and innovations. *Annals of the New York Academy of Sciences* **1321** (1): 1–19.
Zlatanović, S., Kalušević, A., Micić, D. et al. (2019). Functionality and storability of cookies fortified at the industrial scale with up to 75% of apple pomace flour produced by dehydration. *Foods* **8** (11): 561.

9

Post-Processing of 3D Printed Foods

CHAPTER MENU

9.1 Introduction

3D printing of foods is considered as a future disruptive technology of digitalized food manufacturing process that involves direct solid freeform fabrication (SFF) of food constructs in a layered fashion (Baiano 2020). As an additive processing method, 3D printing offers several benefits in terms of multi-material food construct, simplified food supply chain, process automation, design novelty, food structuring, efficient usage of material resources, diverse product range, and customized foods at low cost (Liu and Zhang 2019). The freedom of food product designing depends on material selection and its composition. The significance of the macro constituents of the food system and its role in the enhancement of printability has been discussed in the earlier chapter on food components and their role in printability of the present book. 3D printing of foods is not as easy as polymer printing because food contains a diverse range of attributes. Hence, food 3D printing involves a cautious deposition of material and a safe delivery of the 3D printed foods until consumption (Tran 2019).

3D Printing of Foods, First Edition. C. Anandharamakrishnan, Jeyan A. Moses, and T. Anukiruthika.

Food is a perishable commodity and the presence of moisture serves as a good culture medium for microbes. Therefore, the task of 3D printing does not get stopped with the delivery of printed construct but until the safe delivery of finished 3D printed foods for consumption. Some material supplies such as chocolate spread, jam, jellies, hummus, and cheese frostings do not require a post-process cooking step (Sun et al. 2018). These materials may sometimes require a minimal cold setting for maintaining the structural stability of 3D constructs. However, most of the dough-based material supplies require additional post-processing, i.e. cooking immediately after the 3D printing process (Yang et al. 2019).

Traditional cooking methods such as boiling, baking, steaming, and frying are often employed for the post-processing of 3D printed products (Caporizzi et al. 2019). There occurs a sequence of multiple chemical reactions (starch gelatinization, protein denaturation, and reduction in moisture content) and physical transformations (changes in colour, texture, taste, and volume of product) during the cooking of 3D constructs (Nachal et al. 2019). The associated transformations during the cooking process greatly depend on the method of post-processing employed. Traditionally, the cooking processes are categorized as wet heat and dry heat methods. In the former type of cooking method, water serves as a cooking medium while the latter uses oil as a medium for cooking. When speaking about post-processing, it must be adequate that the printed 3D construct must withstand the cooking conditions without shape deformation (Feng et al. 2018). Hence a successfully developed 3D printed food involves not only the optimization of printability but also the optimization of the post-processing method. The choice of a suitable post-processing method is important for preserving the shape and print fidelity of the 3D printed product and the material's ability to withstand the rigorous cooking step. Although food printing has gained attention in recent days, most of the 3D printing studies focused on material formulation and optimization for the printing process (Dankar et al. 2018; Gholamipour-Shirazi et al. 2019; Liu et al. 2019c). Only a handful of reports were available on post-processing characteristics of the 3D printed construct. From the consumer's point of view, post-processing of the developed 3D construct is equally important as that of the printing process for ensuring the end applications of the developed product, consumer acceptance, edibility, and safety assurance. An overview of the 3D food printing process from raw material formulation to post-processing is presented in Figure 9.1. Hence, the present chapter focused on the scientific principles involved in the post-processing of 3D construct. Further, the significance and various approaches used for the development of 3D printed foods are discussed. Thus, the present chapter would provide broad knowledge on the selection of suitable post-processing methods based on the nature of material supplies without compromising the print fidelity of 3D printed construct. Also, this chapter on the post-processing of 3D printed foods would deliver valuable insights on emerging novel post-processing methods which can be used for 3D printed constructs that helps in preserving both the sensory and nutritional attributes of 3D printed foods.

9.2 Material Supply Requirements for Food 3D Printing

The three criteria that were considered to be significant in translating the requirements of the material supplies are printability, applicability, and post-processing in the 3D printing process (Godoi et al. 2016).

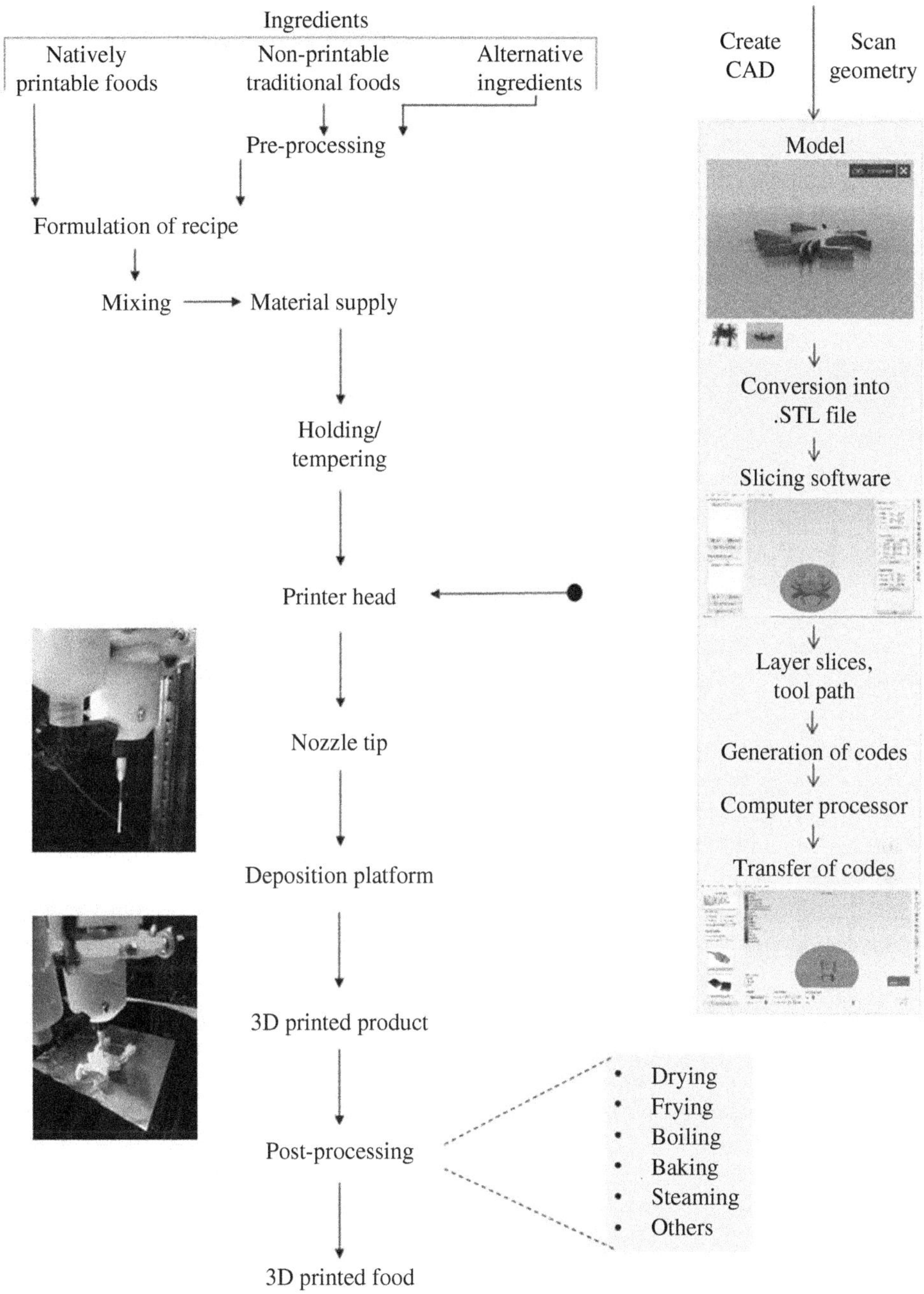

Figure 9.1 Overview of the 3D food printing process. *Source:* From Nachal et al. (2019), Figure 01 [p. 003] / With permission of Springer Nature. DOI-https://doi.org/10.1007/s12393-019-09199-8.

- Printability refers to the dimensional stability and the material's handling properties that impart adhesiveness and are self-supportive to hold the printed layers (Kim et al. 2018). As a foremost requirement of any material supply that has been selected for 3D printing must possess good printability for constructing a 3D structure. The fabrication of complex intricate designing of the 3D structures further depends on pre-processing and incorporation of food additives (He et al. 2019). 3D printing not only allows the use of traditional food materials such as staple food crops, fruits, and vegetables, and meat products but also allows to use of novel unfamiliar and uncommon food sources such as insects, microalgae, and pseudo-cereals (Lupton and Turner 2018; Tuorila and Hartmann 2020). The printability of material supplies relies on various aspects of material properties such as physiochemical, rheological, textural, mechanical, frictional, and thermal properties (Nachal et al. 2019). Mainly the printability was closely correlated with the rheological properties and the various printability tests based on the characterization of material supplies were detailly presented in the earlier chapters.
- The next requirement of material supply towards 3D printing is its applicability. The term applicability denotes a broad range of feasibilities that exists with the 3D printing of foods. The most attractive fascinating application of 3D printing is the creation of customized foods based on individual likeness, shape, colour, and nutritional needs (Sun et al. 2015). Further, the available food printing technologies are more specific towards the printing of particular material supplies. For instance, broadly it has been stated that highly viscous semi-solid materials best suits extrusion printing; low viscous liquid-based food inks are printed using inkjet printing and binder jetting; and powder-based dry materials are 3D printed based on the sintering process (Liu et al. 2017). Due to versatility and simplicity, 3D printing by extrusion method is most commonly used where advanced designing aids in the fabrication of dual materials through coaxial printing (Uribe-Wandurraga et al. 2020). Also, multi-material is feasible with extrusion technique. With advancements of technology 3D printing has been well utilized for the textural modification of foods for serving peoples with swallowing difficulty (dysphagia) (Dick et al. 2020). With innovations based on scientific principles, 3D printing is recently focused on developing foods with reduced salt, sugar, and calorie intake (Prakash et al. 2019). More recently, scientists are putting efforts into the fabrication of plant-based 3D printed meat analogues as a view of a sustainable approach possessing both environmental and economic benefits.
- The third significant attribute of material supply towards 3D printing is the material's ability to withstand post-processing conditions (Yang et al. 2019). The consumer's acceptability of 3D-printed food is highly influenced by the post-processing quality. The common chemical reactions that occur during post-processing include Maillard browning, hydrogel-forming and gelation, and protein denaturation (He et al. 2019). The complex interactions of macro and microconstituents of foods under post-processing conditions determine the mechanical strength of the 3D printed product (Jonkers et al. 2020). The product hardness is adequate that imparts printed construct to withstand handling, packaging, and transportation of the final product to the consumers. Some of the conventional post-processing methods that were used for 3D printed construct include oven drying, frying, boiling, baking, and freezing (Liu et al. 2020a; Nachal et al. 2019). To preserve the heat-labile components of the 3D printed construct, attempts were made on the application of non-thermal processing methods such as freeze-drying and conductive

hydro-drying for post-processing of 3D printed constructs (Anukiruthika et al. 2020a; Lille et al. 2018). A detailed review of the concepts of various post-processing methods are presented in the following sections.

9.3 Post-Processing Methods

9.3.1 Drying

Hot air drying is one of the simpler methods used for drying high moisture foods such as fruits, vegetables, and meat products. The working principle of oven drying is based on the convective heat transfer where the hot and dry air is allowed to circulate over the drying chamber that removes the moisture from wet foods (Anukiruthika et al. 2020b). Based on the initiation of motion of the fluid, the convectional currents can be of the natural and forced types. In natural convection, the less dense hot air rises and gets circulated mainly due to the effects of gravity and buoyancy. On the other hand, a blower is provided for the circulation of the hot air in the forced convection (Mohana et al. 2020). Since 3D printing involves a layered deposition of food, the material density and its distribution are not as same as that of the conventional moulded samples. Hence, the 3D printed samples require only a short duration of post-processing (Liu et al. 2020a). The 3D printed protein and fibre-based material supply were post-processed using two different drying approaches: oven drying and freeze-drying (Lille et al. 2018). Results showed that the layer details were well retained with the freeze-drying than the oven drying. This was due to the dehydration of the samples at their frozen state that aids in maintaining the structural integrity of the samples. The basic principle of freeze-drying is sublimation where the moisture in the form of ice is directly converted into gaseous vapour (Vishali et al. 2019). Since the material is in its frozen solid state, the structure has more ability to resist the associated stresses undergone during the process. On the other hand, product shrinkage was more obvious with the oven-dried 3D printed samples. The samples of 15% starch and 35% oat protein concentrate (OPC) showed the spreading of the material due to the reduction in the viscosity of the material supplies during the heating process of oven drying (Figure 9.2). However, the samples of both the post-processing methods were comparable when the initial dry matter content was $\geq$45% (Lille et al. 2018). This was attributed due to the lesser removal of water content and higher solids content that reflects the structural strength of the materials.

In another study on the development of cereal-based 3D construct, a conventional oven was used for cooking the 3D printed samples (Severini et al. 2016). The effect of the infill density and layer height on the printing performance of cereal dough was examined. It was observed that a greater amount of material was extruded with a higher infill percentage. This leads to the creation of new pores due to the evaporation of moisture during cooking. Also, the study inferred that the porosity of the cooked samples was positively correlated with layer height and the infill level of the 3D construct (Figure 9.3).

A similar approach of post-processing was adapted for the hot-melt extruded cereal-based insect flour (Alam et al. 2019). The extrudates were oven dried at 40 °C for 24 hours and used for further physiochemical analysis. In this study, oven drying was performed to achieve a

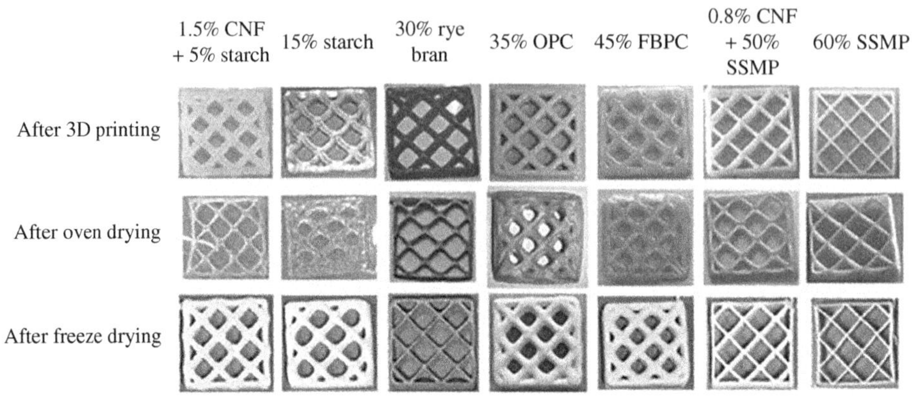

Figure 9.2 Effect of different post-processing methods on appearance of 3D printed samples. *Source:* From Lille et al. (2018), Figure 04 [p. 007] / With permission of Elsevier. DOI-https://doi.org/10.1016/j.jfoodeng.2017.04.034.

Layer height (mm)	Infill (%)	Calculated solid fraction (%)	Designed objects	Printed raw objects	Printed cooked objects
0.3	10	74.8			
0.4	15	76.2			
0.5	20	77.6			

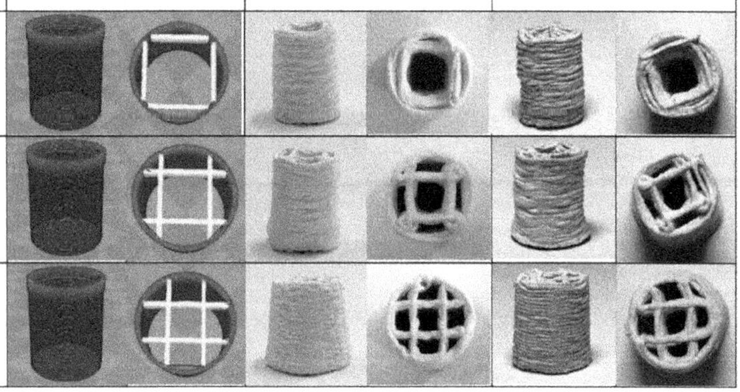

Figure 9.3 3D structures of 3D printed and post-processed 3D constructs. *Source:* From Severini et al. (2016), Figure 01[p.026] / With permission of ELSEVIER. DOI-https://doi.org/10.1016/j.ifset.2016.10.001

stable product for determining the printing behaviour of material supply. 3D printing is a promising technology not only for the fabrication of customized foods in terms of sensorial attributes but also as an efficient technology for the delivery of personalized foods in terms of nutritional requirements with specific needs (Kumar et al. 2020). In consideration of the development of a personalized functional diet, yam paste was 3D printed with edible rose pollen (Feng et al. 2020b). The shape stability, colour retention, bioactive content, and hardness were explored for different post-processing methods such as hot air oven drying, freeze-drying, and microwave vacuum drying. Results showed that freeze-dried samples possessed the best post-processing stability and hot air-dried samples had the worst post-processing stability. Although microwave vacuum dried samples were less scored than freeze-drying in appearance and colour, the flavour of microwave vacuum dried 3D construct was found to be superior to hot air-dried samples. Also, microwave vacuum drying showed greater efficiency of drying with a reduction of 84% drying time. Thus, this study inferred that combination of 3D printing with microwave vacuum drying could result in high valuable 3D printed foods especially applied for bioactive compounds in terms of both process efficiency and product quality.

9.3.2 Frying

Deep frying commonly known as frying is the cooking process of immersion of the food in a medium of hot oil at a very high temperature of around 150–190 °C (Choe and Min 2007). The sudden contact of food with hot oil imparts a desired crispy outer layer for fried products. The frying oil acts as a medium for heat transfer that allows simultaneous heat and mass transfer contributing to the unique texture and flavour profile of fried products. Factors such as oil temperature, duration of frying, the surface area of foods, material composition, moisture content, and type of frying oil significantly affect the oil absorption behaviour of the fried snacks (Asokapandian et al. 2020). An optimally fried product must be completely cooked and possess a golden-brown colour, crispy crust, and optimal oil absorption. The post-processing characteristics and the sensory quality of the 3D printed composite flour snacks were evaluated using different post-processing methods (Krishnaraj et al. 2019). In this study, three different post-processing methods such as deep-frying, microwave drying, and hot air drying followed by frying (HAD-DF) was performed. Results showed that the deep-fried 3D constructs at 165 °C had a bulged structure due to the puffing effects at the centre. This was due to the exertion of high pressure during the frying of 3D printed samples that leads to the sudden transition of moisture into steam. In contrast, the oil absorption of the HAD-DF samples was comparatively lower than the deep-fried samples. In HAD-DF, the hot air drying of 3D constructs before the frying process significantly reduced the exchange of moisture and oil. Further, the drying process assists in the retention of the shape and 3D structure of the printed samples during deep frying. On the other hand, microwave drying had the highest drying rates. Microwave drying is based on volumetric heating that generates heat directly inside the samples associated with rapid internal heating and drying. Results showed that the development of the outer crust layer was due to sudden moisture loss that prevents the removal of moisture from the centre of the sample (Krishnaraj et al. 2019). This significantly restricts the physical expansion of microwaved samples; however, a small extent of visible shrinkage was observed (Figure 9.4). Although the microwaved samples retained the details of the print quality, the HAD-DF samples had the highest overall acceptability.

In another study, the 3D printed scallop puree with transglutaminase as an additive was deep-dried (Lipton et al. 2010). The printed truncated 'L' and space shuttle from scallop puree demonstrated the material's ability to retain the fine details of the curve, edges, and smooth surface during 3D printing and post-processing. However, the layer definitions of deep-fried samples were not much clear that require further investigation. More recently, 3D printed chicken nuggets were developed with added refined wheat flour (Wilson et al. 2020). The post-processing of the optimized sample (GC: RWF – 2 : 1) followed an initial hot air oven drying followed by deep-frying. The initial processing of hot air oven drying helps

Raw 3D printed sample	Deep fried sample	HAD-DF sample	Microwave dried sample
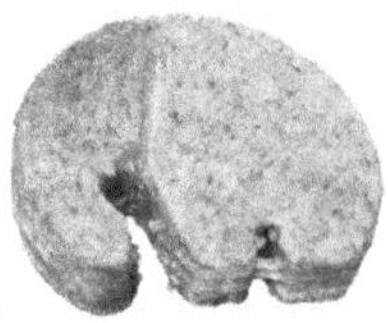			
Length: 2.82 cm Breadth: 3.93 cm Height: 0.63 cm	Length: 3.04 cm Breadth: 4.12 cm Height: 0.93 cm	Length: 2.89 cm Breadth: 3.95 cm Height: 0.61 cm	Length: 2.66 cm Breadth: 3.72 cm Height: 0.61 cm

Figure 9.4 Post-processed 3D printed snack from composite flour. *Source:* From Krishnaraj et al. (2019), Figure 03 [p. 008] / With permission of Springer Nature. DOI-https://doi.org/10.1007/s11947-019-02336-5.

to reduce the moisture level that in turn significantly influenced the oil uptake during deep-frying (Figure 9.5). The associated taste and characteristic flavour of the 3D printed nuggets were attributed to the fat coalescence and drip loss with simultaneous heat and mass transfer during deep frying.

9.3.3 Baking

Baking is another cooking method that utilizes dry heat generally done in an enclosed space such as a baking oven. A greater range of temperature up to 315 °C is used for baking that is much higher than the oven drying. During baking, heat is transferred from the outer surface to the core of the product (Ureta et al. 2019). Various physiochemical changes that occur

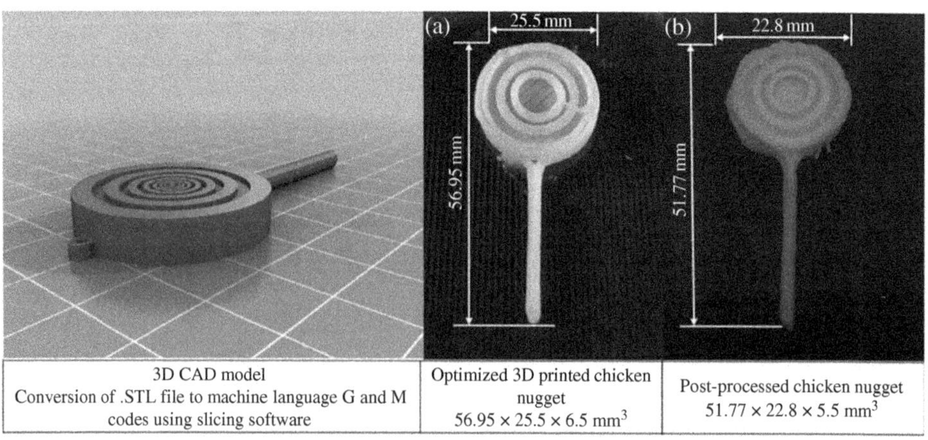

3D CAD model Conversion of .STL file to machine language G and M codes using slicing software	Optimized 3D printed chicken nugget $56.95 \times 25.5 \times 6.5$ mm^3	Post-processed chicken nugget $51.77 \times 22.8 \times 5.5$ mm^3

Figure 9.5 3D printed chicken nugget. (a) 3D printed chicken nugget, and (b) post-processed 3D printed chicken nugget. *Source:* From Wilson et al. (2020), Figure 05 [p. 14] / With permission of Springer Nature. DOI-https://doi.org/10.1007/s11947-020-02537-3.

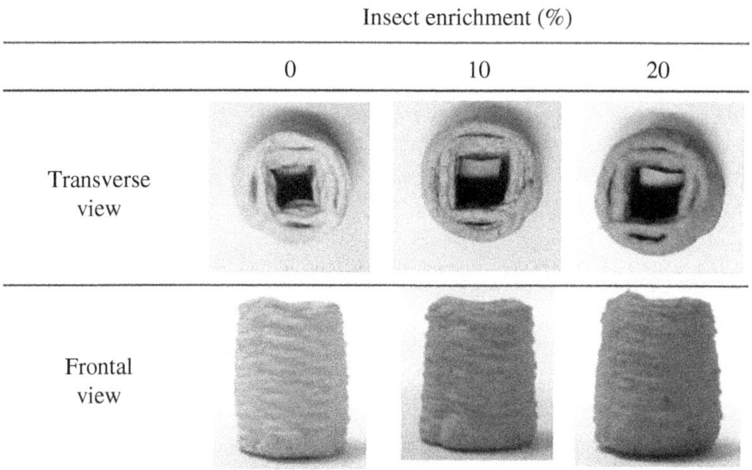

Figure 9.6 3D printed snacks with different levels of insect enrichment baked at 200 °C for 22 minutes. *Source:* From Severini et al. (2018a), Figure 03 [p. 671] / With permission of Elsevier. DOI-https://doi.org/10.1016/j.foodres.2018.01.034.

during baking include melting of fat, formation, and entrapment of gases, starch gelatinization, and protein coagulation, the caramelization of sugars, water evaporation, and volume expansion (Arepally et al. 2020). It was reported that the traditional cookie recipes were not suited for 3D printing (Lipton et al. 2010; Sun et al. 2015). A modified cookie recipe with novel formulations aids in the fabrication of cookies with complex internal structures using a 3D printing approach. The level of protein enrichment on 3D printing of cereal-based dough with insect flour was reported (Severini et al. 2018a). Results showed that there was a significant reduction in the size of the baked cookie mainly due to the associated water loss (Figure 9.6). Thus, proper optimization of baking conditions had positive effects on the nutritional and sensory properties of the 3D printed snacks.

In view of analysing the post-processing baking conditions, the 3D printing conditions of dough were optimized (Liu et al. 2019b). The material supply was formulated with wheat flour, olive oil, freeze-dried mango powder, and water, and the material's response to the printing performance of the dough was studied. Results inferred when the mass fraction of the material composition of the printing formula exceeded the critical limit, the material's extrudability was significantly affected resulting in poor printing characteristics. A similar study on the impact of post-processing on the accuracy of the 3D printed shape of the baking dough was reported (Yang et al. 2019). In this study, a post-process fast cooling (PPFC) at −65 °C was done prior to baking of 3D printed samples for ensuring the dimensional stability of the samples during baking. It was found that with an increase in PPFC time, the baked 3D printed samples retained 3D shape and structure. These observations were confirmed from the results of the creep recovery test.

The components affecting the printing and baking performance of cookie dough were studied. It was stated that the reduction in sugar levels must be adequate for retaining a stable 3D printed structure after baking. Further, the reduced levels of the milk content in the dough lead to the surface drying of samples during printing which significantly affects the sample deposition (Yang et al. 2018b). These results inferred that modifying the levels of printing composition would yield a better 3D construct with stable post-printing characteristics. The optimization of baking conditions of 3D printed dough remained to be a challenge as the material becomes ductile at high temperatures that affect the shape retention (Yang et al. 2018b). In a more recent study, the cookie dough was fortified with micro algae (*Arthrospira platensis*) and post-processed by baking (Vieira et al. 2020). Different approaches of extraction and incorporation (biomass, free extract, and encapsulation) of micro algae into the dough were examined. Results showed that upon baking, the thickness of the cookie was varied from 3.37% (biomass cookies) up to 18.22% (encapsulated extract cookies). The increased dimensions of the cookies were due to the evaporation of moisture, gas production, and higher dough elasticity. In contrast, the diameter of the cookies undergone a smaller change after baking that resulted in shrinkage of the encapsulated extract cookies (7%) due to high loss of water during post-processing. While the baking temperature had a positive influence over the remaining formulations of cookies. The applied heat during baking melts down the fat present in the cookie dough that imparts plasticity resulting in the initial spreading of deposited dough followed by dimensional retraction of samples.

The viability of incorporated probiotics of baked 3D printed cereal-based structures was reported (Zhang et al. 2018). The 3D printed constructs with incorporated probiotics (*Lactobacillus plantarum*) were baked at two different processing conditions at 145 °C for about 16 minutes and 175 or 205 °C for about 8 minutes. Results showed with respect to the baking time, no significant difference was evident with the survival rate of the probiotics. Considering the infill pattern, the residual viability of the probiotics was found to be higher in honeycomb structure rather than in concentric structure at constant moisture content for baking temperatures of 145 or 175 °C but not for the baking temperature of 205 °C. Further results inferred that the higher surface area of the 3D construct accelerated the drying rate during the baking process. The 3D constructs with honeycomb structure baked at 145 °C for about six minutes showed a minimal microbial count greater than $>10^6$ CFU g^{-1}.

9.3.4 Microwave Cooking

Microwave heating involves a complex interaction of electromagnetic radiations with dielectric food materials that leads to heat generation and a sudden rise in temperature (Guo et al. 2017). This results in the removal of moisture from food materials. The product attributes such as mass, density, polarity, specific heat, and dielectric properties significantly affect the drying rate of microwave heating (Taheri et al. 2018). It was reported that microwave heating could be conveniently used for post-processing of 3D printed constructs as it has fewer deterioration effects on the print fidelity of the post-processed samples with lesser processing time (Krishnaraj et al. 2019). A fibre-enriched 3D printed snack was developed from wheat flour with different levels of mushroom (Keerthana et al. 2020). Results of printability showed that 20% of mushroom incorporated samples had better printing performance and the printed samples were post-processed using the microwave. Microwave processing resulted in the denaturation of wheat gluten and mushroom proteins that imparts hardness to the 3D constructs. While lipid fractions of the material supply helped in retaining the crispness of the final product. Among the tested microwave power levels, samples exposed to 800 W for 10 minutes better retained the shape and structural integrity

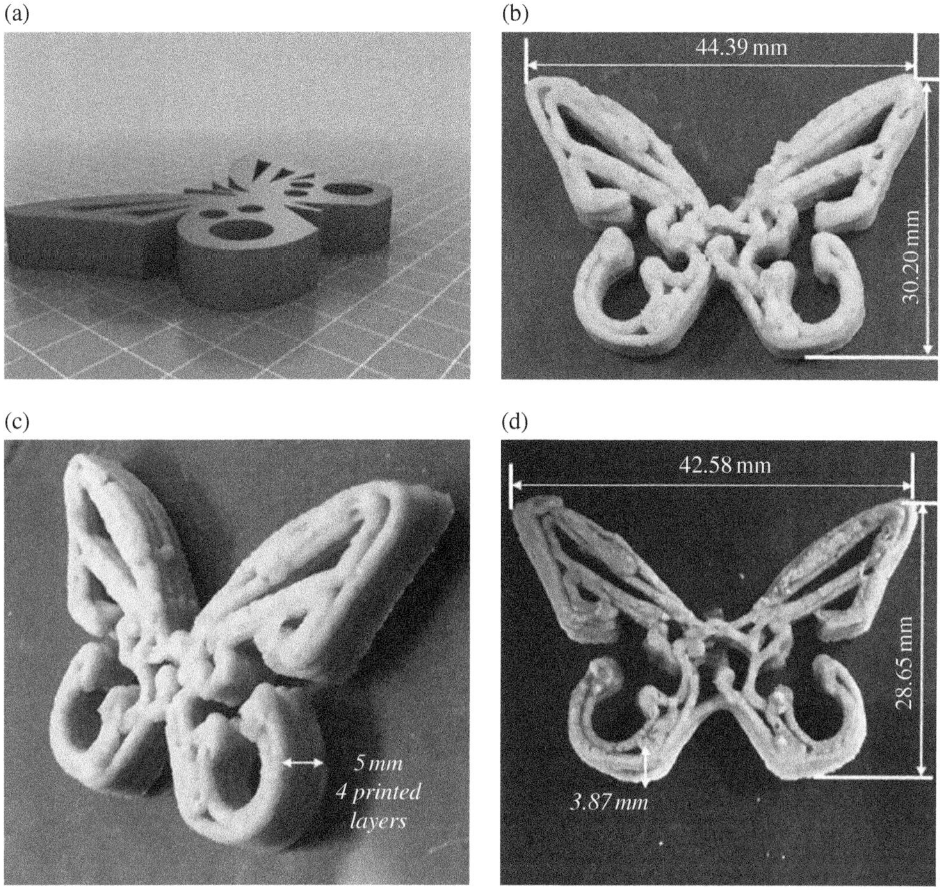

Figure 9.7 Optimized conditions of 3D printed mushroom constructs. (a) 3D model, (b) top view of 3D printed sample, (c) side view of 3D printed sample, and (d) post-processed sample. *Source:* From Keerthana et al. (2020), Figure 03 [p. 07] / With permission of Elsevier. DOI-https://doi.org/10.1016/j.jfoodeng.2020.110116.

of the samples (Figure 9.7). Although microwave processing showed a 5% shrinkage of 3D construct, the associated changes were insignificant.

Interestingly, the effect of printing parameters and microwave heating was employed for triggering the colour/ aroma enhancement of 3D printed buckwheat with yellow peach incorporated dough with complex coacervates of gelatin-gum Arabic (Guo et al. 2021). Results showed that the inclusion of microcapsules greatly improved the printing performance of the dough (Figure 9.8). The morphological characteristics of the microcapsules revealed that microwave heating gradually destroyed the structure of the capsules. This structural breakdown acts as a trigger that resulted in the release of the capsulated pigment and oil from the printed 3D constructs. Thus, the combination of 3D printing with microwave heating allows to design a 3D construct with dissemination of colour and aroma upon post-processing.

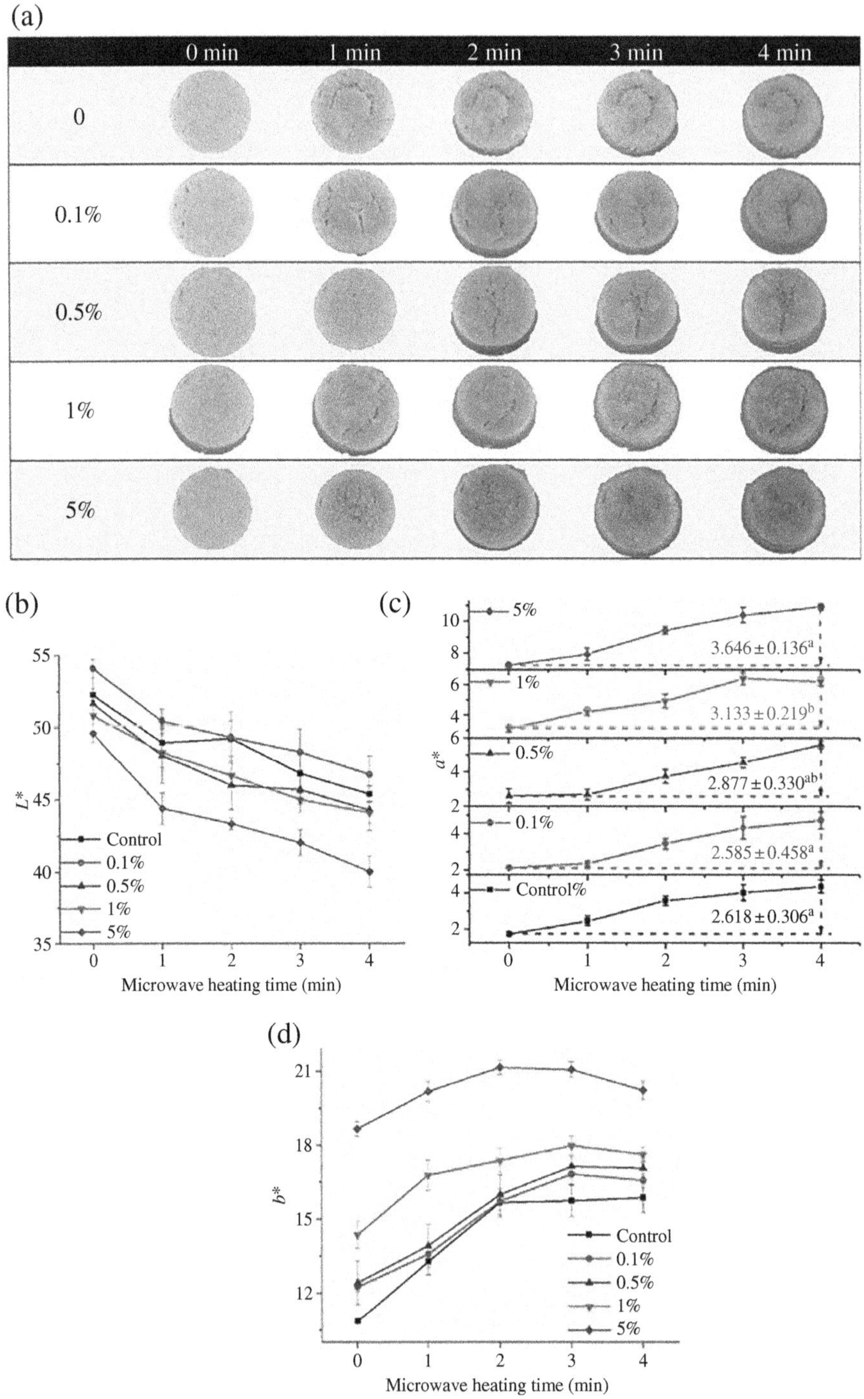

Figure 9.8 3D printed buckwheat dough. (a) With different period of microwave treatment, (b) L^* colour scale, (c) a^* colour scale, and (d) b^*colour scale after microwave treatment. *Source:* From Guo et al. (2021), Figure 05[p.08] / With permission of ELSEVIER. DOI-https://doi.org/10.1016/j.foodhyd.2020.106358

9.3.5 Sous Vide Cooking

Sous vide cooking is one of the cooking methods for food that involves the slow heating of food ingredients at a low temperature for a long duration of time (Baldwin 2012). The food is cooked in the hot water well below its boiling point that helps in the preservation of nutritional and functional properties of food materials. Often the cooking process involves the heating of food material in a vacuum-sealed package in hot water (He et al. 2019). When comes to 3D printed constructs, cooking of the printed material in vacuum sealing would deteriorate its structure and shape. Hence, it was reported to cook the 3D printed constructs in a 100% humid environment in a water bath under a controlled temperature. The unit used for sous vide cooking is known as CVap (controlled vapour) oven which produces a humid environment that surrounds the food product with vapour thereby preventing the loss or gain of moisture (Church and Parsons 2000). Understanding of wet and dry bulb temperature would help in better controlling of humidity and temperature of the cooking process. In a study reported by Lipton et al. (2010) turkey puree was 3D printed and post-processed using the sous vide technique. Results showed that post-processing caused the inward contraction of the printed construct that significantly reduced product dimensions. Despite the sous vide cooking produced a better taste and proper texture with complete cooking of the printed construct, it prevents the volatilization of flavour and inhibits oxidation of food.

A composite multi-layer 3D printed beef-lard mixture was 3D printed and post-processed by the sous vide method (Dick et al. 2019b). The effect of infill density and fat content on the post-processing conditions was studied. Results showed that the higher the fat percentage of the samples, the greater the cooking loss and material shrinkage of the cooked 3D printed construct (Figure 9.9). This was attributed to the unstable nature of fat molecules that resulted in the displacement of the melted fat from the inner structure. Ultimately, this resulted in the loss of the self-supporting ability of the cooked 3D printed construct. Considering the infill density, the shrinkage of the post-processed samples was lower. However, the shrinkage was significantly affected by the combined effects of infill density that interacts with fat content. There was a significant reduction in shrinkage with increased infill density. This was due to higher infill density where the meat proteins got aggregated with heat treatment that resulted in the formation of a network with entrapped water thereby forming a stable emulsion imparting strength to the post-processed product. Similar results

Figure 9.9 3D printed sous-vide cooked beef sample at varying fat content. (a) 0 fat (no lard), (b) low fat (1 lard layer), (c) medium fat (2 lard layers), and (d) high fat (3 lard layers). *Source:* From Dick et al. (2019b), Figure 05 [p. 14] / With permission of Elsevier. DOI-https://doi.org/10.1016/j.meatsci.2019.02.024.

were visualized for cooking loss as influenced by infill density. However, a greater mass loss during cooking was pronounced with an increased infill percentage from 50 to 75%. This was mainly due to the highly porous structure of less infilled samples that had a lower heat transfer rate. Thus, complex interactions of 3D printing parameters on the post-processing conditions were studied.

9.3.6 Low-Temperature Processing

As far as 3D printing is concerned, the post-processing capacity of the material supply must be adequate. The post-processing capacity of the material supplies refers to the material's ability to withstand post-processing conditions and retaining the 3D shape without structural collapse and dimensional deviation (Pulatsu et al. 2020). Most of the thermal processing method employs high heat that results in deterioration of print quality as well as affects the nutritional and sensory attributes of 3D printed products. Considering the development of functional foods, preservation of thermo-labile food components is most important. In a study on 3D printing of probiotics (*Bifidobacterium animalis*), the viability of probiotics was examined by storing the 3D constructs under refrigeration conditions (5 °C) (Liu et al. 2020a). This study aimed at developing a ready-to-eat (RTE) cold serving food by incorporating probiotics in the mashed potato. Results showed that the temperature of printing higher than 55 °C significantly reduced the viability. Considering the storage of probiotic incorporated 3D printed mashed potato, the viability remained at higher levels of greater than 9.773 log CFU g^{-1} with no significant difference. Further, the study inferred that the fabrication of a 3D printed construct with an internal microstructure devoid of oxygen is adequate for avoiding the adverse effects of oxygen on probiotics.

The shape of 3D printed pastry dough was found to be highly retained after rapid cooling that aids in maintaining the integrity of the microstructure even after post-processing (Lipton et al. 2010). A similar approach was used for baking dough using ultra freezer at −65 °C for 0, 5, and 10 minutes (Yang et al. 2018b). Results confirmed the shape accuracy of the baked sample as that of model design when subjected to fast freezing for 10 minutes. In another study, the hot-melt extruded, and cold extruded chocolates were refrigerated to retain the deposited 3D shape and structural integrity for further analysis (Mantihal et al. 2017; Mantihal et al. 2019a, 2019b). A similar approach can be used for maintaining the structure of hydrocolloid-based and hydrogel-based 3D constructs. Recently, oleogels are gaining research potential as a fat replacer with fortified bioactive compounds such as curcumin and resveratrol (Gomez-Estaca et al. 2020; Qiu et al. 2021; Shi et al. 2021). Although 3D printing remains a promising approach in the fabrication of thermal-sensitive food materials, the post-processing of the developed construct is critical. Since these type of material supplies is more prone to oxidation and thermal degradation. Hence, the vast post-processing opportunities that exist with low-temperature processing must be explored. This would open up a new dimensional range of novel bioactive and functional 3D constructs for the delivery of nutraceuticals using 3D food printing.

9.3.7 Other Post-Processing Methods

Other post-processing methods include boiling, steaming, laser heating, infrared cooking, cryo-cooking, cook-chilling, and cook frozen techniques (Aguilera 2018). In a more recent study, the waste streams of potato processing (PP) line, i.e. potato peel powder (0, 20, 40, 60, 80, and 100%) was incorporated into whole wheat flour for developing 3D printed fibre-rich ready-to-cook (RTC) noodles (Madhumitha et al. 2021). The material supply of the printing mixture consists of potato peel powder, whole wheat flour, guar gum, salt, and oil. Results showed that the optimal printing conditions for all formulations were found to be at 600 mm $minute^{-1}$ printing speed, 600 rpm extrusion motor speed, 6 bar pressure with an extrusion rate of 7.72 mm^3 $second^{-1}$. The 3D printed noodles were subjected to steaming for about 5 minutes followed by drying in a hot air oven at 68 °C for 2.5 hours (Figure 9.10). Among the tested formulations, 3D printed noodles containing 40% potato peel powder possessed adequate strength and exhibited higher post-processing stability which was comparable to commercial market samples (Madhumitha et al. 2021).

Recently, gastronomists and culinary professionals are working towards digitalized cooking. The digital gastronomy projects employ a multipurpose robotic arm where the first arm integrates with a 3D printer, the second arm assists in the positioning of the readymade food elements over the 3D printed substrate, and the third arm heats the prepared recipe through a laser-based heating element that cooks a part of the dish (Zoran 2019). This kind of multi-capable food manufacturing workstations is feasible only with the adaptation of the novel 3D printing technology. However, the complexity involves with controlling and coordination of multiple integrated robotic arms. This emphasizes the need for the development of high-tech software assistance for managing and prioritizing the task of the digital cooking robots for parallel food production (Mizrahi et al. 2016). A multi-material composed of sesame paste, shrimp paste, and chicken paste was subjected to 3D printing with *in-situ* IR heating (Hertafeld et al. 2019). The employed food 3D printer has the ability to simultaneous printing and cooking via IR sources. Through this approach, it was feasible to heat foods selectively and uniformly in a layered pattern of a portion of food. Thus, this study proved that IR spot was highly energy efficient in delivering the heat than the conventional oven heating. A hybrid cooking methodology was used for exploring the programmable flavour through selective heating (Mizrahi et al. 2016). The effect of laser on the biochemical reactions of meringue was analysed and results showed that laser cooking reduces charring of foods that has positive effects on the sensorial and nutritional attributes. This type of cooking would have precise control over the aroma, flavour, texture, and colour of the cooked food. Taking into the advancement of food transformations upon external stimuli, the 3D printed Italian pasta (ravioli) was post-processed by boiling method that imparts inflation and floating of pasta in boiling water (Wang et al. 2017). The external heat stimuli induce starch hydration along with vapour generation that resulted in the shape transformation of the original 3D printed construct. The associated transformation of the original 3D printed construct to another new variation (shape/ colour) is being referred to as 4D printing. Common external stimuli that induce physical and chemical transformations in 3D printed food include heat, pH variation (acid/ base), light, and magnetic response (Ghazal et al. 2019; He et al. 2020a, 2020b; Phuhongsung et al. 2020a). All these post-treatments used for shape and colour transformation of the 3D printed foods are also meant to be post-processing. Hence, an appropriate post-processing is selected carefully based on the applicability of material supplies used for 3D printing.

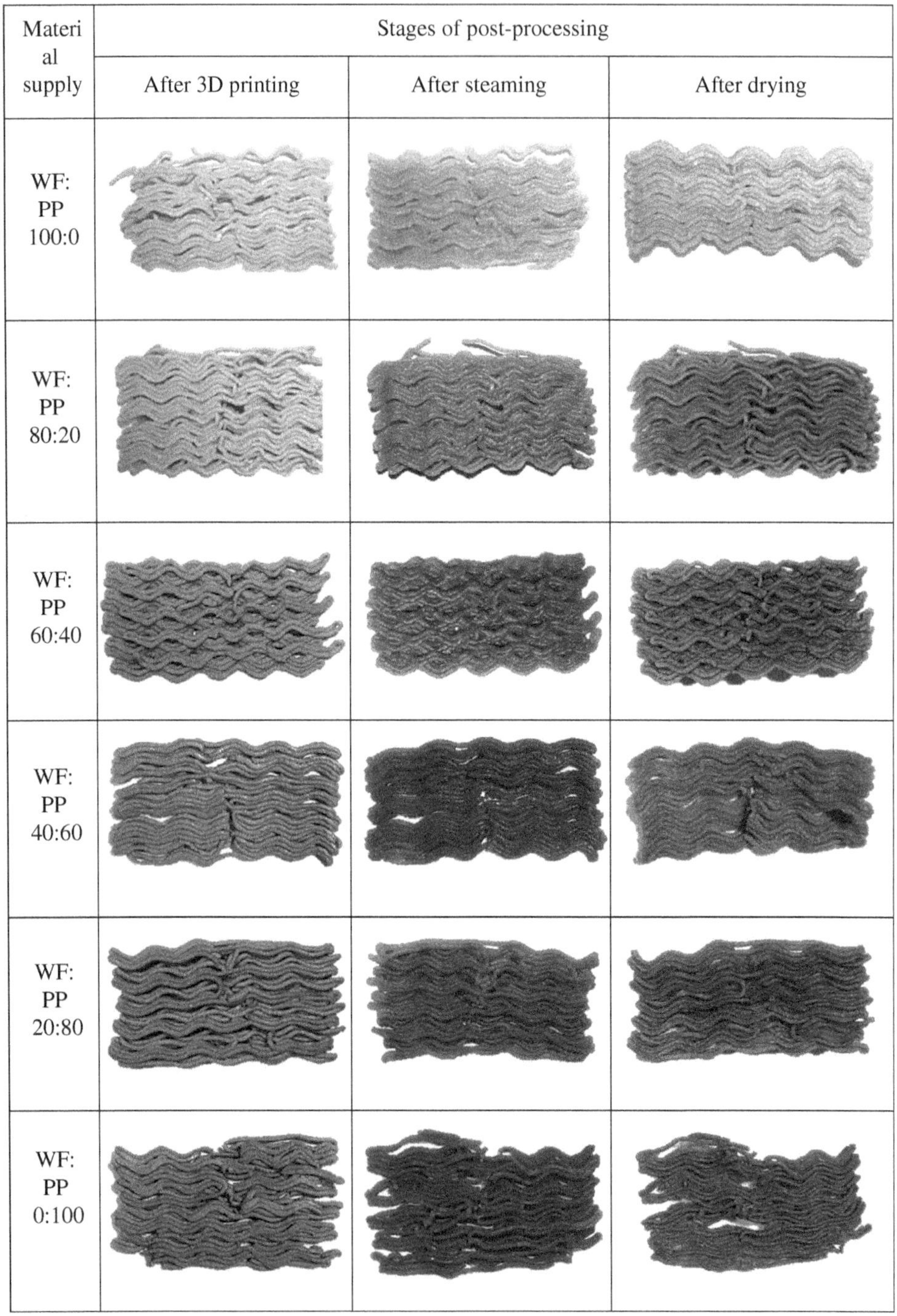

Figure 9.10 Post-processing of 3D printed noodles of different formulations of fine fractions of potato peel powder. *Source:* From Madhumitha et al. (2021) / With permission of Springer nature.

9.4 Novel Post-Processing Methods

A major strength of 3D food printing is the fabrication of novel foods with precise control over nutrition. With rising consumer awareness about a healthy diet, researchers are now focusing on various food processing approaches and techniques for reduced intake of salt, sugar, and fat (Prakash et al. 2019). Taking this into account, the textural modification of potato-based 3D printed snacks using air-frying was analysed (Liu et al. 2020b). The study examined the effects of infill density on reduced uptake of oil content during air frying. A combination of potato flakes and wheat flour was added into the boiling water of 1 : 2 ratio served as a material supply for 3D printing. Initially, the bovine fat was mixed with the boiling water and then added into the powdered materials. The bovine fat serves as a cooking medium enhancing post-processing characteristics and also aids in tailoring the flowability of materials during extrusion of material supply. The 3D printed potato constructs of varied infill density (30, 50, and 70%) and infill pattern (rectilinear, cubic, and honeycomb) were air-fried at 140 °C immediately after removing from storage at −18 °C. As expected with increased infill density, the time required for frying also increased. Results showed that percentage of oil uptake in the final air-fried 3D printed sample was 13.47% that was lesser than conventional potato chips of 39%. Further, comparison of 3D printed, and casted samples showed the significance of infill levels in fried snacks. Apart from modifying texture, the lesser infills with a more porous structure allowed the escape of force-generated steam due to moisture dehydration during the frying process. Thus, the 100% infill sample had lost its original shape due to the severity of the puffing phenomenon while 50 and 70% of infill samples hardly retained its designed geometry. Thus, the varying infill density and infill pattern had a significant impact on reduced oil uptake thereby providing a new concept of developing low-fat fried texture modified 3D printed foods. The mixture of yam powder and the by-product of the PP line was used for 3D printing at various infill levels and structures and its effect on air-frying was studied (Feng et al. 2020a). The effect of infill support on the mass and porosity of the 3D constructs were related to the infill percentage (Figure 9.11). Results showed that the degree of deformation was significantly affected by the amount of incorporation of PP to yam powder. Further, the hardness of the developed 3D printed air-fried product decreased with increased porosity. Thus, the printing variables greatly affect the physical and mechanical properties of the developed 3D printed product.

The post-processing of food mainly focused on the removal of water to achieve a shelf-stable product with extended shelf-life (Moses et al. 2014). From this point of view, drying seems to be a simpler cost-effective approach for post-processing of 3D printed constructs. However, thermal damage of the 3D printed structures is more witnessed in conventional drying approaches. Hence, modern emerging drying techniques must be explored for achieving a stable 3D printed post-processed product without structural collapse. In this regard, the effect of microwave vacuum drying on the post-processing of 3D printed mango juice gel was examined (Yang et al. 2018a). Microwave vacuum drying is a combined drying approach based on rapid volumetric heating along with low-temperature moisture evaporation at reduced energy consumption. The 3D printed mango juice gel was dried by setting process conditions of upper chamber air temperature 35 °C; microwave power level at 150 W; and time duration of 0, 2, and 4 minutes. Among the tested samples, the 3D construct microwave vacuum dried at four minutes showed a compact gel network with better

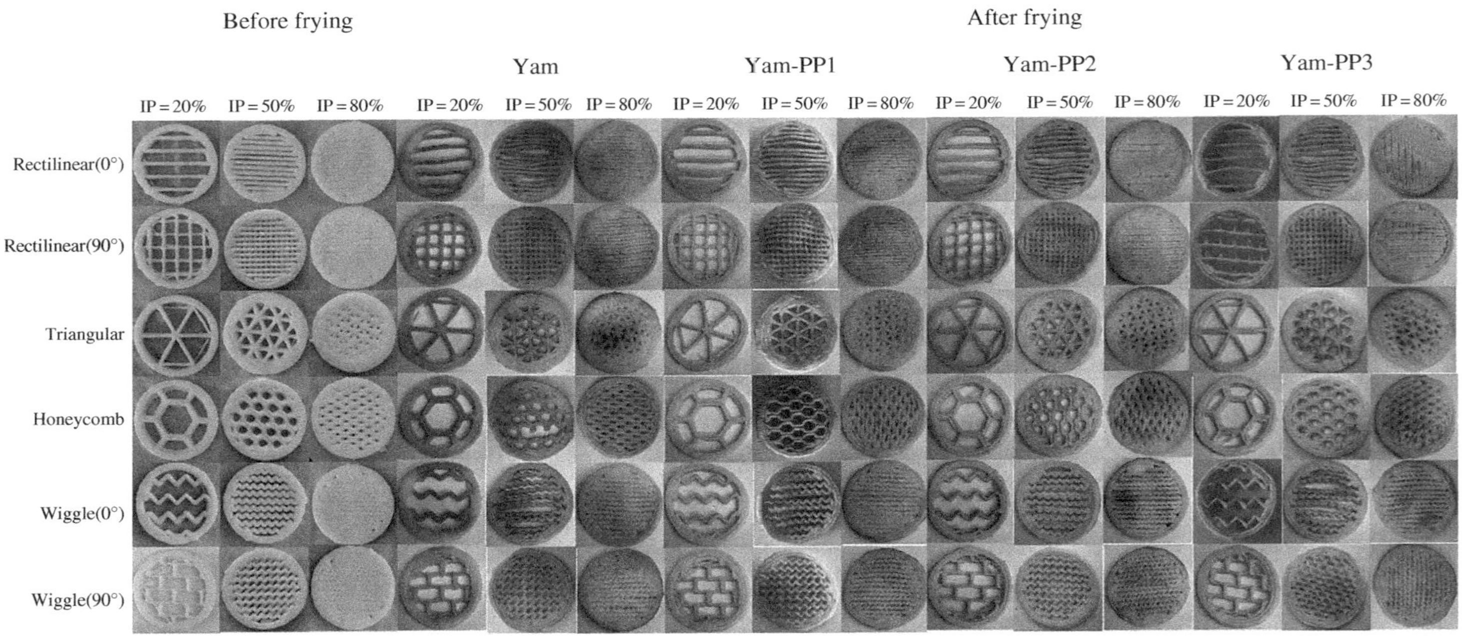

Figure 9.11 Effect of air-frying on the 3D printed samples of different infill levels and structures for different proportions of yam to potato processing by-product (PP) (Yam-PP1 = 9 : 1, Yam-PP1 = 8 : 2; Yam-PP1 = 7 : 3). *Source:* From Feng et al. (2020a), Figure 06 [p. 06] / With permission of Elsevier. DOI-https://doi.org/10.1016/j.lwt.2020.109265.

shape accuracy. Thus, results confirmed that microwave vacuum drying can be potentially used as an energy-efficient approach of post-processing that can be conveniently used for preventing the collapse of post-processed 3D constructs.

In another study on 3D printing of egg fractions, the egg white and egg yolk powders were prepared by refractance window drying (Anukiruthika et al. 2020a). This study emphasized the pre-processing of the material supply conducive for 3D printing applications without thermal degradation of egg fractions. In another study, a similar approach based on conductive hydro drying was reported as an alternative efficient method for the production of egg white powder (Preethi et al. 2020a). The conductive hydro drying, a variant of refractance window drying is a non-thermal approach that involves all the three modes of heat transfer: conduction, convection, and radiation through a thin transparent film (Raghavi et al. 2018). Above this transparent film, the drying material was spread as a thin layer. The advantageous features of this drying technique are safe retention of nutritional and sensory components, efficient drying, and no risk of overheating which makes it suitable for drying of thermal sensitive foods. Using this approach, the beetroot pulp was dried that confirmed the application of this technique for preservation of food colourant pigments (Preethi et al. 2020b). Recently, this approach was used for drying of probiotics (*Lactobacillus plantarum*) (Yoha et al. 2020). Hence, conductive hydro drying could be promisingly used for post-processing 3D printed constructs. More recently, a 3D printed food construct from symbiotic-composite flour was developed using different post-processing approaches such as freeze-drying, hot air drying, microwave drying, and refractance window drying (Yoha et al. 2021). Various encapsulation methods such as spray drying, spray freeze-drying, freeze-drying, and refractance window drying were employed for encapsulation of probiotics (*Lactiplantibacillus plantarum*). Results showed that freeze-drying and spray freeze-drying synbiotics incorporated 3D constructs followed by freeze-drying had retained the highest viability of 8.23 $\log_{10}$ and 8.18 $\log_{10}$ CFU g^{-1}, respectively (Figure 9.12). In addition, the freeze-dried synbiotics incorporated freeze-dried post-processed samples showed less viability of 6.12 CFU ml^{-1} than spray-freeze-dried synbiotics incorporated freeze-dried post-processed sample of 6.43 CFU ml^{-1} during *in-vitro* digestion studies. Further, the storage studies showed that the encapsulated probiotics possessed average viability of 94% against the 83% viability of the free cells. Thus, the study provides valuable insights for 3D printing applications for the development of novel functional probiotic 3D printed foods. Various novel thermal drying approaches used in the food industry and their advantages in terms of production, processing, and cost economics were reported by Moses et al. (2014). Based on the nature of material supply and end applications, these novel non-thermal drying approaches can be explored in 3D printing studies for achieving a stable 3D printed food product.

Modern cooking techniques based on combined information technology and mechatronics help in imparting a novel taste and flavour to foods. Unlike traditional cooking techniques, modern technologies assist in localized cooking of food in short times that provides a unique flavour to the food (Fukuchi et al. 2012; He et al. 2019). One such cooking technique is a laser-based technology that works based on localized heating of food thereby producing a new taste, mouthfeel, and eating experience. A laser cutter integrated with image processing software was tested for analysing the captured images to detect the fat portion in the bacon for localized heating. The acquired data were processed to develop a heat pattern that

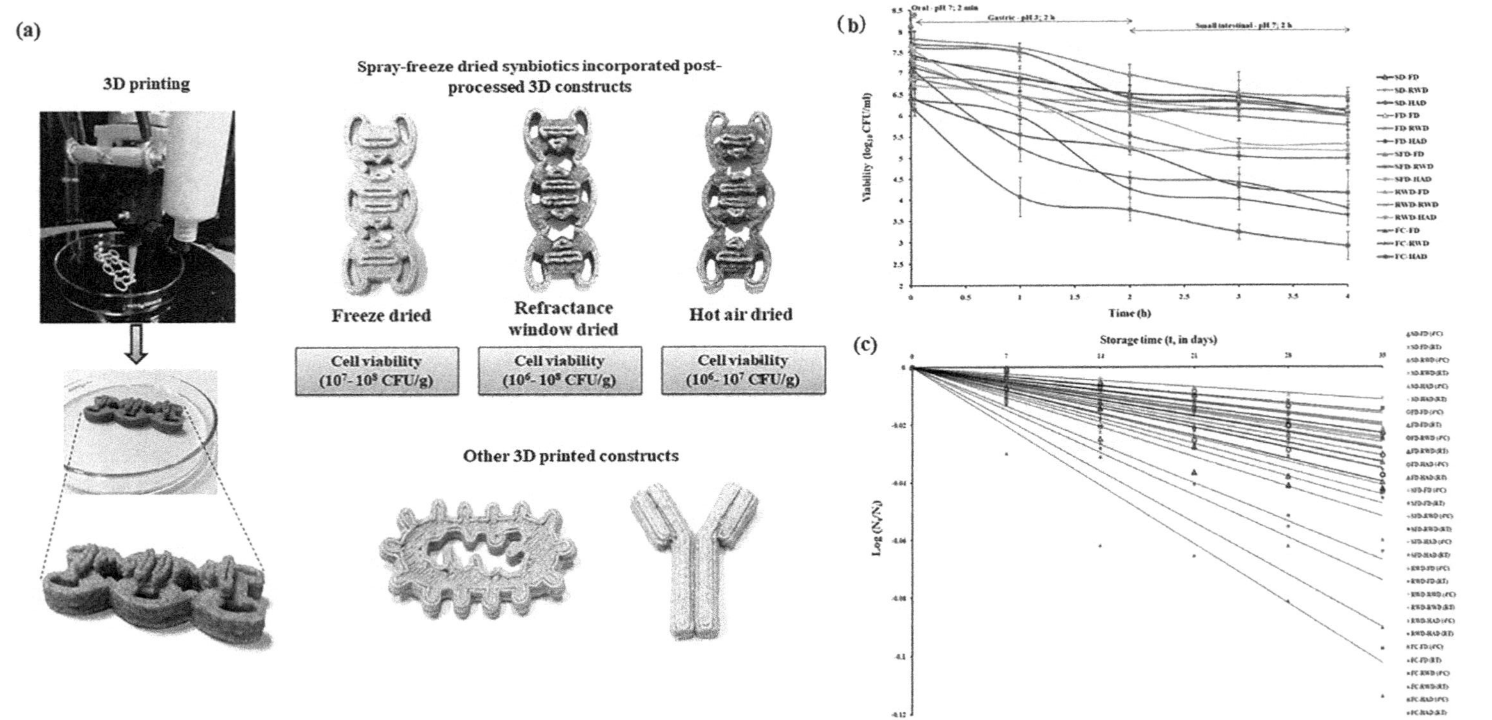

Figure 9.12 3D printing of symbiotic incorporated composite flour. (a) 3D printed probiotic constructs, (b) survival profile of probiotics in post-processed 3D printed symbiotic-composite flour constructs under oral-gastro-intestinal cond tions, and (c) inactivation kinetics of probiotics in post-processed samples during storage. (SD, spray drying; FD, freeze-drying; SFD, spray-freeze-drying; RWD, refractance window drying; HAD, hot air drying; FC, free cells. All the combinations define encapsulation technique – post-processing technique). *Source:* From Suppiramaniam et al. (2021), Figure 02 / With permission of ELSEVIER. DOI-https://doi.org/10.1016/j.lwt.2021.111461

allows melting of fat portion of bacon and leaving the lean meat. This produced a 'melt-fat raw bacon' with a combination of two kinds of flavours of cooked fat and uncooked bacon (Fukuchi et al. 2012). 3D printing can be conveniently combined with laser technology as like the sintering process that is based on laser curing of powdered materials. In a similar approach, laser cooking can be integrated with 3D printing unit as like conveyor that must be directed carrying printed 3D construct to laser post-processing unit. Otherwise, the laser unit is attached within the 3D printing chamber that enables 360° rotation of laser beam to ensure the maximum possible surface contact for post-processing of 3D construct. Recently, researchers of Columbia University employed two servo-driven mirrors for reflecting the emitted laser beam from laser source onto the 3D printed food (Cameron 2018). This allowed the cooking of 3D-printed food in an intricate pattern with precise control over the temperature. The 3D printed dough-based cuboids were post-processed using baking and laser cooking (Blutinger et al. 2018). Structurally both laser cooked samples and oven-baked samples had a comparable nutritional level with similar starch swelling characteristics. However, the laser cooking resulted in precise control over the texture of the final product imparting a new mouthfeel that was more acceptable to consumers. Thus, post-processing using laser is a promissory approach to impart new flavour with higher shape accuracy. Likewise, more research needs are required in designing a compatible heat source integrated with 3D printer for immediate processing of 3D printed constructs that have a great scope of commercialization in near future. Besides the optimization of 3D printing process, the post-processing of the developed 3D construct is crucial for ensuring the success of 3D food printing.

9.5 Assessment of Post-Processing Characteristics

The print fidelity of the 3D printed construct can be assessed by determining the printing precision, accuracy, and dimensional deviation (Liu et al. 2019a; Pulatsu et al. 2020; Vieira et al. 2020). When considering the post-processing characteristics, the final quality of the post-processed product greatly depends on printing parameters such as infill density and infill pattern, sample weight, and thickness (Dick et al. 2019a). Different attributes that describe the post-processing characteristics of the material supply are cooking loss, moisture content, fat retention, shrinkage, hardness, texture, colour, taste, and appearance. The material supply is one of the most significant factors that directly influence the quality of the post-processed 3D printed product (Feng et al. 2018). Hence the selection of post-processing must be based on the nature of the material supply and end application (functionality) of the 3D printed product. In view of customization of a healthy diet, Dick et al. (2019b) reported a study on the development of a lean meat-lard composite layer 3D printed beef. The texture of the 3D-printed beef was significantly affected by infill density and fat percentage. Results showed that post-processed products retained the internal as well as external structure. The major determinants of quality of post-processed 3D printed beef are cooking loss (Eq. 9.1), shrinkage (Eq. 9.2), moisture retention (Eq. 9.3), and fat retention (Eq. 9.4).

$$\text{Cooking loss (\%)} = \frac{W_r - W_c}{W_r} \times 100 \tag{9.1}$$

$$\text{Shrinkage (\%)} = \frac{(L_r - L_c) + (B_r - B_c) + (T_r - T_c)}{L_r + B_r + T_r} \times 100 \tag{9.2}$$

$$\text{Moisture retention (\%)} = \frac{W_c \times \text{percent moisture in cooked sample}}{W_r \times \text{percent moisture in raw sample}} \times 100 \tag{9.3}$$

$$\text{Fat retention (\%)} = \frac{W_c \times \text{percent fat in cooked sample}}{W_r \times \text{percent fat in raw sample}} \times 100 \tag{9.4}$$

where, W_r is the weight of raw 3D construct, W_c is the weight of the cooked 3D construct (g), L_r is the length of raw 3D construct, L_c is the cooked 3D construct (mm) respectively, B_r is the breadth of raw 3D construct, B_c is the cooked 3D construct (mm), T_r is the thickness of the raw 3D construct, and T_c is the thickness cooked 3D construct (mm). Results showed that infill density of composite beef construct was positively correlated with hardness, chewiness, and moisture retention while negatively correlated with shrinkage of material and cohesiveness with no effects on fat retention of the post-processed product (Figure 9.13). On the other side, the fat content of the material supply had positive effects

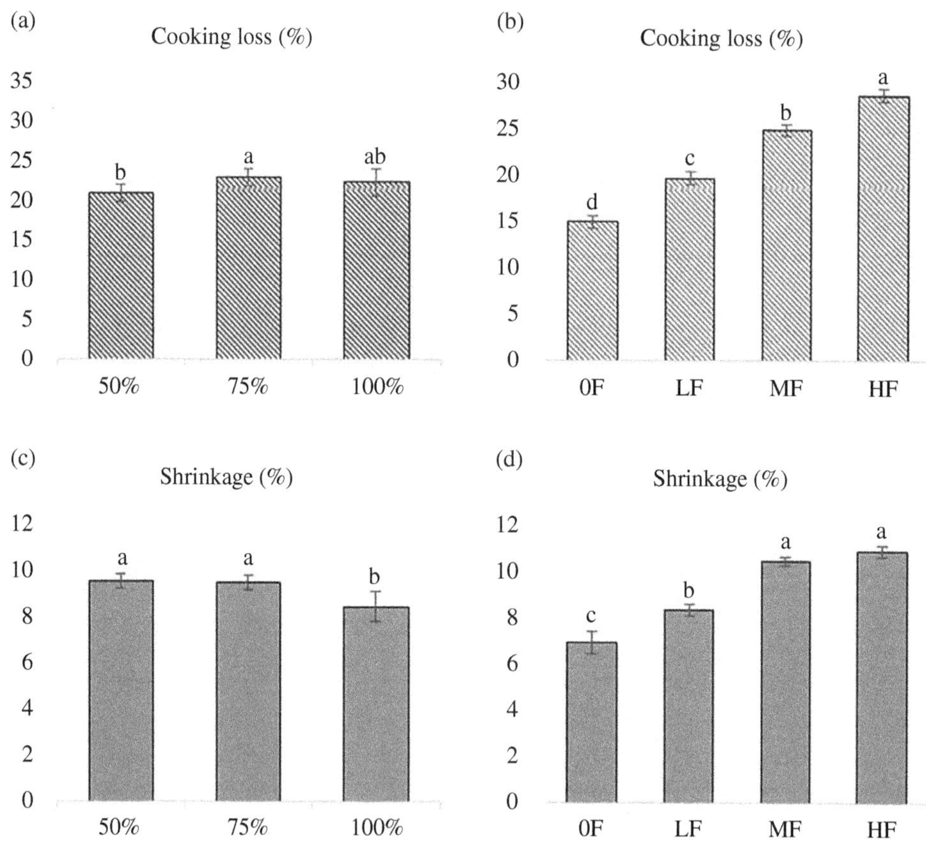

Figure 9.13 Post-processing characteristics of 3D printed beef. (a) Cooking loss vs. infill density, (b) cooking loss vs. fat content, (c) shrinkage vs. infill density, and (d) shrinkage vs. fat content. *Source:* Dick et al. (2019b) / With permission of Elsevier.

on cohesiveness, shrinkage, and cooking loss while negatively related to hardness, chewiness, moisture, and fat retention.

The evaluation of post-processing quality of microwave processed mushroom-based material supply showed hardness and fracturability values of 657 N and 34.32 mm, respectively (Keerthana et al. 2020). These attributes impart hardened texture and crispy crust to the post-processed snack. Although a shrinkage of 5.43% was evident there were no visible effects on the appearance of the post-processed snacks. The water activity of the microwave processed sample was found to be 0.6 that clearly indicates the shelf stability of the developed snack suitable for safe storage at ambient conditions. The effect of post-processing on the dry matter and material hardness was studied on protein-rich and fibre-based material supplies (Lille et al. 2018). Complete cooking of samples was not achieved with oven drying due to the limited post-processing time. Under such conditions, the highest hardness of 43 N was observed for the semi-skim milk powder (SSMP) samples. The partial replacement of milk powder with cellulose nanofibers gradually declined the hardness due to lesser dry matter content. Further, the added cellulose nanofibers significantly reduced the drying rate because of their high-water binding capacity. Comparing the material supply, a lesser amount of water was removed from the faba bean protein concentrate (FBPC) than the OPC sample that was evident with the soft texture of the FBPC samples. On the other hand, the highest dry matter content of 89–96% was evident with all the freeze-dried samples. Compared to SSMP, the effect of freeze-drying was more effective with FBPC and OPC-based samples (Lille et al. 2018). Considering the hardness, the hardness of the printed construct decreased in the following order FBPC>SSMP>OPC. Thus, the study inferred that hardness and drying rate was significantly affected by the drying technique and the material composition of the printing mixture.

In another study on the formulation of fruit and vegetable blend for 3D printing, the storage stability of the printed 3D construct was examined using air storage and modified atmospheric packaging (MAP) (Severini et al. 2018b). The gaseous component analysis showed the absence of the respiration rate of the printed sample. The antioxidant potential and the total phenolic content of the printed constructs remained constant until the first six and eight days, respectively. Thereafter both the antioxidant potential and the total phenolic content were reduced with storage time. Microbial studies showed that there was no significant increase in microbial concentration until eight days after which the count of psychrophilic bacteria was reduced to 3 logs CFU g^{-1} in air storage and 3.72 log CFU g^{-1} MAP. The sensory results of material composition and 3D printed constructs confirmed that 3D printing has positive effects on the visual aspects that drive the marketability of the product. Although nutritionally the microwaved 3D construct of composite flour was superior, taste-wise the HAD-DF samples scored higher (Krishnaraj et al. 2019). These results indicate the necessity to have an optimal balance among the sensory and nutritional aspects of the post-processed 3D constructs.

Although the addition of microbial transglutaminase aids in better printability of the surimi gel, the cooking loss showed an increasing trend with an increased proportion of transglutaminase (Figure 9.14) (Dong et al. 2020). The addition of transglutaminase enhanced the interaction among the salt and solubilized proteins that has dissolved out of muscle proteins during the cooking process. This leads to a substantial decrease in the amount of water molecules available for binding with protein network higher cooking loss due to the

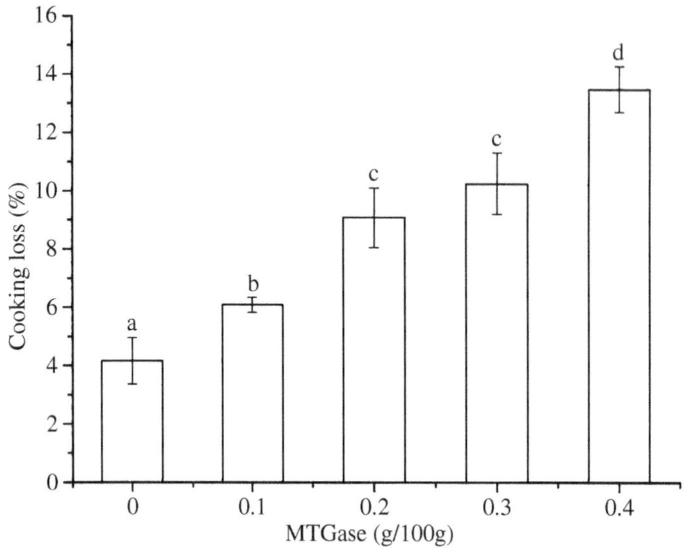

Figure 9.14 Cooking loss of 3D printed surimi gels containing different levels of microbial transglutaminase (MTGase). *Source:* Dong et al. (2020) / With permission of Elsevier.

decreased water-binding capacity. Thus, the higher proportion of added transglutaminase possesses a greater degree of cross-linking that increased water loss.

Interestingly, the 3D printing was integrated with the microwave power for exploring its synergistic effects on the self-gelation of fish surimi with and without transglutaminase (Zhao et al. 2021). Results showed that transglutaminase was activated by microwaves thereby promoted the self-gelation process (Figure 9.15). The moulding quality of the 3D printed samples was mainly attributed to hydrogen bonds and ε-(γ-Glu)-Lys forces. From the observation, it was inferred that samples with transglutaminase treated at 40 and 50 $W\,g^{-1}$ resulted in a firmer solid gel with better shape fidelity and larger protein aggregation (Figure 9.16). Thus, this study dictates the feasibility of combined 3D printing-post-processing for personalized manufacturing of foods.

In a study on the effects of process variables on the printability of cereal-based foods, the printing performance of wheat dough with and without cooking was analysed (Severini et al. 2016). Results showed the shape accuracy of the 3D construct was not highly affected by the cooking process. The solid fraction of uncooked and cooked samples was directly related to infill density. The microstructural analysis revealed that the mismatch of the printed layers due to random filling of void spaces for samples with 0.3 mm layer height while the rest of the samples possessed good print fidelity. Thus, the study emphasized the significance of process variables on the mechanical strength of the developed 3D printed snack. With an increase in the levels of insect proteins, the baking of the 3D printed constructs significantly affected the microstructure, induced non-enzymatic browning, altered mechanical strength, and protein digestibility of post-processed samples (Severini et al. 2018a). Results of the desirability function for optimizing baking conditions showed that

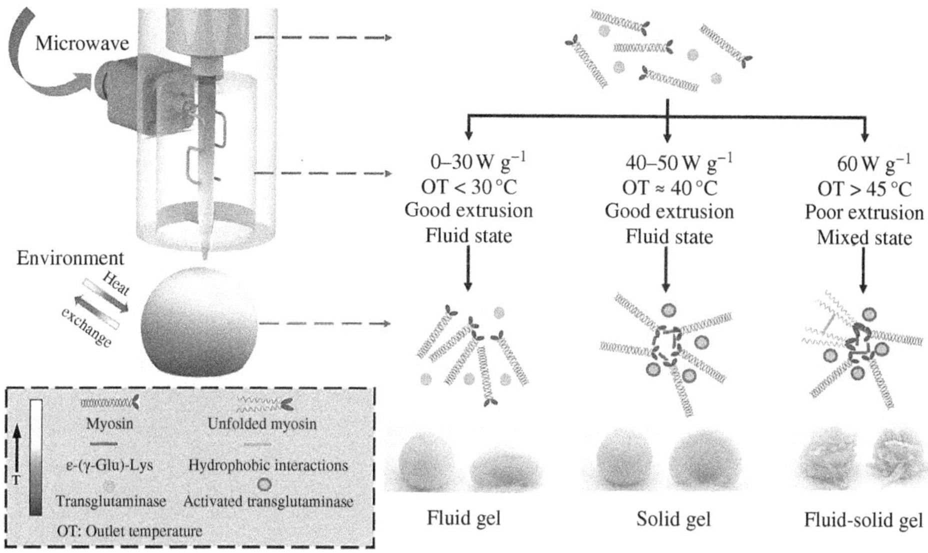

Figure 9.15 Self-gelation mechanism of synergism of microwave and 3D printing with added transglutaminase. *Source:* From Zhao et al. (2021), [p.28] / With permission of ELSEVIER. DOI-https://doi.org/10.1016/j.ifset.2020.102546

the samples subjected to baking at 200 °C for about 22 minutes had maximum desirability of 0.693 with a hardness of 38.3 N, moisture loss of 92.8%, and solid fraction of 4.9%. The amino acid profile showed that the mass fraction of essential amino acids increased from 32.5 to 38.2% and 41.3% for 10 and 20% insect enrichment of flour. Further, the true protein digestibility values were found to be 84.2, 82.2, and 80.3% for 0, 10, and 20% insect enriched 3D printed snacks, respectively. Thus, this study proved that yellow mealworm insects combined with baking flour can be used for fabricating a protein-enriched stable 3D printed snack. Thus, the various process attributes that affect the post-processing characteristics of the 3D printed product were discussed.

9.6 Sensorial Characterization

9.6.1 Qualitative Analyses

The end quality of the 3D printed foods was significantly influenced by the extent of pre-processing, 3D printing technology, and the post-processing approach (He et al. 2019). Various approaches to the sensorial characterization of the material supply and 3D printed materials have been reported. The print fidelity of the material supply was assessed through a visual appearance by a group of sensory panel members. The print quality of the 3D printed rice starch was assessed based on dimensional stability, layer definition, texture, shape, surface finishing, thread quality, binding ability, appearance, and overall acceptability (Theagarajan et al. 2020). In context with the post-processing, the effects of post-processing methods on the final attributes of the developed 3D printed food were examined

Sample ($W\ g^{-1}$)	Without TGase				With TGase			
	Height (mm)	After printing	After cutting	After 3 min	Height (mm)	After printing	After cutting	After 3 min
0	23.57 ± 1.61^{Aa}				27.32 ± 0.24^{Ab}			
10	24.64 ± 1.54^{ABa}				27.84 ± 0.18^{Bb}			
20	25.85 ± 0.51^{BCa}				28.60 ± 0.12^{Cb}			
30	26.54 ± 0.58^{CDa}				29.05 ± 0.17^{Db}			
40	27.49 ± 0.26^{CDa}				29.15 ± 0.05^{Db}			
50	27.87 ± 0.34^{Da}				29.29 ± 0.09^{Db}			
60	—		—	—	—		—	—

Note: Uppercase letters indicate significant difference ($p < 0.05$) between different micrcwave powers, lowercase letters indicate the difference between surimi with and without TGase.

Figure 9.16 Images of printed products at different microwave powers (No shape at 60 $W\ g^{-1}$): after printing, after cutting, and after standing for 3 minutes. *Source:* From Zhao et al. (2021), Figure 09 [p. 42] / With permission of Elsevier. DOI-https://doi.org/10.1016/j.ifset.2020.102546.

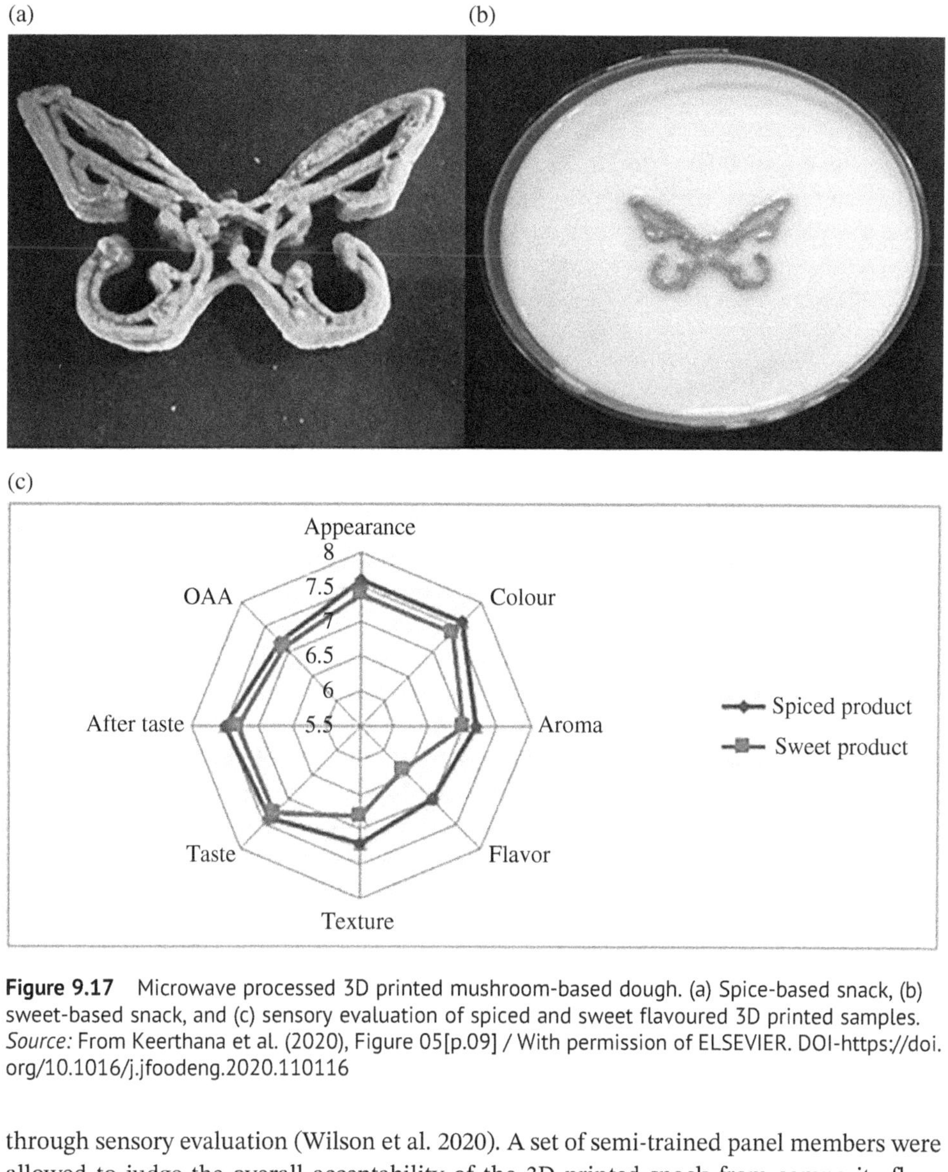

Figure 9.17 Microwave processed 3D printed mushroom-based dough. (a) Spice-based snack, (b) sweet-based snack, and (c) sensory evaluation of spiced and sweet flavoured 3D printed samples. *Source:* From Keerthana et al. (2020), Figure 05[p.09] / With permission of ELSEVIER. DOI-https://doi.org/10.1016/j.jfoodeng.2020.110116

through sensory evaluation (Wilson et al. 2020). A set of semi-trained panel members were allowed to judge the overall acceptability of the 3D printed snack from composite flour (Krishnaraj et al. 2019) and 3D printed fibrous rich mushroom (Keerthana et al. 2020). Various sensory attributes such as colour, aroma, off-odour, taste, after taste, flavour, off-flavour, texture, mouthfeel, and appearance were considered during the sensory evaluation of 3D printed food. In a study on 3D printing of mushroom-based snacks, two different flavours (sweet and spicy) of 3D printed snacks were kept for sensory (Figure 9.17). Among these, the spiced snack was more acceptable among the consumers that showed the applicability of mushroom for the development of spiced 3D printed products. A similar approach was followed for comparing the sensory attributes of the food from fruit and vegetable blend formula and its 3D printed construct (Severini et al. 2018b).

9.6.2 Quantitative Analyses

Apart from human sensorial perception, instruments are also used for analysing the sensorial characteristics of the 3D structures. The colour analysis of 3D-printed foods determines the final acceptability of the developed product. Most commonly, colourimeter and spectrophotometer are used for quantification of the visual perception of colour. The measured colour values are described in terms of L^*, a^*, and b^*values where L^* represents lightness (100) to darkness (0), a^* represents redness (+ve) to greenness (−ve), and b^* represents yellowness (+ve) to blueness (−ve) (Anukiruthika et al. 2020a; Ghazal et al. 2019; He et al. 2020b). The chroma properties of the pre-treated soy protein isolate (SPI) and strawberry mixture with different levels of microwave were reported (Fan et al. 2020). The microwave pre-treatment had negative effects on colour of both salted and unsalted samples. Compared to the unsalted sample, the salted sample had a smaller L^* value while a similar result was observed with a^* and b^* values. This indicates the dominance of green and blue colour that led to a bad visual sense. The decreased a^* and b^* values might be because of the anthocyanin pigment deterioration of strawberry powder. Although the sample colour became deep with microwave treatment, the low microwave power imparts minor change in the colour with improved performance of printing. The effect of the pH levels on colour change of 3D printed composite mixture was studied (Phuhongsung et al. 2020b). The total colour difference ΔE was determined using Eq. 9.5.

$$\Delta E = \sqrt{(\Delta L^*)^2 + (\Delta a^*)^2 + (\Delta b^*)^2} \quad (9.5)$$

Results showed that the total colour change values were in the range of 0.3 to 0.2. Among which the samples that were treated with low pH solution exhibited lower ΔE values, while those samples that were treated with higher pH exhibited a higher ΔE values. These results indicated the degradation of betanin under acidic conditions of pH 2 to 6 with the formation of betalamic acid, betaxanthins, and neobatacianins. Beyond pH value of 6, the colour of the printed samples became highly stable at neutral and alkaline conditions of pH range of 6–10. This caused the alterations of the functional groups such as amine groups into a more stable molecule. These results were supported by the Fourier transform infrared spectroscopy (FTIR) analysis of the samples.

In a more recent study, the effects of microwave irradiation on the flavour changes of the 3D printed SPI samples treated with different microwave power levels (0, 50, 80, and 110 W) were quantified using an electronic nose (e-nose) (Phuhongsung et al. 2020a). To ensure the reliability of the results, prior to the start of the experiments the gaseous pathway of the e-nose was cleaned for about 50 minutes with reference air of flow rate 1 l minute^{-1}. The analysis followed an array of 14 sensors to detect the biological compounds and functional groups such as aromatic compounds (S1); nitrogenous and oxygen compounds (S2); sulfides, mercaptans, and thioether (S3); organic acid ester terpene (S5); lanthionine (S6); fatty acid derivatives (S7), helium (S8), hydrogen (S9); hydrocarbons (S10), volatile compounds (S11); sulfides (S12); ethylene (S13); and volatile gases (S14). The analysis for a single test took about 120 seconds and the recovery time was about 200 seconds. Results indicated that e-nose was not able to distinguish the odours evolved at the different levels of heat treatment. However, the samples with vanilla flavour showed a distinguished flavour than samples without vanilla. This clearly indicates the thermal sensitiveness and volatility of vanillin. From the radar plot, it was clear that samples treated with greater microwave power levels showed a higher response indicating the impact of heat on flavour changes of the samples. Further with respect to storage time, there was a significant difference observed in the volatility. Thus, the samples after heating at 20 minutes with different power levels showed a significant difference, hence it was taken forward for the determination of volatile compounds using GC-MS. In the same study (Phuhongsung et al. 2020a), the electronic tongue (e-tongue) was used for the classification of eight basic senses that includes saltiness, bitterness, astringency, sourness, aftertaste-A, aftertaste-B, umami, and richness, of the 3D printed products treated with different power levels (0, 50, 80, and 110 W). The sensory quantification started with the stabilization of five taste sensors. The sensory characteristics of 3D printed products with and without vanilla flavour were assessed using e-tongue. Results showed that with increase in temperature, the perception of umami, richness, and saltiness was intensified among the samples without vanilla while the sourness and bitterness was not affected. On the other hand, the added vanilla flavour showed intensified umami, astringency, richness, saltiness, and bitterness due to its thermal sensitivity than the original taste of SPI.

In another study, the flavour compounds of microwave-salt synergistic pretreated SPI-strawberry mixture were assessed using e-nose (Fan et al. 2020). A 14 e-nose sensor was used in this study and results showed that e-tongue was not significantly distinguished the flavours of pre-treated samples. The corresponding signals of each sensor were significantly increased and then reduced with increased power levels. The salted samples had a larger response than the unsalted sample at the same power level. These results inferred that lesser microwave power levels enhance volatilization of flavour compounds while higher temperature could significantly deteriorate the aromatic compounds thereby decreases the flavour quality. Thus, qualitative and quantitative measurements of sensory attributes would provide information about the consumer's acceptability of the developed 3D printed food. A comparative assessment of human sensory assessment as well as instrumentation values would remain as a valuable source of database that helps in the formulation and development of future 3D printed food based on consumer's perception and likeness.

9.7 Requisites, Challenges, and Future Trends

The adaptability of the 3D printed construct with traditional cooking is adequate for commercialization and consumer acceptability of the 3D printed foods (Nachal et al. 2019). Common cooking methods used in conventional food manufacturing for mass production of foods are baking, steaming, and frying. However, the maintenance of the 3D printed shape to withstand the post-processing condition remains to be challenging (Lipton et al. 2010). Since material composition is a significant attribute in the development of the desired 3D printed food, proper material formulation and optimization are essential for maintaining the shape of the 3D construct (Godoi et al. 2016; Perez et al. 2019). A deeper understanding of material chemistry and biophysics associated with post-processing must be explored for developing a mechanically stable and sensorially acceptable 3D printed food. The basic mechanisms associated with the binding ability of 3D printed layers are crosslinking and gelation (He et al. 2019). Considering post-processing, the material's response towards thermal treatment and associated chemical transformation such as starch gelatinization, protein denaturation, and dehydration during cooking should be studied. The cooking of 3D printed product is not as same as conventional moulded products. As 3D printing involves, a layer-by-layer deposition of material, the extent of processing is significantly varied. As expected, the 3D printed constructs require very less cooking time due to the porous structure and the higher surface-to-volume ratio accelerates the cooking process (Liu et al. 2020a).

Approaches that can be adapted for solving the post-processing collapse of the 3D printed constructs are control over formulation and incorporation of additives (Pulatsu et al. 2020). Modification of traditional recipes for achieving printability through proper optimization of food constituents helps in preserving the shape of post-processed 3D constructs. Based on the first approach, a cookie recipe was formulated with flour, egg yolk, sugar, and butter for 3D printing (Lipton et al. 2010). In addition to the reduction of the proportion of fat in the printing mixture, the oriented 3D construct was chilled prior to baking (Yang et al. 2019). Results showed that baked 3D printed samples retained the shape without collapse due to alteration of formulation as well as a chilling treatment prior to baking. This low-temperature treatment helped in preventing the increased surface temperature of the sample that would otherwise lead to material charring and structural collapse. The latter approach involves the addition of food hydrocolloids for improving the mechanical integrity of the 3D construct to withstand the post-processing conditions. The effect of the addition of methylcellulose and xanthan gum on dimensional stability during baking was studied (Kim et al. 2019). Although the inclusion of hydrocolloids improved the rheological properties (storage modulus and loss modulus) of material supply, 3D constructs with methylcellulose resulted in high dimensional error. In contrast, the xanthan gum added material supply showed better printability and shape retention with less dimensional error. Thus, the material supply with xanthan gum (0.5/100 g) baked at 170 °C for 15 minutes yielded cooked 3D printed cookies without any deformations. When dealing with printability, rheological properties of material supply are most significant in achieving higher printing resolution. However, with the post-processing, the mechanical properties of the material supply are crucial in achieving less deformed stable 3D printed food. Better understanding of the effect of infill structure with material composition on post-processing conditions is required for overcoming the structural collapse of post-processed 3D constructs.

Methods like intermittent post-processing seem to be beneficial in preserving the nutritional and sensorial properties and the shelf stability of the 3D printed constructs. For instance, traditional preservation techniques like drying, freezing can also be utilized for extending the shelf-life of the printed product that requires an additional post-cooking process before consumption. This approach portraits the synergism of traditional processing with 3D printing. Further intermittent post-processing of 3D printed constructs assists in the serving of two kinds of foods: hot plate and cold plate. The former type of hot serving involves partial post-processing immediately after the 3D printing which helps in extending the shelf-life of the product. However, the 3D constructs of this method require further complete processing prior to consumption. For instance, the 3D printed breakfast cereals are based on this concept that requires immediate post-processing like drying followed by the cooking step in a cooking medium before consumption. The advantage of intermittent processing is the prevention of thermal damage and structural collapse of the post-processed 3D printed foods. While the latter type of cold serving of food involves pre-processing of material supply, the complete process of 3D printing followed by storage that helps in achieving inactivation of microbial activity. However, the processed 3D printed construct based on this principle requires continued storage at controlled conditions generally low-temperature storage. Most of the 3D printed soft foods, such as mashed potato, lemon juice gel, and fish surimi gel, are based on this concept that well suits for cold serving. Irrespective of the way of processing, the shelf stability of the 3D construct is achieved by reducing the water activity of the printed matrix. Most of the studies reported were based on immediate post-processing of 3D printed food that serves as ready-to-eat foods (Feng et al. 2020a; Keerthana et al. 2020; Krishnaraj et al. 2019; Lipton et al. 2010; Vieira et al. 2020; Wilson et al. 2020). Techniques such as drying, frying, baking, refrigeration, and freezing are commonly used for the delivery of RTE 3D printed foods. Taking into advantage of the on-demand production of 3D printing, future studies must be directed to adopting the intermittent method of post-processing. Another benefit of intermittent post-processing is efficient utilization of material and energy resources that can be conveniently handled during packaging and transportation of 3D printed foods.

9.8 Conclusion

The major requirements of the printing material supply are printability, applicability, and post-processing. Considering these aspects, the success of the developed food formulation greatly depends on post-processing to ensure the edibility of the developed 3D printed food. The post-processing ability is the capacity of the material to withstand the post-processing conditions without structural collapse. Although traditional cooking methods are applied for the post-processing of foods, the extent of post-processing varied based on the design attributes of the 3D construct. The significance of post-processing of 3D printed foods was discussed elaborately along with conventional and novel post-processing approaches. Various factors that influence the post-processing characteristics of the 3D printed product have been presented. Apart from these process parameters, the sensorial assessment of raw material, 3D printed construct, and post-processed samples are essential for determining the consumer's acceptability of the 3D printed foods. Different sensory attributes that influence the consumer's perception are colour, taste, flavour, texture, mouthfeel, and eating experience. With advanced quantification methods, e-nose and e-tongue are useful in quantifying sensorial attributes. Thus, a combination of qualitative and quantitative sensory assessment aids in novel food formulation of 3D printed foods to cover a wide range of people's choices based on their taste preference. Also, it would help in the transformation of lab-scale printing recipes into large-scale production with immense scope for marketability and commercialization. In addition, the present chapter highlights the challenges associated with the post-processing of the printed construct and various approaches to overcome the post-processing stability issues. It has been forecasted that 3D printers would revolutionize food manufacturing from domestic kitchens to large-scale industrial production. With this viewpoint, understanding the post-processing of 3D constructs is equally important as that of the optimization of the 3D printing process. In this regard, the present chapter provides valuable insights on the scientific prerequisites for achieving a stable post-processed 3D printed food.

References

Aguilera, J.M. (2018). Relating food engineering to cooking and gastronomy. *Comprehensive Reviews in Food Science and Food Safety* **17** (4): 1021–1039.

Alam, M.R., Scampicchio, M., Angeli, S., and Ferrentino, G. (2019). Effect of hot melt extrusion on physical and functional properties of insect based extruded products. *Journal of Food Engineering* **259**: 44–51.

Anukiruthika, T., Moses, J.A., and Anandharamakrishnan, C. (2020a). 3D printing of egg yolk and white with rice flour blends. *Journal of Food Engineering* **265**: 109691. https://doi.org/10.1016/j.jfoodeng.2019.109691.

Anukiruthika, T., Moses, J.A., and Anandharamakrishnan, C. (2020b). Electrohydrodynamic drying of foods: principle, applications, and prospects. *Journal of Food Engineering* **295**: 110449.

Arepally, D., Reddy, R.S., Goswami, T.K., and Datta, A.K. (2020). Biscuit baking: a review. *LWT* **131**: 109726.

Asokapandian, S., Swamy, G.J., and Hajjul, H. (2020). Deep fat frying of foods: a critical review on process and product parameters. *Critical Reviews in Food Science and Nutrition* **60** (20): 3400–3413.

Baiano, A. (2020). 3D printed foods: a comprehensive review on technologies, nutritional value, safety, consumer attitude, regulatory framework, and economic and sustainability issues. *Food Reviews International*: 1–31.

Baldwin, D.E. (2012). Sous vide cooking: a review. *International Journal of Gastronomy and Food Science* **1** (1): 15–30.

Blutinger, J.D., Meijers, Y., Chen, P.Y. et al. (2018). Characterization of dough baked via blue laser. *Journal of Food Engineering* **232**: 56–64.

Cameron (2018). 3D printing food: lab creates 3D food printer that cooks its prints with a laser. http://www.3ders.org/articles/20181231-columbia-university-3d-food-printer-that-cooks-its-prints-with-a-laser.html (accessed 27 January 2021).

Caporizzi, R., Derossi, A., and Severini, C. (2019). Cereal-based and insect-enriched printable food: from formulation to postprocessing treatments. Status and perspectives. In: *Fundamentals of 3D Food Printing and Applications.* Elsevier Inc https://doi.org/10.1016/B978-0-12-814564-7.00004-3.

Choe, E. and Min, D.B. (2007). Chemistry of deep-fat frying oils. *Journal of Food Science* **72** (5): R77–R86.

Church, I.J. and Parsons, A.L. (2000). The sensory quality of chicken and potato products prepared using cook–chill and sous vide methods. *International Journal of Food Science and Technology* **35** (2): 155–162.

Dankar, I., Pujolà, M., El Omar, F. et al. (2018). Impact of mechanical and microstructural properties of potato puree-food additive complexes on extrusion-based 3D printing. *Food and Bioprocess Technology* **11** (11): 2021–2031. https://doi.org/10.1007/s11947-018-2159-5.

Dick, A., Bhandari, B., Dong, X., and Prakash, S. (2020). Feasibility study of hydrocolloid incorporated 3D printed pork as dysphagia food. *Food Hydrocolloids* **107**: 105940.

Dick, A., Bhandari, B., and Prakash, S. (2019a). 3D printing of meat. In: *Meat Science*, vol. **153**, 35–44. Elsevier Ltd https://doi.org/10.1016/j.meatsci.2019.03.005.

Dick, A., Bhandari, B., and Prakash, S. (2019b). Post-processing feasibility of composite-layer 3D printed beef. *Meat Science* **153**: 9–18. https://doi.org/10.1016/j.meatsci.2019.02.024.

Dong, X., Pan, Y., Zhao, W. et al. (2020). Impact of microbial transglutaminase on 3D printing quality of *Scomberomorus niphonius* surimi. *LWT* **124**: 109123.

Fan, H., Zhang, M., Liu, Z., and Ye, Y. (2020). Effect of microwave-salt synergetic pre-treatment on the 3D printing performance of SPI-strawberry ink system. *LWT* **122**: 109004.

Feng, C., Zhang, M., and Bhandari, B. (2018). Materials properties of printable edible inks and printing parameters optimization during 3D printing: a review. *Critical Reviews in Food Science and Nutrition* **8398**: 1–8. https://doi.org/10.1080/10408398.2018.1481823.

Feng, C., Zhang, M., Bhandari, B., and Ye, Y. (2020a). Use of potato processing by-product: effects on the 3D printing characteristics of the yam and the texture of air-fried yam snacks. *LWT* **125**: 109265.

Feng, C., Zhang, M., Liu, Z. et al. (2020b). Effect of drying method on post-processing stability and quality of 3D printed rose-yam paste. *Drying Technology* **39**: 1–9.

Fukuchi, K., Jo, K., Tomiyama, A., and Takao, S. (2012). Laser cooking: a novel culinary technique for dry heating using a laser cutter and vision technology. Proceedings of the ACM Multimedia 2012 Workshop on Multimedia for Cooking and Eating Activities, pp. 55–58.

Ghazal, A.F., Zhang, M., and Liu, Z. (2019). Spontaneous color change of 3D printed healthy food product over time after printing as a novel application for 4D food printing. *Food and Bioprocess Technology* **12** (10): 1627–1645.

Gholamipour-Shirazi, A., Norton, I.T., and Mills, T. (2019). Designing hydrocolloid based food-ink formulations for extrusion 3D printing. *Food Hydrocolloids* **95**: 161–167. https://doi.org/10.1016/j.foodhyd.2019.04.011.

Godoi, F.C., Prakash, S., and Bhandari, B.R. (2016). 3d printing technologies applied for food design: status and prospects. *Journal of Food Engineering* **179**: 44–54.

Gomez-Estaca, J., Pintado, T., Jimenez-Colmenero, F., and Cofrades, S. (2020). The effect of household storage and cooking practices on quality attributes of pork burgers formulated with PUFA-and curcumin-loaded oleogels as healthy fat substitutes. *LWT* **119**: 108909.

Guo, C., Zhang, M., and Devahastin, S. (2021). Color/aroma changes of 3D-printed buckwheat dough with yellow flesh peach as triggered by microwave heating of gelatin-gum Arabic complex coacervates. *Food Hydrocolloids* **112**: 106358.

Guo, Q., Sun, D.-W., Cheng, J.-H., and Han, Z. (2017). Microwave processing techniques and their recent applications in the food industry. *Trends in Food Science and Technology* **67**: 236–247.

He, C., Zhang, M., and Devahastin, S. (2020a). Investigation on spontaneous shape change of 4D printed starch-based purees from purple sweet potatoes as induced by microwave dehydration. *ACS Applied Materials and Interfaces* **12** (34): 37896–37905.

He, C., Zhang, M., and Fang, Z. (2019). 3D printing of food: pretreatment and post-treatment of materials. *Critical Reviews in Food Science and Nutrition* **60** (14): 2379–2392.

He, C., Zhang, M., and Guo, C. (2020b). 4D printing of mashed potato/purple sweet potato puree with spontaneous color change. *Innovative Food Science and Emerging Technologies* **59**: 102250.

Hertafeld, E., Zhang, C., Jin, Z. et al. (2019). Multi-material three-dimensional food printing with simultaneous infrared cooking. *3D Printing and Additive Manufacturing* **6** (1): 13–19.

Jonkers, N., van Dommelen, J.A.W., and Geers, M.G.D. (2020). Experimental characterization and modeling of the mechanical behavior of brittle 3D printed food. *Journal of Food Engineering* **278**: 109941.

Keerthana, K., Anukiruthika, T., Moses, J.A., and Anandharamakrishnan, C. (2020). Development of fiber-enriched 3D printed snacks from alternative foods: a study on button mushroom. *Journal of Food Engineering*. https://doi.org/10.1016/j.jfoodeng.2020.110116.

Kim, H.W., Bae, H., and Park, H.J. (2018). Classification of the printability of selected food for 3D printing: development of an assessment method using hydrocolloids as reference material. *Journal of Food Engineering* **215**: 23–32.

Kim, H.W., Lee, I.J., Park, S.M. et al. (2019). Effect of hydrocolloid addition on dimensional stability in post-processing of 3D printable cookie dough. *LWT* **101**: 69–75.

Krishnaraj, P., Anukiruthika, T., Choudhary, P. et al. (2019). 3D extrusion printing and post-processing of fibre-rich snack from indigenous composite flour. *Food and Bioprocess Technology* **12** (10): 1776–1786. https://doi.org/10.1007/s11947-019-02336-5.

Kumar, S., Anukiruthika, T., Dutta, S. et al. (2020). Iron deficiency anemia: a comprehensive review on iron absorption, bioavailability and emerging food fortification approaches. *Trends in Food Science and Technology* https://doi.org/10.1016/j.tifs.2020.02.021.

Lille, M., Nurmela, A., Nordlund, E. et al. (2018). Applicability of protein and fiber-rich food materials in extrusion-based 3D printing. *Journal of Food Engineering* **220**: 20–27.

Lipton, J., Arnold, D., Nigl, F. et al. (2010). Multi-material food printing with complex internal structure suitable for conventional post-processing. Solid Freeform Fabrication Symposium, pp. 809–815.

Liu, Y., Yu, Y., Liu, C. et al. (2019a). Rheological and mechanical behavior of milk protein composite gel for extrusion-based 3D food printing. *LWT* **102**: 338–346. https://doi.org/10.1016/j.lwt.2018.12.053.

Liu, Y., Liang, X., Saeed, A. et al. (2019b). Properties of 3D printed dough and optimization of printing parameters. *Innovative Food Science and Emerging Technologies* https://doi.org/10.1016/J.IFSET.2019.03.008.

Liu, Z., Bhandari, B., Prakash, S. et al. (2019c). Linking rheology and printability of a multicomponent gel system of carrageenan-xanthan-starch in extrusion based additive manufacturing. *Food Hydrocolloids* **87**: 413–424. https://doi.org/10.1016/j.foodhyd.2018.08.026.

Liu, Z., Bhandari, B., and Zhang, M. (2020a). Incorporation of probiotics (*Bifidobacterium animalis* subsp. Lactis) into 3D printed mashed potatoes: effects of variables on the viability. *Food Research International* **128**: 108795.

Liu, Z., Dick, A., Prakash, S. et al. (2020b). Texture modification of 3D printed air-fried potato snack by varying its internal structure with the potential to reduce oil content. *Food and Bioprocess Technology* **13** (3): 564–576.

Liu, Z. and Zhang, M. (2019). 3D food printing technologies and factors affecting printing precision. In: *Fundamentals of 3D Food Printing and Applications*. Elsevier Inc https://doi.org/10.1016/b978-0-12-814564-7.00002-x.

Liu, Z., Zhang, M., Bhandari, B., and Wang, Y. (2017). 3D printing: printing precision and application in food sector. *Trends in Food Science and Technology* **69**: 83–94.

Lupton, D. and Turner, B. (2018). Food of the future? Consumer responses to the idea of 3D-printed meat and insect-based foods. *Food and Foodways* **26** (4): 269–289.

Madhumitha, M., Abinash, V., Rahul, T. et al. (2021). Valorization of food industry waste streams using 3D printing: a study on noodles prepared from potato peel waste. *Food and Bioprocess Technology* **4**: 100036.

Mantihal, S., Prakash, S., and Bhandari, B. (2019a). Texture-modified 3D printed dark chocolate: sensory evaluation and consumer perception study. *Journal of Texture Studies* **50** (5): 386–399.

Mantihal, S., Prakash, S., Godoi, F.C., and Bhandari, B. (2017). Optimization of chocolate 3D printing by correlating thermal and flow properties with 3D structure modeling. *Innovative Food Science and Emerging Technologies* **44**: 21–29. https://doi.org/10.1016/j.ifset.2017.09.012.

Mantihal, S., Prakash, S., Godoi, F.C., and Bhandari, B. (2019b). Effect of additives on thermal, rheological and tribological properties of 3D printed dark chocolate. *Food Research International* **119**: 161–169.

Mizrahi, M., Golan, A., Mizrahi, A.B. et al. (2016). Digital gastronomy: methods & recipes for hybrid cooking. Proceedings of the 29th Annual Symposium on User Interface Software and Technology, pp. 541–552.

Mohana, Y., Mohanapriya, R., Anukiruthika, T. et al. (2020). Solar dryers for food applications: concepts, designs, and recent advances. *Solar Energy* **208**: 321–344.

Moses, J.A., Norton, T., Alagusundaram, K., and Tiwari, B.K. (2014). Novel drying techniques for the food industry. *Food Engineering Reviews* **6** (3): 43–55.

Nachal, N., Moses, J.A., Karthik, P., and Anandharamakrishnan, C. (2019). Applications of 3D printing in food processing. *Food Engineering Reviews* **11** (3): 123–141. https://doi.org/10.1007/s12393-019-09199-8.

Perez, B., Nykvist, H., Brogger, A.F. et al. (2019). Impact of macronutrients printability and 3D-printer parameters on 3D-food printing: a review. *Food Chemistry* **287** (October 2018): 249–257. https://doi.org/10.1016/j.foodchem.2019.02.090.

Phuhongsung, P., Zhang, M., and Bhandari, B. (2020a). 4D printing of products based on soy protein isolate via microwave heating for flavor development. *Food Research International* **137**: 109605.

Phuhongsung, P., Zhang, M., and Devahastin, S. (2020b). Influence of surface pH on color, texture and flavor of 3D printed composite mixture of soy protein isolate, pumpkin, and beetroot. *Food and Bioprocess Technology* **13** (9): 1600–1610.

Prakash, S., Bhandari, B.R., Godoi, F.C., and Zhang, M. (2019). Future outlook of 3D food printing. In: *Fundamentals of 3D Food Printing and Applications* (eds. F.C. Godoi, B.R. Bhandari, S. Prakash and M. Zhang), 373–381. United Kingdom: Academic Press.

Preethi, R., Shweta, D., Moses, J.A., and Anandharamakrishnan, C. (2020a). Conductive hydro drying as an alternative method for egg white powder production. *Drying Technology* **39**: 1–13.

Preethi, R., Shweta, D., Moses, J.A., and Anandharamakrishnan, C. (2020b). Conductive hydro drying of beetroot (*Beta vulgaris* L) pulp: insights for natural food colorant applications. *Journal of Food Process Engineering* **43** (12): e13557.

Pulatsu, E., Su, J.-W., Lin, J., and Lin, M. (2020). Factors affecting 3D printing and post-processing capacity of cookie dough. *Innovative Food Science and Emerging Technologies* **61**: 102316.

Qiu, H., Qiu, Z., Chen, Z. et al. (2021). Antioxidant properties of blueberry extract in different oleogel systems. *LWT* **137**: 110364.

Raghavi, L.M., Moses, J.A., and Anandharamakrishnan, C. (2018). Refractance window drying of foods: a review. *Journal of Food Engineering* **222**: 267–275.

Severini, C., Azzollini, D., Albenzio, M., and Derossi, A. (2018a). On printability, quality and nutritional properties of 3D printed cereal based snacks enriched with edible insects. *Food Research International* **106** (November 2017): 666–676. https://doi.org/10.1016/j.foodres.2018.01.034.

Severini, C., Derossi, A., and Azzollini, D. (2016). Variables affecting the printability of foods: preliminary tests on cereal-based products. *Innovative Food Science and Emerging Technologies* **38**: 281–291. https://doi.org/10.1016/j.ifset.2016.10.001.

Severini, C., Derossi, A., Ricci, I. et al. (2018b). Printing a blend of fruit and vegetables. New advances on critical variables and shelf life of 3D edible objects. *Journal of Food Engineering* **220**: 89–100. https://doi.org/10.1016/j.jfoodeng.2017.08.025.

Shi, Y., Zhang, M., and Bhandari, B. (2021). Effect of addition of beeswax based oleogel on 3D printing of potato starch-protein system. *Food Structure* **27**: 100176.

Sun, J., Peng, Z., Yan, L. et al. (2015). 3D food printing—an innovative way of mass customization in food fabrication. *International Journal of Bioprinting* **1** (1): 27–38.

Sun, J., Zhou, W., Huang, D., and Yan, L. (2018). 3D food printing: perspectives. In: *Polymers for Food Applications* (ed. T.J. Gutierrez), 725–755. Argentina: Springer International Publishing.

Taheri, S., Brodie, G., Jacob, M.V., and Antunes, E. (2018). Dielectric properties of chickpea, red and green lentil in the microwave frequency range as a function of temperature and moisture content. *Journal of Microwave Power and Electromagnetic Energy* **52** (3): 198–214.

Theagarajan, R., Moses, J.A., and Anandharamakrishnan, C. (2020). 3D extrusion printability of rice starch and optimization of process variables. *Food and Bioprocess Technology*. https://doi.org/10.1007/s11947-020-02453-6.

Tran, J.L. (2019). Safety and labelling of 3D printed food. In: *Fundamentals of 3D Food Printing and Applications* (eds. F.C. Godoi, B.R. Bhandari, S. Prakash and M. Zhang), 355–371. United Kingdom: Academic Press.

Tuorila, H. and Hartmann, C. (2020). Consumer responses to novel and unfamiliar foods. *Current Opinion in Food Science* **33**: 1–8.

Ureta, M.M., Diascorn, Y., Cambert, M. et al. (2019). Water transport during bread baking: impact of the baking temperature and the baking time. *Food Science and Technology International* **25** (3): 187–197.

Uribe-Wandurraga, Z.N., Zhang, L., Noort, M.W.J. et al. (2020). Printability and physicochemical properties of microalgae-enriched 3D-printed snacks. *Food and Bioprocess Technology* **13** (11): 2029–2042.

Vieira, M.V., Oliveira, S.M., Amado, I.R. et al. (2020). 3D printed functional cookies fortified with *Arthrospira platensis*: evaluation of its antioxidant potential and physical-chemical characterization. *Food Hydrocolloids* **107**: 105893.

Vishali, D.A., Monisha, J., Sundari, S.S.K. et al. (2019). Spray freeze drying: emerging applications in drug delivery. *Journal of Controlled Release* **300**: 93–101.

Wang, W., Yao, L., Zhang, T. et al. (2017). Transformative appetite: shape-changing food transforms from 2D to 3D by water interaction through cooking. Proceedings of the 2017 CHI Conference on Human Factors in Computing Systems, pp. 6123–6132.

Wilson, A., Anukiruthika, T., Moses, J.A., and Anandharamakrishnan, C. (2020). Customized shapes for chicken meat–based products: feasibility study on 3D-printed nuggets. *Food and Bioprocess Technology* **13**: 1968–1983.

Yang, F., Zhang, M., Fang, Z., and Liu, Y. (2019). Impact of processing parameters and post-treatment on the shape accuracy of 3D-printed baking dough. *International Journal of Food Science and Technology* **54** (1): 68–74. https://doi.org/10.1111/ijfs.13904.

Yang, F., Zhang, M., and Liu, Y. (2018a). Effect of post-treatment microwave vacuum drying on the quality of 3D-printed mango juice gel. *Drying Technology*: 1–9. https://doi.org/10.1080/07373937.2018.1536884.

Yang, F., Zhang, M., Prakash, S., and Liu, Y. (2018b). Physical properties of 3D printed baking dough as affected by different compositions. In: *Innovative Food Science and Emerging Technologies* (vol. 49, iss. 2017). Elsevier Ltd https://doi.org/10.1016/j.ifset.2018.01.001.

Yoha, K. S., Anukiruthika, T., Anila, W., Moses, J. A., & Anandharamakrishnan, C. (2021). 3D printing of encapsulated probiotics: Effect of different post-processing methods on the stability of *Lactiplantibacillus plantarum* (NCIM2083) under static in-vitro digestion conditions and during storage. *LWT*, **146**, 111461.

Yoha, K.S., Moses, J.A., and Anandharamakrishnan, C. (2020). Conductive hydro drying through refractance window drying–an alternative technique for drying of *Lactobacillus plantarum* (NCIM 2083). *Drying Technology* **38** (5–6): 610–620.

Zhang, L., Lou, Y., and Schutyser, M.A.I. (2018). 3D printing of cereal-based food structures containing probiotics. *Food Structure* **18**: 14–22.

Zhao, Z., Wang, Q., Yan, B. et al. (2021). Synergistic effect of microwave 3D print and transglutaminase on the self-gelation of surimi during printing. *Innovative Food Science and Emerging Technologies* **67**: 102546.

Zoran, A. (2019). Cooking with computers: the vision of digital gastronomy [point of view]. *Proceedings of the IEEE* **107** (8): 1467–1473.

10

4D Printing Technology

CHAPTER MENU

3D Printing of Foods, First Edition. C. Anandharamakrishnan, Jeyan A. Moses, and T. Anukiruthika.

10.1 Introduction

The current research on additive manufacturing transformed the industrial sector with the emergence of four-dimensional (4D) printing that steps forward to the next array of smart manufacturing practices. The 4D printing technology can be described as the fabrication of a physical object from a stimuli-responsive smart material through 3D printing with additional functionalities (Momeni et al. 2017). Thus, the 3D printed constructs exerting the desired characteristic response under external or internal stimuli showing a sequentially programmed change with time. The materials that are capable enough to respond to changing environment with the fourth dimension (time) are known as smart materials (Pei and Loh 2018). The 4D printing technology brings about new opportunities and research potential in the designing of novel metamaterials. Various materials that are being used as smart materials are shape memory metals, alloys, ceramics, polymers, composites, and hydrogels (Choi et al. 2015). Although the 4D printed engineered segment is time-dependent, the resulting transformation is predictable. The transformation in terms of shape, volume, or colour is evident with the fluctuations of the external environment due to different kinds of stimuli such as temperature, pressure, concentration difference, light, pH change, magnetic effects, and so on (Li et al. 2017). Mainly the 4D concept relies on the printing method, nature of stimulus material, type of stimulus response of the smart material, interaction mechanism, and its modelling. Since additive manufacturing is a free-form fabrication technique capable of designing a 3D object through digitally controlled process, most of the 3D printing technologies can be conveniently used for 4D printing applications. Only criteria that must be considered is the selection of stimuli responsive programming materials that must be compatible with 3D printer.

The growing research interest in 4D printing kindled the development of engineered programmable smart structures and materials especially in the design of soft robotics and actuators. These kinds of materials are having a great scope in smart textiles, biomedical devices, defence and military, aerospace, and automobiles (Mitchell et al. 2018). The constraint of the limited bed space of the print bed in 3D printing can be easily surpassed with the concept of self-assembly of 4D printing. The 4D printing allows to 3D print a simple flat structure which can be conveniently transformed into a complex structure through predefined programming at the post-fabrication stage. Thus, the 3D printed parts can be self-assembled during the stage of its end application. So, the concept of self-assembly in 4D printing paves a way for the development of flat packaging that greatly reduces the volume required for materials at the initial storage (Pei and Loh 2018). Hence the individual parts can easily be printed using small additive manufacturing systems which can be conveniently self-assembled at a later stage. This characteristic feature of 4D printing differentiates its potentiality from 3D printing that found its applications in the design of biomedical and electromechanical devices (Zhou et al. 2015). For instance, the engineered nano-scale parts can be employed for treatment of ailments at the targeted sites through self-assembling of the delivered parts at the desired location; development of structures and parts that can be assembled in a predetermined environment without any external power source or human intervention in space exploration; design of self-healing pipes and fabrication of self-healing hydrogels. Overall, 4D printing is moving towards the development of miniatured smart robotics that is capable enough to replace the bulky, energy consuming, and complex machine parts of conventional approaches (De Marco et al. 2019). With these smart applications, food scientists are seeking the potential applicability of 4D printing in the food sector. The applicability of 4D printing in the food sector seems to be an iteration of the concepts of 2D and 3D printing with smart programmable biomaterials with dynamic functionalities (Wang et al. 2017). The concept of 4D printing would give rise to the development of functional foods in a customized and personalized way thereby grabbing the consumer's interest and marketability. The present chapter provides an overview of 4D printing with concepts, characteristics of smart materials, self-assembling potential, and applicability. Understanding the basic principles of 4D printing would provide insights into the new research opportunities in the food sector. Hence, the advances, applicability, and recent trends of 4D printing is also presented along with the challenges and limitations. Thus, the present chapter put forth insights on the novel concept of 4D printing in spontaneous physiochemical transformations of foods, future prospective, and challenges.

10.2 4D Printing: Concept and Functionality

The fabrication of a physical entity from a composite material or multi-material with transforming properties that exhibits a visible physiochemical change in shape, volume, or colour over time due to environmental stimuli from natural causes or through human intervention is termed as 4D printing (Mu et al. 2018). In contrast to the conventional 3D printing technologies that are focused on the fabrication of static structures from a single or composite material, 4D printing could be able to produce a dynamic structure with programmed functionalities such as self-folding packaging, soft grippers for surgeries, and adaptive wind turbines to name a few (Zhang et al. 2019). The idea of 4D technology was first put forward by Tibbit during Technology, Entertainment, and Design (TED) talks (Tibbits 2013). Later the definition of 4D printing has been refined over the years as the 3D structures that exhibit a change in property or shape transformation upon environmental stimuli with respect to time is known to be 4D printing. With this key idea, researchers had been focused on the simple basic shape transformations of 4D structures based on bending, twisting, elongating, and corrugating mechanisms (Figure 10.1). These material behaviours are utilized for programming in the development of microtubes, lockers, lifters, sensors, actuators, and toys. The pre-programmed directional shape change of the 4D structures is possible by making use of the advantage of the material's anisotropy property (van Manen et al. 2018). Research in 4D printing is correlated to the principle and methods of 3D printing with an additional concept of smart materials usage that led to 4D technology. Thus, 4D printers are considered as a subset of 3D printers. It is worth noting that not all the 3D printing techniques are applicable for 4D printing. Some of the printing approaches used in 4D printing are fused deposition modelling, stereolithography, inkjet curing, direct ink writing, selective laser melting, and laser-assisted bioprinting (Subash and Kandasubramanian 2020). Careful selection of smart materials and understanding of its response are crucial for success of 4D structures.

The stimulus-responsive materials are highly dynamic that has been considered as a key element in determining the capability of the self-transformation and functionality. This

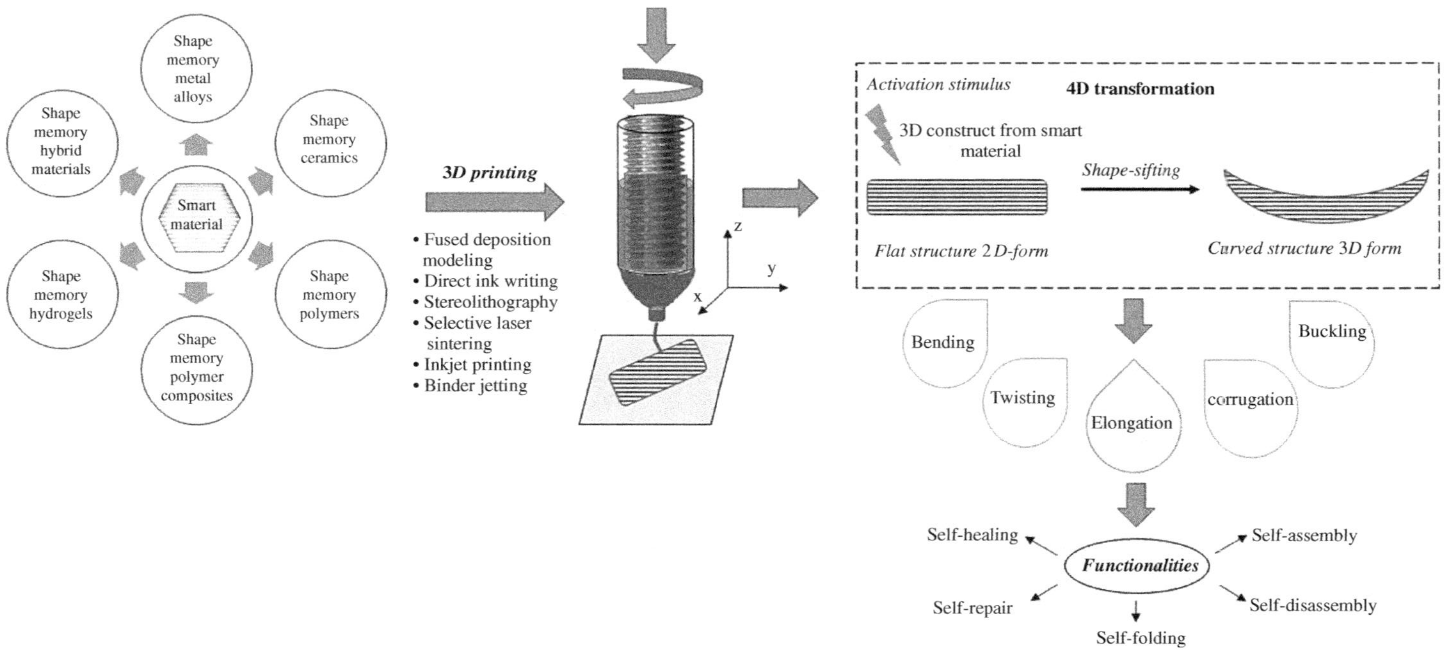

Figure 10.1 *Overview* of 4D printing process.

transformative change in the stimuli-responsive properties of smart materials allows coupling phenomenon i.e., the energy conversion among the physical domains (Pei and Loh 2018). There are two ways of coupling energy: direct and indirect. The former type describes the mechanical response due to field-induced Eigen strain in the smart entities while the latter type describes the mechanical response due to field-induced variation in mechanical properties like stiffness. Hence the type of stimulus-responsive smart materials that can change their physical properties is categorized as shape memory material (SMM) and shape change material. The associated stimulus-induced behaviour of the shape change material is termed as shape-change effect (SCE) where the material transforms and respond instantly to a temporary change in shape followed by returning to the original permanent shape with the removal of stimulus (Zhang et al. 2019). On the other hand, shape memory polymers (SMPs) have the inherent ability to memorize and recover to programmed shape from temporary shape on the application of a stimulus. The associated changes undergone by the SMPs are referred to as the shape memory effect (SME) (Patil and Song 2017). The three common kinds of stimuli for which the smart material transforms spontaneously are physical, chemical, and biological stimuli. Among which the physical stimuli such as temperature, moisture, light, magnetic energy, electricity, and ultraviolet light are the most commonly used in 4D transformations. While the chemical stimuli include oxidants, reductants, ionic strength, and pH change. On the other hand, substances that are used as biological stimuli are enzymes and glucose (Pei and Loh 2018). Most experimental studies reported in literature explored the applicability of water, heat, combination of both i.e., hydration, and combination of heat and light i.e., curing as stimuli mechanisms. These stimuli often result in a primary phase-shifting response with visible folding, curling, twisting, shrinkage, swelling, contraction, and expansion of programmed materials.

The process of 4D transformation follows a series of pre-defined mechanism that exhibits a pre-planned shape-shifting behaviour in response to stimuli over a certain period of time and appropriate process conditions (Figure 10.2). The fixation of phases of the smart materials can be done through physical mechanism manipulation techniques or by mechanical loading for designing the deformation and programming the structure at the temporary shape on exposure to stimuli (Li et al. 2017). Often mathematical modelling is used to manipulate and modulate these transformations for planning the time required and sequence of stimulus that acted on stimulus-responsive materials. The mathematical modelling of a single composite material to be 4D print involves the design of orientation and distribution of material which is useful in the determination of contraction or expansion rates during the self-assembling of designed parts. The series of transformative changes can be thoroughly monitored by using mathematical modelling in conjunction with 3D model programming using finite-element analysis (FEA) and computer-aided design (CAD) (Ding et al. 2017). In addition to the benefits and merits of conventional 3D printing, the 4D printing technology offers greater freedom and versatility in product design with only a very little additional cost for design complexity. Further, 4D printing allows designing of smart structures that are used and modified to utilize the environment's free energy i.e. the applicability of temperature, altitude, moisture, pressure, sound, humidity, as potential stimuli (Pei and Loh 2018). Thus, the design of 4D printed structures involves fully pre-programmed transformations to account for the anticipated time-dependent deformations of a physical entity. The applicability of smart materials allows making use of the advantage

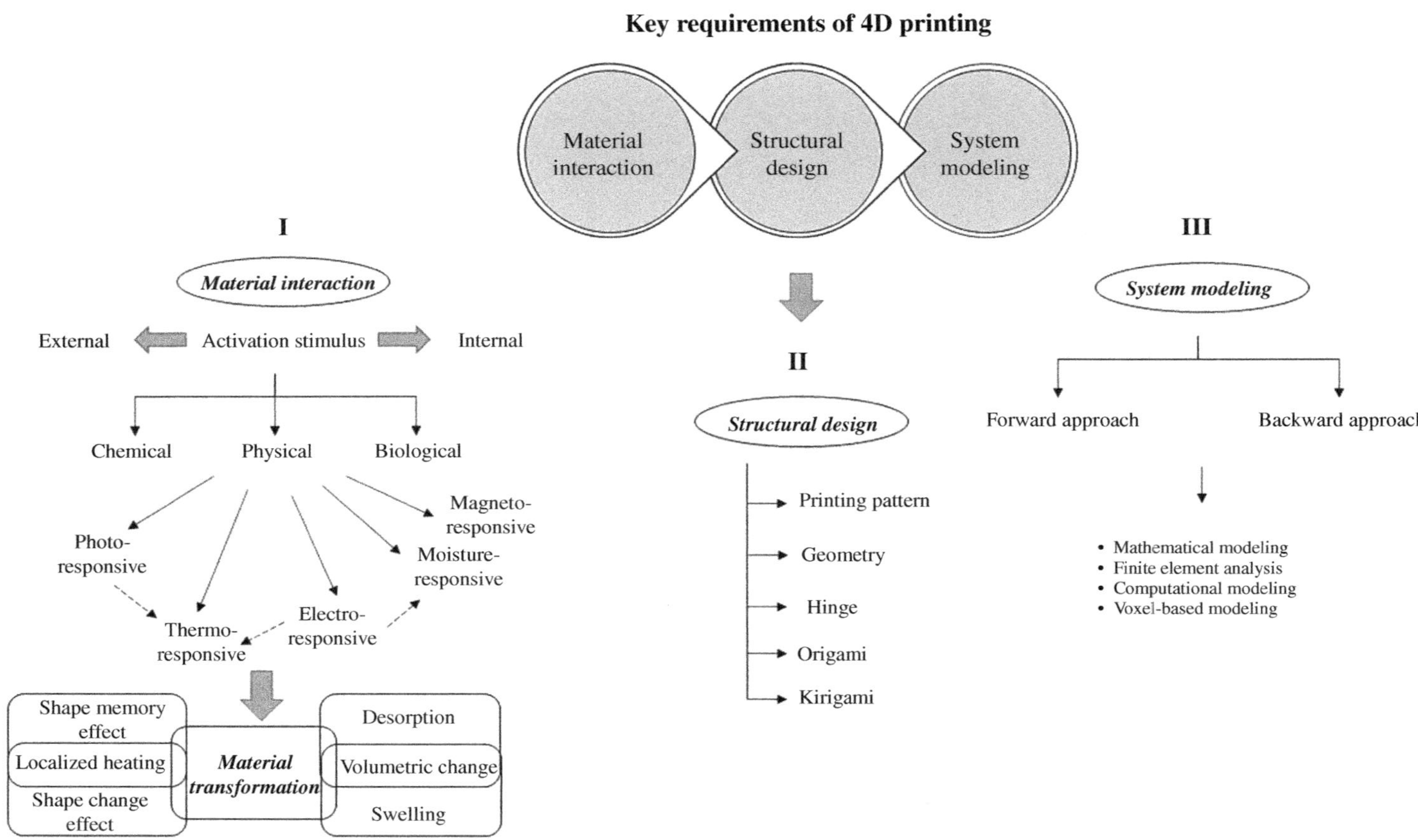

Figure 10.2 Key requirements of 4D printing of smart materials.

of material properties that are flexible, expandable, and deformable in response to the applied stimulus. Hence the implementation of 4D structures with the combined design of smart materials would pave a way for the emergence of novel 4D printer concepts and the development of 4D structures by improving the existing highly functionalized 3D printers. Undoubtedly 4D printing is a reforming technology with the inherent capability to use programmable smart materials based on a self-assembling approach (Zhou et al. 2015). This aids in the development of 3D functional structures exhibiting programmable transformation only under response to stimuli at the targeted location for the defined time period.

10.3 Smart Materials for 4D Printing

As stated earlier, the stimuli-responsive materials exert a spontaneous transformation in response to changing environment. This would lead to an interaction or coupling effect resulting in a mechanical response due to variation in stiffness or field-induced Eigenstrain of the smart material (Pei and Loh 2018). Here the Eigenstrain refers to those deformations without the influence of external forces of thermal expansion or phase change. These programmable materials can be categorized into two types: shape-changing materials and SMMs. Materials that come under shape memory smart materials are SMPs, shape memory alloys (SMAs), shape memory metals, shape memory ceramics (SMCs), shape memory gels (SMGs), and shape memory hybrid (SMH) materials (Zhang et al. 2019). When a stimulus is applied, the SMMs can regain their original shape from a temporary state due to the SME whereas the shape-changing materials can morph the transformation in response to stimulus and regaining the original shape upon removal of stimulus as a result of SCE (Li et al. 2017). Various characteristic transformations of smart materials are shape memory, self-sensing, self-repair, self-adaptability, responsiveness, and multifunctionality. The present section provides an overview of the shape-memory mechanism of the most commonly used smart materials. Understanding the governing factors that affect the mechanical behaviour of the smart materials would assist in gaining deeper knowledge on the design and fabrication of the smart materials, programming of a pre-defined pattern of transformations, and aids in modelling of the material behaviour.

10.3.1 Shape Memory Alloys

The SMAs is another kind of shape-memory materials where these smart materials exert a diffusion-less phase transformation through a well-coordinated shear dominant atomic displacement (Tsuchiya 2011). The SMAs are also referred to as memory metal, smart metal, smart alloy, and memory alloy. This mechanism of phase transformation occurs between the austenite and martensite state of the material. The austenite of the alloys refers to the material state that exists at a high-temperature phase with a face-centred cubic structure while the martensite is the state where the material experiences a low-temperature phase with a low crystallographic symmetry and body-centred tetragonal structure (Wen et al. 2018). This sequence of the material transformation of the SMAs is evident because of SME and superelasticity (SE) or pseudoelasticity (PE) behaviours. This can be explained by the gradual recovery of the original shape of the smart materials through heating of

SME and subsequent unloading of the SE. Due to which a reversible martensite transformation would be induced by the effects of temperature (thermal) and stress (mechanical) for SME and SE behaviours, respectively (Li et al. 2017). Thus, a deformed alloy recovers to its original shape by heating in SME while in SE state, the material undergoes severe deformation and recover its real shape on removal of applied mechanical load. For instance, when the material under its austenite phase was subjected to stress, it would be mechanically deformed due to stress effects and induces a martensitic transformation. This transformed martensite phase would completely get back to its austenite state upon removal of the applied load (Khoo et al. 2015). The associated recovery of the materials to their original shape was mainly due to its thermodynamic instability of the transformed state of the material (martensite) at that temperature. The SMAs are widely used in bioengineering, civil engineering, automotive applications, fabrics, and textiles (Jani et al. 2014). Recently, the applications of SMAs are being exploited for their applicability in biomedical engineering that provides new insights on the development of cardiovascular stents, guided wires, and organ frame retractors (Wen et al. 2018). The SMA are having a great scope to be used in the manufacturing of smart and intelligent food packaging containers i.e. self-folding packages (Zhang et al. 2019). However, more studies are required in this area for exploring its food applications in terms of convenience, safety, and sustainability.

10.3.2 Shape Memory Polymers

The next class of smart memory materials that are widely applied in 4D printing are SMPs. The SMPs are capable of programming to get temporarily fixed at its deformed state and subsequently recovered to the initial state when exposed to external stimuli like heat, pH, moisture, light, pressure, electricity, magnetic field, radiation, laser heating, and chemical solvent or vapours (Meng and Li 2013). In contrast to the martensitic transformation of SMAs, the mechanism of SMPs involves a combination of a fixity phase and a reversible phase. Here the stabilization of the original shape is attributed by fixity phase (net point) that are formed by the supramolecular bonding or physical/ chemical cross-linking while the reversible phase (switching unit) was accountable for the material response and recovery of shape upon different stimuli. The reversible phase of the SMPs is associated with crystallization, light-induced glass transition, reversible bonding, and supramolecular hydrogen bonding (Jiang et al. 2017). Typically, an SMP consists of a net-point and a switching unit that assists in spontaneous transformation. Based on the nature of the fixity phase, i.e. net point, the SMP can be of either physical cross-linking or chemical cross-linking. On the other hand, based on the nature of external stimuli, the SMPs are classified as thermo-responsive, electro-responsive, and solvent-induced. Further, the SMPs are subdivided as glass transition temperature (T_g)-type and melting temperature (T_m)-type SMPs based on the activation of stimuli by T_g and T_m, respectively (Wen et al. 2018). The SMEs are achievable through the programming and designing of smart polymers with special molecular structure and varied transition segments. As stated, the fixity segment assists in attaining permanent shape while the switching segment assists in gaining of temporary phase (Miao et al. 2017). With rapid advancements in material science, a different range of shape memory programming methods has been developed. This includes triple SME, multi-functional SME, multi-stimuli SME, one- and two-way SME, and temperature memory polymer. Among which the temperature-induced SMPs are the widely used smart materials that are receiving greater attention in recent days. Apart from simplicity and ease of processing, the SMPs exhibits 200% strain that are much larger than the strain of SMAs (Pei and Loh 2018). Various advantages of SMPs are larger strain, wider range of tailorable temperature, lower densities, biocompatibility, biodegradation, lower cost, and convenience of processing (Khoo et al. 2015). Due to these potential features, SMPs are most widely used in various industrial sectors in designing of smart actuators and sensors that found its applications in aerospace, automobiles, textiles, biomedicine, and food industries. The latter half of the present chapter discusses in detail about the various methods of environmental stimulus methods of SMPs for food applications in 4D printing.

10.3.3 Shape Memory Composites

Shape memory polymer composites (SMPCs) are variants of SMPs that have been developed to overcome the limitations that exist with the engineering aspects of SMPs in the reinforcement of mechanical properties, improvement of driving methods, designing, and programming of distortions and material formation with multi functionalities (Khoo et al. 2015). The combination of SMPs with diverse fillers gives rise to the SMPCs that overcome the drawbacks with SMPs like less modulus of elasticity, mechanical integrity, stiffness, stress recovery, low electrical, and thermal conductivity. Further, the incorporation of conductive fillers into SMPs would create composite structures with the potential ability that can respond under the magnetic or electrical field instead of heating it directly (Madbouly and Lendlein 2009). The main purpose of the production of SMPCs is to improve the thermomechanical properties of the SMPs and make use of the advantage of composite layers in the discovery of new stimulus with novel applications in material science. SMPCs with active filler materials for biomedical applications aids in enhancing the temperature of the specific/ local pointe through indirect heating in the human body that can be used as an alternative to direct heating or increasing temperature of the environment (Mu et al. 2018). The common and appropriate way of triggering SMPs is to directly increase the surrounding temperature. However, this mode of exogenous heating is not suitable to be used for some applications especially in biomedicine that requires the direct generation of heat from the interior of the material. This sort of requirement leads to the development of novel SMPCs through the embedment of micro and nano inert fillers into the polymeric matrix that converts the chemical, optical, magnetic, acoustic, electrical energy into heat energy (Mitchell et al., 2018). Thus, the fabrication of novel SMPCs is a facile way of tailoring the smart materials by combining the advantages of polymers and the functional properties of micro/ nano fillers. The role of microstructures in the development of the smart behaviour of polymer composites are detailed in a recent report (Panahi-Sarmad et al. 2019). It was inferred that the silicate and carbon-based micro/ nano- and hybrid particles in SMPCs aids in controlling the recovery state and shape, improve the electrical and thermal conductivity of the filler, enhances the modulus of elasticity, maintains the crystallinity of the composite, maximizes the fixity ratio and shape recovery of the composite material. All these added features lead to higher performance of the SMPCs.

10.4 Mechanism of Shape Memory Polymers

The SMPs not only employed at the bulk level but can also be used as shape memory surface, porous shape-memory materials, shape-memory microparticles, and shape memory hydrogels (Panahi-Sarmad et al. 2019). The polymers used for programming the shape memory are categorized as thermoplastic and thermoset polymers. Both these classes of polymers have their own merits as thermoset polymers could store a large amount of energy and tolerate a greater number of shape transforming cycles because of the cross-linked nature of the bonds. While the thermoplastic polymers are simpler and follow a facile method of processing (Xie et al. 2016). Hence the end-use and intended applications of SMMs determine the selection of thermoset and thermoplastic polymers. The process of shape transformation in SMPs usually occurs in two steps of shape fixing and recovery stages. The first step of shape fixing involves the attainment of transition temperature (T_{trans}) from permanent state where the material reaches a temporary shape and is capable of programming (Ge et al. 2016). This would augment the mobility of polymeric chains by imposing an external force. During which the SMP is tuned to be in its programmed interim shape known as shape fixity as the entropy of the system got declined under the quenching process. This process of shape fixation stabilizes the SMP at its temporary state. The entropy of the system increases with reduction in the free energy when the temperature becomes higher than T_{trans}. This would result in the shape recovery of the material from transition state to its permanent state.

According to the type of polymers employed, the T_{trans} can be the glass transition temperature (T_g) in amorphous polymers; melting temperature in the crystalline polymers (T_m); and liquid crystal isotropic transition temperature (T_i) in semi-crystalline polymers (Mu et al. 2018). Thus, the polymeric chains get unlocked with the application of a stimulus and get back to their original shape. The memory performance of the SMPs is determined by shape fixity and recovery parameters of the fixing ratio (R_f) and recovery ratio (R_{rec}) (Eqs 10.1 and 10.2) (Zhao et al. 2015). From these equations, it is clear that the viscoelastic nature of the polymers leads to the recovery of strain instantly after the removal of the applied external force.

$$R_f = \frac{\varepsilon}{\varepsilon_{load}} \times 100 \tag{10.1}$$

$$R_{rec} = \frac{\varepsilon - \varepsilon_{rec}}{\varepsilon} \times 100 \tag{10.2}$$

where, ε refers to the fixed strain observed after the removal of external force below T_{trans}, ε_{load} is the interim strain under the external force at permanent state and ε_{rec} is the permanent recovery strain observed after heating the system above T_{trans}.

10.5 Shape Memory Effect in 4D Printing

The potential feature of the shape memory smart materials is their inherent ability to transform from a temporary shape to their pre-determined programmed shape when subjected to stimuli. This attribute is described as SME that comprise of programming process and shape

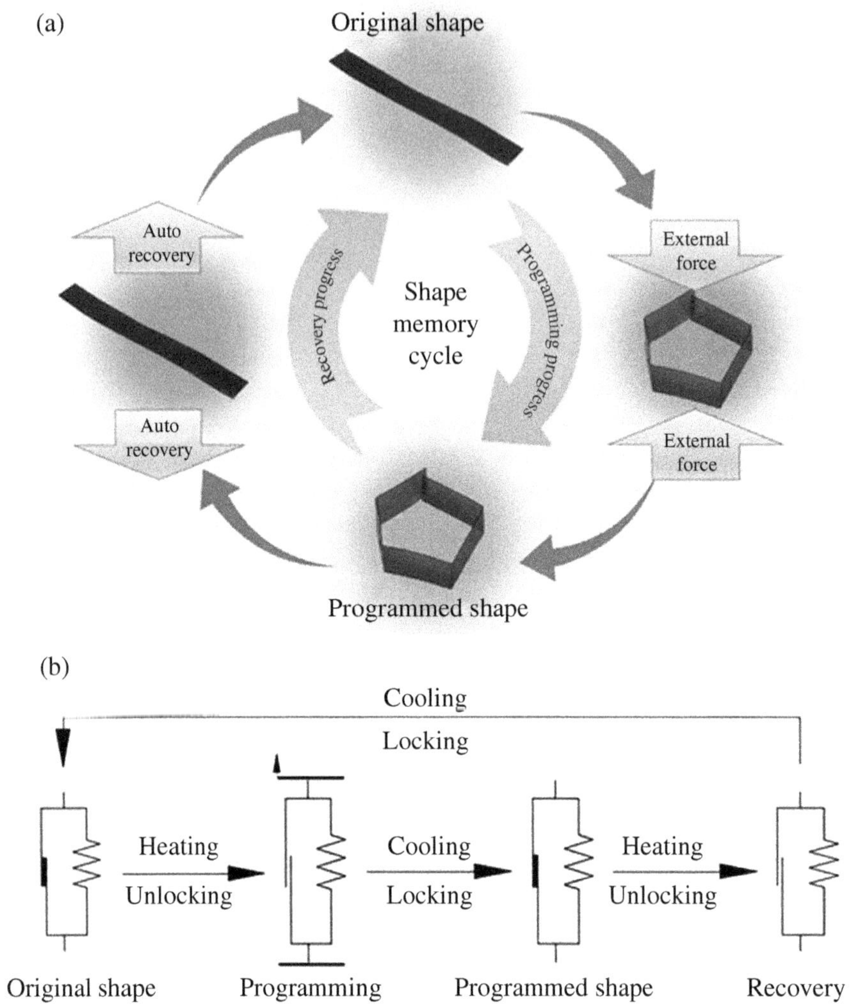

Figure 10.3 Concept of SME. (a) shape memory cycle, and (b) analogy of SME. *Source:* Mu et al. (2018) / With permission of Elsevier.

recovery process to complete one shape memory cycle (Figure 10.3) (Mu et al. 2018). The programming process involves the deformation of the material into a temporary shape followed by recovery of shape. The SMMs would stand in their interim shape until affected or triggered by the optimal stimulus at the right time that initiates the process of shape recovery.

The quick response of the geometric transformation from interim shape relies on the material's response, programming strategies, and structural design of the intended geometry (van Manen et al. 2018). The elasticity of these materials regulates the 'memory' of geometric structures as desired. The SME of materials is determined by strain recovery rate (R_{rec}) and strain fixity rate (R_f) (Zhao et al. 2015). The capability of smart material to realize

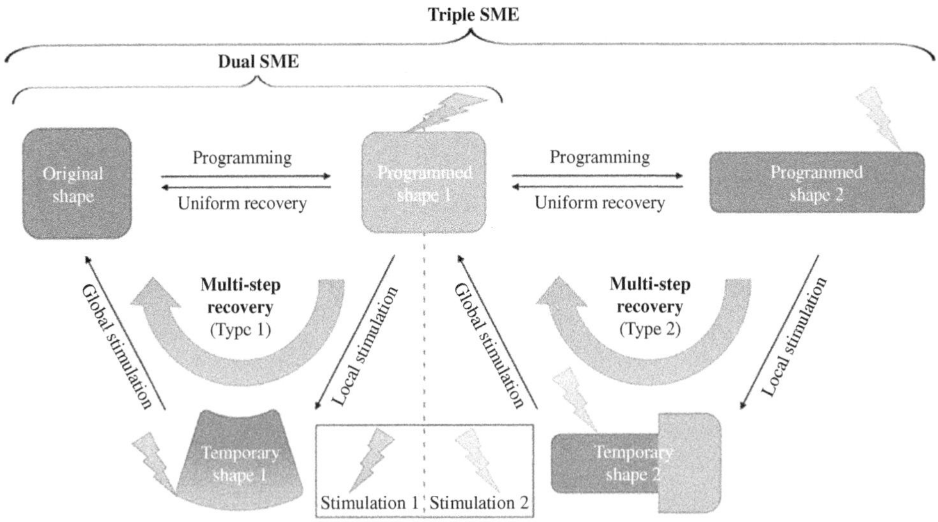

Figure 10.4 Working concept of dual and triple shape memory effect (SME) of smart materials. *Source:* Mu et al. (2018) / With permission of Elsevier.

its permanent shape is termed as strain recovery rate while the material's ability to switch the segment within the mechanical deformation is termed as strain fixity rate. Both the R_f and R_{rec} should be added up to 100% to be considered as an effective SMP (Pei and Loh 2018). Based on the SMEs, the SMPs are classified as one-way, two-way, and three-way SMEs (Panahi-Sarmad et al. 2019). Both one- and two-way SMEs are apt to get programmed into dual shapes, triple shapes, and multiple interim shapes which are distinct from each other by the process of reversibility in each effect (Figure 10.4). Various types of SMEs as employed for SMPs are discussed briefly in the present section.

10.5.1 One-Way SME

The one-way SME is the commonly used process for programming the SMPs. The one-way SME is irreversible that involves the deformation of the temporary shape into a permanent shape on the application of the external stimulus (Mu et al. 2018). In the first step of the process of programming smart materials, the SMP is heated to a temperature higher than T_{trans} in order to soothe the material to endure the deformational force (loading) which could be applied to the original geometry. This pre-deformed material is then cooled under the subjected load to get fixed with the interim shape. The fixed interim shape of the material on exposure to the optimal external stimuli such as heat would transform the material to regain its original shape (Li et al. 2017). Different kinds of thermal transition temperatures that are related to the switching domains are glass transition temperature, melting transition temperature, and liquid crystalline transition temperature. Although these temperatures are varied according to the type of polymers used, most of the melting temperature and liquid crystalline transition temperatures of polymers are nearly very small. For smart materials like SME, the glass transition temperature is mostly considered. Overall, the action of one-way SME follows a sequence of steps: deformation of the structure at temperature $>T_{trans}$ through mechanical loading; phase fixation where the temperature is reduced well below the T_{trans} without removing the mechanical load; phase transformation occurs with the removal of load to attain a desired temporary programmed shape; and recovery of shape from temporary to permanent shape by reheating the geometry till reaching the T_{trans} (Pei and Loh 2018).

10.5.2 Two-Way SME

The two-way SME is capable of remembering two different geometrical shapes when subjected to external stimuli. It undergoes a transformation from the interim shape and returns to its permanent shape where the shape transition is reversible. This kind of reversible behaviour of SMPs is considered neither to be structurally nor mechanically constrained which allows for a number of switchings between the encoded shapes without subjecting them to external force (Zhou et al. 2014). The photo-actuation deformed polymers and the liquid state crystalline elastomers exert two-way SME (Basit et al. 2013).

10.5.3 Three-Way SME

The three-way SME is similar to a one-way SME with additional interim geometry existing between the permanent and temporary shapes. Thus, a three-way SME involves more than one interim shapes (Erkecouglu et al. 2016). Hence this kind of material behaviour is also termed as multiple SME that can be attained by combining multiple two-way SMPs of different T_g or by heating a tuned programmable SMP to temperature greater than T_g at first followed by increasing the temperature above temperature of melting of the switching domains (Bothe et al. 2012). Various methods used for modulating the triple SMEs of polymers are blocking of copolymers, blending, grafting, hybrid SMP, and other polymer composite laminates (Li et al. 2015). In contrast to the dual-SME that results in one permanent shape and one temporary shape, the triple-SME materials possess a single stationary shape and two interim shapes. A variant of SME exhibiting a dual shape mechanism that can be achieved by organizing two different components of soft and hard segments within a single unit. Such a designed dual-component system possess the capability to transform into a ductile state on attaining T_g followed by recovering to its original shape. This mechanism is based on making use of advantage of glass transition material where the polymer exists in a rubbery state (Pei and Loh 2018). The material is soft enough and can be easily transformed at this rubbery state. Upon cooling the polymeric material to a temperature below the glass transition temperature, the polymer is in its glass state and becomes hard. Thus, the original shape of the material can be recovered upon heating to above T_g where cross-linked chains aids in storing the elastic energy that acts as a driving force for shape recovery (Yang et al. 2014). Therefore, the dual component mechanism (DCM) refers to the assembly of different components of varied forms of soft and hard partitions within a single unit. During which the hard portion possess its elasticity within the working range of temperature and subsequently the soft portions become ductile upon heating that leads to material deformation at high temperatures (Monzon et al. 2017).

10.6 Stimuli-Responsive Systems

Recent developments in material science have brought up the novel applications of composite polymers in 4D printing. The applications of 3D printing in fabrication and the development of 4D printed smart materials allow for precise material deposition, flexible process, and controlled programming (Miao et al. 2017). The selection of materials with smart capabilities to transform its structure upon external stimulus is the most significant aspect of 4D printing. The response of the smart materials to various environmental stimuli determines the end-application of 4D printing (Khoo et al. 2015). Various environmental temporal stimuli that are commonly employed in designing smart materials and 4D printing are discussed as follows.

10.6.1 Thermo-Responsive

Among the various stimuli, the temperature is the most commonly used stimulus for SMPs. The deformation behaviour of SMPs is based on two mechanisms: the SME and SCE. The SMM based on SME are further classified as SMPs, SMAs, SMPCs, SMCs, SMHs, and SMGs (Zhang et al. 2019). Among which SMPs have great research potential due to their versatility, simplicity, and ease of 3D printing. The SMPs have a T_g that is higher than its operational temperatures. By utilizing this property, the SMPs are programmed to respond to the specific mechanical and thermal treatments above their T_g. Upon cooling the material get fixed to its interim geometry independent of external loading. Later material responds by recovering to its stationary state when the temperature is increased above the T_g (Hager et al. 2015). The SMPs works based on the SME mechanism that usually has two or more discrete states with no intermediate stable shapes. Contrarily, the SCE is proportional to the applied treatment of stimulus that remains continuously changing between its extreme states. The SCE occurs in bilayer structures of thermo-responsive materials that possess an adequate difference in its coefficient of thermal expansion (Hu et al. 2015). The presence of a constant interfacial area between the layers would incur stress fields resulting in bending of the structures.

10.6.2 Moisture-Responsive

Next to the thermo-responsive materials, hydrophilic/ hydrophobic or moisture-responsive materials possess higher research attention especially for food and biomedical applications. Hydrogels are water-loving materials that remain excellent moisture-responsive material (Zhang et al. 2019). The shape memory hydrogels are capable of expanding to 200% from their original volume. In addition, hydrogels are polymers that exhibit good printability, biocompatibility, and highly suitable for the direct method of ink writing (Zhang et al. 2017b). Despite these advantages, the applications of SMHs are limited because of the long process time for drying and material shrinking. The process of programming of hydrogels is endowed with mechanisms of anisotropy to swelling. The response of hydrogels is evident through material swelling when the hydrogels are submerged into an aqueous environment (Wang et al. 2017). The materials would stop absorbing water on reaching its saturation level that hinders the intermediary control of the hydrogels. However, the swelling behaviour of hydrogels can be directed by modulating the temperature of the aqueous medium.

10.6.3 Photo-Responsive

The temperature- and moisture-response of smart materials come under the class of direct stimuli in 4D printing. In contrast, light is the indirect stimulus where the exposed part/ surface of the photo-responsive materials absorbs light and converts it into heat energy. A sequentially controlled self-folding structure was demonstrated for its photo-responsive stimuli (Liu et al. 2017). The photo energy was absorbed by the joints in the form of heat and the absorption rate can be assessed through visual colour difference among the surface joint and light source. In another study, the photo-responsive chromophores were infiltrated into polymer gel dispersed all over the surface. The infiltrated chromophores would block a portion of hydrogel and only those locations with chromophores are exposed to the light (Kuksenok and Balazs 2016). The applicability and flexibility of light as a stimulus in the designing of patterns are well known. Hence, by projecting the weaker UV light onto a liquid resin that would create a gradient crosslink reachable to the depth initiating the bending of 4D printed structure based on anisotropy (van Manen et al. 2018).

10.6.4 Electro-Responsive

The electrical energy can be used as indirect stimuli in designing the 4D structures. The applicability of electro-responsiveness of silicone elastomer and ethanol composite actuator was reported. When electric energy is applied, the thermal energy is induced due to resistance heating that aids in the evaporation of ethanol. The associated shifting of phases from liquid to gas increased the volume of gas resulting in the expansion of the whole matrix (Miriyev et al. 2017). In another study, electrical energy was used for controlling the water absorption and desorption in polypyrrole films. In this report, the polypyrrole film was applied to an origami robot with a foot designed of special geometries that possess less resistance when moved forward (Okuzaki et al. 2013). Under a humid environment, it was observed that the voltage drives the head forward followed by the tail due to absorption and desorption of the moisture. The moisture gets desorbed with the absence or decrease of the applied voltage.

10.6.5 Magneto-Responsive

The 4D smart geometries responding to the applied magnetic field is known as magneto-responsive materials. A microgripper was 3D printed from hydrogels incorporated with magnetic nanoparticles. Here the nanoparticles were embedded during pre-processing of material where powdered ferric oxide was blended into solution (Breger et al. 2015). The printed structure has been remotely controlled through varied magnetic fields. The magneto-responsive materials can be applied for polymer and metal printing. However, the size of the print restricts its applicability as the sufficiently lightweight prints are less affected by the magnetic fields.

10.7 Programming Strategies

4D technology of printing dealt with the science of programming 3D printed structure to undergo spontaneous shape-shifting with respect to time when exposed to appropriate external stimuli. Recently, the shape-shifting of smart materials have gained greater research potential that holds a lot of scope in industrial sectors for developing a complex material with unmatchable properties and functionalities (Mitchell et al. 2018). Although 4D printing is an emerging research area, the concept is never new as many shape-shifting phenomena exist as common in nature. These shape-shifting behaviours are originated from the spatial arrangement of the structured elements and their varied responses under diverse environmental stimuli. For instance, the dimensional transition of the smart elements in regard to the application of external stimuli would result in activation of shape changing of materials like leaves of pine cone and *Mimosa pudica* (Studart 2015; Volkov et al. 2010). The natural phenomena of this response of leaves are due to the evolutionary changes that occurred in the extracellular systems that encourage the idea of the 4D transformation of synthetic materials. Thus, inspired by nature, studies are focused on the design of complex shape transformations of synthetic materials as triggered by certain stimuli. In addition to the surface-induced functionalities, the shape-changing materials offer several unique functionalities that found their applications in diverse fields of the industrial sector. The main three groups of stimuli-responsive smart polymers employed in shape-shifting are liquid crystalline elastomers, polymeric gel systems, and SMPs. As discussed earlier, the SMPs are most commonly used materials for shape transformation.

Consider a smart material of flat slender structure of elastic body incorporated with passive and active elements. The structural deformation of such an element is approximated based on assumptions of Kirchoff–Love theory (Efrati et al. 2013). Here the deformation of a thin object is viewed by the induced changes due to deformation at the mid-plane in the dimensions and curvature of the object. The energy of elasticity of the flat structure is a sum of bending and stretching. When a smart material is triggered by an activation stimulus, axial forces and bending moments are generated. These actions lead to out-of-plane buckling and out-of-plane bending (van Manen et al. 2018). The creation of these two forces is the main driving mechanism of shape transformation. The out-of-plane bending moment along with the axial forces of compression induces the out-of-plane deformation resulting in the geometrical transformation to a complex 3D shape from the flatten structure. The required moment of bending can be achieved by developing a gradient in planar stresses along the dimensional thickness of the structure upon activation by external stimuli. Various strategies are employed for programming the smart material in regard to the activation stimulus (Figure 10.5). The smart memory material will usually exhibit pure stretching due to the axial forces upon activation. For a thin construct, a bending deformation is more desirable than pure stretching in terms of its molecular energy (Sharon and Efrati 2010). This was mainly because of the energy dependency of deformations as bending energy possess cubic dependency on dimensional thickness as against the linear dependency of molecular stretching. In this regard, it is essential to know about the various strategies based on which the smart material can be programmed in a pre-defined way exerting desired characteristics. Understanding the deformation mechanisms for shape transformation of smart materials provides a deeper insight on the rational design of shape-shifting materials.

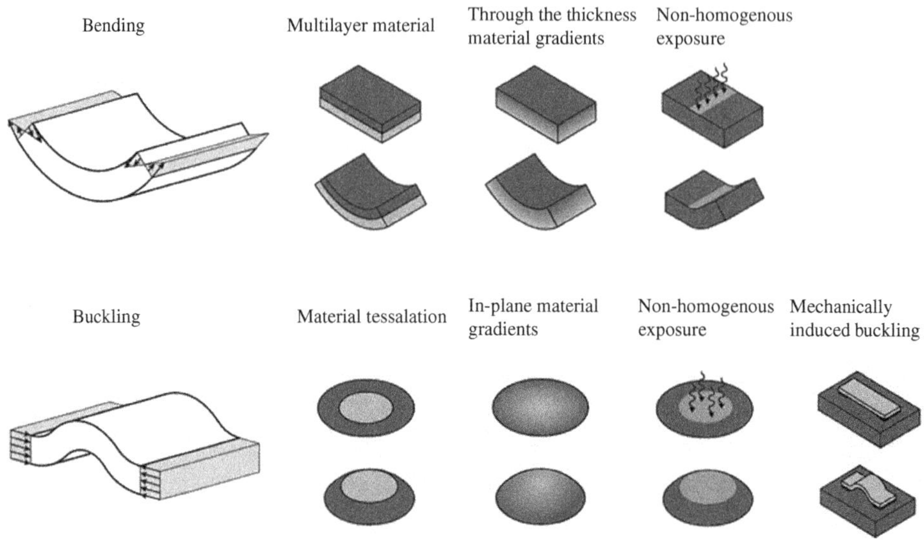

Figure 10.5 Programming strategies of bending and buckling. *Source:* van Manen et al. (2018) / With permission of Elsevier.

In this regard, various programming strategies that are employed for transformation of flat structure to 3D structure is discussed that could be in par with the shape-shifting of 3D printed materials for 4D food printing applications.

10.7.1 Bending Strategy

One of the common strategies that are evident during 4D shape-shifting is the bending strategy where gradients of stress act by the side of the material's thickness. Various approaches that are employed for the creation of stress gradients for different stimuli-responsive materials are the multilayer approach, material gradient approach and localized activation approach which are explained in the following sections.

10.7.1.1 Multilayer Approach

The multilayer approach is the most widely used strategy for generating the bending moment where two layers are combined as a single bilayer construct to exert shape-shifting. The combination of different behaviour of surface in response to active stimuli would reflect a mismatch of swelling/ shrinkage. The mismatch of material properties when sustaining a strain at the interface yielding different kinds of deformations. Based on the induced strain and thickness of the material, the shrinkage and bending can be programmed (Cendula et al. 2009). For instance, the surface wrinkling can be generated by the forces of compression when applied over a stiff slender top layer by a thick-soft substrate while the bending is evident with comparable layer thickness and stiffness considering the bending from out-of-plane. Upon activation by external stimuli, the main straining layer undergoes deformation and experiencing stress is hindered by the second layer. This would be reflected as a tension load on the first layer while the compressive forces are evident in the second layer.

The developed bending moment results in the bending of the printed construct by the limited release of the internal stresses.

A bilayer comprising of two different materials with a varying coefficient of thermal expansion would yield a uniform curvature that is inversely related to the thickness. The normalized curvature can be described by the general theory of bilayers (Eq. 10.3) (Timoshenko 1925).

$$\tilde{\kappa} = \frac{\kappa h}{\Delta\alpha} = \frac{6(1+m)^2}{3(1+m)^2 + (1+mn)\left(m^2 + \frac{1}{mn}\right)} \tag{10.3}$$

where, $\tilde{\kappa}$ is the normalized curvature, h is the layer thickness, κ is the curvature of the bilayer, $\Delta\alpha$ is the deviation in coefficients of thermal expansion, n is the stiffness ratio of both the materials and m is thickness ratio of both the layers. Although the general theory of bilayers was derived by considering the bilayer metallic strips with a very small thickness ($h \ll L$), this theory could be applicable for predicting the properties of the SMP with different mechanisms of activation (Liu et al. 2016). Based on the properties of the material, the multilayer construct can be of two types: isotropic multilayers and anisotropic multilayers.

10.7.1.1.1 Isotropic Multilayers

It was reported that isotropic de-swelling and swelling of smart polymers were the activation mechanism that acts as a major driving force for bending of bilayer structures (Egunov et al. 2016). The designed curvature of bilayer strips can be controlled based on the swelling ratio of the different polymer compositions of the bilayer. It was reported that a larger curvature could be generated by increasing the swelling ratio as a result of a larger bending moment. Based on Timoshenko's bilayer theory and results reported by Guan et al. (2005), a linear relation was drawn between the swelling ratio and radius of curvature. Apart from the thickness and swelling ratio, the stiffness ratio also influences the radius of curvature by adjusting the bilayer resistance against deformation (Guo et al. 2013). Comparatively, the thickness ratio would strongly affect the radius of curvature of bilayer construct than the stiffness ratio. In addition to the factors considered in the 2D model of bilayer theory, other factors related to the 3D structure of the smart materials must also be included during the evaluation of shape transformation in 4D printing.

A study on isotropic bilayer elastic polymer stated that a flat material could adopt a double-curved structure when there exists a small mismatch of strain (Lin et al. 2016). In case of a large extent of strain mismatch, the outcoming shape could be of a single curve since in-plane stretching is formidably high for a double-curved shape. The strain mismatch is witnessed during wrapping of a sphere with a sheet of plastic as rolling along one direction become energetically beneficial for mismatch of strains that exceeds the threshold level (Egunov et al. 2016). The associated bifurcation of double to a single curvature model relies on geometrical attributes, layer thickness, and material properties that significantly affects the balance between the energies of bending and stretching. A bilayer rectangular strip with an aspect ratio of 1 and larger width could adopt a spherical shape while the larger samples with uniform thickness would result in cylinder form. In the case of an unconstrained strip of the bilayer, the structure preferred to curve by the side of the shortest direction (Alben et al. 2011). Thus, the geometry of the flat bilayers highly influences the curvature direction of the shape-shifting materials. It was reported that an isotropic swelling of polymer in an aqueous medium in response to variations in temperature would result in the bending of the star-shaped arms of the polymeric bilayer. The incorporation of active and passive elements in a star-shaped bilayer is used for restricting the diffusion directions (Stoychev et al. 2011). As a result, the water diffuses through the active layer by the side of the edges inducing deformation. As against unconstrained bilayers, the edge-activation and the polymer adhesion to the substrate cause the bilayer to curve by the side of longest direction (Stoychev et al. 2012). Additionally, the 3D printing of multi-material allows greater degree of freedom to tune the polymeric material through spatial alignment of passive and active elements. This had led to the development of 4D printed origami structures at microscale with multiple functionalities.

10.7.1.1.2 Anisotropic Multilayers

The programming of a more complex shape transformation includes an additional programming parameter in the form of anisotropic properties of the materials (van Manen et al. 2018). Anisotropic is direction-dependent, hence it can be induced by generating stress in desired directions, shape-shift transformations are evident. This includes self-helixing elements, self-bending, self-twisting strips of soft matters. The self-twisting strips are developed by nature-inspired opening of chiral seed pods by a combination of two anisotropic layers with strain directions perpendicular to one another. The helixing or twisting would be dominant in anisotropic bilayer as it relies on the dimensionless width of the strip ($\widetilde{w}$) (Eq. 10.4) (Guo et al. 2014).

$$\widetilde{w} = w\sqrt{\kappa_0/t} \tag{10.4}$$

where, w is the width of the strip, κ_0 is the reference curvature, and t is the thickness of the bilayer. Here, deformation of twisting is dominant because of the in-plane stretching and bending is dominant in the case of helixing of bilayer strips. The stretching energy would be rapidly increased with an increase in the width of the self-twisting samples while the energy of bending in self-helixing constructs are proportional to the width of the sample. Hence, twisting is energetically beneficial for samples with a small width. In contrast, the self-helixing would occur when $\widetilde{w} \gg 1$ (Forterre and Dumais 2011). The self-twisting and self-helixing constructs could be defined by varying the radius (r) and pitch (p) based on angle (φ) and width of the bilayer (w). Various methods were described for encompassing anisotropy within the multilayer sample among which uniaxial stretching is the most commonly used method for programming of SMPs (Ge et al. 2014). With the application of a passive layer of elastomer, a self-helixing and self-rolling behaviour SMP could be designed. The uniaxial arrangement of liquid crystal in a polymeric network imparts another strategy for the enhancement of mechanical properties of the anisotropic active layer (Zhou and Sheiko 2016). Using this approach, a layer of liquid crystal elastomer was combined with isotropic polymer for shape-changing with different shapes when submerged in the water (Boothby and Ware 2017). Thus, multiple shapes can be defined within the same bilayer through activation of temperature resulting in layer shrinkage of liquid crystal elastomer along the direction of molecules and expansion in the orthogonal direction. Another strategy is the direct 3D printing of swellable polymeric gel by imprinting onto the cellulose fibrils resulting in localized programming of anisotropy and get aligned along the direction of extrusion. The 3D printed structures exhibit anisotropic swelling when immersed into the water that could be used for self-rolling, self-twisting, and a combination of different modes of shape-shifting. Recently, SMP material has been 4D printed based on the fused deposition modeling (FDM) approach for simultaneous fabrication and programming of SMP constructs (van Manen et al. 2017). Using this method, a directional strain till 40% can be defined inside the SMP along the direction of printing that enables shape-shifting according to the pattern of printing. This kind of novel 4D printing approach is useful in the printing of monolayer passive panels and bilayer bending actuators that can be conveniently integrated to form origami constructs.

10.7.1.2 Material Gradients

Another approach for generating gradients of stress along the material's thickness is by making use of gradient in the properties of monolayer material. This strategy of material gradients offers two main advantages over the bilayer constructs. This includes the elimination of issues of delamination or sliding of interfacial layers since the high shear stresses are avoided at the bilayer interface; and the elimination of complex process steps that are required in multilayer structures. On the basis of this strategy, different approaches are being employed for shape-shifting applications. The polymerization of photocurable resin is one of the simpler approaches for the generation of self-rolling structures (Zhao et al. 2017). This method follows the development of gradients of light along with the material's thickness when exposed to light during the polymerization process that has been partially absorbed by the polymer. Bidirectional curving was achieved from the top and bottom sides when exposed to light. A similar technique based on the laser was employed for a partial relaxation of pre-strained polymer layers for programming stress gradients (Danielson et al. 2017). The release of pre-strain after the programming of stress relaxation had resulted in a curvature of the monolayer specimen. The desired stress gradient can be produced by modulating the irradiation duration and light intensity while non-irradiated passive panels could also be developed using a photomask. Another approach is the flattening of 3D polymer above its T_g for generation of desired stress gradient that has been saved within the material by reducing the temperature (Behl et al. 2010). This would result in a spontaneous curving of the layer after releasing from the material's substrate. This technique is quite useful in self-assembling origami structures at the macro scale.

Stress gradients could also be generated through the creation of particle gradient by the side of the thickness of smart polymer for localized reduction of swelling/ shrinkage ratio. The magnetic field was used to generate magnetic gradients of nanoparticles in the swellable polymer (Liu et al. 2012). In this case, flat material starts to curve in one direction followed by the curving in a contrasting direction when the deswelling rate increases with increased particle concentration as a result of a reduction in deswelling ratio. The orientation and intensity of the applied magnetic stimulus were useful in locally controlling material curvature upon deswelling. Another interesting strategy useful in modulating the moment of bending in a monolayer is described as ionoprinting (Lee and Konst 2014). The metallic ions got embedded within the hydrogel under the potential gradient field. The ionic discharge from the electrode and bonding of gel network enhances the crosslinking at local pointe stiffen the material and causes deswelling. The subsequent drying of hydrogels may induce a second shape-shifting as further deswelling would be inhibited at a greater amount of cross-linking. In response to variations in humidity and temperature, a gradient developed in the nematic direction of liquid crystal material is utilized for induction of self-helixing and self-rolling of strips (Lee et al. 2017). Here, the main straining direction is varied along with the material thickness. The self-helixing and self-twisting could be pre-defined by modulating the width of the strip and its angle (φ) as like bilayer strips with anisotropic layers.

10.7.1.3 Localized Activation

Another strategy for the generation of stress gradients is the localized activation of stimulus by the side of the thickness of monolayered material based on non-homogeneous exposure of the material. A dynamic gradient in shrinkage/ swelling can be induced from the inside of the material as the stimulus applied on the surface would take a longer diffusion time. The diffusion would get completed when the material is exposed continuously to activation stimulus yielding a uniform shrinkage/ swelling through the material's thickness which inhibits bending from out-of-plane. A study was reported on single side activation of flat structure in response to humidity (Holmes et al. 2011). This approach was based on placing a desired droplet of solvent at one side of a thin planar structure. This induces swelling of material at local pointe that in turn depends on solvent's viscosity. Results showed that the time for maximum curvature has been varied with two orders of magnitude for a beam with 1 mm thickness. After the activation, the normalized curve has attained a maximum value of about 0.33 that was found to be only 22% of the theoretical limit for bilayer materials. Another way of programming the shape-shifting materials is the application of light with high spatial resolution and temporal resolutions (Tomatsu et al. 2011). Based on this approach, a monolayer hydrogel was embedded with nanoparticles on one side. The incorporation of nanoparticles converts the irradiated light into thermal energy that causes deswelling of the hydrogel. The microstructural properties determine bending direction and the extent of deformation was influenced by power density of exposed light and its time of exposure (Wang et al. 2013). As a variant of this approach, one-sided irradiation of material with laser would locally heats the black ink that in turn warm up the underlying layer to a temperature above its T_g. This series of process leads to strain recovery which could be used for programming a desired shape-shifting pattern by direct ink writing onto the surface of the material using laser (Liu et al. 2014).

Another way of programming is the use of global activation stimulus making use of sensitivity of the material instead of localized stimulus (Davis et al. 2015). An optically transparent pre-strained polymer was patterned with black lines that transform the exposure to infrared light into the temperature rise at the local pointe. With a long exposure time, there occurs a uniform shrinkage with the transformation of the structure curve back to a flat state. Using a similar approach, bidirectional folding could be programmed through patterning of black lines on both sides of transparent material. This kind of bidirectional folding is useful in the development of self-assembling origami and kirigami constructs (Zhang et al. 2017a). The patterning of hinge regions with black ink would create a similar construct with the desired direction of bending. Thus, localized activation of stimulus would provide a deeper insight into the development of dynamically active 4D printed construct through spatial tuning based on the diffusion of active stimulus.

10.7.2 Buckling Strategy

Buckling strategy involves programming based on the in-plane configuration of variable passive and active elements which on activation generates a desired compressive stress. In addition, externally induced forces of compression are utilized for inducing out-of-plane buckling in a passive layer of material. These methods are employed for the development of a broad range of materials with diverse shape-shifting behaviour (van Manen et al. 2018). The local space among the adjoining points on the surface of the material is represented by metric tensor g. By pre-defining of changes of in-plane dimension, a target metric $\bar{g}$ is directed that eventually generates compressive stresses. As the structure attains a 3D shape, the in-plane stresses would be relieved thereby minimizing the elastic energy. The declining energy of elasticity can be expressed as the surface integration of the energy of elasticity in density function w (Eq. 10.5) (Efrati et al. 2009).

$$w = \frac{1}{8}Et(g - \bar{g})^2 + \frac{1}{24}Et^3\kappa^2 \tag{10.5}$$

where, E is the stiffness of the tensor material, t is material thickness, and κ is surface curvature. The energy of elasticity is consisting of a stretching term as a function of on-plane bending and stretching term relating to curvature of the surface. The increase in material's thickness would lead to the balance of bending and stretching energies resulting in variations from $\bar{g}$. The buckling strategy can be viewed in the natural phenomenon of differentially growing soft matter tissues. A simple in-plane development would impart the shaping mechanism of thin membranes and leaves.

10.7.2.1 Material Tessellation

Material tessellation is another programming strategy where the materials are tessellated in a construct of a single layer instead of stacking multiple layers together. Photopolymerization is one of the most commonly used methods for producing material tessellations from hydrogels. By controlling the exposure to light, the degree of cross-linking and transition of monomers are improved. This process would further increase the stiffness of the material and assists in shape-shifting. The immersion of hydrogel in an aqueous solution influences density of cross-linking by affecting the swelling ratio. The modulation of pattern and the light exposure would create the regions of hydrogel with a different swelling ratio that in turn generates in-plane compressive stresses upon activation. A single layer constructs with the parallel orientation of a dual strip of hydrogel comprised of higher and lower swelling ratios is an example of swellable material tessellation (Kim et al. 2012b). The material swells on exposure to an aqueous medium resulting in self-rolling around the breadth of the dual strip allowing the strip with the high swelling capacity to adopt a longer length than the strip with lower swelling capacity thereby minimizes the material's strain. The balance energies of bending and stretching at the material's interface of dual strip lead to transition that in turn depends on the strip width and thickness (Byun et al. 2013). Another method of material tessellation is the side-by-side arrangement of multistrip alternatively with lower and higher swelling ratios (Wu et al. 2013). The generation of self-rolling and self-helixing structures depend on the strip orientation along the longitudinal direction of the specimen. The same concept can be extended to produce constructs with strips along the direction of width and length that responds to different stimuli. Another method of shape-shifting process is the axisymmetric tessellation of material to produce 3D geometries with double curvature into flat structures (Therien-Aubin et al. 2015). Both curved surface and cylindrical structures could be developed through patterning of single-layered structure using material tessellation. However, the explained methods were restricted with simple geometries. Hence future studies must be directed in programming of 3D complex structures with desired transformations.

10.7.2.2 In-Plane Material Gradients

The programming strategy of in-plane material gradients employs gentle gradients in shrinkage ratio/ swelling ratio to develop a stress-free 3D structure. Using this approach, a pre-defined gradient in the forms of waves and dome shapes were created over the material's surface. The developed axisymmetric configuration in dome shapes slightly varied from the desired stretch-free state as the small thickness of the polymer restricts the bending energy (Klein et al. 2007). Another approach of in-plane material gradient programming is the use of grey scale gradients of photomask during photopolymerization resulting in polymer cross-linking (Na et al. 2016). This kind of localized swelling can also be programmed through lithography technique by patterning extremely cross-linked dots in swellable polymer with a lower cross-linked unit (Kim et al. 2012a). The average swelling ratio can be programmed by monitoring the amount of dots in its dimensions that are highly cross-linked. The programming gradients with a swelling ratio of 2:8 would impart a spherical shape to the flat structure. Potentially this strategy is used for designing and programming complex shapes with regions of positive and negative Gaussian curvatures. Since the buckling direction is determined by small imperfections, even the similarly programmed flat materials could result in different surfaces on activation by stimulus. Hence, the creation of complex structures with non-constant Gaussian curvature is merely difficult. More research on analysing the material characteristics would explore different approaches of in-plane material gradients for shape-shifting.

10.7.2.3 Non-Homogenous Exposure

A non-homogenous material's exposure to active stimuli can be employed for defining a program to undergo a large variety of shape transitions. This kind of strategy helps to program a large variety of configurations as it relies on the pattern of activation instead of programming shape-shifting from inside of the material. The incorporation of nanoparticles into a sheet of hydrogel transforms the illuminated light into a thermal source resulting in the deswelling of the gel based on non-homogenous exposure. This approach can be employed for programming a complex surface with double curvature based on the exposure pattern to light (Hauser et al. 2015). By adopting a similar approach, direct light writing on the material can be done to program shape-shifting of localized relaxation of the pre-strained polymer through laser heating (Chen et al. 2016). The attachment of flexible micro-electrodes onto a hydrogel sheet allows for localized temperature control based on resistive heating that enables the tuning of materials with different swelling/ deswelling ratios (Yu et al. 2013). This combined structure can then be programmed to demonstrate multiple transformations in a single construct through modulating the input energy of the attached electrode.

10.7.2.4 Mechanically Induced Buckling

The compression of a stiff layer at the top of the soft substrate due to induced dimensional changes cause surface wrinkling resulting in shape-shifting (Liu et al. 2016). Compared to compression, the mechanical buckling over the layer at the top was energetically more beneficial in inducing the wrinkles to pop up. In addition, compression leads to delamination at the interface that would result in a buckle-delamination rather than wrinkling of surfaces depending on defects at interface and stiffness ratio. The spatial programming result in the adhesion of layers through a controlled delamination process thereby develops a strategy for out-of-plane shape shifting. A pre-strained soft substrate bonded to semiconductor nanoribbons was used for demonstrating controlled out-of-plane buckling (Sun et al. 2006). The use of a very thin top layer exerts a low bending stiffness as a semiconductor material is more flexible. A slender top layer would be subjected to forces of compression upon releasing the substrate thereby forms an out-of-plane buckling. The spatial changes in the top layer thickness would lead to the development of more complex structures in defining smart kirigami and origami constructs (Yan et al. 2017). In another approach, the inclusion of multiple overlapping layers at the top would form a multiplex topology (Yan et al. 2016). Shape-shifting through buckling is an interesting programming strategy that enables the use of a different range of active substrate allowing spatial changes in the compression thereby imparts feasibility to pre-define an SMP with a more complex 3D topology.

10.7.3 Sequential Shape-Shifting

The sequential shape-shifting of the smart materials refers to the programming of the shape transformations along with the time depending deformations. An illustration of this strategy is the sequential folding of origami panel to transform into a complex structure without engaging the self-deforming collisions. The incorporation of desired locking mechanisms into the folding process improves the mechanical and structural integrity of the desired configuration. The localized application of active stimulus is another approach of sequential shape-shifting that can be applied in a pre-stressed patterned strip bilayer with a stiff polymer (Laflin et al. 2012). Such configuration imparts mechanical strength and prevents the curving of the construct. On exposure to the laser, the polymer layer becomes soft due to heating and allows the strip of the bilayer to bend. Thus, programmed series of folding can be controlled by modulating the pattern of exposure of the laser light. A similar approach based on localized activation by Joule's heating instead of laser light can also be employed for sequential shape-shifting (Felton et al. 2015). Here the hinges get activated by managing the input energy delivered at different electrodes thereby forms self-assembly of a complex structure through a sequential deformation. Another approach of sequential folding is through tuning the response of the material global stimulus with respect to time by localized programming of material's properties (Yu et al. 2015). An example of this approach is the 3D printing of the origami structure containing hinges from SMPs of varied T_g that enables the programming of the shape-shifting in a sequence of folding. Based on the glass transition temperature, the hinges started to fold one by one in response to the gradual increase of the temperature. This temperature response as well as difference in glass transition temperature are the driving force in programming of sequential folding. Similar approach can also be applied for swellable polymers where the temperature response of the material can be tuned based on polymer composition (So and Hayward 2017). This would result in the shape transformation through fall or rise of the temperature with time. Another form of sequential folding/ unfolding approach is relying on the variations in the activation stimulus using ionoprinting of hydrogels employing different electrodes. The ionoprinting method allows to imprint portions with different cations thereby the structure exhibits different swelling behaviour based on the hydrophilic property and pH of the solvent (Baker et al. 2016).

Another approach of sequential shape-shifting is the spatial programming of the diffusion rate through the global stimulus of the active material. An example of this approach is the line patterning with grey scales on one side of the transparent pre-strained SMP sheet through activation on exposure to UV light (Lee et al. 2015). The absorption of UV light through the patterned regions allows a localized temperature rise. Depending on the extent of greyscale patterning, a varied amount of light gets converted into heat energy that allows for predefined onset time of folding which starts recovery of shape as temperature reaches T_g. As the patterned gray scale lines were applied on one side, the induction of recovery of shape resulting in out-of-plane curving. This kind of approach was adopted in employing inks of different colours with ability to absorb different spectra of light that allows higher freedom in controlling the shape-shift (Liu et al. 2017). The tuning of the thickness and void space of the bilayer 3D construct SMP allows for modulating the rate of heat transfer from the interior of SMP material (van Manen et al. 2017). As material stiffness is large at low temperature, the recovery process will be initiated only when the entire element reaches its glass transition temperature. This kind of programming strategy allows the exploration of a different range of designs based on sequential folding.

10.8 Spontaneous Transformation in Foods

4D printing is a recently emerged concept that integrates the 3D printing technology along with the fourth dimension, i.e. transformation over time in response to environmental stimuli (Chen et al. 2021a). Different types of stimulus used for transformations in 4D food printing are temperature, moisture, humidity, light, UV, pH, electrical field, and magnetic field (Ghazal et al. 2021). These stimuli induce changes in the 3D construct in terms of shape, dimensions, property, and functionality in a controlled pre-defined manner. With rising interest in 4D printing, the application of which is extended to diverse industrial sectors such as material science, civil engineering, computer science, medicine, and bioengineering (Teng et al. 2021). However, research works are in progress in identifying the potential applications of 4D printing in the food industry. As with 3D printing of foods, 4D food printing is a novel science dealing with combined concepts of biophysics, food chemistry, digital gastronomy, and 3D printing. The 4D printing of foods is still in its infancy that has a great scope in the future food industry. To realize the potential benefits of 4D printing, the basic principle of material transformation in response to stimuli over time should be well understood. The advancements of printing technology from 3D printing to 4D imparts certain technological and functional advantages (Zhao et al. 2020). The 3D food printing is limited with the customization of food material till the process of printing.

Despite 4D printing helps in overcoming this limitation of 3D printing which allows customizing the food material even after the printing through desired transformations over time. Thus, 4D printed foods possess the inherent ability to change their characteristics and functional performance in a controlled pre-determined way. The breakthrough of 4D printing technology is the advantage of material's response to external stimuli that reflects desired changes in shape, colour, flavour, and taste of foods (He et al. 2020b). These changes not only impart a desired sensory attribute to 4D printed foods but also eating pleasure and fun that enhances the intake of healthy meal with tailored functionality. This remains as a promising technology for ensuring nutritional diet to children, adults, and people with poor appetite. Thus, 4D printing offers a range of options for innovation of the foods that stimulates the imagination of food technologist, engineers, and gastronomists in implementing innovative ideas for developing novel food products.

The word '4D printing' was put forward by Skylar Tibbits during his TED talk on 3D printing (Tibbits 2013). Unlike 3D printing, 4D printing depends on the applicability of the smart behaviour of the materials in the delivery of customized products with improved functionalities. In this regard, 4D food printing can be defined as the 3D printed constructs with the inherent potential to vary with time under different stimuli such as moisture, temperature, pressure, wind, the difference in material concentration, light, pH, magnetic, and electric fields (Ghazal et al. 2019). The major characteristics of 4D printing are a high degrees of design freedom, design variants, versatility, self-assembly, and self-healing that enables the development of products with predictable changes that are time-dependent and printer independent (Wang et al. 2017). Inspired by nature as like response to pinecones and the opening of pea pods, the concept of 4D printing gains the attention of food scientists and researchers in developing novel food products. These plants could sense the variations in ambient humidity and drives the deformation that led to the concept of design of the novel structure and smart properties for 4D printing (Chen et al. 2021b). It was reported that the outer and inner layers of pea pods possess a fibre with an angle of 45° in an orthogonal dimension where the layers are perpendicular to one another. With these structural features, the peapod absorbs water at the surface during high humid conditions. This phenomenon would cause deformation by inward bending due to directional expansion. While in the dry conditions, the two layers of pea pod will get twisted to the outside that helps in releasing seeds through the opening of pods thereby completes the seeding cycle. In the case of a pinecone, the mechanism of bending deformation depending on the orientation of the cellulose microfibrils and controls the hygroscopicity at the outer and inner layers of cells. On the basis of the bilayer asymmetric structure of pinecone, a synthetic double-layer paper film was designed by Reyssat and Mahadevan (2009). Results showed that the absorption of moisture by the matrix produce expansion forces forming the bending moment thereby inducing the bending of the overall 3D structure. In another study, a double-layered temperature-sensitive hydrogel structure was developed to demonstrate the bending strategy (Naficy et al. 2017). Results showed that absorption of water by the top layer causes the geometry to bend downwards when the water temperature was kept at 20 °C while a gradual rise in temperature to 60 °C causes water loss shrinkage making the structure bend upwards. Thus, the design of geometry and the properties of materials are the most significant aspects in designing 4D structures. Considering the food applications, 4D transformations are feasible through spontaneous change in geometric and sensorial attributes of

3D printed foods. The success of 4D food printing depends on nature of smart biopolymers, stimulants, printability, material response, and post-processing transformations.

10.9 Recent Advancements in 4D Food Printing

The food application of 4D printing was first reported by MIT researchers (Wang et al. 2017). This study demonstrated the concept of 2D to 3D transformations based on hybrid fabrication strategy and activation stimulus. An edible 2D film was developed from a composite mixture of starch, cellulose, and protein that when immersed in the water get transformed to its 3D forms. This concept was found to have the following distinct application techniques as hydration induced wrapping, 2D to 3D folding, and temperature-induced self-fragmentation resulting in novel geometry, texture, and material interactions improving the taste and flavour of foods. Based on this approach several novel food cuisines could be created that demonstrates the futuristic dining experience based on the design of material interactions. 4D printing allows for the personalized morphing of food that enhances the conventional foods with novel human-interaction capabilities to structurally design and tailor the functional properties of foods. Tao et al. (2019) reported an integrated design strategy for the special mechanism of transition of kneaded and sheeted dough with two different cooking methods (baking and boiling). This programming strategy allows for self-wrapping, self-folding, self-chopping, and self-assembly that would certainly possess a great reach among the consumers due to its attractive applications such as improvement of the aesthetic appeal of foods, information display through shape-shifting, personalized pasta/ noodles in flat-packages, and customized multiple flavours and dishes. The mechanisms associated with cooking such as hydration stimulus and dehydration stimulus were employed for shape transformation of flat structures from 2D to 3D forms. Thus, this study put forth the concept of hybrid cooking techniques that envisioned to have a significant impact on consumer's food decisions and possess a great scope for real-world applications in near future (Tao et al. 2019). In this regard, the present section describes the various innovative strategies used for the spontaneous 4D transformation of foods based on various external and internal stimuli to impart changes in geometry and sensory attributes of the 3D constructs.

10.9.1 pH-Triggered Colour Transformation

A healthy 4D printed food product was developed through 3D printing of anthocyanin-potato starch gel with spontaneous colour change (Ghazal et al. 2019). This study highlights the influence of colour change in terms of aesthetic appearance and food choices of consumers. In view of consumer's health and well-being, the study utilized a response of phytochemical anthocyanin in response to different pH for colour transformation. The combination of potato starch gel with anthocyanin enhances printability. This study demonstrated the spontaneous colour change based on external and internal pH stimulus. In external pH stimulus, the effect of concentrations of potato starch on 3D printability and colour variations of anthocyanin-potato starch gel was evaluated by external spraying of different levels of pH solutions. On the other hand, the internal pH stimulus was induced by combining two portions of food product with one portion containing smart anthocyanin and the other portion containing external pH stimulus. Here, the smart material was anthocyanin-potato starch gel while the lemon juice gel was used as stimulus material. In addition, the intensity of colour transition of anthocyanin-potato starch 3D construct was assessed after spraying the pH solutions at varying levels and different percentage of anthocyanin. Results showed that potency of colour variations of 3D printed anthocyanin potato starch gel with and without pH treatment decreased with higher concentrations of potato starch. On the other hand, the potency of colour variations of 3D printed anthocyanin potato starch with and without pH treatment increased with increasing anthocyanin concentrations. Regardless of anthocyanin and potato starch concentration, the change in colour intensity was more distinct for samples at pH of 2, 2.5, 3, and 3.5 while there was no significant difference in the colour change at higher levels of pH (4, 4.5, and 5). Thus, this study demonstrates the potential of 4D printing in the development of a colourful healthy product with consumer satisfaction and acceptability (Ghazal et al. 2019).

More recently, sago starch was used for demonstrating the 4D colour transformation based on pH change (Shanthamma et al. 2021). The transformation in the colour of the 3D printed sago starch was in response to the external stimuli that significantly resulted in a visual colour change. The 3D printing material supply was a mixture of sago flour with different percentage of turmeric powder (0, 0.5, 1.5, 2.5%, w/w) that in turn subjected to gelatinization with 1 : 3 ratio (w/v) of water at 72 °C for 90 seconds. The used turmeric powder contains about 7% of the bioactive functional compound curcumin. The optimized 3D printing conditions of sago material supply with 2.5% turmeric powder at 300 rpm extrusion motor speed, 2000 mm minute^{-1} printing speed, 3 bar extrusion pressure, printed using 0.84 mm nozzle was immersed in 10 ml of different levels of sodium bicarbonate solutions (1, 3, and 5%). With sudden variation in the pH of the turmeric powder incorporated sago 3D construct, there occurs a visual colour transition from yellow to orange-red (Figure 10.6). This was mainly because of the keto-enol tautomerism of curcumin present in turmeric powder. The concentration of the bioactive compound curcumin is a significant factor responsible for the colour change that in turn closely associated with the levels of sodium bicarbonate.

Results showed that the increasing trend of chroma values clearly depicts the molecular transformation of curcumin from keto (in acidic conditions) to enol form (in alkaline conditions) (Figure 10.7). Further, the curcumin intensity at higher levels of sodium bicarbonate increases the darkness of the 3D construct resulting in a higher redness value in

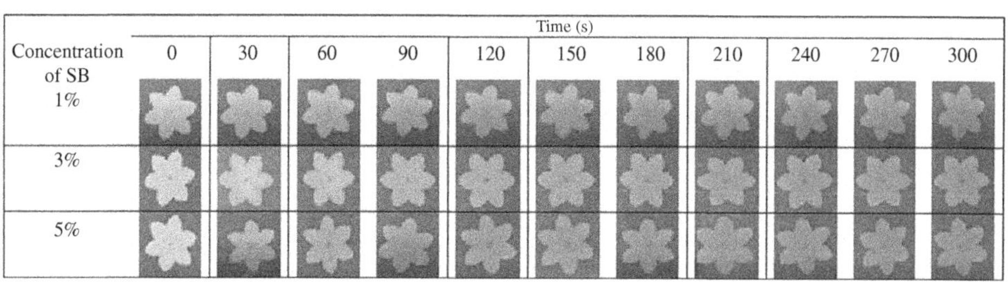

Figure 10.6 Images of pH-triggered colour transition of the 3D constructs of turmeric sago blends. *Source:* From Shanthamma et al. (2021) / With permission of American Chemical Society.

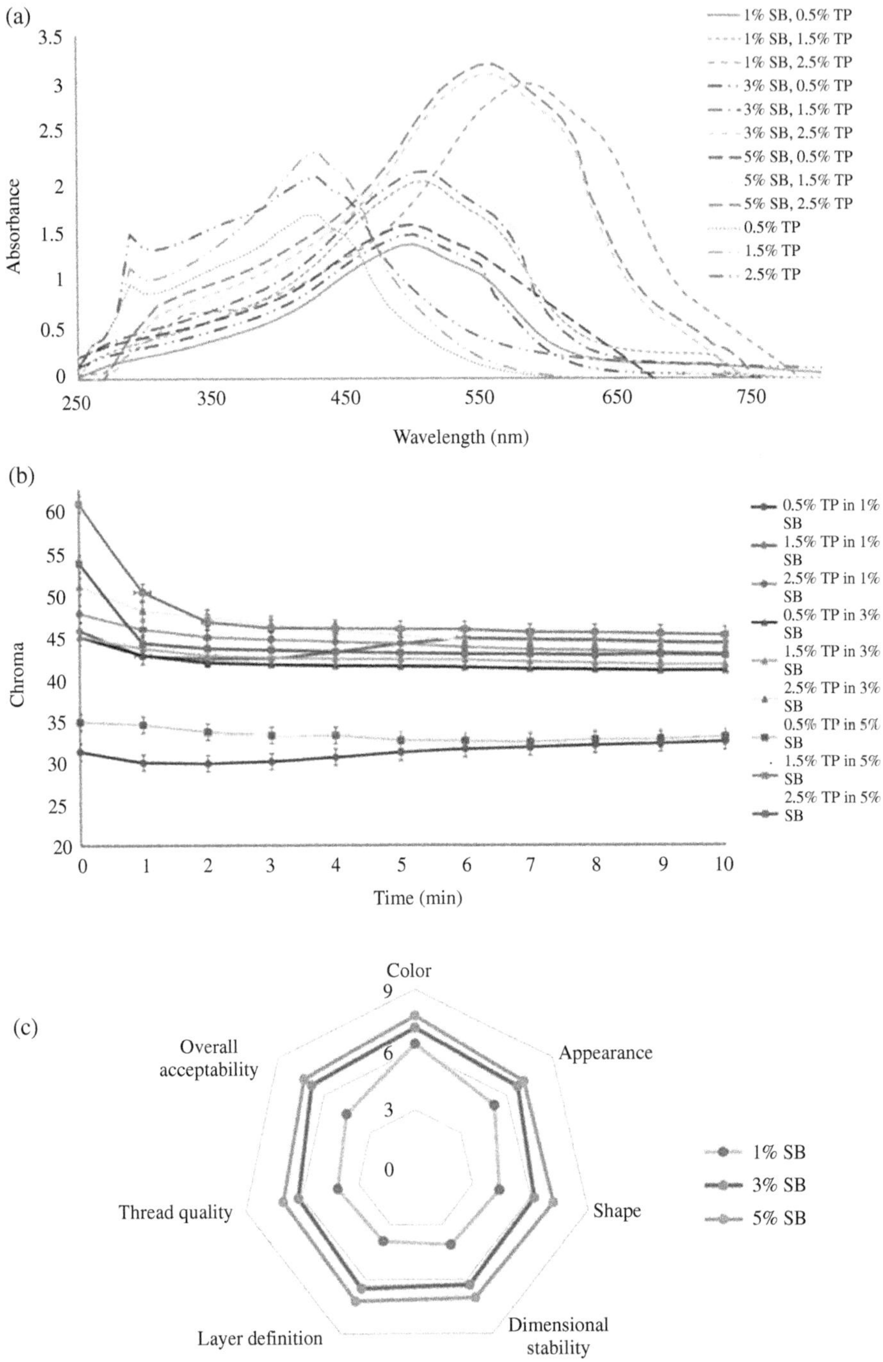

Figure 10.7 4D printed turmeric-sago constructs. (a) UV absorbance of different 3D printed constructs in different concentrations of sodium bicarbonate solutions, (b) changes in chroma values with different formulations during the colour transformation, and (c) appearance evaluation of 4D printed constructs. *Source:* Shanthamma et al. (2021).

alkaline conditions. These results are consistent with the hedonic scale values of the appearance evaluation parameters. Therefore, colour preference significantly affects the acceptability and marketability of any food product (Shanthamma et al. 2021). Thus, this study provides useful insights on developing attractive and fascinating colour-changing 4D food products from underutilized ingredient tapioca with natural colourant turmeric powder possessing therapeutic value.

A similar concept of pH-triggered colour transition was demonstrated from a 3D construct made of purple sweet potato puree and mashed potato (He et al. 2020b). The effect of different pH levels (2.5, 6.5, and 7.8) and potato flakes (15, 19, 23, and 27%) on the printability of mashed potato was studied. Results showed that the inclusion of citric acid enhanced material viscosity, storage modulus, gumminess, and hardness of 3D construct while in the meantime the increase in the concentration of potato flakes enhanced the shape fidelity. A dual-extrusion printing was used for demonstrating the 4D colour transformation by incorporating anthocyanin-rich purple sweet potato into mashed potato. It was inferred that both pH and potato flake content had significant effects on colour transformation of mashed potato (Figure 10.8). Thus, this study provides insights into the fabrication of multi-material constructs with potential colour transformations through 4D printing (He et al. 2020b).

In a more recent study, a spontaneous colour and flavour transformation in response to internal and external pH was reported (Ghazal et al. 2021). The formulation of the 3D construct comprised of multiple smart materials combined together to respond to pH stimulus. This includes cabbage juice from red coloured variety (RCJ), starch from potato (PS), fruit juices (lemon, orange, and apple), and vanillin powder (V). Here the red cabbage juice and vanillin were the smart materials responsible for colour and flavour transformation, respectively. Results showed that colour of the printed construct was changed from blue colour to red, purple, violet, blue, bluish-green, and greenish-yellow when sprayed with solutions with pH values 2, 3–4, 5–6, 7, 8–9, and 10, respectively. Further, prominent distinct taste and aroma profiles were observed among samples with different pH. In addition, the freeze-dried 4D printed construct retained anthocyanin content and its colour during the storage period of three weeks at ambient temperature. Thus, this study successfully demonstrated that the colour variation of 3D constructs is a function of pH and anthocyanin stability during storage (Figure 10.9) (Ghazal et al. 2021). Therefore, pH can be used as an excellent stimulus in developing attractive 4D printed healthy foods.

10.9.2 Dehydration-Triggered Colour and Flavour Transformation

The effect of the colour change triggered by microwaves on 3D printed curcumin incorporated lotus root powder gel was studied. It was observed that the inclusion of curcumin in its emulsion forms enhanced the loss modulus, storage modulus, and the viscosity of printing mixture. Here, bioactive component curcumin (smart material) was added in emulsion form (curcumin, medium-chain triglycerides, and Tween 80). The printed gel system appears yellow before microwave treatment; the colour of gel was changed to red after microwave treatment (Figure 10.10). Thus, the heat generated by the microwaves triggered the colour change through the deprotonation of curcumin. In addition, the texture of the lotus gel was significantly improved by the microwave treatment. Thus, this study

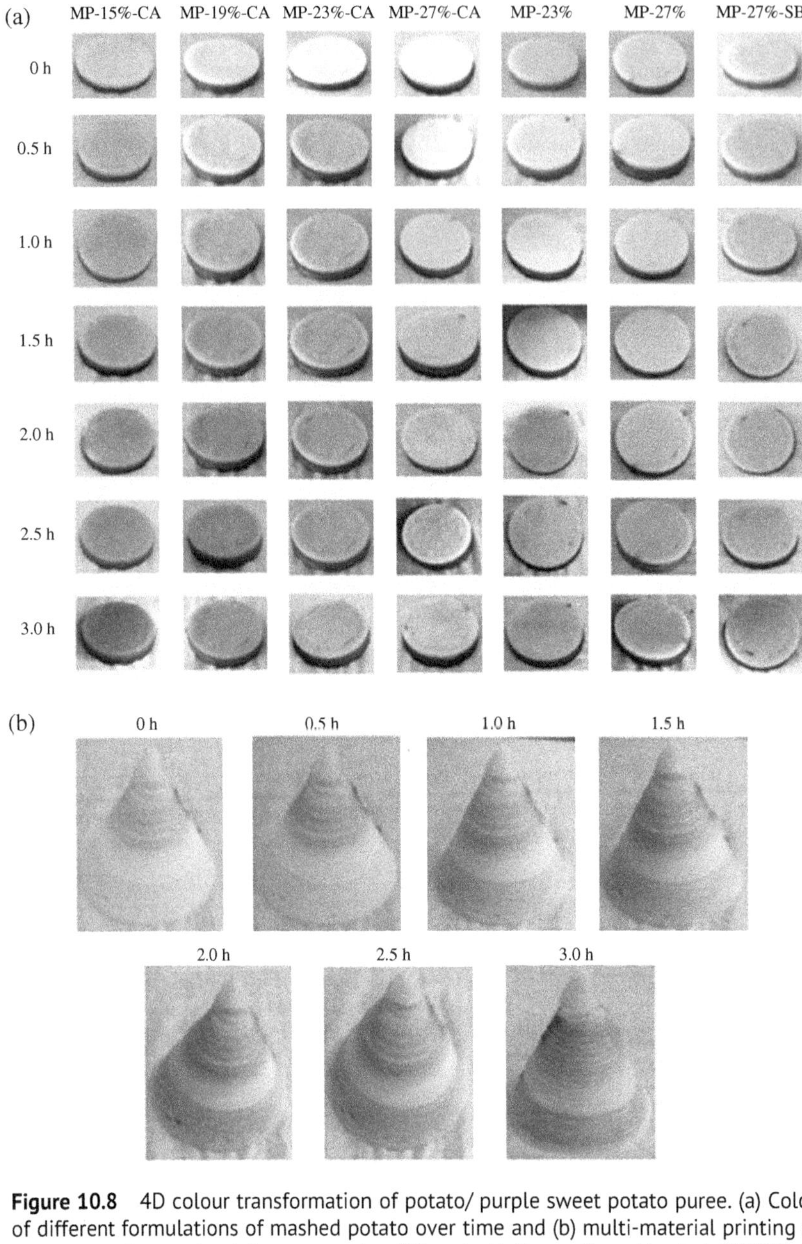

Figure 10.8 4D colour transformation of potato/ purple sweet potato puree. (a) Colour transition of different formulations of mashed potato over time and (b) multi-material printing of mashed potato- purple sweet potato puree under different pH. *Source:* From He et al. (2020b), Figure 09 & 13 [p. 10 & 13] / With permission of Elsevier. DOI-https://doi.org/10.1016/j.ifset.2019.102250.

successfully demonstrates the use of microwaves as external stimuli in colour transformation of curcumin incorporated lotus gel where the consumers could experience two different sensory attributes before and after microwave treatment that offers a new method of 4D printing through heat stimuli (Chen et al. 2021a).

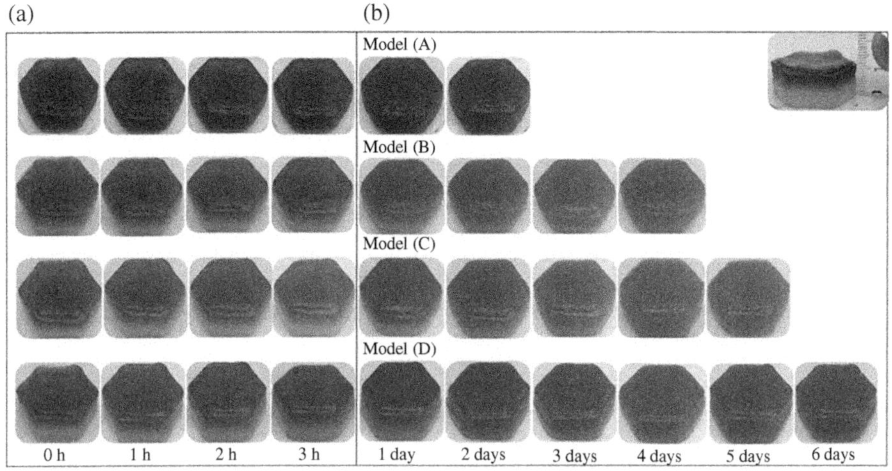

Figure 10.9 (a) Spontaneous colour transition of the 3D-printed RCJ-V-PS-Gel over storage time (0, 1, 2, 3 hours) and (b) colour stability when stored under refrigeration at different multi-material models (two parts of RCJ-V-PS-Gel (control) – Model (A), two parts: upper part of RCJ-V-PS-Gel and apple juice gel in lower part – Model (B), two parts: upper part of RCJ-V-PS-G and orange juice gel in lower part – Model (C), and two parts: upper part of RCJ-V-PS-Gel and lemon juice gel in lower part – Model (D)). *Source:* From Ghazal et al. (2021), Figure 10 / With permission of Elsevier. DOI-https://doi.org/10.1016/j.foodres.2021.110215.

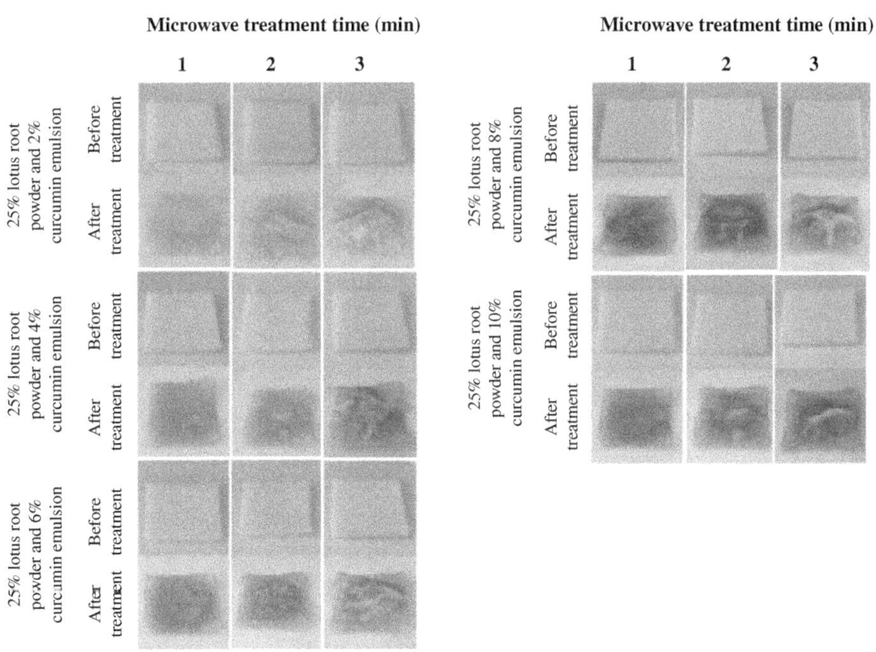

Figure 10.10 Spontaneous colour transformation of different combinations of lotus root powder and curcumin emulsion for different microwave time. *Source:* From Chen et al. (2021a) / With permission of Elsevier.

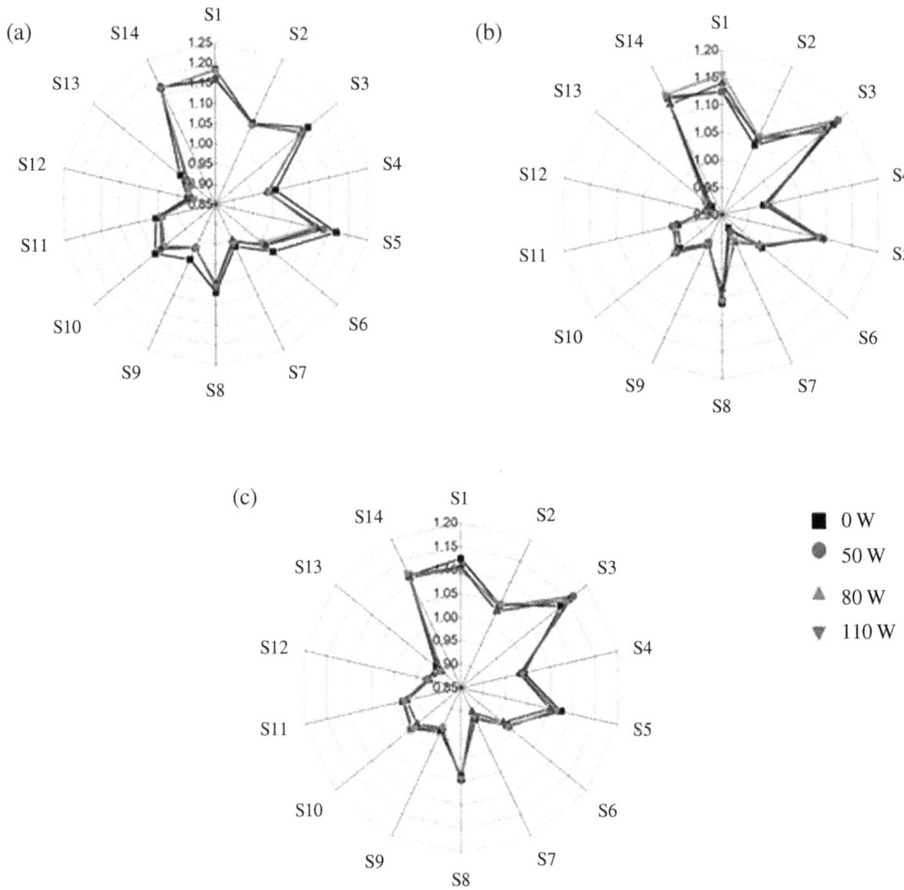

Figure 10.11 Radar chart showing changes in aroma of SPI-3% carrageenan-0.5% vanillin after microwave treatment of different power levels under varied time. (a) 20 minute, (b) 40 minute, and (c) 60 minute. *Source:* Phuhongsung et al. (2020)/ With permission of Elsevier.

Another interesting application of 4D printing in the development of food products with microwave triggered flavour transformation by an external thermal stimulus. The 3D printing material supply composed of soy protein isolate, vanilla flavour, and *k*-carrageenan that forms a gel system. Among which vanillin is the smart material whose quality is significantly influenced by moisture and temperature. Hence, in this study, the 3D constructs with incorporated vanillin were subjected to varying levels of heat treatment that causes the precursors of vanillin flavour to undergo different chemical reactions thereby producing diverse distinct flavours (Figure 10.11). Results showed that the presence of carrageenan and post-processing by microwaves enhanced the printability as well as flavour of the soy protein isolate-based gel systems. Thus, this study delivers key insights on a novel approach of 4D printing with flavour transformation induced by microwaves (Phuhongsung et al. 2020). A study was reported in exploring the applicability of microwaves for the aroma and colour variations of 3D printed buckwheat flour through microwave heating of complex moisture of gelatin-gum Arabic coacervates (Guo et al. 2021). Thus, this study provides insights on the iteration of 3D printing to 4D printing where the developed gelatin-gum Arabic-oil microcapsules can be used as an efficient smart material in inducing the spontaneous aroma and colour changes.

10.9.3 Dehydration-Triggered Shape Transformation

Microwaves are not only imparting sensorial changes but also the geometrical transition of 3D printed products. The effect of microwave dehydration as external stimuli in the shape transition of the 3D printed purple sweet potato puree was reported (He et al. 2020a). Various formulations of printing mixture are the gelatinized purple sweet potato gel with and without salt/ sugar. The 3D printing process involves the printing of a bilayer structure of paper/ sweet potato puree for which a food-grade plastic sheet of 0.07 mm thickness was attached to the printing platform. The extruded purple sweet potato threads were deposited over the frosted sheet and the printed sample was cut along the outer contour to obtain a bilayer structure. The designed bilayer structure was subjected to microwave treatment that eventually dehydrates the gel and detached from paper. It was observed that the inclusion of salt decreased viscosity, yield stress, loss and storage modulus but increased the time of relaxation of sweet potato puree. These results were vice versa with the inclusion of fructose syrup into sweet potato puree. However, with the inclusion of both syrup and salt, the dielectric constant was decreased with dielectric loss. In context with shape transformation, the higher microwave power and the salt concentration enhanced the dehydration rate and structural deformation thereby decreasing the maximal deformation degree. Even the addition of syrup reduced the maximal deformation degree. The study inferred that variation of material composition along with manipulation of infill patterns aids in achieving desired deformation pattern. Thus, the study successfully demonstrated the shape transition of 3D printed 2D flower planar design and butterfly constructs into 3D forms through appropriate material design and optimization of material properties. This approach can be conveniently adopted for inducing shape change of starch-based 3D printed samples through a dehydration mechanism that leads to novel and fascinating 4D printed foods (He et al. 2020a).

A similar approach of spontaneous dehydration triggered a shape change of the 3D construct from potato flakes/ potato starch was reported (Liu et al. 2021). A bilayer structure of food structure was 3D printed over the top of the plastic film of thickness 0.2 mm. The developed structure when subjected to two different dehydration methods (microwave, infrared, and air drying), the heat energy was utilized to evaporate the water thereby dehydrates the top layer of the gel matrix while the volume of the plastic layer remains constant that forms the bottom layer. This associated shrinkage between food material and plastic layer induces bending in an upward direction resulting in shape-shifting of material (Figure 10.12). Results inferred that degree of adhesion of printing material is quite adequate for programming the shape-shifting where the concentrations of added soybean oil significantly influence the bending angle. It was found that the bending angle (degree) was directly proportional to the shrinkage ratio and amount of evaporation of water; and inversely proportional to the dielectric loss factor. On comparing dehydration methods, air dehydration resulted in a greater shrinkage ratio and improved bending degree than dielectric

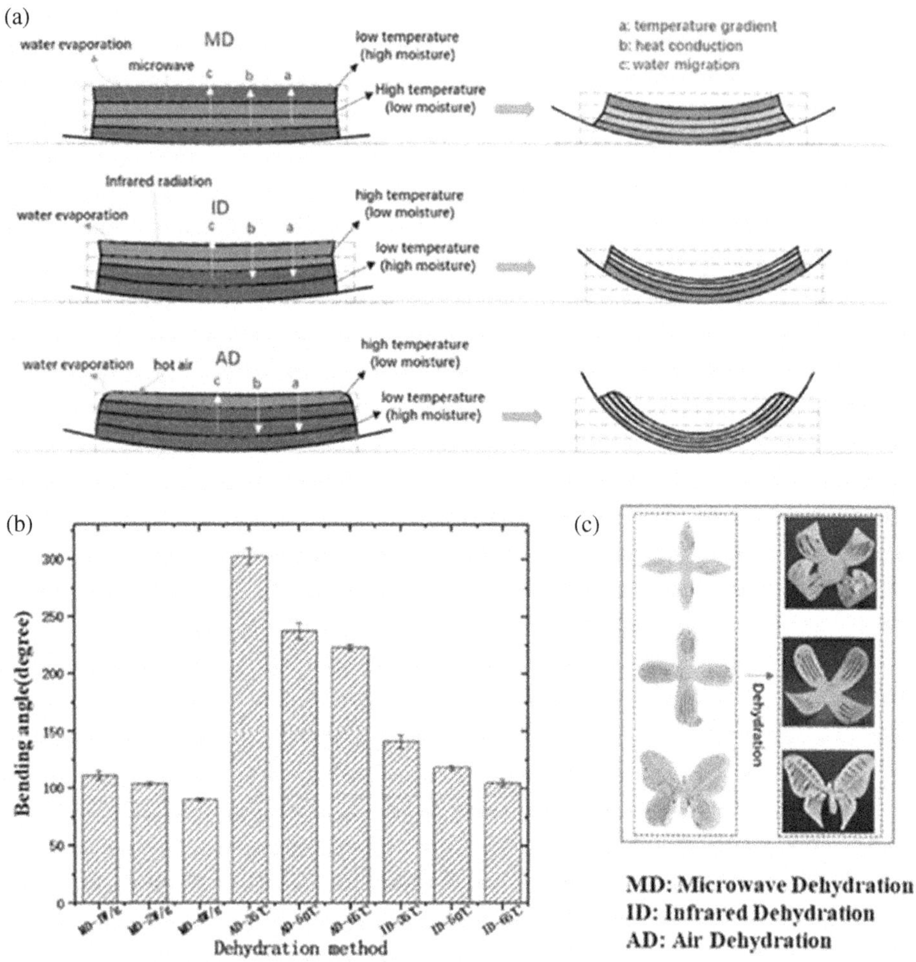

Figure 10.12 Dehydration-triggered shape transition of 4D printed gel. (a) Dehydration mechanism of different drying methods, (b) degree of bending of 3D constructs at different process conditions, and (c) shape transformation of 3D printed constructs. *Source:* From Liu et al. (2021), Figure 01 / With permission of ELSEVIER. DOI-https://doi.org/10.1016/j.foodhyd.2021.106608

microwave dehydration. This was due to the internal heating and rapid dehydration rate of the microwaves. Also, the higher bending degree was evident in air dehydration because the accumulating effects of bending developed from the surface causes inward dehydration thereby enhances the bending degree of samples. In the case of infrared dehydration, the increase in temperature resulted in an increased shape transformation rate; however, the angle of bending of the sample at 50 or 65 °C was significantly lower than that of samples at 35 °C. Thus, the study explores the novel concept of 4D printing through shape transformation by triggered dehydration mechanism that paves a way for delivery of healthy foods (Liu et al. 2021).

10.9.4 Temperature-Triggered Shape Transformation

A 4D shape transformation of the double-layered structure of pumpkin/ paper in response to activation stimulus (temperature) was reported (Chen et al. 2021b). In this study, the anisotropy of bilayer construct and direction of layers were used as a programming strategy for shape deformation. Both paper and pumpkin possess diverse properties, so the 3D printed bilayer construct in response to air drying (25 °C and 65% RH) lost water. The evaporation of water resulted in shrinkage of the pumpkin layer while the layer of paper remains unchanged exhibiting anisotropic properties in the direction of dimensional thickness. The model path of printing imparts directional transitions on the pumpkin layer through managing the porosity that in turn help in controlling the deformation of a sample of equal-sized print. The central angle corresponding to the arc was considered for characterizing the angle of bending that assists in its measurement between the radial and axial direction of the 3D printed pumpkin layer (Figure 10.13). Therefore, the characteristic moisture loss of the pumpkin layer acts as a driving force for deformation making the bilayer twist towards the side of the pumpkin puree. The total time of drying, the path of printing and thickness are major factors that aid in controlling the deformation degree. Thus, this study provides novel strategies based on anisotropy characteristics of materials to program the 4D shape transformation of 3D printed foods.

10.10 Future Trends and Challenges

The previous section on the advancements of 4D printing of foods gives a broad overview of the current trends and concept of food transformations based on hybrid cooking approaches and human-computer interactions in the manufacturing of novel foods. The future of printing technology relies on the integrated concepts of 3D printing with smart materials that paves a way for a novel technique '4D printing'. The selection of the smart materials and its response to various activation stimuli are the most significant criteria that influence the programming strategy and targeted transformations in 4D printing (Tao et al. 2019). The emergence of 4D printing provides advantages over 3D printing in designing novel foods with transformations during post-printing operations. The shape, colour, flavour, aroma, and taste changes of 3D constructs are achieved through appropriate programming based on tuning the material properties based on activation stimuli of temperature, moisture, and pH. The 4D transformations employed in food materials are sensorial change, swelling, shrinkage, and bending that are associated with pH change, hydration, dehydration, and anisotropic properties (Chen et al. 2021b). Two basic components of 4D food printing are the smart material and stimulants. It was reported the potential ability of phytochemicals and flavours as smart materials due to their changing physicochemical properties with activation stimulus. Thus, 4D printing allows designing a healthy diet with the incorporation of bioactive substances such as anthocyanins, betanidins, carotenoids, flavonoids, curcumin, lycopene, and xanthophylls (Chen et al. 2021a; Ghazal et al. 2021). The addition of these pigments would yield novel appealing foods with colour transformations as colour is one of the significant criteria in the food industry for consumer's acceptance. These desired sensory characteristics impart a higher intake of a healthy diet among children, elders, and people with the physiological disorder.

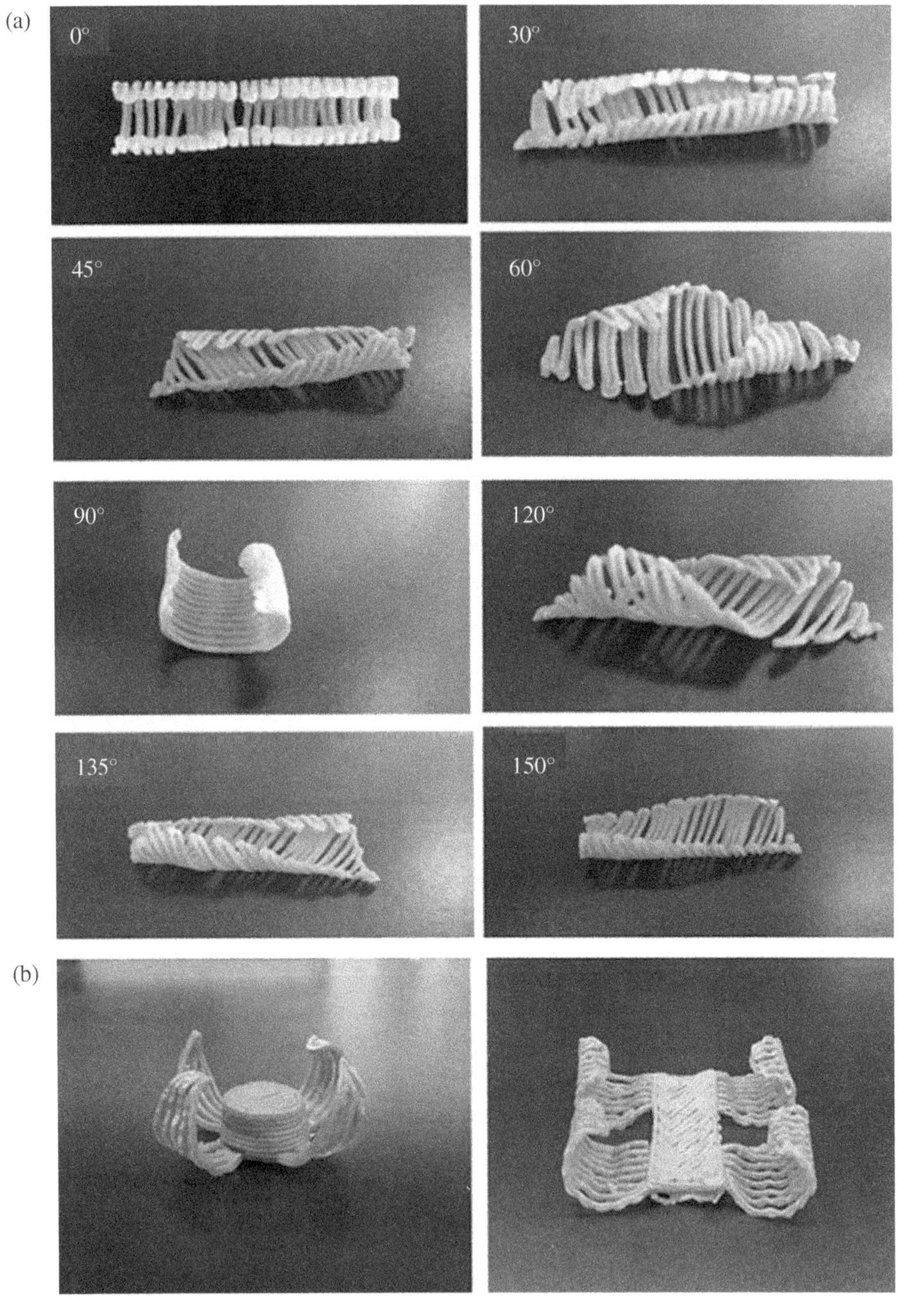

Figure 10.13 4D shape transformation of bi-layered pumpkin/ paper 3D construct. (a) Deformation process of a flat bilayer structure, and (b) complex deformation of bi-layered structure. *Source:* From Chen et al. (2021b), Figure 04 & 06 [p. 23 & 25] / With permission of Elsevier. DOI-https://doi.org/10.1016/j.foostr.2020.100168.

The determination of the endpoint of cooking was mostly based on colour change of foods; however, the complex interactions of materials impart darkness to some of the foods making them difficult to judge the endpoint of cooking (Chen et al. 2021b). This challenge associated with the determination of the endpoint of cooking based on colour can be overcome with the visible shape transformations. The programming strategies for shape-shifting not only determines the endpoint of cooking but also provides a pleasant and fun eating experience with fascinating shape transformation. Further, the shape transformations of foods are the recent topic of interest in the design of flat packages of foods that save time, cost, material, and energy. Hence, the major breakthrough of 4D printing over 3D printing is its ability to develop products with changing properties and material characteristics from the environmental stimulus with respect to time (Zhao et al. 2020). These advancements would make feasible the design of foods with smart capabilities of desired transformations at targeted site and right time; creation of novel and innovative shapes with improved functionalities; and self-assembly structures with health promotion properties. Although the progress of 4D printing is quite active and is at its infant stage in the food sector, the worldwide 4D printing market is anticipated to rise with 42.95% CAGR during 2015 and 2025 (Market Research Report 2015). Considering the market potential, the trend of 4D printing looks forward to reaching $64.2 million in 2025 that is 8.38 times higher than that in 2019 (Tibbits 2013). Thus, 4D printing serves as the best technological platform to establish new business strategies in reducing capital investments and inventory space that helps in process and product efficiency.

Despite these advantages, the challenges, and limitations of 4D printing should be addressed to enjoy the benefits of 4D food printing. Although 4D printing is similar in concept to 3D printing, the addition of the fourth dimension adds complexity in assessing the material properties and programming strategy. The associated transformation is linked with the post-processing operations; hence it is most important to select appropriate stimulus that not only impart transformations but also ensures edibility. In this regard, it is quite important to assure the quality and retention of the bioactive compounds as smart material used in 4D food printing are thermal and chemical sensitive. In addition to the optimization of printability, the programming strategy and material response to activation stimulus greatly influence the success of 4D transformations. More studies are required in assessing the production efficiency, process time, processing cost, safety, storability, and practical implementation of 4D printing of foods. Future research must be directed in simulation and validation of the experimental results associated with shape-shifting that would save time in the testing of existing materials as well as in exploring new intelligent smart food materials.

10.11 Conclusion

The changing consumer attitude and food decisions made food technologist to explore novel advanced food technologies. Although 3D printing is a breakthrough technology in the food sector, the recent advancements of printing remain as a quantum leap that paves for the emergence of a new technique called '4D printing'. The 4D printing is a fascinating science that deals with the transformations of the physical construct with respect to the fourth dimension time in response to environmental stimuli. In the food sector, 4D printing remains under progress that mainly associated with shape transformations and changes in sensorial attributes such as colour, flavour, aroma, and taste. At present, the programming strategies are limited with few material properties (pH, moisture, temperature, anisotropy). Hence, it is adequate to investigate material properties and stimulus mechanism to explore more about 4D transformations in the food sector. From this perspective, the present chapter discussed the advancements of 4D printing in industrial sectors, smart materials, material responsive systems, and programming strategies used for synthetic materials. This clearly indicates the future benefits of 4D printing where the structural design and programming strategies are significant aspects determining the desired and targeted transformations. In this regard, various studies attempted in exploring the potential of 4D printing of foods were detailed. Recent trends and challenges associated with 4D food printing were briefed that put forth the need for future research to be directed in linking the food structure with activation stimulus, innovative changes in material properties, printing and programming strategies, optimization of post-processing, evaluation, and validation of the process. By addressing these aspects, 4D printing allows for the production of novel healthy food with enhanced functional and sensory properties that not only saves resources and energy but also satisfies the consumer. Thus, consumer satisfaction is the most significant criteria that most food scientists and industries are working on in exploring novel food processing technologies. Hence 4D food printing remains as the futuristic technology of the food industry that meets the nutritional and sensorial needs apart from satisfying the consumers.

References

Alben, S., Balakrisnan, B., and Smela, E. (2011). Edge effects determine the direction of bilayer bending. *Nano Letters* **11** (6): 2280–2285.

Baker, A.B., Wass, D.F., and Trask, R.S. (2016). 4D sequential actuation: combining ionoprinting and redox chemistry in hydrogels. *Smart Materials and Structures* **25** (10): 10LT02.

Basit, A., L'Hostis, G., Pac, M.J., and Durand, B. (2013). Thermally activated composite with two-way and multi-shape memory effects. *Materials* **6** (9): 4031–4045.

Behl, M., Razzaq, M.Y., and Lendlein, A. (2010). Multifunctional shape-memory polymers. *Advanced Materials* **22** (31): 3388–3410.

Boothby, J.M. and Ware, T.H. (2017). Dual-responsive, shape-switching bilayers enabled by liquid crystal elastomers. *Soft Matter* **13** (24): 4349–4356.

Bothe, M., Mya, K.Y., Lin, E.M.J. et al. (2012). Triple-shape properties of star-shaped POSS-polycaprolactone polyurethane networks. *Soft Matter* **8** (4): 965–972.

Breger, J.C., Yoon, C., Xiao, R. et al. (2015). Self-folding thermo-magnetically responsive soft microgrippers. *ACS Applied Materials & Interfaces* **7** (5): 3398–3405.

Byun, M., Santangelo, C.D., and Hayward, R.C. (2013). Swelling-driven rolling and anisotropic expansion of striped gel sheets. *Soft Matter* **9** (34): 8264–8273.

Cendula, P., Kiravittaya, S., Mei, Y.F. et al. (2009). Bending and wrinkling as competing relaxation pathways for strained free-hanging films. *Physical Review B* **79** (8): 85429.

Chen, C., Zhang, M., Guo, C., and Chen, H. (2021a). 4D printing of lotus root powder gel: color change induced by microwave. *Innovative Food Science & Emerging Technologies* **68**: 102605.

Chen, F., Zhang, M., Liu, Z., and Bhandari, B. (2021b). 4D deformation based on double-layer structure of the pumpkin/paper. *Food Structure* **27**: 100168.

Chen, T., Li, H., Li, Z. et al. (2016). A "writing" strategy for shape transition with infinitely adjustable shaping sequences and *in situ* tunable 3D structures. *Materials Horizons* **3** (6): 581–587.

Choi, J., Kwon, O.-C., Jo, W. et al. (2015). 4D printing technology: a review. *3D Printing and Additive Manufacturing* **2** (4): 159–167.

Danielson, C., Mehrnezhad, A., YekrangSafakar, A., and Park, K. (2017). Fabrication and characterization of self-folding thermoplastic sheets using unbalanced thermal shrinkage. *Soft Matter* **13** (23): 4224–4230.

Davis, D., Mailen, R., Genzer, J., and Dickey, M.D. (2015). Self-folding of polymer sheets using microwaves and graphene ink. *Rsc Advances* **5** (108): 89254–89261.

De Marco, C., Alcântara, C.C.J., Kim, S. et al. (2019). Indirect 3D and 4D printing of soft robotic microstructures. *Advanced Materials Technologies* **4** (9): 1900332.

Ding, Z., Yuan, C., Peng, X. et al. (2017). Direct 4D printing via active composite materials. *Science Advances* **3** (4): e1602890.

Efrati, E., Sharon, E., and Kupferman, R. (2009). Elastic theory of unconstrained non-Euclidean plates. *Journal of the Mechanics and Physics of Solids* **57** (4): 762–775.

Efrati, E., Sharon, E., and Kupferman, R. (2013). The metric description of elasticity in residually stressed soft materials. *Soft Matter* **9** (34): 8187–8197.

Egunov, A.I., Korvink, J.G., and Luchnikov, V.A. (2016). Polydimethylsiloxane bilayer films with an embedded spontaneous curvature. *Soft Matter* **12** (1): 45–52.

Erkecouglu, S., Sezer, A.D., and Bucak, S. (2016). Smart delivery systems with shape memory and self-folding polymers. *Smart Drug Delivery System* **1**: 1–30.

Felton, S.M., Becker, K.P., Aukes, D.M., and Wood, R.J. (2015). Self-folding with shape memory composites at the millimeter scale. *Journal of Micromechanics and Microengineering* **25** (8): 85004.

Forterre, Y. and Dumais, J. (2011). Generating helices in nature. *Science* **333** (6050): 1715–1716.

Ge, Q., Dunn, C.K., Qi, H.J., and Dunn, M.L. (2014). Active origami by 4D printing. *Smart Materials and Structures* **23** (9): 94007.

Ge, Q., Sakhaei, A.H., Lee, H. et al. (2016). Multimaterial 4D printing with tailorable shape memory polymers. *Scientific Reports* **6** (1): 1–11.

Ghazal, A.F., Zhang, M., Bhandari, B., and Chen, H. (2021). Investigation on spontaneous 4D changes in color and flavor of healthy 3D printed food materials over time in response to external or internal pH stimulus. *Food Research International* **142**: 110215.

Ghazal, A.F., Zhang, M., and Liu, Z. (2019). Spontaneous color change of 3D printed healthy food product over time after printing as a novel application for 4D food printing. *Food and Bioprocess Technology* **12** (10): 1627–1645.

Guan, J., He, H., Hansford, D.J., and Lee, L.J. (2005). Self-folding of three-dimensional hydrogel microstructures. *The Journal of Physical Chemistry B* **109** (49): 23134–23137.

Guo, C., Zhang, M., and Devahastin, S. (2021). Color/aroma changes of 3D-Printed buckwheat dough with yellow flesh peach as triggered by microwave heating of gelatin-gum Arabic complex coacervates. *Food Hydrocolloids* **112**: 106358.

Guo, Q., Mehta, A.K., Grover, M.A. et al. (2014). Shape selection and multi-stability in helical ribbons. *Applied Physics Letters* **104** (21): 211901.

Guo, W., Li, M., and Zhou, J. (2013). Modeling programmable deformation of self-folding all-polymer structures with temperature-sensitive hydrogels. *Smart Materials and Structures* **22** (11): 115028.

Hager, M.D., Bode, S., Weber, C., and Schubert, U.S. (2015). Shape memory polymers: past, present and future developments. *Progress in Polymer Science* **49**: 3–33.

Hauser, A.W., Evans, A.A., Na, J.-H., and Hayward, R.C. (2015). Photothermally reprogrammable buckling of nanocomposite gel sheets. *Angewandte Chemie* **127** (18): 5524–5527.

He, C., Zhang, M., and Devahastin, S. (2020a). Investigation on spontaneous shape change of 4D printed starch-based purees from purple sweet potatoes as induced by microwave dehydration. *ACS Applied Materials & Interfaces* **12** (34): 37896–37905.

He, C., Zhang, M., and Guo, C. (2020b). 4D printing of mashed potato/purple sweet potato puree with spontaneous color change. *Innovative Food Science & Emerging Technologies* **59**: 102250.

Holmes, D.P., Roché, M., Sinha, T., and Stone, H.A. (2011). Bending and twisting of soft materials by non-homogenous swelling. *Soft Matter* **7** (11): 5188–5193.

Hu, Y., Wu, G., Lan, T. et al. (2015). A graphene-based bimorph structure for design of high performance photoactuators. *Advanced Materials* **27** (47): 7867–7873.

Jani, J.M., Leary, M., Subic, A., and Gibson, M.A. (2014). A review of shape memory alloy research, applications and opportunities. *Materials & Design (1980–2015)* **56**: 1078–1113.

Jiang, Z.-C., Xiao, Y.-Y., Kang, Y. et al. (2017). Shape memory polymers based on supramolecular interactions. *ACS Applied Materials & Interfaces* **9** (24): 20276–20293.

Khoo, Z.X., Teoh, J.E.M., Liu, Y. et al. (2015). 3D printing of smart materials: a review on recent progresses in 4D printing. *Virtual and Physical Prototyping* **10** (3): 103–122.

Kim, J., Hanna, J.A., Byun, M. et al. (2012a). Designing responsive buckled surfaces by halftone gel lithography. *Science* **335** (6073): 1201–1205.

Kim, J., Hanna, J.A., Hayward, R.C., and Santangelo, C.D. (2012b). Thermally responsive rolling of thin gel strips with discrete variations in swelling. *Soft Matter* **8** (8): 2375–2381.

Klein, Y., Efrati, E., and Sharon, E. (2007). Shaping of elastic sheets by prescription of non-Euclidean metrics. *Science* **315** (5815): 1116–1120.

Kuksenok, O. and Balazs, A.C. (2016). Stimuli-responsive behavior of composites integrating thermo-responsive gels with photo-responsive fibers. *Materials Horizons* **3** (1): 53–62.

Laflin, K.E., Morris, C.J., Muqeem, T., and Gracias, D.H. (2012). Laser triggered sequential folding of microstructures. *Applied Physics Letters* **101** (13): 131901.

Lee, B.P. and Konst, S. (2014). Novel hydrogel actuator inspired by reversible mussel adhesive protein chemistry. *Advanced Materials* **26** (21): 3415–3419.

Lee, D.-W., Phadikar, J., and Shankar, M.R. (2017). Multiplicity of shape selection in functionally graded liquid crystalline polymers. *RSC Advances* **7** (37): 23046–23054.

Lee, Y., Lee, H., Hwang, T. et al. (2015). Sequential folding using light-activated polystyrene sheet. *Scientific Reports* **5** (1): 1–9.

Li, X., Shang, J., and Wang, Z. (2017). Intelligent materials: a review of applications in 4D printing. *Assembly Automation* **37** (2): 170–185.

Li, X., Zhu, Y., Dong, Y. et al. (2015). Epoxy resin composite bilayers with triple-shape memory effect. *Journal of Nanomaterials* **2015**: 1–8.

Lin, S., Xie, Y.M., Li, Q. et al. (2016). On the shape transformation of cone scales. *Soft Matter* **12** (48): 9797–9802.

Liu, Y., Takafuji, M., Ihara, H. et al. (2012). Programmable responsive shaping behavior induced by visible multi-dimensional gradients of magnetic nanoparticles. *Soft Matter* **8** (12): 3295–3299.

Liu, Y., Genzer, J., and Dickey, M.D. (2016). "2D or not 2D": shape-programming polymer sheets. *Progress in Polymer Science* **52**: 79–106.

Liu, Y., Miskiewicz, M., Escuti, M.J. et al. (2014). Three-dimensional folding of pre-strained polymer sheets via absorption of laser light. *Journal of Applied Physics* **115** (20): 204911.

Liu, Y., Shaw, B., Dickey, M.D., and Genzer, J. (2017). Sequential self-folding of polymer sheets. *Science Advances* **3** (3): e1602417.

Liu, Z., He, C., Guo, C. et al. (2021). Dehydration-triggered shape transformation of 4D printed edible gel structure affected by material property and heating mechanism. *Food Hydrocolloids* **115**: 106608.

Madbouly, S.A. and Lendlein, A. (2009). Shape-memory polymer composites. In: *Shape-Memory Polymers. Shape-Memory Polymers. Advances in Polymer Science*, vol. **226** (ed. A. Lendlein), 41–95. Berlin, Heidelberg: Springer. https://doi.org/10.1007/12_2009_28.

Market Research Report (2015). 4D printing market by material (programmable carbon fiber, programmable wood - custom printed wood grain, programmable textiles), end user (aerospace, automotive, clothing, construction, defense, healthcare & utility) & geography - global trends & forecas. //www.marketsandmarkets.com/Market-Reports/4d-printing-market-3084180.html (accessed 29 March 2021).

Meng, H. and Li, G. (2013). A review of stimuli-responsive shape memory polymer composites. *Polymer* **54** (9): 2199–2221.

Miao, S., Castro, N., Nowicki, M. et al. (2017). 4D printing of polymeric materials for tissue and organ regeneration. *Materials Today* **20** (10): 577–591.

Miriyev, A., Stack, K., and Lipson, H. (2017). Soft material for soft actuators. *Nature Communications* **8** (1): 1–8.

Mitchell, A., Lafont, U., Hołyńska, M., and Semprimoschnig, C. (2018). Additive manufacturing — a review of 4D printing and future applications. *Additive Manufacturing* **24**: 606–626.

Momeni, F., Liu, X., Ni, J., and others (2017). A review of 4D printing. *Materials & Design* **122**: 42–79.

Monzon, M.D., Paz, R., Pei, E. et al. (2017). 4D printing: processability and measurement of recovery force in shape memory polymers. *The International Journal of Advanced Manufacturing Technology* **89** (5–8): 1827–1836.

Mu, T., Liu, L., Lan, X. et al. (2018). Shape memory polymers for composites. *Composites Science and Technology* **160**: 169–198.

Na, J.-H., Bende, N.P., Bae, J. et al. (2016). Grayscale gel lithography for programmed buckling of non-Euclidean hydrogel plates. *Soft Matter* **12** (22): 4985–4990.

Naficy, S., Gately, R., Gorkin, R. III et al. (2017). 4D printing of reversible shape morphing hydrogel structures. *Macromolecular Materials and Engineering* **302** (1): 1600212.

Okuzaki, H., Kuwabara, T., Funasaka, K., and Saido, T. (2013). Humidity-sensitive polypyrrole films for electro-active polymer actuators. *Advanced Functional Materials* **23** (36): 4400–4407.

Panahi-Sarmad, M., Abrisham, M., Noroozi, M. et al. (2019). Deep focusing on the role of microstructures in shape memory properties of polymer composites: a critical review. *European Polymer Journal* **117**: 280–303.

Patil, D. and Song, G. (2017). A review of shape memory material's applications in the offshore oil and gas industry. *Smart Materials and Structures* **26** (9): 93002.

Pei, E. and Loh, G.H. (2018). Technological considerations for 4D printing: an overview. *Progress in Additive Manufacturing* **3** (1): 95–107.

Phuhongsung, P., Zhang, M., and Bhandari, B. (2020). 4D printing of products based on soy protein isolate via microwave heating for flavor development. *Food Research International* **137**: 109605.

Reyssat, E. and Mahadevan, L. (2009). Hygromorphs: from pine cones to biomimetic bilayers. *Journal of the Royal Society Interface* **6** (39): 951–957.

Shanthamma, S., Preethi, R., Moses, J.A., and Anandharamakrishnan, C. (2021). 4D printing of sago starch with turmeric blends: a study on pH-triggered spontaneous color transformation. *ACS Food Science and Technology* https://doi.org/10.1021/acsfoodscitech.0c00151.

Sharon, E. and Efrati, E. (2010). The mechanics of non-Euclidean plates. *Soft Matter* **6** (22): 5693–5704.

So, S. and Hayward, R.C. (2017). Tunable upper critical solution temperature of poly (*N*-isopropylacrylamide) in ionic liquids for sequential and reversible self-folding. *ACS Applied Materials & Interfaces* **9** (18): 15785–15790.

Stoychev, G., Puretskiy, N., and Ionov, L. (2011). Self-folding all-polymer thermoresponsive microcapsules. *Soft Matter* **7** (7): 3277–3279.

Stoychev, G., Zakharchenko, S., Turcaud, S. et al. (2012). Shape-programmed folding of stimuli-responsive polymer bilayers. *ACS Nano* **6** (5): 3925–3934.

Studart, A.R. (2015). Biologically inspired dynamic material systems. *Angewandte Chemie International Edition* **54** (11): 3400–3416.

Subash, A. and Kandasubramanian, B. (2020). 4D printing of shape memory polymers. *European Polymer Journal* **134**: 109771.

Sun, Y., Choi, W.M., Jiang, H. et al. (2006). Controlled buckling of semiconductor nanoribbons for stretchable electronics. *Nature Nanotechnology* **1** (3): 201–207.

Tao, Y., Do, Y., Yang, H. et al. (2019). Morphlour: personalized flour-based morphing food induced by dehydration or hydration method. Proceedings of the 32nd Annual ACM Symposium on User Interface Software and Technology, 329–340.

Teng, X., Zhang, M., and Mujumdar, A.S. (2021). 4D printing: recent advances and proposals in the food sector. *Trends in Food Science & Technology* **110**: 349–363.

Therien-Aubin, H., Moshe, M., Sharon, E., and Kumacheva, E. (2015). Shape transformations of soft matter governed by bi-axial stresses. *Soft Matter* **11** (23): 4600–4605.

Tibbits, S. (2013). The emergence of "4D printing". TED Conference. //www.ted.com/talks/skylar_tibbits_the_emergence_of_4d_printing (accessed 29 March 2021).

Timoshenko, S. (1925). Analysis of bi-metal thermostats. *Josa* **11** (3): 233–255.

Tomatsu, I., Peng, K., and Kros, A. (2011). Photoresponsive hydrogels for biomedical applications. *Advanced Drug Delivery Reviews* **63** (14–15): 1257–1266.

Tsuchiya, K. (2011). Mechanisms and properties of shape memory effect and superelasticity in alloys and other materials: a practical guide. In: *Shape Memory and Superelastic Alloys* (eds. K. Yamauchi, I. Ohkata, K. Tsuchiya and S. Miyazaki), 3–14. United Kingdom: Woodhead publishing.

van Manen, T., Janbaz, S., and Zadpoor, A.A. (2017). Programming 2D/3D shape-shifting with hobbyist 3D printers. *Materials Horizons* **4** (6): 1064–1069.

van Manen, T., Janbaz, S., and Zadpoor, A.A. (2018). Programming the shape-shifting of flat soft matter. *Materials Today* **21** (2): 144–163.

Volkov, A.G., Foster, J.C., Ashby, T.A. et al. (2010). *Mimosa pudica*: electrical and mechanical stimulation of plant movements. *Plant, Cell & Environment* **33** (2): 163–173.

Wang, E., Desai, M.S., and Lee, S.-W. (2013). Light-controlled graphene-elastin composite hydrogel actuators. *Nano Letters* **13** (6): 2826–2830.

Wang, W., Yao, L., Zhang, T. et al. (2017). Transformative appetite: shape-changing food transforms from 2D to 3D by water interaction through cooking. Proceedings of the 2017 CHI Conference on Human Factors in Computing Systems, 6123–6132.

Wen, C., Yu, X., Zeng, W. et al. (2018). Mechanical behaviors and biomedical applications of shape memory materials: a review. *AIMS Materials Science* **5** (4): 559–590.

Wu, Z.L., Moshe, M., Greener, J. et al. (2013). Three-dimensional shape transformations of hydrogel sheets induced by small-scale modulation of internal stresses. *Nature Communications* **4** (1): 1–7.

Xie, F., Huang, L., Leng, J., and Liu, Y. (2016). Thermoset shape memory polymers and their composites. *Journal of Intelligent Material Systems and Structures* **27** (18): 2433–2455.

Yan, Z., Zhang, F., Liu, F. et al. (2016). Mechanical assembly of complex, 3D mesostructures from releasable multilayers of advanced materials. *Science Advances* **2** (9): e1601014.

Yan, Z., Zhang, F., Wang, J. et al. (2017). Correction to: controlled mechanical buckling for origami-inspired construction of 3D microstructures in advanced materials (Advanced Functional Materials, (2016), 26, 16,(2629–2639), 10.1002/adfm.201504901). *Advanced Functional Materials* **27** (22): 2629–2639. https://doi.org/10.1002/adfm.201702339.

Yang, W.G., Lu, H., Huang, W.M. et al. (2014). Advanced shape memory technology to reshape product design, manufacturing and recycling. *Polymers* **6** (8): 2287–2308.

Yu, C., Duan, Z., Yuan, P. et al. (2013). Electronically programmable, reversible shape change in two-and three-dimensional hydrogel structures. *Advanced Materials* **25** (11): 1541–1546.

Yu, K., Ritchie, A., Mao, Y. et al. (2015). Controlled sequential shape changing components by 3D printing of shape memory polymer multimaterials. *Procedia Iutam* **12**: 193–203.

Zhang, Q., Wommer, J., O'Rourke, C. et al. (2017a). Origami and kirigami inspired self-folding for programming three-dimensional shape shifting of polymer sheets with light. *Extreme Mechanics Letters* **11**: 111–120.

Zhang, Y.S., Yue, K., Aleman, J. et al. (2017b). 3D bioprinting for tissue and organ fabrication. *Annals of Biomedical Engineering* **45** (1): 148–163.

Zhang, Z., Demir, K.G., and Gu, G.X. (2019). Developments in 4D-printing: a review on current smart materials, technologies, and applications. *International Journal of Smart and Nano Materials* **10** (3): 205–224.

Zhao, L., Zhang, M., Chitrakar, B., and Adhikari, B. (2020). Recent advances in functional 3D printing of foods: a review of functions of ingredients and internal structures. *Critical Reviews in Food Science and Nutrition*: 1–15.

Zhao, Q., Qi, H.J., and Xie, T. (2015). Recent progress in shape memory polymer: new behavior, enabling materials, and mechanistic understanding. *Progress in Polymer Science* **49**: 79–120.

Zhao, Z., Wu, J., Mu, X. et al. (2017). Desolvation induced origami of photocurable polymers by digit light processing. *Macromolecular Rapid Communications* **38** (13): 1600625.

Zhou, J. and Sheiko, S.S. (2016). Reversible shape-shifting in polymeric materials. *Journal of Polymer Science Part B: Polymer Physics* **54** (14): 1365–1380.

Zhou, J., Turner, S.A., Brosnan, S.M. et al. (2014). Shapeshifting: reversible shape memory in semicrystalline elastomers. *Macromolecules* **47** (5): 1768–1776.

Zhou, Y., Huang, W.M., Kang, S.F. et al. (2015). From 3D to 4D printing: approaches and typical applications. *Journal of Mechanical Science and Technology* **29** (10): 4281–4288.

11

Applications of Food 3D Printing Technology

CHAPTER MENU

11.1 Introduction

The consumer's perception of the dietary pattern is gradually changing with industrialization and evolving lifestyles. The five major criteria that determine the consumer's food decisions and product marketability are taste, nutrition, cost, eating experience, and convenience (Sun et al. 2018). Physical fitness and healthy diets are the potential driving claims for the emerging market trends on functional foods, nutraceuticals, and personal care products. Despite the effects of metabolism, the nutritional and dietary requirements of each individual are varied based on health status, age, and specific functional needs (Flint et al. 2012). Considering all these attributes, the urge to satisfy individual needs and likeness drives the food industry crazy. Food scientists work towards the concepts of customization and personalization of foods (Johnson et al. 2021). In the conventional food manufacturing process, customization can be achieved at the implication of higher productional cost. However, none of the conventional food processing techniques claims to satisfy all the aforementioned criteria of food customization and personalization (Derossi et al. 2020b). Focusing on the individual needs and on-demand production, food

3D Printing of Foods, First Edition. C. Anandharamakrishnan, Jeyan A. Moses, and T. Anukiruthika.

3D printing is forecasted to be a revolutionizing technology that would rule digitalized food market in the next 10 years. The concept of food 3D printing relies on precise material deposition with control over macro- and micronutrients. The so-called food printing is an integration of 3D printing and gastronomy that involves the digitalized fabrication of complex food structures in a layer-by-layer manner with intricate design and unique functionality (Anukiruthika et al. 2020). 3D food printing is a novel emerging technology that allows for customization of foods in terms of size, shape, portion, calorie whilst maintains sensorial and nutritional attributes. Despite its nascent growing stage in the food sector, 3D printed foods are slowly capturing the food market with their fascinating designing approach (Jagtap et al. 2021).

Although applications of 3D printing are well known in allied industrial sectors such as construction, automobiles, fashion and textiles, aerospace, and medicine, its food applications are obscure (Nida et al. 2020). Only in the past decade, 3D printing of foods gained research interest and has been explored for the hidden opportunities and functionalities. 3D printing is a single stand-alone technology that combines a sequence of unit operations of conventional food processing such as weighing, mixing, and slicing into a single step thereby saves time and energy (Nachal et al. 2019). Nevertheless, 3D printing seems to reduce complexity and simplifies the food supply chain. From a broad perspective, 3D printing significantly reduces human intervention through process automation in addition to customized designing of foods (Sun et al. 2015a). Further, 3D printing is claimed to be sustainable technology that aids in tackling climate change through minimal and efficient utilization of material resources and energy with fewer carbon footprints (Rezek Jambrak et al. 2021). Further, 3D printing provides opportunities to explore uncommon alternative food ingredients such as insects, macro and microalgae, seaweeds, fungi, and lupin seeds that are listed to be the future foods (Derossi et al. 2020b). As an additive process, food printing utilizes less amount of raw material that promotes its potentiality in the conversion of 'waste into wealth', i.e. food processing waste streams can be conveniently processed into a value-added product through 3D printing (Prakash et al. 2019). With technological advancements, 3D printing can be integrated with other food processing technologies such as encapsulation and electrospinning to produce functional 3D printed foods (Sun et al. 2018). Above all the features, the key advantage of 3D printing is the possibility of reduced usage of chemical preservatives that best suits for on-demand production of foods. Thus, with adequate pre- and post-processing of material supplies, 3D printing significantly reduces the multiple handling steps involved in food production that significantly prevents food loss, reduces processing cost, and saves energy (Nachal et al. 2019). Various opportunities exist with the 3D printing of foods are presented in Figure 11.1. The present chapter summarizes the different applications of 3D printing of foods. In addition, the future prospects of food printing have been discussed in context with customization and personalization of foods. Also, the feasibility and applicability of 3D printing as a modern digitalized technique in the production of future sustainable foods are presented. Thus, the present chapter provides valuable insights on the potential research opportunities in the designing of customized and personalized foods based on individual likeness and nutritional requirements.

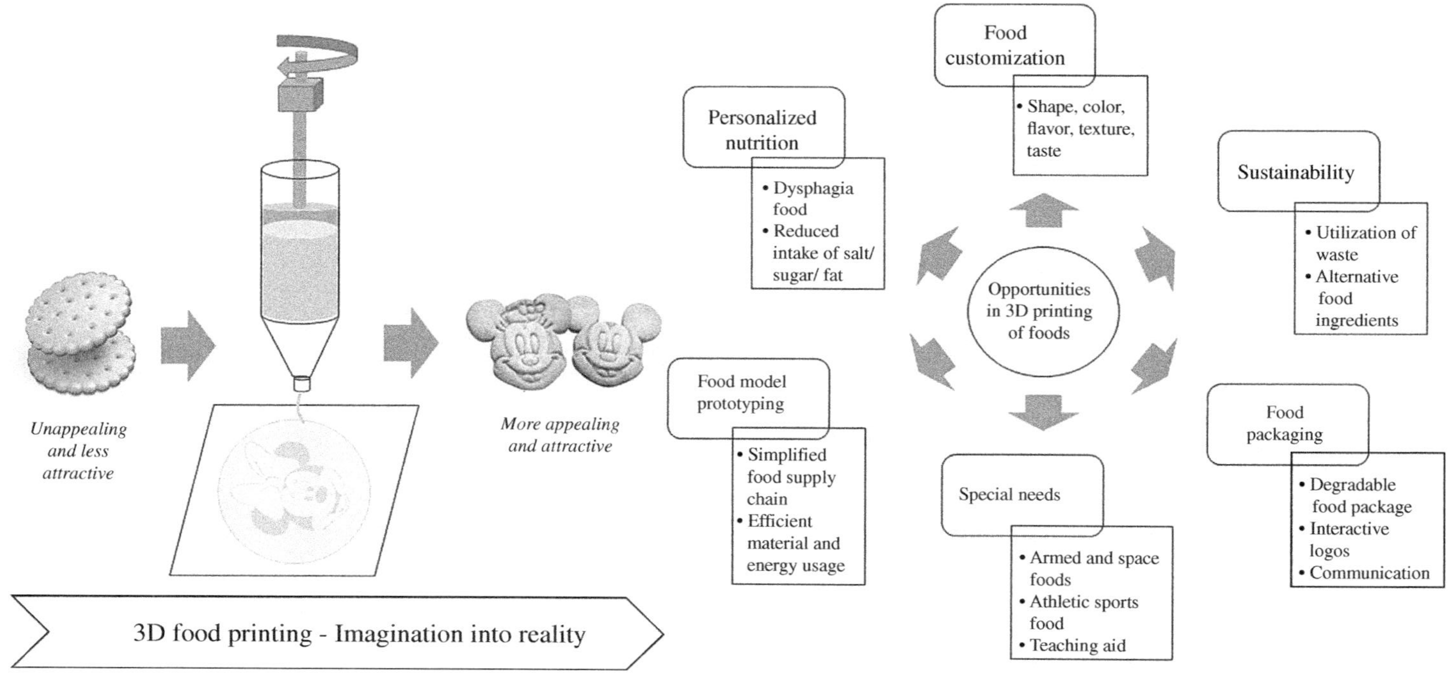

Figure 11.1 Various opportunities exist with 3D printing of foods.

11.2 Applications of 3D Food Printing

11.2.1 Food Customization

The most exploited application of 3D printing is the customized design of foods. The fabrication of complex 3D constructs with internal microstructure is highly impossible with existing conventional technologies. The development of customized foods with intricate complex designs using conventional approaches is limited to material selection, processing, sensorial attributes, nutritional functionality, creativity, automation, and skilled personnel (Dankar et al. 2018). Fortunately, 3D printing could overcome all the above-mentioned limitations through customizing the shape, size, colour, taste, flavour, and texture of the food. A well-known feature of 3D printing is its scope and potential for mass customization and commercialization. With the existing conventional technologies, bulk production of foods is possible. However, the cost of production increases with the increasingly complex design of fabrication objects. In contrast, additive manufacturing can be conveniently used for the mass production of foods with complex internal designs at less productional cost (Dev et al. 2020). A variety of 3D food printing technologies are available that aids in designing foods with novel texture, flavour, and taste. As discussed in earlier chapters, extrusion-based 3D printing is most commonly used for printing of semi-solid foods; sintering technique is employed for the printing of dry powders; and inkjet printing and binder technologies are applied for less viscous food materials (Yang et al. 2017). All these technologies dictate the design opportunities existing with the food materials that cover a broad range of material consistencies. Based on the consumer's choice and preferences, food materials are selected for 3D printing that allows for the development of attractive foods even from unappealing and unappetizing food materials. The nutrients dense cereals and pulses can be conveniently processed into printable form for the development of protein and fibrous enriched snacks (Krishnaraj et al. 2019). Interestingly, the buckwheat flour has been combined with yellow peach powder for the development of colour and flavour triggered 3D construct (Guo et al. 2021). The most lavishly used materials for 3D printing are chocolates, starch, hydrogels, hydrocolloids, cheese, hummus, and cake frostings that exclusively belong to the category of natively printable foods (Sun et al. 2015b). Although most of the 3D printing studies are under lab-scale research, some of the commercial 3D printers are utilized for printing the chocolates and sugar confectioneries focusing on customized designs. Researchers had developed a jelly-like food capsules in desired shape and flavours (Molitch-Hou 2014). The surface fillings and designing on cookies, biscuits, pizza toppings, and cake carvings are also possible with 3D printing technology.

Focusing on the development of customized foods with altered texture, the infill pattern and internal structure of the 3D constructs are varied. By varying the infill level, the material supplies such as chocolates and cereal dough were 3D printed (Lanaro et al. 2019; Zhang et al. 2018). More recently, the air-fried texture modified potato-based 3D construct was developed (Liu et al. 2020). In another study, a cuboid-shaped 3D printed wheat-based snack was developed with programmed texture modification (Figure 11.2) (Derossi et al. 2020a).

All these studies open up interesting perspectives on the application of 3D printing in the creation of novel foods with unique tastes and eating experiences. The 3D printed mashed potato samples were compared with the casted samples and the results showed that porous

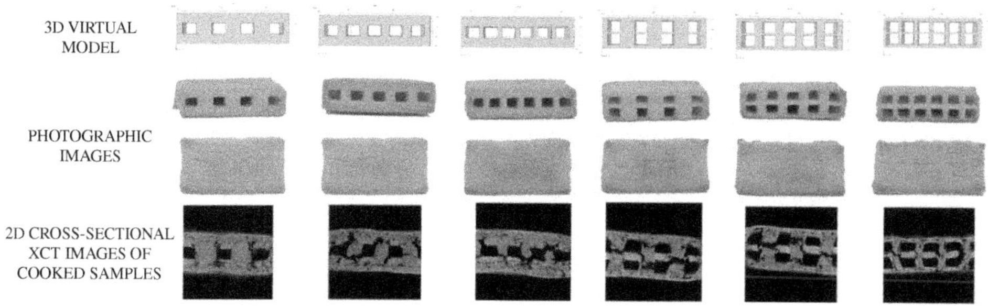

Figure 11.2 Cross-sectional image showing the programmable texture modified 3D printed snack. *Source:* From Derossi et al. (2020a), Figure 03 [p. 005.] / With permission of Elsevier. DOI-https://doi.org/10.1016/j.jfoodeng.2020.110160.

microstructure of 3D printed samples were less harder even with the 100% infill samples (Liu et al. 2018b). The incorporation of additives not only tailors the rheological properties but also improves the mechanical stability of the post-deposited 3D printed constructs. The lemon juice was successfully 3D printed with potato starch as an additive (Figure 11.3) (Yang et al. 2018). Similarly, the sweet potato starch was added to improve the printability of the surimi gel (Dong et al. 2019).

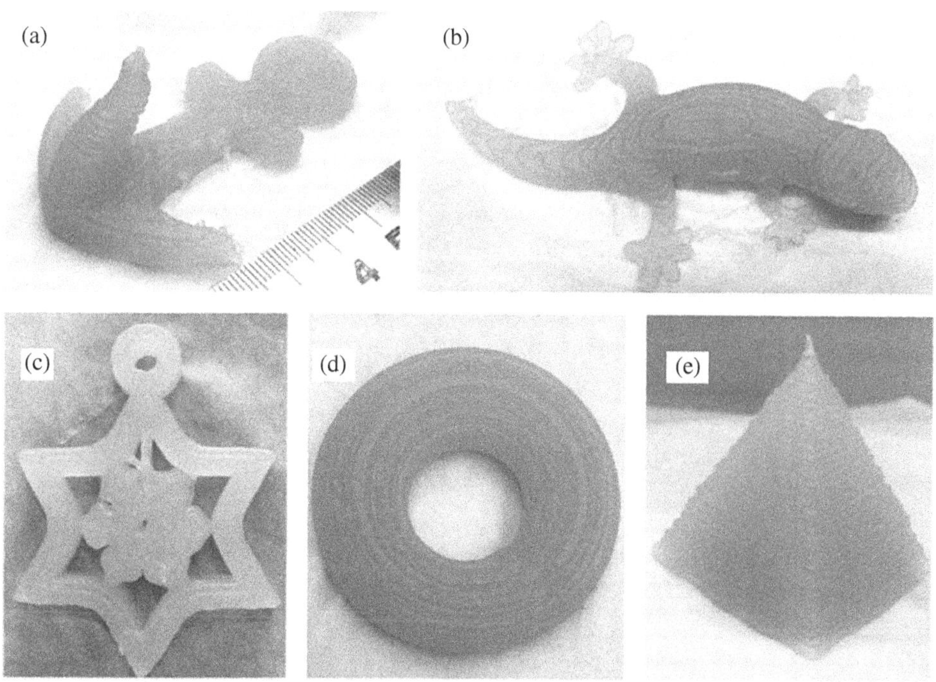

Figure 11.3 Lemon juice gel 3D printed in different customized shapes. (a) anchor, (b) gecko, (c) snowflake, (d) ring, and (e) tetrahedron. *Source:* From Yang et al. (2018), Figure 10 [p. 038] / With permission of Elsevier. DOI- https://doi.org/10.1016/j.lwt.2017.08.054.

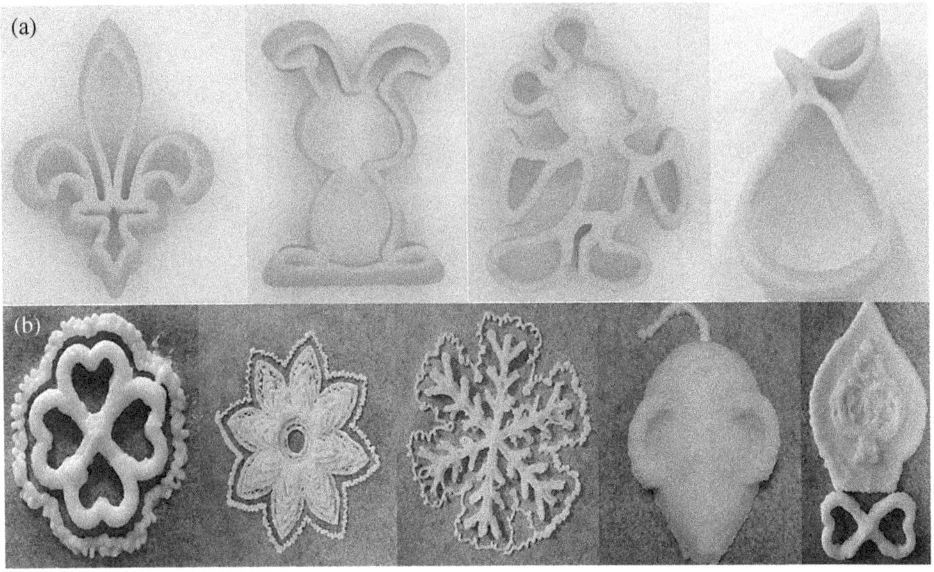

Figure 11.4 Customized shapes of 3D constructs from different material supplies. (a) Soy protein isolate mixture. *Source:* From Chen et al. (2019b), Figure 07 [p. 011] / With permission of Elsevier. DOI- https://doi.org/10.1016/j.jfoodeng.2019.03.016 and (b) milk protein food simulant. *Source:* From Liu et al. (2018a), Figure 07 [p. 39] / With permission of Elsevier. DOI- https://doi.org/10.1016/j.ifset.2018.07.018.

More recently, 3D printed nuggets were prepared into desired shapes from chicken meat in a customized manner (Wilson et al. 2020). Each of the deposited layers of the printed construct can be customized with desired taste, flavour, and texture through multi-material printing. The mashed potatoes and the strawberry juice gel were used as printing mixture for the development of attractive customized foods in desired shape and colour using dual extrusion-based 3D printing (Liu et al. 2018c). Similar studies were reported on the fabrication of colourful 3D constructs with multiple materials (Dick et al. 2019a; Lipton et al. 2010). Focusing on the delivery of nutritious foods, protein, and fibrous-rich material supplies were 3D printed and post-processed using oven drying and freeze-drying methods (Lille et al. 2018). In view of analyzing the printability of different protein-based material supplies, soy protein isolate (Chen et al. 2019b), milk protein (Liu et al. 2018a), and pea protein (Oyinloye and Yoon 2020) were successfully printed into customized 3D shapes (Figure 11.4).

Hydrogels are the cross-linked polymeric 3D network that can be 3D printed through the sol-gel transition process. These kinds of soft materials can be conveniently 3D printed into desired 3D structures. Recently, the effect of pre-processing on the 3D printing of starch was analyzed from different material sources such as wheat, rice, corn, and cassava (Chen et al. 2019a; Maniglia et al. 2020). In a more recent study, the effect of hybrid gelator system (beeswax-carrageenan-xanthan) on the printability of litchi was reported (Figure 11.5) (Tian et al. 2021). Results showed that the hybrid gelator system aids in improving the stability, precision, and formability of 3D printed structure that could withstand the layer height comprised of 85 layers with an inclination angle of about 53.5°. Thus, 3D printing of foods is mainly used for the fabrication of customized foods with desired shape, taste, flavour, colour, modified texture, and tailored nutrition.

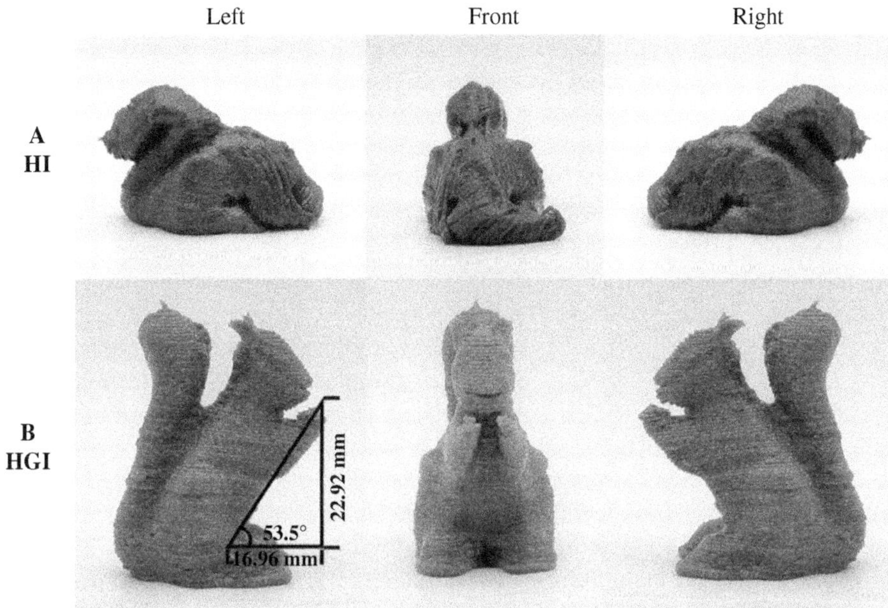

Figure 11.5 3D printing evaluation of litchi into squirrel shape with a hybrid gelator system (HI - hydrogelator ink; HGI - hybrid gelator ink). *Source:* From Tian et al. (2021), Figure 07 [p. 36] / With permission of Elsevier. DOI- https://doi.org/10.1016/j.foodhyd.2020.106482.

11.2.2 Personalized Foods and Digitalized Nutrition Control

Another most important application of 3D printing is the personalization of diets. The nutritional and dietary requirements would vary for each individual based on metabolic rate, physical activity, age, and gender. By considering all these factors, 3D printing can be promisingly used for the development of a customized food formula with the right proportions of functional ingredients and nutrients as required (Nachal et al. 2019). One such application of food printing is the texture modification of foods for elderly people suffering from swallowing disorder (dysphagia). Based on this concept, the researchers of Biozoon Food Innovation had prepared texture-modified cookies from insect-derived flours for people suffering from swallowing disorders (Pinna et al. 2016). Dysphagia is a medical condition related to the difficulty in swallowing foods experienced by elderly people. There are several soft-textured foods available in the market are thickened drinks, soft bite moist foods, and pureed fruits and vegetables. The provision of soft foods with the right appropriate material consistency is crucial otherwise it would result in choking and aspiration problems in dysphagia patients (Prakash et al. 2019). Considering all these aspects, solid foods are subjected to adequate processing by transforming them into a pourable liquid consistency. However, intense processing leads to a less appealing appearance, taste, and texture. And also, the continual supply of these texture-modified foods creates aversion and boredom to people suffering from dysphagia that significantly impacts dietary daily intake. These issues can be overcome through 3D printing of soft food matrices into attractive shapes, tastes, colour, flavour, and texture with better appearance (Portanguen et al. 2019). In addition, the digital deposition of a layered pattern of foods allows to supplement the essential macro- and micronutrients in between the layers whilst maintains proper consistency. Thus, the applicability of 3D printing in the fabrication of dysphagia foods ensures the proper delivery of nutrients, appetizing texture modified foods, and correct times of meal intake in a day that leads to a better quality of life. The scientists of Netherlands Organization for Applied Scientific Research (TNO) had 3D printed food purees with customized soft food formulas for elderly people with mastication problems (Nachal et al. 2019).

Recently, the effect of material formulation on the printability of texture-modified pork paste was reported (Dick et al. 2020). The rheological, textural, and microstructural properties were examined for enhancement of printability. Results showed that the samples with the incorporated hydrocolloids showed a visible extended network with heterogeneous pores and high-water retention after post-processing. Thus, the developed pork-based food formula was categorized as a transitional food within the International Dysphagia Diet Standardization Initiative (IDDSI) framework which can be consumed by elderly people suffering from dysphagia. In the formulation of dysphagia diets, the desired texture could be achieved through appropriate material characteristics such as gel resistance and post-processing conditions. The printability and textural attributes of the cooked beef paste were analyzed with different proportions of hydrocolloids (xanthan gum, guar gum, *k*-carrageenan, and locust bean gum) for dysphagia diet (Dick et al. 2021). Results confirmed the applicability of different formulations of beef paste as texture-modified foods that comes under levels of 5, 6, and 7 of the IDDSI framework (Figure 11.6).

In another study, the effect of hydrocolloids on three different categories of vegetables (garden peas, bok choy, and carrot) in the formulation of a dysphagia diet was reported (Pant et al. 2021). The IDDSI tests showed that the optimized food ink formulation exhibited excellent printability, minimal water seepage, the minimal addition of

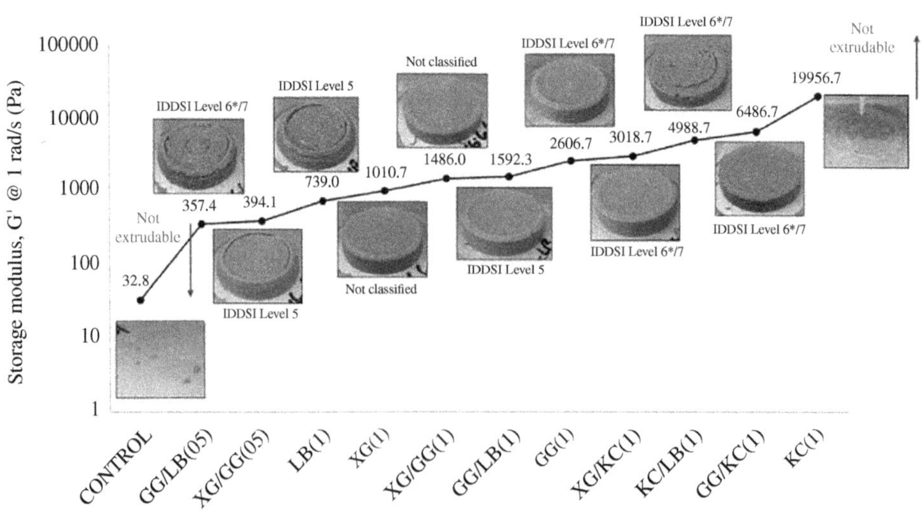

Figure 11.6 Correlation of rheology with the 3D printing of cooked beef paste showing different levels of dysphagia IDDSI framework. *Source:* From Dick et al. (2021), Figure 01 / CC BY 4.0 / With permission of Elsevier. DOI- https://doi.org/10.1016/j.fufo.2020.100006.

hydrocolloids, and dense microstructure (Figure 11.7). The results confirmed the use of fresh vegetables instead of freeze-dried foods for the formulation of dysphagia food using 3D printing that could potentially preserve the flavour and nutrition profile of the real food.

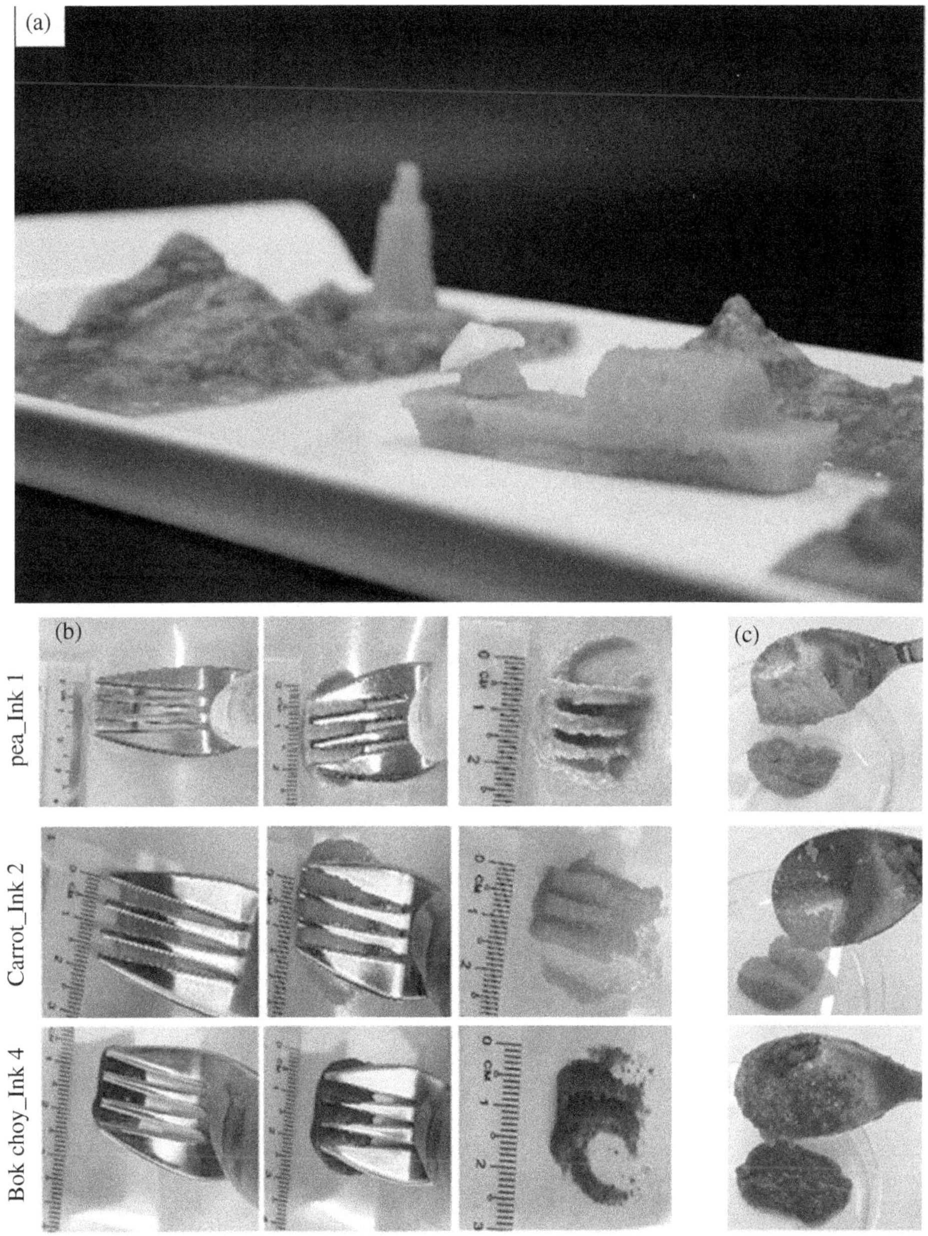

Figure 11.7 3D printed fresh vegetables for dysphagia patients. (a) 3D printed designs of garden peas, carrot, and corn, (b) IDDSI fork pressure test, and (c) IDDSI spoon tilt test of material supplies. *Source:* From Pant et al. (2021), Figure 02 & 06 [p. 29 & 33] / With permission of Elsevier. DOI-https://doi.org/10.1016/j.foodhyd.2020.106546.

A fruit-based snack was 3D printed based on reviewing the nutritional requirement of children of age group of 3–10 years (Derossi et al. 2018). Thus, the developed 3D printed snack could satisfy about 5–10% of the recommended dietary allowance (RDA) of energy, calcium, iron, and vitamin D. Similarly, a composite flour was formulated from barnyard millet, green gram, fried gram, and ajwain seeds (Krishnaraj et al. 2019). The formulated printing mixture was successfully 3D printed into a fibrous and protein-rich snack suitable for people of all age groups. In general, children used to show a strong aversion towards eating of green leafy vegetables that severely affect physical growth and may lead to malnourishment. In such case, the vegetables can be conveniently processed into printable form and can be 3D printed into attractive shapes. Thus, the food aversion can be transformed into product likeness using 3D printing. The effect of particle size on 3D printing performance of cellular materials (spinach) was analyzed and results showed that the increased particle size had positive effects on improving the mechanical strength of the 3D constructs (Lee et al. 2019). In another study, the addition of hydrocolloids on rheological properties and printability of vegetable ink was analyzed. Among the tested samples, food ink (spinach/ broccoli/ carrot) with xanthan gum exhibited smooth extrusion that yielded a 3D construct with higher printing resolution (Kim et al. 2018). Research on the free radical scavenging capacity of polyphenols and bioactive compounds confirmed the significance of these ingredients in a regular diet for the prevention of lifestyle disorders. Edible rose petals are a good source of anthocyanin which possess higher free radical scavenging activity. More recently, the printability of edible rose petals was analyzed with the sodium alginate gel model system (Feng et al. 2021a). In this study, the effect of size and quantity of rose petals on 3D printing behaviour was investigated. Results showed that the rose petals with a plane diameter of <0.5 cm and material quantity of 6% (w/w) yielded more appealing 3D constructs with rich rose-fragrance. In a similar approach, the viability of probiotics was analyzed with varied infill densities and structures. Thus, 3D printing is proved to be a promising technology in the fabrication of functional probiotic foods with appropriate optimization and adequate post-processing (Zhang et al. 2018). With rising awareness of the healthy diet, 3D printing can be used for the fabrication of personalized foods with the desired proportion of nutrients as required. Thus, the amount and density of nutrients, texture, shape, colour, and flavour can be tailored as per the individual needs (elderly, athlete, pregnant woman, adolescent, and infants) using 3D printing. Thus, it is possible to print the vegetable purees in any shapes for instance carrot puree can be either printed as in the shape of carrot or apple without affecting the original nutritional profile of the food material.

11.2.3 Delivery of Specific Foods with Unique Functionality

The 3D printed food products are more appealing in shape and size that could be related to human emotions. Other than customization and personalization, 3D-printed foods serve as a medium of communication (Gayler et al. 2020). The 3D printed constructs can be used as a teaching aid for early pre-school children. 3D printing is a potential education tool that enhances the knowledge of school children through practical learning by printing edible foods in attractive shapes and colours. Vegemite and marmite were used as material supplies in surface designing of the bread substrate that is more attractive and could be used as a teaching aid for school kids (Figure 11.8) (Hamilton et al. 2018).

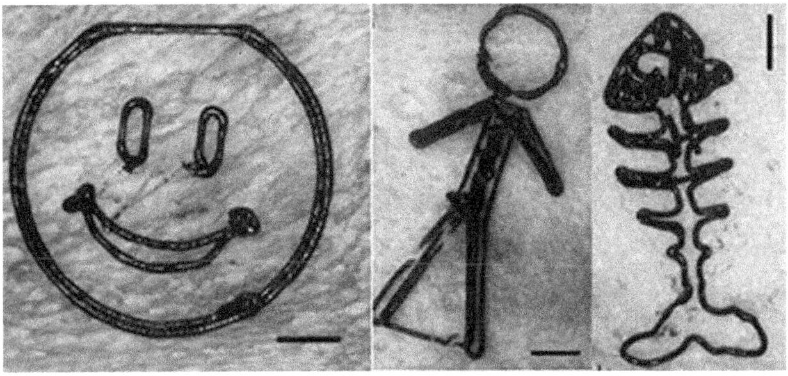

Figure 11.8 3D printed designs of vegemite and marmite over the toasted bread substrate. *Source:* From Hamilton et al. (2018), Figure 04 [p. 13] / With permission of Elsevier. DOI- https://doi.org/10.1016/j.jfoodeng.2017.01.008.

For instance, the 3D-printed alphabets from a mixture of fruits and vegetables would not only provide nourishment but also act as a teaching aid. This way of interaction fascinates school children that results in a more effective acquirement of knowledge and keeps them engaged without boredom (Figure 11.9). 3D food printers have been evaluated as a teaching

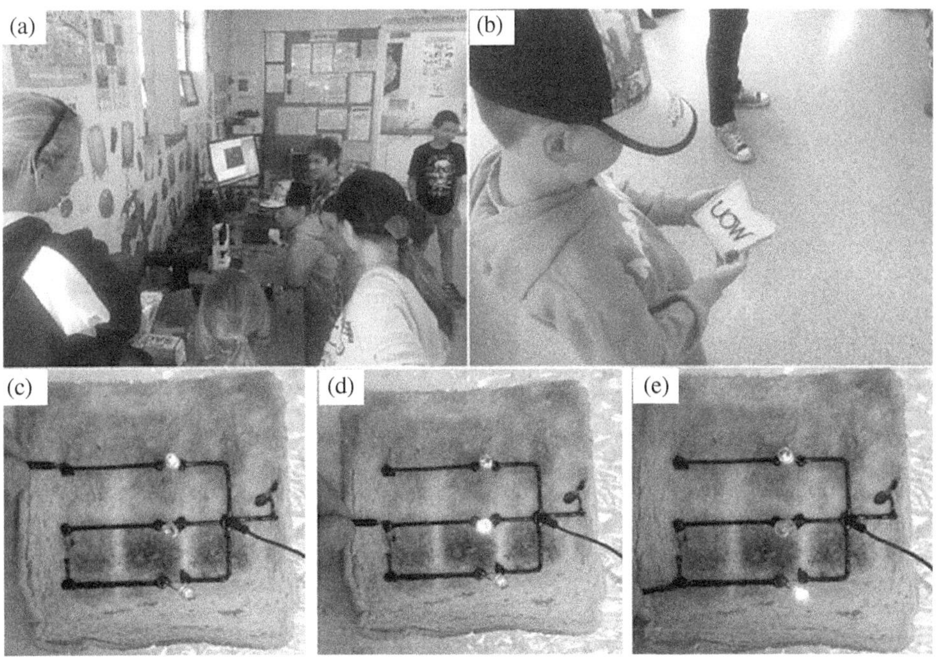

Figure 11.9 Illustration of 3D printing as an educational aid. (a) and (b) children looking at the designs of 3D printed vegemite; (c), (d), and (e) vegemite electronic circuit 3D printed on bread substrate. *Source:* From Hamilton et al. (2018), Figure 05 & 06 [p. 12 & 13] / With permission of Elsevier. DOI-https://doi.org/10.1016/j.jfoodeng.2017.01.008.

tool among dieticians, teachers, and students in disseminating information about food and nutrition (Gosine et al. 2021). Initially, the participants worried about learning of 3D printing for the printing of foods and had negative thoughts that 3D printing could lead to confusion and misunderstanding of concepts. However, later the audience realized that observing the 3D printing process grabbed attention and developed more interest on the subject that helps in better understanding of the teaching concepts through fabrication of appealing foods.

As a convenient way of delivering the digitalized supply of nutritional and energy requirements, 3D printing can be promisingly used for manufacturing foods for specific population groups based on their convenience. With the advancements of technology, 3D printing can be conveniently used for fabrication of the military and space foods (Nachal et al. 2019). For instance, the specific nutritional needs of the soldiers can be tailored by delivering 3D printed foods through on-demand production. Similarly, the convenience and safety aspects make 3D printing more appropriate for the fabrication of space foods that remains still challenging with conventional techniques in meeting the long shelf-life requirements of food products (Lin 2015). The United States army shows a great deal of interest in the application of 3D printing in delivery of the military foods. Research experimentation is still under progress in exploring the application of ultrasonic waves with 3D printing for the fusion of materials through ultrasonic agglomeration (Jennifer 2014). This would offer a variety of meals with customized sensorial and nutritional attributes in soldier's foods. Trials were being conducted for the fabrication of 3D units that could transform the forage plant materials into edible foods of attractive shapes and sizes (Davide and Xavier 2015). Similarly, National Aeronautics and Space Administration (NASA)-funded projects are in active progress in developing a 3D printed space food for long space mission program. With the least amount of spacecraft resources, NASA tried to implement a 3D printer for space foods by meeting all the safety, nutritional, and sensory requirements. With the currently available resources, the traditional processing methods are not suitable for long mission space programs as micronutrients tends to decline with increasing storage time. Hence, 3D printing could be the best option in meeting the personalized nutritional requirements of the astronauts. Based on the proposed NASA-funded project of Systems and Materials Research Corporation (SMRC), extrusion-based 3D printing could be used for delivering macronutrients with tailored food structure and texture while inkjet printing could be used for delivering the micronutrients with desired flavour and aroma in the space foods (Liu et al. 2017). These approaches must follow a strict protocol and regulations for ensuring the nutritional delivery, customized shapes, modified texture, long-term storage, dietary needs, sustenance, and improving the pleasure of eating.

11.2.4 Food Model Prototyping

The changing consumer's perception on food products demands the need for changing the existing food manufacturing approaches in consideration with the food safety, nutritional value, convenience, shelf life, and reduced food wastage (Augustin et al. 2016). These aspects insist on the restructuring and redevelopment of current conventional products to cope up with the ever-changing consumer's attitude towards food. However, these gradual changes may lead to a large amount of product failure in satisfying consumer needs. In order to avoid product failure and material loss, there must be proper communication among the food scientists, design engineers, marketing people, distributors, and consumers that pave a way for alteration of ingredient formulation and design innovation at the product development stage (Sun et al. 2018). This dictates the need for prototyping the food model at the small-batch production for which a 3D printer seems to be a promising cost-effective tool for material testing and product development. Thus, 3D printing aids in a broad understanding of the technical requirements such as material mixing, ingredient combination, consumer preference, feasibility and product handling, distribution, and marketing prior to the implementation of developed formula at large-scale mass production (Liu et al. 2017). This significantly saves time, energy, material usage, and cost in eliminating the food designs of non-interest to the consumers. Further, it allows for testing of printability of novel food ingredients at the small-batch level and feasibility of which for mass production in consideration with consumers acceptability (Tuorila and Hartmann 2020). 3D food printers and printing technology would allow for a sustainable supply of foods in satisfying the global hunger in an efficient way (Garcia-Oliveira et al. 2020). To date, none of the conventional food technologies match the features of 3D printing. In brief, 3D printing is an innovative technology with vast research opportunities that have the potential to alleviate hidden malnourishment. However, more awareness about 3D food printing to consumers is adequate and necessary in determining the success of the technology.

11.2.5 Sustainable Approach for Conversion of Waste into Wealth

Urbanization and industrial revolution lead to deterioration of renewable resources that are evident with unpredictable climate change and extinction of fauna and flora (Morgado and Vieira 2020). This gives an alarm for the development of sustainable productional practices using renewable resources. The nutritional and food security can be conserved by exploring the efficient utilization of available material resources exploring the uncommon, unexplored, and under-utilized food sources (Nachal et al. 2019). Insects are recognized as a sustainable food of the future by the European Union and World food Organization to cope up with the problem of global hunger (Stamer 2015). The advent of 3D food printing not only provides the nutrient-dense food matrix but also makes use of exotic foods such as insects and microbial proteins for the creation of novel food products to feed global hunger. 3D printing is considered a sustainable technology that can convert low-value waste materials into value-added 3D-printed food (Nachal et al. 2019). As an additive process, 3D printing generally requires a minimal adequate amount of raw materials that seem to be advantageous in utilizing the waste stream and by-products of the food processing line (Prakash et al. 2019). The generated waste from the main food processing line for instance potato fines from the chips processing line, meat trimmings from meat processing, distorted fruits, and vegetable remains from the juice processing line can be conveniently used as raw materials for 3D printing. Thus, these materials must require appropriate pre-processing for converting them into the printable form which would otherwise discard as waste and dumped into the environment. As a sustainable approach, 3D printing paves a way for cleaner productional practices that have both economic as well as environmental benefits (Baiano 2020). Recently, there is a growing interest in the use of novel uncommon, unfamiliar, and under-utilized food ingredients as future foods. Materials of this category include the applicability of renewable food sources such as insects, seaweeds, algae, fungi, lupin seeds, pseudocereals, grass, and beet leaf in the fabrication of customized sensorially acceptable foods (Lupton and Turner 2018). Some of the unappealing food ingredients such as insects and algae are potentially transformed into an attractive shape with a more appetizing taste using 3D printing. Recently, a mushroom-based material supply (*Agaricus bisporus*) has been incorporated into wheat flour for the development of 3D printed snacks (Keerthana et al. 2020). Results showed that the fibrous nature of mushrooms has negative effects on the extrudability of material supply. However, with adequate pre-processing and appropriate proportion, mushroom can be successfully 3D printed. In a similar approach, 3D printed functional cookies were developed with the incorporation of *Arthrospira platensis* (microalgae) (Vieira et al. 2020). Due to the high value of proteins, insects are added to cereal-based products for protein enrichment. Interestingly, conventional hot-melt extrusion was used for the production of pellets from a mixture of cornflour enriched with insect protein (Alam et al. 2019). Various opportunities exist with the fabrication of novel cereal-based 3D constructs with insect proteins and the process factors that affect the printability were discussed by Caporizzi et al. (2019). The effect of the different levels of insect enrichment on the quality and nutritional properties of 3D printed snacks was reported. In this study, the ground larvae of yellow mealworm (*Tenebrio molitor*) of 20 g/100 g were added into the wheat flour and 3D printed into the desired shape (Severini et al. 2018). The printed constructs were post-processed by baking at 200 °C for 22 minutes that yielded a 3D snack with a maximum desirability value of 0.693. Results obtained from these studies imply that 3D printing can be used for the innovative fabrication of foods from a novel protein source of insects. However, the associated risks of using alternative ingredients should not be ignored as not all insects and seaweeds are edible (Baiano 2020). Hence, the procedures followed for the fabrication of 3D constructs from alternative ingredients must be standardized and regulated to avoid the risk of toxin intake from uncommon food sources.

Another class of uncommon food ingredients is the food processing waste from the main processing line. The applicability of 3D printing in the conversion of 'waste into wealth' has been realized in recent days. More recently, functional cookies were 3D printed through the valorization of grape pomace (Jagadiswaran et al. 2021). The food processing by-product grape pomace was collected, processed, and dried into powder for the efficient replacement of flour in conventional cookies. Also, the broken wheat was considered in the study as an effective means of the utilization of milling fractions. The printing process was optimized for printing a mixture with different levels of grape pomace powder (0, 4, 6, and 8%) in broken wheat flour (Figure 11.10). Results showed that the optimized conditions of 6% grape pomace incorporated 3D construct was post-processed by baking at 130 °C for about 12 minutes. While the results of the sensory assessment revealed that 3D printed cookies of 4 and 6% formulations were sensorially acceptable than 8% grape pomace cookies as it tasted to be slightly bitter (Figure 11.11). Thus, the developed cookies remain as a rich source of essential proteins, dietary fibre, and functional bioactive compounds that provides valuable insights for future studies on the conversion of waste into a value-added product.

A 3D printed texture modified air-fried yam snack was prepared by utilizing the potato processing by-product (Feng et al. 2020). Although this study focused on the optimization of the printability of material supply and post-processing characteristics, the study provides a conceptual idea of blending uncommon ingredients from the waste streams of the main processing line that could be converted into valuable 3D printed customized food. In a similar

Nozzle size (mm)	Print speed (mm/min)	Control			4% Formulation			6% Formulation			8% Formulation		
		Motor speed (rpm)			Motor speed (rpm)			Motor speed (rpm)			Motor speed (rpm)		
		360	480	600	360	480	600	360	480	600	360	480	600
0.84	400	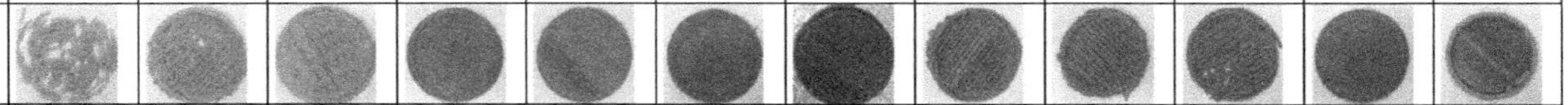											
	Observation	Under-extrusion		Slight under-extrusion	Desired extrusion			Over-extrusion	Slight under-extrusion	Under-extrusion	Under-extrusion	Desired extrusion	
	600	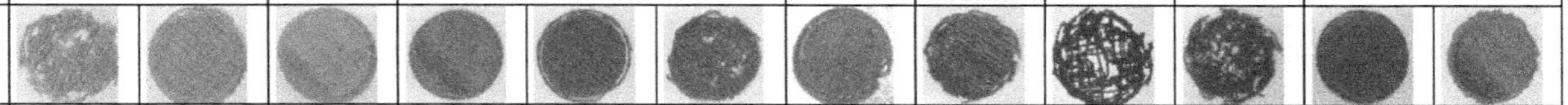											
	Observation	Under-extrusion	Slight under-extrusion	Desired extrusion	Desired extrusion	Slight under-extrusion	Under-extrusion	Over-extrusion	Under-extrusion		Under-extrusion	Desired extrusion	Slight over-extrusion
	800												
	Observation	Under-extrusion		Desired extrusion	Slight under-extrusion	Under-extrusion		Under-extrusion			Under-extrusion		Slight under-extrusion
1.28	400	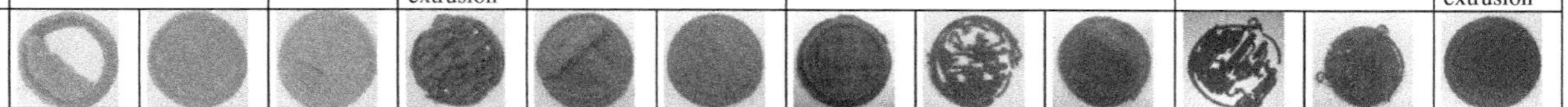											
	Observation	Over-extrusion		Desired extrusion	Under-extrusion	Over-extrusion	Desired extrusion	Slight over-extrusion	Under-extrusion	Desired extrusion	Under-extrusion		Desired extrusion
	600	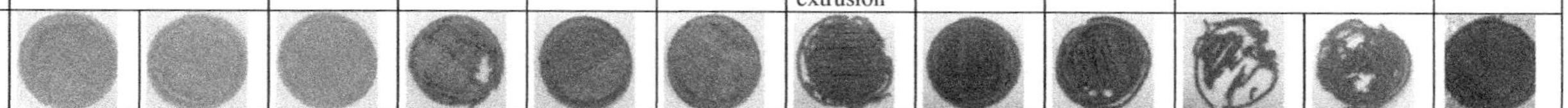											
	Observation	Desired extrusion	Slight over-extrusion		Desired extrusion	Under-extrusion	Desired extrusion	Under-extrusion	Slight over-extrusion	Under-extrusion	Under-extrusion		Slight over-extrusion
	800	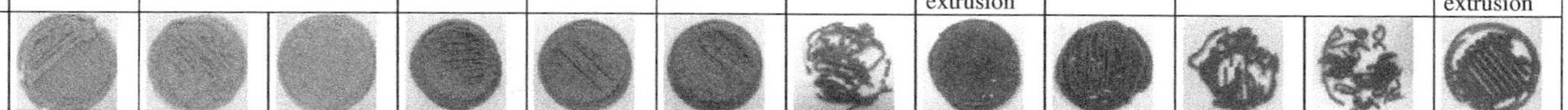											
	Observation	Under-extrusion		Desired extrusion	Under-extrusion	Slight under-extrusion		Under-extrusion			Under-extrusion		

Figure 11.10 Optimization of 3D printing of grape pomace incorporated broken wheat-based material supply. *Source:* From Jagadiswaran et al. (2021), Table 02 / CC BY 4.0 / With permission of Elsevier. DOI- https://doi.org/10.1016/j.fufo.2021.100036.

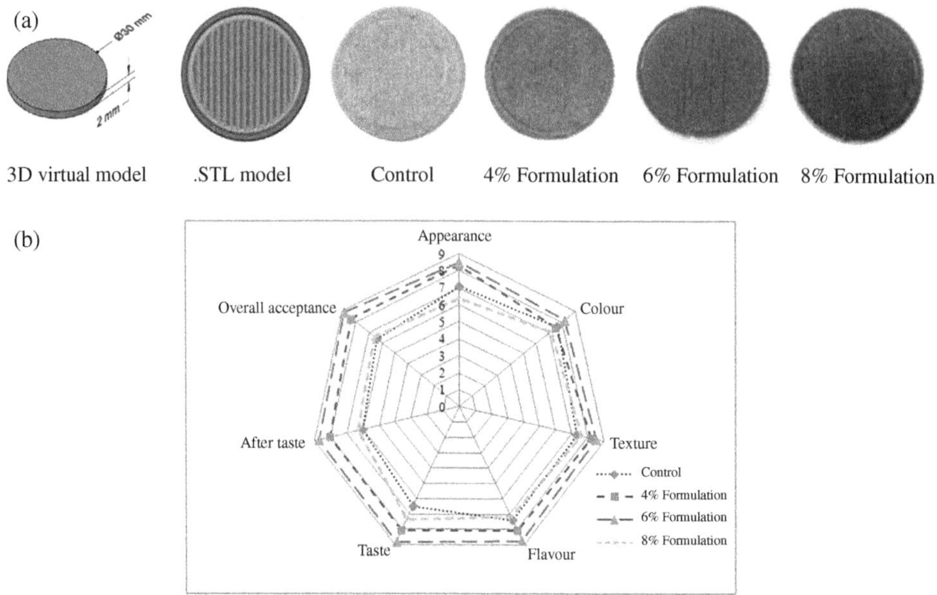

Figure 11.11 3D printing of grape pomace functional cookies. (a) 3D printed functional cookies with different levels of grape pomace. *Source:* From Jagadiswaran et al. (2021), Figure 05 / CC BY 4.0 / With permission of Elsevier. DOI-https://doi.org/10.1016/j.fufo.2021.100036 and (b) Sensory analysis of the developed 3D printed cookies.

approach, a study on 3D printing of chicken nuggets emphasized the feasibilities and opportunities in the fabrication of value-added 3D printed meat products from meat off-cuts (Wilson et al. 2020, 2021). More importantly, these ingredients serve as a good source of nutrients that could be possibly reused and utilized as human food otherwise it is less valued as animal feed and fertilizers. These approaches significantly reduce the percentage of food waste that is being dumped into landfills. As a revolutionary technique, food printing aids in the development of innovative food products from uncommon and underutilized materials in such a way of customizing the foods based on the individual likeness (Derossi et al. 2020b). However, consumer perception towards new foods in terms of the acceptance of alternatives foods, taste preferences, neophobia, customs, and social norms must be addressed for exploring the ample scope that exists with the development of sustainable 3D printed foods (Possidónio et al. 2020).

11.2.6 Food Packaging Designs

In a similar approach of using 3D printing as a communication tool, the 3D printed messages are more closely correlated with the emotions and moods of people of all ages. With the ability to customize the text, logos, and designs, 3D printing can be used as a communication tool in expressing greetings and messages in a personalized manner (Khot et al. 2015). Thus, 3D printing can be combined with the intelligent and active packaging concepts for the development of 3D printed self-communicable packages ensuring food safety

Printing speed (mm/min)	Printing rate (g/min)	3D printed MRH + GG	Observations	Nozzle height, h_0 (mm)
1200	0.624 ± 0.002	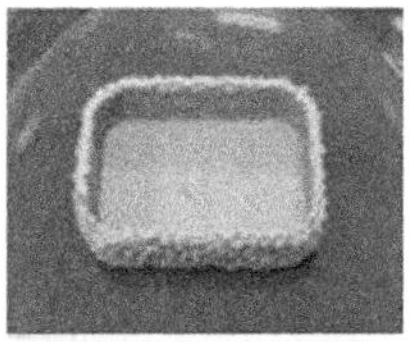	Over extrusion with curved unstable layers	0.262
1500	0.528 ± 0.005	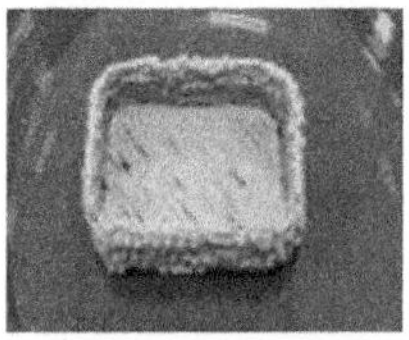	Over extrusion with curved unstable layers	0.165
1800	0.535 ± 0.002		Desired extrusion with curved unstable layers	0.147
2100	0.678 ± 0.003	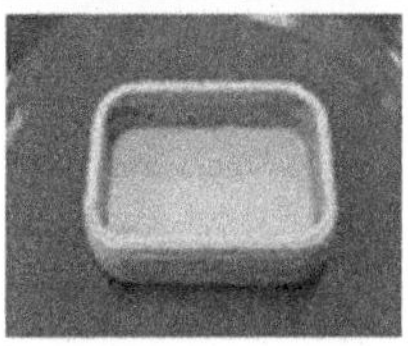	Desired extrusion with stable layers	0.161
2400	0.596 ± 0.002		Desired extrusion with curved unstable layers	0.126
2700	0.803 ± 0.005	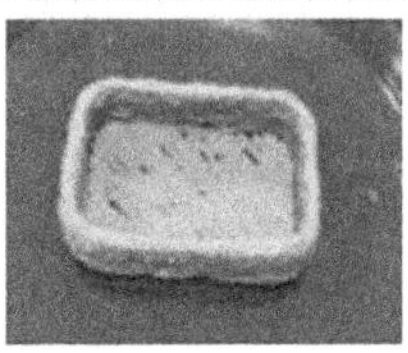	Slightly under extrusion with discontinuous layers	0.141

Figure 11.12 Prototype of 3D printed food casing developed from agricultural biomass. *Source:* From Nida et al. (2020), Table 03 [p. 07] / With permission of Springer Nature. DOI-https://doi.org/10.1007/s12649-020-01000-w

and extended shelf-life (Wu et al. 2018). The depletion of renewable resources and unpredictable climate change provides a warning alarm to transform conventional processing into a sustainable approach. This urges the need for reducing the usage of non-renewable petroleum-based plastics (Nida et al. 2020). Taking this into account, researchers are focusing on the development of degradable environmentally safe technologies. 3D printing is not only used for the printing of foods but also for the design and construction of food packages.

3D printing is well known for the construction of 3D structures from plastics, metals, and ceramics. Moreover, the applicability of 3D printing in the development of food packages is more interesting in delivering the package with customized colour, text, logos, and designs (Nachal et al. 2019). In this way, consumers would have a choice of selecting their preferred initials and photographs on the package that serves as a customized gift. Focusing on sustainability, 3D printing can be used as a promising tool in the fabrication and designing of customized biodegradable food packages from agricultural biomass. As a proof-of-concept, a 3D printed food casing was developed from rice husk (Figure 11.12) (Nida et al. 2020). Results showed that rice husk can be used as a promising ingredient for extrusion-based 3D printing with appropriate pre-processing for the development of biodegradable eco-friendly food packages. However, research on aspects of mechanical strength, structural integrity, compatibility with food, impact of handling and degradability studies must be carried out for exploring the vast benefits of 3D printing of food packages. This could even make the concept of 'edible food and edible plate' possible using 3D printing implying the fabrication of both food as well as package based on 3D printing approach that allows us to consume the food along with the package with no remains left as waste.

11.3 Future Outlook of 3D Food Printing

11.3.1 Healthy Dietary Practice

With the advent of industrialization, people are becoming more workaholics that severely impacts the general well-being and leads to lifestyle disorders such as obesity, diabetes, cardiovascular disease, hypertension, and cancer (Pinheiro et al. 2017). The only way to address this rising concern is to implement dietary changes other than medical supplementation. For which 3D printing is an opt technology that customizes the food based on the specific needs of the individuals. Thus, 3D food printing is capable of delivering foods with reduced-fat or salt, or sugar (Kistler et al. 2021). Food sensory and perception plays a significant role in cutting down the intake of fat, sugar, and salt. 3D printing allows for the localized use of the food ingredients between the layers without mixing them to the whole material supply. This approach works based on the sudden burst of sensation of salt/ sugar/ fat with the first bite of the 3D printed food that subsequently stays for a while than the actual period as oral processing continues (Prakash et al. 2019). Some of the conventional processing approaches such as size reduction, structural modification, and encapsulation are used for modifying salt perception and reduced intake of salt (Vinitha et al. 2021). In cohort to these approaches, 3D printing can be synergistically combined with these techniques that would result in altering the sensory perception by tailoring the reduced intake of salt/ sugar/ fat than actual amount in current practice. An illustration for the entrapment of the localized food ingredients depends on playing around the structural modification and solvation chemistry of food ingredients by adapting specific shapes based on its dissolving characteristics. By taking the advantage of structural morphing, the basic cuboid salt crystals were transformed into a pyramid shape that resulted in a greater rush of the salty flavour due to greater surface area with less mass (Arumugam 2011). When this pyramid-shaped salt is used as a localized ingredient in 3D printing between the layers that would further reduce the amount of salt intake. Appropriate material formulation and pre-processing of material supply would promisingly result in 3D constructs with precise control over the sensory perception of food ingredients such as salt, sugar, and fat. Also, 3D printing allows the use of seaweed as food ingredient that is considered as a potential salt replacer (Liu et al. 2019). The unpleasant aroma and flavour of the seaweeds could be effectively masked using 3D printing by incorporating seaweeds in inner layers of the 3D printed constructs that also avoid oxidation damage.

Lipids are one of the major food constituents that impart flavour and enhances the taste of the food. However, lipids are more prone and susceptible to oxidation which significantly deteriorates the sensory properties, nutritional profile, and reduces the shelf-life of the food products. The additive manufacturing of foods not only allows for the cutting down of fat but also prevents oxidation (Portanguen et al. 2019). Since the localized use of ingredients allows fat to be hidden between the composite layers that protect the 3D printed food from oxidation by preventing the direct exposure of fat to light and air. Thus, 3D printing opens up endless opportunities for the functional delivery of food ingredients that are more susceptible to oxidation. Dick et al. (2019a,b) reported a study on the fabrication of composite layer 3D printed beef with alternate layers of lard fat. This study aimed at replacing the beef fat with lard in developing a composite layer of lean meat and lard as alternate layers. Thus, 3D printing can be used for localized use of desired ingredients in the fabrication of customized foods. Traditional processing of sweets is more popular in the Middle East and Asian countries. The conventional processing method is time-consuming, energy-intensive, and laborious. Due to which the sweets prepared in one set of batches of production do not resemble the exact shape, size, and taste of the previous batch. Sugar being the major ingredient for the preparation of sweet recipes, 3D printing can be conveniently used for the fabrication of sweets as sugar is a unique material suitable for 3D printing (Prakash et al. 2019). By analyzing the melting and crystalline property of different sources of sugar, it is possible to fabricate attractive fascinating colourful 3D sugar cubes with tailored solvation characteristics. Hence, the proper optimization of the particle size, temperature, printing variables would result in texture modified, mechanically more stable, uniform shaped 3D printed sweet delicacies. Also, 3D printing provides ample scope of opportunities for utilization of different sources of sugars such as fruits, vegetables, and crops with incorporation of the functional ingredients that essentially tailors the sweet perception with reduced intake of sugars. In a similar approach, the proportions of fat and sugar content of dairy products can be customized through 3D printing. Thus, 3D printing can be used as nutrient monitoring tool that satisfies the individual needs without compromising the nutritional and sensorial attributes. However, more studies on the material formulation, structural designing, and pre-processing are required to explore the above-mentioned applications of 3D printing for a healthy dietary practice.

11.3.2 Complementing Existing Food Processing Practices

The delivery of sustainable nutrition and appealing food is the need of the hour to feed the ever-growing human population. In view of sustainability, food practices that have a low environmental impact are most required in the food industry that satisfies global hunger and preserves nutritional security. Due to higher CO_2 emissions, researchers are focusing on plant-based diets (Ramachandraiah 2021). It has been reported that carbon footprints in the rearing of animals for meat production would be higher for beef than pork and chicken. Recent research trends are being focused on the development of plant-based foods that mimics the nutritional profile and texture of meat products. Plant-based meat analogues are becoming more popular nowadays which resembles the texture of meat protein such as chicken, pork, and beef. The common vegetable proteins that are used for the production of meat analogues include soy proteins (tofu, tempeh), gluten proteins, pea proteins, and mycoproteins (Sha and Xiong 2020). Studies were reported on the extraction of carbohydrates, proteins, and nutrients from alternative food ingredients such as algae, fungi, and insects to 3D print the chicken and steak with less energy and labour (Smetana et al. 2015). Thus, 3D printing is considered to be sustainable technology in terms of cleaner productional practices, digital supply chains, resource utilization, less wastage, CO_2 emissions, and overall product life cycles that complement the existing food processing approaches.

The insufficient intake of essential macro and micronutrients leads to nutritional deficiencies. Various conventional approaches of food supplementation and fortification are in practice. Although these conventional approaches of nutritional supplements help overcome nutritional deficiency, improper absorption, and bioavailability pattern would result in a nutrient loss (Gharibzahedi and Jafari 2017). This issue could be addressed through the targeted delivery of nutrients by adopting the techniques such as encapsulation and micronization. The integration of these approaches with 3D printing would result in a highly efficient way of delivering the micronutrients and bioactive compounds (Nachal et al. 2019). This area of research is still under progress that requires an improvement in the existing designs of 3D printers and a deeper understanding of synergistic interactions of food vehicle and bioactive compounds and their effects on 3D printing. As an initial attempt, the effect of particle size distribution on the carotenoids release of carrot pulp and its printing characteristics was studied (Feng et al. 2021b). When speaking about personalized nutrition through 3D printing, the bioavailability of the functional targeted compound is the most significant aspect in addition to the physiochemical properties, rheological properties, mechanical properties, and printing parameters. The material formulation with added potato starch and xanthan gum revealed the elastic dominance of the material supply that aids in printability. Considering the particle size, the printability, rheological properties, and release of carotenoids were found to be higher with an increased particle size of the carrot pulp. Thus, this study demonstrated the correlation among the physiochemical properties, printability, and the release pattern of bioactive compounds. In this cohort, more 3D printing studies are required in analyzing the fate of the incorporated bioactive compounds and its bioavailability.

With advancements in biotechnology, it is possible to deliver nutrients based on the gene expression of the individuals that are referred to as 'nutriepigenetics'. In line with this, 'metagenomics' refers to the biochemical interactions among the diet, gut microbiota, and the associated metabolism. These approaches comes under a broad category of precision nutrition that implies a unique targeted delivery of the nutrients to the group of individuals (Kumar et al. 2020). Recently, precision nutrition has been emerging as a revolutionary approach in alleviating nutritional deficiency. In contrast to the conventional fortification approaches, the major strength of precision nutrition is addressing the nutritional status at the individual level based on their health status. Precision nutrition is differing from personalized nutrition where the people are stratified into different subgroups based on the biomarkers and therapeutic efficacy (Zeisel 2020). In a broader sense, 3D printing also requires a database of the nutritional dietary pattern of the people to address the nutritional deficiency. Thus, 3D printing is potential technology that offers both precision nutrition as well as personalized nutrition. Hence the convergence of 3D food printing with omics technologies could built a robust and sustainable technology that paves a way for precision nutrition 4.0 (Ozdemir and Kolker 2016). Thus, the advanced biotechnological approach in combination with 3D printing seems to be the futuristic technology that governs the nutraceutical and functional food market in near future.

11.3.3 Kitchens with Food 3D Printers?

3D food printing is well known for its simplicity in the automation of the process and higher product quality highlighting lower resistance to enter into the food market (Godoi et al. 2019). As a consumer-friendly technology, 3D printers could be used as a domestic kitchen appliance for the creation of fascinating foods in households. With the use of open-source software, the users can customize the printing recipes based on the fresh blending of food ingredients. Additionally, the 3D printer is also referred to as a 'digital cooker' that could combine both printing and cooking of foods (Sun et al. 2018). More research needs are required in the design aspects integrating post-processing with the printing process without deterioration of nutritional and sensory attributes of foods. The implementation of 3D printing technology in small-scale food production unit like restaurants, cafes, and bakeries would replace the traditional artistic surface designing with digitalized process (Lipton et al. 2015). This significantly reduces the labour requirement and decreases the overall productional cost. However, the successful establishment of 3D printing at the small-scale level requires a multi-head 3D printing unit with a large reservoir for the continuous supply of materials (Mantihal et al. 2020). 3D food printing reduces the complexities of the food supply chain associated with low inventory risk and process efficiency (Ben-Ner and Siemsen 2017). A study on the feasibility of the supply chain strategy on 3D printing of chocolates forecasted that food printing was a manufacturer and retailer dominant business model that has inheritable profitability with fewer impacts on the environment (Jia et al. 2016). Although 3D food printing has a vast potentiality, there is limited information on the socioeconomic aspects. Further, the technology is yet to be streamlined for food safety and quality regulations, usage of the online designs, authenticity, piracy, and ethical concerns to be adapted for mass production of customized foods at industrial level (Nachal et al. 2019).

The intrusion of robotic systems in food manufacturing leads to the development of kitchen 3D printers. These systems are equipped with complex sensors, actuators, microprocessors, controllers, and processing software. The robotic systems are designed to interact and repeat the motions of cooking chefs for producing the traditional cuisines in a novel approach using 3D printing. With the current advanced technology, the available robot chefs are highly expensive that cost several million dollars. Hence, food scientists, gastronomists, and culinary professionals are working towards the development of affordable 3D printers. An example of a robotic chef is the PR2 Robot developed by the Technical University of Munich. It is a custom-built robotic assembly designed to interact with the kitchen appliances such as refrigerator and griddle to produce a pancake (Kunze et al. 2012). Another machine designed by the researcher of MIT's CSAIL is the Bakebot PR2 robot that has been programmed to gather and mix the ingredients for cookies followed by placing the mixture into the oven for baking (Bollini et al. 2011). Although both these robotic systems complete the programmed task flawlessly, it takes a much longer time in hours which could otherwise get completed in minutes by humans. Commercial players are being attempted to develop a viable robotic chef. An example of this type of robotic system is developed by Moley Robotics system which is the first robotic multifunctional kitchen-based cooking platform (Bogue 2016). These systems are well furnished and capable of replaying a prerecorded motion of human chefs in a precise manner under a controlled environment. Although it looks simple and easy, the reproduction of inherent variability associated with each recipe in a small cooking batch is difficult. Closer to this approach is the development of tailor-made machines that are capable of preparing each type of dish. This approach has been utilized by Momentum Machines to produce burgers where the designed machine is capable of automatically slices veggies and grinds meat thereby producing a customized burger based on the demand in marketplace such as shops and retail units (Hughes 2016). However, the main technical challenge associated with this strategy is instead of having a single machine that could produce 100 items, there would be 100 machines to produce each of the preferred dishes in a perfect manner. This is economically infeasible in the longer run, hence more research needs are required in designing food machines based on 3D printing approaches that could use a wide diverse range of food ingredients efficiently. The 3D printing-based systems that are available in market are the kitchen-ready 3D printers developed by 3D Systems specifically to print chocolates and sugars; and multi-food 3D printer Foodini developed by Natural Machines. Some of the other 3D food printers available in the market are Chefjet and Cocojet (Lipton 2017). Another global hub for experimentation of culinary skills is Robots in Gastronomy that focuses on research and innovations in interactions of technology with gastronomy. Another 3D printing firm is XYZ Printing whose printers are capable to print dough-based foods for producing 3D printed cookies, cakes, and pastries (Sher and Tuto 2015). With these advancements, more startups and research groups are currently working on the development of customized food 3D printers with multifunctionality to be adopted for a diverse range of foods.

11.4 Conclusion

The present chapter on applications of 3D printing summarizes the various opportunities with the practical implementation of 3D printing technology. In addition to material formulation and printability, the end-use and the targeted application of 3D printed foods determines the growth of this flourishing technology in the food sector. 3D printing as a single standalone technology has the capability to revolutionize the food industry that offers several socio-economic benefits. Various potential applications of 3D printing are food customization, personalized nutrition, food packaging, communication, and teaching aid. With more specific needs, 3D printing is being used for the fabrication of armed and space foods with unique functionalities. Moreover, as a sustainable approach, 3D printing has been forecasted to be futuristic technology that counteracts the negative effects of climate change, the use of petroleum-based products, and CO_2 emissions. By utilizing the less valued food by-products, 3D printing could convert them into value-added 3D printed food. Thus, 3D printing is an innovative technology that utilizes the uncommon, unexplored, and under-utilized food sources thereby provide foods in customized shape, colour, taste, flavour, and texture without compromising the nutritional properties. With advancements in technology, 3D printing could be used as a potential tool for the fabrication of personalized foods with reduced intake of salt, sugar, and fat. Considering the physiological needs and metabolic rates, 3D printing tailors the individual dietary requirements based on age, gender, physical activity, special needs such as pregnant woman, athletic person, and people suffering from dysphagia. In conjunction with advanced biotechnological approaches, 3D printing can be used for the delivery of targeted micronutrients and bioactive compounds that seems to be far more efficient than conventional techniques of food supplementation. Along with precision nutrition, 3D printing could evolve as a more robust sustenance technique that aids in feeding global hunger and alleviating the hidden hunger 'malnourishment'. Thus, 3D printing being a disruptive technology plays a significant role in food customization and personalization from domestic to industrial levels.

References

Alam, M.R., Scampicchio, M., Angeli, S., and Ferrentino, G. (2019). Effect of hot melt extrusion on physical and functional properties of insect based extruded products. *Journal of Food Engineering* **259**: 44–51.

Anukiruthika, T., Moses, J.A., and Anandharamakrishnan, C. (2020). 3D printing of egg yolk and white with rice flour blends. *Journal of Food Engineering* **265**: 109691. https://doi.org/10.1016/j.jfoodeng.2019.109691.

Arumugam, N. (2011). Tasty ways manufacturers can reduce salt but keep the flavor. *Forbes* https://www.forbes.com/sites/nadiaarumugam/2011/07/28/tasty-ways-manufacturers-can-reduce-salt-but-keep-the-flavor/?sh=660aa92a4602 (

Augustin, M.A., Riley, M., Stockmann, R. et al. (2016). Role of food processing in food and nutrition security. *Trends in Food Science & Technology* **56**: 115–125.

Baiano, A. (2020). 3D printed foods: a comprehensive review on technologies, nutritional value, safety, consumer attitude, regulatory framework, and economic and sustainability issues. *Food Reviews International*: 1–31.

Ben-Ner, A. and Siemsen, E. (2017). Decentralization and localization of production: the organizational and economic consequences of additive manufacturing (3D printing). *California Management Review* **59** (2): 5–23.

Bogue, R. (2016). Flexible and soft robotic grippers: the key to new markets? *Industrial Robot* **43** (3): 258–263. https://doi.org/10.1108/IR-01-2016-0027.

Bollini, M., Barry, J., and Rus, D. (2011). Bakebot: baking cookies with the PR2. The PR2 Workshop: Results, Challenges and Lessons Learned in Advancing Robots with a Common Platform, IROS.

Caporizzi, R., Derossi, A., and Severini, C. (2019). Cereal-based and insect-enriched printable food: from formulation to postprocessing treatments. Status and perspectives. In:

Fundamentals of 3D Food Printing and Applications. Elsevier Inc. https://doi.org/10.1016/B978-0-12-814564-7.00004-3.

Chen, H., Xie, F., Chen, L., and Zheng, B. (2019a). Effect of rheological properties of potato, rice and corn starches on their hot-extrusion 3D printing behaviors. *Journal of Food Engineering* **244**: 150–158. https://doi.org/10.1016/j.jfoodeng.2018.09.011.

Chen, J., Mu, T., Goffin, D. et al. (2019b). Application of soy protein isolate and hydrocolloids based mixtures as promising food material in 3D food printing. *Journal of Food Engineering* https://doi.org/10.1016/j.jfoodeng.2019.03.016.

Dankar, I., Haddarah, A., Omar, F.E.L. et al. (2018). 3D printing technology: the new era for food customization and elaboration. *Trends in Food Science and Technology* **75**: 231–242. https://doi.org/10.1016/j.tifs.2018.03.018.

Davide, S. and Xavier, T. (2015). Review of 3D food printing.

Derossi, A., Caporizzi, R., Azzollini, D., and Severini, C. (2018). Application of 3D printing for customized food. A case on the development of a fruit-based snack for children. *Journal of Food Engineering* **220**: 65–75.

Derossi, A., Caporizzi, R., Paolillo, M., and Severini, C. (2020a). Programmable texture properties of cereal-based snack mediated by 3D printing technology. *Journal of Food Engineering* **289**: 110160.

Derossi, A., Husain, A., Caporizzi, R., and Severini, C. (2020b). Manufacturing personalized food for people uniqueness. An overview from traditional to emerging technologies. *Critical Reviews in Food Science and Nutrition* **60** (7): 1141–1159.

Dev, N.K., Shankar, R., and Qaiser, F.H. (2020). Industry 4.0 and circular economy: operational excellence for sustainable reverse supply chain performance. *Resources, Conservation and Recycling* **153**: 104583.

Dick, A., Bhandari, B., and Prakash, S. (2019a). 3D printing of meat. *Meat Science* **153**: 35–44. https://doi.org/10.1016/j.meatsci.2019.03.005.

Dick, A., Bhandari, B., and Prakash, S. (2019b). Post-processing feasibility of composite-layer 3D printed beef. *Meat Science* **153**: 9–18. https://doi.org/10.1016/j.meatsci.2019.02.024.

Dick, A., Bhandari, B., Dong, X., and Prakash, S. (2020). Feasibility study of hydrocolloid incorporated 3D printed pork as dysphagia food. *Food Hydrocolloids* **107**: 105940.

Dick, A., Bhandari, B., and Prakash, S. (2021). Printability and textural assessment of modified-texture cooked beef pastes for dysphagia patients. *Future Foods* **3**: 100006.

Dong, X., Huang, Y., Pan, Y. et al. (2019). Investigation of sweet potato starch as structural enhancer for 3D printing of Scomberomorus niphonius surimi. *Journal of Texture Studies*: 1–24. https://doi.org/10.1111/jtxs.12398.

Feng, C., Zhang, M., Bhandari, B., and Ye, Y. (2020). Use of potato processing by-product: effects on the 3D printing characteristics of the yam and the texture of air-fried yam snacks. *LWT* **125**: 109265.

Feng, C., Zhang, M., Bhandari, B. et al. (2021a). Improvement of 3D printing properties of rose-sodium alginate heterogeneous gel by adjusting rose material. *Journal of Food Process Engineering* **44** (1): e13583.

Feng, L., Wu, J., Song, J. et al. (2021b). Effect of particle size distribution on the carotenoids release, physicochemical properties and 3D printing characteristics of carrot pulp. *LWT* **139**: 110576.

Flint, H.J., Scott, K.P., Louis, P., and Duncan, S.H. (2012). The role of the gut microbiota in nutrition and health. *Nature Reviews Gastroenterology & Hepatology* **9** (10): 577.

Garcia-Oliveira, P., Fraga-Corral, M., Pereira, A.G. et al. (2020). Solutions for the sustainability of the food production and consumption system. *Critical Reviews in Food Science and Nutrition*: 1–17.

Gayler, T., Sas, C., and Kalnikaite, V. (2020). Material food probe: personalized 3D printed flavors for emotional communication in intimate relationships. Proceedings of the 2020 ACM Designing Interactive Systems Conference, 965–978.

Gharibzahedi, S.M.T. and Jafari, S.M. (2017). The importance of minerals in human nutrition: bioavailability, food fortification, processing effects and nanoencapsulation. *Trends in Food Science & Technology* **62**: 119–132.

Godoi, F.C., Bhandari, B.R., Prakash, S., and Zhang, M. (2019). An introduction to the principles of 3D food printing. In: *Fundamentals of 3D Food Printing and Applications* (eds. F.C. Godoi, B. R. Bhandari, S. Prakash and M. Zhang), 1–18. Elsevier.

Gosine, L., Kean, B., Parsons, C., and McSweeney, M.B. (2021). Using a 3D food printer as a teaching tool: focus groups with dietitians, teachers, and nutrition students. *Journal of Food Science Education* **20** (1): 18–25.

Guo, C., Zhang, M., and Devahastin, S. (2021). Color/aroma changes of 3D-Printed buckwheat dough with yellow flesh peach as triggered by microwave heating of gelatin-gum Arabic complex coacervates. *Food Hydrocolloids* **112**: 106358.

Hamilton, C.A., Alici, G., and Panhuis, M. (2018). 3D printing vegemite and marmite: redefining "breadboards.". *Journal of Food Engineering* **220**: 83–88.

Hughes, B. (2016). Iron chef: San Francisco getting world's first robot-powered burger bar. In: *International Business Times*. https://www.ibtimes.co.uk/worlds-first-robot-burger-restaurant-open-san-francisco-momentum-machines-1568470 (

Jagadiswaran, B., Alagarasan, V., Palanivelu, P. et al. (2021). Valorization of food industry waste and by-products using 3D printing: a study on the development of value-added functional cookies. *Future Foods*: 100036.

Jagtap, S., Bader, F., Garcia-Garcia, G. et al. (2021). Food logistics 4.0: opportunities and challenges. *Logistics* **5** (1): 2.

Jennifer, K.P. (2014). US army looks to 3D print food for soldiers.

Jia, F., Wang, X., Mustafee, N., and Hao, L. (2016). Investigating the feasibility of supply chain-centric business models in 3D chocolate printing: a simulation study. *Technological Forecasting and Social Change* **102**: 202–213.

Johnson, M., Barlow, R., and Ghuman, P. (2021). Psychological mechanisms underlying design thinking's impact on gustatory perception: implications for food experiences and well-being. In: *Design Thinking for Food Well-Being* (ed. W. Batat), 83–98. Springer.

Keerthana, K., Anukiruthika, T., Moses, J.A., and Anandharamakrishnan, C. (2020). Development of fiber-enriched 3D printed snacks from alternative foods: a study on button mushroom. *Journal of Food Engineering* https://doi.org/10.1016/j.jfoodeng.2020.110116.

Khot, R.A., Pennings, R., and Mueller, F. (2015). EdiPulse: supporting physical activity with chocolate printed messages. Proceedings of the 33rd Annual ACM Conference Extended Abstracts on Human Factors in Computing Systems, 1391–1396.

Kim, H.W., Lee, J.H., Park, S.M. et al. (2018). Effect of hydrocolloids on rheological properties and printability of vegetable inks for 3D food Printing. *Journal of Food Science* **83** (12): 2923–2932.

Kistler, T., Pridal, A., Bourcet, C., and Denkel, C. (2021). Modulation of sweetness perception in confectionary applications. *Food Quality and Preference* **88**: 104087.

Krishnaraj, P., Anukiruthika, T., Choudhary, P. et al. (2019). 3D extrusion printing and post-processing of fibre-rich snack from indigenous composite flour. *Food and Bioprocess Technology* **12** (10): 1776–1786. https://doi.org/10.1007/s11947-019-02336-5.

Kumar, S., Anukiruthika, T., Dutta, S. et al. (2020). Iron deficiency anemia: a comprehensive review on iron absorption, bioavailability and emerging food fortification approaches. *Trends in Food Science & Technology* https://doi.org/10.1016/j.tifs.2020.02.021.

Kunze, L., Haidu, A., and Beetz, M. (2012). Making virtual pancakes-acquiring and analyzing data of everyday manipulation tasks through interactive physics-based simulations. 35th German Conference on Artificial Intelligence September 24–27, 2012 Saarbrucken, Germany, 140.

Lanaro, M., Desselle, M.R., and Woodruff, M.A. (2019). 3D printing chocolate: properties of formulations for extrusion, sintering, binding and ink jetting. In: *Fundamentals of 3D Food Printing and Applications* (eds. F.C. Godoi, B.R. Bhandari, S. Prakash and M. Zhang), 151–173. Elsevier.

Lee, J.H., Won, D.J., Kim, H.W., and Park, H.J. (2019). Effect of particle size on 3D printing performance of the food-ink system with cellular food materials. *Journal of Food Engineering* **256**: 1–8. https://doi.org/10.1016/j.jfoodeng.2019.03.014.

Lille, M., Nurmela, A., Nordlund, E. et al. (2018). Applicability of protein and fiber-rich food materials in extrusion-based 3D printing. *Journal of Food Engineering* **220**: 20–27.

Lin, C. (2015). 3D food printing: a taste of the future. *Journal of Food Science Education* **14** (3): 86–87.

Lipton, J.I. (2017). Printable food: the technology and its application in human health. In: *Current Opinion in Biotechnology*, vol. **44**, 198–201. Elsevier Ltd. https://doi.org/10.1016/j.copbio.2016.11.015.

Lipton, J., Arnold, D., Nigl, F. et al., (2010). Multi-material food printing with complex internal structure suitable for conventional post-processing. Solid Freeform Fabrication Symposium, 809–815.

Lipton, J.I., Cutler, M., Nigl, F. et al. (2015). Additive manufacturing for the food industry. *Trends in Food Science and Technology* **43** (1): 114–123. https://doi.org/10.1016/j.tifs.2015.02.004.

Liu, Z., Zhang, M., Bhandari, B., and Wang, Y. (2017). 3D printing: printing precision and application in food sector. *Trends in Food Science & Technology* **69**: 83–94.

Liu, Y., Liu, D., Wei, G. et al. (2018a). 3D printed milk protein food simulant: improving the printing performance of milk protein concentration by incorporating whey protein isolate. *Innovative Food Science and Emerging Technologies* **49**: 116–126. https://doi.org/10.1016/j.ifset.2018.07.018.

Liu, Z., Zhang, M., Bhandari, B., and Yang, C. (2018b). Impact of rheological properties of mashed potatoes on 3D printing. *Journal of Food Engineering* **220**: 76–82. https://doi.org/10.1016/j.jfoodeng.2017.04.017.

Liu, Z., Zhang, M., and Yang, C. (2018c). Dual extrusion 3D printing of mashed potatoes/strawberry juice gel. *LWT* **96**: 589–596.

Liu, J., Sun, L., Xu, W. et al. (2019). Current advances and future perspectives of 3D printing natural-derived biopolymers. *Carbohydrate Polymers* **207**: 297–316.

Liu, Z., Dick, A., Prakash, S. et al. (2020). Texture modification of 3D printed air-fried potato snack by varying its internal structure with the potential to reduce oil content. *Food and Bioprocess Technology* **13** (3): 564–576.

Lupton, D. and Turner, B. (2018). Food of the future? Consumer responses to the idea of 3D-printed meat and insect-based foods. *Food and Foodways* **26** (4): 269–289.

Maniglia, B.C., Lima, D.C., da Matta Júnior, M. et al. (2020). Dry heating treatment: a potential tool to improve the wheat starch properties for 3D food printing application. *Food Research International* **137**: 109731.

Mantihal, S., Kobun, R., and Lee, B.-B. (2020). 3D food printing of as the new way of preparing food: a review. *International Journal of Gastronomy and Food Science*: 100260.

Molitch-Hou, M. (2014). The 3D fruit printer and the raspberry that tasted like a strawberry. https://3dprintingindustry.com/news/3d-fruit-printer-raspberry-tasted-like-strawberry27713/ (accessed 1 September 2020).

Morgado, F. and Vieira, L.R. (2020). Biodiversity and biogeography of zooplankton: implications of climate change. In: *Climate Action. Encyclopedia of the UN Sustainable Development Goals* (eds. W. Leal Filho, A.M. Azul, L. Brandli, et al.), 53–65. Cham: Springer https://doi.org/10.1007/978-3-319-95885-9_119.

Nachal, N., Moses, J.A., Karthik, P., and Anandharamakrishnan, C. (2019). Applications of 3D printing in food processing. *Food Engineering Reviews* **11** (3): 123–141. https://doi.org/10.1007/s12393-019-09199-8.

Nida, S., Anukiruthika, T., Moses, J.A., and Anandharamakrishnan, C. (2020). 3D printing of grinding and milling fractions of rice husk. *Waste and Biomass Valorization*: 1–10. https://doi.org/10.1007/s12649-020-01000-w.

Vieira, M.V., Oliveira, S.M., Amado, I.R. et al. (2020). 3D printed functional cookies fortified with *Arthrospira platensis*: evaluation of its antioxidant potential and physical-chemical characterization. *Food Hydrocolloids* **107**: 105893.

Oyinloye, T.M. and Yoon, W.B. (2020). Stability of 3D printing using a mixture of pea protein and alginate: precision and application of additive layer manufacturing simulation approach for stress distribution. *Journal of Food Engineering* **288**: 110127.

Ozdemir, V. and Kolker, E. (2016). Precision nutrition 4.0: a big data and ethics foresight analysis—convergence of agrigenomics, nutrigenomics, nutriproteomics, and nutrimetabolomics. *Omics: A Journal of Integrative Biology* **20** (2): 69–75.

Pant, A., Lee, A.Y., Karyappa, R. et al. (2021). 3D food printing of fresh vegetables using food hydrocolloids for dysphagic patients. *Food Hydrocolloids* **114**: 106546.

Pinheiro, M., Ivandic, I., and Razzouk, D. (2017). The economic impact of mental disorders and mental health problems in the workplace. In: *Mental Health Economics* (ed. D. Razzouk), 415–430. Cham: Springer https://doi.org/10.1007/978-3-319-55266-8_28.

Pinna, C., Ramundo, L., Sisca, F.G. et al. (2016). Additive Manufacturing applications within Food industry: an actual overview and future opportunities. 21st Summer School Francesco Turco 2016, 18–24.

Portanguen, S., Tournayre, P., Sicard, J. et al. (2019). Toward the design of functional foods and biobased products by 3D printing: a review. *Trends in Food Science & Technology* **86**: 188–198.

Possidónio, C., Prada, M., Graça, J., and Piazza, J. (2020). Consumer perceptions of conventional and alternative protein sources: a mixed-methods approach with meal and product framing. *Appetite* **156**: 104860.

Prakash, S., Bhandari, B.R., Godoi, F.C., and Zhang, M. (2019). Future outlook of 3D food printing. In: *Fundamentals of 3D Food Printing and Applications* (eds. F.C. Godoi, B.R. Bhandari, S. Prakash and M. Zhang), 373–381. Elsevier.

Ramachandraiah, K. (2021). Potential development of sustainable 3D-printed meat analogues: a review. *Sustainability* **13** (2): 938.

Rezek Jambrak, A., Nutrizio, M., Djekic, I. et al. (2021). Internet of nonthermal food processing technologies (IoNTP): food industry 4.0 and sustainability. *Applied Sciences* **11** (2): 686.

Severini, C., Azzollini, D., Albenzio, M., and Derossi, A. (2018). On printability, quality and nutritional properties of 3D printed cereal based snacks enriched with edible insects. *Food Research International* **106** (November 2017): 666–676. https://doi.org/10.1016/j.foodres.2018.01.034.

Sha, L. and Xiong, Y.L. (2020). Plant protein-based alternatives of reconstructed meat: science, technology, and challenges. *Trends in Food Science & Technology* **102**: 51–61.

Sher, D. and Tuto, X. (2015). Review of 3D food printing. *Temes de Disseny* **31**: 104–117.

Smetana, S., Mathys, A., Knoch, A., and Heinz, V. (2015). Meat alternatives: life cycle assessment of most known meat substitutes. *The International Journal of Life Cycle Assessment* **20** (9): 1254–1267.

Stamer, A. (2015). Insect proteins—a new source for animal feed: the use of insect larvae to recycle food waste in high-quality protein for livestock and aquaculture feeds is held back largely owing to regulatory hurdles. *EMBO Reports* **16** (6): 676–680.

Sun, J., Peng, Z., Yan, L. et al. (2015a). 3D food printing—an innovative way of mass customization in food fabrication. *International Journal of Bioprinting* **1** (1): 27–38.

Sun, J., Zhou, W., Huang, D. et al. (2015b). An overview of 3D printing technologies for food fabrication. *Food and Bioprocess Technology* **8** (8): 1605–1615.

Sun, J., Zhou, W., Huang, D., and Yan, L. (2018). 3D food printing: perspectives. In: *Polymers for Food Applications* (ed. T.J. Gutierrez), 725–755. Springer.

Tian, H., Wang, K., Lan, H. et al. (2021). Effect of hybrid gelator systems of beeswax-carrageenan-xanthan on rheological properties and printability of litchi inks for 3D food printing. *Food Hydrocolloids* **113**: 106482.

Tuorila, H. and Hartmann, C. (2020). Consumer responses to novel and unfamiliar foods. *Current Opinion in Food Science* **33**: 1–8.

Vinitha, K., Leena, M.M., Moses, J.A., and Anandharamakrishnan, C. (2021). Size-dependent enhancement in salt perception: spraying approaches to reduce sodium content in foods. *Powder Technology* **378**: 237–245.

Wilson, A., Anukiruthika, T., Moses, J.A., and Anandharamakrishnan, C. (2020). Customized shapes for chicken meat–based products: feasibility study on 3D-printed nuggets. *Food and Bioprocess Technology* **13**: 1968–1983. https://doi.org/10.1007/s11947-020-02537-3.

Wilson, A., Anukiruthika, T., Moses, J.A., and Anandharamakrishnan, C. (2021). Preparation of fiber-enriched chicken meat constructs using 3D printing. *Journal of Culinary Science & Technology*: 1–12.

Wu, S., Wang, W., Yan, K. et al. (2018). Electrochemical writing on edible polysaccharide films for intelligent food packaging. *Carbohydrate Polymers* **186**: 236–242.

Yang, F., Zhang, M., and Bhandari, B. (2017). Recent development in 3D food printing. *Critical Reviews in Food Science and Nutrition* **57** (14): 3145–3153.

Yang, F., Zhang, M., Bhandari, B., and Liu, Y. (2018). Investigation on lemon juice gel as food material for 3D printing and optimization of printing parameters. *LWT - Food Science and Technology* **87**: 67–76. https://doi.org/10.1016/j.lwt.2017.08.054.

Zeisel, S.H. (2020). Precision (personalized) nutrition: understanding metabolic heterogeneity. *Annual Review of Food Science and Technology* **11**: 71–92.

Zhang, L., Lou, Y., and Schutyser, M.A.I. (2018). 3D printing of cereal-based food structures containing probiotics. *Food Structure* **18**: 14–22.

12

Integrating Encapsulation Technique with 3D Food Printing

CHAPTER MENU

12.1 Introduction

The twenty-first century is witnessing tremendous growth in globalization and industrialization that has both pros and cons in the food sector (Jansen et al. 2020). The advent of communication technology and urbanization demands the need for convenient foods. With changing lifestyles, the consumer's food decisions are evolving towards superior foods in terms of nutritional and sensory attributes (Manstan and McSweeney 2020). On the other

3D Printing of Foods, First Edition. C. Anandharamakrishnan, Jeyan A. Moses, and T. Anukiruthika.

side, rising global population, food demand, nutritional security, environmental impact, climate change, animal grazing, and land pressure, depletion of fossil fuels, CO_2 emissions, and carbon footprints are the pressing needs of the present era (Derossi et al. 2020). This gives a red alert that draws the attention of scientists and ecologists for the development of sustainable technologies. In addition, sedentary works and business activities in most professions have increased the incidence of the lifestyle disorders such as cardiovascular disease (CVD), diabetes, cancer, chronic respiratory syndromes, hypertension, and obesity. All these scenarios could be addressed through the adoption of well-prepared dietary plan and eating practices. In a recent survey, the global nutraceutical market size was expected to grow with compound annual growth rate (CAGR) of 8.3% from USD 382.51 billion during the forecasted period of 2020–2027 (Market Analysis Report 2020). This highlights the rising market demand for the functional foods that seems to likely increase the dietary supplements to combat the lifestyle disorders. In this view, food technologists, scholars, and scientists are aiming towards the production of functional foods and nutraceuticals. Although, numerous ranges of nutraceuticals and food supplements are available as on today, the question is on 'how efficient they are?'. This rises the need for the interrogation on the efficiency of the developed food delivery system in terms of bioavailability and targeted delivery. Focusing on these aspects, encapsulation has emerged as a potential food technology in the fabrication of functional foods with improved bioavailability, better storability, controlled release, and targeted delivery (Bhushani and Anandharamakrishnan 2014). On the other hand, the desire for enjoying food and a pleasant eating experience would lead to the emergence of food printing that allows for the fabrication of customized and personalized foods (Nachal et al. 2019). Personalized nutrition is a hot topic of the present decade in the delivery of nutrients based on the physiological conditions and nutritional status of individuals. 3D printing of foods is a well-known technology for the delivery of personalized foods based on individual likeness. Several advantageous features of 3D food printing technology and its applications have been discussed in earlier chapters. With one step forward, 3D printing offers several health benefits in combination with other food processing technologies, especially encapsulation (Kumar et al. 2020).

Food encapsulation is the concept of entrapping the desired functional bioactive compounds in a biopolymeric matrix that ensures the targeted delivery as well as protects it from gastro environmental/ external degradation (Ezhilarasi et al. 2013b). Based on the encapsulated particle size, encapsulation can be of three types: macro encapsulation, microencapsulation, and nanoencapsulation. With the reduction in physical size, the surface-to-volume ratio becomes larger that enhances the bioavailability of the encapsulated bioactive compounds (Bharathi et al. 2018). Considering this, micro and nano encapsulation are the most investigated area of research in the delivery of micronutrients. Therefore, the micro and nano encapsulation on integration with 3D printing would open up a new window with an array of research innovations and opportunities in personalized nutrition making it possible to satisfy the needs of both individuals as well as specific population groups. Although both 3D printing and encapsulation are emerging novel food technologies, the integration of both these approaches is not yet explored. Upon integration with encapsulation techniques, 3D printing would offer much more benefits that gives rise to a nôvel 3D printer designs, food ingredient formulations, and functional modifications of foods. On realizing the promissory features of these combined food technologies, it is adequate to analyze the feasibilities existing with this combined technology for the delivery of 3D printed functional foods. Hence, the present chapter deals with the concepts of 3D printing and food encapsulation, common food delivery systems used for encapsulation, integration of 3D printing with encapsulation providing insights on design needs, and encapsulation methods. Thus, the present chapter throws a spotlight on the future research prospects and innovations in the delivery of personalized foods through integrated food technologies that aids in the alleviation of hidden hunger 'malnourishment' and preserves nutritional security.

12.2 Integration of 3D Printing and Encapsulation

The exploration of encapsulation in providing 3D printed foods is the need of the hour that ensures the targeted micronutrient delivery with enhanced bioavailability. In this regard, micro encapsulation in the range 10^{-6} m and nano encapsulation in the range of 10^{-9} m are promising technologies that could be combined with 3D food printing. Encapsulation is the process of entrapment of bioactive compounds (known as core material) surrounded by a biopolymeric matrix (known as wall material) (Bhushani and Anandharamakrishnan 2014). The combined concept of 3D printing and encapsulation refers to the possible combinations of coaxial extrusion of bioactive core material surrounded by the printing material supply during the 3D printing; and the addition of encapsulated functional food ingredients directly into material supply that can be taken for 3D printing. Both these approaches should consider the need for the protection of encapsulated compounds from functional degradation suitable for targeted delivery at the intestinal gut for enhanced bioavailability. In this cohort, it is essential to understand the basic principles of encapsulation as well as 3D printing. The common biopolymers such as polysaccharides, cellulose and its derivatives, starch and its derivatives, proteins, dietary fibres, lipids, plant extracts and exudates, animal extracts, and marine extracts are used as outer shell/ wall material for encapsulation (Preethi et al. 2020a). Similarly, the synthetic polymers and their blends are also used as outer shell/ wall material for encapsulation that includes collagen, fibrin, gelatin, elastin, polyvinyl acetate, polyacrylic acids, polyalkylene oxides, polyethylene glycol, polyvinyl alcohol, polyvinyl pyrrolidone, polypropylene, polystyrene, polycaprolactic acid, polylactic acid, poly glycolic acid, polybutadiene, and so on (Sionkowska 2011). However, natural biopolymers are more suitable for food applications due to their compatibility, degradability, solubility, physiochemical modifications, retention of volatiles, low viscosity at higher solid fractions, and suitability for high process conditions. Realizing the potential of natural polymers, the prospects of pectin hydrogels in food applications as wall material for the delivery of bioactive and as the material supply for 3D printing was summarized in a recent report (Ishwarya and Nisha 2021).

On the other hand, food flavors, pigments, polyphenols, colourants, preservatives, antioxidants, sweeteners, proteins, lipids, acids, bases, buffer, minerals, vitamins, and probiotics are the commonly used core materials (Anandharamakrishnan and Ishwarya 2015). The properties of the core material (solid/ liquid/ gas) determines the nature of the final product. The final resulting encapsulates can be referred to as mononuclear with single continuous core material implanted over the shell material, or polynuclear with multiple units of core material implanted over the wall matrix, or matrix-type capsule with the core that has been uniformly distributed in an ordered pattern in a wall matrix (Bharathi et al. 2018). The difference in the size of the encapsulate differentiates the micro- and nano-encapsulation from macro encapsulation. The various advantageous feature of the food encapsulation is controlled release of active compounds, targeted delivery at the intestinal gut, the improved storage stability of encapsulated ingredients, longer shelf-life as it protects the bioactive compounds from thermal and photo-oxidation, masking of unpleasant taste and smell, and enhanced bioavailability of micronutrients (Leena et al. 2020b). All these features are linked with the surface-to-volume ratio and particle size. Various top-down approaches and the challenges involved in it were summarized by Katakam et al. (2015). This review broadly described the integration of encapsulation with 3D printing techniques and provides future prospects of this emerging area for food applications. In another study, the capability of stereolithography (SLA) printing was used for the fabrication of solid dosage loaded with ascorbic acid (Karakurt et al. 2020) (Figure 12.1). The technique utilizes the chemistry of the novel photopolymer resin combined with riboflavin as the photoinitiator.

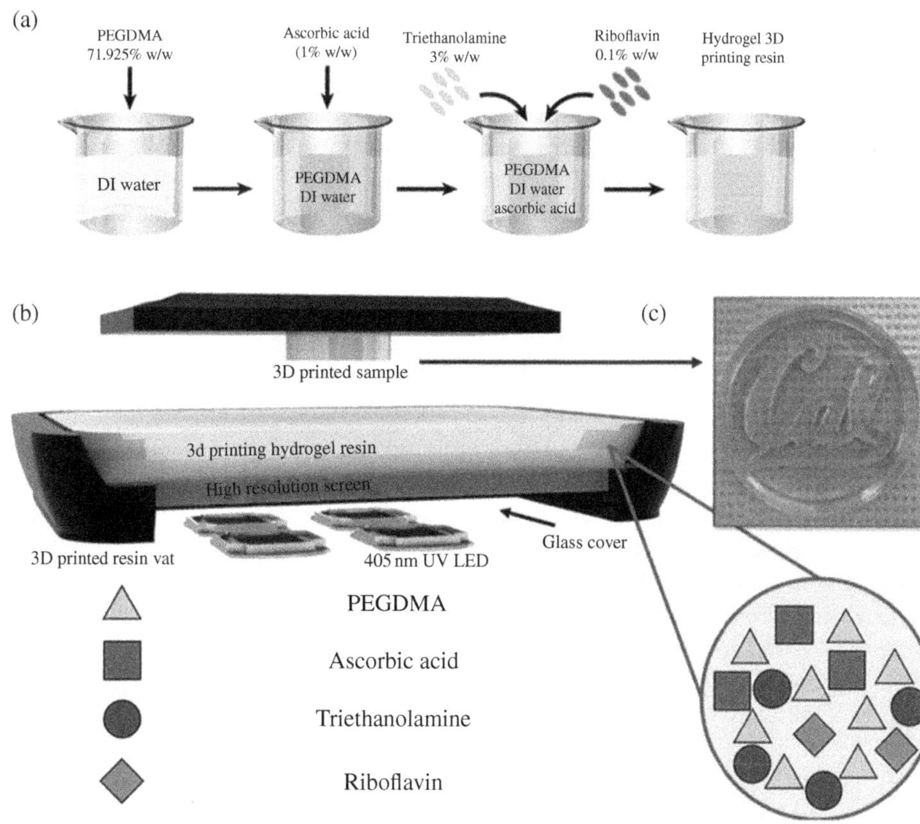

Figure 12.1 3D printing of ascorbic loaded hydrogel. (a) Method of preparation of 3D printing resin, (b) SLA-based 3D printer setup, and (c) 3D printed hydrogel showing University of California Berkeley logo. *Source:* From Karakurt et al. (2020) / With permission of Elsevier.

Thus, ascorbic acid was successfully encapsulated, and 3D printed hydrogel drugs showed sustained and controlled release of water-soluble bioactives. Thus, this study remains as a proof-of-concept emphasizing the advancements of material science and layered design fabrication that could lead to advanced drug delivery systems for formulation of personalized medicines using 3D printing. An in-depth understanding of the encapsulation would help in the functional product development upon integration with 3D printing and design innovations in 3D food printer for simultaneous encapsulation during 3D printing.

12.2.1 Encapsulation Followed by 3D Printing

The simpler approach of integration of encapsulation with 3D printing is the micro/ nano encapsulation followed by the incorporation of encapsulates into the printing mixture. This mixture is further taken for the regular 3D printing process by the extrusion principle. The success of this approach depends on the type of the delivery system as well as the process variables of the encapsulation technique employed. Various kinds of micro and nano structured delivery systems used in the preparation of feed solution are liposomes, emulsions, emulsification-solvent evaporation, solid lipid nanoparticle, nanostructured lipid carriers, nanoprecipitation, inclusion complexation coacervation, gelation, and use of supercritical techniques (Bharathi et al. 2018). The composition, polymeric concentration, and molecular weight of the polymers are the significant attributes that determine the nature of the resultant feed solution. The selection of the core and shell/ wall materials is based on their compatibility and physiochemical properties that influence the mechanical strength and storage stability of the encapsulated particles. Various types of delivery systems used for the preparation of micro and nano encapsulates as well as the principle of encapsulation techniques are discussed in the subsequent sections of this chapter. All these process steps from the preparation of the feed solution to the obtained encapsulated bioactive compounds comprise the first stage of processing. The second stage follows a well-defined sequential process involving the incorporation of the developed encapsulates with desired bioactive compounds into the printing material supply usually cereal-based dough, jam/ jellies, hydrogels, and so on. Later this material supply is taken for the usual printing process with the printing variables optimization. This kind of integrated approach is more applicable for extrusion-based printing rather than other food printing technologies (Jia et al. 2016). Advantageously, the first half of the integrated approach aids in the development of a stable micro/ nano encapsulated material in powder form that can be conveniently stored until it is used for 3D printing.

The challenge associated with this encapsulation process is the development of a well-optimized formulation of bioactive compounds (polyphenols, antioxidants, probiotics, proteins, vitamins, minerals, and enzymes) that are shelf-stable without degradation due to light, heat, and oxygen. Later, in the second stage of processing, the material response (encapsulates) for pre-processing steps of 3D printing such as mixing, agitation, and thermal treatments for gelation are most crucial that should not deteriorate the functional properties of the encapsulated bioactive compounds (Feng et al. 2018). The material supply acts as a base where the selection of it as well the post-processing method to be used after 3D printing of bioactive incorporated mixture must be selected appropriately. The printing material supplies such as dough-based materials often require high thermal post-processing treatments such as baking, hence the final bioavailability of the encapsulates must be in the acceptable range (He et al. 2019). For instance, the probiotic incorporated 3D printed foods should possess minimal cell viability in the range of 10^6–10^7 CFU g^{-1} (Zhang et al. 2018). A cereal-based 3D printed probiotic construct was developed in two different internal structures (honeycomb and concentric infill patterns) using baking as a post-processing method. Interestingly, the residual viability of the *Lactobacillus plantarum* in the 3D printed honeycomb structure was higher than that of the concentric structures at baked conditions of 145 °C for 6 minutes (viable counts > 10^6 CFU g^{-1}). This was owing to the fact of high surface-to-volume ratio of the honeycomb infill structure of the 3D printed baked foods. Although this study followed the direct addition of probiotics into material supply without encapsulation, the results provide insights for the integration of 3D printing with encapsulation. Further, this study assessed the printability of high protein flour and low protein flour with calcium caseinate as additive. More recently, the 3D printing of encapsulated probiotics (*Lactiplantibacillus plantarum*) using composite flour formulated by Krishnaraj et al. (2019) was reported (Yoha et al. 2021). This study has analyzed the 3D printing of the encapsulated probiotics by the following encapsulation techniques such as freeze-drying, spray drying, spray-freeze-drying, and refractance window drying (RWD) subjected to different post-processing methods (freeze-drying, hot air drying, microwave drying, and RWD). Results inferred that the printing process has no effects on the survival rate of the encapsulated probiotics in the printing mixture with the retention of survival rate of 98–99% after printing process. Further, the spray-freeze-dried synbiotics incorporated sample followed by freeze-drying post-processing method was observed to be the best among the examined encapsulation methods and post-processing techniques.

In a similar way, future studies can utilize fortified food materials based on encapsulation and micronization for the production of functional 3D printed foods (Kumar et al. 2020). Similarly, ready-to-eat materials such as hydrogels, jams/ jellies, hydrocolloid mixtures, mashed potatoes, fermented milk, fruit juices, yoghurt, cheese, and ice creams often require minimal post-processing and low-temperature treatments such as refrigeration and freezing. These types of ready-to-eat materials are more suitable for thermal labile bioactive compounds that can be used as an excellent substrate for the incorporation of bioactive compounds. The possibility of the incorporation of probiotics *Bifidobacterium animalis* into mashed potato was successfully 3D printed (Liu et al. 2020). The probiotic culture was directly added into mashed potato without encapsulation and taken for printing. Results showed that the smaller-sized nozzle reduced the viable cell count from 9.93 to 9.74 log CFU g^{-1}. Also, there was a significant reduction in probiotic count from 10.07 to 7.99 log CFU g^{-1} when the temperature of the heating nozzle of feed supply was kept at 55 °C for 45 minutes. Further, the study confirmed the minimal cell viability of 3D printed probiotic mashed potato when stored in refrigeration temperature at 5 °C for 12 days. Recent reports were focused on 3D printing of more complex structures with multiple constituents of hydrocolloids: vitamin D enrichment of orange concentrate with material properties (Azam et al. 2018a,b) and fabrication of 3D constructs from sodium caseinate dispersions incorporated with olive oil droplets (Figure 12.2) (Schutyser et al. 2018).

The synergistic effect among the encapsulate, printed layers, infill percentage, time, and temperature of baking was reported for 3D printed functional cookies (Oliveira et al. 2020). In this study, the encapsulating matrix was prepared from grape pomace with maltodextrin

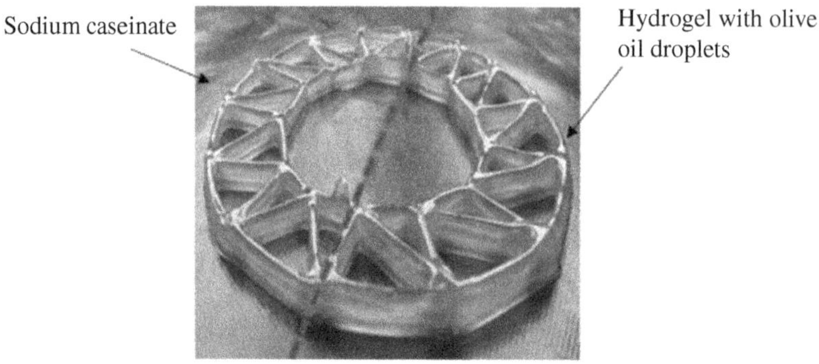

Figure 12.2 3D printed sodium caseinate dispersions with loaded olive oil (left side – sodium caseinate and right side – hydrogel with olive oil droplets). *Source:* From Schutyser et al. (2018), [p. 30] / With permission of Elsevier.

using a spray-drying approach and the designed particles were in the size of 10 and 60 μm. The cereal-based dough material supply was prepared with free extract (free pomace extract 2.5% and powdered sugar 19.5%) and encapsulated extract (encapsulated pomace extract 5% and powdered sugar 17%). The prepared dough was kept in tightly covered package and stored at lower temperature of 4 °C for 48 hours that aids in complete hydration of the added hydrocolloid xanthan gum. Results showed that printed 3D construct with four layers and 30% infill percentage retained the highest bioactivity (115%) and total phenolic content (173%) of the developed cookie when baked at 180 °C for about 10 minutes (Figure 12.3). This study confirmed the advantage of the presence of interconnected pores in 3D constructs in accelerating the baking and browning while at the same time reduced the thermal degradation of functional cookies (Oliveira et al. 2020).

In another study, three different incorporation approaches (dried biomass, freeze-dried extract, and encapsulated extract) were used for the fortification of cookies with *Arthrospira platensis* (Vieira et al. 2020). Results showed that fortified 3D printed cookies with encapsulated extract retained improved antioxidant activities, colour stability, and good mechanical properties than other formulations. 3D printing also provides a sustainable solution for the rising demand for animal protein. Interestingly, through the concept of bioprinting, 3D printing allows to print alive cells, i.e., it enables the culture of artificially grown meat tissues. Based on this concept, interestingly the 3D printing was used for the encapsulation of the alive cellular plant cells into a pectin-based printing mixture as a food model (Vancauwenberghe et al. 2019). This study successfully demonstrated the use of land-plant cells (*Valerianella locusta*) embedded onto the bio-ink low methoxy matrix (Figure 12.4). Results showed that increased pectin concentration reduced the cell viability of encapsulated lettuce cells due to its high viscosity. As a result, the encapsulated cells of the 3D printed bio-ink showed a decreased mechanical strength and structural properties. This study throws light on the applicability of 3D printing for encapsulation of cells in a bio-ink matrix; however, it needs further research on the response of cell cultures on the mechanical integrity of the 3D printed construct.

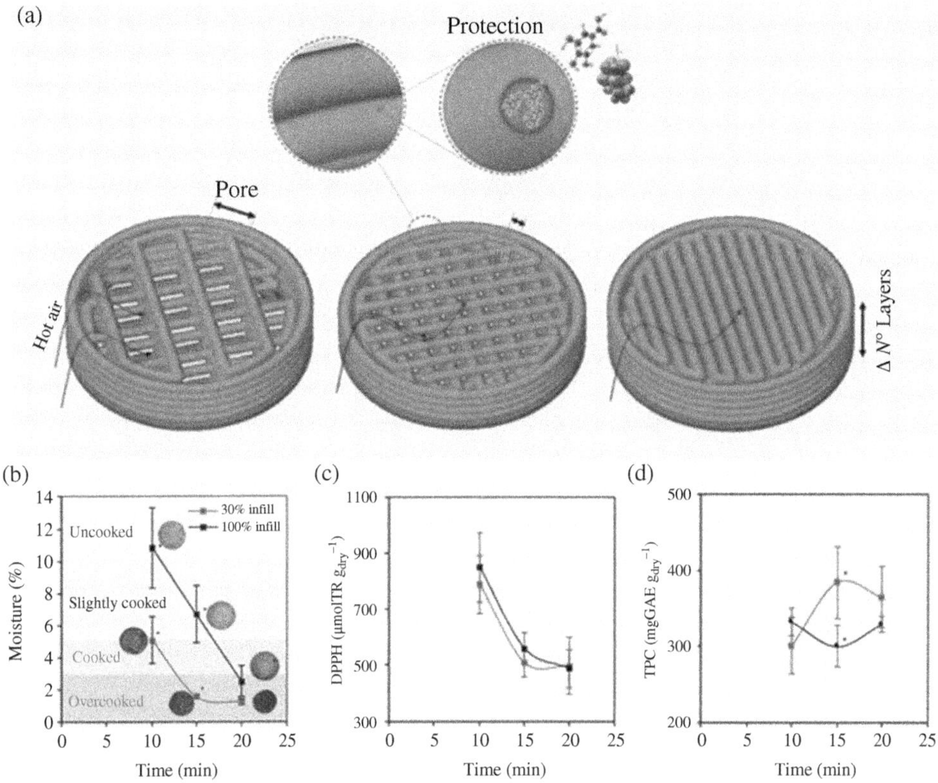

Figure 12.3 Fabrication of 3D printed baked functional cookie. (a) multiscale strategy adopted to model functional cookies enriched with grape-skin derived polyphenols, (b) evaluation of moisture content, (c) DPPH evaluation, and (d) evaluation of total phenolic content of developed cookies. *Source:* From Oliveira et al. (2020) / With permission of Elsevier.

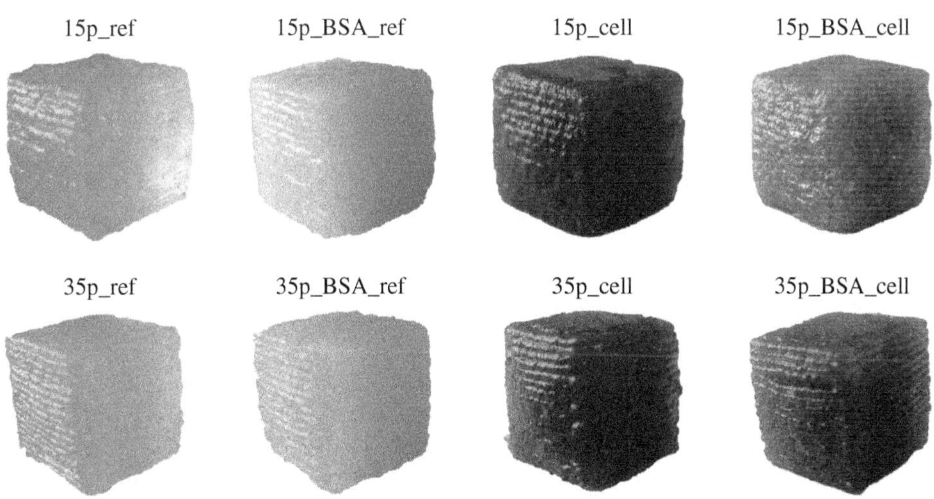

Figure 12.4 3D printed cubes of different formulations of pectin-based bio-ink. *Source:* From Vancauwenberghe et al. (2019), [p. 34] / With permission of Elsevier.

For which a thorough understanding of the cell's response towards rheological behaviour, mechanical strength, and physiochemical properties must be adequate. As a proof-of-concept, a similar approach could be adopted for exploring the material formulation through different kinds of filler materials such as meat cell cultures, plant/ animal/ microbial extracts, and synthetically developed micro/ nano encapsulates. Following a similar experimental procedure, callus-based food ink was prepared through varied composition of callus (carrot) dispersion with 4% alginate at ratio of 1 : 2, 1 : 1, and 2 : 1 (w/w) (Park et al. 2020). Results showed that the cells implanted in the hydrogels formed into clusters on incubation that exhibit unique textural characteristics of artificial plant tissues. Further, this concept provides a valuable insight for the simulation of artificial plant tissue for the development of texture-modified 3D printed foods from plant cell-laden hydrogels that are similar to the texture of the real foods. Although both the studies on lettuce cells and carrot callus had no encapsulation at micro level, the entrapment of plant cells in a food matrix made it to be referred as encapsulation at macro levels. Similar approaches with fine precision and particle size accuracy assist in the fabrication of 3D printed food matrix encapsulated with bioactive compounds.

12.2.2 Simultaneous Encapsulation and 3D Printing

The second approach of simultaneous encapsulation and 3D printing refers to the direct encapsulation of the active bioingredients during 3D printing. This could be possible through design modifications of the existing 3D food printers with coaxial extrusion of the material supplies. The core material may contain active ingredients such as probiotics, enzymes, and micronutrients while the outer ring forms the food wall matrix for encapsulation. Since 3D food printing is in its initial nascent stage, most of the studies utilize the commercial polymer fused deposition modelling (FDM) printers with slight design modifications suitable for food printing. However, the concept of simultaneous encapsulation and 3D printing could be feasible with design innovations and modifications as that of coaxial extrusion at micro and nano levels. The bioprinting of materials through the micro extrusion approach is gaining attention in recent days for the fabrication of human tissues and organs. This concept of micro extrusion can be positively applied for food applications in developing an encapsulated 3D printed food. In addition to the existing procedure of 3D printing, intense pre-processing must be required to assure the printability of micro particles with desired functionality using micro-sized printing nozzles. This area of research is not yet explored for food applications, hence more reports investigating the material characteristics and printability of micro and nano encapsulated 3D printed foods are needed. Cell encapsulation is a well-explored area in bioprinting due to its several advantages it has found in medical applications especially customized fabrication of vascular tissues and organ printing (Seoane-Viano et al. 2021). The droplet formation and its deposition have been deeply investigated for the past decade. The cell encapsulation follows different methods for entrapping the biomaterials that include T junction, fluid focusing, cross junction, and many other methods (Liu et al. 2017). In addition, different strategies are adopted for desired size droplets of bio-ink that are based on the microfluidic approach. A similar kind of encapsulation can be envisioned with food inks as well as through appropriate material formulation and optimization of the 3D printing process variables. A study was reported on direct 3D bioprinting of the vascular constructs using a blend of bio-ink mixture through a multilayer encapsulation approach (Jia et al. 2016). This study utilized a coaxial nozzle device containing two injection channels with different-sized needles. The top cross-sectional view of the setup shows a concentric pattern as it was fixed with each other using epoxy resin that enables the multilayer encapsulation of the 3D construct. This coaxial nozzle device in turn is attached with the commercial bioprinter with motor-driven *X-Y-Z* axes. The cells were mixed with polymeric ink for bioprinting of cell-laden ink. Immediately after the printing process, the printed 3D constructs were exposed to ultraviolet (UV) light at predefined time points for allowing covalent cross-linking of the material followed by the immersion of 3D construct in ethylenediamine tetraacetic acid (EDTA) solution for 5 minutes in order to remove the alginate (substrate) matrix (Figure 12.5). Thus, this study successfully demonstrated the design of a multilayer coaxial nozzle with concentric channels for continuous generation and simultaneous encapsulation of the 3D construct in a single step. This approach has a clear distinct advantage in reducing the complexity of the conventional microfabrication of scaffolds and sacrificial templating approaches. Hence, this kind of approach for material fabrication can be conveniently adopted for fabrication of functional multilayer encapsulated 3D printed foods. The food printing using dual extrusion was already reported for fabrication of mashed potato and strawberry juice gel (Liu et al.

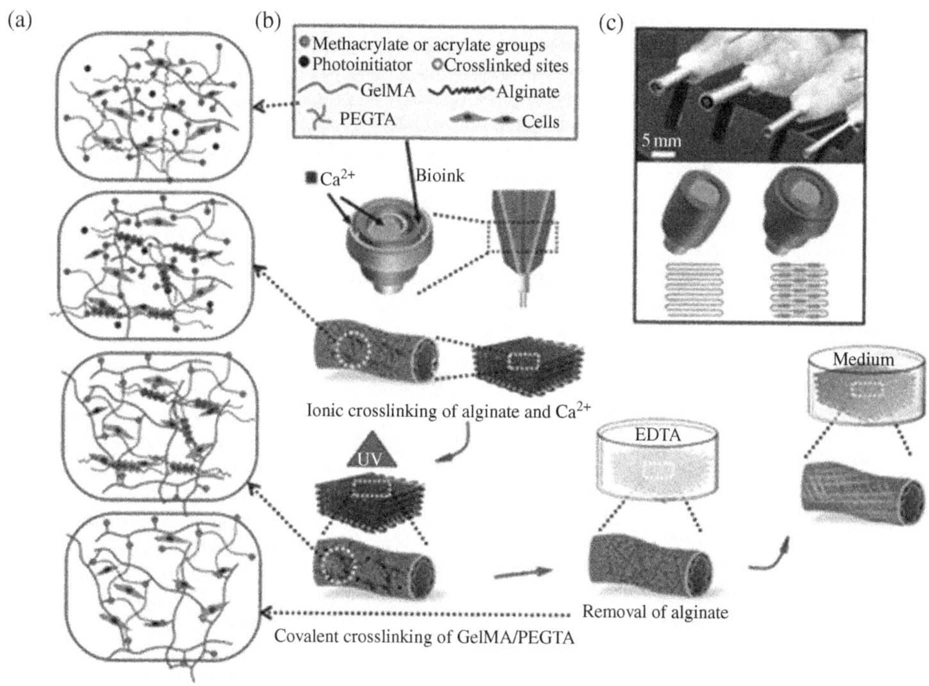

Figure 12.5 3D bioprinting of vascular construct. (a) illustration of cross-linking process of bio-ink, (b) bioprinting of perfusable hollow tubes with cell encapsulated blend of bio-ink, and (c) designed multilayer coaxial nozzle used for fabrication of vascular construct. *Source:* From Jia et al. (2016) / With permission of Elsevier.

2018); and micro-algae (*Chlorella vulgaris* and *Arthrospira platensis*) fortified 3D printed snacks (Uribe-Wandurraga et al. 2020; Vieira et al. 2020). Considering this, the integration of encapsulation of food ingredients with 3D printing as one single step would also be feasible. With the integration of 3D printing, the localized use of functional ingredients such as micronutrients, enzymes, proteins, lipids, salt, sugar, pigments, and probiotics are possible and can be encapsulated in its active form directly during 3D printing. This offers a wide range of functional and health benefits along with a pleasant eating experience and desired sensory attributes that lead to the design of novel innovative personalized 3D printed foods.

12.3 Structure Modified Delivery Systems

The delivery system refers to the material formulation or device employed for the introduction of bioactive compounds in their active functional forms for consumption with improved efficacy, controlled and targeted release of bioactive compounds (Hosseini et al. 2021). The food-based carrier systems are based on polysaccharides, proteins, and lipids. Various approaches and preparation methods adopted for targeted delivery of bioactive compounds in the drug delivery system using melt-extrusion were reported (Chauhan et al. 2020) (Figure 12.6). In the context of 3D printing, pre-processing of the material supply is a critical step for achieving the desired printability (Anukiruthika et al. 2020a). Hence, the process of pre-processing is a bit complicated when integrated with encapsulation. For a successful fabrication of functional 3D printed foods, a basic understanding of the existing conventional delivery systems used for micro and nano encapsulation is essential.

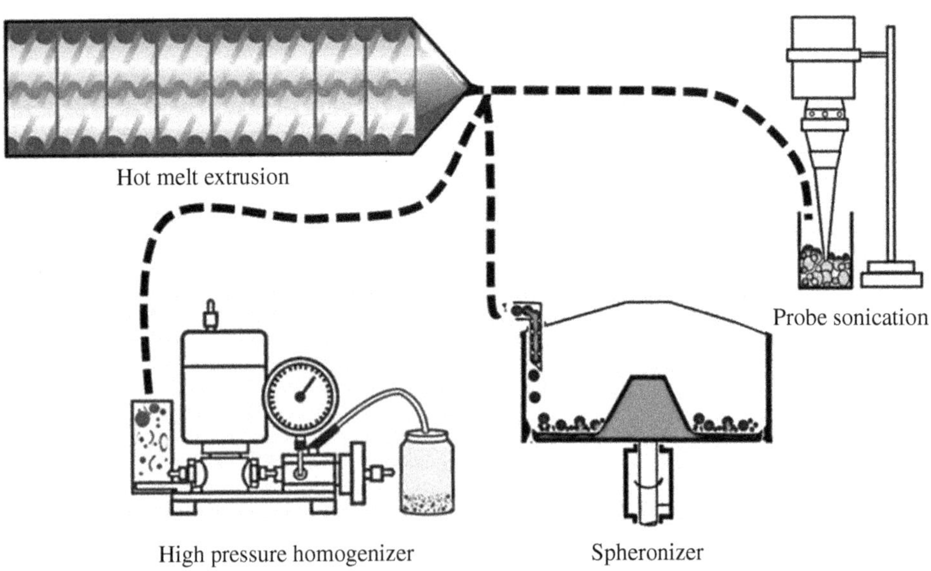

Figure 12.6 Schematic representation of post-hot-melt extrusion used for lipid-based drug delivery system. *Source:* Chauhan et al. (2020) / With permission of Elsevier.

Therefore, the present section discusses the various food delivery systems of encapsulation with further insights on the applicability of these delivery systems in the integrated approach of 3D printing and encapsulation.

12.3.1 Micro and Nano Emulsions

Emulsions are the colloidal dispersions of a mixture of two or more immiscible liquids where one of the liquids is dispersed (dispersed phase) over the other one (continuous phase). Based on the size range of the formulated liquid particles, emulsions can be either microemulsions or nanoemulsions. The reduction in the size of droplets provides stability against sedimentation and creaming (Karthik et al. 2017). Based on the relative spatial arrangement of water and oil phases, the emulsions are of two types: oil-in-water (O/W) and water-in-oil (W/O). The O/W emulsion represents dispersed oil droplets over water phase dispersion and vice versa is referred to as W/O emulsion. The micro and nano emulsions offer great potential for encapsulating the higher concentrations of lipophilic bioactive compounds such as essential fatty acids (ω3 and ω6 fatty acids), β carotene, carotenoids, and phytosterols through O/W emulsion. On the other side, hydrophilic bioactive compounds such as polyphenols can be delivered through W/O emulsion (Anandharamakrishnan 2014d). The main characteristic attribute of the delivery system is the protection of the bioactive compounds against physical, chemical, and biological degradation. Hence appropriate processing in achieving an emulsion, specific for desired end application is the most crucial step. The formation of emulsion is not a spontaneous process that requires energy inputs from external or internal sources. Apart from the applied energy sources, the final quality of the emulsion is influenced by the method of preparation, order of ingredient addition, chemical behaviour of the continuous and the dispersed phase, surfactant type, and the concentration of surfactants used. Based on the applied energy source, micro and nano emulsions can be prepared using high and low energy approaches (Kumar et al. 2019). The former type of high energy sources involves the sudden disruption of the lipid and aqueous phases into a fine droplet with the assistance of mechanical instruments such as high-pressure homogenizers, microfluidizers, and powered sonicators. While the latter type of low-energy approach was based on the phase transition effects that occurred during emulsification with changing process conditions such as temperature and material composition (Ezhilarasi et al. 2013b). Various challenges associated with the stability of the food-grade emulsions in terms of flocculation, Ostwald ripening, coalescence, creaming, phase separation, and sedimentation were detailly explained by Karthik et al. (2017). Commonly used low-energy approaches for emulsion formation are self-emulsification, phase inversion, and emulsion inversion point principles. Since emulsions are in a flowable consistency as such they cannot be used in 3D printing. Recently, food scientists had successfully developed jelly-like capsules that resemble raspberry but taste like strawberry by combining the concept of spherification and 3D printing (Molitch-Hou 2014). In a similar approach, the micro and nano emulsions could be encapsulated in the form of capsules. More research needs are required in analyzing the capsule formulation, filming properties of wall material, stability of developed 3D printed micro and nano food capsules. This would lead to the development of range of novel functionalized foods based on the combined concepts of molecular gastronomy, encapsulation, spherification, and 3D printing.

12.3.2 Lipid-Based Delivery Systems

A lipid-based delivery system is capable of entrapping materials with different solubilities which can be developed from the naturally available lipid-based food ingredients (Anandharamakrishnan 2014a). This kind of encapsulation protects the sensitive and heat-labile bioactive compounds from structural transformation and degradation thereby enhancing the encapsulation efficiency and bioavailability. The advantage of the lipid-based delivery systems is the delivery of bioactive compounds at the targeted sites through active and passive mechanisms (Mozafari 2005). Further, this approach imparts mechanical stability to aqueous soluble materials and can also utilize fat-soluble agents providing combined benefits of both the ingredients. Therefore, the formulation of lipid-based delivery systems consists of emulsifiers like lecithin, polysorbate 20, and poloxamer 188; lipids like steroids and triglycerides; and bilayer lipids like phospholipids that include mixtures of the monoglycerides, diglycerides, and triglycerides (Attama et al. 2012). The application of lipid-based drug delivery systems is gaining research potential due to the advent of precision medicine in recent days. In order to provide individualized dosages and personalized treatments, a novel approach of dropwise inkjet printing was used for amorphous and self-emulsifying drug delivery based on the precise deposition of melt-based formulations onto inert tablet substrate (Icten et al. 2017). This approach yields promising results in the development of solid oral dosages from the formulation using active pharmaceutical ingredients and the surfactant (Figure 12.7). This approach enhances the solubility of lipophilic drugs that enables the production of precise deposition with adjustable sizing and higher accuracy. Various potential opportunities of the 3D printing technologies for personalized lipid-based drug delivery systems were reviewed by Vithani et al. (2019b). In a similar manner, the bioactive food ingredients can be possibly fabricated using lipid-based delivery system with enhanced bioavailability and efficacy. The lipid-based drug delivery system

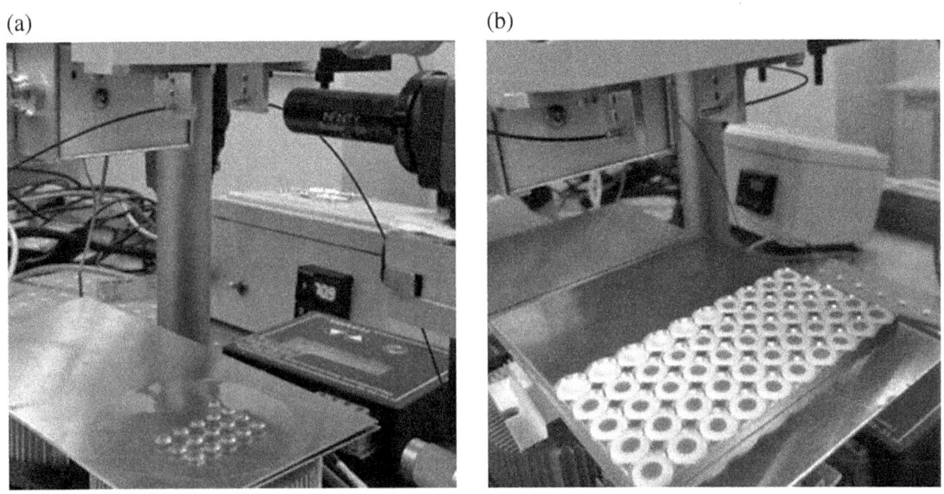

Figure 12.7 Dropwise additive manufacturing of pharmaceutical melt-based formulations during material deposition on (a) polymeric films and (b) placebo tablets. *Source:* From Icten et al. (2017), Figure 03 [p. 29] / With permission of Elsevier.

is proven platforms for enhancing the stability of engineered nano and micro particles, to impart higher bioavailability, and biological efficacies with different range of bioactive compounds. These features showcase the fruitful prospects of these delivery systems in direct food fabrication using 3D printing. Although the design features of drug delivery systems using encapsulation are quite clear, an in-depth understanding of the *in-vivo* biological efficacies and bioavailability of bioactive food ingredients are adequate. Hence future research must be focused in these directions on delivery of functional foods through integrated approach of 3D food printing and encapsulation.

12.3.3 Solid Lipid Nanoparticles

The structure of the solid lipid nanoparticles comprises an aqueous colloidal dispersion (core) within a solid lipid matrix (shell). The solid lipid nanoparticles range in the submicron size that is comprised of solid lipids such as triacylglycerol, waxes, and paraffin (da Silva Santos et al. 2019). This differs from the emulsions with the replacement of liquid lipid (oils) with solid lipid. Compare to other delivery systems such as nano emulsions and liposomes, the solid lipid nanoparticles possess advantageous features of possibility for large-scale production; slower degradation rate; sustained release; avoidance of organic solvents; and high encapsulation efficiency (Saupe and Rades 2006). The solid lipid nanoparticles can be fabricated by high-pressure homogenization by passing the liquid through a narrow gap in the micron range under the pressure of 100–2000 bar. During which the fluid accelerates over a short distance at higher velocities greater than 1000 km h^{-1}. The associated shear stress and the cavitation forces disrupt the particles that lead to size reduction in a desired submicron level. Using solid lipid nanoparticles, high concentrations of lipids up to 40% can be homogenized to lipid nano dispersions (Lippacher et al. 2000). Conventionally, the fabrication process of solid lipid nanoparticles follows either hot or cold homogenization. Initially, the bioactive ingredient is incorporated into lipid bulk either through dissolving or dispersing it using the lipid melt technique. The former type of hot homogenization process involves the melting of the lipid by about 10 °C higher than the melting point of the lipid followed by the addition of this melted mixture into the surfactant at the same temperature. The formed pre-emulsion is then homogenized under the controlled conditions of temperature and pressure with predetermined number of flow cycles. This would result in the nano emulsion that is allowed to recrystallize upon cooling to room temperature forming nanostructured lipid carrier and solid lipid nanoparticle (Anandharamakrishnan 2014a). On the other hand, cold homogenization involves the immediate exposure of the melted lipid and bioactive mixture to the liquid nitrogen thereby forming solid lipid microparticles. Later the formed solid lipid microparticle is homogenized with a cold surfactant solution and subjected to high-pressure homogenization under predetermined conditions well below the room temperature. This would result in the formation of solid lipid nanoparticles and nanostructured lipid carriers (Lippacher et al. 2002). The cold homogenization method is most suitable for heat sensitive and hydrophilic compounds; even though the hot homogenization method is still applicable for thermal sensitive compounds because of relatively short exposure of the bioactive compounds to high temperatures. Sometimes, the solid lipid nanoparticles can be prepared by combining ultrasonication with highspeed homogenization techniques. Various top-down and bottom-up approaches of the printing technologies for the delivery of drugs were summarized by Katakam et al. (2015). The applicability of solid lipid nanoparticles in 3D printing has not yet been explored. In a recent study, 3D printing was employed for preparing solid self-micro emulsifying drug delivery systems with a precise surface area to volume ratio (Vithani et al. 2019a). Results showed that the printing process was not significantly influenced by the formulated drug. Thus, this study remains a proof-of-concept for the printing of solid lipid-based drug delivery systems with controlled dispersion kinetics. More detailed studies are required in the integration of 3D printing with micro and nano delivery systems for the fabrication of customized bioactive food supplements and nutraceuticals.

12.3.4 Nanoliposomes

Liposomes are spherical vesicles fabricated at micro and nano levels with at least one lipid bilayer that is used for the encapsulation of different materials from natural to synthetic polymers and active compounds (Mohammadabadi and Mozafari 2018). Liposomes are the drug delivery system comprised of membrane-like phospholipid bilayers surrounded by an aqueous medium. Due to the structural organization that composed of aqueous and lipid phase, the liposomes are commonly used for the encapsulation of water-soluble, lipid-soluble, and amphiphilic materials (Mozafari and Khosravi-Darani 2007). Based on the size and the count of lamellae, liposomes are categorized as small unilamellar vesicles (<30 nm) and large unilamellar vesicles (30–100 nm); multilamellar vesicles; and multivesicular vesicles. Here, the unilamellar vesicles refer to liposomes with a single bilayer membrane while multilamellar vesicles are composed of more than two concentric bilayers. On the other hand, the liposomes with many tiny non-concentric vesicles present within a single bilayer of lipids is known as multivesicular vesicles. Both multilamellar vesicles and multivesicular vesicles are more conducive for the encapsulation of the fat-soluble materials while the water-soluble compounds are entrapped using unilamellar vesicles (Lasekan et al. 2016). Other than the core material, liposome formation can be greatly influenced by hydrophilic–hydrophobic interactions between the aqueous and lipid phases. The nanoliposomes are liposomes with submicron sizes at nanometric levels. With the reduction in the physical size, the efficacy of the delivery system is improved with enhanced storage stability. A slight variant of the liposomes is the bifunctional liposomes that are capable of entrapment and simultaneous release of dual materials with different solubilities (Anandharamakrishnan 2014a). The amount of energy input to the liposomes affects the molecular arrangement of lipids thereby resulting in the formation of bilayer vesicles through attaining a thermodynamic equilibrium in the aqueous phase. Some of the common mechanical techniques applied for the liposome's formation include extrusion, high-pressure homogenization, membrane homogenization, high-intensity ultrasonication, and microfluidization. Other non-mechanical methods include evaporation by reverse phase, freeze-drying cum rehydration, and removal of detergents from lipid micelles (Taylor et al. 2005). Recently, researchers made efforts in encapsulating gels into the liposomes based on the droplet-shooting method. In this study, the gelatin and the shape memory gel have been used as the inner core for reinforcement of liposomes in order to impart mechanical strength to the fabricating material (Adachi et al. 2020). Thus, the study looked into the effect of the gel on liposome formation and came up with a design of a 3D printer suitable for liposome-gel composite mixtures to be used as a novel way of drug delivery system. This study throws light on the integration of 3D printing and encapsulation to be used as a potent vehicle for bioactive compounds and nutraceuticals delivery. In another study, the bioavailability of curcumin has been improved by encapsulating curcumin into liposomes followed by incorporation onto 3D printed calcium phosphate scaffolds (Sarkar and Bose 2019). The concept of designing the customized shape and interconnected porosity allows for personalized treatment with patient-specific implants. Results showed that fabricated 3D printed scaffolds with the tailored release of the liposomal curcumin showed significant cytotoxicity against *in-vitro* osteosarcoma cells and promoted cell viability and proliferation of osteoblast cells. This study would open a new dimension of advanced manufacturing technology in designing alternative personalized medicines. A similar concept could be applied for the delivery of micronutrients through 3D printed foods as a fortification approach to alleviating nutritional deficiencies.

12.3.5 Nanostructured Lipid Carriers

Nanostructured lipid carriers are known to be the second generation of lipid nanoparticles that are characterized by a solid lipid core along with a combination of solid and liquid lipids with average particle size at nano-level (Muller et al. 2016). The nanostructured lipid carriers are the colloidal carriers composed of a lipid matrix with a specialized nanostructure that remained to be in a solid state by controlling the liquid lipid through controlling the quantity of liquid lipid included in the formulation. Due to this, nanostructured lipid carriers possess good stability and enhanced loading capacity that has found its applications for the delivery of drugs in the pharmaceutical field (Khosa et al. 2018). The nanostructured lipid carriers possess similar advantages as that of the lipid nanoparticles that include the controlled rate of release, low toxicity, biodegradability, and protection of bioactive compounds. Other than these features, the nanostructured lipid carriers have a high drug payload, show minimal drug expulsion, and enhanced physiochemical stability. Hence, the nanostructured lipid carriers are often known to be the new generation of drug delivery systems with promising colloidal carriers. They can be produced either by mixing solid and liquid lipids or by mixing optimal formulations of liquid and solid lipids thereby forming an amorphous solid. Here, the surfactants have a significant effect in achieving a stable formulation and the characterization of the nanostructure lipid carriers. The production method includes hot or cold high-pressure homogenization, microemulsion technique, precipitation, and dispersion with combined ultrasonication (Anandharamakrishnan 2014a). The nanostructured lipid carriers show a stronger capacity to immobilize the drugs and prevent particle coalescence than the emulsions. Further, the mobility of the incorporated drugs shows a drastic reduction in the solid phase. The nanostructured lipid carriers possess a higher drug payload capacity than solid lipid nanoparticles mainly because of the presence of liquid lipids in the solid matrix that would increase the release of drugs in a sustained manner (Hosseini et al. 2021). The whey protein isolate (WPI) and sodium alginate (ALG) nanocomplexes were used for stabilizing bioactive curcumin (CUR) in a fat-free model beverage (Yi et al. 2021). The higher loading quantity of CUR in the nanocomplex CUR-WPI-ALG was found to be 15.26 $\mu g\, mg^{-1}$. The developed nanocomplex showed greater stability with no obvious precipitation under simulated conditions of processing and storage. This was due to the spontaneous interaction among the nanocomplex CUR and WPI-ALG mainly characterized by hydrogen bonding and hydrophobic interactions. Further, the chemical stability and DPPH (2,2-Diphenyl-1-picryl-hydrazyl) scavenging capacities of CUR in the developed nanocomplex were found to be greater than the free curcumin. Although the potential of nanostructured lipid carriers is a well-proven promissory drug delivery system, the integration of nanostructured lipid carriers with 3D printing is obscure that must be explored for designing of novel ingredient and material formulations for delivery of bioactive compounds through food.

12.4 Techniques and Methods for Micro and Nanoencapsulation

The functionality of the developed encapsulate is not only based on physiochemical properties of core and shell materials but also techniques adopted for its fabrication. Hence it is essential to select appropriate encapsulation techniques based on the nature of core and wall materials, the desired size of the final encapsulate, the efficiency of encapsulation, rate of release, etc. In this regard, the microencapsulation techniques can be of a physical and chemical process (Bharathi et al. 2018). The former type of physical fabrication includes physicochemical processes such as sol–gel mechanism, coacervation, supercritical CO_2 assisted encapsulation; and physicomechanical processes such as fluid bed coating, vacuum encapsulation, centrifugal encapsulation, and electrostatic-based encapsulation. On the other hand, techniques like *in-situ* and interfacial polymerization, polycondensation comes under the chemical process of microencapsulation process. Based on the medium of suspension, microencapsulation techniques are categorized as liquid suspension systems (coacervation, evaporation by emulsification-solvent interaction, *in-situ* and interfacial polymerization) and gas suspension systems (spray drying, fluidized bed coating, and spray cooling) (Ezhilarasi et al. 2013b).

The method used for the fabrication of nano encapsulates is quite complex than microcapsules. The challenge associated with the difficulty in attaining the complex miniatured morphology with the particle size of around 10–1000 nm and controlled rate of release. The fabrication of nanoencapsulation is broadly classified as bottom-up (based on self-assembly and self-arrangement of the molecules) and top-down (application of efficient mechanical tools for the desired reduction in particle size and engineered shape of nanoparticles) approaches. Techniques such as inclusion complexation, coacervation, supercritical fluid techniques, nanoprecipitation are grouped as bottom-up approaches. On the other hand, the emulsification and evaporation by emulsification-solvent interaction are categorized as top-down approaches (Ezhilarasi et al. 2013b). The pH, concentration, temperature, ionic strength, molecular weight, mechanical stress, electrical force, and magnetic force associated with these techniques influence the end quality of the nanoparticles. In another context, nanoencapsulation can be classified as liquid-based, drying techniques, and electrohydrodynamic processes (Bharathi et al. 2018). The present section discusses the first two classes of liquid-based techniques (inclusion complex, coacervation, evaporation by emulsification-solvent interaction, supercritical fluid techniques, nanoprecipitation) and drying techniques (freeze-drying, spray drying, and spray-freeze-drying). The electrohydrodynamic processes refer to the electrostatic-based methods such as electrospraying and electrospinning. The latter group of electrohydrodynamic processes for nanoencapsulation of materials are discussed in the subsequent chapter of the present book.

12.4.1 Polymer-Lipid Based Encapsulation Techniques

12.4.1.1 Nanoprecipitation

Nanoprecipitation also known as antisolvent precipitation or solvent displacement method that is based on the polymer solubility with change in the quality of the solvent. This technique follows a series of steps that starts with the polymer precipitation from an organic solution followed by the organic solvent diffusion through an aqueous medium (Hosseini et al. 2021). This technique is based on the principle of spontaneous emulsification of the organic inner phase that contains a mixture of dissolved polymer, bioactive ingredient, and organic solvent into an aqueous external phase. Based on the nature of displacement of solvent, the resulting encapsulates can be of nanospheres and nano capsules in size of 100–300 nm (Barreras-Urbina et al. 2016). Both synthetic polymers and biopolymers can be used in the nanoprecipitation technique. The resulting nanodispersion showed higher stability against particle coalescence and Ostwald ripening. The applicability of this approach is restricted to the water-miscible solvents since the rate of diffusion is well enough for the production of spontaneous emulsification (Anandharamakrishnan 2014c). Recently, flash precipitation techniques are employed for nanoprecipitation of particles that involves the rapid mixing of the polymer solution with highly efficient mixers to obtain a uniform-sized particle of around 50–150 nm (Barreras-Urbina et al. 2016). Commonly used polymers in nanoprecipitation are polylactic acid, polyalkyl cyanoacrylate, polylactide-co-glycolide, and copolymers of polyethylene glycol. Based on the nanoprecipitation approach, nanocapsules of a size of even below 100 nm can be produced. Thus, the nanoprecipitation technique is efficiently used for producing uniform-sized nanoparticles with high drug load capacity, and encapsulation efficiency at the low processing temperature (Hosseini et al. 2021). Further, the final attributes of the formed nanoparticles can be optimized by controlling the nucleation and growth of particles in polymeric solutions. Several studies were reported on the development of stable nanoparticles using the nanoprecipitation technique for the delivery of carotenoids, polyphenols, and resveratrol using chitosan, casein, and zein (Davidov-Pardo et al. 2015; Hu and McClements 2014; Li et al. 2018; Wang et al. 2018). One major drawback of the conventional nanoprecipitation technique is the use of toxic chemical solvents. Hence nowadays, studies are focusing on the development of nanoparticles with an ecofriendly approach and less toxicity. Recently, the use of supercritical CO_2 is explored as a potential solvent for nanoparticle formation (Kaga et al. 2018). The efficacy of the developed particles and their fate on consumption should be investigated for exploring the potential benefits for the delivery of bioactive compounds by integrating with 3D printing.

12.4.1.2 Emulsification-Solvent Evaporation

As the name indicates, this technique of fabrication of nanoparticles starts with emulsification followed by the evaporation of the solvent. It involves homogenization of organic polymeric solution with aqueous phase followed by evaporation of solvent thereby precipitating polymer into nanospheres (Hosseini et al. 2021). High-pressure homogenization and sonication approaches are used for sample preparation. Often, the O/W emulsions are used rather than W/O emulsions since it utilizes water as the anti-solvent agent that enhances the process economy and sustainability and minimizes particle aggregation. The resulting nanoparticle size is greatly affected by the rate of stirring; nature, type, and the quantity of the dispersing agent; process temperature; and viscosities of aqueous and organic phases (Anandharamakrishnan 2014c). Most commonly synthetic polymers are used in emulsification-solvent evaporation; however, natural polymers are also employed. Polymers such as polylactic acid, cellulose acetate phthalate, ethyl cellulose, polycaprolactone, polylactic-co-glycolic acid, and β-hydroxybutyrate are majorly used in this technique (Bharathi et al. 2018). The emulsification-solvent preparation technique is one of the efficient techniques used for producing nanoparticles with a size below 100 nm. The nanocapsules developed by this technique are characterized by a spherical shape, higher drug loading capacity, higher efficiency of encapsulation (75–95%), sustained-release, and enhanced bioavailability (Anandharamakrishnan 2014c). Further, the nanodispersions and nanoemulsions that are prepared based on this approach exhibit good mechanical stability, although it depends on the methods used like microfluidization, high-pressure and high-speed homogenization, and drying technique adopted for formation of nanocapsules.

12.4.1.3 Inclusion Complexation

The process of encapsulation based on the supramolecular association of ligand of the core into a cavity containing the substrate, i.e., wall material through various intermolecular interactions such as entropy-driven hydrophobic interactions, van der Waals interactions, stearic interactions, and hydrogen bonds are termed as inclusion complexation (Ezhilarasi et al. 2013b). Mainly this technique is used for encapsulating volatile organic molecules like essential oils, minerals, vitamins, and polyphenols. Conveniently this method can be used for converting these volatile molecules into a stable free-flowing powder forms. Considering the food applications, inclusion complexation is useful in masking the off-odour and off-flavour in addition to preserving the bioactive compound from thermal and oxidative degradation. Further, the encapsulated particles shows an improved solubility, antioxidant and antimicrobial activity, sustained release, and reduced volatility. Although this technique provides greater encapsulation efficiency and higher material stability, only a few of the compounds like β-lactoglobulin and β-cyclodextrin are well suited for this method of encapsulation (Anandharamakrishnan 2014c). With advancements in material science, self-healing hydrogels are gaining recent attention in the biomedical field through the spontaneous self-recovery of host-guest interactions. The synergistic benefits of self-healing hydrogels especially in drug delivery, cell encapsulation, tissue engineering, and 3D printing were summarized (Jin et al. 2019). A combination of antiviral and antitumor drugs has been inkjet printed onto the adhesive polymeric film through an inclusion complexation encapsulation approach for localized treatment of cervical cancer (Varan et al. 2017). Results showed that the printed film exhibited the highest anticancer efficiency due to the synergistic effects of combined drugs (Figure 12.8). Thus, inclusion complexation can be combined with 3D printing for the delivery of functional bioactive compounds. Further, this approach could be applied for the 3D printing of active/ intelligent packaging films with

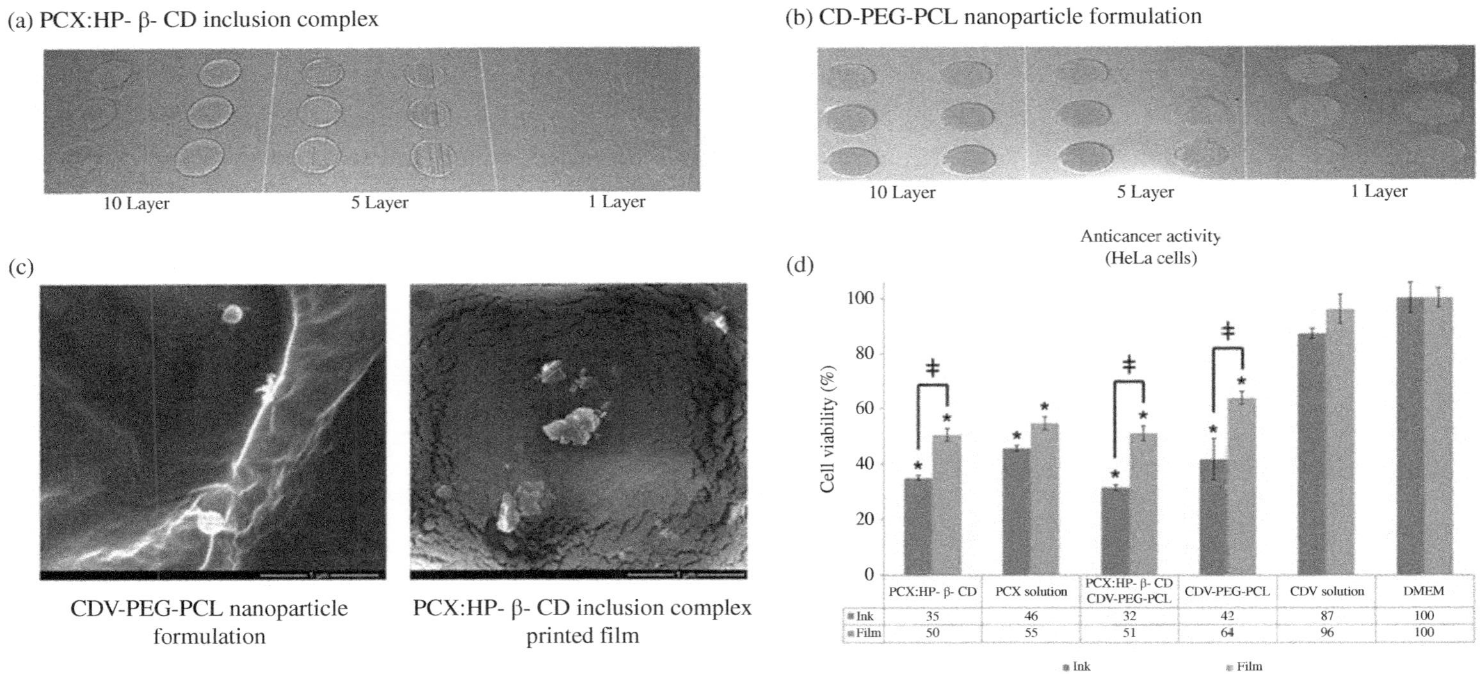

Figure 12.8 Inkjet printing of antiviral polycaprolactone (PCX) nanoparticles and antitumor cyclodextrin (CDV) inclusion on bio-adhesive film. (a), (b) 3D printed film, (c) images of SEM morphology, (d) anticancer effect of developed film. *Source:* From Varan et al. (2017) / With permission of Elsevier.

tailored functionality. More studies on the application of integration of complex coacervation with 3D printing for development of functional foods and food packages are required in unwinding the research scope in this area.

12.4.1.4 Coacervation

Coacervation is also known as ionic gelation is the process of suspending the core material in the polymeric solution followed by the deposition of the shell that surrounds the core after the separation of phases (Ezhilarasi et al. 2013b). Coacervation can be of simple coacervation or complex coacervation based on the induction of phase separation. The simple coacervation involves the use of desolvating coacervating agents to induce phase separation while for complex coacervation, the phased separation is induced by the oppositely charged polymers (cations and anions). Later the formed particles are stabilized through gelation by either thermal treatment or using suitable cross-linking agents (Bharathi et al. 2018). Thus, coacervation refers to the process of electrostatic interaction between the two aqueous media that undergoes sol–gel transition under normal conditions. Other than electrostatic interactions, hydrogen bonding and interactions of hydrophobic groups also contribute to the complex formation. Hence the nature of the polymers and their interactions significantly affects the size of the formed nanocapsules. The resulting nanoparticles would be of 100–600 nm; however, the type of process depends on the selection of apt drying techniques like vacuum or freeze-drying for further processing. Other factors like molar mass and biopolymer charges, pH, concentration, ionic strength, and the ratio of biopolymers added would influence the strength of the chemical interaction between the biopolymers and the resulting complex mixture (Anandharamakrishnan 2014c). The major force involved in the formation of coacervation is the reduced electrostatic free energy due to the association of the oppositely charged polymers. Also, the stirring and gelation rates have a significant effect in controlling the particle size. Biodegradable polymers such as sodium alginate, gum acacia, chitosan, and gelatin are commonly used for the formation of nanoparticulate systems (Jana and Jana 2017). This method of encapsulation has very high payloads up to 99% and encapsulation efficiency (90%) at low process temperatures without the use of organic or toxic solvents suitable for scalability and renders low processing costs. Other advantageous features include the formed nanocapsules are highly resistant to heat and exhibit controlled sustained release of bioactive compounds on applied mechanical stress and temperature (Bharathi et al. 2018). The morphology of the nanocapsules (size, dispersion, and particle distribution) was affected by the treatment with tannic acid. Further, the thermal stability and the melting temperature of the nanocapsules are raised with addition of glutaraldehyde. However, the major limitation of this approach to be used for applications in food is the use of the cross-linking chemical agent glutaraldehyde and its applicability is restricted only to hydrophobic compounds (Ezhilarasi et al. 2013b). Instead of glutaraldehyde, the natural cross-linking agents such as proteins and anionic polysaccharides; and enzymes like transglutaminase and laccase are used for food applications. Also, the formed coacervates are more susceptible to pH change and ionic strength as they are formed within a narrow range of pH. Therefore, it is difficult to protect the low pH sensitive compounds during complex coacervation (Hosseini et al. 2021). Thus, coacervation can be conveniently used for the fabrication of micro and nano encapsulates with the required amount of cross-linking agents and shell materials for the efficient delivery of thermo-labile

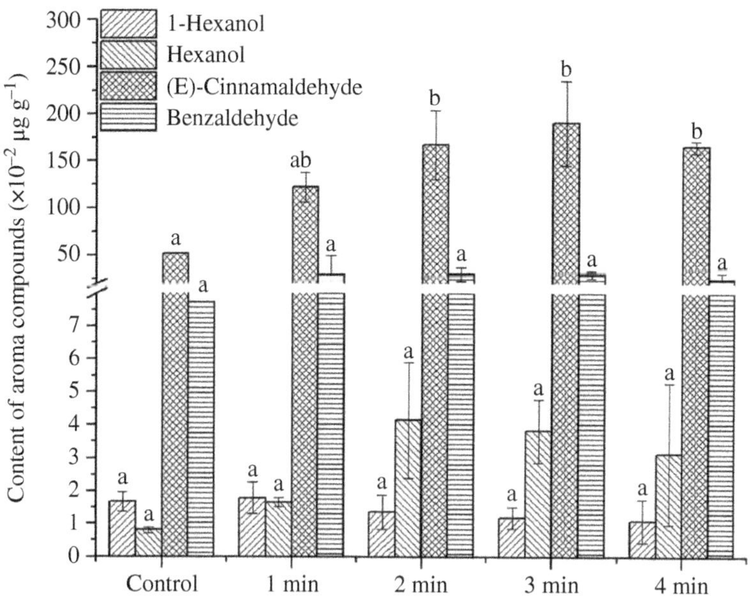

Figure 12.9 Concentration of selected aromatic compounds of 3D constructs from yellow peach and buckwheat dough triggered by microwaves. *Source:* Guo et al. (2021) / With permission of Elsevier.

bioactive materials. Different kinds of coacervation methods used for encapsulating essential oil and its effect on thermal stability and release properties were summarized in a recent report (Muhoza et al. 2020). The effect of microwave post-processing on the colour and aroma changes of 3D printed buck wheat dough incorporated with yellow peach was reported (Guo et al. 2021). A complex coacervate microcapsules was formulated prior to printing using gelatin and gum Arabic that exhibited spherical multicore structure with pigments and flavoured oil (8 : 1 : 1, (v/v/v) – corn oil: pepper red pigment: cinnamaldehyde essential oil) (Guo et al. 2021). Upon microwave treatment, the aroma of the developed 3D construct was enhanced by two-fold (Figure 12.9). Further, the gelatin-Arabic gum-oil complex coacervate microcapsules were confirmed to be effective in inducing colour/ aroma changes in 3D printed foods. Thus, this study provides insights on the applicability of complex coacervation for 3D printing that open up a new dimension of research on 4D printing of foods. In this cohort, complex coacervation can be used for the incorporation of functionalized micro and nano capsules into the 3D printed foods that acts as an efficient carrier vehicle for bioactive compounds and micronutrients.

12.4.1.5 Supercritical Fluid Technique

Supercritical fluid techniques are becoming more popular due to its greener approach of processing the materials. Recently, supercritical fluid techniques are more explored for the synthesis of colloidal particles (Vo et al. 2018). The supercritical fluids exert intermediate properties between the gases and liquids. These properties include less density, less viscosity, high solvation power, high diffusivities, and mass transfer that are well above the critical point (temperature and pressure) of the fluid (carbon dioxide, water, nitrogen, propane, and so on) (Anukiruthika et al. 2019). Supercritical fluids can be conveniently applied for similarly encapsulating thermolabile bioactive compounds as that of the spray drying process. This encapsulation approach follows the solubilization of the bioactive compounds and polymer into a supercritical fluid and the solution is expanded by allowing it to flow through the nozzle. During the spraying process, the supercritical fluid is evaporated and the solute particles get precipitated (Anandharamakrishnan 2014c). This is the most common method used for encapsulation at low critical temperatures, minimal use of organic solvents, tunable solvent polarity, ease of the removal of solvents, less toxicity of solvents, and less processing temperatures. Due to these advantages, this method is apt for heat-sensitive hydrophobic bioactive materials (Hosseini et al. 2021). Among the supercritical fluids, supercritical CO_2 is an excellent solvent that is suitable for both water-soluble and lipid-soluble bioactive substances. The strong polar agents are sometimes difficult to be dissolved in supercritical CO_2 due to the quadrupole moment. In such cases, the solubility can be enhanced through the addition of surfactants or cosolvents in the form of microemulsions (Pisoschi et al. 2018).

Various techniques based on supercritical fluids are supercritical fluid extraction of emulsion (SFEE), aerosol solvent extraction system (ASES), the rapid expansion of supercritical solutions (RESS), particles from gas-saturated solution (PGSS), gas or supercritical fluid antisolvent (GAS/ SAS), rapid expansion of supercritical solutions into solvent (liquid) chamber (RESOLVE), supercritical antisolvent precipitation, supercritical fluid solution-enhanced dispersion (SEDS), and precipitation with compressed antisolvent fluid (Bharathi et al. 2018). The bioactive ingredients loaded nanoparticles can be developed using PGSS technique where the core material and wall material become saturated with CO_2 under a supercritical environment. This process upon immediate depressurization through nozzle creates a rapid cooling effect that enables core and coating materials to be precipitated resulting in the encapsulation of the bioactive substances. Another advantageous feature of PGSS is the ability of supercritical CO_2 to reduce the melting temperature of the wall material thereby decreases the overall operating temperature of the process (Vo et al. 2018). The SFEE is another technique used as a food-based delivery vehicle. In this technique, the O/W emulsion is put into contact with the supercritical fluid that aids in the extraction of solvents from the organic phase and promotes the precipitation of bioactive molecules in micro and nano forms. In contrast, the bioactive substances can be dissolved into the organic phase thereby promoting the entrapment of the encapsulation inside the polymer (Hosseini et al. 2021). Although the supercritical technique has seemed to be environmentally friendly technology, it requires higher capital investment for handling high-pressure equipment that limits its application in the food industry.

12.4.1.6 Fluid Bed Coating

Fluid bed coating is another technique employed for the uniform entrapment of solid particulates, powders, and granules. The process involves the spray coating of the polymeric shell material onto the fluidized core (bioactive substances). This technique is also referred to as the Wurster process named after the initiation of this process firstly by Dr. Dale E. Wurster who developed this technique (Bharathi et al. 2018). Fluid bed coating can be categorized into several types: based on the entry point of feed, the process can be of tangential spray (rotary head and static head), bottom spray, and top spray; based on spray direction to the fluidized particles, the process can be of concurrent, counter current and tangential flow (Frey 2014). The static tangential spray has differed from the rotary tangential spray provided the base plate for fluidization of the core. The static type possesses perforations in a diagonally cut manner for allowing the air to pass through angularly. On the other side, the rotary type contains a non-perforated plate that rotates and allows the air to pass only through the outer edge of the plate. The process efficiency of the fluid bed coating is affected by temperature, relative humidity, the temperature of the air at the inlet, the temperature of atomized air, rate of the feed solution, volume of atomized air, and volume of fluidized air (Gupta et al. 2016). Using this approach, the bioactive curcumin was coated over the rice kernel that had a significant effect on colour, antioxidant capacity, and total phenolic content. In addition, fluid bed coating can be employed as the post-treatment after the spray-dried encapsulated particles. Since the spray-dried particles could leave a partially exposed bioactive core to the ambient conditions causing a structural collapse of the shell due to heat employed during the process of spray drying (Bharathi et al. 2018). Considering the scalability of the operation, the tapered or conical fluidized bed coating setup with top-spray inlet entry is most preferred in the food industry due to its higher batch size and versatility that aids in easy dismantling and cleaning process (Depypere et al. 2009). With proper knowledge on the optimization of process parameters, the fluid bed coating method can be efficiently used for the development of uniform encapsulates with higher encapsulation efficiency that could overcome the drawbacks of higher processing costs associated with freeze-drying and spray drying. Recently, a study was reported on the applicability of 3D printing for fluid bed coating of the conventionally prepared tablets (Tsintavi et al. 2020). The model drugs used in this study were hydrophilic Melevodopa and lipophilic Acyclovir with glyceryl distearate as coating material. The tablet coating was performed using a custom-designed FDM-based 3D printer (Figure 12.10). Results confirmed the controlled release rate of the two selected drugs without modifying its core formulation based on the fluid coating method using a 3D printer. Both the selected hydrophilic as well as lipophilic active pharmaceutical ingredients exhibited different dissolution profiles through fine-tuning of three process parameters: surface area coverage, number of the coated layers, and number of coated sides. The non-Fickian release of the drug was evident and the dissolution power was mainly influenced by the levels of the coating parameters instead of the composition of active pharmaceutical ingredients.

In another study, the potential of 3D printing in the fabrication of the drug with customized filaments from polymers of different solubility was reported (Melocchi et al. 2016). The developed drug showed an immediate or modified release profile based on the coating layers at tablets or in whole capsules (Figure 12.11). Thus, partial coating based on a 3D printing approach was applied for drugs loaded with active pharmaceutical ingredients showing a sustained release profile with enhanced drug efficiency.

A cylindrical-based implant of active pharmaceutical ingredients over the polymeric wall matrix was 3D printed using inkjet technology for tailored drug release profiles (Huang et al. 2007). Thus, 3D printing technology could be promisingly used for custom designing of drugs with a unique release profile as per the needs of the patient or based on the localized demand of the drugs at the production points. In a similar approach, functionalized

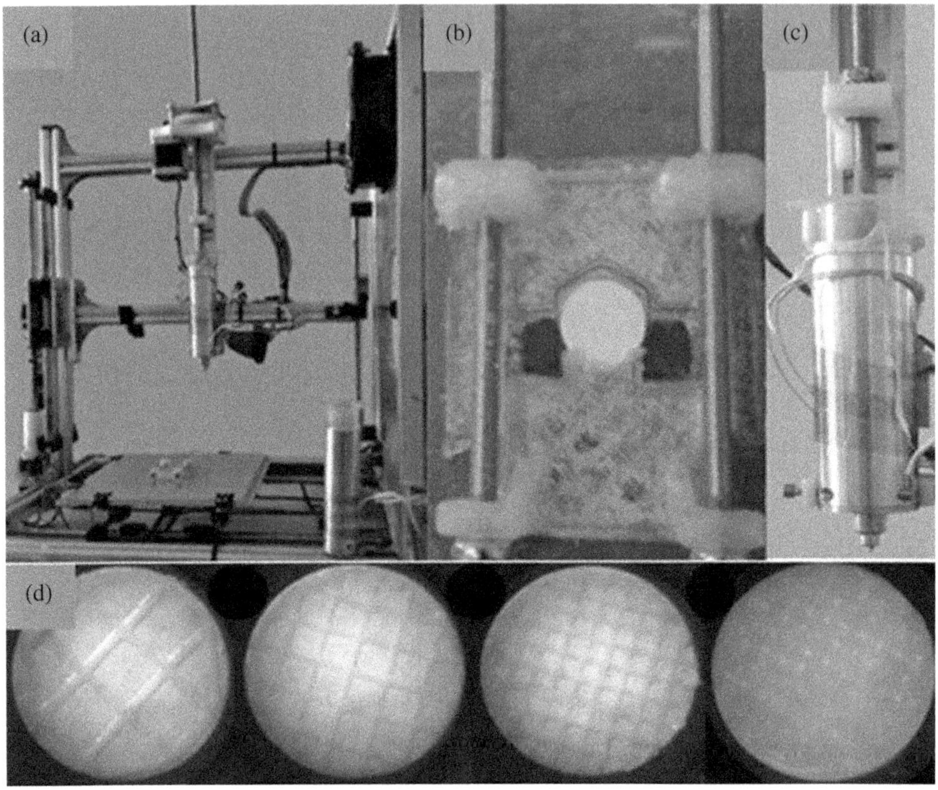

Figure 12.10 3D printing for partial tablet coating. (a) experimental set-up used for 3D printing, (b) support system used to hold tablet in position, (c) heating assembly used to keep material supply in liquid form, and (d) coated samples of different levels of surface coverage and thickness. *Source:* From Tsintavi et al. (2020), Figure 01 & 03 [p. 02 & 03] / With permission of Elsevier.

food capsules could be developed with bioactive nutraceutical ingredients for the development of novel functional foods with enhanced health benefits.

12.4.2 Drying Techniques for Micro and Nanoencapsulation

Encapsulation is commonly employed for the production of bioactive suspensions entrapped within the shell in either liquid or dried forms. Although liquid-based encapsulation showed good efficacy, it possesses few limitations in terms of irreversible aggregation of particles, chemical instability, and seepage of encapsulates (Ezhilarasi et al. 2013b). Hence the formulated micro/ nano suspensions are often transformed into dried forms that impart greater stability of encapsulates. Also, drying of micro/nano particles would be easy for handling and storage with higher solubility that can be conveniently reconstituted into liquid forms (Nakagawa et al. 2011). Although the drying imparts additional stress on the dried powders, these encapsulate are capable of exhibiting sustained and controlled release of bioactives. The present section discusses the various emerging drying techniques that are applied for the encapsulation of bioactives.

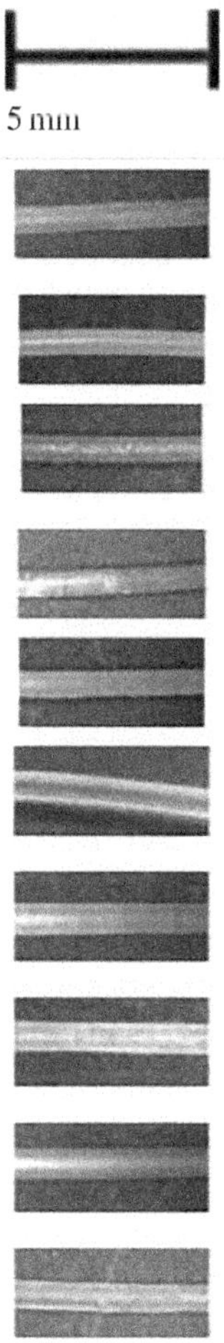

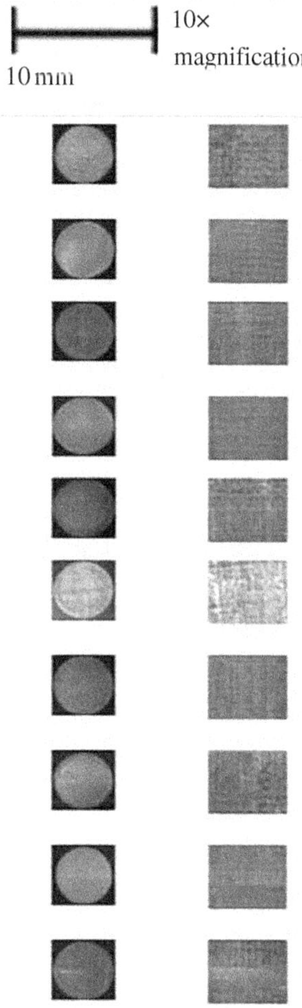

Formulation	HME				FDM	
	T (°C)	Screw speed (rpm)	Torque (N × cm)	Product	T (°C)	Product
KIR + 12% GLY	160	100	80		180	
PEO	65	100	100		160	
HPMC + 5% PEG 400	160	70	70		200	
HPC	165	80	40		180	
PVA + 5% GLY	190	70	80		225	
SLP + 10% PEG 400	120	80	80		200	
HPMCAS + 5% PEG 8000	180	100	100		200	
EDR L + 20% TEC	160	80	120		160	
EDR RL + 15% TEC	120	95	60		160	
EC + 10% TEC	160	100	100		200	

Figure 12.11 Formulations, process parameters, and printed samples of different pharmaceutical grade polymers. *Source:* From Melocchi et al. (2016) / With permission of Elsevier.

12.4.2.1 Spray Drying

Spray drying transforms the feed solution (micro/ nano suspensions) into a dried particulates through spraying feed solution into a hot drying medium. This process yields fine particles in less processing time that are more suitable for continuous large-scale industrial applications at economical cost (Anandharamakrishnan 2014b). Thus, spray drying not only allows encapsulation of particles but also aids in the drying of encapsulates. The process involves dispersing bioactive core in an aqueous medium that contains shell material followed by spraying of this mixture into a hot drying chamber leads to encapsulate formation (Figure 12.12).

Spray drying follows a series of process steps that includes the transformation of micro/nano suspension into powder through liquid atomization; contact of atomized spray droplets with hot air; and evaporation of the moisture with particle separation (Bharathi et al. 2018). Common types of atomizer used in spray drying process include centrifugal (rotary) atomizer, two-fluid nozzle atomizer, and pressure nozzle atomizer. More recently, three fluid nozzle has been designed for the co-encapsulation of curcumin and resveratrol using spray drying technique (Leena et al. 2020a). Results showed that the loading capacity, as well as the encapsulation efficiency of micro-sized bioactives, was 9% higher in a three-fluid nozzle (Figure 12.13). The controlled release of the encapsulated resveratrol from a three-fluid nozzle is comparable to that of the conventionally used two-fluid nozzle. Thus, the obtained release kinetics showed the best fit with the Korsmeyer-Peppas model that explains the underlying diffusion mechanisms (resveratrol exerts pseudo-Fickian diffusion from the core and curcumin exerts anomalous diffusion from the shell). Thus, this study explains the feasibility of co-delivery of bioactive compounds through a novel design of a three-fluid nozzle using a spray drying approach.

In another study, a dual bioactive compound (quercetin and ferulic acid) loaded microcapsules were developed through cross-linking of chitosan with sodium alginate and tripolyphosphate by three-fluid nozzle atomization (Figure 12.14). The obtained spray-dried microparticles showed a sustained release pattern of 85–88% for about 24 hours and had higher potency of DPPH activity. Thus, the fabrication of core-shell structured microparticles using a three-fluid nozzle system remains to be a promising method for delivering the encapsulated bioactive with synergistic antioxidant activity. Based on the contact of spray and gas medium, the flow can be of either co-current (parallel) or counter-current flow (Anandharamakrishnan 2014b). Both the flow process has its own advantages where the parallel flow aids in quick evaporation of spray droplets, reduced time of drying, and less thermal degradation. On the other hand, the product temperature at the outlet is greater than the air temperature of exhaust and is nearly the same as that of the feed-air contact temperature for counter-current flow. Hence, counter-current flow is commonly employed for thermally insensitive foods. Another variant is the mixed flow where the dryer utilizes both parallel and counter-current flows based on its design. Mixed flow arrangements are employed for drying coarse free-flowable materials however process efficiency is limited by imparting higher product temperature that deteriorates its quality. Commercially available spray dryers can produce the encapsulates only at the micro-sized level. However, a slight modification of the equipment design such as the use of laminar airflow instead of turbulent flow; replacement of atomizer with vibrating mesh piezoelectric actuator; and the use of electrostatic particle collector rather than cyclone separator would help in the production of nano-sized particles (Bharathi et al. 2018).

Spray drying can be applied for encapsulating both hydrophilic as well as lipophilic materials. The optimization of feed formulation must consider the selection of core and shell materials, viscosity, molecular weight, and ionic strength as well as optimization of drying parameters like inlet and outlet temperature of the air, the flow rate of the feed solution, and airflow rate would significantly influence the powder characteristics of obtained

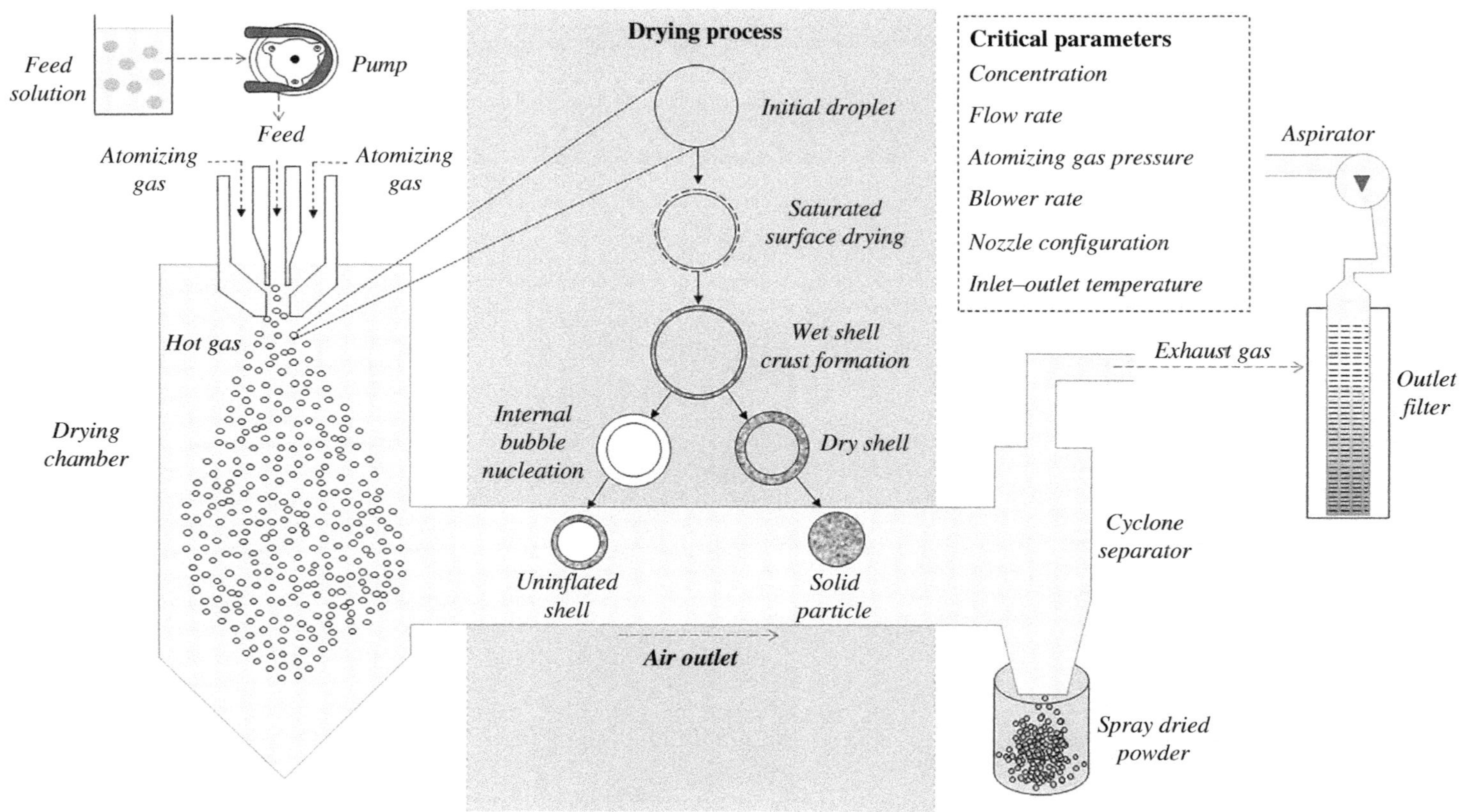

Figure 12.12 Illustration of spray drying process. *Source:* Yoha et al. (2020b) / With permission of Elsevier.

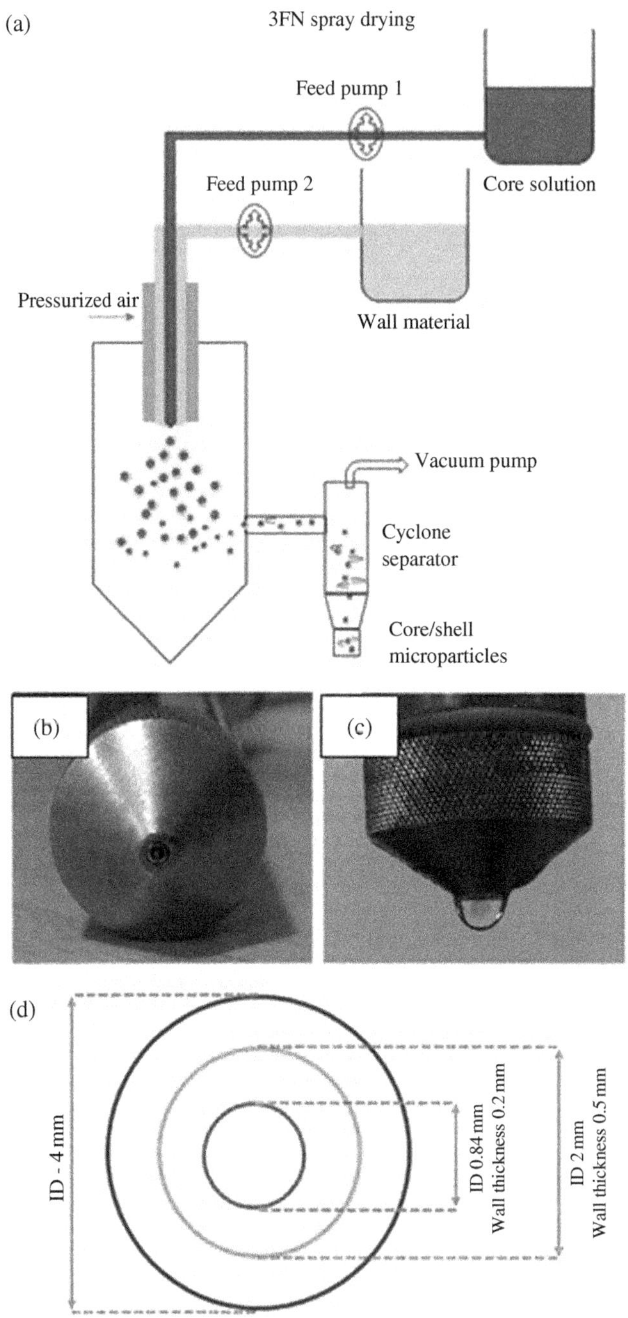

Figure 12.13 Formation of core-shell microparticles. (a) Schematic process of spray drying using 3-fluid nozzle, (b) front-view of 3-fluid nozzle, (c) formation of core/shell droplet at tip of nozzle, and (d) dimensions of designed concentric nozzle. (b) (c) *Source:* From Leena, Antoniraj, et al. (2020), Figure 01[p.02] / With permission of ELSEVIER.

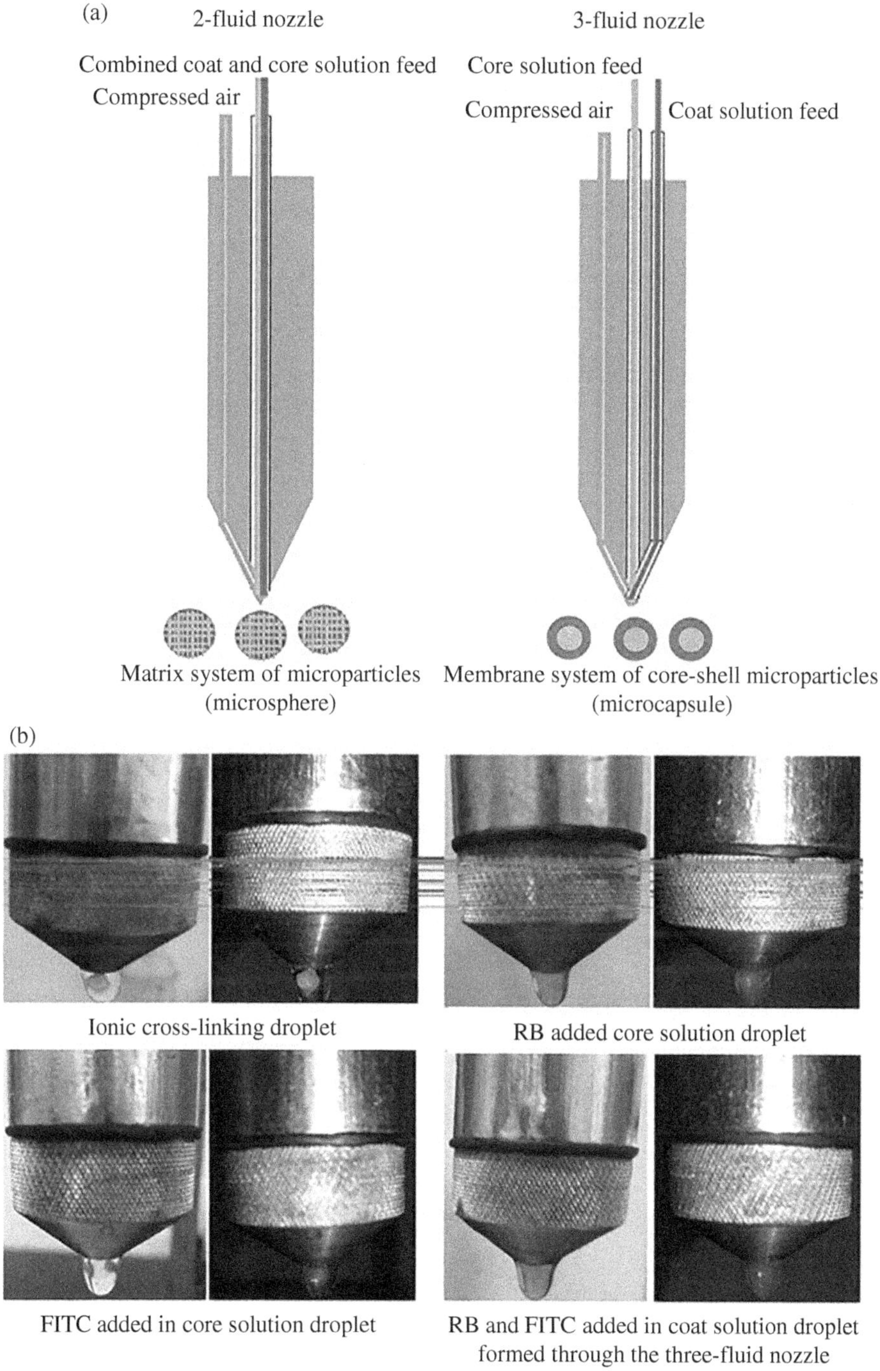

Figure 12.14 Modified three-fluid nozzle for production of cross-linked chitosan microparticles. a) comparison of schematic view of two- and three-fluid nozzle, and b) images of morphology of droplet formation. *Source:* From Antoniraj et al. (2020) / With permission of Elsevier.

encapsulates (Hosseini et al. 2021). Various processing conditions of spray drying for microencapsulation of chia oil and fish oil were reported (Lavanya et al. 2019). The study concluded that spraying of oil at 55 °C resulted in maximum efficiency of encapsulation and minimum peroxide values. These results are consistent with the Fourier transform infrared spectroscopy (FTIR) spectra of the obtained microcapsules that clearly indicate the presence of polyunsaturated fatty acids in the encapsulated samples. Morphological studies showed the presence of blowholes confirming the smooth surface of shell matrix with the particle size of 5.04 μm for chia oil powder and 3.89 μm for fish oil powder. Further, this study confirmed the maximum release of bioactives in gastrointestinal conditions indicating the stability of the microcapsules under acidic conditions with maximum absorption of oil would be at the intestine. In another study, the spray nozzle of the conventional spray dryer was replaced with a nebulizer for producing inhalable β-carotene formulations through a modified-spray drying process (Figure 12.15) (Lavanya et al. 2020a). The slow and sustained release of β-carotene based on diffusion mechanism was confirmed through *in-vitro* studies. Thus, the developed method is a novel way of delivering the bioactive in dry forms for direct inhalation applications.

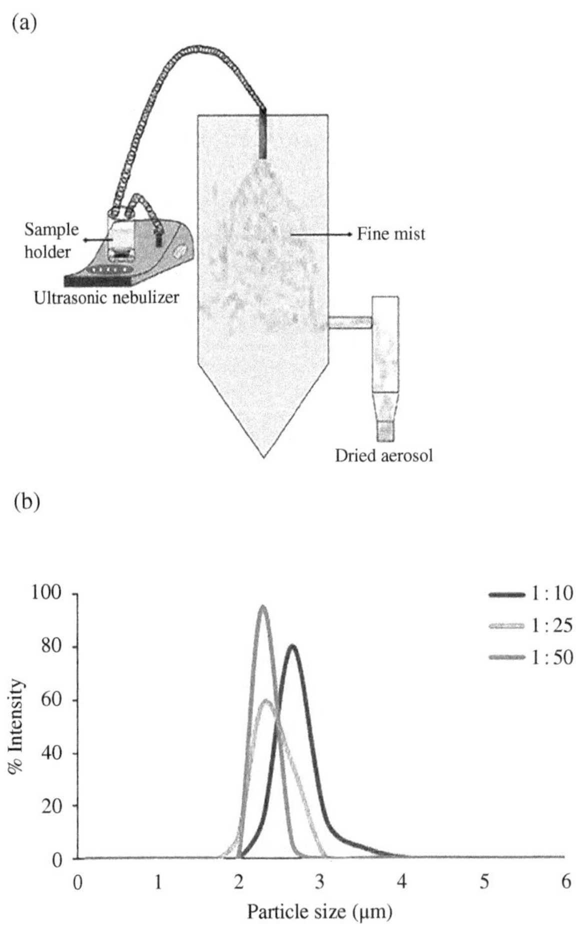

Figure 12.15 Modified spray dryer for aerosol formulations. (a) illustration of modified spray drying process, and (b) particle size distribution of β-carotene powder of different core to wall ratios. *Source:* Unpublished images from Lavanya et al. (2020a).

Spray drying is seemed to be an efficient drying approach for encapsulating thermal sensitive bioactive compounds. The advantageous features of spray drying rely on its ability to produce a broad range of size and shape of the micro/ nanoparticles through optimization of drying process parameters, material properties, and preparation method; and short exposure time of hot air that preserves the functionality of encapsulated bioactive compounds from thermal degradation (Ezhilarasi et al. 2013b). Other than the spray drying technique, processes like spray chilling and freeze-drying are also used for the recovery of the powder particles after encapsulation through the conversion of micro/ nano suspensions into capsules. All these processes contribute to maintaining storage storability, easy handling, and controlled release suitable for targeted delivery (Assadpour and Jafari 2019). Spray drying has been used for a long time for encapsulating a broad range of bioactive ingredients such as vitamins, flavours, minerals, colours, oils, fats, enzymes, probiotics, and so on. The common shell materials used in the encapsulation process using spray drying include maltodextrin, modified starch, guar gum, gum acacia, alginate, carboxymethyl cellulose, whey proteins (WPs), sodium caseinate, soy proteins, etc. Thus, spray drying is a one-step method that produces uniform spherical nanoparticles and can be used for the fabrication of functional beverages and foods for the delivery of bioactives.

12.4.2.2 Freeze-Drying

Freeze-drying also termed lyophilization is a multi-stage process where the water present in the foodstuff is initially frozen to form ice then it is removed through direct sublimation followed by desorption. During the first stage of freezing, the crystallization rate depends on the rate of cooling, nuclei and crystal growth, concentration, and final cooling temperature (Anandharamakrishnan et al. 2010). Most of the food components exist in an amorphous glassy stage that is not crystallized but water present in it gets crystallized. The freezing is followed by the primary drying where the formed ice crystals are removed through sublimation at low pressure and temperature conditions. While in the next stage of secondary drying, the unfrozen water present in the food material is removed through desorption. This stage of freeze-drying seems to consume about one-third of the drying time that results in lowering the moisture content up to 2–10% and 0.1–3% for food materials and biological products, respectively (Anandharamakrishnan 2014b). In order to preserve the products after drying, often, the drying column is filled with inert nitrogen or argon gas as a final treatment of freeze-dried products to enhance storage stability. Recently, freeze-drying is used as an encapsulation approach in which the aqueous solution or dispersion containing the wall material, core material, and bioactive ingredient is dried through subsequent stages of freezing followed by drying. During which the incorporated core material and bioactive ingredient result in co-lyophilization that would result in a porous, non-shrunken, and complex matrix (Fang and Bhandari 2012). Freeze-drying is highly useful to produce heat-sensitive encapsulates from bioactive as it significantly minimizes the thermal degradation than other drying techniques. The resulted porous structure of the freeze-dried matrix allows the partial exposure of the bioactive to the surrounding atmosphere that would result in the loss of bioactives. This can be avoided through a post-process treatment of the dried product to the inert atmosphere. The obtained freeze-dried product particles are more stabilized in dried form, showing high solubility in aqueous medium, preserve colour and nutritional value. Hence this method of drying is frequently used for commercial production of encapsulates from thermal labile compounds such as flavour, aroma, pigments, oils, probiotics, and so on (Karthik and Anandharamakrishnan 2013). Freeze-drying seems to be a stable process that aids in enhancing the physicochemical stability of the encapsulates suitable for long-term storage. However, the drawbacks of this technique are its higher processing cost, energy-intensive process, long drying time (even last for 20 hours), loss of bioactive due to its porous structure, and poor flowability of final encapsulated particles (Ishwarya et al. 2015).

In a more recent review, the various synergistic benefits of freeze-drying assisted with other drying techniques were reported in reducing cost and drying time associated with freeze-drying (Waghmare et al. 2021). Despite these limitations, freeze-drying is generally applied for separating nanoparticles that are produced through other encapsulation techniques. Freeze-drying imparts porous structure to the encapsulates that might result in loss of bioactive. Hence this process of freeze-drying for encapsulates is not purely meant as encapsulation as some portion of bioactive is exposed to the atmosphere results in its degradation. Therefore, freeze-drying is employed for the removal of water from encapsulates without altering their shape and structure. Thus, freeze-drying can improve the stability of core material during storage against deterioration with an encapsulation efficiency above 70% (Anandharamakrishnan 2014b). The final quality attributes of the obtained freeze-dried nanoparticles depend on high-energy intensification method used for the formulation of the delivery system and the encapsulation technique adopted for a breakdown of the microdroplets to the nano-sized forms. Often freeze-drying requires cryoprotectants for preserving the particle size and to prevent the material from aggregation. The commonly used cryoprotectants in freeze-drying are mannitol, sucrose, trehalose, and lactose. The freezing temperature used in the process also influences the size of the nanocapsules. Different range of freezing temperatures on the size of the developed nano encapsulates was reported by Ezhilarasi et al. (2013b). Thus, freeze-drying is an efficient process in stabilizing and retaining the particle in the nano-level even after the drying. The stability of the polyphenols was enhanced through microencapsulation of green tea extracts using spray drying and freeze-drying (Pasrija et al. 2015). Results showed that the freeze-dried micro encapsulates showed 60–76% encapsulation efficiency and antioxidant activity with IC_{50} value of 54.77–60.26 μg mL^{-1} that was found to be significantly higher than the spray-dried microencapsulate. Further, the incorporation of developed encapsulates into bread dough improves the volume of bread and crumb firmness with significant difference on total phenolic content (Figure 12.16). Similar approach of freeze-drying was adopted for microencapsulation of Garcinia fruit extract for enrichment of bread with hydroxycitric acid (HCA) (Ezhilarasi et al. 2013a). Results showed that bread incorporated with WPI encapsulates exhibited higher volume and possess soft crumb with the sensory attributes possessing higher free HCA concentration. In a more recent study, the tuna fish oil was microencapsulated using starch from arrowroot and maltodextrin using freeze-drying (Charles et al. 2020). Results indicated that freeze-dried microcapsules had higher encapsulation efficiency of 80.5–86.4% with a mean particle size of 26.72–103.14 μm. Further, the developed microcapsules had lower peroxide value (4.8–9.2 mEq kg^{-1} oil) and acid value (1.46–2.24% free fatty acid (FFA)) well below the industry-set maximum limits for refined fish oils. In addition, the obtained microcapsules exhibited the highest oxidation stability during storage period of 90 days. The arrowroot starch acts as a cryoprotectant as well as shell for the

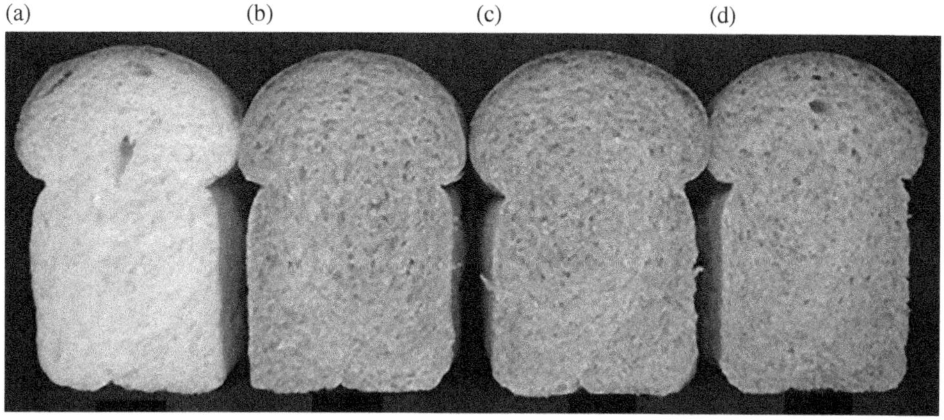

Figure 12.16 Images of microencapsulated green tea polyphenols incorporated bread. (a) control, (b) bread with free green tea extract, (c) bread with freeze-dried maltodextrin-green tea extract encapsulates, and (d) bread with spray-dried maltodextrin-green tea extract encapsulates. *Source:* From Pasrija et al. (2015), Figure 04 [p. 295] / With permission of Elsevier.

microencapsulation of oil that could be used as potential food carrier for fortification. A study was performed on the microencapsulation of the lycopene (tomato peels) by freeze-drying and complex coacervation approach in combination with acacia gum and WPI (Gheonea et al. 2021). The developed double microencapsulates had an average particle size of 378.8 nm to 3.56 μm with 83.6% encapsulation efficiency. Around 63% of lycopene retention was observed during 14-day storage period at 4 °C. Further, the obtained microparticles were used in the dressing samples exhibiting higher antioxidant activity and functionality. Thus, this study highlighted the efficacy of the encapsulated lycopene microparticles in terms of phytochemical profile, chemical stability, and its applicability in food systems.

12.4.2.3 Spray-Freeze-Drying

Spray-freeze-drying is a combined technique used for providing synergistic benefits of both freeze-drying and spray drying. The limitations of freeze-drying can be overviewed by adopting combined technology of spray-freeze-drying that would significantly result in reducing the pore size and process time (Karthik and Anandharamakrishnan 2013). The spray-freeze-drying involves the atomization of the aqueous dispersion with wall material and core material followed by freezing and subsequent drying. The combined spray-freezing can be done by the following methods: spray-freezing into either liquid or vapour, and spray-freezing into vapour over the liquid (Ishwarya et al. 2015). The resultant spray-freeze-dried powder possesses free flowability, spherical shape, and porous structure that exhibit higher rehydration ability. Thus, spray-freeze-drying can be used for developing encapsulates without any loss of bioactive whilst maintaining good sensorial and nutritional attributes. Further, spray-freeze-drying significantly reduces the drying time with precise control over the particle size. Nevertheless, the initial cost, processing cost, and complexity of the combined process that restrict its commercial applications for large-scale

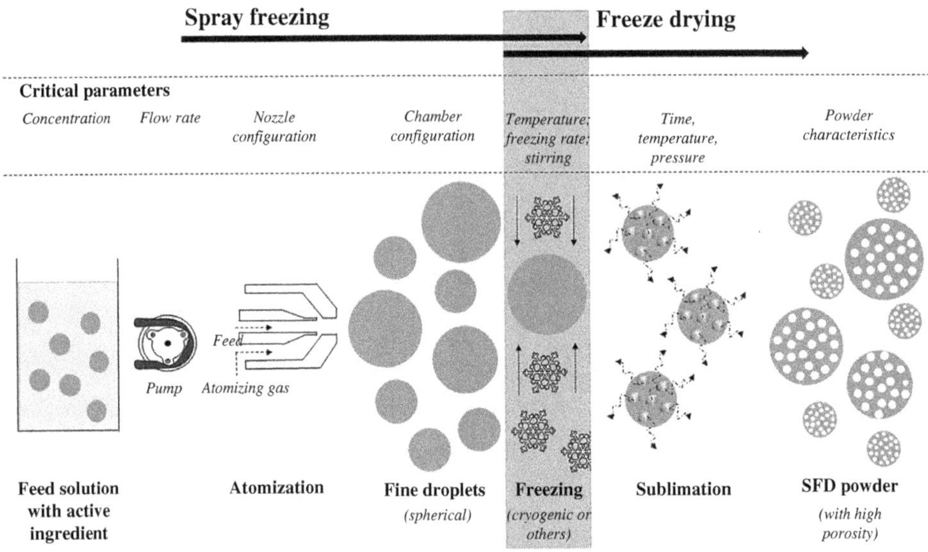

Figure 12.17 Illustration of working principle of spray-freeze-drying. *Source:* Vishali et al. (2019) / With permission of Elsevier.

production of high-value products. The mechanism of spray-freeze-drying, process variables, and various approaches employed for the targeted delivery of pharmaceutical drugs was explained by Vishali et al. (2019) (Figure 12.17).

The effect of spray-freeze-drying on microencapsulation of vitamin E was compared with other drying approaches such as spray drying and freeze-drying (Parthasarathi and Anandharamakrishnan 2016). The encapsulation efficiency of vitamin E for spray-freeze-drying (89.3%) was on par with the spray drying (89.6%), while the encapsulation efficiency obtained for freeze-drying (86.1%) was comparable to that of both of the other techniques. Results of oral bioavailability studies showed a maximum area under the curve (AUC) and highest plasma concentration (C_{max}) for the obtained microcapsules through a combined approach than freeze-dried and spray-dried microcapsules. Thus, the spray-freeze-dried vitamin E microcapsules increased the bioavailability by 1.13 and 1.19-fold compared to encapsulates obtained from individual processes. The spray drying and spray-freeze-drying were compared to be used for encapsulation of probiotics *Lactiplantibacillus plantarum* (NCIM 2083). The fructooligosaccharide (FOS), maltodextrin (MD), and WP was used as prebiotics for the matrix-type encapsulation of probiotics (Yoha et al. 2020b). Among the tested formulations, the FOS: WP : MD of 2 : 0.5 : 0.5 ratio showed greater encapsulation efficiencies of 89.21% for spray drying and 96.16% for combined spray-freeze-drying. The spray-freeze-dried synbiotics showed greater survivability during *in-vitro* digestion while the spray-dried synbiotics showed greater retention of cell viability of 0.95-fold during storage studies. The foam stability of the instant coffee was enhanced through the spray-freeze-drying approach (Deotale et al. 2020). Results showed that coffee powder obtained from spray-freeze-drying produced a stable foam in the size ranging from 100–200 nm that was found to remain intact up to 40 minutes (Figure 12.18). Thus, spray-freeze-drying is

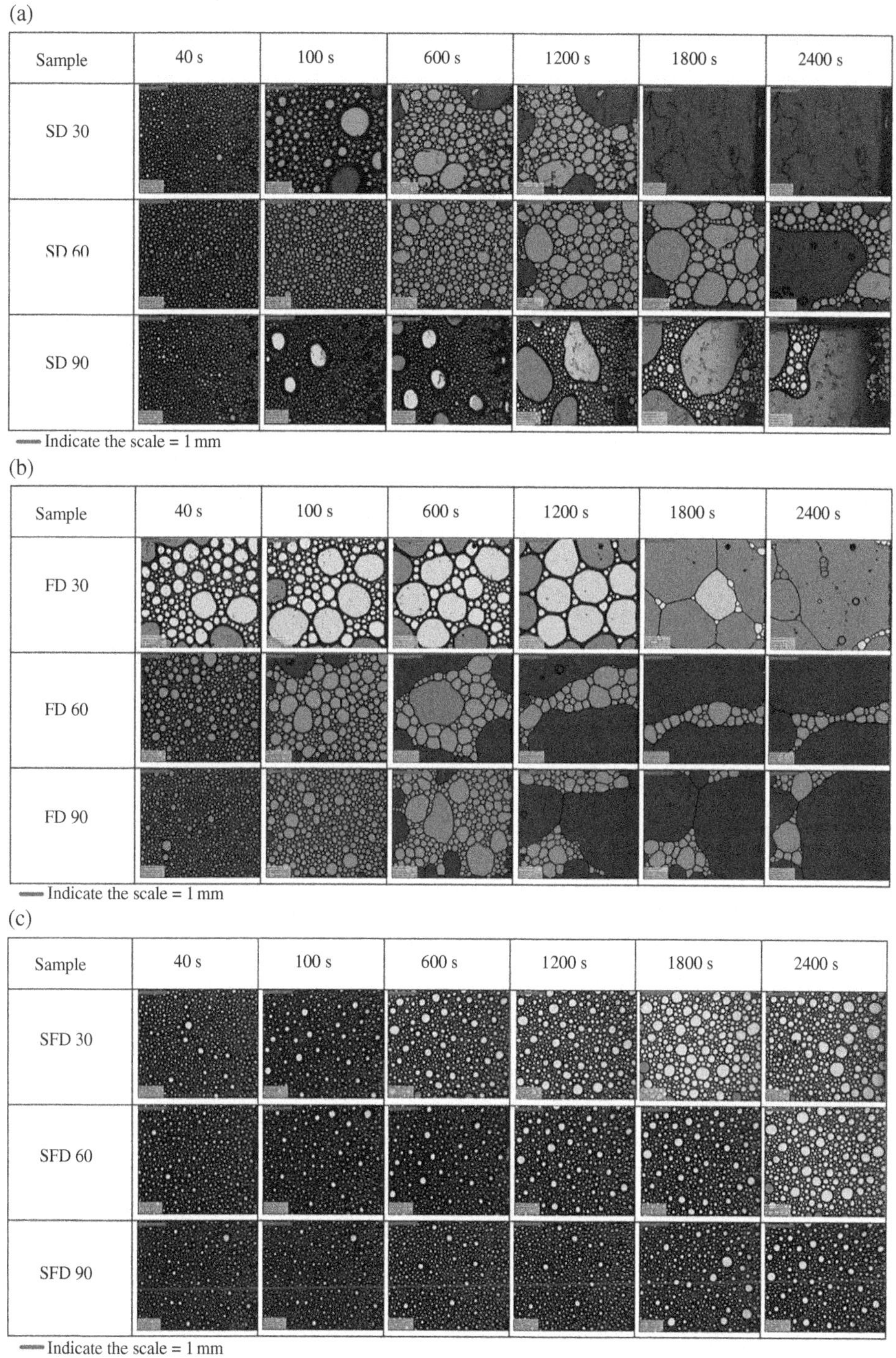

Figure 12.18 Effect of temperature on foaming ability and structure of bubbles of instant coffee obtained by (a) spray drying (SD), (b) freeze-drying (FD), and (c) spray-freeze-drying (SFD). *Source:* From Deotale et al. (2020), Figure 02 [p. 10] / With permission of Springer Nature.

a novel way of enhancing the foam stability of instant coffee powders. T spray-freeze-drying has been also used for the production of bromelain aerosols for pulmonary supplementation (Lavanya et al. 2020b). The total bromelain concentration of 413.73–462 mg g^{-1} and functional activity of 333.22–404.64 casein digestion $units^{-1}$ were observed in the obtained aerosols. Further, the *in-vitro* release study showed a steady bromelain release of 46–55% for 12 hours. Thus, this study confirmed the development of pulmonary supplementation through a spray-freeze-drying approach with higher retention and bioactivity of the bioactive compounds. Various synergistic benefits of this unified approach applied for encapsulation of volatile compounds, flavours, lipids and fatty acids, probiotics, vitamins, enzymes, and nucleic acids were comprehensively reviewed (Dutta et al. 2018). The above studies clearly dictate the advantageous features of the spray-freeze-drying in enhancing bioavailability, enhanced storage stability, ease of handling, quality retention, sustained release, and targeted delivery. Hence, the fabricated spray-freeze-dried encapsulates is characterized by a matrix type structure with higher reconstitution capacity which on integration with 3D printing would provide a novel way of delivering the food bioactives in a more convenient, attractive, and well-accepted manner based on individual preferences.

12.4.2.4 Conductive-Hydro Drying

Conductive-hydro drying is a novel drying method based on the formation of a 'window' that aids in heat and mass transfer. Here drying occurs through all the modes of heat transfer: conduction, convection, and radiation. The heat is being transferred from the hot water medium through the transparent plastic film (food-grade polyester) interface above which the products to be dried are placed uniformly in a thin layer (Preethi et al. 2020b). The conductive-hydro drying is a variant of RWD with differences that exist with the type of plastic film used. It involves the use of Mylar plastic sheet as a film interface between the hot medium and food material (Figure 12.19). Other than the type of plastic film, the mechanism of drying and the overall approach remains the same for both these drying methods.

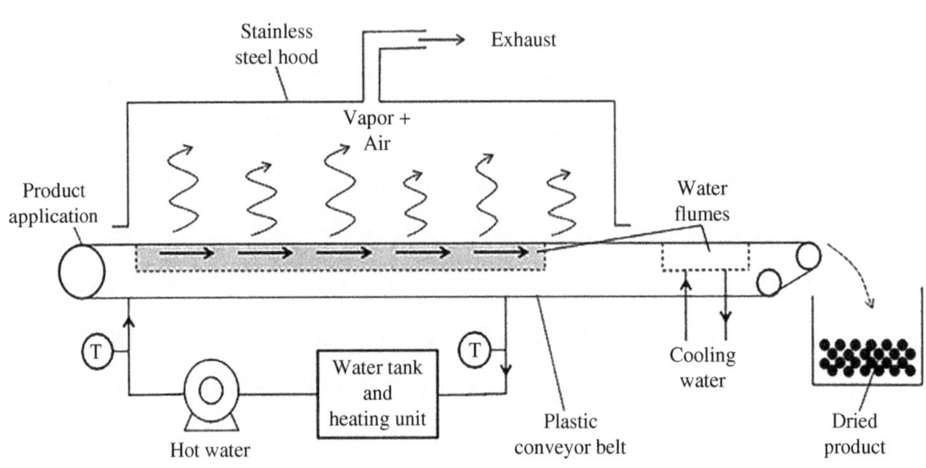

Figure 12.19 Schematic diagram of a RW™ drying system. *Source:* Moses et al. (2014).

The system components and working principle of this approach with energy consumption and product quality have been detailed by Raghavi et al. (2018). Owing to the high transport mechanism, the drying rate of conductive-hydro drying is higher that significantly reduces the drying time. Further, the product temperature at any point of drying is less than that of the hot water temperature because of the effects of evaporative cooling. Due to which this method of drying has been referred to as a non-thermal approach and is widely used for drying thermal sensitive compounds like proteins, enzymes, probiotics, and pigments (Moses et al. 2014).

Factors such as drying temperature, hot water temperature, moisture content of the product, product thickness, and optical properties of the plastic film significantly influence the drying characteristics and final end-quality of the powders/ flakes. The dried product is characterized by a smooth surface and possesses porous structure. Hence, it has a higher rehydration capacity suitable for production of functional fruit juice powders. More recently, this approach has been used for drying of beetroot pulp (Preethi et al. 2020d). Results showed that the dried beet-root pulp has a smooth flaky structure with higher porosity, superior colour, higher reconstitution, retention of polyphenolics, betalain pigment and possess a higher antioxidant capacity. It provides valuable insights on the preservation of natural food colourants of higher value through conductive-hydro drying approach. In another study, the same approach was used for production of egg white powder (Preethi et al. 2020c). As against spray drying and freeze-drying, the conductive-hydro-dried egg white powder showed superior product quality in terms of higher protein content, powder characteristics, foaming properties, and solubility. These results confirm the potentiality of conductive-hydro drying in achieving a superior product quality that is on par with conventional drying approaches of spray drying and freeze-drying.

The RWD would result in a matrix encapsulation that implies the embedment of bioactive material onto the wall material. The survival and the drying kinetics of *Lactobacillus fermentum* K73 during encapsulation of probiotics were reported based on the RWD technique (Aragon-Rojas et al. 2019). The obtained results of the viability and thin layer drying kinetics showed that the best fit with the modified Gompertz model and Midilli model. The best process conditions of drying temperature and time for RWD were found to be at 80 °C and 41 minutes, respectively. Thus, at these conditions, the obtained product had a cell viability of 9.1 log CFU g^{-1} that proved the survivability of probiotics in the encapsulated wall matrix. In another study, conductive-hydro drying has been explored for the encapsulation of probiotics *Lactiplantibacillus plantarum* (NCIM 2083) (Yoha et al. 2020a). The effect of encapsulation techniques (RWD and freeze-drying) on the cell viability of probiotics was reported. Among these two approaches, the RWD of probiotics had a cell viability of 9.18 $\log_{10}$ CFU g^{-1} with an encapsulation efficiency of 93%. These results confirmed that the RWD has a comparable efficiency as that of freeze-drying. Hence, conductive-hydro drying can be promisingly used for encapsulation of the aqueous dispersions of wall and core materials. Compare to other drying techniques, this approach is quite simple, requires less processing time, and has a great scope for commercialization (Yoha et al. 2019). Hence, conductive-hydro drying can be promisingly used as an encapsulation approach for the delivery of functional foods. In a 3D printing study on egg yolk and egg white fractions, the material supply has been prepared based on this drying method (Anukiruthika et al. 2020a). Using a similar approach, the aqueous dispersion with functional bioactive compounds can be used to obtain encapsulates in dried forms which can be further used as printing material supply for the delivery of personalized functional 3D printed foods.

12.5 Future Outlook and Prospects of Synergistic Approaches

Undoubtedly, 3D printing is the futuristic technology that would rule the food and nutraceutical markets. Different approaches that are employed for the fabrication methods of functional bioactive compounds were discussed in the earlier sections. As stated, 3D printing can be integrated with these delivery systems either by modifying the design of a 3D printer for simultaneous encapsulation during 3D printing or the developed bioactive compounds can be incorporated into the material supply used for 3D printing. The possible research opportunities that exist with the integrated technologies of 3D printing and encapsulation are not yet explored (Nachal et al. 2019). From the literature survey, it is evident that these two technologies have a great scope and can be integrated into a single process through design innovations of 3D food printers. More mechanistic studies are required in the proper understanding of the material response of bioactive incorporated material supply for the delivery of novel personalized functional foods. The combination of 3D printing technologies and encapsulation remains an active area of research in the biomedical field. An ample amount of studies are available on this combined technique for targeted delivery of bioactives especially in tissue engineering, drug formulations, and pharmaceuticals delivery (Adachi et al. 2020; Icten et al. 2017; Vithani et al. 2019a). Some of the key illustrations of these combined technologies in the biomedical field are detailed in this section for a better understanding of key elements and strategies adopted for integrating 3D printing with encapsulation. A similar approach can be applied for food applications in the delivery of functional customized foods. A structured review on the applications of encapsulation in the 3D printing of drugs was reported (Ursan et al. 2013). This review summarizes the techniques of 3D printing in combination with encapsulation for targeted release and formulation of drug delivery systems. A research study was reported on the encapsulation of cells in gelatin bio-ink based on extrusion printing technique where the effect of material properties on the print quality was examined (Schwartz et al. 2020). The preparation of bio-ink involves the dilution of culture to cell density as desired by the addition of a pre-warmed solution of gelatin. Later the samples were continuously agitated until achieving uniform dispersion of cell pellets through gel solutions and the formed hydrogel is taken for bioprinting. Results of the rheological properties of the material supply showed a 50% reduction in the steady shear viscosity with a minimum required force for the onset of the material flow of about 30%. These observations justify the obtained lower resolution of the 3D-printed hydrogel. In another study, the inkjet printing technology was used for the fabrication of antiviral (cidofovir)/ antitumor (paclitaxel) drug for mucosal application (Varan et al. 2017). In this study, the cidofovir-polyethylene glycol-polycaprolactone and paclitaxel: hydroxypropyl-β-cyclodextrin containing ink formulations prepared using double emulsion technique was 3D printed on the hydroxypropyl cellulose films using a piezoelectric inkjet printer through a nozzle size of 50 μm. The prepared inclusion complex had a particle size below 200 nm that is conducive for passing through the printing nozzle without clogging. Thus, the developed film showed highest anticancer efficacy with synergistic benefits in combined delivery of both the drugs in a sustained manner. More interestingly, the capability of 3D printing in producing a versatile personalized anti-acne drug incorporated with salicylic acid through and SLA and FDM techniques was investigated (Goyanes et al. 2016). On comparing the printing techniques, the SLA 3D printed drug showed higher printing resolution and higher drug loading of 1.9% (w/w) than the FDM approach. This was mainly because of the degradation of drugs during the hot-melt extrusion process in FDM printing. Thus, this study inferred that the SLA technique was more appropriate in 3D printing of personalized anti-acne device with salicylic acid. This study provides insights on not only the drug dosage based on individual requirement but also design and development of personalized drug-loaded device as per shape and size requirements of organ parts of the individual patients.

A novel method for the fabrication of dosage formulations containing drug (deflazacort) loaded in nanocapsules was proposed using combined technologies 3D printing and nanotechnology (Beck et al. 2017). Initially, the drug delivery carrier was prepared from poly (ε-caprolactone) and Eudragit® filaments with and without mannitol as channelling agent (Figure 12.20). The 3D printed tablets were immersed in deflazacort-loaded nanocapsules of 138 nm for the fabrication of printed tablets referred to as 'printlets'. The presence of the channelling agent significantly improved the drug loading capacity and showed a linear correlation with soaking time ($R^2 = 0.9739$). Further, the tablets printed with 50% infill percentage had exhibited a greater drug loading capacity of 0.27% (w/w) and rapid release of drugs. Thus, this study throws light on the conversion of nanosuspensions into a solid dosage through 3D printing for the development of a novel drug delivery medium for personalized medicines.

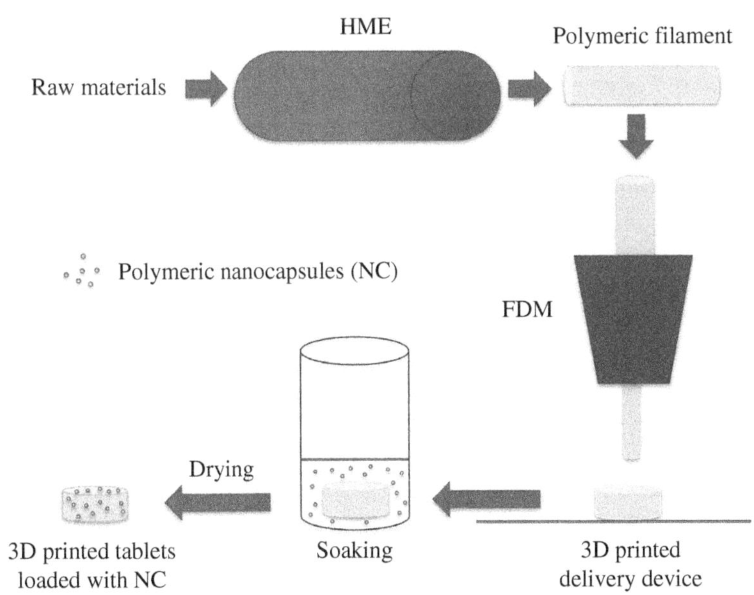

Figure 12.20 Illustration of the hot-melt extrusion process used for production of 3D printed tablets loaded with nanocapsules. *Source:* Beck et al. (2017) / With permission of Elsevier.

A similar approach could be conveniently applied for the delivery of food bioactives by integrating 3D printing with encapsulation. Attention must be paid to the formulation of food ink by considering the material properties and encapsulation technique. Also, the effect of pre-processing and post-processing on incorporated bioactives should be studied in designing a suitable food delivery system. The pre- and post-treatments should not reduce the biological activity and the encapsulated capsules must be stable until its delivery at the targeted site. The combined processes significantly increase the complexity. More process variables such as the selection of wall and core materials, encapsulation conditions, preparation of material supply with stable encapsulates, pre- and post-processing of 3D constructs, temperature, storage stability, release kinetics of the bioactives, absorption, and bioavailability would come into the picture. All these factors complicate the optimization of the 3D printing process. Most often, it requires modification of system design components for achieving the compatibility of the combined techniques encapsulation and 3D printing under a single process (Antoniraj et al. 2020; Leena et al. 2020a). More research needs on physicochemical properties, textural properties, rheological properties, thermal properties, solubility, diffusion mechanism, and solvation chemistry of biopolymers matrix, bioactive ingredients, and 3D printing ink must be studied for a better understanding of this integrated approach. Addressing these attributes would bridge up the knowledge gap that exists in combining 3D printing and encapsulation that would open up a new dimension of research on delivery of 3D printed nutraceutical and functional foods. This novel active area of research would provide a promising solution in alleviating malnourishment and global hunger by using a synergistic combination of a single combined process (3D printing and encapsulation) in addressing the nutritional needs at both individual level and population group without compromising the pleasure of eating and sensory properties of foods.

12.6 Barriers and Research Constraints

3D printing and encapsulation are two different technologies that have their own advantages and limitations. This in combination would provide not only the synergistic benefits for the development of personalized functional foods but also add on to the complications associated with the process (Nachal et al. 2019). The potential benefits of integrated technologies such as 3D printing and encapsulation are realizable from the advancements encountered in the modern biomedical world. A better understanding of the frontiers of this novel combined technology would lead to many innovations in the state-of-the-art of printing. This includes the development of the 3D printed double chamber or tablet-in-tablet dosage forms (Duo Tablet), drugs delivered in customized colour and shapes, five-in-one polypill, oral capsules with two compartments and two different colours, buccal wafers, oral dispersible films, customized beads and catheters, topical medications, transdermal microneedles, vaginal rings, intrauterine devices, medical implants, and personalized inhalers (Wallis et al. 2020). All these advancements would transform the conventional approach of drug delivery that leads to precision medicine. In this regard, the prospects of the combined 3D printing and encapsulation seem to be enormous;

however, this status is available only at the lab-scale prototype model which has yet to be explored for its large-scale practical applications. The integration of 3D printing of foods with encapsulation not only aids in the delivery of functionalized food but also provides insights for advanced quality sensing and preserving applications in food packaging (active and smart packaging) (Kalpana et al. 2019). The progress of multi-layer food packaging would remain a solid base for realizing the opportunities that exist in integrating 3D printing and encapsulation for the fabrication of functional food packages (Anukiruthika et al. 2020b). Nevertheless, food 3D printing and encapsulation would provide a promising solution as a unified technology in developing personalized functional foods that pave the way towards 'precision nutrition'. With advancements in biotechnology, the foods could be delivered based on analyzing the genetic pool of the individuals that provide nutritional food delivery as well as nutraceutical delivery in treating the ailment under combined process (Kumar et al. 2020). However, this takes a long time for practical implementation. Despite these aspects the process simplifications/ modifications/ progressions of the combined processes would increase the challenges. The feasibility of 3D printing for consumer's acceptance as regular food in their diet is questionable. Often, food neophobia limits the applicability of 3D printing. Also, consumer's psychological beliefs that food printing is not a natural way of producing food, safety, and quality regulations remain as a metal barrier for exploring the benefits of 3D printed foods (Manstan and McSweeney 2020). Some of the technological barriers of 3D printing and encapsulation at an individual basis are fabrication speed, selection of compatible printing material, wall and core material, complex physiochemical interactions with macro and micronutrients, design requirements, process optimization, post-processing, edibility, and safety. Food is a perishable commodity consists of a complex combination of macro and micronutrients, enzymes, flavours, aroma, beneficial microbes, and so on. The reduction in the molecular size will significantly change the behaviour and response of the food material. The reduction in the size of particles will increase the surface area to volume ratio that remains chemically and biologically more active at the micro and nanoscale of food constituent than at the macro level. The nano foods are already facing constraints concerning toxicity and safety (Bhushani and Anandharamakrishnan 2014). So, the integration of 3D printing with encapsulation (micro and nano) would remain to be debatable. As of today's date, there is no proper protocol and standard practices available for the fabrication of food using 3D printing. Also, legal frameworks related to nano foods are yet to be regulated. In this context, government organizations and statutory bodies like Food and Drug Administration (FDA) should regulate the current protocol of good manufacturing practices (GMP) to comply with the standards by thoroughly analyzing these integrated approaches. The declaration of these materials as generally recognized as safe (GRAS) must require more detailed studies on the better understanding of material interactions and physiological response of these foods upon consumption (Baiano 2020). Addressing all the challenges would reform the food manufacturing approaches with future advancements in design of multi-head printers and multi-material printing for the delivery of functional bioactives through integration of 3D printing with encapsulation. Advancements of material design, 3D model scanning, data acquisition, material chemistry, biophysics, software configurations, printing technologies, encapsulation techniques, gastronomy, and delivery systems would transform the 'balanced diet' into a 'digitalized diet' that provides both personalized nutrition as well as precise nutrition.

12.7 Conclusion

The emergence of 3D printing often incurs innovations and numerous benefits in the designing of customized and personalized foods. The changing consumer decisions and desires for fascinating foods with pleasant eating experiences would nurture the idea of integrated concepts of combined technologies. In this cohort, 3D printing and encapsulation are two novel technologies that offer several health benefits that on integration would undoubtedly yield promising solutions in the delivery of customized and personalized functional foods. The synergistic benefits of these technologies not only satisfy global hunger but also protects global food security. However, the practical implementation of these integrated technologies is questionable as most of these concepts are lab-scale proven technology. Hence, the present chapter focused on the various benefits of combining 3D printing with encapsulation. The principle and underlying mechanism of each of these techniques along with several approaches used for the delivery of functional bioactive at the targeted sites are discussed in detail. Various prospects of the combined technology with possible solutions and innovations in the delivery of functional foods and nutraceuticals are described along with the challenges encountered in its implementation. Hence, a thorough understanding of the material response in terms of allergy, toxicity, safety, and edibility must be required. The short- and long-term effects associated with the consumption of foods developed by these combined technologies should be examined for the safe administration of functionalized foods. Apart from these constraints, the challenges associated with process optimization, standard protocols, and safe dosage are yet to be regulated. Further, the risk associated with piracy and software trading adds to its complexity. Thus, the present chapter aimed at unwinding the research scope and forecasted the potential futuristic opportunities that exist with this promising combined technology of 3D printing and encapsulation. Certainly, the present work remains a valuable resource base for futuristic advancements and provides useful insights on material formulations, system design and components, process efficacy, encapsulation efficiency, bioavailability, release profile and targeted delivery of the 3D printed functional foods with micro and nano encapsulates.

References

Adachi, K., Yoshida, K., Makino, M. et al. (2020). Formation of liposomes containing Pre-Gel solution and 3D-printing applications by droplet-shooting method. *ECS Transactions* **98** (13): 85.

Anandharamakrishnan, C. (2014a). Bioactive entrapment using lipid-based nanocarrier technology. In: *Techniques for Nanoencapsulation of Food Ingredients* (ed. C. Anandharamakrishnan), pp. 17–28. Springer.

Anandharamakrishnan, C. (2014b). Drying techniques for nanoencapsulation. In: *Techniques for Nanoencapsulation of Food Ingredients* (ed. C. Anandharamakrishnan), pp. 51–60. Springer.

Anandharamakrishnan, C. (2014c). Liquid-based nanoencapsulation techniques. In: *Techniques for Nanoencapsulation of Food Ingredients* (ed. C. Anandharamakrishnan), pp. 29–41. Springer.

Anandharamakrishnan, C. (2014d). Techniques for formation of nanoemulsions. In: *Techniques for Nanoencapsulation of Food Ingredients* (ed. C. Anandharamakrishnan), pp. 7–16. Springer.

Anandharamakrishnan, C. and Ishwarya, P.S. (2015). *Spray Drying Techniques for Food Ingredient Encapsulation*. Wiley.

Anandharamakrishnan, C., Rielly, C.D., and Stapley, A.G.F. (2010). Spray-freeze-drying of whey proteins at sub-atmospheric pressures. *Dairy Science & Technology* **90** (2): 321–334.

Antoniraj, M.G., Leena, M.M., Moses, J.A., and Anandharamakrishnan, C. (2020). Cross-linked chitosan microparticles preparation by modified three fluid nozzle spray drying approach. *International Journal of Biological Macromolecules* **147**: 1268–1277.

Anukiruthika, T., Dutta, S., Moses, J.A. et al. (2019). Modern applications of supercritical fluids extraction in food toxicology. In: *Innovative Food Processing Technologies: A Comprehensive Review*, vol. **2** (eds. K. Knoerzer and K. Muthukumarappan), pp. 640–659. Elsevier doi: B978-0-08-100596-5.22939-9.

Anukiruthika, T., Moses, J.A., and Anandharamakrishnan, C. (2020a). 3D printing of egg yolk and white with rice flour blends. *Journal of Food Engineering* **265**: 109691. https://doi.org/10.1016/j.jfoodeng.2019.109691.

Anukiruthika, T., Sethupathy, P., Wilson, A. et al. (2020b). Multilayer packaging: advances in preparation techniques and emerging food applications. *Comprehensive Reviews in Food Science and Food Safety* **19**: 1156–1186.

Aragon-Rojas, S., Quintanilla-Carvajal, M.X., Hernandez-Sanchez, H. et al. (2019). Encapsulation of *Lactobacillus fermentum* K73 by refractance window drying. *Scientific Reports* **9** (1): 1–15.

Assadpour, E. and Jafari, S.M. (2019). Advances in spray-drying encapsulation of food bioactive ingredients: from microcapsules to nanocapsules. *Annual Review of Food Science and Technology* **10**: 103–131.

Attama, A.A., Momoh, M.A., and Builders, P.F. (2012). Lipid nanoparticulate drug delivery systems: a revolution in dosage form design and development. *Recent Advances in Novel Drug Carrier Systems* **5**: 107–140.

Azam, R.S.M., Zhang, M., Bhandari, B., and Yang, C. (2018a). Effect of different gums on features of 3D printed object based on vitamin-D enriched orange concentrate. *Food Biophysics* **13** (3): 250–262.

Azam, S.M.R., Zhang, M., Mujumdar, A.S., and Yang, C. (2018b). Study on 3D printing of orange concentrate and material characteristics. *Journal of Food Process Engineering* **41** (5): 1–10. https://doi.org/10.1111/jfpe.12689.

Baiano, A. (2020). 3D printed foods: a comprehensive review on technologies, nutritional value, safety, consumer attitude, regulatory framework, and economic and sustainability issues. *Food Reviews International*, pp. 1–31.

Barreras-Urbina, C.G., Ramirez-Wong, B., Lopez-Ahumada, G.A. et al. (2016). Nano-and micro-particles by nanoprecipitation: possible application in the food and agricultural industries. *International Journal of Food Properties* **19** (9): 1912–1923.

Beck, R.C.R., Chaves, P.S., Goyanes, A. et al. (2017). 3D printed tablets loaded with polymeric nanocapsules: an innovative approach to produce customized drug delivery systems. *International Journal of Pharmaceutics* **528** (1–2): 268–279.

Bharathi, S.K.V., Moses, J.A., and Anandharamakrishnan, C. (2018). Nano and microencapsulation using food grade polymers. In: *Polymers for Food Applications* (ed. T.J. Gutierrez), pp. 357–400. Springer.

Bhushani, J.A. and Anandharamakrishnan, C. (2014). Electrospinning and electrospraying techniques: potential food based applications. *Trends in Food Science & Technology* **38** (1): 21–33.

Charles, A.L., Abdillah, A.A., Saraswati, Y.R. et al. (2020). Characterization of freeze-dried microencapsulation tuna fish oil with arrowroot starch and maltodextrin. *Food Hydrocolloids* **112**: 106281.

Chauhan, G., Shaik, A.A., Kulkarni, N.S., and Gupta, V. (2020). The preparation of lipid-based drug delivery system using melt extrusion. *Drug Discovery Today* **25** (11): 1930–1943.

Davidov-Pardo, G., Joye, I.J., and McClements, D.J. (2015). Encapsulation of resveratrol in biopolymer particles produced using liquid antisolvent precipitation. Part 1: preparation and characterization. *Food Hydrocolloids* **45**: 309–316.

Deotale, S.M., Dutta, S., Moses, J.A., and Anandharamakrishnan, C. (2020). Stability of instant coffee foam by nanobubbles using spray-freeze drying technique. *Food and Bioprocess Technology* **13** (11): 1866–1877.

Depypere, F., Pieters, J.G., and Dewettinck, K. (2009). PEPT visualisation of particle motion in a tapered fluidised bed coater. *Journal of Food Engineering* **93** (3): 324–336.

Derossi, A., Husain, A., Caporizzi, R., and Severini, C. (2020). Manufacturing personalized food for people uniqueness. An overview from traditional to emerging technologies. *Critical Reviews in Food Science and Nutrition* **60** (7): 1141–1159.

Dutta, S., Moses, J.A., and Anandharamakrishnan, C. (2018). Modern frontiers and applications of spray-freeze-drying in design of food and biological supplements. *Journal of Food Process Engineering* **41** (8): e12881.

Ezhilarasi, P.N., Indrani, D., Jena, B.S., and Anandharamakrishnan, C. (2013a). Freeze drying technique for microencapsulation of Garcinia fruit extract and its effect on bread quality. *Journal of Food Engineering* **117** (4): 513–520.

Ezhilarasi, P.N., Karthik, P., Chhanwal, N., and Anandharamakrishnan, C. (2013b). Nanoencapsulation techniques for food bioactive components: a review. *Food and Bioprocess Technology* **6** (3): 628–647. https://doi.org/10.1007/s11947-012-0944-0.

Fang, Z. and Bhandari, B. (2012). Encapsulation techniques for food ingredient systems. In: *Food Materials Science and Engineering* (eds. B. Bhandari and Y.H. Roos), 320–348. Wiley-blackwell https://doi.org/10.1002/9781118373903.

Feng, C., Zhang, M., and Bhandari, B. (2018). Materials properties of printable edible inks and printing parameters optimization during 3D printing: a review. *Critical Reviews in Food Science and Nutrition* **8398**: 1–8. https://doi.org/10.1080/10408398.2018.1481823.

Frey, C. (2014). Fluid bed coating-based microencapsulation. In: *Microencapsulation in the Food Industry*, pp. 65–79. Elsevier.

Gheonea, I., Aprodu, I., Circiumaru, A. et al. (2021). Microencapsulation of lycopene from tomatoes peels by complex coacervation and freeze-drying: evidences on phytochemical profile, stability and food applications. *Journal of Food Engineering* **288**: 110166.

Goyanes, A., Det-Amornrat, U., Wang, J. et al. (2016). 3D scanning and 3D printing as innovative technologies for fabricating personalized topical drug delivery systems. *Journal of Controlled Release* **234**: 41–48.

Guo, C., Zhang, M., and Devahastin, S. (2021). Color/aroma changes of 3D-printed buckwheat dough with yellow flesh peach as triggered by microwave heating of gelatin-gum Arabic complex coacervates. *Food Hydrocolloids* **112**: 106358.

Gupta, S., Khan, S., Muzafar, M. et al. (2016). Encapsulation: entrapping essential oil/flavors/ aromas in food. In: *Encapsulations: Nanotechnology in the Agri-Food Industry*, vol. **2** (ed. A.M. Grumezescu), 229, 229–268, 268. Academic Press https://doi.org/10.1016/B978-0-12-804307-3.00006-5.

He, C., Zhang, M., and Fang, Z. (2019). 3D printing of food: pretreatment and post-treatment of materials. *Critical Reviews in Food Science and Nutrition* **60** (14): 1–14.

Hosseini, S.F., Ramezanzade, L., and McClements, D.J. (2021). Recent advances in nanoencapsulation of hydrophobic marine bioactives: bioavailability, safety, and sensory attributes of nano-fortified functional foods. *Trends in Food Science & Technology* Elsevier **109**: 322–339.

Hu, K. and McClements, D.J. (2014). Fabrication of surfactant-stabilized zein nanoparticles: a pH modulated antisolvent precipitation method. *Food Research International* **64**: 329–335.

Huang, W., Zheng, Q., Sun, W. et al. (2007). Levofloxacin implants with predefined microstructure fabricated by three-dimensional printing technique. *International Journal of Pharmaceutics* **339** (1–2): 33–38.

Icten, E., Purohit, H.S., Wallace, C. et al. (2017). Dropwise additive manufacturing of pharmaceutical products for amorphous and self emulsifying drug delivery systems. *International Journal of Pharmaceutics* **524** (1–2): 424–432.

Ishwarya, P.S. and Nisha, P. (2021). Advances and prospects in the food applications of pectin hydrogels. *Critical Reviews in Food Science and Nutrition*, pp. 1–25.

Ishwarya, P.S., Anandharamakrishnan, C., and Stapley, A.G.F. (2015). Spray-freeze-drying: a novel process for the drying of foods and bioproducts. *Trends in Food Science & Technology* **41** (2): 161–181.

Jana, S. and Jana, S. (2017). Natural polymeric biodegradable nanoblend for macromolecules delivery. In: *Recent Developments in Polymer Macro, Micro and Nano Blends* (eds. P.M. Visakh, G. Markovic and D. Pasquini), pp. 289–312. Woodhead publishing https://doi.org/10.1016/C2014-0-03998-6.

Jansen, M., Guariguata, M.R., Raneri, J.E. et al. (2020). Food for thought: the underutilized potential of tropical tree-sourced foods for 21st century sustainable food systems. *People and Nature* **2** (4): 1006–1020.

Jia, W., Gungor-Ozkerim, P.S., Zhang, Y.S. et al. (2016). Direct 3D bioprinting of perfusable vascular constructs using a blend bioink. *Biomaterials* **106**: 58–68.

Jin, J., Cai, L., Jia, Y.-G. et al. (2019). Progress in self-healing hydrogels assembled by host—guest interactions: preparation and biomedical applications. *Journal of Materials Chemistry B* **7** (10): 1637–1651.

Kaga, K., Honda, M., Adachi, T. et al. (2018). Nanoparticle formation of PVP/astaxanthin inclusion complex by solution-enhanced dispersion by supercritical fluids (SEDS): effect of PVP and astaxanthin Z-isomer content. *The Journal of Supercritical Fluids* **136**: 44–51.

Kalpana, S., Priyadarshini, S.R., Leena, M.M. et al. (2019). Intelligent packaging: trends and applications in food systems. *Trends in Food Science & Technology* **93** (September): 145–157. https://doi.org/10.1016/j.tifs.2019.09.008.

Karakurt, I., Aydougdu, A., Cikrikci, S. et al. (2020). Stereolithography (SLA) 3D printing of ascorbic acid loaded hydrogels: a controlled release study. *International Journal of Pharmaceutics* **584**: 119428.

Karthik, P. and Anandharamakrishnan, C. (2013). Microencapsulation of docosahexaenoic acid by spray-freeze-drying method and comparison of its stability with spray-drying and freeze-drying methods. *Food and Bioprocess Technology* **6** (10): 2780–2790.

Karthik, P., Ezhilarasi, P.N., and Anandharamakrishnan, C. (2017). Challenges associated in stability of food grade nanoemulsions. *Critical Reviews in Food Science and Nutrition* **57** (7): 1435–1450.

Katakam, P., Dey, B., Assaleh, F.H. et al. (2015). Top-down and bottom-up approaches in 3D printing technologies for drug delivery challenges. *Critical Reviews™ in Therapeutic Drug Carrier Systems* **32**, 32 (1, 1).

Khosa, A., Reddi, S., and Saha, R.N. (2018). Nanostructured lipid carriers for site-specific drug delivery. *Biomedicine & Pharmacotherapy* **103**: 598–613.

Krishnaraj, P., Anukiruthika, T., Choudhary, P. et al. (2019). 3D extrusion printing and post-processing of fibre-rich snack from indigenous composite flour. *Food and Bioprocess Technology* **12** (10): 1776–1786. https://doi.org/10.1007/s11947-019-02336-5.

Kumar, M., Bishnoi, R.S., Shukla, A.K., and Jain, C.P. (2019). Techniques for formulation of nanoemulsion drug delivery system: a review. *Preventive Nutrition and Food Science* **24** (3): 225.

Kumar, S., Anukiruthika, T., Dutta, S. et al. (2020). Iron deficiency anemia: a comprehensive review on iron absorption, bioavailability and emerging food fortification approaches. *Trends in Food Science & Technology* **99**: 58–75. https://doi.org/10.1016/j.tifs.2020.02.021.

Lasekan, O., Ng, S., Teoh, L. et al. (2016). Current trends in nano encapsulation of flavours and aromas. *Journal of Food Bioengineering and Nanoprocessing* **1** (1–2): 1–3.

Lavanya, M.N., Kathiravan, T., Moses, J.A., and Anandharamakrishnan, C. (2019). Influence of spray-drying conditions on microencapsulation of fish oil and chia oil. *Drying Technology* **38** (3): 1–14.

Lavanya, M.N., Dutta, S., Moses, J.A., and Anandharamakrishnan, C. (2020a). Development of β-carotene aerosol formulations using a modified spray dryer. *Journal of Food Process Engineering* **43** (3): e13233.

Lavanya, M.N., Preethi, R., Moses, J.A., and Anandharamakrishnan, C. (2020b). Production of bromelain aerosols using spray-freeze-drying technique for pulmonary supplementation. *Drying Technology* **39** (3): 1–13: Special Issue for the 10th Asia Pacific Drying Conference (ADC 2019).

Leena, M.M., Antoniraj, M.G., Moses, J.A., and Anandharamakrishnan, C. (2020a). Three fluid nozzle spray drying for co-encapsulation and controlled release of curcumin and resveratrol. *Journal of Drug Delivery Science and Technology* **57**: 101678.

Leena, M.M., Bharathi, S.K.V., Moses, J.A., and Anandharamakrishnan, C. (2020b). Potential applications of nanofibers in beverage industry. In: *Nanoengineering in the Beverage Industry* (eds. A.M. Grumezescu and A.M. Holban), pp. 333–368. Elsevier.

Li, H., Xu, Y., Sun, X. et al. (2018). Stability, bioactivity, and bioaccessibility of fucoxanthin in zein-caseinate composite nanoparticles fabricated at neutral pH by antisolvent precipitation. *Food Hydrocolloids* **84**: 379–388.

Lippacher, A., Müller, R.H., and Mäder, K. (2000). Investigation on the viscoelastic properties of lipid based colloidal drug carriers. *International Journal of Pharmaceutics* **196** (2): 227–230.

Lippacher, A., Müller, R.H., and Mäder, K. (2002). Semisolid SLN™ dispersions for topical application: influence of formulation and production parameters on viscoelastic properties. *European Journal of Pharmaceutics and Biopharmaceutics* **53** (2): 155–160.

Liu, H., Zhou, H., Lan, H., and Liu, T. (2017). Organ regeneration: integration application of cell encapsulation and 3D bioprinting. *Virtual and Physical Prototyping* **12** (4): 279–289.

Liu, Z., Zhang, M., and Yang, C. (2018). Dual extrusion 3D printing of mashed potatoes/strawberry juice gel. *LWT* **96**: 589–596.

Liu, Z., Bhandari, B., and Zhang, M. (2020). Incorporation of probiotics (*Bifidobacterium animalis* subsp. Lactis) into 3D printed mashed potatoes: effects of variables on the viability. *Food Research International* **128**: 108795.

Manstan, T. and McSweeney, M.B. (2020). Consumers' attitudes towards and acceptance of 3D printed foods in comparison with conventional food products. *International Journal of Food Science & Technology* **55** (1): 323–331.

Market Analysis Report. (2020). Nutraceutical market size, share & trends analysis report by product (dietary supplements, functional foods, functional beverages), by region, and segment forecasts, pp. 2020–2027. grand view research. https://www.grandviewresearch.com/industry-analysis/nutraceuticals-market (accessed 6 March 2021).

Melocchi, A., Parietti, F., Maroni, A. et al. (2016). Hot-melt extruded filaments based on pharmaceutical grade polymers for 3D printing by fused deposition modeling. *International Journal of Pharmaceutics* **509** (1–2): 255–263.

Mohammadabadi, M.R. and Mozafari, M.R. (2018). Enhanced efficacy and bioavailability of thymoquinone using nanoliposomal dosage form. *Journal of Drug Delivery Science and Technology* **47**: 445–453.

Molitch-Hou, M. (2014). The 3D fruit printer and the raspberry that tasted like a strawberry. https://3dprintingindustry.com/news/3d-fruit-printer-raspberry-tasted-like-strawberry-27713/ (1 September 2020).

Moses, J.A., Norton, T., Alagusundaram, K., and Tiwari, B.K. (2014). Novel drying techniques for the food industry. *Food Engineering Reviews* **6** (3): 43–55.

Mozafari, M.R. (2005). *Nanoliposomes: From Fundamentals to Recent Developments*. Trafford.

Mozafari, M.R. and Khosravi-Darani, K. (2007). An overview of liposome-derived nanocarrier technologies. In: *Nanomaterials and Nanosystems for Biomedical Applications* (ed. M.R. Mozafari). Dordrecht: Springer https://doi.org/10.1007/978-1-4020-6289-6_7.

Muhoza, B., Xia, S., Wang, X. et al. (2020). Microencapsulation of essential oils by complex coacervation method: preparation, thermal stability, release properties and applications. *Critical Reviews in Food Science and Nutrition*, pp. 1–20.

Muller, R.H., Alexiev, U., Sinambela, P., and Keck, C.M. (2016). Nanostructured lipid carriers (NLC): the second generation of solid lipid nanoparticles. In: *Percutaneous Penetration Enhancers Chemical Methods in Penetration Enhancement*, 161–185. Springer.

Nachal, N., Moses, J.A., Karthik, P., and Anandharamakrishnan, C. (2019). Applications of 3D printing in food processing. *Food Engineering Reviews* **11** (3): 123–141. https://doi.org/10.1007/s12393-019-09199-8.

Nakagawa, K., Surassmo, S., Min, S.-G., and Choi, M.-J. (2011). Dispersibility of freeze-dried poly (epsilon-caprolactone) nanocapsules stabilized by gelatin and the effect of freezing. *Journal of Food Engineering* **102** (2): 177–188.

Oliveira, S.M., Gruppi, A., Vieira, M.V. et al. (2020). How additive manufacturing can boost the bioactivity of baked functional foods. *Journal of Food Engineering* **294**: 110394.

Park, S.M., Kim, H.W., and Park, H.J. (2020). Callus-based 3D printing for food exemplified with carrot tissues and its potential for innovative food production. *Journal of Food Engineering* **271**: 109781.

Parthasarathi, S. and Anandharamakrishnan, C. (2016). Enhancement of oral bioavailability of vitamin E by spray-freeze drying of whey protein microcapsules. *Food and Bioproducts Processing* **100**: 469–476.

Pasrija, D., Ezhilarasi, P.N., Indrani, D., and Anandharamakrishnan, C. (2015). Microencapsulation of green tea polyphenols and its effect on incorporated bread quality. *LWT- Food Science and Technology* **64** (1): 289–296.

Pisoschi, A.M., Pop, A., Cimpeanu, C. et al. (2018). Nanoencapsulation techniques for compounds and products with antioxidant and antimicrobial activity-a critical view. *European Journal of Medicinal Chemistry* **157**: 1326–1345.

Preethi, R., Leena, M.M., Moses, J.A., and Anandharamakrishnan, C. (2020a). Biopolymer nanocomposites and its application in food processing. In: *Green Nanomaterials* (eds. S. Ahmed and W. Ali), pp. 283–317. Springer.

Preethi, R., Moses, J.A., and Anandharamakrishnan, C. (2020b). Effect of conductive hydro-drying on physiochemical and functional properties of two pulse protein extracts: green gram (*Vigna radiata*) and black gram (*Vigna mungo*). *Food Chemistry* **343**: 128551.

Preethi, R., Shweta, D., Moses, J.A., and Anandharamakrishnan, C. (2020c). Conductive hydro drying as an alternative method for egg white powder production. *Drying Technology* **39** (3): 1–13.

Preethi, R., Shweta, D., Moses, J.A., and Anandharamakrishnan, C. (2020d). Conductive hydro drying of beetroot (*Beta vulgaris* L) pulp: insights for natural food colorant applications. *Journal of Food Process Engineering* **43** (12): e13557.

Raghavi, L.M., Moses, J.A., and Anandharamakrishnan, C. (2018). Refractance window drying of foods: a review. *Journal of Food Engineering* **222**: 267–275.

Sarkar, N. and Bose, S. (2019). Liposome-encapsulated curcumin-loaded 3D printed scaffold for bone tissue engineering. *ACS Applied Materials & Interfaces* **11** (19): 17184–17192.

Saupe, A. and Rades, T. (2006). Solid lipid nanoparticles. In: *Nanocarrier Technologies*, pp. 41–50. Springer.

Schutyser, M.A.I., Houlder, S., de Wit, M. et al. (2018). Fused deposition modelling of sodium caseinate dispersions. *Journal of Food Engineering* **220**: 49–55.

Schwartz, R., Malpica, M., Thompson, G.L., and Miri, A.K. (2020). Cell encapsulation in gelatin bioink impairs 3D bioprinting resolution. *Journal of the Mechanical Behavior of Biomedical Materials* **103**: 103524.

Seoane-Viano, I., Januskaite, P., Alvarez-Lorenzo, C. et al. (2021). Semi-solid extrusion 3D printing in drug delivery and biomedicine: personalised solutions for healthcare challenges. *Journal of Controlled Release* **332**: 367–389.

da Silva Santos, V., Ribeiro, A.P.B., and Santana, M.H.A. (2019). Solid lipid nanoparticles as carriers for lipophilic compounds for applications in foods. *Food Research International* **122**: 610–626.

Sionkowska, A. (2011). Current research on the blends of natural and synthetic polymers as new biomaterials. *Progress in Polymer Science* **36** (9): 1254–1276.

Yoha, K.S., Anukiruthika, T., Anila, W. et al. (2021). 3D printing of encapsulated probiotics: effect of different post-processing methods on the stability of *Lactiplantibacillus plantarum* (NCIM 2083) under static *in-vitro* digestion conditions and during storage. *LWT* **146**: 111461.

Taylor, T.M., Weiss, J., Davidson, P.M., and Bruce, B.D. (2005). Liposomal nanocapsules in food science and agriculture. *Critical Reviews in Food Science and Nutrition* **45** (7–8): 587–605.

Tsintavi, E., Rekkas, D.M., and Bettini, R. (2020). Partial tablet coating by 3D printing. *International Journal of Pharmaceutics* **581**: 119298.

Uribe-Wandurraga, Z.N., Zhang, L., Noort, M.W.J. et al. (2020). Printability and physicochemical properties of microalgae-enriched 3D-printed snacks. *Food and Bioprocess Technology* **13** (11): 2029–2042.

Ursan, I.D., Chiu, L., and Pierce, A. (2013). Three-dimensional drug printing: a structured review. *Journal of the American Pharmacists Association* **53** (2): 136–144.

Vancauwenberghe, V., Mbong, V.B.M., Vanstreels, E. et al. (2019). 3D printing of plant tissue for innovative food manufacturing: encapsulation of alive plant cells into pectin based bio-ink. *Journal of Food Engineering* **263**: 454–464.

Varan, C., Wickstrom, H., Sandler, N. et al. (2017). Inkjet printing of antiviral PCL nanoparticles and anticancer cyclodextrin inclusion complexes on bioadhesive film for cervical administration. *International Journal of Pharmaceutics* **531** (2): 701–713.

Vieira, M.V., Oliveira, S.M., Amado, I.R. et al. (2020). 3D printed functional cookies fortified with *Arthrospira platensis*: evaluation of its antioxidant potential and physical-chemical characterization. *Food Hydrocolloids* **107**: 105893.

Vishali, D.A., Monisha, J., Sundari, S.S.K. et al. (2019). Spray freeze drying: emerging applications in drug delivery. *Journal of Controlled Release* **300**: 93–101.

Vithani, K., Goyanes, A., Jannin, V. et al. (2019a). A proof of concept for 3D printing of solid lipid-based formulations of poorly water-soluble drugs to control formulation dispersion kinetics. *Pharmaceutical Research* **36** (7): 1–13.

Vithani, K., Goyanes, A., Jannin, V. et al. (2019b). An overview of 3D printing technologies for soft materials and potential opportunities for lipid-based drug delivery systems. *Pharmaceutical Research* **36** (1): 4.

Vo, D.T., Saravana, P.S., Woo, H.-C., and Chun, B.-S. (2018). Fucoxanthin-rich oil encapsulation using biodegradable polyethylene glycol and particles from gas-saturated solutions technique. *Journal of CO_2 Utilization* **26**: 359–369.

Waghmare, R.B., Choudhary, P., Moses, J.A. et al. (2021). Trends in approaches to assist freeze-drying of food: a cohort study on innovations. *Food Reviews International*: 1–22.

Wallis, M., Al-Dulimi, Z., Tan, D.K. et al. (2020). 3D printing for enhanced drug delivery: current state-of-the-art and challenges. *Drug Development and Industrial Pharmacy* **46** (9): 1385–1401.

Wang, M., Fu, Y., Chen, G. et al. (2018). Fabrication and characterization of carboxymethyl chitosan and tea polyphenols coating on zein nanoparticles to encapsulate β-carotene by anti-solvent precipitation method. *Food Hydrocolloids* **77**: 577–587.

Yi, J., Peng, G., Zheng, S. et al. (2021). Fabrication of whey protein isolate-sodium alginate nanocomplex for curcumin solubilization and stabilization in a model fat-free beverage. *Food Chemistry* **348**: 129102.

Yoha, K.S., Priyadarshini, S.R., Moses, J.A., and Anandharamakrishnan, C. (2019). *Refractance Window Drying and Its Applications in Food Processing*. New York: Apple Academic Press, Taylor & Francis.

Yoha, K.S., Moses, J.A., and Anandharamakrishnan, C. (2020a). Conductive hydro drying through refractance window drying—an alternative technique for drying of *Lactobacillus plantarum* (NCIM 2083). *Drying Technology* **38** (5–6): 610–620.

Yoha, K.S., Moses, J.A., and Anandharamakrishnan, C. (2020b). Effect of encapsulation methods on the physicochemical properties and the stability of *Lactobacillus plantarum* (NCIM, 2083) in synbiotic powders and in-vitro digestion conditions. *Journal of Food Engineering* **283**: 110033.

Zhang, L., Lou, Y., and Schutyser, M.A.I. (2018). 3D printing of cereal-based food structures containing probiotics. *Food Structure* **18**: 14–22.

13

Integrating Electrohydrodynamic Processes with Food 3D Printing

CHAPTER MENU

13.1 Introduction

3D printing is rolling out to be the technology of the day that reforms almost all of the industrial sectors. As an additive technology, 3D printing not only contributes to the growth of society but also ensures environmental benefits (Nachal et al. 2019). With recent innovations in a digitalized world, 3D printing is seemed to be well flourishing in the biomedical field especially in tissue engineering, scaffold development, organ transplantation, and so on (Adachi et al. 2020; Icten et al. 2017; Vithani et al. 2019). In this cohort, 3D printing is a promising technology to be explored in the food sector in the delivery of the personalized foods. Nowadays, consumers are becoming more cautious and had a concern about healthy diets like functional foods and bioactive compounds (Portanguen et al. 2019). This upholds the old maxim of 'prevention is better than cure' stating the significance of the food that can be used as medicine. Various emerging food technologies such as encapsulation, fortification, micronization, and nanotechnology remains to be the pillars of the future nutraceutical industry (Kumar et al. 2020). Nevertheless, 3D printing is one such technology that has numerous health benefits in delivering the personalized foods in a customized way (Keerthana et al. 2020). The ability to print foods is a glimpse to modern food processing

3D Printing of Foods, First Edition. C. Anandharamakrishnan, Jeyan A. Moses, and T. Anukiruthika.

that in combination with electrospraying and electrospinning certainly would lead to a new dimensional array of research. The earlier chapter of the present book explained the applicability of 3D printing for the delivery of functional 3D printed foods through integration with encapsulation. Several encapsulation methods and approaches used for production of the functional foods are detailed. In this regard, the electrospraying and electrospinning are the electrostatic-based nanoencapsulation techniques. Both electrospraying and electrospinning are an active area of research in the past decades that has several food applications such as encapsulation, food packaging, food coating, filtration, and enzyme immobilization to name a few (Bhushani and Anandharamakrishnan 2014). Electrospraying and electrospinning are known as sister technologies that are based on the electrohydrodynamic principle. Here, the solution of polymer can be sprayed or spun by applying high voltage electrical forces that result in the production of powders or fibres at micro and nanoscale (Anandharamakrishnan 2014). In contrast, 3D printing is a robotic food manufacturing process that involves a layered deposition of food material in a pre-determined way. The synergistic benefits of the combined technologies are a multi-faceted approach that targets the delivery of functional foods in a pleasing appearance through integrating 3D printing with high electrotechnologies: electrospraying and electrospinning. The advantages of this combined food processing technologies were discussed in several reports (Deotale et al. 2020; Jayan et al. 2019; Lavanya et al. 2020; Vimala Bharathi et al. 2019); however, the practical implementation of which is still obscure.

In a broader view, each of these technologies has its own advantages and limitations in terms of process efficiency, production capacity, cost, and suffers from scale-up issues. To date, there is no clear knowledge of the practical feasibilities of 3D printing with other food technologies. These factors leave a great research gap that cannot be neglected when talking about the merits of individual food processing technologies. The long-run sustenance of any novel emerging food technology must be determined based on the social, economic, cultural, and environmental benefits (Lupton and Turner 2016). Hence analysis of the success of any proposed approach requires a solid knowledge on basic understanding of the scientific principle and the state-of-the-art of future developments. In this regard, the present chapter is an initial attempt that discusses various merits and challenges associated with the integration of 3D printing and electrospraying/ electrospinning. Although the combination of the 3D printing of foods with electrospraying and electrospinning is forecasted to provide several health benefits, it also complicates the risks associated with the targeted delivery of functional bioactive foods at the micro and nano level. The present chapter focused on exploring the hidden research opportunities with this integrated multidisciplinary approach. Certainly the present chapter gives foresight on the novel strategies for the delivery of functional foods in a sustainable way.

13.2 Encapsulation Techniques Involving Electrohydrodynamic Process

Recently, the applications of the electrotechnologies in food processing have received widespread attention of researchers from all over the world due to its advantageous features such as cost-effectiveness, non-thermal approach, greener technology, and sustainable processing (Moses et al. 2014). In this context, electrospraying and electrospinning are electrostatic-based encapsulation techniques based on the application of high potential alternating current/ direct current (AC/DC) voltage in the fabrication of micro/ nano sized particles or/ and fibres. The major application of these technologies in the food industry is the production of the functional bioactive used for drug delivery (Leena et al. 2019). Both electrospraying and electrospinning are very cost-effective technologies that apply a uniform electrohydrodynamic force to break down the charged polymeric solution into fine jets. Various advantages of electrospraying and electrospinning are high encapsulation efficiency, controlled and sustained release of bioactive, higher storage stability, and protection against light and heat (Waghmare et al. 2021). The bioactive compounds encapsulated in electrospray powders and electrospun fibres possess enhanced stability with a sustained release rate at the targeted sites. These kinds of the encapsulated powders and fibres could be conveniently added as food ingredients into the processed foods. The active ingredients that are commonly entrapped through electrospraying and electrospinning include antioxidants, polyphenols, pigments, vitamins, minerals, essential fatty acids, and probiotics. Various biopolymers such as polysaccharides, carbohydrates, proteins, and lipids are most commonly used wall materials for entrapment of the core (Bharathi et al. 2018). Compared to other encapsulation techniques, electrospraying and electrospinning are simpler one-step process to produce encapsulate in dried forms at micro and nanoscale. Importantly, the electrospraying and electrospinning are mostly used for encapsulation of thermolabile compounds because these technologies are operated at ambient temperature and atmospheric pressure conditions (Anukiruthika et al. 2020a). Thus, electrospraying and electrospinning are promissory encapsulation techniques used for the delivery of bioactive compounds via functional foods. Therefore, the present section discusses about the basic system components, process parameters, and working principle of electrospraying and electrospinning that gives an overview of system design which could be adopted for its integration with 3D printing.

13.2.1 System Components and Process Parameters

Electrospraying and electrospinning are based on the electrohydrodynamic principle where the polymeric solution can be sprayed or spun into fine jets through the application of high electric voltage (Anandharamakrishnan 2014). Although both these technologies are based on a similar principle, there exist differences in the charging mechanism for the production of powdered particles or spun fibres. An ideal setup used for electrospraying and electrospinning consists of four major components: high voltage source that ranges from 1 to 30 kV, blunt-ended stainless-steel needle or capillary, feed pump/ syringe, and grounded flat plate or rotating drum (Figure 13.1) (Bhushani and Anandharamakrishnan 2014). Although the high voltage source can work in either DC or AC mode, often the electrospraying and electrospinning systems are operated in DC mode. The needle or capillary placed at the top acts as an emitter while the flat plate at the bottom acts as a collector. When an electric voltage is applied to the pair of electrodes, the free electric charges are induced in the emitter through which the polymer solution is dripped. The two main forces that are acting on the charged polymer solution at the tip of the capillary are electrostatic repulsive force and the Coulombic force (Bharathi et al. 2018). Due to these electrostatic forces,

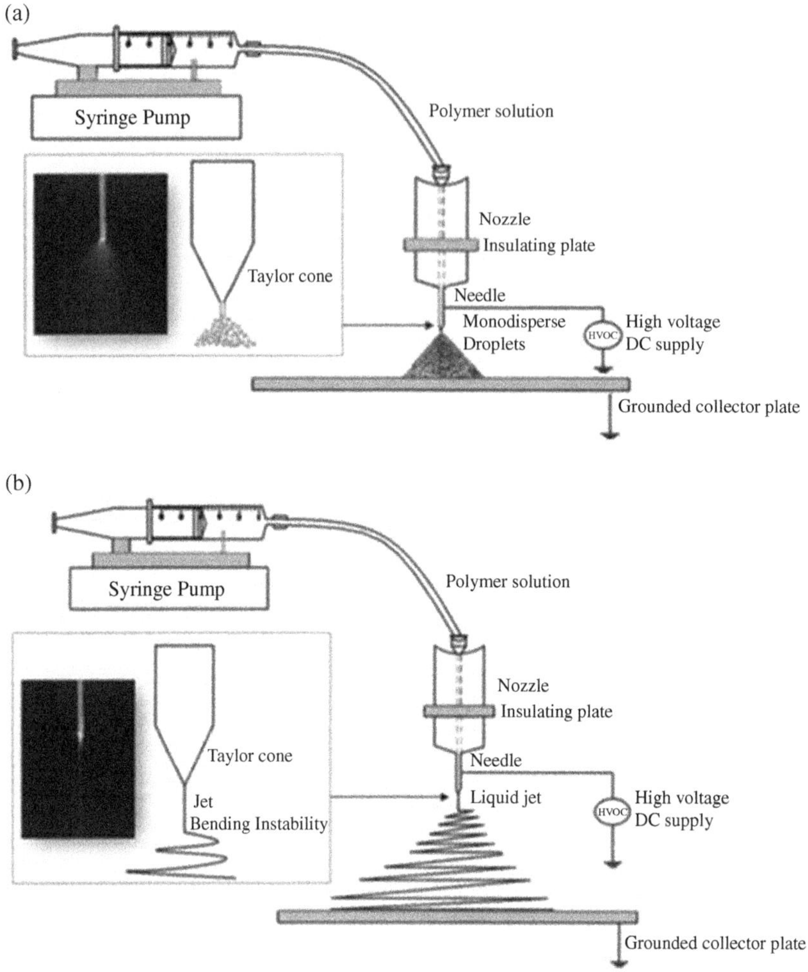

Figure 13.1 System and design components of electrohydrodynamic processes. (a) electrospraying setup and (b) electrospinning setup. *Source:* From Bhushani and Anandharamakrishnan (2014) / With permission of Elsevier.

the hemispherical liquid droplet of polymer solution gets distorted into a cone termed as Taylor cone. When the applied electrostatic force overcomes the surface tension of the solution of polymer, a charged fine jet of polymer solution is ejected from the nozzle tip of the Taylor cone towards the collector (Leena et al. 2020c). The unevenly distributed electrical charges tend to cause whipping motion of bending of the ejecting jet as it gets driven towards the ground electrode. During which the ejected jet gets elongated and rapid evaporation of the solvent takes place resulting in the deposition of the solid thin fibre over the ground electrode, i.e. collector plate in a randomly oriented and non-woven mat-like fashion (Leena et al. 2020b). In contrast, electrospraying is the atomization of liquid polymer solution with applied electrical forces that results in the deposition of the solid particles in dried powder forms (Jayan et al. 2019). The difference between these two technologies is the concentration of the polymer used for the process. When the concentration of the polymer in the solution is high, the ejected jet remains stable due to the formation of the Taylor cone and elongation occurs subsequently as a result of the whipping instability (Zhang et al. 2018). This series of processes result in the production of the fibres. On the other hand, when the concentration of the polymer solution is low, the ejected jet of the viscoelastic polymer is destabilized by the varicose instability and results in the formation of fine droplets. The formed droplets are highly self-dispersed that prevents the coagulation and agglomeration of the droplet thereby yields a fine powdered particle (Anandharamakrishnan 2014). Further evaporation during the electrospraying process would impart contraction and solidification of the droplets resulting in the deposition of the fine powdered particle on the ground collector plate.

The functionality of the electrospray and the electrospun fibers are determined by several variants of the process parameters. The governing factors that manipulate the size of the final product are grouped into system parameters, polymer/ solution parameters, instrumental parameters, and ambient parameters (Bhushani and Anandharamakrishnan 2014). The molecular weight of the polymer and its microstructural features are grouped as system parameters; while the polymer type and its concentration and the solvent are used to determine the solution parameters such as pH, surface tension, conductivity, and viscosity. On the other hand, the instrumental parameters including the liquid feed rate, applied electrical voltage, the distance between the needle end or capillary and collector plate, the nature, and type of material used as collector. Other factors like the temperature, relative humidity, and the velocity of ambient air inside the process chamber are grouped as ambient parameters that determine the evaporation rate of the solvent from the electrospray or electrospun fibres. Technically, the advantages of these technologies are categorized as structural-based and functional-based. The structural advantages of these technologies include the production of the final product in sub-micron and nanoscale, high porosity, tailored morphology, large surface area to volume ratio, and intertwined fibrous structure (Coelho et al. 2020). While the functional advantages of these high voltage technologies are process efficacy, sustained, controlled, and targeted release of bioactive, non-thermal process, reduced denaturation and molecular degradation, higher storage stability, efficient encapsulation, and most importantly use of food-grade polymers (Bharathi et al. 2018). These advantageous features would remain as a promissory approach for opportunities for the integration of electrospraying and electrospinning. The obtained electrosprayed particle and the electrospun fibers ranged in microscale (>1 μm) and submicron/ nanoscale (10–1000 nm). Both electrospraying and electrospinning are well-established technologies in the fabrication of nanodevices, sensors, and material characterization like mass spectrometry. However, the food applications of which in the delivery of food bioactive and enzymes through micro and nanoencapsulation are still under progress. Apart from encapsulation, applications of electrospraying and electrospinning in food science include the development of food packages through the incorporation of functional materials, food coating, and filtration which are less explored (Preethi et al. 2020). The major constraint for scaling-up of these technologies includes high capital investment, less production capacity, and long process time. These limitations can be overcome through design modifications and arrangement of system configuration such as the use of multiple needles in a single setup would greatly reduce process time and operational cost thereby increases the productivity (Anukiruthika et al. 2020b). Despite of these advantageous features of electrostatic encapsulation technologies, the applicability of the developed micron and submicron sized bioactive compounds are debatable in terms of toxicity, allergenicity, and safety. Hence more research studies are required in the direction of the determining the efficacy of the developed drug delivery system and its validation are most important in proving its practical implementation of this novel technologies. Further, understanding of the material response and physiological behavior is most adequate when speaking about the synergistic benefits of integration of these technologies with the 3D printing of foods.

13.2.2 Encapsulation via Electrospraying

The operation of electrospraying is similar to that of electrospinning with the only difference that exists in the concentration of the polymer solution used. The use of the low viscous polymeric solution imparts a lesser degree of cohesion thereby results in the formation of fine liquid droplets (Bhushani and Anandharamakrishnan 2014). The final characteristic attribute of the formed particle is determined by the solution flow rate, viscosity, surface tension, conductivity, ionic charges, density, applied voltage, ambient temperature, and humidity. A relationship was drawn between the electrical pressure resulting from the induced electrical charges on a spherical droplet. It was stated that the natural phenomenon of the quadrupolar oscillation of the liquid droplet due to a field-free environment becomes weak when the electric charges exceed its critical limit (Anandharamakrishnan 2014). This phenomenon of instability of the liquid droplet is reached either by the evaporation of the solvent or when the applied charges exceed the critical limit that is often known as the Rayleigh limit. Hence the size of the formed droplets depends on flow rate and induced charges due to applied voltage (Eq. (13.1)) (Wang et al. 2020). Therefore, the electrospraying technique is often used for developing monodispersed particles at micron and submicron levels.

$$q_R = 8\pi\varepsilon^{1/2}\sigma^{1/2}r^{3/2} \tag{13.1}$$

where q_R is the electrostatic repulsion of the induced charges (q) required to overcome surface tension, r is the induced droplet radius, σ is the surface tension of the droplet, and ε is the electric permittivity of the medium. It was reported that the combination of higher applied voltage, larger gap/ distance between the electrodes, and lower flow rate resulted in a small spherical shaped particle. The self-dispersing nature of the ejected jet yields non-agglomerated dried powders. The higher encapsulation efficiencies of electrosprayed particles clearly showed that most of the core material got entrapped by the wall material with less loss (Jayan et al. 2019). This suggests that encapsulation via electrospraying is an efficient electrohydrodynamic technique used for preserving the thermo-labile bioactive ingredients.

A comparative study was reported on the encapsulation of the β-carotene (a lipophilic bioactive compound) in a zein protein matrix using spray drying and electrospraying with different core to wall ratios. As against spray-dried microparticles, electrospraying yields higher encapsulation efficiency up to 81% and the resulted nanoparticles showed excellent dissolution behaviour due to size reduction and larger surface area (Mahalakshmi et al. 2020). In another study, the antioxidant resveratrol was encapsulated through electrospraying using zein (Jayan et al. 2019). The produced nanoparticles ranged in the size of 230–330 nm and showed higher encapsulation efficiency of about 68.49% with a resveratrol to zein ratio of

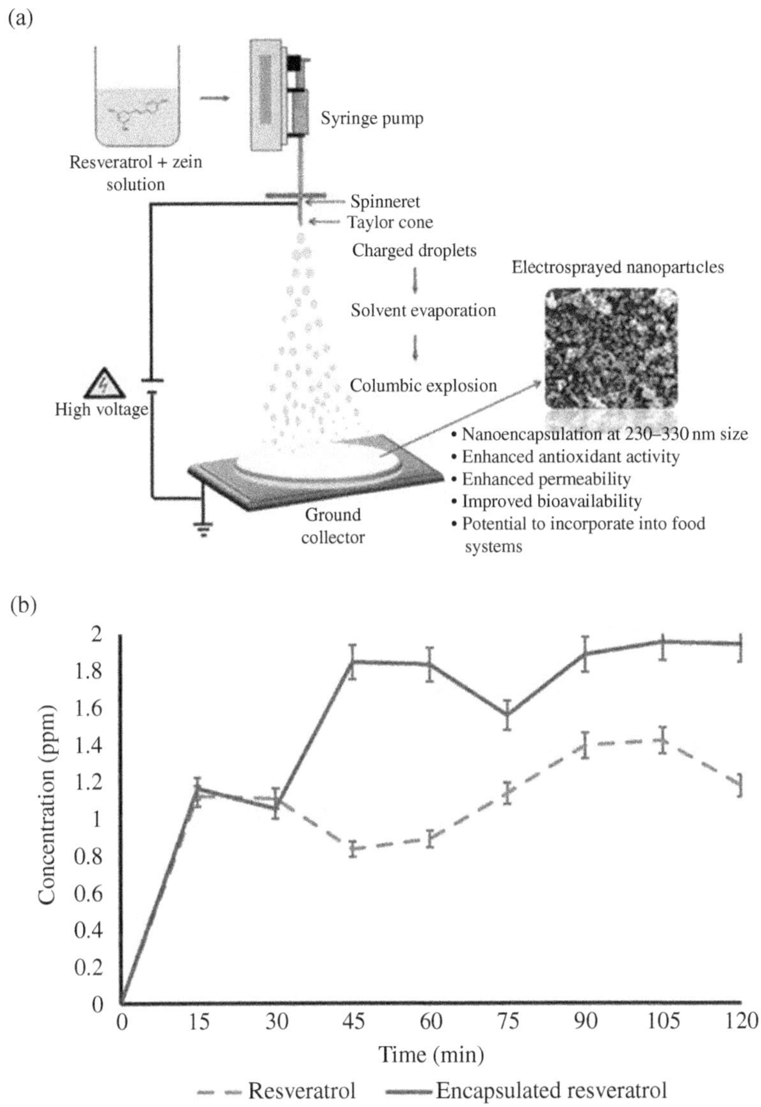

Figure 13.2 Improvement of bioavailability of resveratrol through encapsulation in zein. (a) schematic diagram of experimental setup and (b) intestinal absorption of nanoencapsulated (1 : 50) and unencapsulated resveratrol. *Source:* Jayan et al. (2019) / With permission of Elsevier.

1 : 50 (%, w/w). The nanoencapsulated resveratrol exhibited greater stability and controlled release with 1.15-fold permeability under *ex-vivo* conditions of dynamic engineered small intestine system that in turn showed improved bioavailability (Figure 13.2). Thus, the encapsulated resveratrol nanoparticles can be conveniently applied for oral administration by incorporating into beverages and foods.

Using a similar approach, zein was used for the nanoencapsulation of the green tea catechins (Bhushani et al. 2017). Among the varied concentrations of zein (1–40%, w/w), 5%

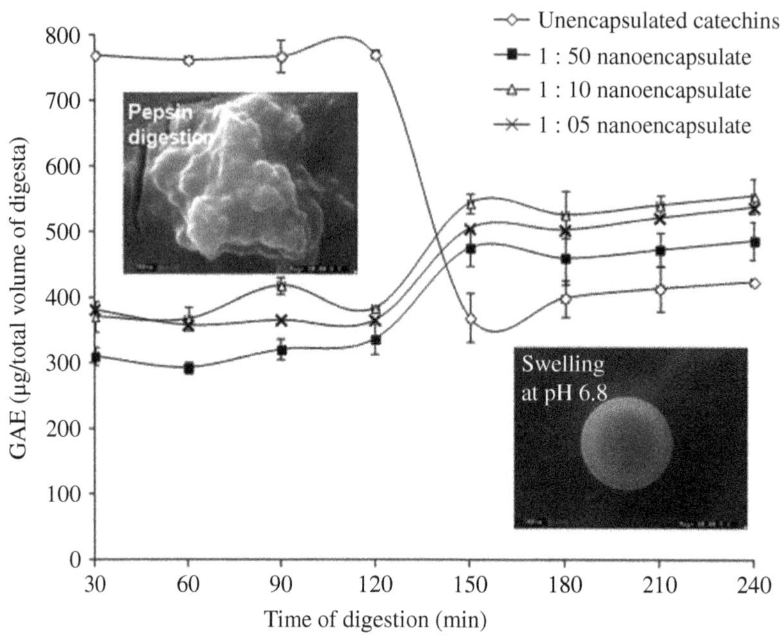

Figure 13.3 Release profile and stability of nanoencapsulated and free polyphenols under simulated gastrointestinal conditions. *Source:* From Bhushani et al. (2017) / With permission of Elsevier.

zein solution (w/w) resulted in a spherical monodispersed nanoparticle with the size of 157 nm. On comparing, the sample with 1 : 50 core to wall ratio exhibited the highest encapsulation efficiency. Further, the nanoencapsulated catechins showed improved stability and higher permeability than unencapsulated catechins under gastrointestinal conditions and Caco-2 cell studies, respectively (Figure 13.3). Thus, electrospraying is proved to be a one-step method used for generation of functional nanoencapsulates with enhanced permeability, controlled release and targeted delivery of bioactives.

13.2.3 Encapsulation via Electrospinning

Electrospinning involves the electrostatic deposition of the electrospun fibres on the ground electrode by applying a high electrical voltage to the droplets of the polymer solution (Leena et al. 2020d). The resulting submicron-sized polymeric fibres are in the range of diameter <1 μm. The process of fiber production involves the transformation of the Taylor cone into a fine jet when the induced charges counteract the droplet's surface tension. The morphology and functionality of developed fibers depends on the chemical properties of polymer solution, feed rate, conductivity, surface tension, viscosity, volatility, and electrode distance (Hosseini et al. 2021). Generally, a set distance of approximately 15 cm is maintained between the electrodes. The formed nanofibers ejected from liquid droplet through the tip of the needle is subjected to a combination of radial stretch and the longitudinal stretch as a result of repulsion between the like charges and higher strength of attraction towards collector plate, respectively (Bhushani and Anandharamakrishnan 2014). The electrospinning can be performed either through coaxial electrospinning where multiple core-sheath layers are formed from different polymers under a single-step process or through direct inclusion of the bioactive into the polymer feed solution (Anukiruthika et al. 2020b; Perez-Masia et al. 2013). Electrospinning has a wider range of applications in biomedicine especially in tissue engineering, tissue scaffolds, vasculature, bone, neural, and tendon repair. However, the potentiality of this technology in the targeted controlled release of bioactive strengthens its application in food and nutraceutical industries for the delivery of bioactives. Another variant of the electrospinning used for encapsulation is two-phase electrospinning. In this process, a biphasic suspension is developed through appropriate mixing of the aqueous solution with the organic polymer solution (Figure 13.4) (Wang et al. 2017). Later, the formed stable delivery system is electrospun resulting in the nanoencapsulation of the aqueous phase within a polymeric fibre. The encapsulation of functional peptides including growth factors, hormones, and cytochrome C into the biopolymeric matrix is based on a two-phase electrospinning approach (Dong et al. 2009).

The end-quality of developed electrospun fibers can be assessed by process parameters and solution characteristics. The following observations are drawn by Chen (2007); Sill and Von Recum (2008) that helps in the basic understanding of the interaction of process variables and solution properties during the electrospinning process. Generally, the fiber diameter is inversely related to the electrode distance when the applied voltage, flow rate,

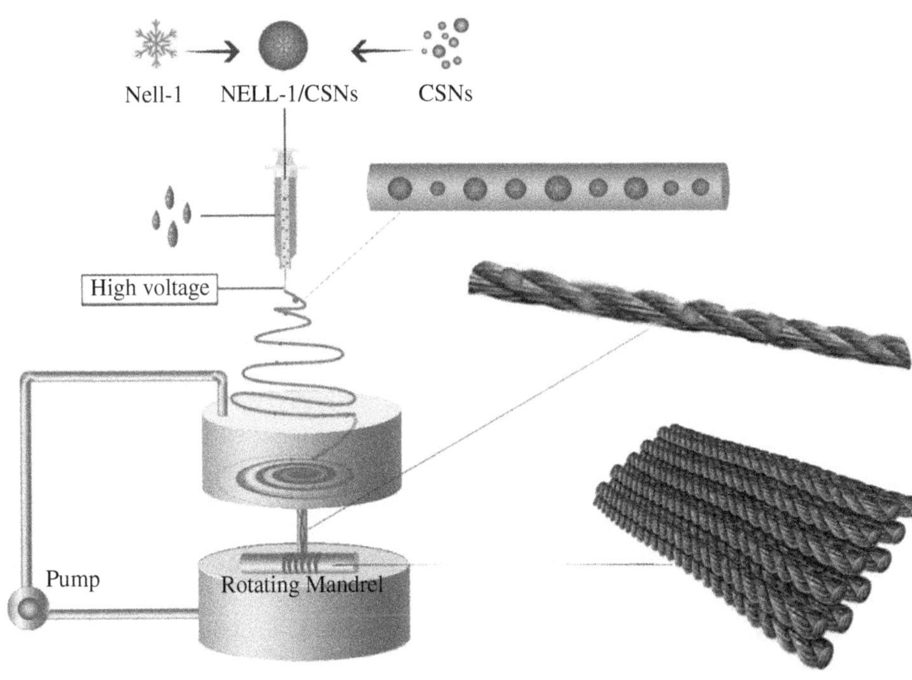

Figure 13.4 Schematic illustration of the two-phase electrospinning for electrospun fibrous scaffold prepared from incorporated chitosan nanoparticle with loaded growth factors. *Source:* Wang et al. (2017) / With permission of Elsevier.

and the concentration of polymer solution are kept constant; and the fiber diameter is inversely related to the applied voltage when the electrode distance, flow rate, and the concentration of polymer solution are kept constant. Contrarily the fiber diameter is directly related to (i) flow rate when concentration is constant and (ii) concentration when the flow rate is constant under the conditions of constant applied voltage and electrode distance. Considering the properties of the polymer solution, the following conclusions were drawn. The concentration of polymer is directly proportional to surface tension and viscosity of solution; the choice of the solvent used is another determinant that influences the formation of the fibre. Hence volatility of the solvent must be adequate that eases the evaporation of the solvent and influences the separation of phase during the travelling of fibres between the electrodes. Though the conductivity of the solution is seemed to have less influence over the process, it directly affects the mechanical strength of the developed fibres (Anandharamakrishnan 2014). Thus, fibre jets that are drawn from high conductivity polymer solutions possess greater tensile strength. Usually, to obtain better results, the polymer solution of higher concentration is used for electrospinning applications. The electrospun fibres produced through needleless electrospinning approach from whey protein isolate and maltodextrin at different mass ratios (5 : 95, 10 : 90, 20 : 80, and 25 : 75 w/w) serves as the promising material for the production of glycated whey proteins (Kutzli et al. 2018). This work explained the applicability of electrospinning for the glycation of the whey protein isolates. The increasing concentrations of whey protein isolate decreased the binding ratio of maltodextrin from 2.1 to 1.2. The fabricated nanofibers upon heating resulted in a protein-specific reaction yield up to 44.52% (w/w) that confirmed the glycoconjugate formation.

A mixture of whey protein isolate (7–12%, w/w) and guar gum (0.7–0.9%, w/w) in the acidic conditions was demonstrated to have excellent spinnability at 20 kV with a flow rate of 0.6 ml min^{-1} and capillary collector distance of 8 cm (Ramazani et al. 2019). With the increase in the concentration of guar gum from 0.8 to 0.9% (w/w) at a fixed concentration of whey protein isolate (7%, w/w), a visible morphological transformation from nanofibers with beads into bead-free nanofiber was observed. The electrospinning efficiency was highly influenced by the concentration ratio and the viscosity of the biopolymer solution. Further, the thermal stability of developed nanofibers (466 nm) showed that the combination of guar gum and whey protein isolate fibres had greater stability at higher temperatures against native guar gum and whey protein isolate. The encapsulation of cinnamon essential oil (CEO) in electrospun polyvinyl alcohol (PVA)/ β-cyclodextrin (CD) matrix was investigated for the antibacterial activity (Wen et al. 2016). The characterization of the developed film clearly indicated the encapsulation of CEO into the β-CD with complex interaction among the PVA, CEO, and the β-CD that enhanced the thermal stability of the CEO. The antimicrobial studies showed that developed PVA/β-CD/CEO film had a minimum inhibitory concentration of 0.9–1 mg ml^{-1} against *Staphylococcus aureus* and *Escherichia coli*. These results showed that electrospinning is a mild process favorable for the encapsulation of CEO for the fabrication of nanofibrous film with antimicrobial activity. In addition, the developed antimicrobial nanofibrous film was successfully used as an active packaging for the extending the shelf-life of the strawberry (Figure 13.5).

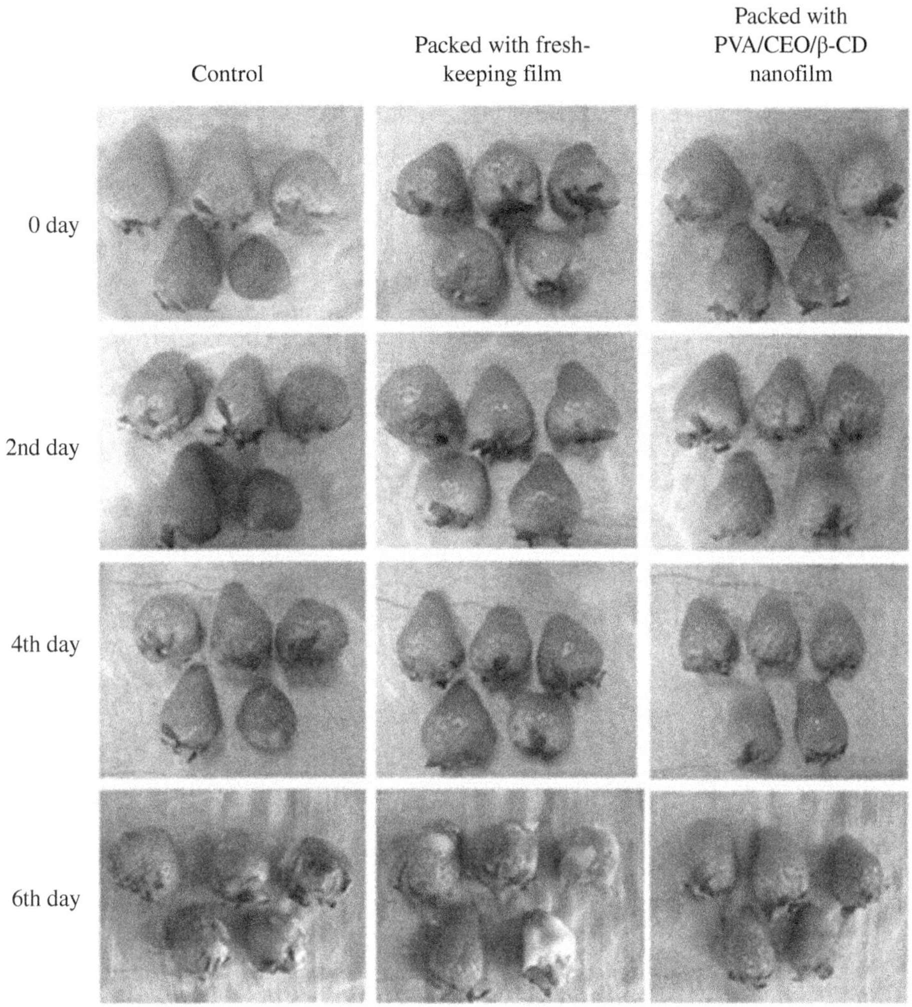

Figure 13.5 Quality changes in the appearance of strawberries stored at 21 °C. *Source:* From Wen et al. (2016), Figure 07 [p. 374] / With permission of Elsevier.

13.3 Applications in the Food Industry

13.3.1 Encapsulation of Bioactives and Probiotics

Encapsulation through electrospraying is a simple, cost-effective, and highly efficient technique used for the delivery of bioactives. The multidisciplinary facets of this approach along with the advancements of nanotechnology found its application in designing functional foods, active food packages, efficient filtration aid, and surface coatings with functional bioactive materials (Leena et al. 2019). Encapsulation is a booming area of research with broader applications focused on protection and bioactive delivery. Although several techniques are available for encapsulation, electrospraying, and encapsulation are a one-step process used for direct entrapment of core material within the polymer matrix at the micro to the nanoscale (Bhushani and Anandharamakrishnan 2014). Since the process is aimed at the delivery of functional foods, the polymer and solvent used should be food-grade that are generally recognized as safe (GRAS). In this regard, biopolymers such as polysaccharides (starches, dextrin, cellulose, and chitin), hydrocolloids (sodium alginate, gum acacia, guar gum, carboxymethyl cellulose, hydroxyethylcellulose, methylcellulose, hydroxypropyl cellulose, and gellan gum), and proteins (zein, casein, albumen, whey protein, soy protein, elastin, keratin, and collagen) are used as wall material to deliver the core materials at the right place with appropriate release rate (Preethi et al. 2020). Apart from the bioactive food ingredients, the interest of food scientists has brought up the idea of encapsulation of live cells such as probiotic bacteria and bacteriocins to impart enhanced stability during storage. Compare to electrospinning, the method of electrohydrodynamic atomization is a new concept used for the production of encapsulated bioactive in its powdered forms. Recently, low-molecular-weight carbohydrates such as maltodextrin and resistant starch can be used as wall materials with a suitable combination of surfactants like Tween 80, Tween 40, Tween 20, Span 20, and lecithin (Bharathi et al. 2018). Thus, the protection of bioactive by encapsulation via electrospray technique holds a range of promising research opportunities in the development of functional foods and beverages. The advantages of electrospraying and electrospinning are high efficacy, sustained release of bioactives, greater protection of bioactives against oxidation, enhanced storage during processing, storage, and handling (Waghmare et al. 2021). In this regard, the electrosprayed particles can be directly incorporated into the food preparation while the electrospun fibres must require appropriate physical modification to ease its incorporation into food since the fibres resemble a non-woven mat-like structure with an intertwined network (Bhushani and Anandharamakrishnan 2014). Although spray drying has been used as a common encapsulation technique for the bioactive, the emerging concept of encapsulation based on electrohydrodynamic processes is most preferred due to its high encapsulation efficiency, improved storage stability, enhanced protection for thermal and oxidation sensitive active ingredients (Drosou et al. 2017).

The effect of the morphology on improving the storability and the retention of the resveratrol was studied using zein-chitosan composite nanoparticles (Khan et al. 2021). In this study, hollow zein encapsulates were prepared by chitosan coating for fabricating a composite zein-chitosan nanoparticle. The hollow zein and the composite zein-chitosan nanoparticles were studied to encapsulate polyphenol and resveratrol. Results showed that zein-chitosan nanoparticles possessed positive charges and had a bigger particle size as compared to the hollow zein nanoparticles. This was mainly because of the entrapment of the resveratrol that subsequently increased the apparent size of the particles. The fabrication of hollow particles has advantages over its solid counterparts in terms of lower density, higher surface area, and sustained release of bioactive. The composite zein-chitosan nanoparticles had 91% encapsulation efficiency and 14% of loading capacity. Further, the added chitosan coating helped in improving its storability and digestion stability of the resveratrol-loaded hollow zein matrix (Figure 13.6). Thus, the developed hollow composite

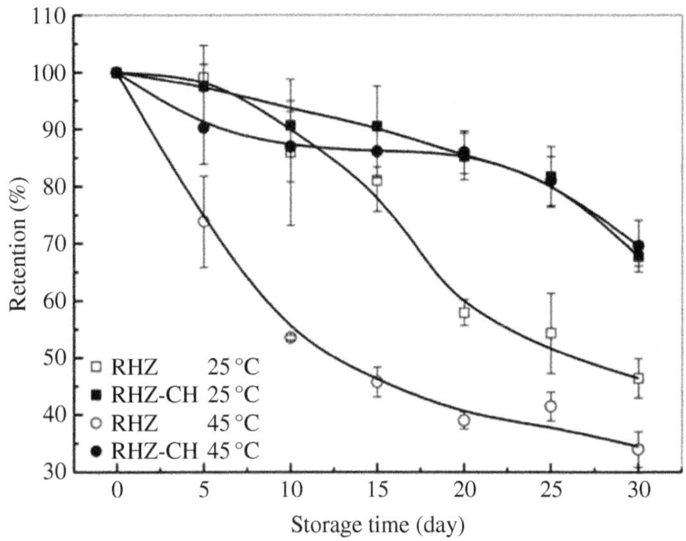

Figure 13.6 Retention of resveratrol in resveratrol-loaded hollow zein (RHZ) and zein-chitosan (RHZ-CH) particles during storage at 25 and 45 °C. *Source:* Khan et al. (2021) / With permission of Elsevier.

zein-chitosan nanoparticles could be used as a potential drug delivery system that can be easily incorporated into foods.

The effect of protein type and formulation ratio on the fabrication and morphology of the nanofibers was studied using electrospinning of whey protein and soy protein along with maltodextrin (Kutzli et al. 2019). It was stated that higher protein content would generally increase the electrical conductivity and viscosity of polymer solution. Results showed that soy protein isolate significantly reduced the surface tension but had increased the electrical conductivity and apparent viscosity of the spinning dispersions than whey protein isolate. However, its higher viscosity limited the production yield of maltodextrin-soy protein isolate nanofibers that were found to be 0.5 g $hour^{-1}$ with a particle diameter of 4.74 μm compared to whey protein isolate with production yield and particle size of about 0.98 g $hour^{-1}$ and 2.85 μm, respectively (Figure 13.7). This can be overcome with the removal of the insoluble fraction of soy protein isolate and electrospinning of dissolvable soy protein isolate with maltodextrin resulting in lower viscous dispersions and better spinning with a yield of 0.5 g $hour^{-1}$.

A novel approach of electrospray coating atomization was used for enhancing the viability of probiotic strains (Libran et al. 2017). In this study, the freeze-dried probiotic strain of *Bifidobacterium longum* subsp. *infantis* (10 log CFU g^{-1}) was uniformly electrosprayed with different food hydrocolloids like whey protein concentrate, zein, and maltodextrin. The water-soluble polymer-derived polyvinylpyrrolidone was employed as reference material under permissible limits as pharma-approved encapsulation material. Results showed a significant enhancement in the survivability of probiotics in encapsulated forms

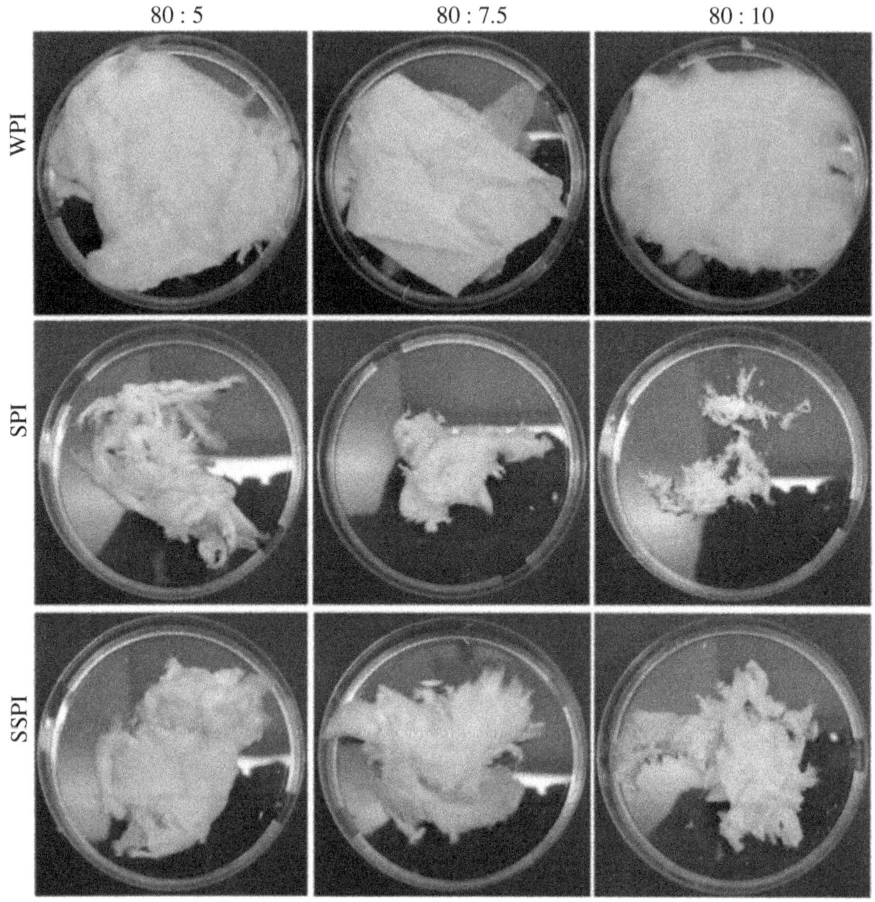

Figure 13.7 Images showing the electrospun fibres from maltodextrin (MD)-protein dispersions with whey protein isolate (WPI), soy protein isolate (SPI), and soluble soy protein isolate (SSPI) at different MD to protein ratios. *Source:* From Kutzli et al. (2019), Figure 04 [p. 210] / With permission of Elsevier.

than in free state during the storage period of more than one year at room temperature and 23% RH for all the formulations with whey protein isolate, resistant maltodextrin, and reference material polyvinylpyrrolidone (Figure 13.8). Thus, electrospray coating atomization could be used as potential encapsulation method for the protection of probiotics in the dried forms.

13.3.2 Enzyme Immobilization

Both electrospraying and electrospinning are also found their applications for the immobilization of enzymes (Coelho et al. 2020). Enzymes are the natural catalysts with higher specificity and efficiency due to which it has several industrial applications. However, enzymes are highly sensitive to temperature, pH, and other chemical reactions that imparts difficulty in their handling (Haider et al. 2018). Furthermore, there are many chances for the

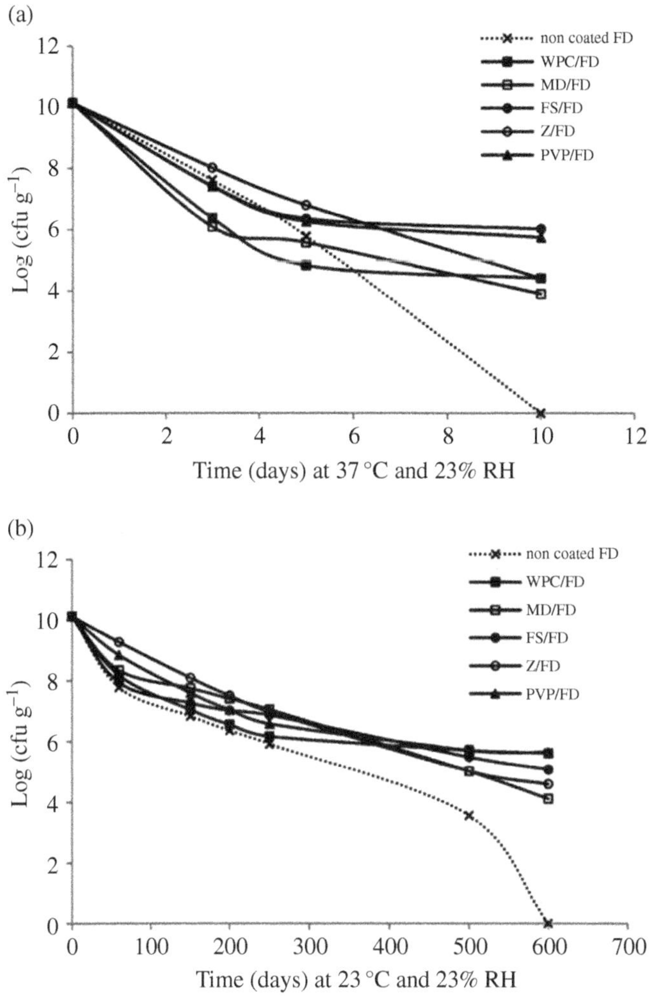

Figure 13.8 Viability of uncoated freeze-dried and electrospray coating atomization (ECA) *Bifidobacterium longum*. (a) stored at 37 °C and 23% RH and (b) stored at 23 °C and 23% RH. *Source:* Libran et al. (2017) / With permission of Elsevier.

inactivation of enzymes during processing by a range of physical and chemical stimuli that limit its commercial usage. These difficulties can be surpassed through immobilization technique that refers to the process of confinement of the enzymes on physical support or entrapment of enzymes in a solid matrix which aids in the retention of the biochemical activities. The applicability of the electrospraying for the immobilization of enzymes is not well explored; however, the electrospinning is well known for immobilization of enzymes (Wang et al. 2009). The co-electrospinning method and coaxial electrospinning method are highly adopted for this purpose. In addition to the encapsulation, the enzymes are covalently attached or physically adsorbed to the electrospun fibres. This process is also known as a surface attachment where the spun fibres act as physical support for the immobilized enzymes. The beneficial outcome of this process of cross-linking of enzymes is the enhancement of residual activity due to the increased porosity and surface area (Bhushani and Anandharamakrishnan 2014). Hence, the catalytic activity of the immobilized enzymes is far superior to free enzymes. This is mainly because of the reduction in the size of the carrier materials that aid in improving the efficacy of the immobilized enzymes. The immobilized enzymes have a different range of applications in the food industry from active food packaging to biosensing of molecules that ensure food safety and quality of the developed product (Homaei et al. 2013). Most of the works on enzyme immobilization are demonstrated based on lab-scale electrospinning setup. A proper channelling and sharing of knowledge among the researchers and manufacturers would lead to the applicability of the electrospinning at a large-scale commercial basis for the production of nanofibers. Some of the manufacturers dealing with the fabrication of nanofibers at an industrial scale are Elmarco, Electrospunra, Asian Nanostructures Technology Company, and eSpin (Bhushani and Anandharamakrishnan 2014). More research studies on the investigation of practical implementation of electrospinning technique would pave a way for the feasibility of a wide range of applications in the food sector.

Immobilized enzymes are more robust and characterized by greater resistance to environmental changes. Moreover, the system heterogeneity of the immobilized enzyme allows for easy recovery of products, reuse of enzymes, rapid termination of chemical reactions, designing of a variety of bioreactors, and are suitable for continuous operations (Homaei et al. 2013). More interestingly, a novel method of 'in cellulo' encapsulation technique was being used for the immobilization of protein, CotA laccase (Pose-Boirazian et al. 2021). Encapsulation is a cost-effective technology that facilitates the one-step production used for the immobilization of enzymes of industrial interest. Results showed that the immobilized enzyme through encapsulation of protein in micro/ nanospheres significantly increased the pH working range and exhibited higher resistance towards chemical inactivation. Further results of the thermal treatments indicated that the enzyme activity remained unaffected when incubated at 90 °C for 30 minutes and when kept at boiling temperature of 95 °C for about 15 minutes. In another study, the *Lactobacillus rhamnosus* cells were encapsulated in polyvinyl alcohol in order to explore its feasibility for the low-cost production of L-(+)-lactic acid fermentation (Radosavljevic et al. 2021). The obtained PVA-based biocatalyst was applied for batch fermentation and repeated batch fermentation with brewer's spent grain and hydrolysates from malt rootlets. The consecutive batch fermentations with developed biocatalyst showed greater mechanical integrity with minimal loss of bacterial activity. This process resulted in productivity of 2.1 $g\,l^{-1}$ $hour^{-1}$ with a lactic acid yield of 97%. The study successfully demonstrated the use of immobilized *L. rhamnosus* in the polyvinyl alcohol cryogel for batch fermentation as low-cost fermentation media with greater fermentation capability. A sol–gel encapsulation approach was used for co-immobilizing glucose oxidase and β-fructofuranosidase in overcoming the inhibition of by-products and to enhance the efficiency of transfructosylation (Long et al. 2019). This study investigated the effect of co-immobilization of enzymes on lactosucrose production efficiency. Results showed that the yield of lactosucrose was found to 160.8 $g\,l^{-1}$ by using the prepared co-immobilized enzymes that had retained about 85.39% of the initial activity of the enzyme at a concentration of 1.47 $mg\,g^{-1}$ sol during the process. Thus, this study demonstrated the application of co-immobilization of glucose and β-fructofuranosidase oxidase for a continuous low-cost method of lactosucrose production with higher purity. The advantages of the immobilization of enzymes for industrial applications are its higher stability and convenience of reusability.

13.3.3 Functional Food Packages

Food packaging is the state-of-the-art of preserving and protecting the foods from abiotic and biotic factors that ensures the quality, wholesomeness, integrity, edibility, and safety of the product. Food packaging is an intermittent unit operation that coordinates the preparation of food for safe delivery of food to the consumers through the entire food supply and distribution chain (Pal et al. 2019). Various determinants that influence the marketability and acceptability of food products are quality, safety, and shelf-life. These aspects depend on temperature, gas concentration, relative humidity, moisture content, chemical reactions, pH, microbial contamination, and respiration of the produce (Biji et al. 2015). Thus, the food package is the physical entity that must fulfil the four basic functions of food packaging: containment, protection, convenience, and communication. Recent awareness of food quality and safety among consumers has reformed food packaging techniques through the emergence of novel techniques such as active and intelligent packaging (Preethi et al. 2020). Unlike the conventional food package that remains inert, active packaging, and intelligent packaging are innovative approaches that maintain and prolong the shelf-life of the product through appropriate communication. The packing system that allows the physical, chemical, and biological interactions among the food, food package, and its headspace for assuring the quality, safety, and freshness of the product is known as active packaging (Vimala Bharathi et al. 2019). It involves the use of absorbers/ scavengers (moisture absorber, oxygen absorber, and ethylene absorber) and emitters (carbon dioxide emitters, ethanol emitters, antioxidant emitters, and flavour emitters) for enhancing the shelf-life of the produce. On the other hand, intelligent packaging is a smart packaging system that continuously monitors and senses the quality of packed food and communicates to the outside environment until reached the consumers (Yildirim et al. 2018). The shelf-life, microbial safety, and other physiochemical properties of the food are communicated through the use of quality indicators that are the devices used for conveying the information to the consumers. It works based on specific chemical and biological responses with the presence or absence of active substances and change in the concentration of chemical species through irreversible visible colour changes (Kalpana et al. 2019). Broadly, the quality indicators are classified into the external indicators (time–temperature indicators) and internal indicators (freshness, gas, and microbial indicator). In addition to the quality indicators, the other components of intelligent packaging are data carriers (radio-frequency identification (RFID) and bar codes) and sensors. In this regard, electrospinning is a potential technology used for designing active packages. Another novel designing of food package with recent advancements in material science are multilayer packaging that amalgamates the unique functional attributes of different polymers into a single package. Various emerging approaches for the fabrication of multilayer food packages based on nanostructured multilayered films were reported (Ajitha et al. 2016; Olmo et al. 2019). One such emerging approach is the application of the electrohydrodynamic-based approach for the development of a multilayer package with several functionalities. Different variants of electrospinning systems used for fabrication of multilayer food package include single needle setup, needleless system, multijet setup, vibration electrospinning, bubble electrospinning, edge electrospinning, roller electrospinning, disc electrospinning, ball electrospinning, and cone electrospinning (Anukiruthika et al. 2020b). These variants are classified based on the modification of design components in terms of functionality and productivity of the nanofibers. The production of electrospun fibres in micron and submicron range with larger surface area to volume ratio extends its application in active food packaging (Leena et al. 2020e). The produced fibres are highly sensitive to changing environmental conditions and more suitable for encapsulation of thermolabile food components encouraging its applicability in the development of functional food packages. As far as food packaging is concerned, studies were reported on the electrospinning of polymer with antimicrobial agents, antioxidants, and aroma compounds for humidity triggered release (Bhushani and Anandharamakrishnan 2014). The requirement of single package with multiple functionalities such as oxygen barrier, thermal management, and antimicrobial properties is possible through multilayer packaging via electrospinning. The temperature of the frozen foods should be maintained at the desired range using appropriate phase change material like dodecane. A study was reported on the encapsulation of dodecane using zein nanofibers with the aim of the development of bio-packaging (Perez-Masia et al. 2013). Results showed that coaxial electrospinning had a higher encapsulation efficiency than uniaxial electrospinning. Zein nanofibers are also used as the strengthening agents for improving the mechanical integrity of the multilayer polymeric structure. Thus, electrospinning is an effective simpler approach for the fabrication of novel food packages with an enhanced barrier, antimicrobial, and mechanical properties along with feasibility of quality tracing and monitoring of foods during transportation and storage (Anukiruthika et al. 2020b).

A bilayer active packaging film with enhanced antioxidant activity was developed using electrospinning zein over solvent-casted chitosan film (Bharathi et al. 2020). As against solvent casted zein film, the electrospun zein nanofibers showed increased antioxidant activity. However, the coating of zein nanofiber over the solvent-casted chitosan film highly increased antioxidant activity from 12.41 to 44.17%. Further, the developed functional bilayer film was assessed for its anti-browning capacity by packing the cut apple slices. Results confirmed the applicability of the developed functional bilayer film to be used as an anti-browning package for minimally processed food. The characteristic anti-browning ability of the developed functional bilayer film was mainly due to the larger surface-to-volume ratio and the greater affinity of the electrospun zein nanofibers. In another study, the yerba-mate extract was encapsulated in zein nanofibers to be used as active packaging material with antioxidant release capacity (Pinheiro et al. 2020). Results showed that the developed nanofiber package exhibited higher thermal stability and the zein nanofibers with 5% yerba-mate extract showed greater release of extract of about 49% within 50 hours when kept in contact with hydrophilic food stimulant. Thus, the yerba-mate extract encapsulated zein nanofibers could be used as active packaging material with the capability of antioxidant release during storage. In a more recent study, the active packaging material was developed from allyl isothiocyanate through electrospinning in the zein ultrafine fibres for analyzing the post-harvest quality of the strawberry (Colussi et al. 2021). Among the examined concentrations of allyl isothiocyanate (4 and 8%), the zein fibres with a lower

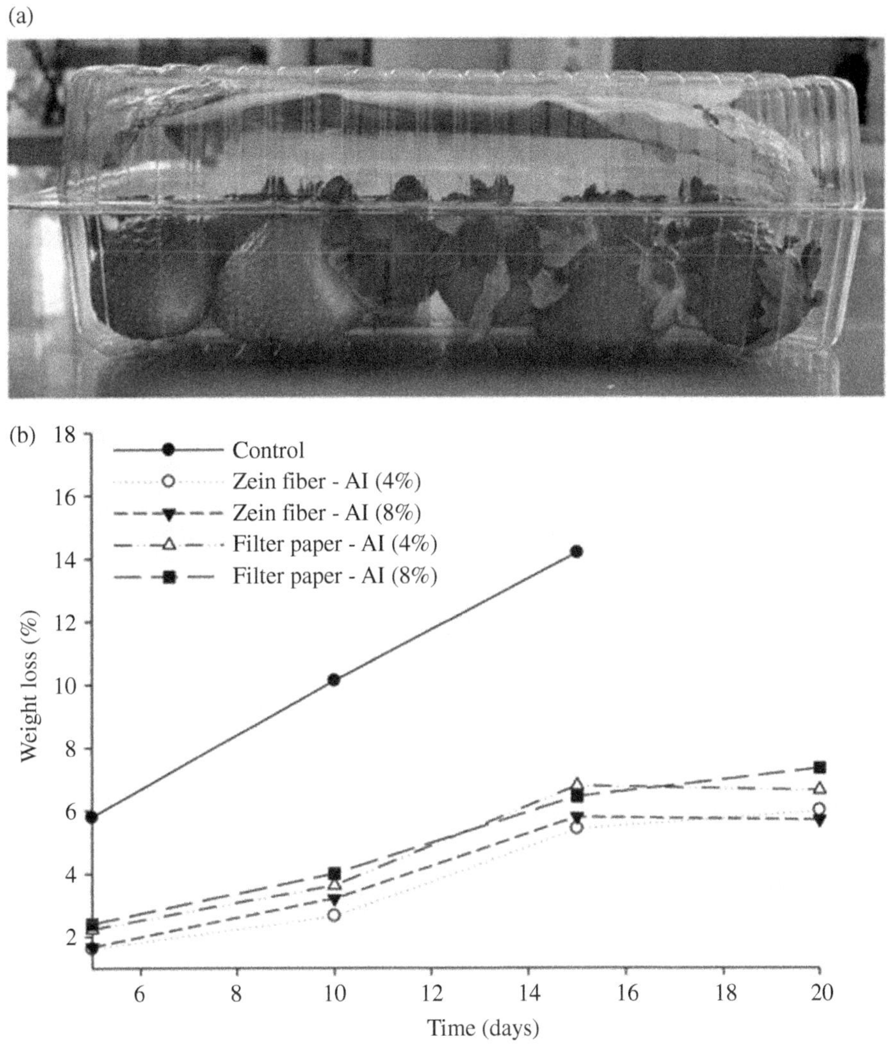

Figure 13.9 Allyl isothiocyanate loaded electrospun zein ultrafine fibres. (a) Package used for applying the ultrafine zein fibres with encapsulated allyl isothiocyanate and (b) weight loss of strawberries stored using encapsulated and unencapsulated allyl isothiocyanate. *Source:* From Colussi et al. (2021) / With permission of Elsevier.

concentration of 4% allyl isothiocyanate were found to be more efficient than the higher concentration (8%). It was observed that higher concentrations of allyl isothiocyanate significantly reduced the firmness, redness (colour quality), and suppressed the anthocyanin and total phenolic contents of the strawberry. Results confirmed that the use of allyl isothiocyanate encapsulated in zein ultrafine fibre at low concentrations could yield a favourable outcome in extending the shelf-life and maintaining strawberry quality (Figure 13.9). Thus, electrotechnologies have a great scope in designing of functional food packages.

13.3.4 Food Coatings

Another interesting application of the electrohydrodynamic process is electrospray coating based on atomization of polymer solution by applied electrical forces. Edible coatings of fresh fruits and vegetables are receiving greater attention due to consumer's preference for minimally processed foods (Vimala Bharathi et al. 2019). Apart from the use of food packages to impart barrier properties, the edible coating is another approach for the enhancement of shelf-life extension of food products. The conventionally used method for edible coating is the spray coating process. The use of electrical forces for spray coating seems to be more efficient than the normal spray coating process. Thus, the electrospray coating aids in the uniform deposition and coating of the biopolymer over the large area of food material with a controlled rate of deposition and layer thickness (Leena et al. 2020c). The size of the droplet formation can be controlled by varying the applied voltage, rate of feed flow, and the concentration of polymer solution. The distribution of the monodisperse particles in micron and nano scale through electrospray coating significantly reduces the count and the size of the cracks and voids over the deposited film coating (Bhushani and Anandharamakrishnan 2014). Studies were reported on the electrospray coating of starch films and hydrophobic films through spraying of sunflower oil, soybean oil, and medium-chain fatty acids (Khan et al. 2012; Zhang et al. 2013). An effective strategy was developed for the oral delivery of resveratrol using electrospinning method (Leena et al. 2020d). In this study, resveratrol of varying concentrations (2, 5 and 10%, w/w) was encapsulated using zein based on electrospinning technique. The resulted nanofibers had a ribbon like morphology with a mean diameter of 404 nm. The electrospun resveratrol-loaded zein nanofibers showed higher encapsulation efficiency (96.9%) and higher radical scavenging activity (66.2%) (Figure 13.10). Further, release profile under simulated *in-vitro* gastrointestinal conditions exhibited greater protection of encapsulated resveratrol in gastric conditions and showed a sustained release in intestinal conditions. In addition, the applicability of the developed nanofiber was evaluated by directly electrospinning over the cut apple slices placed above the collector plate. Results showed that the nanocoating of the cut apple slices significantly reduced moisture loss and prevented browning reactions during the period of study of about 6 hours. Thus, electrospinning can be effectively used for minimal processing of fresh-cut fruits and vegetables.

In another study, the bioactive curcumin was loaded into electrospun fibres of gelatin and zein to be used as edible coating for food packaging (Alehosseini et al. 2019). Additionally, the green tea extract was also added into the formulations of gelatin and zein to impart storage stability and controlled release of curcumin from the nanofibers (Figure 13.11). Results showed that both electrospun gelatin coating and zein coating had higher encapsulation efficiency. The former coating of gelatin electrospray showed promising results of release profile that is more suitable to be used with fatty acid simulant while the electrospun zein coating showed best results for packaging of high moisture foods.

A study was reported on the encapsulation of antioxidant β-carotene in a mixture of soy protein isolate and polyvinyl alcohol emulsion using a soybean oil (Figure 13.12) (Bruni et al. 2020). This mixture is directly electrospun onto the polyhydroxybutyrate-*co*-valerate film. The adhesion of the electrospun mat onto the polymeric packaging film was improved through an annealing process that aids in modulating the release profile of the β-carotene.

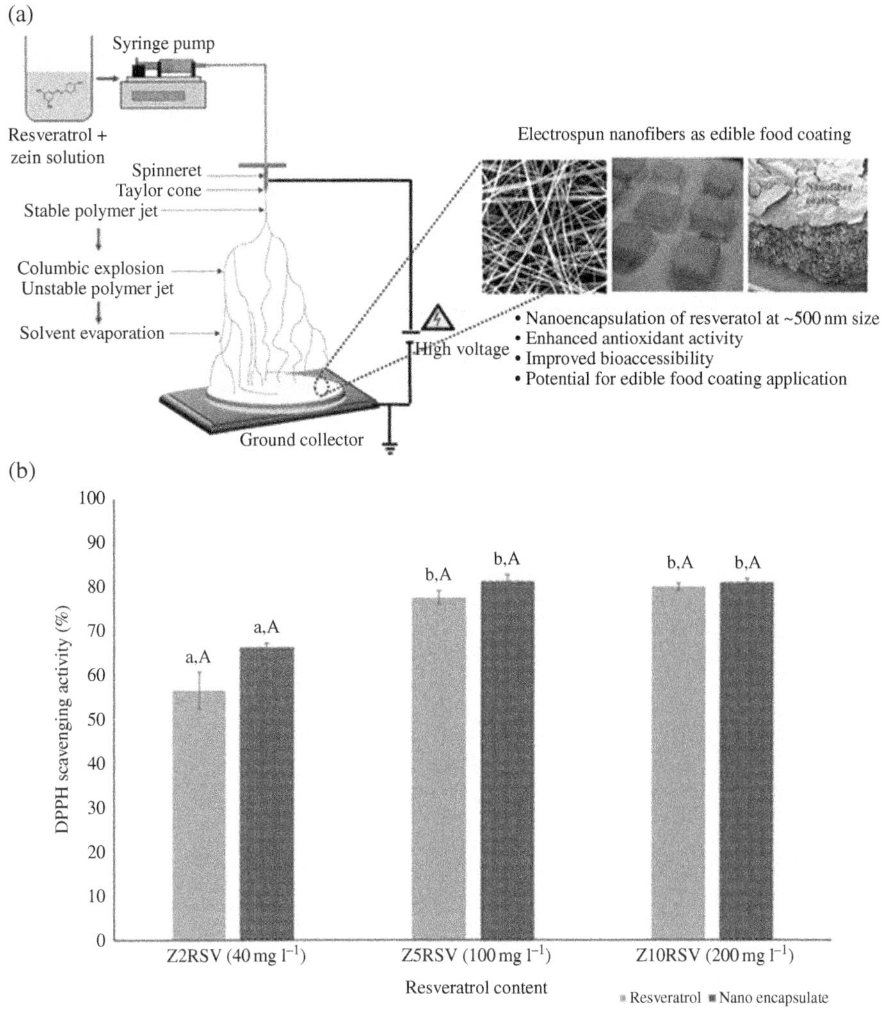

Figure 13.10 Edible coating with resveratrol loaded electrospun zein nanofibers. (a) Illustration of electrospinning process for edible coating and (b) antioxidant activity of resveratrol before and after electrospinning. *Source:* From M. M. Leena, Yoha, et al. (2020a), Figure 06[p.08] / With permission of ELSEVIER.

Results showed that the encapsulation efficiency of the electrospun nanofibers was found to be 65% with about 51.4% encapsulated into the core. Further, the *in-vitro* release profile of antioxidant β-carotene in fatty food simulant (soybean oil) exhibited a slow and sustained release of bioactives. Thus, the emulsion electrospinning can also be used for the fabrication of bioactive food coatings as active packaging material. Thus, electrospraying is a simple and efficient process used for edible coating of foods at low energy and less processing cost that is capable enough for the production of uniform and thin coating (Leena et al. 2020d). This process of electrospray is an effectual process employed for the delivery of foods with unique characteristics. Another interesting application is the chocolate coating used for the production of sweets and confectioneries. The application of the electrospray process for the

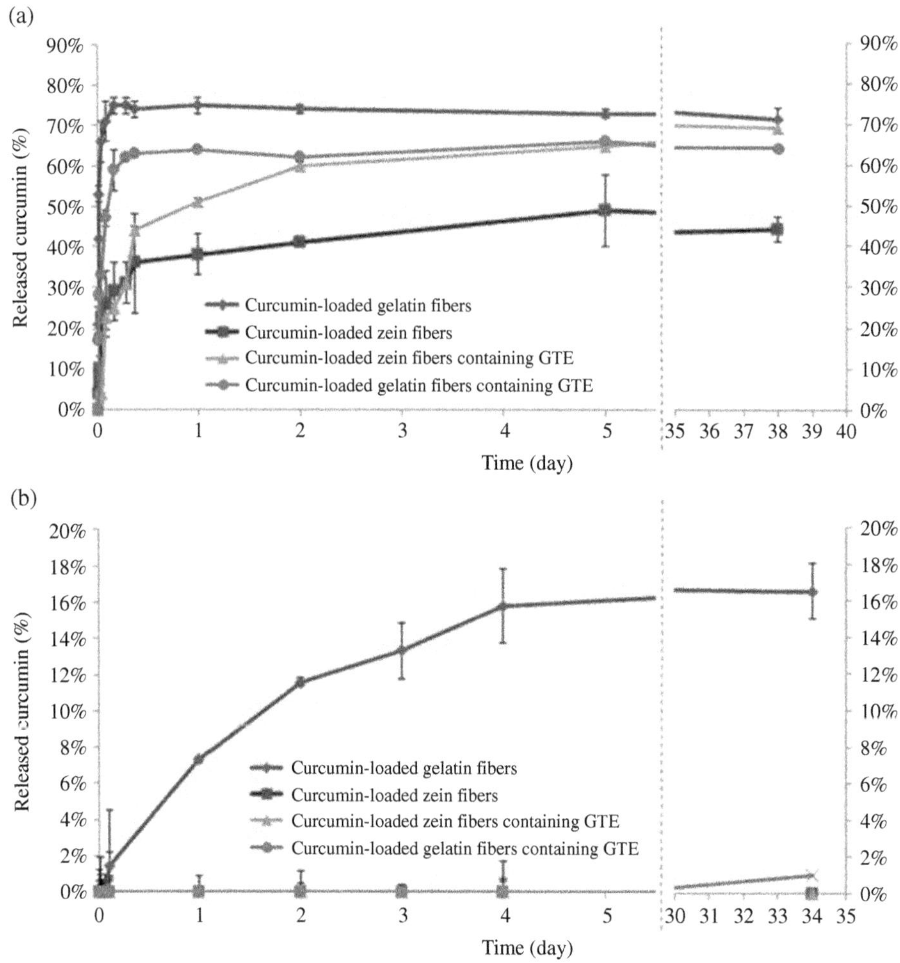

Figure 13.11 Curcumin release profiles from electrospun coatings of gelatin nanofiber. (a) In 50% ethanol (v/v) and (b) in soybean oil. *Source:* Alehosseini et al. (2019) / With permission of Elsevier.

chocolate coating has been investigated (Marthina and Barringer 2012). Process parameters such as viscosity, electrical resistivity, yield stress, and flowability are found to influence the spray quality of chocolate. The reduction in these process attributes reduced the size of the droplet thereby resulting in the uniform coating. A further material composition such as fat content and percentage of lecithin content influence the spray process. The highly efficient electrospraying process can be achieved for chocolate slurries with a higher percentage of fat and lesser concentration of surfactant of below 1.5% lecithin (Gorty and Barringer 2011). Other factors such as process temperature, sugar concentration, and the presence of electrolytes also influence the size of droplet formation. The monodispersing of a particle through electrospraying seems to be a cost-effective process in the confectionery industry that has the potential to replace the conventional spray coating process (Bhushani and Anandharamakrishnan 2014). The electrosprayed chocolates possessed unique organoleptic

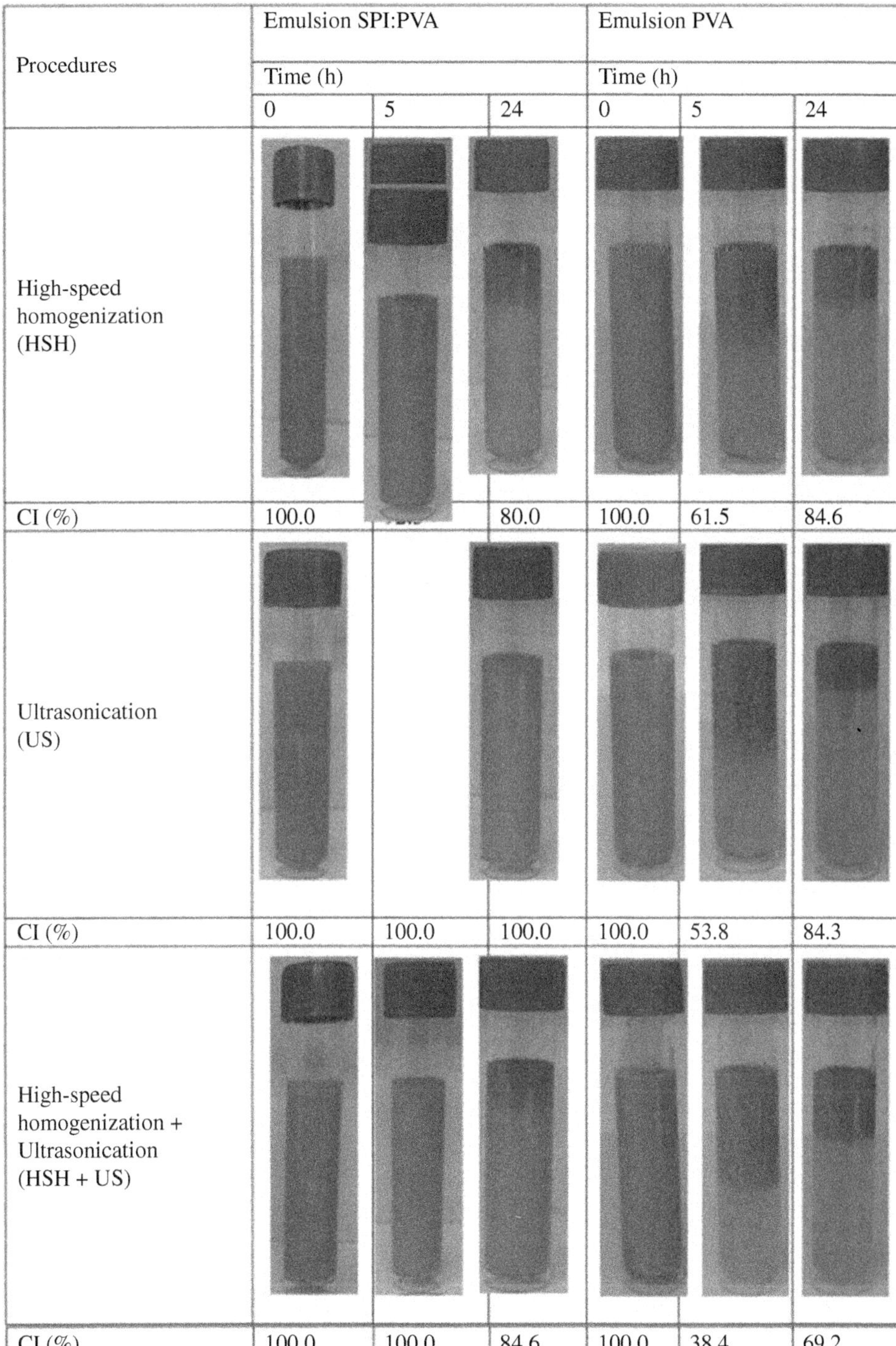

Procedures	Emulsion SPI:PVA			Emulsion PVA		
	Time (h)			Time (h)		
	0	5	24	0	5	24
High-speed homogenization (HSH)						
CI (%)	100.0	[illegible]	80.0	100.0	61.5	84.6
Ultrasonication (US)						
CI (%)	100.0	100.0	100.0	100.0	53.8	84.3
High-speed homogenization + Ultrasonication (HSH + US)						
CI (%)	100.0	100.0	84.6	100.0	38.4	69.2

Figure 13.12 Gravitational stability of different formulations of prepared emulsion during 24-h storage period with their creaming index values. *Source:* From Bruni et al. (2020), Table 02 [p. 06] / With permission of Elsevier.

properties with modified and mouthfeel that were sensorially superior to the bulk chocolates. Thus, the electrospray coating is a cutting-edge technology with simple and minimal processing that imparts texture modification in the confectionery items and enhances the shelf-life of the perishable commodities through incorporation of functional bioactives.

13.4 Integrating 3D Printing with Electrospraying/ Electrospinning

Compare to other encapsulation approaches, the electrohydrodynamic process (electrospraying and electrospinning) is a simple cost-effective technique that can be conveniently integrated with 3D printing through slight modifications of the system design. Electrospraying and electrospinning are a single-step process used for the production of the encapsulates in dried forms. The combination of these techniques with 3D printing allows us to fabricate the 3D printed foods with simultaneous encapsulation of bioactives during the printing process (Leena et al. 2022). The major applications of electrospraying and electrospinning are highlighted in the earlier section that also found to have a great scope for integration of 3D printing (Nachal et al. 2019). In this regard, the combination of electrospraying and electrospinning with 3D printing would provide endless opportunities in the design of novel foods. Various approaches that can be adopted for integration of 3D printing with electrospraying and electrospinning are as follows:

- It is well known that dual extrusion of material is possible with 3D printing that allows us to fabricate a single 3D construct with two different materials (Liu et al. 2018; Uribe-Wandurraga et al. 2020). In a similar manner, coaxial electrospinning allows us to encapsulate multiple cores of different materials (Perez-Masia et al. 2013). These two concepts are seemed to be similar in principle that can be integrated for the simultaneous printing of foods along with encapsulation of bioactive ingredients at micron and submicron range (Figure 13.13).
- Multi-material and multi-head 3D printers are the area of scientific interest that food technologists are actively working on to enhance the productivity of 3D printed foods and to overcome the limitation of low printing speed and low productivity with a single head 3D printer (Nachal et al. 2019). A similar concept goes well with electrospraying and electrospinning where the limitation for scaling up of these techniques is its slow processing speed (Anukiruthika et al. 2020b). Hence 3D printing can be integrated with electrospraying/ electrospinning in such a way with the modification of print heads with multiple spinnerets that enables the fabrication of functional bioactive enriched 3D printed foods (Figure 13.14).
- Another design variant of the electrohydrodynamic process is the co-spinning approach where the materials of interest are electrosprayed and electrospun together under the same process that yields a product with coated powdered particles as well as electrospun fibers with encapsulated bioactive ingredients. Integration of this concept into 3D printing allows us to fabricate 3D printed foods with each layer designed with alternative electrosprayed particle and electrospun fibers for developing a sandwich-type of 3D printed

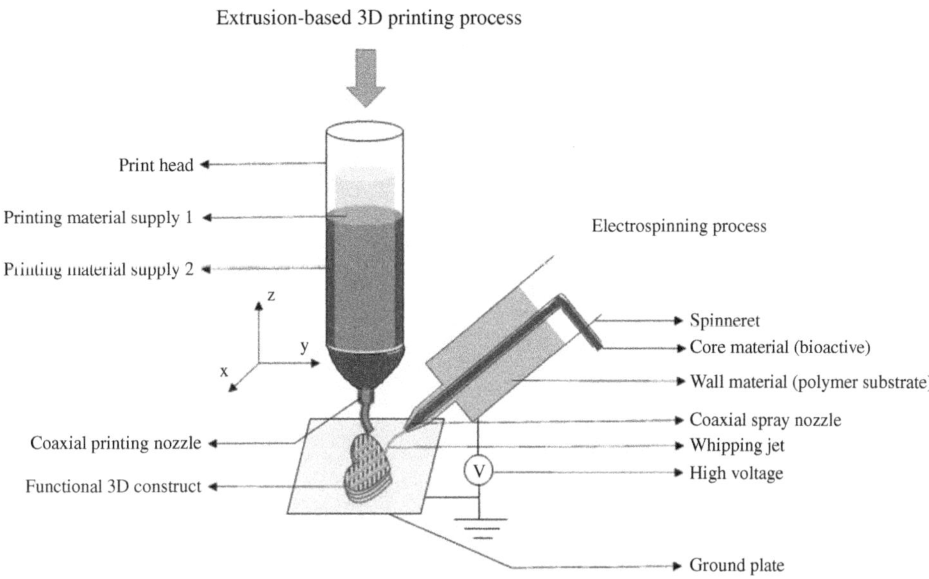

Figure 13.13 Schematic diagram of integration of electrospinning with 3D printing.

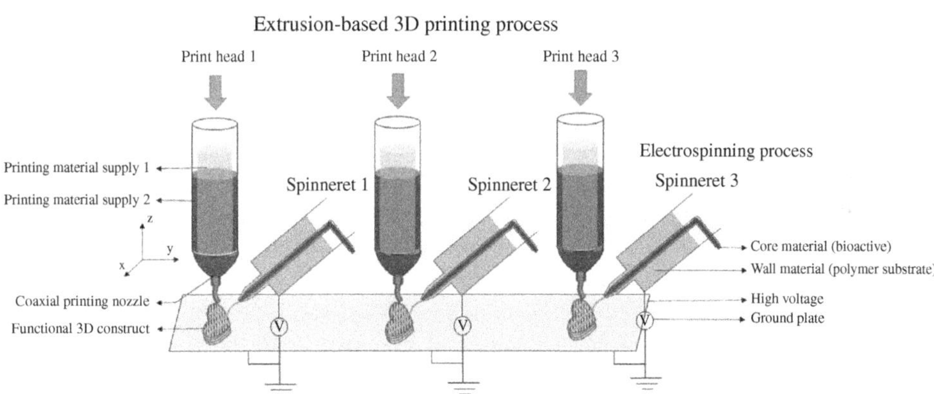

Figure 13.14 Schematic diagram of integration of multijet electrospinning with multi-head 3D printing process.

food with multiple bioactive compounds that help in targeted delivery as well as localized use of functional ingredients (Figure 13.15).

- Recently, encapsulation of bioactive core materials with a novel design of a three-fluid spray nozzle had gained attention. Studies were reported on proving the efficacy of the sustained and controlled release and delivery of bioactive ingredients using a three-fluid spray nozzle (Antoniraj et al. 2020; Leena et al. 2020a). Results showed that the encapsulation efficiency of bioactive compounds using a three-fluid nozzle was higher than the two-fluid nozzle. This kind of spray nozzles could be adopted for the 3D printing of foods

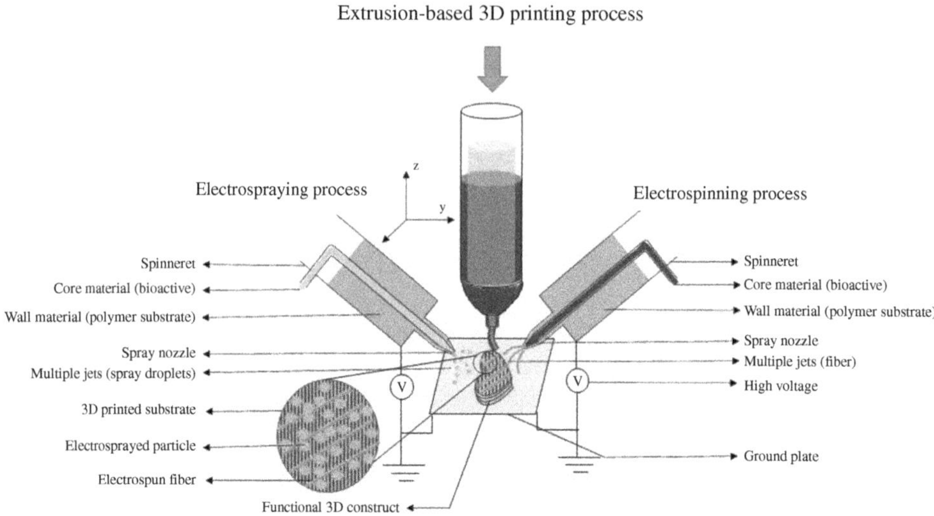

Figure 13.15 Schematic diagram of integration of co-spinning with 3D printing.

in integration with electrospraying/electrospinning that enables the delivery of simultaneous complex coacervation of encapsulates in a single process.

The aforementioned design modifications of the experimental setup of the 3D printer and electrospraying/ electrospinning unit allow us to integrate all the technologies under a one-step process. This integrated approach has numerous advantages in terms of the design of food structure with multiple functionalities. The synergistic benefits of integrated approaches have found its applications in the design of novel innovative 3D printed foods, functionally active and intelligent food packages, minimal processing through electrospray coating of 3D printed foods, enzyme immobilization, and encapsulation of live probiotic cells to enhance its stability during handling and processing (Sun et al. 2018). All these applications focus on the design engineering of food structures and food packages that allows the controlled, sustained, and targeted release of functional ingredients, multifunctional food packages that monitors and senses food quality during storage, protection of sensitive biomolecules from environmental degradation. Upon integration with 3D printing technology, the foods can be delivered with enhanced levels of customization and personalization of foods. Overall, the integration of electrospraying and electrospinning with 3D printing makes 3D printers serve as a technique for food fortification that allows delivery of foods with precise control over the nutrition. Further, the advancements of biotechnology assist in the development of a gene pool database that guides in the eradication of malnourishment among the group of targeted audience by using 3D printing technology for personalized and precision nutrition (Kumar et al. 2020). Although additional design modifications and integration would complicate the system design, working mechanism and increases the cost, the integrated approach seemed to have a great scope that rules the future food and nutraceutical industries. Since 3D printing and electrospraying/ electrospinning are versatile technologies that allow for many innovations in the design of food at the macro and micro level. The utter benefits of the forecasted approach of integration of 3D printing with electrospraying/ electrospinning has a long way to go that requires more research needs in the optimization of design and process variables, studies on the efficacy of the 3D printed functional food, fate of the encapsulated bioactive compounds and its delivery through 3D printed material supply, feasibility for scale-up and commercialization and similar other debatable questions. Despite of all these interrogations 3D printing is a sunrising technology in food sector with immense scope and research opportunities for design and fabrication of functional foods.

13.5 Future Perspectives and Challenges

3D printing is a fascinating technology that allows for mass customization and personalization of foods. The integration of 3D printing with encapsulation techniques particularly electrospraying/ electrospinning allows designing complex food structures with intricate designs and multiple functionalities. The combined approach would consolidate and put forth the advantages of both technologies together that can be achievable in a single process step. Various benefits of this integrated approach are the protection of heat-labile and chemical sensitive active compounds, entrapment of biomolecules of interest, control over the release of bioactive, enhanced absorption, increased bioavailability, and greater efficacy (Jia et al. 2016; Leena et al. 2020e). As a single standalone technology, this integrated approach can yield promising results in alleviation of malnutrition, decrease the incidence of lifestyle disorders, address the specific nutritional needs of elderly people with dysphagia, infants, adolescents, pregnant women with precise delivery of the calculated proportions of calories and nutrients (Keerthana et al. 2020; Krishnaraj et al. 2019). Various prospective techniques that could be integrated include the multi-head print heads with multiple spinnerets; print head with coaxial electrospinning; print head with co-spinning; and multi-material printing with two-fluid/ three-fluid spray nozzles (Antoniraj et al. 2020; Hertafeld et al. 2019; Leena et al. 2020a). This combination would not only increase productivity but also has a great scope for commercialization. However, the integration of these electro-technologies with 3D printing complicates the optimization process where material properties and printing variables contribute significantly in achieving higher printability and post-processing stability of the printed 3D construct. Several printing technologies that are adopted for food applications include extrusion-based printing, selective sintering, inkjet printing, binder jetting, and bioprinting. Among which extrusion is the most commonly used technique for the printing of semi-solid food materials. In this regard, extrusion printing is more suitable for integration with electrospraying/ electrospinning through coaxial design and multiple print heads. While other printing techniques such as inkjet printing and binder jetting can be explored for surface filling and decoration through incorporating the active ingredients into material supply. In contrast, the powder bed fusion technique, i.e., the sintering process could be applied for the design of intricate 3D structures with higher porosity through electrospray coating of bioactive compounds. Although these combinations show a promising scope for the design of functional foods, the actual implementation of technical feasibility is not yet explored even at the lab-scale level.

The prospects of these techniques can be realized by a thorough understanding of the merits and the limitations of the integrated approach. Though the integration of 3D printing with electrostatic-based encapsulation techniques will add on the merits, it also poses complications and limitations in the actual practice. This includes increased complexity in the design of system components at micro and nano level; difficulty in optimization of process variables; severity of pre-treatments of the material supply to ensure smooth flowability through micron and submicron-sized printing nozzles; coordination of movement of print heads and axes of spinneret; integration of slicing software with 3D printer unit; selection of suitable post-processing method that should not degrade the encapsulates; testing the efficacy and bioavailability of the encapsulated active ingredients; choice of core and wall materials, nature, type, properties, and compatibility of core and wall materials; safety, edibility, allergenicity, and toxicity of developed functional foods at the molecular level. Most importantly, both electrospraying and electrospinning are high voltage technologies that require the use of mechanically stable and durable food-grade materials in designing the components of a 3D printer. The accessories and other components of the 3D printer should withstand high voltages when integrated with electrospraying and electrospinning in a single process chamber. The integration of technologies would surpass the limitation of lower productivity and long processing time indeed the scale of operation and commercialization are still questionable. Although the integrated approach has a flourishing future in the functional and nutraceutical industry, it requires more research inputs in addressing these complications and limitations. This can be addressed through channelling the regulations, proper communications, and knowledge transfer among the experts of diverse fields such as food engineers, nanotechnologists, material and design engineers, gastronomists, information technology (IT) professionals, and industrial firms. Hence, the integration of 3D printing with electrospraying and electrospinning would open up a new dimension of research opportunities in the area of functional foods and nutraceutical delivery.

13.6 Conclusion

3D printing on integration with the electrospraying and electrospinning augment the growth of the functional foods and nutraceuticals market. The integration of the 3D printing with electrostatic-based encapsulation techniques such as electrospraying and electrospinning would significantly reduce the processing steps that ensure the direct encapsulation of active ingredient in a single step process. This integrated approach has a great scope and prospects not only in designing functional foods and bioactive delivery but also in developing functional smart food packages and food coating for protection of the foods and extend its shelf-life. In this regard, the present chapter mainly discussed various research opportunities and challenges associated with the scale-up of the integrated approach. The basic working principle, system components, and process parameters of electrospraying and electrospinning techniques have been elaborated. Further, the discussion was extended to cover the emerging trends on the applications of electrospraying and electrospinning in the food sector. Various insights and future outlook on the integration of electrospraying and electrospinning with 3D printing were summarized in terms of design innovations, end-use of the functional product, and applications in the food sector. From the literature review, it is clear that electrospraying/ electrospinning and 3D printing have more or less similar design components with few exceptions. This augments the integration of 3D printing and electrospraying/ electrospinning with multi-head print systems combined with multiple spinnerets. Thus, the fabrication of functional foods with localized use of novel active ingredients with alternate layers of electrosprayed particles and electrospun fibers is feasible with this integrated approach. Although the combination of 3D printing and electrospraying/ electrospinning yield promissory results, no reports are available on the integration of these techniques even at the lab-scale level. Hence, more studies are required in addressing the research aspects on system design, process optimization, printability, encapsulation efficiency, and bioavailability of the active ingredients. All these attributes would assist in analyzing the practical feasibility of this integrated approach, large-scale implementation, and commercialization. Further, the permissible limit of usage of the active ingredients, allergenicity, toxicity, storage stability, quality, and other safety aspects requires a detailed investigation. Nevertheless, the statutory bodies and government authorities must intervene to establish standard protocols and regulations on the application of the bioactive ingredients stating the threshold limits of usage. Hence, proper understanding and knowledge transfer among professional experts, food scientists, manufacturers, and industrial firms is adequate to explore the utter benefits of the integrated approach of 3D printing of foods through electrospraying and electrospinning.

References

Adachi, K., Yoshida, K., Makino, M. et al. (2020). Formation of liposomes containing pre-gel solution and 3D-printing applications by droplet-shooting method. *ECS Transactions* **98** (13): 85.

Ajitha, A.R., Aswathi, M.K., Maria, H.J. et al. (2016). Multilayer polymer films. In: *Multicomponent Polymeric Materials* (eds. K. Kim, S. Thomas and P. Saha), 229–258. Springer.

Alehosseini, A., Gomez-Mascaraque, L.G., Martinez-Sanz, M., and Lopez-Rubio, A. (2019). Electrospun curcumin-loaded protein nanofiber mats as active/bioactive coatings for food packaging applications. *Food Hydrocolloids* **87**: 758–771.

Anandharamakrishnan, C. (2014). Electrospraying and electrospinning techniques for nanoencapsulation. In: *Techniques for Nanoencapsulation of Food Ingredients* (ed. C. Anandharamakrishnan), 43–49. Springer.

Antoniraj, M.G., Leena, M.M., Moses, J.A., and Anandharamakrishnan, C. (2020). Cross-linked chitosan microparticles preparation by modified three fluid nozzle spray drying approach. *International Journal of Biological Macromolecules* **147**: 1268–1277.

Anukiruthika, T., Moses, J.A., and Anandharamakrishnan, C. (2020a). Electrohydrodynamic drying of foods: principle, applications, and prospects. *Journal of Food Engineering*, pp. 110449.

Anukiruthika, T., Sethupathy, P., Wilson, A. et al. (2020b). Multilayer packaging: advances in preparation techniques and emerging food applications. *Comprehensive Reviews in Food Science and Food Safety* **19**: 1156–1186.

Bharathi, S.K.V., Moses, J.A., and Anandharamakrishnan, C. (2018). Nano and microencapsulation using food grade polymers. In: *Polymers for Food Applications* (ed. T.J. Gutierrez), 357–400. Springer.

Bharathi, S.K.V., Leena, M.M., Moses, J.A., and Anandharamakrishnan, C. (2020). Nanofiber-based bilayer biopolymer films: enhancement of anti-oxidant activity and potential for food packaging application. *International Journal of Food Science & Technology* **55**: 1477–1484.

Bhushani, J.A. and Anandharamakrishnan, C. (2014). Electrospinning and electrospraying techniques: potential food based applications. *Trends in Food Science & Technology* **38** (1): 21–33.

Bhushani, J.A., Kurrey, N.K., and Anandharamakrishnan, C. (2017). Nanoencapsulation of green tea catechins by electrospraying technique and its effect on controlled release and in-vitro permeability. *Journal of Food Engineering* **199**: 82–92.

Biji, K.B., Ravishankar, C.N., Mohan, C.O., and Gopal, T.K.S. (2015). Smart packaging systems for food applications: a review. *Journal of Food Science and Technology* **52** (10): 6125–6135.

Bruni, G.P., de Oliveira, J.P., Gómez-Mascaraque, L.G. et al. (2020). Electrospun β-carotene-loaded SPI: PVA fiber mats produced by emulsion-electrospinning as bioactive coatings for food packaging. *Food Packaging and Shelf Life* **23**: 100426.

Chen, M.-J. (2007). Development and parametric studies of carbon nanotube dispersion using electrospraying http://purl.flvc.org/fsu/fd/FSU_migr_etd-3849 (accessed 7 November 2020).

Coelho, S.C., Estevinho, B.N., and Rocha, F. (2020). Encapsulation in food industry with emerging electrohydrodynamic techniques: electrospinning and electrospraying - a review. *Food Chemistry*, p. 127850.

Colussi, R., da Silva, W.M.F., Biduski, B. et al. (2021). Postharvest quality and antioxidant activity extension of strawberry fruit using allyl isothiocyanate encapsulated by electrospun zein ultrafine fibers. *LWT* **143**: 111087.

Deotale, S.M., Dutta, S., Moses, J.A., and Anandharamakrishnan, C. (2020). Stability of instant coffee foam by nanobubbles using spray-freeze drying technique. *Food and Bioprocess Technology* **13** (11): 1866–1877.

Dong, B., Smith, M.E., and Wnek, G.E. (2009). Encapsulation of multiple biological compounds within a single electrospun fiber. *Small* **5** (13): 1508–1512.

Drosou, C.G., Krokida, M.K., and Biliaderis, C.G. (2017). Encapsulation of bioactive compounds through electrospinning/electrospraying and spray drying: a comparative assessment of food-related applications. *Drying Technology* **35** (2): 139–162.

Gorty, A.V. and Barringer, S.A. (2011). Electrohydrodynamic spraying of chocolate. *Journal of Food Processing and Preservation* **35** (4): 542–549.

Haider, A., Haider, S., and Kang, I.-K. (2018). A comprehensive review summarizing the effect of electrospinning parameters and potential applications of nanofibers in biomedical and biotechnology. *Arabian Journal of Chemistry* **11** (8): 1165–1188.

Hertafeld, E., Zhang, C., Jin, Z. et al. (2019). Multi-material three-dimensional food printing with simultaneous infrared cooking. *3D Printing and Additive Manufacturing* **6** (1): 13–19.

Homaei, A.A., Sariri, R., Vianello, F., and Stevanato, R. (2013). Enzyme immobilization: an update. *Journal of Chemical Biology* **6** (4): 185–205.

Hosseini, S.F., Ramezanzade, L., and McClements, D.J. (2021). Recent advances in nanoencapsulation of hydrophobic marine bioactives: bioavailability, safety, and sensory attributes of nano-fortified functional foods. *Trends in Food Science & Technology* **109**: 322–339.

Icten, E., Purohit, H.S., Wallace, C. et al. (2017). Dropwise additive manufacturing of pharmaceutical products for amorphous and self emulsifying drug delivery systems. *International Journal of Pharmaceutics* **524** (1–2): 424–432.

Jayan, H., Leena, M.M., Sundari, S.K.S. et al. (2019). Improvement of bioavailability for resveratrol through encapsulation in zein using electrospraying technique. *Journal of Functional Foods* **57**: 417–424.

Jia, W., Gungor-Ozkerim, P.S., Zhang, Y.S. et al. (2016). Direct 3D bioprinting of perfusable vascular constructs using a blend bioink. *Biomaterials* **106**: 58–68.

Kalpana, S., Priyadarshini, S.R., Leena, M.M. et al. (2019). Intelligent packaging: trends and applications in food systems. *Trends in Food Science & Technology* **93** (September): 145–157. https://doi.org/10.1016/j.tifs.2019.09.008.

Keerthana, K., Anukiruthika, T., Moses, J.A., and Anandharamakrishnan, C. (2020). Development of fiber-enriched 3D printed snacks from alternative foods: a study on button mushroom. *Journal of Food Engineering* https://doi.org/10.1016/j.jfoodeng.2020.110116.

Khan, M.K.I., Schutyser, M.A.I., Schroën, K., and Boom, R. (2012). The potential of electrospraying for hydrophobic film coating on foods. *Journal of Food Engineering* **108** (3): 410–416.

Khan, M.A., Chen, L., and Liang, L. (2021). Improvement in storage stability and resveratrol retention by fabrication of hollow zein-chitosan composite particles. *Food Hydrocolloids* **113**: 106477.

Krishnaraj, P., Anukiruthika, T., Choudhary, P. et al. (2019). 3D extrusion printing and post-processing of fibre-rich snack from indigenous composite flour. *Food and Bioprocess Technology* **12** (10): 1776–1786. https://doi.org/10.1007/s11947-019-02336-5.

Kumar, S., Anukiruthika, T., Dutta, S. et al. (2020). Iron deficiency anemia: a comprehensive review on iron absorption, bioavailability and emerging food fortification approaches. *Trends in Food Science & Technology* https://doi.org/10.1016/j.tifs.2020.02.021.

Kutzli, I., Gibis, M., Baier, S.K., and Weiss, J. (2018). Formation of whey protein isolate (WPI)-maltodextrin conjugates in fibers produced by needleless electrospinning. *Journal of Agricultural and Food Chemistry* **66** (39): 10283–10291.

Kutzli, I., Gibis, M., Baier, S.K., and Weiss, J. (2019). Electrospinning of whey and soy protein mixed with maltodextrin-Influence of protein type and ratio on the production and morphology of fibers. *Food Hydrocolloids* **93**: 206–214.

Lavanya, M.N., Preethi, R., Moses, J.A., and Anandharamakrishnan, C. (2020). Production of bromelain aerosols using spray-freeze-drying technique for pulmonary supplementation. *Drying Technology* **39**: 1–13.

Leena, M., Yoha, K.S., Moses, J.A. et al. (2019). Electrospraying and spinning techniques fabrication and its potential applications. In: *Food Nanotechnology: Principles and Applications*. CRC Press.

Leena, M.M., Antoniraj, M.G., Moses, J.A., and Anandharamakrishnan, C. (2020a). Three fluid nozzle spray drying for co-encapsulation and controlled release of curcumin and resveratrol. *Journal of Drug Delivery Science and Technology* **57**: 101678.

Leena, M.M., Bharathi, S.K.V., Moses, J.A., and Anandharamakrishnan, C. (2020b). Potential applications of nanofibers in beverage industry. In: *Nanoengineering in the Beverage Industry*, 333–368. Elsevier.

Leena, M.M., Mahalakshmi, L., Moses, J.A., and Anandharamakrishnan, C. (2020c). Nanoencapsulation of nutraceutical ingredients. In: *Biopolymer-Based Formulations*, p. 311–352. Elsevier.

Leena, M.M., Yoha, K.S., Moses, J.A., and Anandharamakrishnan, C. (2020d). Edible coating with resveratrol loaded electrospun zein nanofibers with enhanced bioaccessibility. *Food Bioscience* **36**: 100669.

Leena, M.M., Yoha, K.S., Moses, J.A., and Anandharamakrishnan, C. (2020e). Nanofibers in food applications. In: *Innovation in Food Nanotechnology, Innovative Food Processing Technologies: A Comprehensive Review* (eds. K. Knoerzer and K. Muthukumarappan), pp. 634–650. Elsevier. doi: B978-0-08-100596-5.22952-1.

Leena, M. M., Anukiruthika, T., Moses, J. A., and Anandharamakrishnan, C. (2022). Co-delivery of curcumin and resveratrol through electrosprayed core-shell nanoparticles in 3D printed hydrogel. *Food Hydrocolloids* **124**, 107200.

Libran, C.M., Castro, S., and Lagaron, J.M. (2017). Encapsulation by electrospray coating atomization of probiotic strains. *Innovative Food Science & Emerging Technologies* **39**: 216–222.

Liu, Z., Zhang, M., and Yang, C. (2018). Dual extrusion 3D printing of mashed potatoes/strawberry juice gel. *LWT* **96**: 589–596.

Long, J., Pan, T., Xie, Z. et al. (2019). Effective production of lactosucrose using β-fructofuranosidase and glucose oxidase co-immobilized by sol-gel encapsulation. *Food Science & Nutrition* **7** (10): 3302–3316.

Lupton, D. and Turner, B. (2016). Both fascinating and disturbing': consumer responses to 3D food printing and implications for food activism. In: *Digital Food Activism* (eds. T. Schneider, K. Eli, C. Dolan and S. Ulijaszek). London: Routledge.

Mahalakshmi, L., Leena, M.M., Moses, J.A., and Anandharamakrishnan, C. (2020). Micro-and nano-encapsulation of β-carotene in zein protein: size-dependent release and absorption behavior. *Food & Function* **11** (2): 1647–1660.

Marthina, K. and Barringer, S.A. (2012). Confectionery coating with an electrohydrodynamic (EHD) system. *Journal of Food Science* **77** (1): E26–E31.

Moses, J.A., Norton, T., Alagusundaram, K., and Tiwari, B.K. (2014). Novel drying techniques for the food industry. *Food Engineering Reviews* **6** (3): 43–55.

Nachal, N., Moses, J.A., Karthik, P., and Anandharamakrishnan, C. (2019). Applications of 3D printing in food processing. *Food Engineering Reviews* **11** (3): 123–141. https://doi.org/10.1007/s12393-019-09199-8.

Olmo, J.A., Pérez-álvarez, L., Hernáez, E., and Ruiz-rubio, L. (2019). Food hydrocolloids antibacterial multilayer of chitosan and (2-carboxyethyl) - β -cyclodextrin onto polylactic acid (PLLA). *Food Hydrocolloids* **88** (October 2018): 228–236. https://doi.org/10.1016/j.foodhyd.2018.10.014.

Pal, M., Devrani, M., and Hadush, A. (2019). Recent developments in food packaging technologies. *Beverage Food World* **46** (1): 21–25.

Perez-Masia, R., Lopez-Rubio, A., and Lagaron, J.M. (2013). Development of zein-based heat-management structures for smart food packaging. *Food Hydrocolloids* **30** (1): 182–191.

Pinheiro Bruni, G., dos Santos Acunha, T., de Oliveira, J. P., Martins Fonseca, L., Tavares da Silva, F., Martins Guimarães, V., and da Rosa Zavareze, E. (2020). Electrospun protein fibers loaded with yerba mate extract for bioactive release in food packaging. *Journal of the Science of Food and Agriculture* **100** (8): 3341–3350.

Portanguen, S., Tournayre, P., Sicard, J. et al. (2019). Toward the design of functional foods and biobased products by 3D printing: a review. *Trends in Food Science & Technology* **86**: 188–198.

Pose-Boirazian, T., Eibes, G., Barreiro-Piñeiro, N. et al. (2021). Chemical and thermal stabilization of CotA laccase via a novel one-step expression and immobilization in muNS-Mi nanospheres. *Scientific Reports* **11** (1): 1–10.

Preethi, R., Leena, M.M., Moses, J.A., and Anandharamakrishnan, C. (2020). Biopolymer nanocomposites and its application in food processing. In: *Green Nanomaterials* (eds. S. Ahmed and W. Ali), 283–317. Springer.

Radosavljevic, M., Levic, S., Belovic, M. et al. (2021). Encapsulation of *Lactobacillus rhamnosus* in polyvinyl alcohol for the production of L-(+)-lactic acid. *Process Biochemistry* **100**: 149–160.

Ramazani, S., Rostami, M., Raeisi, M. et al. (2019). Fabrication of food-grade nanofibers of whey protein Isolate-Guar gum using the electrospinning method. *Food Hydrocolloids* **90**: 99–104.

Sill, T.J. and Von Recum, H.A. (2008). Electrospinning: applications in drug delivery and tissue engineering. *Biomaterials* **29** (13): 1989–2006.

Sun, J., Zhou, W., Huang, D., and Yan, L. (2018). 3D food printing: perspectives. In: *Polymers for Food Applications* (ed. T.J. Gutierrez), 725–755. Springer.

Uribe-Wandurraga, Z.N., Zhang, L., Noort, M.W.J. et al. (2020). Printability and physicochemical properties of microalgae-enriched 3D-printed snacks. *Food and Bioprocess Technology* **13** (11): 2029–2042.

Vimala Bharathi, S.K., Maria Leena, M., Moses, J.A., and Anandharamakrishnan, C. (2019). Zein-based anti-browning cling wraps for fresh-cut apple slices. *International Journal of Food Science & Technology* **55**: 1238–1245.

Vithani, K., Goyanes, A., Jannin, V. et al. (2019). A proof of concept for 3D printing of solid lipid-based formulations of poorly water-soluble drugs to control formulation dispersion kinetics. *Pharmaceutical Research* **36** (7): 1–13.

Waghmare, R., Moses, J.A., and Anandharamakrishnan, C. (2021). Mucilages: sources, extraction methods, and characteristics for their use as encapsulation agents. *Critical Reviews in Food Science and Nutrition*, pp. 1–22.

Wang, Z.-G., Wan, L.-S., Liu, Z.-M. et al. (2009). Enzyme immobilization on electrospun polymer nanofibers: an overview. *Journal of Molecular Catalysis B: Enzymatic* **56** (4): 189–195.

Wang, C., Hou, W., Guo, X. et al. (2017). Two-phase electrospinning to incorporate growth factors loaded chitosan nanoparticles into electrospun fibrous scaffolds for bioactivity retention and cartilage regeneration. *Materials Science and Engineering: C* **79**: 507–515.

Wang, P., Ding, M., Zhang, T. et al. (2020). Electrospraying technique and its recent application advances for biological macromolecule encapsulation of food bioactive substances. *Food Reviews International*, pp. 1–23.

Wen, P., Zhu, D.-H., Wu, H. et al. (2016). Encapsulation of cinnamon essential oil in electrospun nanofibrous film for active food packaging. *Food Control* **59**: 366–376.

Yildirim, S., Rocker, B., Pettersen, M.K. et al. (2018). Active packaging applications for food. *Comprehensive Reviews in Food Science and Food Safety* **17** (1): 165–199. https://doi.org/10.1111/1541-4337.12322.

Zhang, X., Kobayashi, I., Uemura, K., and Nakajima, M. (2013). Direct observation and characterization of the generation of organic solvent droplets with and without triglyceride oil by electrospraying. *Colloids and Surfaces A: Physicochemical and Engineering Aspects* **436**: 937–943.

Zhang, C., Feng, F., and Zhang, H. (2018). Emulsion electrospinning: fundamentals, food applications and prospects. *Trends in Food Science & Technology* **80**: 175–186.

14

Globalization of Printed Foods and Consumer Perception to 3D Printed Foods

CHAPTER MENU

14.1 Introduction

3D printing is an additive technology that adds on innovations in the production of goods in various industrial sectors. The social sustainability of any emerging technology is adequate in concern with supply chain management. The increasing adoption of 3D printing in the food sector is mainly because of its flexibility, digital fabrication, minimal material wastage, reduced complexity of the process, design innovations, cost-effectiveness in low-volume production, localized on-demand production, and reduced transportation cost that offers several socio-economic benefits (Bandyopadhyay and Heer 2018). These advantageous features of 3D printing provide endless opportunities in the fabrication of novel food products. Although the 3D printing of foods is claimed to simplify the food supply chain, the aspects of social and environmental sustainability in manufacturing 3D printed foods remain unexplored. A piece of thorough knowledge on the management of the operations of the food supply chain and energy resources are quite important for maximizing profitability and social well-being (Beltagui et al. 2020). Hence, the socio-economic benefits of 3D food printing must be analysed for determining the sustenance of the technology in the long-term run.

3D Printing of Foods, First Edition. C. Anandharamakrishnan, Jeyan A. Moses, and T. Anukiruthika.

Reports stated that 3D printing is seemed to be competing with the existing conventional technologies (Kapetaniou et al. 2018; Laplume et al. 2016). The unmatching levels of customization and personalization of 3D printing than conventional food technologies add value to the 3D printed products. Its capability and tendency of fabrication of customized foods with complex internal structures and unique functionality remain as the driving potential to capture the food market (Nachal et al. 2019). 3D food printing allows the firms to engage with the customers through digitized layered manufacturing of foods that provides the possibility of designing their own food for consumers. The advantage of sharing the digitalized files for printing of foods helps in improving the communication among the food technologists, culinary professionals, design engineers, production managers, market firms, suppliers, distributors, and consumers. In this regard, the barriers to social sustainability are the lack of awareness about sustainability among the consumers and the insufficient knowledge on the material flow in the food supply and distribution chain (Govindan 2018). These barriers can be overcome by filling the knowledge gap on how 3D printing of foods impacting the production and economy.

Mainly 3D printing is used as a rapid prototyping tool that allows for improving the supply chain efficiencies, shortens the marketing time, transforms mass production into mass customization, and offers environmentally sustainable practices. The advent of 3D printing in the food sector is known to be disruptive innovation and acceleration of conventional food production towards digitalization (Charlebois and Juhasz 2018). Undoubtedly, 3D printing of foods would pave a way for the technological transitions influencing the existing industrial firms through expansion of industries, development of new industries, and can even overthrow the traditional manufacturing practices. Unlike the other technologies, 3D printing allows for the digitalized production of customized foods in both small and large-scale levels that accelerates the cultural shift from ready-made processed foods into do-it-yourself foods based on individual likeness (Garrett 2014). Thus, in near future 3D printers would become a common domestic kitchen appliance for preparing foods in every home. This dictates the necessity of the ease of material access, protocols, and standards for the printing of foods, design handling, and software flexibility in the creation of own 3D design models (Nachal et al. 2019). The convergence of 3D printing with the co-creation of web-based 2.0 technologies would lead towards the globalization of 3D-printed foods (Rayna et al. 2015). With advancements in information technologies, a number of online platforms allow for sharing the design of 3D models that enable designers, industrial firms, and individuals to access the source file for manufacturing the customized foods. Thus, 3D printing could create new competitive dynamics by enforcing the digitalization and democratization of manufacturing foods (Kapetaniou et al. 2018). In this regard, understanding food manufacturing through 3D printing and its economic paradigms are adequate. An overview of the sustainability and the circular economy of 3D printing of foods has been presented. The significance of globalization of printing recipes and its impacts on the supply and distribution chain of foods are discussed. Although many people are now becoming aware of food printing technology, only a few know about the extent to which this disruptive technology could impact the global world. The present chapter explores the implications of the large-scale adoption of 3D food printing with the current trends of scientific, technological, economic, and social developments evolving with this additive process of food manufacturing.

14.2 Circular Economy in Food Printing

The emergence of novel technologies will provide the feasibility of changing the manufacturing practices to a more organized manner. This would have a positive effect on the distribution chain of the manufactured goods by regulating the materials flow with sustainable benefits (Rossi et al. 2019). One such aspect is the transition of productional practice towards the circular economy that radically improves resource efficiency by reducing material wastage. Thus, the transitional shift from the linear model of the economy to a circular economy would not only saves the resources but also reduces the negative environmental impacts (Lewandowski 2016). Hence, circular economy has gained recent attention as a way towards sustainability in terms of the design of materials, new business models, and globalization. In this view, the current applications of 3D printing in the production of foods are based on a circular economy with the efficient usage of materials and re-utilization of waste streams from food processing lines (Rubio and Hurtado 2019). As a sustainable approach, various circular design strategies of 3D printing are customer attachment, resource efficiency, low inventory, and localized production (Zhong and Pearce 2018). 3D food printing creates a strong user-product relationship as the food products are customized in a personalized way based on individual preferences and nutritional needs. Adoption of food printing technology seemed to reduce material wastage, reduces the complexity of assembly lines, product and process efficiency, reduced energy consumption, and less processing and transportation cost. 3D printing provides a greater degree of freedom in designing the customized foods with less inventory risk (Attaran 2017). Moreover, the re-utilization of agro-food industry waste transforms the 'waste into wealth' that significantly contributes to the circular economy using 3D printing. Focusing on sustainability, recently a study was reported on the utilization of potato processing by-products in the development of texture-modified air-fried snacks using 3D printing (Feng et al. 2020). Although this study described the printability of material supply and the mechanical stability of post-processed 3D printed product, the feasibility of incorporation of less valued waste fraction into printing material supply for the development of the value-added product was well documented. By taking the advantage of 3D printing, the waste streams of food processing line such as milling fractions of cereals, pulses, and millets; assorted cut fruits and vegetables; trimmings, and off-cuts of meat could be potentially transformed into a valuable 3D printed food (Prakash et al. 2019). Although these materials have been used for animal feed, fertilizer, and bioethanol production, they are less valued as it possesses a range of valuable macronutrients that could potentially be served as human foods through appropriate pre-processing through 3D printing.

The application of 3D printing in food packaging is also quite remarkable in developing smart and intelligent packaging with customized designs and functionality. The circular economy is a business model based on the efficient and individual use of biological and technical inputs that adds value to the existing material with less exploitation of natural resources (Despeisse et al. 2017). Recently, the reports on the making use of the advantage of a circular economy for packaging reformed the roles of business models and stakeholders that exist with the conventional materials, logistics, legislation, and policies (Ferreira Gregorio et al. 2018; Meherishi et al. 2019). The 4R strategy in developing a package based on circular economy implies recycling of the used materials from package waste; reusing the already developed package through appropriate processing; renewing is the use of biodegradable and renewable sources for developing a package; and rethinking is the novelty and the innovation of designing methods and improving the existing package (Ten Klooster 2002). In cohort to the stated 4R strategy, 3D printing could be promisingly used for developing food packages. Also, the advent of the global e-commerce market would allow the use of novel technologies and design innovations in the packaging (Escursell et al. 2020). Thus, 3D printing can be applied for both designing of food as well as food packages that contributes sustainability in energy consumption, digitalized and simplified supply chains, less CO_2 emissions and carbon footprints, and overall product life cycles (Liu et al. 2019; Nachal et al. 2019). More recently, a study on the fabrication of 3D printed biodegradable food casings from agricultural biomass was reported. This study utilized the rice husk, a waste generated from the milling of rice (Nida et al. 2020). As an initial attempt, the effect of pre-processing (milling and grinding) of material supply was assessed and converted to the printable form with the incorporation of hydrocolloid guar gum. Thus, a biodegradable and mechanically stable 3D printed food casing was developed from rice husk and this study remains as a proof-of-concept for the development of customized environmentally friendly biodegradable food package using 3D printing. Based on this approach research works could be directed in the fabrication of edible food packages through efficient reutilization of waste based on the concept of circular economy through 3D printing. This would significantly help in improving the circular economy through mitigation of green gas emissions, tackling of climate change, energy conservation, and offers economic benefits.

14.3 Globalization of Food 3D Printing Technology

The development of digitalized technologies, electronics, and communication eases the accessibility of 3D model designs. This enables easy transfer of the 3D designs among the designers, food technologists, and culinary professionals that involves a quick download of the source files for printing of foods based on individual desirability (Sun et al. 2015a). Further, the advancements of communication systems enhance the pooling of culinary knowledge and artistic skills from various professionals under a single online platform ready for 3D printing (Berman 2012). Various online web sources allow professionals to upload 3D design models. These models could be easily downloaded by the users at the time of printing of foods based on on-demand requirements. Thus, the advancements of technology could make 3D printing even available at home domestic kitchen apart from small- and large-scale industrial firms. This highlights the necessity of globalization of 3D printing technology that significantly saves time, energy, transportation, and packaging costs (Rogers et al. 2016). The globalization of 3D printed foods would make 3D models available under a single cloud computing platform that enables the users to download as desired. The materials and technology flow in the 3D food printing ecosystem is presented in Figure 14.1. Other than simplifying the food supply chain, 3D printing prevents product failure by considering the consumer's preferences and likeness. In addition, 3D printing acts as a food prototyping tool that tests the desirability of a product before large-scale implementation (Sun et al. 2018a). Ultimately, consumers can easily configure and transact the 3D designs from e-commerce platforms either through the desktop or mobile applications. This could

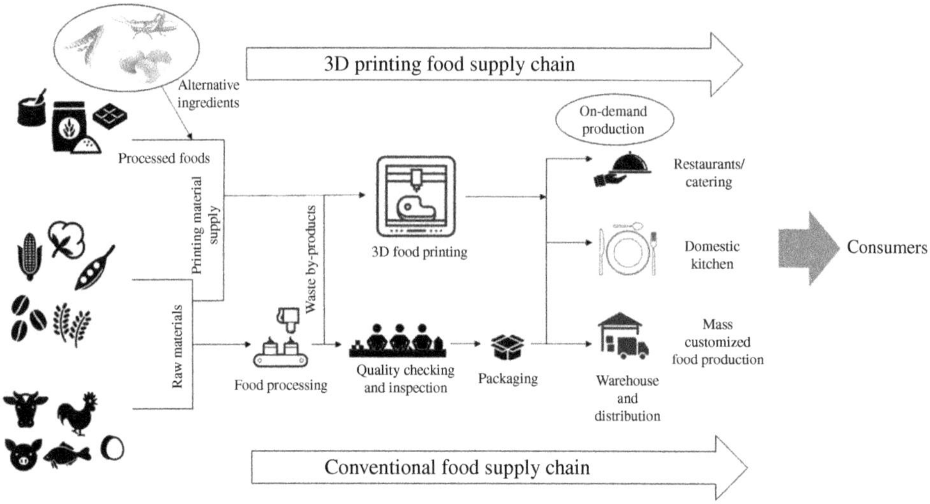

Figure 14.1 Materials and technology flow in 3D food printing ecosystem.

be feasible when the online printing platforms are configured and customized to work under a similar range of compatible software. The compatibility of 3D file is most significant because of the deviations in end quality of the printed 3D construct. Even if the same 3D design is printed with different 3D printers that could result in variations in the quality of the fabricated 3D printed product. This issue can be addressed through standardization of 3D printing technology when the file sharing becomes a business trade.

The growing market trends depict the potentiality of 3D printing to enter into the food trade through online trading and sharing of 3D food printing recipes. This scenario could significantly impact the economy of the current food trade practices that emerge as a rivalry technology in terms of process flexibility, cost, product variety, customization, and personalization of food items. Thus, 3D printing reduces the complexity of the food supply and distribution chain by reducing the manufacturing, transportation, distribution, packaging, and warehousing costs (Keogh et al. 2020). Comparatively, the customized food products are more appealing and reachable to consumers than the conventionally developed foods. With minimal processing time and a simplified supply chain, 3D printing of foods could eliminate the use of chemical preservatives and polymeric food packages for the extended shelf-life of 3D printed foods making it more appropriate for on-demand localized production. Certainly, 3D printing could have a significant impact on the foodservice chain that could alter the organizational structure of the global economy (Nachal et al. 2019). This significantly reduces the gap between the manufacturing and distribution point that is closer and more reachable to the consumers. The 3D printed foods could potentially transform the global economy and trading (import and export) of foods with a newer form of localized production of customized foods. As an overview 3D printing significantly cut off the multiple processing steps such as weighing, mixing, cutting, and dispensing under a single step that seems to offer technical, social, economic, and environmental benefits.

Rapid prototyping comprises a group of additive technologies including 3D printing that widespread in a diverse range of industrial sectors such as automobiles, aircraft, movie props, fashion designing, cosmetics, jewellery, etc. With the advent of 3D printing, all the above-mentioned industrial sectors are evolving towards digitization and global networking. Recently, the emergence of internet technologies had significantly resulted in tracking and traceability of the various developmental and marketing activities such as designing, process planning, manufacturing, customer service, etc., that are distributed at different locations in an integrated environment (Ali et al. 2020). In addition, the global economy is witnessing drastic and dynamic changes related to the 4th industrial revolution. So, the concept of smart manufacturing is nearly adopted in all of the industrial sectors to implement the efficient process with economic operations. The implementation and integration of novel combined technologies in the conventional production approaches would open up opportunities for the setting up of smart factories in order to globalize the production process (Ashima et al. 2021). The deployment of computer-based 3D printing technologies would unleash enormous designing, engineering, and production opportunities that lay down a strong foundation in meeting the custom-built global market demands. The global 3D printing market is predicted to reach 41 587.1 billion dollars in 2027 from 11 223.6 billion dollars in 2019 (Knowles 2019). Only about 16% of firms possess a documented multiyear supply chain and operations strategy as reported by global supply chain institute. However, nevertheless these trends, advancements, and practical feasibility would vary for the food 3D printing applications.

14.4 New Horizons of 3D Food Printing

The adoption of 3D printing in food manufacturing implicates challenges to the existing foodservice chain that would result in the emergence of new strategies, market competitions, and changes in the supply chain which differs with the geographical region (Sasson and Johnson 2016). Various decisive factors that remain as barriers to the globalization of 3D printing of foods are lack of knowledge about 3D printing, insufficient experience with 3D printing systems, high acquisition costs of 3D printers, and technical limitations and challenges associated with the printing of material supplies (Zeleny 2012). The success of globalization of food 3D printing relies on the simplification of services through offering the digital path of design models and the fabrication procedures to the consumers. The major challenge associated with the delivery of the customized food products is its higher cost of production which can be conveniently overcome by increasing the production volume of 3D printing through multi-material and multi-head printing techniques (Sun et al. 2015b). The food supply chain comprises of several complex inter-dependent steps such as pre-harvest, harvest, post-harvest, processing, distribution, retailing, marketing, and consumption. Thus, the intrusion of 3D printing in food supply chain management could reduce the complexity associated with the delivery of the finished foods to the consumers. This involves the transaction of 3D design models from an e-commerce platform that enables the consumers to download the design for printing their own food and also allows the industrial firms for mass customization of foods (Shree et al. 2020). 3D printing is best known for its customization in the fabrication of functional foods in a cost-effective manner with minimal handling steps. This has a significant impact on the materials and energy usage with a transition to a digitalized and localized supply chain. 3D food printing is a revolutionizing technology that offers digitalized solution to multiple problems such as malnourishment, food security, sustainability, and environmental concerns. The present section describes the new novel horizons of 3D printing in food sector.

14.4.1 Strategic Market Foresight

The strategic foresight analysis helps in better understanding the impact of 3D printing on how food manufacturing would be disrupted by this emerging technology. The main focus of food printing is the fabrication and development of innovative foods for a sustainable future in preserving nutritional security. The strategic foresight of the 3D food market acts as a tool that shapes and regulates the business process and planning. Hence it is adequate to know about the innovation economics, business strategies, and policy making to anticipate the future societal implications of 3D food printing. Considering these aspects, Cook et al. (2014) proposed that the market of 3D food printing could be broadened by addressing the existing problems associated with the safety of 3D printed foods; emerging future threats over the conventional food processing technologies; innovative promising research opportunities with the printing of foods; examining the resilience of government policies and schemes; planning the research agenda and outcomes of 3D printing. This should essentially integrate the concepts and insights in addressing the future threats and opportunities with food printing. Hence creative thinking is required in implementing policies and decision making especially in a complex food ecosystem for unwinding the potential applications of 3D food printing (Rasmussen et al. 2010). Thus, strategic foresight is referred to as the point of knowledge transfer that paves a way for the full exploitation of applications of food printing. This includes the set of functions and interface process between the existing knowledge about technology, strategy, business models, and implementation. Hence market strategic foresight of 3D food printing would allow to streamline the existing business model that has significant impacts on trading prospects, employment opportunities, retail pricing, and operational costs (Charlebois and Juhasz 2018). Undoubtedly, 3D food printing has the potential to reform the existing dynamics of production and manufacturing practices at domestic and international levels.

3D food printing offers customized services along with consumer's satisfaction that provides a greater degrees of design freedom in a cost-effective manner. As 3D food printing shortens the foodservice chain, it is essential to review the changes with the design of material flow, supply chain, and logistics for the successful implementation of the 3D printing technology (Jia et al. 2016). The reduction in the number of billing significantly decreases the number of suppliers and traders within the network of the food supply chain. In contrast to conventional technologies, 3D printing eliminates long-distance transportation that reduces handling and transportation costs with less impact on the environment (Petrick and Simpson 2013). Thus, 3D printing is one of the emerging food technologies that should be encouraged and supported by the government through incentives to encourage the development of sustainable and eco-friendly food technologies. In view of mass customization, 3D printing is viewed to download digital models at various production sites simultaneously that has a great scope for ramping up of its production volume. Future research and development on exploring the printability of novel materials would help in designing customized functional foods. Thus, 3D printing of foods has the potential to shift the conventional business models into self-production. In near future, 3D printing allows the users to use open-source materials and mobile applications through sharing the culinary skills, creativity, and knowledge of food experts under a single platform for the printing of foods (Nachal et al. 2019). Understanding all these marketing strategies and business models would drive the customization, food consumption pattern, consumer perceptions, environmental regulations, and implementation of innovative trading policies.

14.4.2 Strategic Shifts and Economic Paradigms

3D printing as a general-purpose technology has an enormous range of applications in food industries such as printing foods, design of spare parts for existing machine units, reduces the intensity of packaging, and eliminating storage warehouses to name a few. 3D printing is viewed as a new industrial revolution that is likely to change the business operations, shift the marketing strategies, change the production site, and shrinks the supply chain thereby altering the global economy (Laplume et al. 2016). Recent attention is paid to the potential of 3D printing in military operations for the printing of missiles, guns, and foods on a forward basis for redesigning the existing manufacturing practices. NASA views 3D printing as a classic disruptive technology that must be incorporated as a part of space exploration plans (Garrett 2014). The printing of foods is now emerging as a topic of interest from bottom-up do-it-yourself hobbyists to top-down businesses and government-sponsored research. 3D food printing offers a range of structural and functional benefits in the food supply chain. This includes increased freedom over product design, complexity at affordable cost, on-demand production, mass customization, food model prototyping, eliminating assembly lines, simplifies the food service chain, instant production, the boost of innovation, and globalization of 3D printed foods (Nachal et al. 2019). Apart from these features, 3D food printing is likely to have positive impacts on the environment through increased resource efficiency, low carbon footprints, zero wastage, elimination of storage and inventory costs, avoids the risk of chemical food preservatives, and use of polymeric packages in the production of 3D printed foods. These attributes of 3D printing would result in the dimensional change of the global economy. The production and distribution of foods could begin to deglobalized with the localized manufacturing of the foods at the consumption point through 3D printing (Campbell et al. 2011). This shift would result in the movement of the finished goods that seems to be reachable to all people around the globe. The decentralization of the manufacturing foods will unleash the opportunities with 3D printing in producing hundreds of customized food items in a single go. The mass customization of foods would ensure the perfect alignment of supply and demand based on on-demand requirements. Moreover, the same 3D printer could be used for producing the customized foods of different intricate designs by simply changing the 3D model file. While the conventional facilities require moulds of different sizes and shapes for the fabrication of foods. Importantly, the requirement of large storage spaces could be possibly eliminated by storing the design files in digitalized forms that can be used only at the time of printing based on requirements (Kubac and Kodym 2017).

3D food printing would create a new range of food processing facilities and professions. Nowadays, research works are focusing on the development of convenient user-friendly 3D printers from individual domestic appliances to industrial levels. The associated shift in global manufacturing of foods using 3D printing technology would involve trillions of dollars of business in coming decades with enhanced customer service, improvements in product, and design engineering (Garrett 2014). 3D printing is less labour-intensive that might be politically destabilizing in some economics that relies on traditional food manufacturing. Most of the developing countries could be benefited by encouraging entrepreneurs for setting up 3D printing facilities for digitalized fabrication of foods for local consumption. This would significantly expand the skilled labor forces in the manufacturing sector based on geographic taste preference, reduce the burden of expensive imports and experience the profits of production. The countries with ageing societies would benefit from the advent of automation of production facilities through 3D food printing that significantly increases the overall productivity (Sun et al. 2018b). Otherwise, these societies would decline with the shift in the ratio of employed to retired workers towards less labour force to support the growing proportion of elderly people. 3D food printing along with the integration of novel food processing technologies such as encapsulation and electrospinning at micro and nano levels could promote the market of functional foods and nutraceuticals (Kumar et al. 2020). Thus, 3D printing of foods is expected to be a major drag on the economic growth of the country in the coming years.

14.4.3 Decentralization and Localization of Production

The advent of 3D printing allows industrial firms to get rid of the standard constraints imposed by the current traditional manufacturing paradigm. This would have a positive effect on product proliferation resulting in the individual and collaborative design with extensive experimentation on the feasibilities of novel food sources for 3D printing (Zuboff 2010). Further, it has the potential to transform the economy of the country from concentrated capitalism to distributed capitalism. With the implementation of reverse engineering, the production of foods can be localized specifically for geographic locations based on their preferences. These changes would severely alter the existing foodservice chain where the production occurs closer to the consumption point. 3D printing of foods is likely to nurture a customer-oriented market rather than a manufacturer dominant system that greatly avoids the risk of product failure (Sun et al. 2018a). This scenario emphasizes the production of food products based on consumer preference that has more direct access to market demand with reduced coordination challenges of the food supply chain. As a result, there occurs a drastic reduction in the complexity associated with the production planning process by forecasting the actual needs through aggregate capacity and materials (Baumers et al. 2016). Therefore, the skill requirements and the occupational demands of the new economy would undergo a change accordingly. The common jobs of the traditional production process such as assembly line, retail sales, packaging, transportation, shipping, and delivery of the food products will be replaced by integrative, creative, analytical, automation, software handling, and design engineering through 3D printing (Ben-Ner and Siemsen 2017). Initially, the adoption of new change with the supply chain seems to be difficult that requires strenuous technical assistance through proper training and education for a better understanding of the workflow of the 3D printing of foods. However, the potential growth of the 3D printed food market will decrease the barriers for the entry of new business firms that allows the establishment of more startups. Since the complexity of food supply and distribution chains is relatively reduced with 3D printing that requires less capital investment with higher profits.

The technological developments in AM (additive manufacturing) result in the efficient use of energy that could lead to the transformation of the economy towards localization and decentralization (Schrape 2019). Improvements in the re-utilization of materials and renewable resources remain as the potential solutions for the depletion of fossil fuels and negative impacts of the environment with the traditional manufacturing approaches (Ramachandraiah 2021). The 3D printing technology allows for the decentralization of business activities by disseminating the basic knowledge about the existing resource utilization for the development of novel 3D printed foods. Hence, the transition from the traditional manufacturing approach into additive technology would result in the drastic reduction in the economies of sales and shortens the food supply chain. However there exists a strong economic and political force that limits the extent of decentralization and localization (Ben-Ner and Siemsen 2017). Like the internet that becomes the necessity of the modern world, 3D printing is also forecasted to become an integral part of the domestic kitchen. Despite the analysis of the food material supply for printability, the exploitation of 3D printing technology in view of material flow in the food supply chain and its impact on the economy must be studied to unleash the potential market opportunities of 3D printed foods.

14.4.4 Role of Industry 4.0

The twenty-first century has been witnessing the 4th industrial revolution known as 'Industry 4.0' that is based on the smart manufacturing approaches (Xu et al. 2018). The vision of this revolution is taking the advantage of digitalization in designing, developing, and delivery of goods that is empowered by the virtual world. The transformation could be assured through interconnecting the major firms of the global supply chain using the internet that enables real-time interaction. The four key design principles underlying Industry 4.0 are interconnection, information transparency, technical assistance, and decentralized decisions (Hermann et al. 2016). This concept covers a wide range of industrial players including manufacturing machine parts, power generation, construction, aerospace, automation, health care, and food sector. The major driving forces of the Industry 4.0 are short development periods, process flexibility, individualization upon demand, decentralization, resource efficiency, mechanization, digitalization, networking, and miniaturization. The cutting-edge technologies coming under Industry 4.0 are the Internet of Things (IoT), cloud computing, cyber-physical system, and cybersecurity, augmented reality, big data analytics, advanced robotics, and 3D printing (Kumar and Nayyar 2020). Smart manufacturing involves efficient utilization of resources through interconnecting the machines, materials, and products (Figure 14.2). The 3D printing technology enables the mass production of customized foods through inventory digitalization that imparts significant cost savings in the food supply chain. With the augment of Industry 4.0, 3D printing of foods has fruitful benefits in the delivery of tailored foods based on individual likeness and nutritional requirements with added functionality. Thus, Industry 4.0 is a promising breakthrough in the manufacturing sector that is aimed at the integration of the reality of the physical world with the digitalization of existing technology. It is being capable of taking forward the global

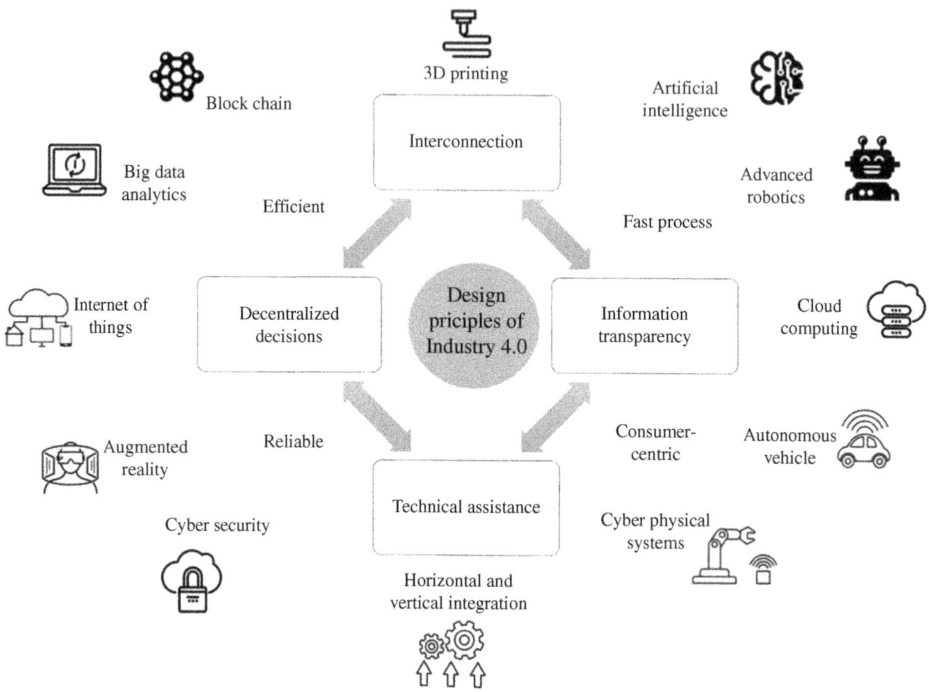

Figure 14.2 Design principles and technologies of Industry 4.0.

manufacturing to the next stage of unforeseen levels of optimization, customization, automation, productivity, and efficiency at lower costs with better management practices (Pramanik et al. 2020). In this cohort, the development of 3D printing in the creation of novel functional foods further requires a lot of research in exploring the feasibility and sustenance of this technology in the food sector. The advent of Industry 4.0 technologies like robotic AM, 3D printing, cloud computing, big data analytics, and IoT drive the globe towards the vision of smart products that remains as promising technologies of the future.

The consumer-driven world focused on the development and delivery of value-added products in a short time to the consumers. This could not be possible with traditional manufacturing as there exist several challenges related to labours and operation skills. So, the automation of the manufacturing process reduces the limitations with labour skills. In contrast to robotic food manufacturing, 3D printing of foods provides automation in addition to customization based on consumer preferences (Sun et al. 2015b). Other than food manufacturing, robotics-based technology has a great scope in packaging. The new generation of robotic technology aids in improving the data flow in packaging by integrating design, production, convenience, distribution, and storage under one single process step. This would result in increasing productivity, ease of handling, transportation, and utility (Nachal et al. 2019). Further, the application of AM technique in food manufacturing results in the correct and accurate dispensing of food materials in all the batches of the production line that ensures efficiency, reliability, and consumer satisfaction. The key elements in the success of 3D printing are food material characteristics, process and product parameters, and post-processing methods. The desirability of the 3D printed foods is determined from printing precision, process efficiency, and end-applications of developed food product. Thus, Industry 4.0 step towards smart manufacturing and intelligent automation of technologies. The inclusion of information technology with AM would play a significant role in the market competitiveness and industrial economy.

14.5 3D Food Printing – A Classic Disruptive Technology

Certainly, 3D printing has transformed the complex slow process of traditional food manufacturing into efficient, fast, and cost-effective with the added advantage of customization and prototyping (Kabir 2019a). With the current research interest on 3D printing of foods, the next decade would be witnessed with the abundance of 3D printed of foods ruling the food sector. 3D printing is a part of the digital manufacturing process that involves the use of CAD 3D designs for the fabrication of foods in three dimensions (Godoi et al. 2016). The overwhelming applicability of 3D printing in the food sector is viewed as a strategic competitive advantage. The state-of-the-art of food printing serves various purposes such as consumer satisfaction, customized foods, personalized nutrition, mass production, education, and communication tool, design artistic skills, and food packaging (Dankar et al. 2018; Sun et al. 2018b). The technology is transforming the food industries towards the delivery of functional and nutraceutical foods by focusing on individual needs (Kumar et al. 2020). Foods such as pizza, cookies, chocolate, cheese, and many other material supplies could be 3D printed based on individual preference. Research works are going on in developing a universal food 3D printer for the fabrication of foods at small and industrial scales with tailored nutrition at the micro level (Tan et al. 2018). 3D printing seems to be an exciting food technology that would significantly impact the economy. The capability of the technique to fabricate food by the consumers instead of buying it will slowly diminish the international trade of some of the other conventional food products (Kabir 2019b). However, expedite innovation and scope for commercialization would overview the limitations and helps in possible technical and cultural collaboration. Hence 3D printing is becoming an integral part of the R&D process in food manufacturing. Various advancements of 3D printing and its applications in the food industries are discussed in the following sections. The schematic representation of the food supply and distribution chain of 3D printing is presented in Figure 14.3.

14.5.1 Food Choice and Consumer Behaviour

Increased awareness of nutrition and health remains at the top of the list of priorities that must be taken into account during food product development and marketing. The consumer's perceptions of personal foods and health care products are greatly influenced by food type, origin, processing, conservation method, and packaging (Graeff-Hönninger and Khajehei 2019). In this context, 3D printing is a print-and-eat technology that involves the fabrication of foods based on consumer's nutritional needs and taste preferences. Recently the healthy diets like 'all-raw', 'vegan', 'super foods', 'gluten-free', 'lactose-free', cutting down of intake of sugars, salt, and fat are more trending in recent days (Falguera et al. 2012). These

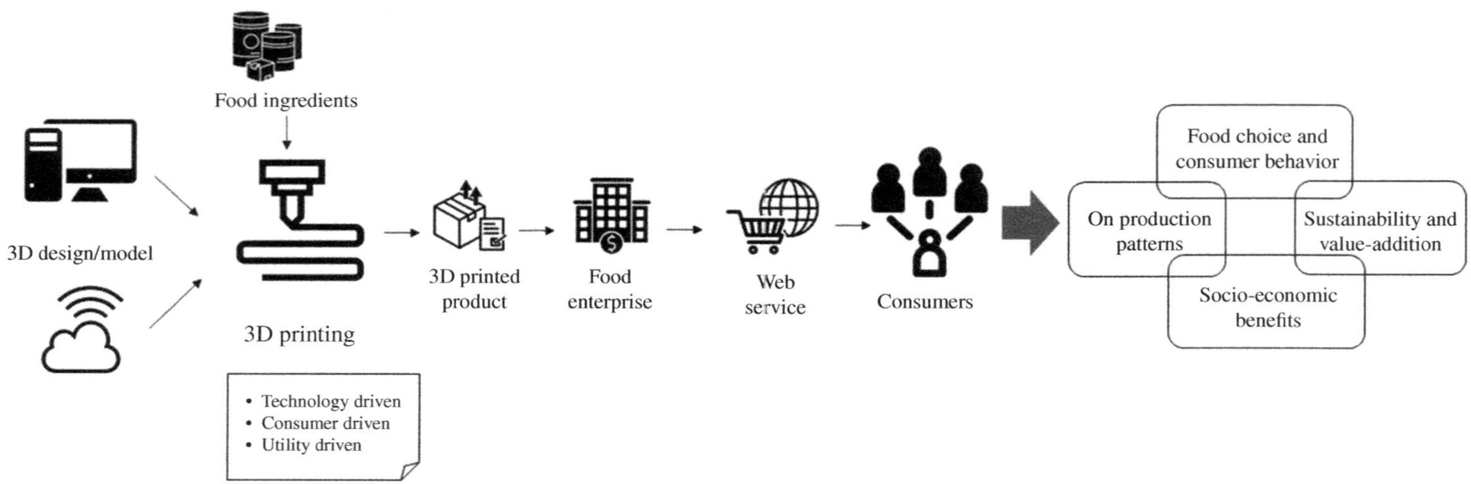

Figure 14.3 Food supply and distribution chain of 3D printing.

food trends show the consumer desires and changing attitudes towards diet. Four major factors that influence the consumer's behaviour and food choice are personal factors, psychological factors, cultural and social factors, and the factors related to the nature of the developed food product. Superfoods are those exotic food that remained as unexplored, uncommon and under-utilized (Meyerding et al. 2018). Foods of this category include moringa leaves, spinach, spirulina, millets, pseudo-cereals, insects, macro and micro algae, and fungi. In context with food printing, these materials are categorized as alternative ingredients (Derossi et al. 2020). Thus, 3D food printing seems to be a versatile technology that is capable of delivering novel foods using alternative ingredients that adds variety and increases food choices to consumers. Also, 3D printing helps in delivery of special foods such as dysphagia food, infant food, elderly food, sports food, space food, and military food with precise control over the proportion of nutrients. The exploitation of food ingredients to be used as alternative sweeteners, salt, and sugar replacers is gaining more research interest in recent days (Prakash et al. 2019). In this regard, 3D printing could be used as a potential vehicle for food fortification by integration with other food technologies such as encapsulation and electrospinning (Kumar et al. 2020; Sun et al. 2015c). The scope of 3D food printing in the localized use of food ingredients such as sugar, salt, and fat are under study for cutting down their daily intake to tackle lifestyle-related diseases such as diabetes, hypertension, and obesity.

3D printing is a flexible technology that can adapt to the changing landscape of the food industry with the need to address the rising global population, changing climate, and new consumer demands. None of the conventional technologies could match the advantages of 3D printing. Further, the integration of multiple unit operations into a single process step reduces the complexity of the food supply chain (Nachal et al. 2019). 3D food printing has a great scope in food niche markets that could satisfy the peoples of all age groups. Based on the automation and influence of IoT, 3D printing could rule the food industry in the coming decades through administering more transparency in a food production line that consumers can print their own food (Keogh et al. 2020). Thus, the technological advancements and biological transformations can evolve as a ground-breaking innovation that assures the possibility of value-added foods through 3D printing. Further, the growth of 3D food printing can be supported through unified communication among food technologists, culinary professionals, design engineers, retailers, and consumers.

14.5.2 On Production Patterns

3D printing is viewed as a solid freeform fabrication (SFF) process that involves the digitalized fabrication of solid 3D constructs without human labour based on a pre-defined printing path (Dankar et al. 2018). Technology and innovation are the two key players determining the success of the developed product in the food industry. The implementation of potential food technology at three market levels such as industrial mass production, individual domestic production, and interconnected businesses influences the success of technology (Pinna et al. 2016). 3D printing is one such technique that offers a range of opportunities from food customization to personalization. Mass production of highly demanding food items such as pizza, cookies, cakes, and confectioneries emphasizes the feasibility of the implementation of 3D printing at restaurants (Portanguen et al. 2019). At the household level, 3D printing can be adopted by stakeholders from start-ups to small companies that make use of printing technology in the domestic kitchen by the implementation of compact digital cooking units '3D printers'. Meanwhile, the industrial applications of 3D printing also have a greater scope in improving the economies of scale. With these potential features, 3D printing has been widely used for the development of customized food products with intricate designs and unique food patterns. Based on the printing technology and the nature of the material supply, the end-quality of frosted 3D printed design patterns will vary. The fused deposition modeling (FDM) based approach and selective sintering process yield 3D constructs of utmost precision than the traditional approaches employed using human labours. The precise cutting and size of the 3D food constructs are the major strength of food printing (Wegrzyn et al. 2012). Similarly, laser-based 3D printing technology (stereolithography) is used for surface pattern and 3D designing from powder-based materials. The sugar-based solid 3D sculptures can be fabricated using inkjet printing and binder jetting technologies.

Considering the nature and type of food materials, 3D printing is well suited for the preparation of both warm as well as frozen foods. The frozen foodstuffs such as ice creams with fascinating art designs can be achieved through a post-process cooling approach using liquid nitrogen. This concept of food printing is often referred to as 'foodForm' which can also be used for dough-based 3D constructs (Skartsaris and Piatti 2019). As an added advantage 3D printing can be used for printing alternative food sources such as insects, lupin seeds, algae, and fungi that are listed to be the future foods. Through appropriate pre-processing of material supplies, a complex 3D food construct can be designed using 3D printing (Anukiruthika et al. 2020). The customer-designed food fabrication can be feasible with high value through printing technology that promises to deliver a range of customized food products with healthy derivatives of the contemporary meals thereby addressing the individual needs and likeness (Nachal et al. 2019). Thus, 3D printing of foods has a great scope for industrial commercialization with impressive cost savings and utmost process efficiency. Consumers can take hold of the food production and design of their own meals based on their personal taste preferences whilst maintain the nutritional needs. Hence 3D printing being referred to as a classic disruptive food technology where each and every step of the food fabrication process can be customizable and tailorable by opening every possibility for online collaboration in the food sector.

14.5.3 Sustainability and Value Addition

The emergence of industrialization and urbanization had significant impacts on dietary change and eating practices resulting in a shift of diet towards animal foods with high fat and refined foods with less fibre (Cohen and Ilieva 2015). The sudden change of nutrition transition has severe health implications resulting in the increased incidence of lifestyle diseases such as obesity, cardiovascular disease, diabetes, and cancer (Burggraf et al. 2015). In addition, the rising demand for food to feed global hunger put forth land pressure for higher food productivity. The increasing land pressure slowly depletes the existing fossil fuels and yields a negative impact on the environment. The enhanced level of processing of food products often generates by-product waste streams (Nida et al. 2020). These generated waste fractions remain a potential source of essential nutrients that are often neglected and dumped in landfills. This poses a serious threat to the environment that puts forth the need for efficient resource utilization. Food wastes are described as the food lost during the processing of foods from upstream to downstream in the food supply chain (Bellemare et al. 2017). The current voluminous materials of generated food wastes demand action by governmental and non-governmental organizations. In order to address these issues, a global-scale societal transition is required in conjunction with ecological and social equity goals (Annosi et al. 2021). Research studies are being focused on the analysis of the food supply chain for the promotion of sustainable food production and consumption (Gheewala et al. 2020; Mak et al. 2020). This emphasizes the application of circular economy rather than the existing linear food supply chain by implementing the principles of reduce, reuse, recycle, and remanufacturing.

3D printing of foods can be a promising solution for the above-mentioned multiple problems. The adoption of digital technologies in food manufacturing simplifies the food supply and distribution chain. The 3D printing could have a great potential to reshape the existing business model, improves technological collaboration, and fosters the development of new business models thereby increasing profitability (Sangwan and Bhatia 2020). The use of digital technology aids in accessing the 3D design files from an e-commerce platform that enables ease of information transfer among food processors, designers, industrial firms, and consumers. This would pave a way for the creation of a digital ecosystem with greater coordination among the industrial firms resulting in the development of novel strategies, process planning, and synchronization of production (Mahapatra 2020). Further, the greater degree of integration of the food service chain allows for efficient utilization of material resources and energy usage offering environmental benefits through less waste generation and carbon footprints. The transitioning of the prevailing food chain to a more sustainable socio-technical system is driven by a range of multiple factors such as consumer's food decisions, transparency of production, and food safety (Manning 2019). The application of 3D printing in the preparation of food satisfies the consumers through enhanced levels of food customization with fewer impacts on the environment. So, the evolving digital technologies remain a socio-economic driver influencing the food supply chain in ensuring the nourishment of the global population. However, a lack of evidence on the implementation of the impact of 3D printing on the food supply chain limits the exploration of its socio-economic benefits. Hence, more case studies are required in analysing the cost benefits, management of 3D printing food supply chain, marketability, profitability, and economic growth.

14.5.4 Anti-Counterfeiting and Food Authentication

Other than food printing and food packaging applications, 3D printing technology is also useful as a novel anticounterfeit measure (Trenfield et al. 2019a). The product authenticity of the personalized medicines is reviewed through 3D printing technology. However, applications for food authentication are not yet explored. Yet the same principle could be successfully adopted for food authentication that ensures the safe delivery of personalized foods. So, the present section summarizes the emerging applications of 3D printing as an authentication tool and its working principle. The track and trace of the personalized drug delivery are based on the combination of 3D printing and 2D printing technology as a single step in the design of 3D printed medicine (Alomari et al. 2018; Awad et al. 2018). The 3D printing was based on the FDM approach where a drug-loaded filament is extruded through a hot printing nozzle in a layer-by-layer manner over the printing platform. The 3D printed substrate was subjected to 2D printing for encoding the tracking information. The inkjet printing technology utilizes liquid-based printing ink for encoding information through precise control over droplet ejection that enables the 2D printing of unique patterns onto the substrate. The advancements in technology allow printing of the quick response (QR) code and data matrices onto the surface of the 3D printed tablets (Chen et al. 2019; Edinger et al. 2018). This enables the use of smartphones to retrieve the tailored information related to the nature of the drug, patient description, and medical prescriber. The designed novel anticounterfeit strategy was based on the deposition of a unique combination of printing ink that could be detected rapidly using Raman spectroscopy (Lawson et al. 2018; Mazivila and Olivieri 2018). The success of the designed strategy was based on the printability of the ink which depends on the material's viscosity, surface tension, and density. In this case, also, post-processing of 3D printed tablets is required for ensuring the proper functionality of tracing.

The compact size of drugs, affordability, and flexibility makes 3D printing to be more suitable for the production of tailored medicines in decentralized locations at the delivery points. This concept enables the production of personalized medicines in the absence of anti-counterfeit packaging (Edinger et al. 2018). This works based on the direct measure of the authentication of developed 3D printed dosage (Figure 14.4). Further, the graphic-designed 2D printed scanning codes onto the 3D printed substrate enables quick scanning using handheld smartphones that retrieve the product information before administration

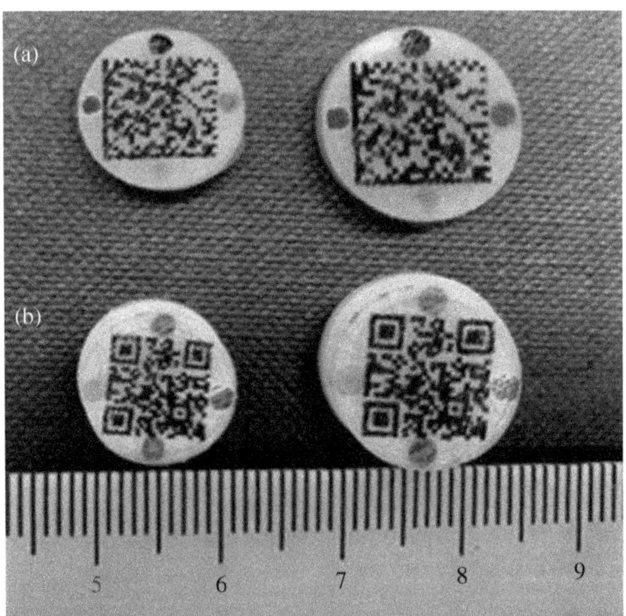

Figure 14.4 3D printed tablets with novel anticounterfeit design. (a) Data matrices with model ink, and (b) QR codes with model ink. *Source:* From Trenfield et al. (2019b), Figure 04 [p. 005] / With permission of Elsevier.

(Trenfield et al. 2019b). The application of the 2D printing of the barcode over the personalized medicines acts as an anticounterfeit measure and improves safety by enabling the digitalized monitoring of medications that are transparent to both physicians and patients. In order to avoid the risk of duplication of the 2D barcodes, a combination of invisible material ink was used for surface patterning. The unique combination of the detectable ingredients is used as material ink that are deposited onto the substrate surface aiding in traceability and eliminates the risk of code duplication. A non-destructive rapid detection technique, Raman spectroscopy is employed for anticounterfeit applications in scanning the barcodes (Kakio et al. 2017). Likewise, the same principle could be applied for the detection and tracing of food authentication. Thus, 3D printing is a versatile technology not only used in the customization of 3D printed foods but also in the safe delivery of personalized foods as an anticounterfeit measure. The advent of digitalized technologies transforms the food supply chain into more transparent that eliminates the adulteration and theft of food products by cultivating consumer's trust through the safe delivery of 3D printed foods. The simplification of the food supply chain enables the localized production and the on-time delivery of personalized foods to consumers by continual digital monitoring of the safety of the developed food product until consumption.

14.6 Technological Barriers and Challenges

The adoption of digital technologies in the food sector would pave a way for sustainable supply chain innovation that involves novelty in the contractual business relationships among the suppliers for the co-creation of value for stakeholders (Beltagui et al. 2020). The impact of 3D printing has significantly impacted the food supply chain management processes such as planning, coordination of activities, a balance between the demand and supply chain, delivery of raw materials, distribution of finished products, after-sale services, and logistics (Janssen et al. 2014). This imposes challenges over the adoption of digitalized technologies that include a shortage of skilled labours, productional cost, complex logistics, and fluctuating product demand (Singh and Agrawal 2021). The good manufacturing practices for AM process like 3D food printing is yet to be well established. In this context, some of the technical challenges associated with 3D printing of foods are a selection of proper raw material, degree of pre-processing, optimization of process and product parameters, post-processing methods, shelf-life, storage stability, packaging, and food safety (Jayaprakash et al. 2019). However, these challenges can be overcome through proper understanding of printability of food materials. Although 3D printing of foods is being tested and implemented at lab-scale level, the present decade is slowly witnessing the success of 3D printed foods. This growing trend depicts the technical innovations of food printing in the coming years. In order to explore the applications of 3D printing at industrial levels some of the management challenges such as reskilling, education of staff, process training, pricing of 3D printed foods, marketing strategies, cybersecurity, and piracy have to be addressed (Singh and Agrawal 2021). Despite the individuals have opportunities in opening a new firm, ease of transferring files, buying of 3D models, and localized printing, it imposes a challenge for the existing food companies due to consumer dominance in food supply chain management (Kapetaniou et al. 2018). Thus, the sudden rapid emergence of advanced digital technologies anticipates the concerns over food regulations that remain as the barrier to the globalization of 3D printing of foods (Savolainen and Collan 2020).

The leading global 3D printing industrial firms such as 3D Systems and Stratasys earn revenue not only from the 3D printers but also from the material services (Charlebois and Juhasz 2018). 3D food printing can augment the capacity of the existing food manufacturing approaches in meeting the continual demand of consumers for the development of novel 3D printed foods. The new research developments on AM techniques could complement and improves the existing food innovation opportunities with the conventional food technologies that impose the applicability of 3D printing for food prototyping and product development. The major driving force for the prospect of 3D printing in the food sector is mass customization that creates a range of innovative food cuisines. The marketing, branding, pricing, and business models of the food supply and distribution chain of 3D printing have to be investigated (Charlebois and Mackay 2010). The ongoing challenge in accessing 3D printed foods relies on the food safety regulations, piracy, manufacturing cost, resource utilization, and energy usage, large-scale commercialization, and rivalry business competitors. The protection of intellectual property rights (IPR) of 3D printing is likely to be a litigious challenge related to the design of 3D models, buying of 3D designs, software piracy, and ownership of developed 3D printers, formulations, and recipes (Garrett 2014). This scenario would probably create a large niche for lawyers and related government officials in the protection of 3D printing intellectual properties. In addition, the AM will present legal regulatory challenges in tracing the origin of 3D printed food sources, operational activities of a decentralized network of 3D printing, process, and product malfunctioning. The abovementioned challenges and barriers must be overcome through a proper understanding of the business model in the globalization of 3D food printing. Researchers, economists, and scholars are started to ponder the legal implications of the food printing process. More case studies on the implementation of 3D food printing and 3D printers at individual, small and large-scale firms would help in a deeper understanding of the barriers and provides insights on mitigation strategies for its commercialization at the global level.

14.7 Conclusion

The impact of 3D printing on food manufacturing, environment, society, geopolitics, and the global economy is likely to occur over the past decade. However, the cumulative impact of digitalized food technologies is forecasted to be disruptive and revolutionary as like 3D food printing. The technological outbreak is evident with the ubiquity of personal computers, the internet, and mobile applications which has been the case with 3D printers in near future. It can be foreseen that 3D food printing would revolutionize eating habits by transforming the balanced diet into a digitalized diet. Although the evolutionary growth of 3D printing technology and its strategic impact is not completely discernible at this point of time, it has significant and large implications on the global food supply chain and distribution in the future. The consumer's behaviours, food decisions, bias, buying habits, and convenience will drive the market of 3D printing of foods. However, gaining the consumer's trust, food safety, and regulation concerns poses a threat to the incentive of 3D printing technology. The business model of 3D printing exists along with a range of spectrum from business firms and individual users of open-source e-commerce platforms to the mobile applications that emphasize the spirit of sharing, creativity among the players of the food supply chain, protection of intellectual property, and piracy to meet the ever-rising demand of consumers in terms of both customization and personalization of foods. The greyscale in between marketing firms of the food supply chain is undoubtedly wide. More research needs and case studies on the management of the food supply distribution chain using 3D printing are required to bridge up the gap and holes in the commercialization and globalization of 3D food printing. The impact of 3D printing can go beyond transforming food production and rebalancing the world economy especially in addressing the current threats of environmental degradation, climate change, and resource scarcity. Thus, 3D printing is beneficial in developing a sustainable global economy, protecting the environment, and preserving nutritional security.

References

Ali, A., Soni, M., Javaid, M., and Haleem, A. (2020). A comparative analysis of different rapid prototyping techniques for making intricately shaped structure. *Journal of Industrial Integration and Management* **5** (03): 393–407.

Alomari, M., Vuddanda, P.R., Trenfield, S.J. et al. (2018). Printing T3 and T4 oral drug combinations as a novel strategy for hypothyroidism. *International Journal of Pharmaceutics* **549** (1–2): 363–369.

Annosi, M.C., Brunetta, F., Bimbo, F., and Kostoula, M. (2021). Digitalization within food supply chains to prevent food waste. Drivers, barriers and collaboration practices. *Industrial Marketing Management* **93**: 208–220.

Anukiruthika, T., Moses, J.A., and Anandharamakrishnan, C. (2020). 3D printing of egg yolk and white with rice flour blends. *Journal of Food Engineering* **265**: 109691. https://doi.org/10.1016/j.jfoodeng.2019.109691.

Ashima, R., Haleem, A., Bahl, S. et al. (2021). Automation and manufacturing of smart materials in additive manufacturing technologies using Internet of Things towards the adoption of Industry 4.0. Materials Today: Proceedings.

Attaran, M. (2017). The rise of 3-D printing: the advantages of additive manufacturing over traditional manufacturing. *Business Horizons* **60** (5): 677–688.

Awad, A., Trenfield, S.J., Gaisford, S., and Basit, A.W. (2018). 3D printed medicines: a new branch of digital healthcare. *International Journal of Pharmaceutics* **548** (1): 586–596.

Bandyopadhyay, A. and Heer, B. (2018). Additive manufacturing of multi-material structures. *Materials Science and Engineering: R: Reports* **129**: 1–16.

Baumers, M., Dickens, P., Tuck, C., and Hague, R. (2016). The cost of additive manufacturing: machine productivity, economies of scale and technology-push. *Technological Forecasting and Social Change* **102**: 193–201.

Bellemare, M.F., Çakir, M., Peterson, H.H. et al. (2017). *On the Measurement of Food Waste*. Wiley Online Library.

Beltagui, A., Kunz, N., and Gold, S. (2020). The role of 3D printing and open design on adoption of socially sustainable supply chain innovation. *International Journal of Production Economics* **221**: 107462.

Ben-Ner, A. and Siemsen, E. (2017). Decentralization and localization of production: the organizational and economic consequences of additive manufacturing (3D printing). *California Management Review* **59** (2): 5–23.

Berman, B. (2012). 3-D printing: the new industrial revolution. *Business Horizons* **55** (2): 155–162.

Burggraf, C., Kuhn, L., Zhao, Q. et al. (2015). Economic growth and nutrition transition: an empirical analysis comparing demand elasticities for foods in China and Russia. *Journal of Integrative Agriculture* **14** (6): 1008–1022.

Campbell, T., Williams, C., Ivanova, O., and Garrett, B. (2011). Could 3D printing change the world? Technologies, Potential, and Implications of Additive Manufacturing, Atlantic Council, Washington, DC, 3.

Charlebois, S. and Juhasz, M. (2018). Food futures and 3D Printing: strategic market foresight and the case of Structur3D. *International Journal on Food System Dynamics* **9** (2): 138–148.

Charlebois, S. and Mackay, G. (2010). Marketing culture through locally-grown products: the case of the fransaskoisie terroir products. *Problems and Perspectives in Management* **8** (4): 91–102.

Chen, F., Luo, Y., Tsoutsos, N.G. et al. (2019). Embedding tracking codes in additive manufactured parts for product authentication. *Advanced Engineering Materials* **21** (4): 1800495.

Cohen, N. and Ilieva, R.T. (2015). Transitioning the food system: a strategic practice management approach for cities. *Environmental Innovation and Societal Transitions* **17**: 199–217.

Cook, C.N., Inayatullah, S., Burgman, M.A. et al. (2014). Strategic foresight: how planning for the unpredictable can improve environmental decision-making. *Trends in Ecology and Evolution* **29** (9): 531–541.

Da Xu, L., Xu, E.L., and Li, L. (2018). Industry 4.0: state of the art and future trends. *International Journal of Production Research* **56** (8): 2941–2962.

Dankar, I., Haddarah, A., Omar, F.E.L. et al. (2018). 3D printing technology: the new era for food customization and elaboration. *Trends in Food Science and Technology* **75**: 231–242. https://doi.org/10.1016/j.tifs.2018.03.018.

Derossi, A., Husain, A., Caporizzi, R., and Severini, C. (2020). Manufacturing personalized food for people uniqueness. An overview from traditional to emerging technologies. *Critical Reviews in Food Science and Nutrition* **60** (7): 1141–1159.

Despeisse, M., Baumers, M., Brown, P. et al. (2017). Unlocking value for a circular economy through 3D printing: a research agenda. *Technological Forecasting and Social Change* **115**: 75–84.

Edinger, M., Bar-Shalom, D., Sandler, N. et al. (2018). QR encoded smart oral dosage forms by inkjet printing. *International Journal of Pharmaceutics* **536** (1): 138–145.

Escursell, S., Llorach, P., and Roncero, M.B. (2020). Sustainability in e-commerce packaging: a review. *Journal of Cleaner Production* **280**: 124314.

Falguera, V., Aliguer, N., and Falguera, M. (2012). An integrated approach to current trends in food consumption: moving toward functional and organic products? *Food Control* **26** (2): 274–281.

Feng, C., Zhang, M., Bhandari, B., and Ye, Y. (2020). Use of potato processing by-product: effects on the 3D printing characteristics of the yam and the texture of air-fried yam snacks. *LWT* **125**: 109265.

Ferreira Gregorio, V., Pié, L., and Terceño, A. (2018). A systematic literature review of bio, green and circular economy trends in publications in the field of economics and business management. *Sustainability* **10** (11): 4232.

Garrett, B. (2014). 3D printing: new economic paradigms and strategic shifts. *Global Policy* **5** (1): 70–75.

Gheewala, S.H., Jungbluth, N., Notarnicola, B. et al. (2020). No simple menu for sustainable food production and consumption. *The International Journal of Life Cycle Assessment* **25** (7): 1175–1182.

Godoi, F.C., Prakash, S., and Bhandari, B.R. (2016). 3d printing technologies applied for food design: status and prospects. *Journal of Food Engineering* **179**: 44–54.

Govindan, K. (2018). Sustainable consumption and production in the food supply chain: a conceptual framework. *International Journal of Production Economics* **195**: 419–431.

Graeff-Hönninger, S. and Khajehei, F. (2019). The demand for superfoods: consumers' desire, production viability and bio-intelligent transition. In: *Food Tech Transitions* (eds. C. Piatti, S. Graeff-Honninger and F. Khajehei), pp. 81–94. Switzerland: Springer Nature.

Hermann, M., Pentek, T., and Otto, B. (2016). Design principles for industrie 4.0 scenarios. 2016 49th Hawaii International Conference on System Sciences (HICSS), pp. 3928–3937.

Janssen, R., Blankers, I., Moolenburgh, E., and Posthumus, B. (2014). The impact of 3-D printing on supply chain management. TNO: Innovation for Life.

Jayaprakash, S., Ituarte, I.F., and Partanen, J. (2019). Prosumer-driven 3D food printing: role of digital platforms in future 3D food printing systems. In: *Fundamentals of 3D Food Printing and Applications* (eds. F.C. Godoi, B.R. Bhandari, S. Prakash and M. Zhang), pp. 331–354. United Kingdom: Academic Press.

Jia, F., Wang, X., Mustafee, N., and Hao, L. (2016). Investigating the feasibility of supply chain-centric business models in 3D chocolate printing: a simulation study. *Technological Forecasting and Social Change* **102**: 202–213.

Kabir, M.N. (2019a). *Knowledge-Based Social Entrepreneurship: Understanding Knowledge Economy, Innovation, and the Future of Social Entrepreneurship*. Springer.

Kabir, M.N. (2019b). Technologies of the future. In: *Knowledge-Based Social Entrepreneurship* (ed. M.N. Kabir), 91–133. New York: Palgrave Macmillan Springer Nature Amerixa Inc.

Kakio, T., Yoshida, N., Macha, S. et al. (2017). Classification and visualization of physical and chemical properties of falsified medicines with handheld Raman spectroscopy and X-ray computed tomography. *The American Journal of Tropical Medicine and Hygiene* **97** (3): 684–689.

Kapetaniou, C., Rieple, A., Pilkington, A. et al. (2018). Building the layers of a new manufacturing taxonomy: how 3D printing is creating a new landscape of production eco-systems and competitive dynamics. *Technological Forecasting and Social Change* **128**: 22–35.

Keogh, J.G., Dube, L., Rejeb, A. et al. (2020). The future food chain: digitization as an enabler of Society 5.0. In: *Building the Future of Food Safety Technology* (ed. D. Detwiler), pp. 11–31. United Kingdom: Academic Press.

Klooster Ten, R. (2002). *Package Design: A Methodical Development and Simulation of the Design Process*. Netherlands: Technische Universiteit Delft.

Knowles, S. (2019). *Five Ways 3D Printing Will Impact the Global Supply Chain*. Maine Pointe. https://www.mainepointe.com/practical-insights/five-ways-3d-printing-will-impact-the-global-supply-chain.

Kubac, L. and Kodym, O. (2017). The impact of 3D printing technology on supply chain. *MATEC Web of Conferences* **134**: 27.

Kumar, A. and Nayyar, A. (2020). si^3-industry: a sustainable, intelligent, innovative, internet-of-things industry. In: *A Roadmap to Industry 4.0: Smart Production, Sharp Business and Sustainable Development* (eds. A. Nayyar and A. Kumar), 1–21. Switzerland: Springer Nature.

Kumar, S., Anukiruthika, T., Dutta, S. et al. (2020). Iron deficiency anemia: a comprehensive review on iron absorption, bioavailability and emerging food fortification approaches. *Trends in Food Science and Technology* https://doi.org/10.1016/j.tifs.2020.02.021.

Laplume, A.O., Petersen, B., and Pearce, J.M. (2016). Global value chains from a 3D printing perspective. *Journal of International Business Studies* **47** (5): 595–609.

Lawson, G., Ogwu, J., and Tanna, S. (2018). Quantitative screening of the pharmaceutical ingredient for the rapid identification of substandard and falsified medicines using reflectance infrared spectroscopy. *PloS One* **13** (8): e0202059.

Lewandowski, M. (2016). Designing the business models for circular economy—towards the conceptual framework. *Sustainability* **8** (1): 43.

Liu, J., Sun, L., Xu, W. et al. (2019). Current advances and future perspectives of 3D printing natural-derived biopolymers. *Carbohydrate Polymers* **207**: 297–316.

Mahapatra, B. (2020). si^3-industry: cloud computing in Industry 4.0. In: *A Roadmap to Industry 4.0: Smart Production, Sharp Business and Sustainable Development* (eds. A. Nayyar and A. Kumar), pp. 113–126. Switzerland: Springer Nature.

Mak, T.M.W., Xiong, X., Tsang, D.C.W. et al. (2020). Sustainable food waste management towards circular bioeconomy: policy review, limitations and opportunities. *Bioresource Technology* **297**: 122497.

Manning, L. (2019). Corporate responsibility in a transitioning food environment: truth-seeking and truth-telling. In: *Food Tech Transitions* (eds. C. Piatti, S. Graeff-Honninger and F. Khajehei), pp. 149–169. Switzerland: Springer Nature.

Mazivila, S.J. and Olivieri, A.C. (2018). Chemometrics coupled to vibrational spectroscopy and spectroscopic imaging for the analysis of solid-phase pharmaceutical products: a brief review on non-destructive analytical methods. *TrAC Trends in Analytical Chemistry* **108**: 74–87.

Meherishi, L., Narayana, S.A., and Ranjani, K.S. (2019). Sustainable packaging for supply chain management in the circular economy: a review. *Journal of Cleaner Production* **237**: 117582.

Meyerding, S.G.H., Kurzdorfer, A., and Gassler, B. (2018). Consumer preferences for superfood ingredients—the case of bread in Germany. *Sustainability* **10** (12): 4667.

Nachal, N., Moses, J.A., Karthik, P., and Anandharamakrishnan, C. (2019). Applications of 3D printing in food processing. *Food Engineering Reviews* **11** (3): 123–141. https://doi.org/10.1007/s12393-019-09199-8.

Nida, S., Anukiruthika, T., Moses, J.A., and Anandharamakrishnan, C. (2020). 3D printing of grinding and milling fractions of rice husk. *Waste and Biomass Valorization*: 1–10. https://doi.org/10.1007/s12649-020-01000-w.

Petrick, I.J. and Simpson, T.W. (2013). 3D printing disrupts manufacturing: how economies of one create new rules of competition. *Research-Technology Management* **56** (6): 12–16.

Pinna, C., Ramundo, L., Sisca, F.G. et al. (2016). Additive manufacturing applications within food industry: an actual overview and future opportunities. 21st Summer School Francesco Turco 2016, pp. 18–24.

Portanguen, S., Tournayre, P., Sicard, J. et al. (2019). Toward the design of functional foods and biobased products by 3D printing: a review. *Trends in Food Science and Technology* **86**: 188–198.

Prakash, S., Bhandari, B.R., Godoi, F.C., and Zhang, M. (2019). Future outlook of 3D food printing. In: *Fundamentals of 3D Food Printing and Applications* (eds. F.C. Godoi, B. R. Bhandari, S. Prakash and M. Zhang), 373–381. United Kingdom: Academic Press.

Pramanik, P.K.D., Mukherjee, B., Pal, S. et al. (2020). Ubiquitous manufacturing in the age of industry 4.0: a state-of-the-art primer. In: *A Roadmap to Industry 4.0: Smart Production, Sharp Business and Sustainable Development* (eds. A. Nayyar and A. Kumar), pp. 73–112. Switzerland: Springer Nature.

Ramachandraiah, K. (2021). Potential development of sustainable 3D-printed meat analogues: a review. *Sustainability* **13** (2): 938.

Rasmussen, B., Andersen, P.D., and Borch, K. (2010). Managing transdisciplinarity in strategic foresight. *Creativity and Innovation Management* **19** (1): 37–46.

Rayna, T., Striukova, L., and Darlington, J. (2015). Co-creation and user innovation: the role of online 3D printing platforms. *Journal of Engineering and Technology Management* **37**: 90–102.

Rogers, H., Baricz, N., and Pawar, K.S. (2016). 3D printing services: classification, supply chain implications and research agenda. *International Journal of Physical Distribution and Logistics Management* **46**: 1–22.

Rossi, E., Di Nicolantonio, M., Barcarolo, P., and Lagatta, J. (2019). Sustainable 3D printing: design opportunities and research perspectives. International Conference on Applied Human Factors and Ergonomics, pp. 3–15.

Rubio, E. and Hurtado, S. (2019). 3D food printing technology at home, domestic application. In: *Fundamentals of 3D Food Printing and Applications* (eds. F.C. Godoi, B.R. Bhandari, S. Prakash and M. Zhang), pp. 289–329. United Kingdom: Academic Press.

Sangwan, S.R. and Bhatia, M.P.S. (2020). Sustainable development in industry 4.0. In: *A Roadmap to Industry 4.0: Smart Production, Sharp Business and Sustainable Development* (eds. A. Nayyar and A. Kumar), pp. 39–56. Switzerland: Springer Naure.

Sasson, A. and Johnson, J.C. (2016). The 3D printing order: variability, supercenters and supply chain reconfigurations. *International Journal of Physical Distribution and Logistics Management* **46**: 82–94.

Savolainen, J. and Collan, M. (2020). How additive manufacturing technology changes business models? – Review of literature. *Additive Manufacturing* **32**: 101070.

Schrape, J.-F. (2019). The promise of technological decentralization. A brief reconstruction. *Society* **56** (1): 31–37.

Shree, M.V., Dhinakaran, V., Rajkumar, V. et al. (2020). Effect of 3D printing on supply chain management. *Materials Today: Proceedings* **21**: 958–963.

Singh, S. and Agrawal, V. (2021). Critical success factors for new horizons in the supply chain of 3-D printed products – a review. Materials Today: Proceedings.

Skartsaris, I. and Piatti, C. (2019). Altering production patterns in the food industry: 3D food printing. In: *Food Tech Transitions* (eds. C. Piatti, S. Graeff-Honninger and F. Khajehei), pp. 97–110. Switzerland: Springer Nature.

Sun, J., Peng, Z., Yan, L. et al. (2015a). 3D food printing—an innovative way of mass customization in food fabrication. *International Journal of Bioprinting* **1** (1): 27–38.

Sun, J., Peng, Z., Zhou, W. et al. (2015b). A review on 3D printing for customized food fabrication. *Procedia Manufacturing* **1**: 308–319.

Sun, J., Zhou, W., Huang, D. et al. (2015c). An overview of 3D printing technologies for food fabrication. *Food and Bioprocess Technology* **8** (8): 1605–1615.

Sun, J., Zhou, W., Huang, D., and Yan, L. (2018a). 3D food printing: perspectives. In: *Polymers for Food Applications* (ed. T.J. Gutierrez), 725–755. Argentina: Springer International Publishing.

Sun, J., Zhou, W., Yan, L. et al. (2018b). Extrusion-based food printing for digitalized food design and nutrition control. *Journal of Food Engineering* **220**: 1–11.

Tan, C., Toh, W.Y., Wong, G., and Lin, L. (2018). Extrusion-based 3D food printing – materials and machines. *International Journal of Bioprinting* **4** (2): 1–13. https://doi.org/10.18063/ijb.v4i2.143.

Trenfield, S.J., Awad, A., Madla, C.M. et al. (2019a). Shaping the future: recent advances of 3D printing in drug delivery and healthcare. *Expert Opinion on Drug Delivery* **16** (10): 1081–1094.

Trenfield, S.J., Tan, H.X., Awad, A. et al. (2019b). Track-and-trace: novel anti-counterfeit measures for 3D printed personalized drug products using smart material inks. *International Journal of Pharmaceutics* **567**: 118443.

Wegrzyn, T.F., Golding, M., and Archer, R.H. (2012). Food layered manufacture: a new process for constructing solid foods. *Trends in Food Science and Technology* **27** (2): 66–72.

Zeleny, M. (2012). High technology and barriers to innovation: from globalization to relocalization. *International Journal of Information Technology and Decision Making* **11** (02): 441–456.

Zhong, S. and Pearce, J.M. (2018). Tightening the loop on the circular economy: coupled distributed recycling and manufacturing with recyclebot and RepRap 3-D printing. *Resources, Conservation and Recycling* **128**: 48–58.

Zuboff, S. (2010). Creating value in the age of distributed capitalism. *McKinsey Quarterly* **12** (1): 1–12.

15

Food Industry Market Trends and Consumer Preferences

CHAPTER MENU

3D Printing of Foods, First Edition. C. Anandharamakrishnan, Jeyan A. Moses, and T. Anukiruthika.

15.1 Introduction

3D printing is an additive technology that utilizes computer-assisted control for the precise deposition of material supply in a pre-defined model path (Theagarajan et al. 2020). With growing research potential 3D food printing has evolved as a novel convenient food processing technique that aids in mass customization and personalized nutrition. Most of the 3D printing reports are focused on the assessment of printability and material characteristics to withstand the rigorous post-processing for ensuring the safety and edibility of 3D printed foods (Krishnaraj et al. 2019; Wilson et al. 2020). Food technologists and research scholars are working towards exploring the potential applications of 3D printing in the food sector. However, the feasibility for implementation, market potential, and acceptance of novel food products are the most important factors in determining the success of food printing. These amalgamate the need for analyzing the consumer's preferences and food decisions (Wilson et al. 2020). Health and well-being are the two significant aspects for the consumers that transform the dietary habits towards a balanced diet. With this as a key point, most of the emerging food processing techniques are focusing on the production of minimally processed foods (Moses et al. 2014). It helps in preserving the naturally present bioactive ingredients and micronutrients. In another dimension, consumer's food choice remains to change with the time that also adds to the desires for experiencing new tastes and eating pleasure (Keerthana et al. 2020). All these aspects remain challenging for conventional food technologies to satisfy the ever-changing consumer food choices. This rises the need for the emergence of novel food processing techniques for satisfying consumers both physiologically and psychologically. In this regard, 3D food printing is a disruptive technology that is being forecasted to be the technology of the future food industry. The most advantageous feature of this technology is the delivery of foods based on individual preference and likeness. Unlike other conventional food processing techniques that automate the food production process, 3D printing allows for the delivery of personalized foods (Nachal et al. 2019). Hence 3D printing is being referred to as prosumption technology that allows the consumers to design and produce their own foods.

3D printing offers socioeconomic and environmental benefits in addition to sustainability. As a versatile technology, 3D printing allows for the fabrication of novel foods from uncommon and unfamiliar food ingredients (Derossi et al. 2020). For instance, it is feasible to print mushrooms through appropriate pre-processing and optimization processes (Keerthana et al. 2020). Likewise, reports were available on the 3D printing of macro and micro algae and insects (Severini et al. 2018; Uribe-Wandurraga et al. 2020). Nevertheless, 3D printing also helps in the efficient utilization of by-products and food wastes generated during main product processing (Nida et al. 2020; Wilson et al. 2020). Although the features of 3D printing of foods fascinate the consumers, it also significantly affects the consumer's food decision. Few aspects such as social and cultural concerns, food neophobia, and thought of 3D printed foods as an unnatural way of processing remains to be metal barriers for the growth of food printing that hinders the attempt of tasting novel foods prepared from uncommon food ingredients (Lupton and Turner 2018b). Despite these limiting factors, 3D food printing has been forecasted to reach a market size of USD 1015.4 million in 2027 (Emergen Research 2021). This clearly shows the rising demand for customized foods with increased adoption for various end-uses in the food sector. However, this market trend of 3D printing of foods could be achievable by overcoming the technological barriers among the consumers. In this regard, consumer's food preferences and choices are most important in determining the sustenance and implementation of food printing technology (Lupton and Turner 2018a). Thus, the present chapter focused on exploring the advantages and limitations from different aspect in consumers' point of view. The consumers' acceptance of novel foods is discussed elaborately in context with 3D printing in comparison with similar other emerging food technologies. The food service chain of 3D printing from production to consumption is analyzed in view of the food business model 'prosumerism'. Various methods and approaches used for analyzing the consumer perception of foods as reported in literature were presented. Further, different measures used for overcoming the challenges and barriers on consumer's attitude towards 3D printing is explained with future trends and research directions. Thus, the present chapter remains as a check point among all the chapters of the present book that provides insights on the consumer's willingness and perception on the adoption of 3D printing of foods and the strategies that could be adopted for overcoming the barriers to this novel technology.

15.2 Food Service Market: Consumption to Prosumption

The emergence of 3D printing allows consumers to prepare foods based on their individual choice and preferences. The 3D food printing is realized as a fascinating technology that has been in a nascent stage in the food sector with very little consumer knowledge (Tuorila and Hartmann 2020). This is the major constraint to the growth of this technology as most of the consumers are not well aware of the fact that food can be printed. However, significant research has to be carried out in this area in exploring the hidden opportunities in the delivery of personalized foods (Kumar et al. 2020). With the recent research potential, 3D printing is forecasted to be the future technology of the food industry. The advancements of technology would lead to the arrival of desktop 3D food printers that will be considered as a common domestic kitchen appliance in near future. The combination of digital e-platforms and marketing will increase the consumer's involvement in food production. With 3D printing technology, the consumers can design their own food with the preferred choice of ingredients in desired shapes and forms (Nachal et al. 2019). Although 3D printing seems to be a competitive rivalry technology to conventional food processing methods, the advent of which would revolutionize the food industry. The synergistic benefits of 3D printing and the advent of information and communication technology are highly anticipated in the food domain that would change consumers into prosumers where the consumers enjoy both the role of consumption as well as production (Jayaprakash et al. 2019). This can be witnessed with current changes in the food business model from centralized production into the localized production of foods. Major reasons for this declination are the lower acceptance rate of boring products and repetitive tasks that rises the consumer's demand for food varieties and choices. Instead of relying on the conventional market, consumers are now becoming more aware of the latest advancements, and are showing more interest in fabricating their own cuisines using 3D printing technology (Dankar et al. 2018). A proportional increment of size of prosumers is also one of the reasons for the growth of additive manufacturing in the food sector.

The latest food business models allow the consumers to participate and contribute to the services of food production and distribution. The productional activities that are closely associated with the consumers would shift the business strategy towards a service-oriented domain with customer dominant system rather than goods-dominant (Jayaprakash et al. 2019). All these efforts would pave a way for experiencing innovation through the amalgamation of products, services, and demand in a user-defined environment. This kind of prosumer dominant market is realizable in the food sector through 3D printing technology. It has been predicted that 3D printing would undergo significant transformative changes during the period of 2015–2025 (Karlgaard 2011). The combination of 3D printing and digital platforms would together take up the power of presumption to the next level that seems to be disruptive to the existing manufacturing paradigm. This business model of co-creation would not only avoid product failure but also builds up a strong relationship with customers (Halassi et al. 2019). Further, the consumers can be able to intervene at any stage of production since all the activities are more transparent. Despite 3D printing has a long way to go for its implementation in the food sector to enjoy the benefits of the prosumer-driven market. In addition to the systematic studies on determining the printability of material supplies, qualitative studies on the acceptance of 3D printed foods must be carried out by considering individual, social, cultural, and geographic factors. The knowledge gathered from this qualitative study would help in gaining the consumer's acceptance and build up trust in the developed 3D printed foods. A panel of experts from academia, research scholars, industrial firms, distributors, start-ups, consultancies, software engineers, and service providers must work together for analyzing the market feasibility and consumer acceptance of 3D printed foods. The changing dietary practices and consumer convenience are also crucial factors in determining the success and sustenance of 3D printing in a long run (Nachal et al. 2019). All these aspects of 3D printing with recent innovations and sustainability would complement its growth as a prosumer-driven food domain.

Recently, digital platforms are transforming the traditional food business from mass production to mass customization. It helps in building a strong co-relation with consumer acceptance through linking consumer food decisions with the developed food product and technology (Laplume et al. 2016). 3D food printing allows for the on-demand production of foods that significantly reduces material wastage. The production process of the current food industry is highly dominated by centralized production and the present level of presumption is relatively very low. More studies must be conducted in analyzing the market segmentation of 3D printing, food value chain, ingredient portfolio, existing design concepts, printing techniques, and post-processing feasibility. The change drivers for the prosumer-driven 3D food printing are broadly grouped into four classes as follows: technology, well-being, market trends, and sustainability (Jayaprakash et al. 2019). Regardless of the geographical discrimination, the advent of information and technology allows ease of transfer of 3D designs and e-recipes. Under the current marketing conditions, entrepreneurs should withstand the competitive market by adopting customer-centric and sustainable business models to differentiate themselves from co-competitors. Other major factors that drive the prosumer-oriented 3D printing market are food wastage, sustainability, climate change, global hunger, and environmental implications. Thus, the acceptance of the 3D printing of foods is not only determined by technological innovations but also on convenience, sustainability, and societal changes (Nachal et al. 2019). The results obtained from the qualitative analysis on consumer perception of food printing can be combined with the advancements in information and technology. This would lead to a new generation of 'modern vending machines' that offers nutritious food in a customized and personalized manner. This kind of 3D printing-based vending machines can be conveniently placed at public places such as railway stations and bus stations, offices, workspaces, hospitals, restaurants, etc. Thus, a 3D food printer coupled with a digital platform has the inherent potential to reform the entire food supply and distribution chain.

15.3 Food Decisions and Consumer Attitude

Consumers can easily reject novel foods and new food technologies without realizing the potential benefits. Various reasons like ethical, social, environmental, and safety concerns are mainly responsible for the reluctance of any novel food processing technique among the consumers (Cifci et al. 2020). This also applies for 3D food printing where the consumers are mindful of the safety and edibility of the 3D printed foods even it provides required nutrition in desired shapes and forms. Most of the innovations provided by 3D printing would be subsided in case of improper understanding and negative consumer perception (La Barbera et al. 2018). Hence it is paramount important to analyze the consumer's food choice and preferences during the earlier stages of the food production process. As food printing deals with the production of foods based on the demand and requirement, it significantly reduces material wastage, decreases the intensity of processing, reduces packaging, storage, and handling cost. In this cohort, 3D printing efficiently utilizes the material and energy resources that provide a sustainable solution for rising food demand in satisfying global hunger (Keerthana et al. 2020; Sun et al. 2018). The present decade is witnessed with a great transformation in food manufacturing due to the advent of 3D printing. Undoubtedly, there are numerous studies reported on the material characteristics, printability, post-processing, and applications of 3D printing of foods (Anukiruthika et al. 2020a; Krishnaraj et al. 2019; Theagarajan et al. 2020; Wilson et al. 2020). However, only a handful of reports were available on the consumer perception towards 3D printing, quality, safety, and regulations of 3D printed foods (Baiano 2020). Despite these are the most significant attributes that decide the sustenance and implementation of 3D printing in the food sector. Hence more research needs are required to bridge up the gap between 3D printed foods and consumer acceptability through qualitative and quantitative assessment. The basic understanding of the consumer's needs, and the perception would help in the identification of the technological and marketing barriers to the consumer's acceptance (Chen et al. 2013). Ultimately this would help in designing and adoption of new market strategies that significantly reduce the reluctance towards 3D printed foods among the consumers (Kim and Jeon 2020; Li et al. 2014). Therefore, the present section discusses about the various factors that affect the consumer's acceptance of the novel 3D printed foods. Various motives for the development of novel foods from consumer and industrial perspectives are presented in Figure 15.1.

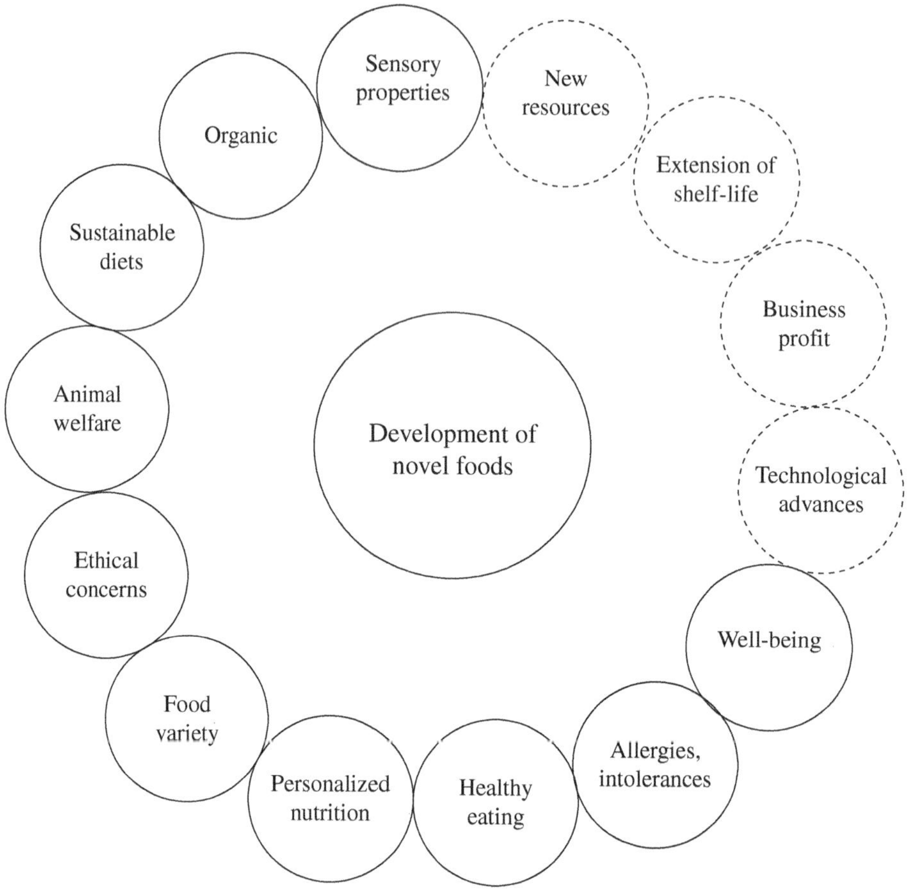

Figure 15.1 Various motives for the development of novel foods from consumer and industrial perspectives (circles with solid line – consumer perspectives; circles with broken line – industrial perspectives). *Source:* Tuorila and Hartmann (2020) / With permission of Elsevier.

15.3.1 Food Neophobia vs Food Neophilia

The success of any introduced food product based on novel food technology say 3D printing depends on the consumer's reactions and acceptability (Siegrist and Hartmann 2020). The modern food industry had realized this fact and consider it significantly by giving more priority to quality testing and screening process. This would significantly avoid the risk of product failure. Unlike the other conventional food processing methods, 3D printing is more unique in the delivery of foods based on the consumer's preferences and choices. Despite the 3D printed foods also gained negative perceptions among consumers due to social, cultural, and ethical reasons (Nachal et al. 2019). Moreover, the acceptance of 3D printed foods is not only based on appearance and taste but also influenced by the psychological behaviour of the consumers. The eating experience and psychological attitude assist in the precise determination of the food behaviour of consumers that in turn assists in the achievement of the target marketability (Guine et al. 2010). Despite the popularity of 3D printing in the fabrication of foods in a novel way, the psychological feeling of food neophobia among the consumers negatively influences the growth of 3D printed foods. Food neophobia is a characteristic feeling of fear that measures an individual's tendency to show food aversion to try novel or unfamiliar foods (Tuorila and Hartmann 2020). The fear of accepting new foods not only reduces marketability through product failure but also decreases the dietary intake and nutritional requirement. In this direction, research studies are now focusing on the dual determinants of the genetic factors and environment on food neophobia (Cooke 2018). It was reported that food neophobia remained at its peak during the early stages of childhood making the consumers more specific with their food choices. The acquirement of food aversion at the early stages of life also has a significant impact at the latter stages of adult since the food preferences and eating habits remained to be constant throughout the life (Jaeger et al. 2021). Thus, food neophobia has been recognized to be a stable trait with the behavioral negative response avoiding the novel foods. The qualitative studies on analyzing consumer's food preferences can be done based on the self-report questionnaires and the behavioural response in measuring food neophobia (Cifci et al. 2020). Even a small variation in the type of foods and eating occasions influences the individual's preferences, the behavioural approach would help in reflecting the mind state of the group of consumers more efficiently than the questionnaire approach (Alley 2018). As like any other sensory scale, the consumer's food choice can be measured using a quantitative scale. Among the various scales available for measuring the food choices, the one proposed by Pliner and Hobden (1992) has been referred as 'gold standard' where a set of 10 statements were given that would be rated on 7-point Likert scales (Cooke 2018). Irrespective of age groups the individual liking and intake were predicted to be correlated with the food neophobia.

Like many other psychological traits, food neophobia exerts a continuum of scale from high to low aversion that may even lead to malnourishment and illness at the extreme case of higher dislikeness of foods (Monteleone et al. 2017). Factors such as age, education, and income were reported to have a significant impact on the food decision-making of individuals. Reports stated that the adults were more neophobic than the younger generation irrespective of gender difference. On the other hand, peoples with a higher level of educational attainment and higher incomes tend to be less neophobic (Jaeger et al. 2019; Soucier et al. 2019). In a broader perspective, the geographical location, demographic factors, socioeconomic reasons, and cultural aspects have a strong argument that adds to the complexity of factors used for measuring food choices (Jaeger et al. 2021). The aforementioned reports were based on analyzing the food preferences for the novel unfamiliar foods prepared based on the conventional food processing techniques. When considering the food printing process, the choice and varieties of the food are quite high. In this regard, there is also the behavioural mood swing for the likeness of the novel foods among a segment of consumers. This stable trait showing the likeness to prefer new foods are referred to as 'food neophilia'. The qualitative and quantitative assessment of the consumer's choice and acceptability of 3D printed foods are clutched by the extreme feelings of food neophobia and food neophilia (Cifci et al. 2020). These aspects would complicate the process of assessment of food choice and consumer attitude towards 3D printed foods. The associated complexity can be reduced through proper understanding and identification of root causes for food aversion. Some of the reasons that were put forth by consumers to strongly resist the 3D printed foods are misconception of the unnaturalness of 3D printed foods, social, and cultural reasons in the use of alternative food ingredients, concerns related to the safety, and edibility of developed product (Baiano 2020). All these factors sought to claim disagreement and remained as a hindrance to the implementation of 3D printing. Therefore, more systematic studies are required in food decision research considering all the above-mentioned factors to predict the food neophobia that in turn gives an idea about the consumer's food behaviour and food choices towards 3D printed foods. To date, there is no report available in correlating the food neophobia and food choice factors in assessing the people's motives for deciding their choice for 3D printed foods. Future studies should be focused on this direction in correlating the food neophobia and food choice questionnaire as a prime importance for determining the success of the 3D food printing and its sustenance based on consumer acceptance.

15.3.2 Food Choice Motives

Understanding the food choice motives helps in determining the role of consumers in the success of the developed foods. Various driving factors that decide the consumer's food choices have been extensively studied in past years using different approaches (Barrena and Sánchez 2013; Cunha et al. 2018; Onwezen et al. 2019). Aspects considering culture, healthy dietary practices (Carrillo et al. 2011) and consumer response for new foods and beverages (Schlup and Brunner 2018) have been reported. Often food choice motives of consumers are measured based on surveys using a questionnaire by considering food-related ideologies and preferences. A basic food choice questionnaire (original) includes nine significant factors that were identified to be influencing the consumer's food decisions. This includes health, mood, convenience, naturalness, sensory appeal, price, familiarity, weight control, and ethical concern (Jaeger et al. 2021). All these nine factors were measured against 36 items that are rated as important in making daily food decisions using a 4-point rating scale where the points 1–4 represent lower to higher preference as follows: not at all important, little important, moderately important, and very important. Later a revised version of the original food choice questionnaire was proposed based on the research carried out in multiple European countries (Onwezen et al. 2019). Compare to the original food choice questionnaire, the newer version is relatively shorter with one question per food choice questionnaire factor thereby reduces the burdensome of the participants. Also, the factor 'ethical concern' in the earlier version has been expanded into three other factors such as environment, animal welfare, and social justice. However, these factors of ethical concerns could be expanded further as per the need and requirement of the specific research topic. Overall, the factors relating the health, price, and sensory appeal are given the utmost importance. While the factors linked with ethical concerns are given the least importance. On the other hand, the familiarity is grouped to the factors with lesser importance while the naturalness/ natural content is placed into the group of factors with higher importance (Markovina et al. 2015; Onwezen et al. 2019).

It is of utmost importance to draw a relation and know how food neophobia associates with the food choice questionnaire. This would help in exploring the correlation of food neophobia with food choice decisions that drive the consumer's motives. Accordingly, the expected association can be positive, negative, or independent of food neophobia (Jaeger et al. 2021). With the increase in neophobia, the food choice questionnaire factors can either gain or lose their importance but do not get systematically changed. This structure of assessment assists in exploring the correlation of the food neophobia in view of how it modulates the food decisions of the consumers is crucial. It was reported that familiarity, convenience, and price would have a positive association between the food neophobia and food choice questionnaire factors (Jaeger et al. 2019). On the other hand, the expected negative associations were found for the health, naturalness, and ethical concerns (environment, animal welfare, and social justice). As expected, the naturalness of the food is directly linked to health-oriented food choices (Roman et al. 2017). Thus, a positive correlation is seemed to be unlikely associated with health and naturalness. Although ethical concerns are ranked to be the least important factors, they are considered of high importance by a certain group of consumers since these factors are linked to working conditions, political, religious, and moral values (Pearcey and Zhan 2018). It was stated that the peoples with high levels of neophobia generally neglects the ethical motives as it is less important compared to the other food choice factors. Other factors such as sensory appeal, weight control, and mood were considered to have non-significant associations between food neophobia and food choice questionnaire factors (Jaeger et al. 2019). Regardless of the degree of neophobia/neophilia, the sensory appeal of foods is always considered with high priority. As expected, consumers with high levels of neophobia least enjoy the varieties of foods than the neophiliac counterparts (Laureati et al. 2018). Studies on the correlation of weight control and its association between food neophobia and food choice questionnaire depicted a weak positive relation and sometimes insignificant correlation. Due to this inconsistent evidence, weight control has been drawn to have a non-significant association. Similarly, there is no significant linkage found between food neophobia and mood as it is not apparent with strong relationship, positive or negative correlations. Hence with the available limited evidence mood has been grouped as non-significant to impart food choice decisions (Jaeger et al. 2021). However, this category of food choice questionnaire factors will be varied when considered for 3D printed foods. Other auxiliary factors that would influence the food decisions of consumers about 3D printing include urbanization, industrialization, sustainability, climate change, and lifestyle. Hence, future studies on determining the significance of the food choice motives for 3D printing of foods must consider all these factors in context with both food neophobia and neophilia. This would provide insights on the changing trends of people's food choice with time, age, and gender that in turn helps in determining the sustenance of 3D printing in the food sector.

15.3.3 Sensorial and Sustainable Claims

The rising global population puts a large pressure on land in producing foods and rearing animals. All these human activities lead to overexploitation of land through depletion of materials and energy resources. One promising solution to combat the demand for food supply is the adoption and implementation of sustainable food practices (Keerthana et al. 2020). A key driver for resource sustainability is the selective consumption of foods. Sustainable strategies can gradually enhance agricultural productivity through the development of alternative novel food ingredients (Derossi et al. 2020). However, determining the consumer's attitude and acceptance is most important for wider usage and commercialization of the food products developed from alternative ingredients. The claim for sustainability in eating practices has been considerably increased among the consumers with higher levels of education. In this regard, the awareness of environmental protection is also increasing among consumers through the use of degradable materials. Hence, a recent decade is witnessed with the emergence of modernized biodegradable food packages with functional entities. Considering these aspects, 3D printing is a prospective technology that not only allows for the fabrication of foods but also food packages by ensuring sustainability (Nida et al. 2020). Regardless of these attractive features, consumer's perception towards the use of alternative ingredients poses a serious threat to the adoption of 3D printing as a sustainable tool. Consumers are more mindful of the health-related concerns and source of origin of food ingredients than environmental awareness in the context of selecting their food choices (Jaeger et al. 2021). Generally, consumers categorize food ingredients in two major dichotomies as natural vs. unnatural and positive effects vs. negative effects that remain as a great barrier for the acceptance of novel alternative foods (Aschemann-Witzel et al. 2019). Although consumers pursue more sustainable food choices, the use of potential new ingredients is limited due to its unfamiliarity and negative perception. Hence it is essential to analyze the various factors that influence food choices both qualitatively and quantitatively. In this regard, 3D food printing is expected to have greater implications in food preferences and dietary practices as it has been forecasted to be the future technology of food manufacturing (Theagarajan et al. 2020). The major breakthrough of this novel technology is the efficient utilization of waste byproducts of food processing line such as cuts of fruits and vegetables, meat trimmings, milling fractions of cereals, and pulses in the development of foods with appealing taste and texture (Ricci et al. 2019; Wilson et al. 2020). Only a few reports were available on the consumer perception of 3D printed foods. In a study reported by Manstan and McSweeney (2020), 3D printed foods were compared with conventionally prepared foods. Surprisingly, the consumers had a positive attitude towards 3D printed foods with the inference that 3D printed foods were believed to be healthier with less intense processing. Contrarily, the reports of Brunner et al. (2018) stated that the participants had a negative attitude towards 3D printed foods considering it to be unnatural. These inconsistent results make it hard to draw inferences about the consumer perception towards 3D printed foods. Hence more studies should be carried out in considering a diverse group of participants in terms of culture, geographic locations, age, gender, and living status.

Recently, a study was reported on consumer perception towards novel food ingredients (Lunden et al. 2020). This study aimed at revealing the consumer's perspectives on a novel and partly traditional but marginally utilized food ingredients against the ingredients that were commonly used in regular cooking by considering the sensory, nutritional, and environmental aspects. Food decisions and preferences are extremely diverse which are influenced by several social, cultural, environmental, and individual factors. Some of the food ingredients that are considered novel and unfamiliar are insects, food wastes and byproducts, wild terrestrial plants, and marine flora (Derossi et al. 2020). Since after the implementation of Novel Food Regulation (Regulation EU 2015/2283), some of the insects are considered edible and are being used as foods in European Union (EU Regulation 2015).

The use of insects as food has gained consumer's acceptance in Western countries. Hence most research scholars are currently working on exploring the potential benefits of utilizing insects in a regular diet as a novel protein source. Although Western consumers are becoming familiar with insects as foods, a portion of consumers from Eastern regions express disgust feelings, and unwillingness to accept insects as foods. Willingness to consume insects was found to be interlinked with cultures and results showed that consumers from Northern Europe expressed a positive attitude towards edible insects than consumers from central Europe (Piha et al. 2018). In this context, the usage of wild traditional plants, herbs, and berries was markedly increased in processed foods in the European culture. However, there was a reluctance in the use of wild plants in regular diet among consumers which was mainly because of the diminishing traditional culinary knowledge and inconvenience in the collection of wild plants as it takes more time for processing (Luczaj et al. 2012). Due to nutritional significance and potential benefits, the use of wild plants in regular cooking is encouraged. The next class of uncommon ingredients that are facing consumer's reluctance includes food wastes and byproducts. Although many reports claim the re-utilization of these materials possessing a rich amount of minerals and fibre in food preparations (Feng et al. 2020), the consumers' mindset of waste to be used as food declines its usage. The most encountered cause for the consumer's reluctance with the introduction of novel alternative food ingredient is food neophobia. Hence, a qualitative and quantitative studies are required in judging the consumer's opinion by taking into account the potential barriers and drivers for these novel food ingredients. Among the considered food ingredients, the plant-based materials were preferred than foods from animal and insect's origin. Neither cultured meat nor 3D printed foods were gained interest without prior experience or knowledge among the consumers (Lunden et al. 2020). Thus, the findings of this study remained as baseline for determining consumer's attitude towards novel food ingredients. Overall, the study concluded that plant-based ingredients have a great potential to be used as future food ingredients. These findings clearly show the need for awareness about 3D printed foods to educate people with potential outcomes and health benefits. Research scholars and food scientists should work along with the professionals of information and communication technology to demonstrate the potential benefits of 3D printing for a wide range of consumers all over the world. Then only the people will think about the use of 3D printers to fabricate their own foods without any reluctance in its acceptance despite sustainability and environmental benefits.

15.4 Approaches and Methods to Assess Consumer Perception

Understanding the consumer's opinions and food preferences are the most important criteria in the success of any new technology (Siegrist and Hartmann 2020). Not only for scientific academic scholars but also the industrial firms are working on exploring the use of sustainable food technologies. Although 3D printing offers several socioeconomic, environmental, and health benefits, the uncertainty and lack of knowledge on perceived risks decline the benefits of 3D printing. This induces hesitation among the consumers to accept novel food technologies like 3D printing (Brunner et al. 2018). Based on the topic of research interest, the approaches used for predicting the consumer acceptance of novel technologies would vary from economists and social scientists in exploring the methods for scientific understanding of consumer's opinions on 3D printing of foods. The present section summarizes the various approaches that are generally used for understanding the consumer perception and associated risks and benefits of novel food technology. Broadly the approaches used for assessing the consumer's opinion on novel food technology are categorized as theoretical and experimental approaches. Certainly, the present section is focused on providing insights on various approaches and methods that are in practice that could be conveniently used in determining the acceptability of 3D printing among consumers.

15.4.1 Theoretical Approaches

15.4.1.1 Quantitative Methods

The quantitative risk assessment involves drawing results based on the models developed using mathematical tools and engineering principles. This would help consumers to understand the technological growth, development, and implications. In another perspective, the quantitative assessment assists in providing insights for industrial firms and food scientists in assessing the implications of novel technology in food service and supply chain management (Castell-Perez and Moreira 2021). Although the social elements are most significant in affecting the technical risk assessments, this approach suffers from the limitation of omitting the social-related elements. In contrast to the mathematical approach, the social science-based studies imply the role of social elements over the acquirement of trust and confidence of the developed product among the consumers (Ronteltap et al. 2007). Other than these methods, several risk assessment models were used for assessing the beneficiaries' and consumer's interest in the implementation of new technology. In this regard, future studies on the assessment of consumer's attitudes towards 3D printing should adopt the combined models considering all aspects of scientific principles and social elements to predict the consumer's preference and barriers for accepting 3D printed foods.

15.4.1.2 Means-end Chain Theory

This theory is based on the consumer's cognitive behaviour of purchasing the newly introduced product. The means-end chain theory is more useful in estimating the success of the developed novel food product and assists in promoting the developed technology through improved advertising of business strategies (Chen et al. 2018). Results of this model provide an overview of product attributes, consumption pattern, outcomes, and personal values which helps in the regulation of the food manufacturing process with the incorporation of novel elements (Castell-Perez and Moreira 2021). Thus, the data gathered from the means-end chain theory have been used for the probing process for determining the linkages of product characteristics from consumer's perspectives through one-on-one interviewing techniques. Undoubtedly, the means-end chain theory helps in improving the marketing strategies through assessing consumers' preferences through their purchasing behaviour.

15.4.1.3 Social Science Models

The social science models are useful in understanding how much the consumers are aware of the newly introduced technology. This method of risk assessment is based on diverse disciplines of sociology, psychology, and anthropology that assist in understanding the consumer's opinion regarding the novel technology. Illustrations based on psychometric approach deals with the personal values and emotional heuristics like fear of the unknown that would significantly affect interpersonal trust (Crowley et al. 2013). Often, the assessments based on heuristic would suffer from individual biasness. Risk assessments using the psychological models are based on emotions and the risk perceptions would be varied from laypeople to experts. This reflects the associated deviation in a wide range of audience and consumer's opinions for the considered technology under study. This kind of behaviour is described as an affect heuristic that infers the overall feeling of individual perceptions of risks and benefits about an object (Erdem and Rigby 2013). On the other hand, the anthropological approach provides an overview of consumer perception in terms of local customs, beliefs, culture, and social organization of the people. It was assumed that diffusion of innovations is acceptable among the consumers endorsed by the trusted organization (Castell-Perez and Moreira 2021). Thus, the consumer's judgement and the perceived risks are influenced by the cognitive and affective process that aids in determining the consumer's response and decisions in accepting new technology.

15.4.1.4 Economic Models

The consumer's risk perception and the food decisions are not only influenced by heuristics but also by intuitive judgements where the behavioural economic models comes into play. Thus, the economic models address the deviation of subjective probability judgements from objective probabilities (Lusk and McCluskey 2018). A traditional economic approach used for risk assessment is based on expected utility theory where the consumers will judge the positive and negative outcomes through observing the expected outcomes multiplied by the utility. For instance, if a consumer associates the introduced food technology with higher probabilities of negative outcomes then obviously the technology would be avoided and eventually failed in the market (Castell-Perez and Moreira 2021). Other types of heuristics that influence the consumer's food decisions in risk assessment are 'representative heuristic' and 'availability heuristic'. The former heuristic involves the rating of the unknown hazards based on perceived similarities between the known and unknown risks while the judgement is derived based on the semantic similarity associated with both known and unknown risks. The latter type of heuristic is based on prospect theory where the consumers assign a probability weighing function to the expected outcome (Lusk et al. 2014). Sometimes, this would lead to overweight the low probability events by the consumers. Hence, a proper understanding of the proposed technology would help in the selection of better risk assessment approaches.

15.4.2 Experimental Approaches

15.4.2.1 Surveys

Experimental approaches for determining the consumer attitude towards food choices are more common. One such experimental approach is the performance of surveys that focus on specific groups with open-ended questions used in assessing consumer's acceptance of new technology (Bolek 2020). The choice of questions would differ based on the type of surveys such as online survey, mailed survey, market survey, and retailer survey. The prepared questionnaire helps in determining how the information drives consumer purchases and product sales. Since the questions are presented as hypotheticals, they do not directly translate the consumer's purchasing behaviour and limit its applicability to assess consumer's food choices (Castell-Perez and Moreira 2021). A psychrometric approach in a questionnaire form including the concepts, awareness, outcomes, labelling, and safety would help in a quick assessment of consumer's knowledge on the novel food technology that in turn useful in the commercialization of the technology.

15.4.2.2 Conjoint Analysis and Choice-Based Conjoint Analysis

The choice-based conjoint analysis is the preferred means for designing experiments for determining the consumer's preference. As a variant of traditional conjoint analysis, the choice-based methodology is depicted through elicited choices of consumer's responses. In other words, the choice-based conjoint analysis allows for rating the multiple elements of the product at a time by choosing the desired attribute from a range of choices (Anabtawi et al. 2020). This scenario simulates the real-time situation of grocery shopping where the consumers have several options and varieties for a single product. Thus, choice-based conjoint analysis helps in knowing the consumer's preference through comparing the same product made out of traditional technology against the novel technology.

15.4.2.3 Heuristics

Studies on understanding how bias affects the food choice when consumers use heuristics for making decisions are most important in relating the sustenance of new technologies in the food industry. Although the heuristic tools are easy to judge the attributes, the food decisions derived out of which often results in biased decisions (Siegrist and Sutterlin 2017). For instance, the perceived naturalness of the product is one of the heuristic attributes that biases the consumer's choice towards a favourable positive response in perceiving the new technologies.

15.5 Consumer's Acceptance of Novel Foods

All the food processing techniques are aimed at the shelf-life extension and maintenance of the fresh-like quality of foods. In the past decades, the food industry witnessed the transformation of traditional processing into a modernized technique with technological advancements (Moses et al. 2014). Unlike other industrial sectors that have undergone the adoption of novel technology with the replacement of older techniques, the food industry is seemed to add on the newer techniques along with the older ones. Because all these food technologies are envisioned to enhance safety, nutritional quality, and sustainability. Thus, it imparts less pressure among the consumers to accept the new innovations and novel food technologies (Siegrist and Hartmann 2020). In contrast to most of the industrial domains that have positive progress with technological advancements, the food domain is slightly different as it has both pros and cons with the emergence of novel food technology.

Factors such as fresh-like foods, minimal processing, and enhanced nutritional quality affect the consumer's choice for the foods. Most importantly the consumer decision influences the market demand and the success of any food processing technology (Lunden et al. 2020). There is still a contradictory thought that prevails among the consumers that food processing technologies negatively affect the wholeness, nutritional, and sensorial attributes of foods. Although the novel emerging food technologies show a promising result in terms of process efficiency and productivity, there exists a strong resistance among a portion of consumers (Anukiruthika et al. 2020b). This misunderstanding was mainly due to a lack of knowledge on the benefits and potential advantages of the process. Hence a thorough understanding of the process is adequate for exploring its benefits in the delivery of nutritional foods. Some of the novel conventional food technologies that suffer from controversy include food irradiation, genetically modified (GM) foods, meat analogues, artificially cultured meat, and nanotechnology (Siegrist and Hartmann 2020). Here the term 'novel' does not imply the time of the invention of technology rather it denotes the consumer's acceptability of the technology all over the world. Individual personal factors such as food aversion, food disgust sensitivity, and food neophobia significantly affect consumer acceptance (Derossi et al. 2020; Manstan and McSweeney 2020). Other than personal factors, socioeconomic factors, cultural factors, political factors, and environmental factors also influence the consumer's reluctance for new technologies (Figure 15.2). Hence, consumer preference and acceptance are very crucial attributes in determining the

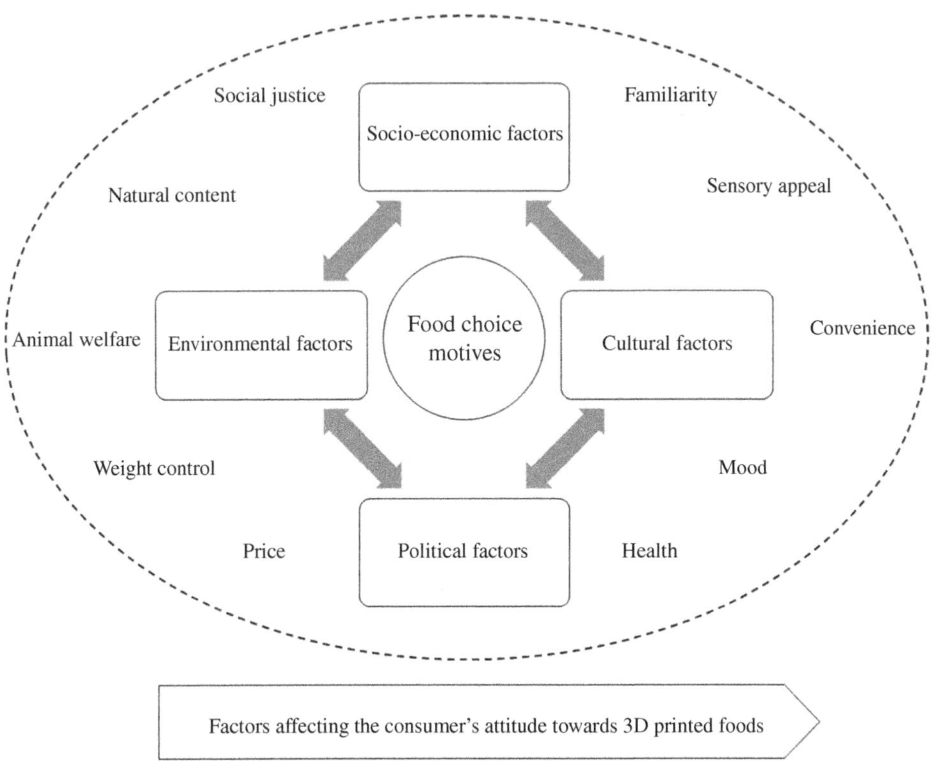

Figure 15.2 Relationship between the consumer's food choice motives and 3D printing.

marketability of the developed food product and the success of the technology. In this regard, the present section provides a comparative analysis of consumer acceptance of conventional novel food technologies and 3D food printing.

15.5.1 Genetically Modified Foods and 3D Printing

GM foods also known as genetically engineered foods are those foods that are derived from plants or animals through gene transfer and modification resulting in changes in their original DNA sequence (Kim et al. 2014). A general view that exists with the acceptance of the GM foods is the unknown long-term illness. Sometimes, the new breed varieties of plants are often confused with GM foods (Zhang et al. 2018). Most people think that seedless grapes and watermelons are GM foods but actually, they are not. The lack of general knowledge and awareness of GM foods often misleads people to rely on trust and naturalness heuristics. Empirical studies conducted in different countries showed that social trust influences the benefits and the risk perceptions of GM foods. Further, the results inferred that there are no tangible benefits, only the perceived unnatural characteristic of GM foods negatively influences consumer acceptance (Ali et al. 2021). Also, the acceptability of GM foods differs widely based on its applications as well as with geographic regions since not all GM foods are perceived in the same way (Kronberger et al. 2014). For instance, inserting the gene seems to be decreasing the naturalness of the food among the consumers than deleting a gene. A report on the acceptability of GM foods in the European Union stated that the cisgenic apples had a positive image while the transgenic apples had a negative perception (Scott and Rozin 2017). These results indicate the oscillating consumer choices towards GM foods. This can be addressed through proper education and awareness among the consumers. Further, the method of evaluation and the factors considered in the survey significantly influences the results of the consumer's acceptance of GM foods. Hence, a standard protocol must be followed in order to avoid individual biases. In context with food printing, people are often told that it was hard to believe the fact that foods are printable (Lupton and Turner 2018b). Although most of the participants showed positive responses for 3D printed foods, a segment of participants strongly resists the acceptability of 3D printed foods as they considered it as unnatural way of manufacturing food (Tuorila and Hartmann 2020). The conceptual framework for food ingredient's naturalness evaluation is presented in Figure 15.3. As like GM foods, the unknown long-term effects on consumption of 3D printed foods hold up the negative perception on 3D printing among the consumers. In this regard, more research needs are required in addressing the safety concerns of the 3D printed foods. Both GM foods and 3D printed foods are in same par facing similar constraints with misconception of the basic concepts of the technology. Enforcement of strict legislations and proper labelling guidelines would considerably help in gaining the trust of the consumers and the acceptability of these novel technologies.

15.5.2 Food Irradiation and 3D Printing

Food irradiation is the process of controlled exposure of food and food packaging to ionizing radiations such as X-rays, gamma rays, and electron beams. Irradiation is an effective process in inhibiting microbial growth, sprouting, and killing of insects in foods. From

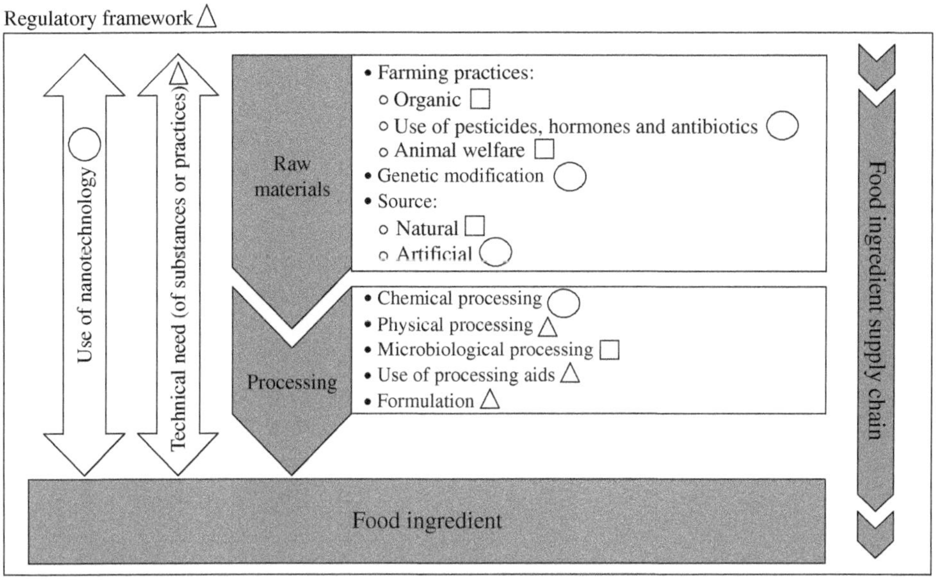

Figure 15.3 Conceptual framework for food ingredient's naturalness evaluation. (The ○ indicate the factors that are expected to negatively influence food ingredient's naturalness perception, while the □ indicate factors that are expected to have a positive influence. The ▽ indicate factors that can have both positive or negative influence, on a case-by-case basis.) *Source:* Battacchi et al. (2020)

the consumer point of view, the name 'irradiation' itself leads to negative perceptions related to nuclear energy (Behrens et al. 2009). Often the acceptability of irradiated foods is severely impacted by individual biasness. Those who had a positive image of irradiation consider it to be a safer technology while some had a negative response in terms of by-products and free radicals' generation. This put forth the perception of the health concerns and risks associated with the consumption of irradiated foods. These aspects remain to be a major reluctance for consumer acceptance (Castell-Perez and Moreira 2021). In this cohort, 3D printing of foods also utilizes a thermal source for the fusion of the deposited layers. Techniques such as selective sintering employ a laser source; and post-process cooking of 3D printed constructs use novel technologies such as microwave heating and infrared heating to enhance the edibility (He et al. 2019). However, the thought of chances of the risk of by-product generation during processing negatively influences the consumer's acceptability of 3D printed foods. Considering these aspects, future research must focus not only on printability but also on the post-processing feasibility with novel cooking technologies. As with the irradiated foods that often bear proper labelling and marking with the 'radura' symbol, 3D printed foods should also be marked with regulated labelling. This could avoid the misconception and reluctance over the acceptability of 3D-printed foods. It was reported that products labelled as 'treated with irradiation' had lower perception and acceptability than the products labelled as 'treated with ionization' (Bearth and Siegrist 2019). Technically both the terms are similar; however, the latter term had surprisingly gained a positive response among the consumers. Despite the misleading labelling terms, this report clearly showed the willingness and acceptability of irradiated foods among the consumers (MacRitchie et al. 2014). Hence a proper standard guideline must be established that would provide a clear idea about the processing technique employed and assist consumers in making their food choices and decisions.

15.5.3 Nanotechnology and 3D Printing

In the past decade, nanotechnology has emerged as an efficient drug delivery tool for the controlled and sustained release of bioactive compounds. The engineered nanoparticles are already in use as additives to enhance the nutritional value, increased absorption, and bioavailability (Leena et al. 2019). Thus, nanotechnology is experiencing tremendous growth in the food sector with a diverse range of applications including functional foods, drug delivery, biosensors, filtration, and active packaging to name a few (Bhushani and Anandharamakrishnan 2014). However, the substantial concern with the applicability of nanoparticles in food systems is the unknown fate of the designed nanoparticles. Yet the possible activation and enhanced reactivity of the same material at the nano level compared to that of the macro size cannot be overviewed (Mahalakshmi et al. 2020). Hence future studies must be directed in understanding the response of the material at different levels of the size reduction process. More than the advantages it provides, consumers are mindful of the negative implications of nanoparticles related to toxicity, allergenicity, and illness. It was reported that a large percentage of the participants unanswered the question related to the risk versus benefits. This clearly indicates the disability of the consumers to assess and differentiate the risk and benefits. Thus, the consumer perception on nanotechnology for food applications is still obscure. Often the negative perception of the nanotechnology is related to that of the GM foods (Duncan 2011). Since the fate of these engineered particles and molecules are unknown and it might cause illness in long-term effects. Despite there exists a distinct difference between these two technologies. The genetic engineering dealt with the addition/ deletion/ insertion of the desired gene thereby modulates the original pattern of genes at the molecular level while nanotechnology includes the physical transformation mainly with the size of the particle from macro to the nano level. Overall, the physical transformation has a lower impact on the perceived naturalness of developed food products compared to the biological transformation (Siegrist and Hartmann 2020). All these understandings clearly indicate the existence of the knowledge gap that resulted in biased opinions among the consumer acceptability. On the other hand, 3D printed foods are also negatively perceived as an unnatural way of manufacturing food by a group of participants in a survey conducted by (Manstan and McSweeney 2020). In this regard, 3D printing also suffers from a similar negative perception as that of GM foods and nano foods. Most importantly during the printing process, the material supply undergoes appropriate phase change for the successful adhesion of printed layers with one another (Nachal et al. 2019). This concept misleads the people's perspectives on 3D printed foods that rises the thought about the chances of generation of by-products during such chemical transitions. Proper education and awareness about the working principle and benefits of 3D printing would overcome this misconception among the consumers.

15.5.4 Stem Cell Technology and 3D Printing

15.5.4.1 *In-Vitro* Cultured Meat and 3D Printing

The process of culturing meat under laboratory conditions sounds weird for most consumers. This dictates their unpleasant state of mind to accept the cultured meat and they strongly resist it as unnatural food (Bonny et al. 2015). Although the meat is produced under *in-vitro* conditions, it undergoes the same biological transformation from culture to matured cells as that of the natural process of raising animals. Consumers are mindful that the meat is grown artificially and express their aversion to cultured meat products (Tomiyama et al. 2020). This could be viewed as the lack of understanding of the technology among the general public. Stem cell technology and tissue engineering are flourishing technologies in the biomedical field. The application of these technologies in the production of animal meat under laboratory conditions has been proposed as a sustainable way of producing foods (Post et al. 2020). Thus, the consumer's perception has been varied with the individuals. Those who considered cultured meat as unnatural showed strong disgust feelings and find the risks associated with it compared to those who accepted cultured meat. As a sustainable approach, researchers are exploring this area as the future technology to combat the rising demand for animal proteins (Derossi et al. 2020). This also reduces animal slaughtering and many other post-process steps that significantly save material and energy resources. However, the consumer's viewpoint of cultured meat as unnatural food has to be changed for the successful implementation of this novel technology. Considering 3D printing which seems to be versatile technology that can be conveniently used for bioprinting applications. Reports were already available on 3D printing of artificial laboratory meat (Handral et al. 2020). In addition to the existing perception of conventional processing of artificial meat as unnatural food when 3D printing is employed the processing conditions and safety concerns are relatively high (Lupton and Turner 2018a). Addressing these negative perceptions with proper processing through adopting standard safety procedures would pave for the driving out of the misconceptions and hesitation among consumers. Even though it reflects the animal welfare and other environmental benefits, this context of argument is relatively less valued when considering the perceived naturalness and other sensory attributes (Siegrist and Hartmann 2020).

15.5.5 Miscellaneous Technologies

15.5.5.1 Alternative Proteins and 3D Printing

The rising global population and the growing demand for protein supply had given pressure on food technologists to look into alternative food sources. The alternative proteins are those protein sources that are uncommon in regular cuisines. This includes insects, macro and micro algae, fungi, and lupine seeds (Sun et al. 2018). In view of sustainability, alternative proteins are the best option that satisfies the protein demand with the energy equivalent as that of animal protein (Keerthana et al. 2020). However, the disgusting feeling of eating insects as food is seemed to be predominant among the consumers. Some of the conventional approaches used for improving sensory attributes of insect and insect-based foods are presented in Figure 15.4. The disgusting feeling eventually hinders the prospective outcomes of alternative proteins as a sustainable food. In regard to 3D printing, based on the printability of the material supply itself, a separate classification named alternative

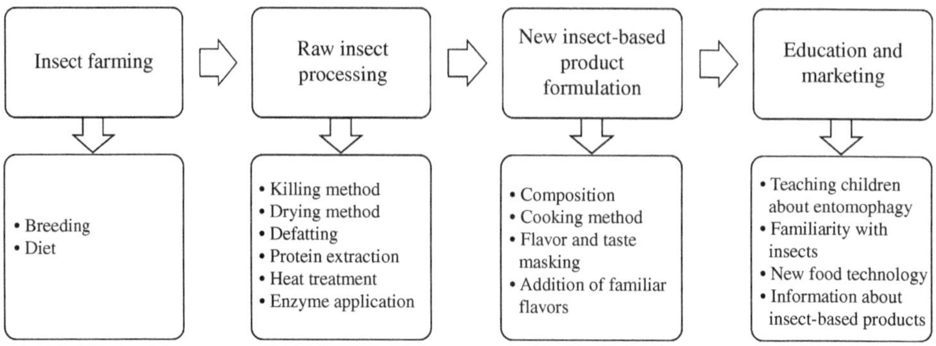

Figure 15.4 Conventional approaches used for improving the sensory attributes of insects and insect-based foods. *Source:* Mishyna et al. (2020) / With permission of Elsevier.

ingredients represents this class of alternative food sources (Nachal et al. 2019). Although 3D printing offers food in customized shape and attractive colour, the origin, and the source of food ingredients from alternative proteins displease the consumers. In addition to disgust feeling, the feeling of food neophobia greatly restricts the consumers to try novel cuisines from new food sources (Derossi et al. 2020). Further, the concerns related to safety, edibility, and allergenicity outrages the negative perception of acceptability of foods from alternative ingredients. More than taste and nutritional aspects, the individual preferences would have significant implications on the developed product out of novel food technology.

15.5.5.2 Meat Analogues and 3D Printing

Considering the hesitance of cultured meat and the applicability of alternative ingredients like insects among the consumers, food scientists had come up with the idea of the development of meat substitutes. The meat substitutes also known as meat analogues are the plant and microbial-based foods that mimic the sensorial and textural attributes of real meat from an animal source (Malav et al. 2015). Foods that are used as meat substitutes are soy, wheat gluten, jackfruit, tofu, tempeh, lentils, beans, chickpeas, and mushrooms (Kyriakopoulou et al. 2019). All these ingredients are successfully demonstrated to mimic the unique texture of animal meat tissues. Interestingly, the consumer perception of meat analogues is totally different with few had liked it but most of them refused to accept meat substitutes. The reluctance of accepting meat analogues showed consumer's food opinion to reject the meat substitutes when there is a choice available for animal-based meat for consumption (Boukid 2020). All these results show diverse perceptions of consumers in accepting cultured meat, alternative proteins, and meat analogues. Regardless all these concepts focused on the efficient utilization of material resources and had a sustainable solution to conserve future food security (Siegrist and Hartmann 2020). In this regard, 3D printing is a print-and-eat technology that utilizes all these three kinds of materials for printing. Different strategies and methods of pre-processing can be adopted to ensure the printability of these novel food ingredients. Of course, the product printed out of these materials also suffers from the diverse food opinions and preferences of consumers. Thus, the success of any novel food technology is highly influenced by the quality and marketability that in turn is decided by the acceptability of consumers.

15.5.6 Presumption and Outcomes of Novel Food Technologies

With increasing consumer demand for novel foods, recent innovations of food science and gastronomy focused on the exploration of novel, non-thermal, sustainable food technologies to deliver nutritious foods without compromising the sensory aspects (Moses et al. 2014). The ultimate aim of any of the proposed agri-food technology is ensuring the delivery of safe and healthy foods with sustainability to preserve nutritional security. All these aforementioned technologies are seemed to be disruptive in the food sector that offers socioeconomic and environmental benefits (Nachal et al. 2019). However, the sustenance of the novel technologies is determined by the consumer's acceptability. The negative perceptions of these technologies totally mask the prospective benefits which it offers. Hence, it is the prime role of food technologists and scientists to duly address the debatable queries related to the negative perception of all these novel technologies. This could be feasible through a thorough understanding of the chemical transitions, associated mechanisms, process outcomes, fate of molecules at gastrointestinal conditions, and bioavailability (Kumar et al. 2020). More clinal trials are yet to be studied for addressing the safety concerns related to short- and long-term effects on consumption of foods prepared from novel techniques. Furthermore, government regulatory and statutory bodies should enforce strict guidelines in the processing of products from these novel technologies with unique labelling that should be transparent to consumers (Baiano 2020). In addition, social media can intervene in promoting awareness and educates the people through the actual process and benefits of novel technologies that could help in changing the consumer's reluctance and misconception about novel food technologies. This implies a significant change in consumer perception, especially to the general public. In this regard, food printing is a nascent emerging technology that has numerous advantages in providing customized food and personalized nutrition in a convenient cost-effective, and sustainable approach. However, more research needs are required to answer the questions related to the safety, edibility, allergenicity, and toxicity of 3D printed foods (Tran 2019). This would significantly open up a newer perception among the consumers that assist in driving out of food neophobia and aversion to the novel foods.

15.6 Intervention Tools for Enhancing Consumer Knowledge

The consumer's eating behaviour and dietary practices must be understood to determine the perceived risks of the novel technology and developed food products. 3D printing is a versatile technology that provides a greater degrees of freedom in the production of foods (Nachal et al. 2019). This is evident with the use of a diverse range of food ingredients from traditional to novel alternative ingredients with appropriate pre-processing that further allows us to deliver it in desired shapes and forms. 3D printing not only remains a sustainable solution for the emerging resources crisis but also helps in satisfying the nutritional and sensory requirements of consumers (Krishnaraj et al. 2019). As expected, 3D printing is a more welcoming technology in the food sector for both manufacturers and consumers; however, it does have few concerns related to safety and other regulations on 3D printed foods. As 3D printing is in its initial stage of development, it has a far way to go for a better understanding of the materials flow in the food supply and distribution chain. Major concerns that are put forth by consumers to resist the 3D printed foods are safety issues in accepting product developed from uncommon food ingredients, food neophobia, long-term side effects, and illness (Tuorila and Hartmann 2020). All these issues impart a negative perception of 3D printing which could be overcome through proper education and spreading awareness among the consumers. In this regard, the social media and public forum platforms should work together with the food scientists and technologists in educating the general public about the merits and prospects of 3D printing technology. This should be supported by proper labelling and regulations that leave the choice of selection of 3D printed products to the consumers (Baiano 2020). For instance, a clear labelling ensures the transparency of ingredients used in the product that has options for consumers to choose either 3D printed food from conventional common food ingredients or 3D printed food from alternative novel ingredients. This could be possible only if the consumers have the ability to distinguish the products for which a prior knowledge is required. Hence the present section gives a brief note on various interventions that could be used as communication tools in creating the awareness among consumers about 3D printing of foods.

15.6.1 Business Schemes and Public Policies

Any novel technology is criticized for favourable and unfavourable outcomes. In this regard, 3D printing has a favourable outcome in terms of customization and personalization; however, the unfavourable outcomes are risks associated with the shelf-life and safety concerns of the 3D-printed foods (Tran 2019). This broadly creates misconceptions among the consumers that can be eliminated by highlighting the fact that 3D printing offers on-demand production and delivery of foods. Further, the existing practice of centralized production can be transformed into a localized production straightaway on the consumption point that greatly reduces the warehouse, transportation, and packaging cost (Laplume et al. 2016). These features would have a significant impact on the existing food supply chain as 3D printing reduces its complexity. Also, 3D printing seems to be a competing technology for the traditional food manufacturing practices since it is a convenient technology dominated by the prosumer-driven food market (Halassi et al. 2019). Accordingly, the analysis of the 3D printing market and consumer acceptability would help in framing new business strategies and marketing policies that in turn assists in gaining consumer's trust over 3D printing and 3D printed foods.

15.6.2 Social Media and Communication

The dissemination of information in social media and other communication platforms would have a great impact on consumer's perception. The messages on benefits and advantages of new technology might be perceived differently by the consumers due to their individual knowledge on science, prior experience, peer review, and other personal factors such as age, gender, and individual interest (Stein 2017). Thus, the disseminated information about the technology will leave a great impression on consumer's food decisions which would severely impact trading relations. As a multifaceted approach, 3D printing encompasses multiple operations in a single step that comprises designing, material processing, printing, and post-processing (Theagarajan et al. 2020). Hence, the consumers must have prior knowledge or proper training for handling the 3D printer is adequate in view of analyzing the consumer's perception towards 3D printed foods. Thus, social media could be used as a potential tool in creating awareness by providing proper training to acquire knowledge about the 3D printing process.

15.6.3 Academia and Scientific Events

One of the interesting applications of 3D printing is the use of 3D food printing itself as a teaching and communication tool in daycares, schools, and colleges. In this regard, 3D printing would have a great impact and draws the attention of the student groups and academic professionals in exploring the advantage of 3D printing (Drury et al. 2019). In promoting these features, food scientists and industrial firms should conduct academic forums and scientific events that encourage people of all age groups from diverse backgrounds to take part in the practical demonstration of working principles and the state-of-the-art of 3D printing of foods. These approaches could reach a wide range of audiences all over the world and assists in exploring the new feasibilities with 3D printing in the delivery of functional foods and nutraceuticals (Stein 2017). Thus, conference proceedings, seminars, and workshops can be powerful platforms in the dissemination of the usefulness of 3D printing of foods to a broad group of audience that has a significant impact on their food decisions.

15.6.4 Internet and e-commerce

With the advent of information and technology, most food services are made easier and reachable to consumers at their doorsteps. Although most of the trading and business policies are becoming online, the lack of its organization complicates in accessing the benefits by the consumers. Recently, the modern digitalized world witnessed the tremendous growth of internet and web services (Rogers et al. 2016). Without much effort, the consumers can now easily browse the required information in an accurate and unbiased manner about the technology and the product. Thus, the food industries and service providers must make use of these e-shopping features by providing all the required information about the product transparently. This could make consumers know the actual fact and reality of both benefits and risks associated with the technology of 3D printing. Further e-commerce and online shopping require less effort by the consumer to access the technology that assists in providing a variety of choices to the consumers to be judged at a time under the comfort of their residence (Jayaprakash et al. 2019). To enjoy these features, efforts must be taken by the food industrial firms in regulating the trading of 3D designs and e-recipes. The government regulatory bodies should enforce strict regulations to prevent piracy and copyright issues related to 3D models and printing recipes (Nachal et al. 2019). Ultimately, proper usage of the internet as a communication tool would help in gaining the consumer's trust over the 3D printed foods thereby the technology becomes more familiar among the general public.

15.7 Trends, Advancements, and Future Directions

Over the past years, the food industry and manufacturing practices have undergone several dimensional changes in view of delivering foods with consumer satisfaction. In this regard, molecular gastronomy is the emerging area in the food sector that implies the science of searching knowledge, i.e. making use of the advantage of physical and chemical transformations and underlying mechanisms during the preparation and consumption of foods (D'Angelo et al. 2016). Molecular gastronomy led to digitalized cooking practices of molecular cuisine and note-by-note cuisine that are based on applied scientific principles of food scientists and culinary chefs (Cifci et al. 2020). The eating experience and taste preference of the consumers will help in determining the acceptability and marketability of the new product. Food printing is an integration of digital gastronomy and 3D printing that has the potential to transform the food production process (Anukiruthika et al. 2020a). With the emergence of alternative ingredients and advancements in food printing, novel 3D printed foods can be fabricated based on molecular and note-by-note cuisines. These approaches are known as science-based cooking that has a great scope in addressing malnourishment and global hunger without compromising the sensorial attributes. Even though science-based cooking proved to provide promising solutions, the greatest hindrance of its implementation is the consumer's fear of the presence of 'chemicals in dish' (This 2013). This misconception of consumers remains a major drawback that can be overcome by educating the general public with the basic concept of 3D food printing. Understanding consumer's attitudes towards novel foods can help to tackle food neophobia and allows for effective marketing of developed food. Most of the novel foods are met with adverse individual reactions such as disgust feeling, unwillingness, and food neophobia that would ultimately lead to product failure (Jaeger et al. 2021). Therefore, consumer's attitude in food choices and preferences are most important and given a high priority in designing food product (Figure 15.5).

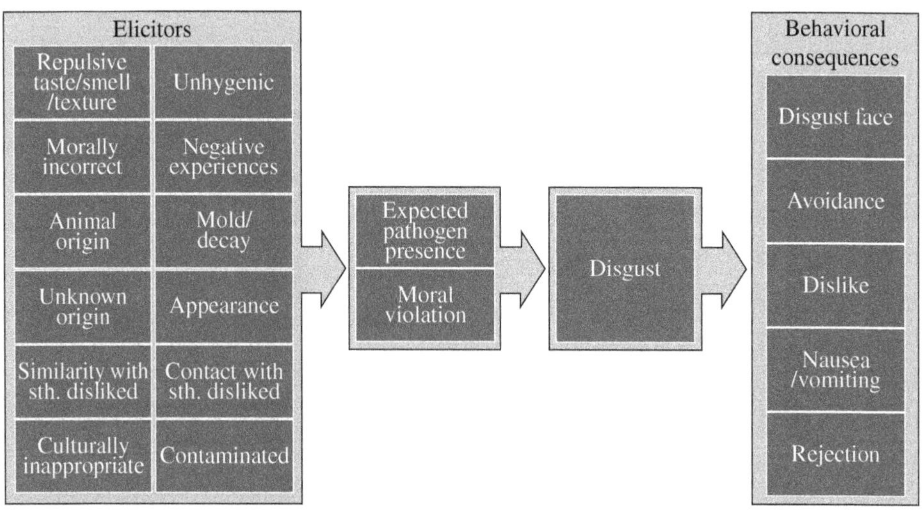

Figure 15.5 Conceptual model showing influence of potential food-related disgust elicitors (left) on behavioural consequences (right). *Source:* Tuorila and Hartmann (2020) / With permission of Elsevier.

Molecular cuisine is based on the application of scientific principles and innovative tools in the fabrication of high-quality foods. Unlike the traditional cooking techniques, it utilizes a technology-based laboratory environment in the material formulation and food preparations (Cifci et al. 2020). Thus, molecular cuisine is defined as the innovative application of scientific knowledge of molecular gastronomy in developing new techniques, methods, and formulations for the fabrication of novel foods (This 2013). While note-by-note cuisine refers to the technique used for the fabrication of food using a pure compound of macro and micro food constituents (Caporaso and Formisano 2016). The science-based cooked foods not only satisfy hunger but also fascinates by spreading positiveness through imparting different eating experience. Thus, 3D printing can be categorized as science-based cooking that not only provides nutrition but also delivers foods in attractive shapes and desired forms. Overall, the digital gastronomy approaches of food preparations improve the existing cuisines, allows for the development of new methods of cooking (like 3D printing), and imparts an entirely different eating experience by developing superior foods (Anukiruthika et al. 2020a). The science-based cooking works are based on the knowledge of scientists and food technologists combined with physicochemical properties of raw materials along with the inputs of culinary professionals, artists, and chefs in developing food recipes. As a futuristic technology, 3D food printing will soon step up to the next level in the fabrication of foods based on the note-by-note cuisine concept (Figure 15.6). The adoption of note-by-note cuisine in 3D printing led to scientific advancement and remains as

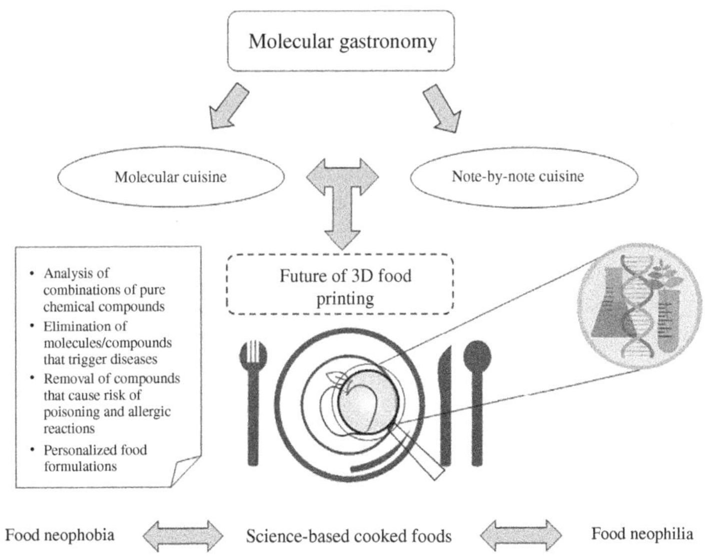

Figure 15.6 Illustration of integration of science-based cooked foods and 3D printed foods.

break-through technology in food biotechnology (Kumar et al. 2020). It allows for the fabrication of functional and personalized 3D printed foods from pure chemical compounds obtained from fractions of plants and animals without using wholesome foods like fruits, vegetables, meat, etc. The main aim is not to mimic or recreate the existing food products but to create new foods with modified textures, new tastes, and pleasant flavours. Various sensorial elements such as texture, colour, taste, odour, pungency, and trigeminal stimulation can be modulated with precise control over macro and micronutrients. Various advantages of these combined concepts are as follows: development of new formulations and flavours by testing various combinations of pure food constituents; elimination and separation of molecules or compounds that cause/ trigger diseases; removal of toxic substances that are responsible for food poisoning and allergic reactions; and development of personalized foods based on metabolic characteristics of the individuals (Cifci et al. 2020). Thus, 3D printing of foods has a numerous benefits and applicability in designing of novel functional foods in customized and personalized manner. Hence, 3D printing is such a potential technology for the future food industry that should not be neglected by consumers through misconception and individual perceptions. It is the role of food technologists in unwinding the scope of 3D printing by conducting a systematic study to relate 3D printing and consumer's attitude towards the acceptability of novel foods.

15.8 Conclusion

3D printing is an emerging novel technology based on science-based cooking techniques that allow for controlled deposition of material with desired calories and nutrients. The success of any novel food technology can be determined by the consumer's acceptability. Hence, the analysis of consumer's attitude towards 3D printing for assessing food choices and individual preferences are most important. The present chapter summarizes the market trends of 3D printing and consumer preferences for 3D printed foods. The consumer reactions to novel technology and food products are discussed briefly in context with 3D printing. Various factors that remain as metal barriers for acceptance of 3D printing from the consumer's viewpoint are food neophobia, unnaturalness, safety, and edibility. These characteristics negatively influence the growth of food 3D printing. The aforementioned misconceptions of 3D printed foods can be overcome by proper education and awareness among the consumers. In this regard, various approaches used for measuring the consumer's food choice and preferences were presented along with the intervention tools for enhancing the knowledge about 3D printing among the consumers. Internet, social media, scientific forums, conferences, seminars, and workshops are some of the powerful intervening tools to reach a wide range of audiences of different backgrounds for creating awareness about the prospects of 3D printing. To date no such studies available on evaluating the consumer's preference and attitude in terms of ethical, social, cultural, environmental, and personal values. Hence future studies must be focused on this direction that would bridge up the gap between the consumer's perception and acceptability of novel food technology (3D printing). Various trends and advancements of molecular gastronomy (molecular cuisine and note-by-cuisine) in context with 3D printing were described in detail. Thus, the success of 3D printing not only relies on technical progress but also on consumer's attitudes and preferences. Hence, qualitative and quantitative studies on assessing consumer's purchasing behaviour and dietary choices would lead to novel business strategies and marketing policies which can be used for evaluating the success rate of 3D printing and provides an overview of the potential of its commercialization in the future.

References

Ali, S., Nawaz, M.A., Ghufran, M. et al. (2021). GM trust shaped by trust determinants with the impact of risk/benefit framework: the contingent role of food technology neophobia. *GM Crops and Food* **12** (1): 170–191.

Alley, T.R. (2018). Conceptualization and measurement of human food neophobia. In: *Food Neophobia* (ed. S. Reilly), pp. 169–192. United Kingdom: Woodhead Publishing.

Anabtawi, O., Swift, J.A., Hemmings, S. et al. (2020). Perceived healthiness of food items and the traffic light front of pack nutrition labelling: choice-based conjoint analysis and cross-sectional survey. *Journal of Human Nutrition and Dietetics* **33** (4): 487–495.

Anukiruthika, T., Moses, J.A., and Anandharamakrishnan, C. (2020a). 3D printing of egg yolk and white with rice flour blends. *Journal of Food Engineering* **265**: 109691. https://doi.org/10.1016/j.jfoodeng.2019.109691.

Anukiruthika, T., Moses, J.A., and Anandharamakrishnan, C. (2020b). Electrohydrodynamic drying of foods: principle, applications, and prospects. *Journal of Food Engineering* **295**: 110449.

Aschemann-Witzel, J., Varela, P., and Peschel, A.O. (2019). Consumers' categorization of food ingredients: do consumers perceive them as 'clean label'producers expect? An exploration with projective mapping. *Food Quality and Preference* **71**: 117–128.

Baiano, A. (2020). 3D printed foods: a comprehensive review on technologies, nutritional value, safety, consumer attitude, regulatory framework, and economic and sustainability issues. *Food Reviews International*, pp. 1–31.

Barrena, R. and Sánchez, M. (2013). Neophobia, personal consumer values and novel food acceptance. *Food Quality and Preference* **27** (1): 72–84.

Battacchi, D., Verkerk, R., Pellegrini, N. et al. (2020). The state of the art of food ingredients' naturalness evaluation: a review of proposed approaches and their relation with consumer trends. *Trends in Food Science and Technology* **106**: 434–444.

Bearth, A. and Siegrist, M. (2019). "As long as it is not irradiated"–Influencing factors of US consumers' acceptance of food irradiation. *Food Quality and Preference* **71**: 141–148.

Behrens, J.H., Barcellos, M.N., Frewer, L.J. et al. (2009). Brazilian consumer views on food irradiation. *Innovative Food Science and Emerging Technologies* **10** (3): 383–389.

Bhushani, J.A. and Anandharamakrishnan, C. (2014). Electrospinning and electrospraying techniques: potential food based applications. *Trends in Food Science and Technology* **38** (1): 21–33.

Bolek, S. (2020). Consumer knowledge, attitudes, and judgments about food safety: a consumer analysis. *Trends in Food Science and Technology* **102**: 242–248.

Bonny, S.P.F., Gardner, G.E., Pethick, D.W., and Hocquette, J.F. (2015). What is artificial meat and what does it mean for the future of the meat industry? *Journal of Integrative Agriculture* **14** (2): 255–263. https://doi.org/10.1016/S2095-3119(14)60888-1.

Boukid, F. (2020). Plant-based meat analogues: from niche to mainstream. *European Food Research and Technology* **247**: 1–12.

Brunner, T.A., Delley, M., and Denkel, C. (2018). Consumers' attitudes and change of attitude toward 3D-printed food. *Food Quality and Preference* **68**: 389–396.

Caporaso, N. and Formisano, D. (2016). Developments, applications, and trends of molecular gastronomy among food scientists and innovative chefs. *Food Reviews International* **32** (4): 417–435.

Carrillo, E., Varela, P., Salvador, A., and Fiszman, S. (2011). Main factors underlying consumers' food chocie: a first step for understanding of attitude towards "Healthy Eating". *Journal of Sensory Studies* **26** (2): 85–95.

Castell-Perez, M.E. and Moreira, R.G. (2021). Irradiation and consumers acceptance. *Innovative Food Processing Technologies*, pp. 122–135.

Chen, C.-H., Wang, C.-L., and Chen, P.-Y. (2018). Performance evaluation of the service industry innovation research program: the application of a means-end chain. *Technology in Society* **54**: 111–119.

Chen, Q., Anders, S., and An, H. (2013). Measuring consumer resistance to a new food technology: a choice experiment in meat packaging. *Food Quality and Preference* **28** (2): 419–428.

Cifci, I., Demirkol, S., Altunel, G.K., and Cifci, H. (2020). Overcoming the food neophobia towards science-based cooked food: the supplier perspective. *International Journal of Gastronomy and Food Science* **22**: 100280.

Cooke, L. (2018). Genetic and environmental influences on food neophobia. In: *Food Neophobia* (ed. S. Reilly), pp. 237–254. United Kingdom: Woodhead Publishing.

Crowley, O.V., Marquette, J., Reddy, D., and Fleming, R. (2013). Factors predicting likelihood of eating irradiated meat. *Journal of Applied Social Psychology* **43** (1): 95–105.

Cunha, L.M., Cabral, D., Moura, A.P., and de Almeida, M.D.V. (2018). Application of the food choice questionnaire across cultures: systematic review of cross-cultural and single country studies. *Food Quality and Preference* **64**: 21–36.

D'Angelo, G., Hansen, H.N., and Hart, A.J. (2016). Molecular gastronomy meets 3D printing: layered construction via reverse spherification. *3D Printing and Additive Manufacturing* **3** (3): 152–159.

Dankar, I., Haddarah, A., Omar, F.E.L. et al. (2018). 3D printing technology: the new era for food customization and elaboration. *Trends in Food Science and Technology* **75**: 231–242. https://doi.org/10.1016/j.tifs.2018.03.018.

Derossi, A., Husain, A., Caporizzi, R., and Severini, C. (2020). Manufacturing personalized food for people uniqueness. An overview from traditional to emerging technologies. *Critical Reviews in Food Science and Nutrition* **60** (7): 1141–1159.

Drury, S.A.M., Rush, R.A., Wilder, S.E., and Wysocki, L.M. (2019). Encouraging bridges: connecting science students to public problem-solving through science communication. In: *Communication in Chemistry* (eds. G.L. Crawford, K.D. Kloepper, J.J. Meyers and R.H. Singiser), pp. 263–280. Wahington, DC: ACS Publications.

Duncan, T.V. (2011). The communication challenges presented by nanofoods. *Nature Nanotechnology* **6** (11): 683–688.

Emergen Research (2021). 3D food printing market size to reach USD 1,015.4 million in 2027. Intrado GlobeNewswire. http://www.globenewswire.com/news-release/2021/03/10/2190536/0/en/3D-Food-Printing-Market-Size-to-Reach-USD-1-015-4-Million-in-2027-Rising-Demand-for-Customized-Food-Products-and-Increasing-Adoption-in-Various-End-Use-Industries-are-Some-Key-Fact.html (accessed 11 December 2020).

Erdem, S. and Rigby, D. (2013). Investigating heterogeneity in the characterization of risks using best worst scaling. *Risk Analysis* **33** (9): 1728–1748.

EU Regulation (2015). Regulation (EU) 2015/2283. https://eur-lex.europa.eu/legal-content/en/TXT/PDF/?uri=CELEX:32015R2283&from=EN (accessed 15 December 2020).

Feng, C., Zhang, M., Bhandari, B., and Ye, Y. (2020). Use of potato processing by-product: effects on the 3D printing characteristics of the yam and the texture of air-fried yam snacks. *LWT* **125**: 109265.

Guine, R., Lima, M.J., Pato, L. et al. (2010). Consumer study and sensorial evaluation of a newly developed spicy strawberry syrup. *International Journal of Academic Research* **2** (3): 173–178.

Halassi, S., Semeijn, J., and Kiratli, N. (2019). From consumer to prosumer: a supply chain revolution in 3D printing. *International Journal of Physical Distribution & Logistics Management* **49**: 200–216.

He, C., Zhang, M., and Fang, Z. (2019). 3D printing of food: pretreatment and post-treatment of materials. *Critical Reviews in Food Science and Nutrition* **60** (14): 2379–2392.

Jaeger, S.R., Roigard, C.M., Hunter, D.C., and Worch, T. (2021). Importance of food choice motives vary with degree of food neophobia. *Appetite* **159**: 105056.

Jaeger, S.R., Roigard, C.M., Le Blond, M. et al. (2019). Perceived situational appropriateness for foods and beverages: consumer segmentation and relationship with stated liking. *Food Quality and Preference* **78**: 103701.

Jayaprakash, S., Ituarte, I.F., and Partanen, J. (2019). Prosumer-driven 3D food printing: role of digital platforms in future 3D food printing systems. In: *Fundamentals of 3D Food Printing and Applications* (eds. F.C. Godoi, B.R. Bhandari, S. Prakash and M. Zhang), pp. 331–354. United Kingdom: Academic Press.

Handral, H.K., Tay, S.H., Chan, W.W., and Choudhury, D. (2020). 3D printing of cultured meat products. *Critical Reviews in Food Science and Nutrition*, pp. 1–10.

Karlgaard, R. (2011). 3D printing will revive american manufacturing. Forbes. https://www.forbes.com/sites/richkarlgaard/2011/06/23/3d-printing-will-revive-american-manufacturing/?sh=4bc5b7e82548 (accessed 20 November 2020).

Keerthana, K., Anukiruthika, T., Moses, J.A., and Anandharamakrishnan, C. (2020). Development of fiber-enriched 3D printed snacks from alternative foods: a study on button mushroom. *Journal of Food Engineering* https://doi.org/10.1016/j.jfoodeng.2020.110116.

Kim, S.-H. and Jeon, H.-M. (2020). Chocolate choice motives and attitudes in foodservice market: fine store product vs. manufactured product consumers. *Journal of Foodservice Business Research* **23** (2): 149–168.

Kim, Y.G., Jang, S.Y., and Kim, A.K. (2014). Application of the theory of planned behavior to genetically modified foods: moderating effects of food technology neophobia. *Food Research International* **62**: 947–954.

Krishnaraj, P., Anukiruthika, T., Choudhary, P. et al. (2019). 3D extrusion printing and post-processing of fibre-rich snack from indigenous composite flour. *Food and Bioprocess Technology* **12** (10): 1776–1786. https://doi.org/10.1007/s11947-019-02336-5.

Kronberger, N., Wagner, W., and Nagata, M. (2014). How natural is "more natural"? The role of method, type of transfer, and familiarity for public perceptions of cisgenic and transgenic modification. *Science Communication* **36** (1): 106–130.

Kumar, S., Anukiruthika, T., Dutta, S. et al. (2020). Iron deficiency anemia: a comprehensive review on iron absorption, bioavailability and emerging food fortification approaches. *Trends in Food Science and Technology* https://doi.org/10.1016/j.tifs.2020.02.021.

Kyriakopoulou, K., Dekkers, B., and van der Goot, A.J. (2019). Plant-based meat analogues. In: *Sustainable Meat Production and Processing*. Elsevier Inc. https://doi.org/10.1016/b978-0-12-814874-7.00006-7.

La Barbera, F., Verneau, F., Amato, M., and Grunert, K. (2018). Understanding Westerners' disgust for the eating of insects: the role of food neophobia and implicit associations. *Food Quality and Preference* **64**: 120–125.

Laplume, A.O., Petersen, B., and Pearce, J.M. (2016). Global value chains from a 3D printing perspective. *Journal of International Business Studies* **47** (5): 595–609.

Laureati, M., Spinelli, S., Monteleone, E. et al. (2018). Associations between food neophobia and responsiveness to "warning" chemosensory sensations in food products in a large population sample. *Food Quality and Preference* **68**: 113–124.

Leena, M., Yoha, K.S., Moses, J.A. et al. (2019). Electrospraying and spinning techniques fabrication and its potential applications. In: *Food Nanotechnology: Principles and Applications* (eds. C. Anandharamakrishnan and S. Parthasarathi), 187–216. New York: CRC Press Taylor & Francis group.

Li, P., Mellor, S., Griffin, J. et al. (2014). Intellectual property and 3D printing: a case study on 3D chocolate printing. *Journal of Intellectual Property Law and Practice* **9** (4): 322–332.

Luczaj, L., Pieroni, A., Tardio, J. et al. (2012). Wild food plant use in 21st century Europe, the disapperance of old traditions and the search for new ciusines involving wild edibles. *Acta Societatis Botanicorum Poloniae* **81** (4): 359–370.

Lunden, S., Hopia, A., Forsman, L., and Sandell, M. (2020). Sensory and conceptual aspects of ingredients of sustainable sources - Finnish consumers' opinion. *Foods* **9** (11): 1669.

Lupton, D. and Turner, B. (2018a). Food of the future? Consumer responses to the idea of 3D-printed meat and insect-based foods. *Food and Foodways* **26** (4): 269–289.

Lupton, D. and Turner, B. (2018b). "I can't get past the fact that it is printed": consumer attitudes to 3D printed food. *Food, Culture and Society* **21** (3): 402–418.

Lusk, J.L. and McCluskey, J. (2018). Understanding the impacts of food consumer choice and food policy outcomes. *Applied Economic Perspectives and Policy* **40** (1): 5–21.

Lusk, J.L., Roosen, J., and Bieberstein, A. (2014). Consumer acceptance of new food technologies: causes and roots of controversies. *Annual Review of Resource Economics* **6** (1): 381–405.

MacRitchie, L.A., Hunter, C.J., and Strachan, N.J.C. (2014). Consumer acceptability of interventions to reduce Campylobacter in the poultry food chain. *Food Control* **35** (1): 260–266.

Mahalakshmi, L., Leena, M.M., Moses, J.A., and Anandharamakrishnan, C. (2020). Micro-and nano-encapsulation of β-carotene in zein protein: size-dependent release and absorption behavior. *Food and Function* **11** (2): 1647–1660.

Malav, O.P., Talukder, S., Gokulakrishnan, P., and Chand, S. (2015). Meat analog: a review. *Critical Reviews in Food Science and Nutrition* **55** (9): 1241–1245. https://doi.org/10.1080/10408398.2012.689381.

Manstan, T. and McSweeney, M.B. (2020). Consumers' attitudes towards and acceptance of 3D printed foods in comparison with conventional food products. *International Journal of Food Science and Technology* **55** (1): 323–331.

Markovina, J., Stewart-Knox, B.J., Rankin, A. et al. (2015). Food4Me study: validity and reliability of Food Choice Questionnaire in 9 European countries. *Food Quality and Preference* **45**: 26–32.

Mishyna, M., Chen, J., and Benjamin, O. (2020). Sensory attributes of edible insects and insect-based foods-Future outlooks for enhancing consumer appeal. *Trends in Food Science and Technology* **95**: 141–148.

Monteleone, E., Spinelli, S., Dinnella, C. et al. (2017). Exploring influences on food choice in a large population sample: the Italian Taste project. *Food Quality and Preference* **59**: 123–140.

Moses, J.A., Norton, T., Alagusundaram, K., and Tiwari, B.K. (2014). Novel drying techniques for the food industry. *Food Engineering Reviews* **6** (3): 43–55.

Nachal, N., Moses, J.A., Karthik, P., and Anandharamakrishnan, C. (2019). Applications of 3D printing in food processing. *Food Engineering Reviews* **11** (3): 123–141. https://doi.org/10.1007/s12393-019-09199-8.

Nida, S., Anukiruthika, T., Moses, J.A., and Anandharamakrishnan, C. (2020). 3D printing of grinding and milling fractions of rice husk. *Waste and Biomass Valorization*: 1–10. https://doi.org/10.1007/s12649-020-01000-w.

Onwezen, M.C., Reinders, M.J., Verain, M.C.D., and Snoek, H.M. (2019). The development of a single-item Food Choice Questionnaire. *Food Quality and Preference* **71**: 34–45.

Pearcey, S.M. and Zhan, G.Q. (2018). A comparative study of American and Chinese college students' motives for food choice. *Appetite* **123**: 325–333.

Piha, S., Pohjanheimo, T., Lahteenmaki-Uutela, A. et al. (2018). The effects of consumer knowledge on the willingness to buy insect food: an exploratory cross-regional study in Northern and Central Europe. *Food Quality and Preference* **70**: 1–10.

Pliner, P. and Hobden, K. (1992). Development of a scale to measure the trait of food neophobia in humans. *Appetite* **19** (2): 105–120.

Post, M.J., Levenberg, S., Kaplan, D.L. et al. (2020). Scientific, sustainability and regulatory challenges of cultured meat. *Nature Food* **1** (7): 403–415.

Ricci, I., Derossi, A., and Severini, C. (2019). 3D printed food from fruits and vegetables. In: *Fundamentals of 3D Food Printing and Applications* (vol. 153, iss. 1), pp. 117–149. Elsevier Ltd https://doi.org/10.1016/b978-0-12-814564-7.00005-5.

Rogers, H., Baricz, N., and Pawar, K.S. (2016). 3D printing services: classification, supply chain implications and research agenda. *International Journal of Physical Distribution and Logistics Management* **46**: 886–907.

Roman, S., Sanchez-Siles, L.M., and Siegrist, M. (2017). The importance of food naturalness for consumers: results of a systematic review. *Trends in Food Science and Technology* **67**: 44–57.

Ronteltap, A., Van Trijp, J.C.M., Renes, R.J., and Frewer, L.J. (2007). Consumer acceptance of technology-based food innovations: lessons for the future of nutrigenomics. *Appetite* **49** (1): 1–17.

Schlup, Y. and Brunner, T. (2018). Prospects for insects as food in Switzerland: a tobit regression. *Food Quality and Preference* **64**: 37–46.

Scott, S.E. and Rozin, P. (2017). Are additives unnatural? Generality and mechanisms of additivity dominance. *Judgment and Decision Making* **12** (6): 572.

Severini, C., Azzollini, D., Albenzio, M., and Derossi, A. (2018). On printability, quality and nutritional properties of 3D printed cereal based snacks enriched with edible insects. *Food Research International* **106** (November 2017): 666–676. https://doi.org/10.1016/j.foodres.2018.01.034.

Siegrist, M. and Hartmann, C. (2020). Consumer acceptance of novel food technologies. *Nature Food* **1** (6): 343–350.

Siegrist, M. and Sutterlin, B. (2017). Importance of perceived naturalness for acceptance of food additives and cultured meat. *Appetite* **113**: 320–326.

Soucier, V.D., Doma, K.M., Farrell, E.L. et al. (2019). An examination of food neophobia in older adults. *Food Quality and Preference* **72**: 143–146.

Stein, J.A. (2017). The political imaginaries of 3D printing: prompting mainstream awareness of design and making. *Design and Culture* **9** (1): 3–27.

Sun, J., Zhou, W., Huang, D., and Yan, L. (2018). 3D food printing: perspectives. In: *Polymers for Food Applications* (ed. T.J. Gutierrez), pp. 725–755. Argentina: Springer International Publishing.

Theagarajan, R., Moses, J.A., and Anandharamakrishnan, C. (2020). 3D extrusion printability of rice starch and optimization of process variables. *Food and Bioprocess Technology*. https://doi.org/10.1007/s11947-020-02453-6.

This, H. (2013). Celebrate chemistry (recent results of molecular gastronomy). *Chemistry* **398** (7): 3139–3153.

Tomiyama, A.J., Kawecki, N.S., Rosenfeld, D.L. et al. (2020). Bridging the gap between the science of cultured meat and public perceptions. *Trends in Food Science and Technology* **104**: 144–152.

Tran, J.L. (2019). Safety and labelling of 3D printed food. In: *Fundamentals of 3D Food Printing and Applications* (eds. F.C. Godoi, B.R. Bhandari, S. Prakash and M. Zhang), pp. 355–371. United Kingdom: Academic Press.

Tuorila, H. and Hartmann, C. (2020). Consumer responses to novel and unfamiliar foods. *Current Opinion in Food Science* **33**: 1–8.

Uribe-Wandurraga, Z.N., Zhang, L., Noort, M.W.J. et al. (2020). Printability and physicochemical properties of microalgae-enriched 3D-printed snacks. *Food and Bioprocess Technology* **13** (11): 2029–2042.

Wilson, A., Anukiruthika, T., Moses, J.A., and Anandharamakrishnan, C. (2020). Customized shapes for chicken meat–based products: feasibility study on 3D-printed nuggets. *Food and Bioprocess Technology* **13** (11): 1968–1983.

Zhang, Y., Jing, L., Bai, Q. et al. (2018). Application of an integrated framework to examine Chinese consumers' purchase intention toward genetically modified food. *Food Quality and Preference* **65**: 118–128.

16

Safety, Challenges, and Research Needs

CHAPTER MENU

16.1 Introduction

3D printing is a fascinating technology where the printing materials are subjected to pre-processing for enhancing the formation of a 3D object in a layer-by-layer manner through appropriate phase transitions and solidification (Liu et al. 2017). The advent of 3D printing is well established in allied industrial sectors such as construction, aerospace, automobiles, fashion designing, and pharmaceuticals (Nida et al. 2020). In this cohort, recently, the potential of 3D printing is more realized in the food sector for customization and personalization of foods. The 3D printing is well known for sustainability and process efficiency that has several socioeconomic benefits. This also applies to the printing of foods and nutraceuticals delivery (Kumar et al. 2020). However, when talking about the edibility of printed products for human consumption, food safety remains as a question. The 3D food printing is a versatile technology that plays with a combination of novel food ingredients to add on varieties in food cuisines. In this regard, 3D printing is successfully used for exploring

3D Printing of Foods, First Edition. C. Anandharamakrishnan, Jeyan A. Moses, and T. Anukiruthika.

the printability of uncommon and under-utilized food sources such as insects, micro and macro algae, fungi, lupin seeds, grasses, herbs, pseudo-cereals, and food processing by-products (Derossi et al. 2020). Hence, the nature and the source of the ingredients should be transparent to the consumers as some people are subjected to allergies and possess personal eating habits or preferences. As like any other conventionally processed foods that are labelled as vegan, vegetarian, non-vegetarian, gluten-free and lactose free, 3D printed foods must be properly labeled mentioning the list of ingredients, additives used and preservatives if any. Despite the 3D printing is mainly used for on-demand production, the safety and edibility of the food is a significant factor that ensures the shelf-life of the 3D printed foods (Nachal et al. 2019). As the printing of foods is in its nascent stage, most research works on 3D printing of foods are focused on analyzing material characteristics and printability (Keerthana et al. 2020; Theagarajan et al. 2020; Wilson et al. 2020). At this moment, the storage stability of the developed 3D printed foods cannot be overviewed which should be considered as equally important to the printability.

The most important key terminologies of food printing are printability, applicability, and post-processing (Godoi et al. 2016). Among these, post-processing is a significant attribute that ensures the edibility of 3D printed foods. Post-processing involves the combination of cooking steps that imparts sensorial attributes to the raw 3D printed foods (Krishnaraj et al. 2019). The post-processing of foods is quite common for conventional foods; however, when comes to 3D printed foods it is entirely different comprising of the digital cooking process. The digital cooking of 3D printed foods is an active area of research that is under progress where researchers are focusing on the development of novel cooking techniques that ensures cooking with minimal impacts on nutritional attributes and structural stability of the printed constructs (He et al. 2019). Modern cooking methods such as microwave cooking, infrared cooking, and laser cooking are nowadays applied in integration with 3D printing for ensuring the edibility of 3D foods with fewer structural deviations. In this context, the physicochemical interactions of food materials and cooking medium must be studied to assure that there is no generation of by-products or free radicals during post-processing. With this background, these aspects are most important from the consumer end as 'right to know the source and origin of the meals'. The retailer and consumer end questions such as 'what is the expiry of 3D printed foods?', 'whether 3D printed food can be reheated before consumption?', 'How to store the 3D printed foods?', 'Is there any foreign materials or free radicals generated during post-processing?', 'what type of package can be used for 3D printed foods?', and many other similar questions have remained unanswered. In this regard, the success of any technology not only depends on its technical development but also on its implementation with societal and economic implications over the food supply chain and distribution in a long run. The present chapter deals with hidden facts and implementation of food 3D printing in context with safety, challenges, and research needs (Figure 16.1). Various measures to be considered in preserving the storability and edibility of 3D printed foods are discussed. Further, the food safety, labelling, and regulation concerns of 3D printed foods are elaborated at different endpoints of the food supply chain. Certainly, the present chapter throws light on the practical challenges associated with the food safety concerns of 3D printed foods and provides insights on the future outlook and measures that can be undertaken for overcoming the issues in the implementation of food 3D printing technology at the commercial level.

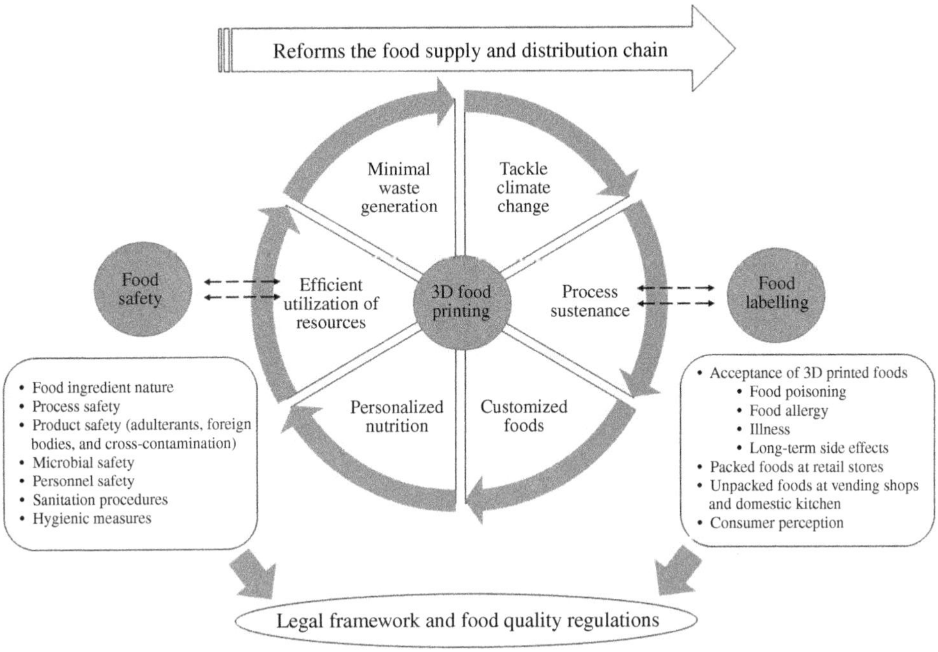

Figure 16.1 Overview of the current status of 3D printing of foods.

16.2 Implications of Food Printing

3D food printing offers several endless possibilities in terms of customization, personalization, and product innovation that solves multiple problems such as food scarcity, climate change, and malnourishment (Nachal et al. 2019). The evolutionary trends of new proteinaceous-based foods in consideration with 3D printing were presented in Figure 16.2. The rearing of animals consumes a larger amount of material resources that imparts greater land pressure. Therefore, the resulting carbon footprints are higher for rearing/ rising beef than pork and chicken (Gerbens-Leenes et al. 2013). Recently, 3D printing has entered into a next dimensional research where researchers successfully have 3D printed meat, i.e. lab-grown artificially cultured meat (Lupton and Turner 2018a). An edible porcine strip of tissue was bio-printed where the droplets of bio-ink were deposited through inkjet technique onto the agarose rods as a substrate for growth (Forgacs et al. 2014). Later, the printed samples were kept inside the bioreactor at controlled conditions to allow for maturation until ready for consumption. The technology of bioprinting could be based on the deposition of biomaterial through micro-extrusion, inkjet printing, and laser-assisted techniques (Salah et al. 2020). When this technology is fully explored, the increased demand of animal protein and environmental impacts would be significantly reduced in near future. In view of sustainability, meat substitutes and meat analogs are becoming popular in recent days. The meat analogues are plant-based and microbial-based food materials that mimic the texture of meat which can be conveniently used as meat substitutes (Malav et al. 2015). With technological advancements, the sensorial properties of animal-based food can be imitated without

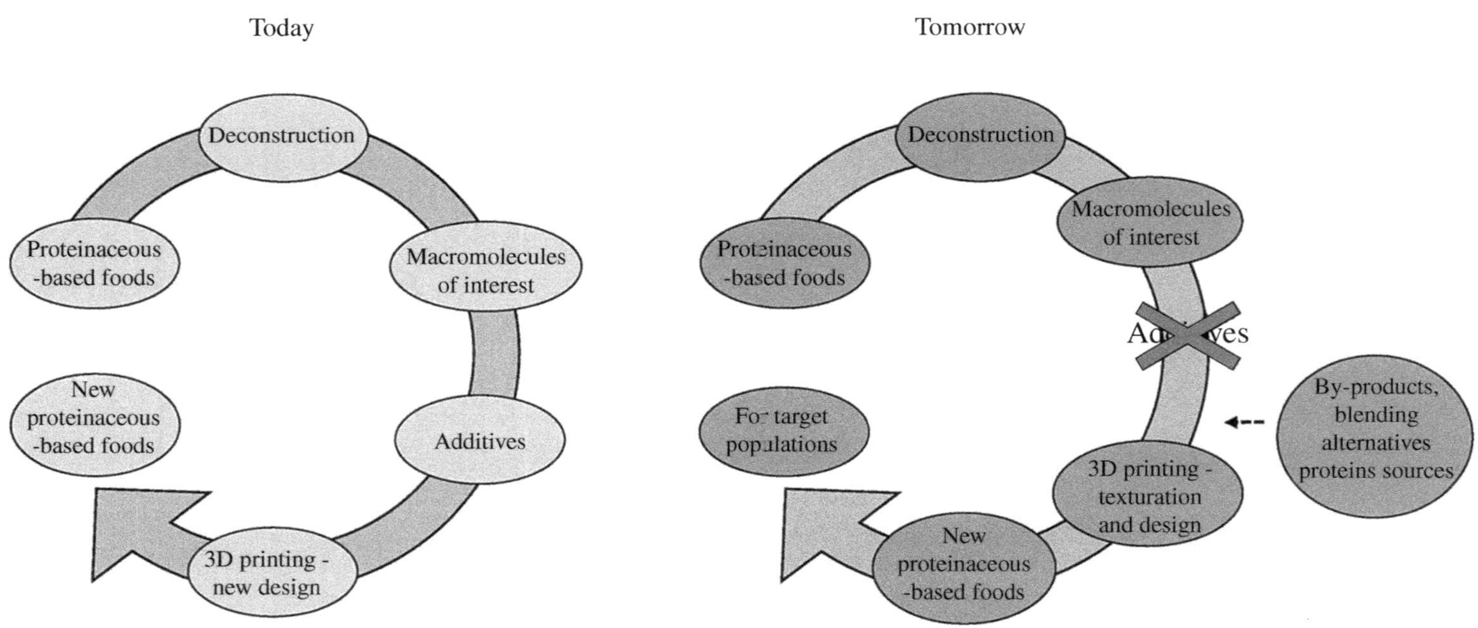

Figure 16.2 Evolutionary trends of new proteinaceous-based foods in consideration with 3D printing. *Source:* Portanguen et al. (2019) / With permission of Elsevier.

compromising the nutritional attributes. Surprisingly, bio-printed meat was relatively accepted by vegetarians as well as those consumers with religious restrictions on meat consumption (Marga 2012). In another report on consumer perception towards cultured meat, most of the people had the mindset that the meat analogues were unnatural and potentially harmful that might cause health implications upon consumption (Lupton and Turner 2018a). 3D printing is well recognized for the preparation of foods from alternative sources that have the potential to significantly reduce environmental impacts, efficient utilization of resources, and overall product life cycle (Nachal et al. 2019). Further, the feeling of neophobia would hinder the development of foods from alternative food sources such as insects, algae, and fungi. Hence, more research needs are required in unwinding the future scope for sustainable foods. Meanwhile, increased awareness about the sustainability and socio-economic benefits of 3D printing among consumers would help in utilizing the maximum technical benefits from 3D food printing technology.

The urbanization and evolving lifestyle influence the consumer perception towards food with increased demand for food varieties, convenient and healthy foods. In this view, 3D printing is a single standalone technology that competes with the existing food processing technologies through its unmatched level of customization (Liu et al. 2017). This would greatly simplify the traditional food supply chain and distribution that gives rise to new business strategies and policies. Further, the advent of Industry 4.0 explores the applications of digitalized technologies in the food sector. 3D printing is a disruptive technology that integrates the multi-step processing of conventional food processing into a single step (Godoi et al. 2016). This greatly reduces material wastage and waste by-product generation during the processing of foods. In contrast, the traditional food processing units generate a considerable amount of by-products that usually goes as waste or less valued to be used for animal feed. Since 3D printing requires minimal usage of raw materials that could be efficiently used for the conversion of these waste by-products into valuable 3D printed foods (Nida et al. 2020). As in this case, the raw material is being recovered from waste, the edibility and microbial safety of the material remains a great concern. This kind of application requires cautious pre-processing to reduce the incidence of microbial load as well as a strenuous printing process to reduce the cross-contamination. Moreover, the thought of a food developed out of waste could have a serious psychological impact on consumers that gives rise to food aversion. Hence more studies on the proper processing of these by-products and the development of safe 3D printed food would help in overcoming the consumer's perception towards food from waste.

Fortunately, food printing is a promissory technology for addressing nutritional deficiency and malnourishment. The most advantageous feature of 3D printing is the delivery of the personalized foods based on individual likeness, preferences, and physiological needs (Theagarajan et al. 2020). Various advantageous and disadvantageous features of 3D food printing technology were presented in Figure 16.3. Several studies were reported on the delivery of nutrients enriched 3D printed foods in customized shapes, tastes, texture, and appearance. Accordingly, food printing can be integrated with encapsulation, electro-spinning, and electro-spraying techniques for the targeted delivery of micronutrients and bioactive compounds (Sun et al. 2018a). In near future, the delivery of personalized foods is possible in addressing nutritional requirements at the individual as well as a large group of populations based on the genetic pool. Advancements in omics technologies such as

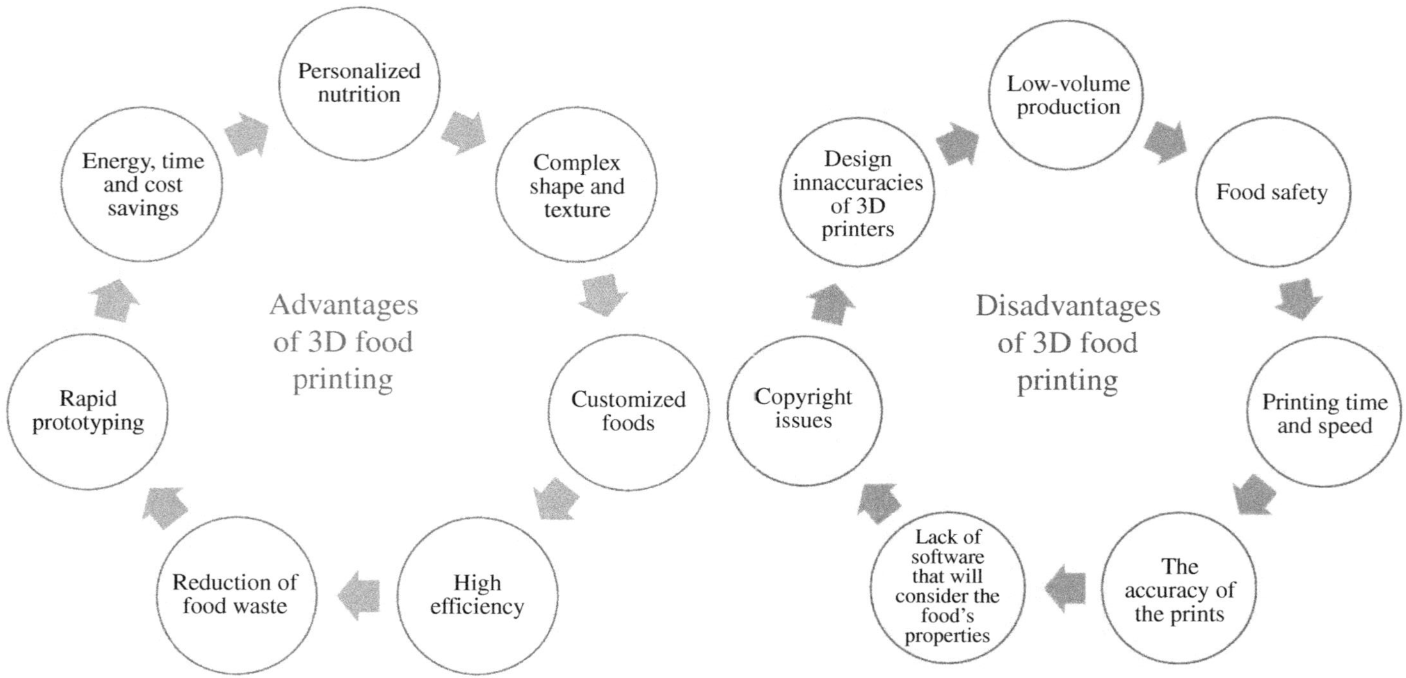

Figure 16.3 Characteristic features of 3D food printing technology. *Source:* Tcmasevic et al. (2021) / With permission of Elsevier.

genomics, metabolomics, transcriptomics, and proteomics will favour the development of new fortification strategies to address micronutrient deficiency (Kumar et al. 2020). Further, 3D printed foods serve as an excellent food vehicle that aids in the development of formulations with key ingredients and enhancers for a healthy diet. Thus, 3D printing remains to be a potentially proven technology that improves societal well-being with a focus on food customization. Already, reports were available for the localized use of food ingredients that helps in reduced intake of sugar, salt, and fat (Dick et al. 2019; Prakash et al. 2019). The knowledge on food structures and solvation chemistry along with printing technology would do wonders in sensory perception and cutting down sugar, salt, and fat. 3D printing can be used for reducing the incidence of lifestyle disorders among people. Taking the advantage of a greater degree of design freedom, 3D food printing is being successfully used for the formulation of dysphagia diets, i.e. a medical condition experienced by elderly people who suffer from swallowing difficulties and choking problems (Hadde and Chen 2019). Traditionally processed dysphagia foods possess an unappetizing taste, unpleasant aroma, and unappealing appearance that significantly reduces the meal intake among dysphagia patients (Dick et al. 2021). 3D food printing is a proven technology for addressing this issue that enables the delivery of textured modified foods in more appealing taste, appearance whilst maintaining the nutritional requirements (Dick et al. 2020). Thus, 3D printing has a great scope for the development of personalized foods. When the delivery of personalized foods becomes a business, food fraud, and adulteration become a great concern that could be addressed through adopting appropriate anti-counterfeit measures (Trenfield et al. 2019).

In a long-term view, when food printing becomes a trade with the emergence of globalization of 3D printed foods, people would start to print their own foods at the comfort of their homes using domestic digital cookers '3D food printers' (Sun et al. 2018a). In this scenario, the percentage of buying habits would be reduced among the consumers where people will no longer go to grocery shops and restaurants if they started to print their own recipes at home. This would significantly cut down the shopping behaviour and affects the traditional business of conventional foods. Further, it would lead to new requirements on the development of standard protocol and procedures for printing foods; selling of preprocessed food ink cartridges that can be brought and loaded into printers; and on the regulation of e-commerce of 3D design models which could be reachable to consumers at doorsteps (Nachal et al. 2019). Providing the ease of ordering and buying food ingredients required for 3D printing would significantly reduce the labour force that transforms the existing process to be automated. The technological advancements of robotic food manufacturing and automation significantly reduces the processing cost of the foods (Sun et al. 2015b). When people become to print their own food with customized designs and creativity, the culinary professionals might go out of business. When the technology is available at doorsteps in a single click to select the desired foods, people would refuse to spend money on the prepared foods that eventually cease to exist. In this scenario, buying foods at marketplaces would become a luxury activity for those who have extra money and time (Koenig 2016). This would have a significant transition in the global economy and sales of food. In another viewpoint, 3D printing seems to be a convenient technology that is more suitable for supplying the 'food on the go' while travelling outside like picnics, trekking, and camping (Tran 2019). During these situations, it can be affordable to carry portable 3D printers that supply food based on demand. A special case is the delivery of foods for astronauts and military personnel who has specific nutritional needs. Hence, the NASA and US military are being worked on the applications of 3D food printing for the delivery of the space and military foods (Liu et al. 2017). The capability of preparing foods on demand would reduce the cost associated with materials, energy, package, warehouse, and distribution costs; eliminates waste generation; and efficient use of available resources that provides sensorially and nutritionally enhanced foods.

16.3 Applicability and Storability

The printability of any material supplies depends on the pre-processing and post-processing methods (He et al. 2019). Few materials such as chocolates, starch gels, and hydrocolloids are adequately subjected to thermal pre-processing before printing and during the printing process, the temperature has been controlled for desired melting and solidification properties. Some other dough-based material supplies require post-processing such as baking, frying, and steaming immediately after printing. The post-process cooking step is adequate in order to ensure safety and to impart desired sensorial attributes to 3D printed foods (Wilson et al. 2020). The most commonly used extrusion-based printing employs a semi-solid dough with a considerable amount of water to impart pasting characteristics. However, the presence of moisture along with other nutrients makes a printed material supply to be an excellent growth medium for microbes (Zhang et al. 2021). Hence, essentially the 3D printed samples must be post-processed without delay to prevent microbial growth. The post-processing can be partial or complete cooking. The former type of partial post-treatment aids in the removal of water to a safe storage level perfect for long-term storage; however, it requires another step of cooking prior to consumption. Most commonly used method for partial cooking is drying which reduces the availability of moisture for biochemical reactions and microbial growth. This type of developed 3D printed products can be categorized as ready-to-cook foods. Several reports were available on employing drying as a post-processing method (Feng et al. 2020). The latter type of complete cooking ensures the edibility by a thorough cooking process of the 3D printed samples. Post-processing methods such as baking, frying, microwave cooking, boiling, broiling, and steaming aids in complete cooking thereby inhibit microbial growth. These post-processing methods are well suited for an immediate serving of 3D printed foods and can be categorized as ready-to-eat foods. Other than microbial safety, post-processing must assure the structural stability and mechanical integrity of the 3D printed samples (Pulatsu et al. 2020). Most of the reports on the post-processing of 3D printed foods focused on the mechanical strength, sensory and nutritional attributes of the developed 3D printed foods (Dick et al. 2019; Feng et al. 2020; Keerthana et al. 2020). Not even a handful of studies reported on storage stability and safety of 3D printed foods. However, this remains to be an important aspect to be considered for the implementation of 3D printing technology.

In a study performed by Severini et al. (2018), a blend of fruits and vegetable formula was 3D printed. In addition to morphological and microstructural properties, the antioxidant, sensory, and microbiological attributes were monitored during the storage period of eight days at 5 °C. During the storage, the total phenolic content was gradually declined from 18.8

to 10.5 mg GAE/100 g while the antioxidant capacity was remained constant at 10.9 mg Trolox/100 g. The concentration of bacteria was found to be 4.28 log CFU g^{-1} after 3D printing that clearly indicates the mandatory sanitation of the whole premise of the workplace, equipment, and other accessories. In this study, the microbial count was determined by a modification of method reported by Martinez-Hernandez et al. (2017). The procedure includes sample dilution followed by the spread-plate count method on agar plates. The plate with the samples was incubated at 30 °C for 24–48 hours and 5 °C for 10 days for enumeration of total aerobic mesophilic bacteria and psychrophilic bacteria, respectively. While a similar procedure was adopted for the enumeration of yeast on potato dextrose agar base at 25 °C for 48 hours. Results of the 3D printed samples stored at air showed a higher value of microbial counts of 4.23, 4.27, and 5.02 log CFU g^{-1} for yeasts, psychrophilic, and mesophilic bacteria, respectively. These results were in accordance with a study on microbial load of fruit-based smoothies. The initial total count of the microbial load was found to be 5.3 and 5.1 log CFU ml^{-1} for mesophilic bacteria and yeasts, respectively. Despite the cleaning and washing of raw materials as well as equipment surface, the contact of printing formula with feed syringe piston, extruder unit, printing head, etc., were responsible for relatively higher microbial load after 3D printing than the initial count. There was no significant increase in the microbial load during the first six days of storage; however, from the eighth day, there occurred a significant reduction in the psychrophilic bacteria to 3 and 3.72 log CFU g^{-1} for the samples stored at air and modified atmospheric packaging (MAP), respectively. While the mesophilic bacteria and yeasts showed no significant difference during the storage time in both air and MAP conditions (Severini et al. 2018). This is the only report available solely on examination of microbial load of 3D printed food during the storage. In another study, the viability of the incorporated probiotic bacteria in the 3D printed baked dough (Zhang et al. 2018) and mashed potato was reported (Liu et al. 2020). These studies indicate the necessity of careful processing and proper sanitation of 3D printing units and raw materials to prevent cross-contamination of probiotic incorporated samples with other microbial species. Hence, cleaning and sanitation are the most important steps in the 3D printing process for assuring food safety and edibility of the developed 3D printed foods. Similar studies on the applicability and storage stability must be analyzed for determining the shelf-life of the developed 3D printed foods. As like the conventional food processing methods, 3D printing should follow a standard protocol of good manufacturing practices (GMPs), standard operating procedures (SOPs), and sanitation standard operating procedures (SSOPs) to assure the delivery of safe foods to the consumers. This could be possible when 3D printing has become a trade business and the 3D printed foods have been globalized. The government organizations would implement a threshold limit for microbial loads of the 3D printed foods that are safe for consumption.

16.4 Food Safety Considerations

Food is a major source of essential macro and micronutrients that critically contribute to physical well-being. With changing lifestyles and climate, consumer's eating experience and food habits are continuously transformed with the timeline (Doppler and Steffen

2020). Recently, the requirement for the most convenient, healthy, and easy to use foods are higher that enforces the food industry to seek for development of acceptable foods with novel ingredients. One such approach that satisfies the needs of all age groups is 3D printing that considers both the sensory and nutritional attributes (Sun et al. 2018b). Apart from providing nourishment, 3D printing provides a pleasant and joyful eating experience with its enhanced level of customization. Talking about food safety, quality control is an integral part of any processing technique. In this context, 3D printing does not deviate from this concept where instrumentation safety, as well as sanitation measures, are given the priority that enforces the need for the hygienic operating standards. With increased awareness about healthy concepts, consumer's concerns about the food safety not only depends on healthy diet but also on cultivational practices, ecology, social customs, and culture (Lupton and Turner 2018b). As an emerging technique, 3D printing of foods has been successfully demonstrated at a lab-scale level. Even though commercial 3D printers are available, the operational procedures are yet to be standardized for large-scale implementation. Since 3D food printing is in the nascent emerging stage, education and awareness about the advantage of printing technology among consumers are highly important. However, the lack of knowledge about food safety concerns from the general public to the various regulatory agencies remains a hindrance for the transition of this technique from lab-scale into a commercial level. More research needs are required in bridging this void on food safety and other related concerns for the successful implementation and sustenance of the 3D food printing technology. The present section summarizes the various safety concerns that must be addressed at different stages of the food supply chain of 3D printing based on the knowledge obtained from the available works of literature. Certainly, this section would provide valuable insights for the commercialization of 3D food printing and highlights the roles of different statuary agencies in framing standard regulations and protocol.

16.4.1 Process and Product Safety

16.4.1.1 Nature of Raw Materials

Like any other food processing technique, the first and foremost step in 3D printing is the washing and cleaning of the raw ingredients for the preparation of material supply. 3D printing is a highly versatile technology that uses a wide range of food sources as raw ingredients. In general, the materials have been categorized as natively printable, non-natively printable, and alternative ingredients (Sun et al. 2015a). Each of the categories has specific attributes that require a different extent of processing for converting into a printable form. Most of the food materials are perishable such as fruits, vegetables, and flesh foods that possess moisture content higher than the safer limit. Further, a few ingredients like cereals, pulses, and legumes require the addition of water while processing for achieving desirable attributes to form a smooth paste-like consistency suitable for extrusion-based printing. In this regard, inherently the material supply has high moisture that makes it conducive for microbial growth. Hence, it is always recommended to 3D print the material immediately after preparation without holding time. Even though the raw ingredients are cleaned and washed thoroughly, food materials are a good source of nutrients that acts as excellent growth media for microbes possessing a considerable amount of initial microbial load (Anukiruthika et al. 2020a). Another fact to be considered is when alternative ingredients such as fungi, algae, and probiotic bacteria are incorporated into the printing mixture, the added ingredients accelerate the microbial growth, and the material is susceptible to safety concerns. In view of sustainability, food processing by-products are conveniently used for the fabrication of value-added 3D printed products. Although this application is successful in lab-scale studies, the implementation of which requires proper hygienic measures as the raw ingredients are recovered from wastes. Practically, there might be great concern with the safety and edibility of these ingredients. Hence, careful process recovery of materials from conventional food processing unit and proper pre-processing must be adequate to keep the material supply microbially safe for further printing applications.

16.4.1.2 Processing and Design Factors

Besides raw material quality the way of pre-processing of these materials also significantly impacts the final quality and contributes to microbial load. Each of the unit operations such as cutting, trimming, slicing, grinding, and mixing would alter the physiochemical properties (Severini et al. 2018). Further, the size reduction imparts a larger surface area of contact of the food materials with equipment surfaces that leads to microbial contamination. Thus, 3D printing requires quick strenuous processing with hygienic practices and sanitation measures. This applies to the whole process of printing starting from material input until the completion of printing. In addition, the equipment and accessories that are used for 3D printing must be in good sanitary condition (Nachal et al. 2019). Especially, design parts such as syringe feed barrel, extruder assembly and print head, material flow path, screw unit, piston assembly, and printing nozzle have direct contact with the material supply. Sometimes, the holding of material and time lag in any of these design parts would significantly contribute to microbial contamination as the printing mixture has a higher amount of moisture as well as enriched with macro and micronutrients. To date, there is no report on the protocol for the sanitation of the 3D printing unit. Future research studies must consider this aspect that would pave for standardization of sanitation procedures. The protocol must include the frequency of cleaning, type of sanitation liquid, the concentration of sanitation liquid, washing, cleaning, and drying cycle. Since the food material is employed that is meant for human consumption, the food-grade sanitation procedures must be adopted. Precautions must be taken in avoiding the residues of the detergents that would otherwise lead to serious health implications when comes in contact with food. Similarly, attention must be paid that the used sanitation agents should not corrode the equipment surface and components of the 3D printing unit.

Most of the commercial 3D printers available for printing foods are based on the FDM approach that is mainly designed for the printing of polymers, plastics, ceramics, and metals (Liu et al. 2018). Hence the component of the printing system is compatible and designed for non-food printing applications. However, when dealing with food printing, food-grade design components are the most required otherwise the material begins to corrode when comes in contact with acidic/ basic material supplies. Since the polymer printer as such was adopted for food printing with slight modifications, there are no standard regulations available on the applications of food-grade materials for the design of 3D printing components. In addition to the study on the material characteristics and printability, design components of different grades of materials used in the design of 3D food printers must be analyzed for determining the best suitable design material for food printing applications.

Conventionally, materials like stainless steel, aluminium, and copper are commonly used for designing the components in a food manufacturing unit (Deshwal and Panjagari 2020). Regarding food printing, design parts such as extruder units, syringe feed barrels, and printing nozzles are available in stainless steel and food-grade plastics. Since plastic materials are more convenient in the fabrication of miniatured 3D printing components and possess less weight, they are more commonly used. However, with the continuous use of the same component and deterioration with tear and wear, there occurs leaching of these plastic materials that might contaminate the printing material supply. Hence, the expected lifetime of usage of these food-grade plastic materials must be monitored.

16.4.1.3 Finished Product Safety

Post-processing is the process that has to be done immediately after 3D printing for achieving post-print stability and ensures the edibility of the developed 3D printed foods. Other than chocolates, cheese, hummus, food spread, and hydrogels, most of 3D printed materials require post-process cooking. Post-processing is one of the most critical steps in the 3D printing process for ensuring safety (Keerthana et al. 2020). During cooking, the moisture content of the material supply is removed and the associated macro component interactions with the thermal source impart thorough cooking, improves texture, adds flavour, and taste to the product. Importantly, post-processing increases the product shelf-life and storage stability of the developed 3D printed food. Based on the applicability and consumer's need, the post-processing method is being selected (Krishnaraj et al. 2019). Due to the increased consumer's choice for minimally processed foods and healthy recipes, researchers are paying attention to novel methods of post-processing of 3D printed constructs such as microwave cooking, infrared cooking, laser heating, air frying, electrohydrodynamic drying, and conductive hydro drying (Anukiruthika et al. 2020b; He et al. 2019; Yoha et al. 2020). Although technically these post-process methods are feasible and have a great scope for implementation, the use of novel methods must require the knowledge of the science of the interaction of heatwaves with the food materials. There is no sufficient information available on the food component interactions with novel thermal sources and by-products free radical generation. A detailed study on these aspects would pave a way for the successful implementation of a cost-effective post-processing method that ensures safe delivery of 3D printed foods. Apart from microbial safety, other concerns related to foreign material contamination and the risk of free radicals must be addressed.

16.4.1.4 Working Premises and Personnel Safety

3D food printer is a compact digital food processing unit that involves layered fabrication of foods. Some of the printing units are integrated with a thermal source to combine the 3D printing and post-processing in one single step. The design components of the printer unit consist of several minute assembly parts, metal screws, and nuts. Precautions must be taken while printing food materials in monitoring the working conditions of system components. Due to the wearing and agitation of the motor assembly, the tightened metal screws might loosen and fall into the 3D printed food. Sometimes, the chemical lubricants used for motors of movement arms and extruder units may cross-contaminate the edible 3D printed material. All these conditions, pose a serious threat to humans as the end application of 3D printed foods is meant for consumption. The working premises, as well as process conditions, must be carefully monitored for assuring product safety (Moerman 2017). In addition, personnel safety cannot be overviewed which is the most significant aspect in any food processing facility. 3D printing is a newer technology that is applied for printing foods in a customized manner. Although 3D printing reduces the complexity of the traditional food supply chain, to get started with this novel technique appropriate personnel training is required. Hence, the workers who are meant for operating a 3D printing unit must possess material, process, and software knowledge in handling the unit. The personnel must be well trained and equipped with a deeper understanding of 3D printing that aids in encountering sudden breakdown issues during the printing process. 3D printing dealt with the handling of sharp metal objects and wear parts such as metal tapered nozzles, extrusion screws, coupling assemblies, and pointed thin metal rods used for cleaning of material residues that might cause personnel injuries if not paid attention during processing. Further, when the 3D printer is integrated with a thermal source for the combined cooking process, the printer would be equipped with the infrared/ laser/ microwave source that must be shielded properly to avoid accidents due to radiation leakage (Vishnyak et al. 2020). Overall, as with any other food processing facility, the workers must wear personal protective equipment (PPE) when 3D food printing is implemented for commercialization at a large-scale level. Various safety aspects in terms of material quality, cleaning, sanitation, hygienic, and safeguard measures from microbes, foreign objects, chemical contaminants, and thermal sources must be considered during the printing process (De Boeck et al. 2015). The statutory body and regulatory agencies must consider all these aspects in framing a standard protocol for safety measures during 3D printing of foods when food printing enters into the global business trade.

16.4.2 Acceptance of 3D Printed Foods

3D printing of foods sounds to be a fascinating technology that delivers food in the desired forms and sensory attributes. However, the acceptance of 3D printed foods relies on several factors such as knowledge of food printing, raw material sources, printing technology, optimization of the process, and post-processing. Although consumers are aware of the benefits of 3D printed foods, the psychological conditions of neophobia among a certain group of populations hinder its acceptability. The sociocultural perspectives of 3D printed foods were examined in assessing the consumer's acceptability (Lupton 2017). The result showed that 3D food printing was viewed as futuristic, healthy, creative, efficient, and sustainable technology that seems to satisfy the food decisions of consumers in terms of time-saving, novel, convenient, nutritional foods with environmental benefits. In another study, the consumer's attitude towards 3D printing was qualitatively studied (Lupton and Turner 2016). During which some aspects of reluctance towards 3D printed foods in terms of safety and edibility were highlighted. Some people stated that the method of printing foods sounds strange and unusual; the use of uncommon food ingredients such as insects and algae considered to be unacceptable for consumption in the western culture; concerns related to chemical preservatives and additives in regard to healthier diets; the possibility of food allergens and food poisoning remained questionable (Lupton and Turner 2018b; Tran 2019). The study on the consumer's acceptance of 3D printed foods in comparison with conventional food products drawn three significant conclusions. Results of the survey categorized the group of participants as interested participants who wanted to know more about food printing and believed that this technology would result in several socioeconomic benefits; moderately interested participants who were excited to give a try for 3D printed foods; and non-interested participants who had the mindset that 3D printed foods were unacceptable and unsafe for consumption (Manstan and McSweeney 2020). Thus, more awareness is required in exploring the advantages of 3D printed foods. With these inputs, future studies must be focused on aspects of safety of 3D printed foods in order to have solid proof of evidence for encouraging 3D printing. The present section discusses the health implications of 3D printed foods in the short and long-term run.

16.4.2.1 Food Poisoning, Food Allergy, and Cross-Contamination

As 3D printing is an emerging technology there are no statutory regulations available as of today's date (Tran 2016). Generally, the legislation would progress slower than the development of the technologies. It is mandatory to know about the implications on the inclusion of 3D printed foods in a regular diet. When 3D printed foods become common, there also comes food contamination, adulteration, and food fraud as in any other conventional food processing technique. The short-term intake of 3D printed foods could rise to the practice of adulteration among the food manufacturers until unless a standard regulation is stated on the tolerance level of deleterious substances by the Food and Drug Administration (FDA). In general, the adulterated food is typically regulated under food safety that comes under the jurisdiction of the US Department of Agriculture and FDA. A food is considered to be adulterated if it possesses poisonous substances that are injurious to health or it contains any poisonous substances that are externally added during processing. On the occurrence of an event of food poisoning, FDA would have all rights to prosecute the food production company that even considers it as criminal prosecution in the case of any deaths (Tran 2019). Considering the case of 3D printing, there is a possibility for the chance of food poisoning at the individual level or a group of the community if the process of 3D printing is not regulated under the law. The first case for the incidence of food poisoning at the individual level implies an individual's response to 3D printed foods as say for example the food allergy. In common practice, the issues on food allergens are addressed through proper labelling of food ingredients used in processed foods. This case is seemed to occur rarely that too based on an individual's allergic response to specific food ingredients. The major food allergens that must be presented in labelling, if the product contains milk, fish, shellfish, egg, peanuts, tree nuts, sesame, wheat, and soybeans (Boden et al. 2005).

Similar practices of food labelling should be enforced for 3D printing as it utilizes common foods, alternative foods, and waste by-products for printing applications. In this way, the labelling of all traces of food ingredients must be mandated so that consumers would be aware and can buy the 3D printed foods at their own risk. Thus, each individual is being responsible for selectively omitting the 3D printed foods that are allergic to them (Baiano 2020). In case of mislabelling, the outbreak of any health issues is wholly from the manufacturer's side that should be punishable under law. In case of food poisoning occurred for a group of population in batches due to the printing ingredients, the parties who distributed the supply, i.e. the service providers are responsible for liability. In such cases, a nationwide recall must be done to retrieve the distributed supply of ingredients. Since the total number of players in 3D food printing are very less in the count, it enables the quick and easy process of recalling the products. Based on the nature of deleterious ingredients, the incidence of food poisoning cases would likely be less or more. If the party of service providers that were responsible for the outbreak remains unidentifiable then the liability must be applied to the product providers like the manufacturers of 3D food printers or material ingredients or the retailers (Tran 2019). Thus, the manufacturers, retailers, and service providers are more prone to legal implications and bear the burden of apportionment of food poisoning rather than the consumers.

16.4.2.2 Long-Term Health Effects and Illness

The physical well-being of humans depends on eating habits and food behaviour. The long-term intake of 3D printed foods would result in changes in the human body. However, the change can be either good or bad that severely impacts the food economy (Baiano 2020). In the case of acceptance of 3D printed foods, in a long-term view, the conventional and traditional food practices will eventually get depleted. In contrast, if the change in eating practices is bad then the rejection of 3D printed foods by the human body would have legal safety issues. On the other hand, when traditional foods are rejected by humans then it indirectly affects 3D printing as the ingredients used for food printing would become scarce (Tran 2019). These questions are unanswerable without the controlled long-term population studies on 3D printed foods. Similar to the testing studies employed for genetically modified (GM) foods using controlled population groups, trials on 3D printed foods must be carried out for answering such questions concerning health implications. For instance, if everyone who consumed 3D printed foods developed health concerns such as cancer development then the root cause of the deleterious substance would be identified easily. Otherwise, the illness on the consumption of 3D printed foods is also influenced by several other process variables and health factors that cause a continual change making it impossible to determine the causation element. Hence, more research on the fate of the food ingredients used in 3D printing of foods must be addressed in relation to safety and legal concerns.

16.5 Legal Framework and Regulations

The significance of framing legal regulations was discussed in regard to 3D printing facilities and personnel safety. Various key steps of the past and the current EU novel food regulations were given in Figure 16.4. The applicability of existing regulations of traditional food manufacturing to 3D printing remains questionable as it might require new additional concerns to address. In the future when the 3D printing technology is globalized, there might be chances for adulteration, or it could contain food allergens in case of improper processing. Such issues are already encountered in traditional food manufacturing processes. Similar food safety and quality control legislations must be enforced at national and international levels (Crevel et al. 2008). Another significant and common legal issue that would occur with 3D printed foods is labelling. The 3D printed foods can come under the broad category of imitation foods that should require a different kind of labelling to distinguish from the rest of processed foods. For instance, if the cost of production of traditional and

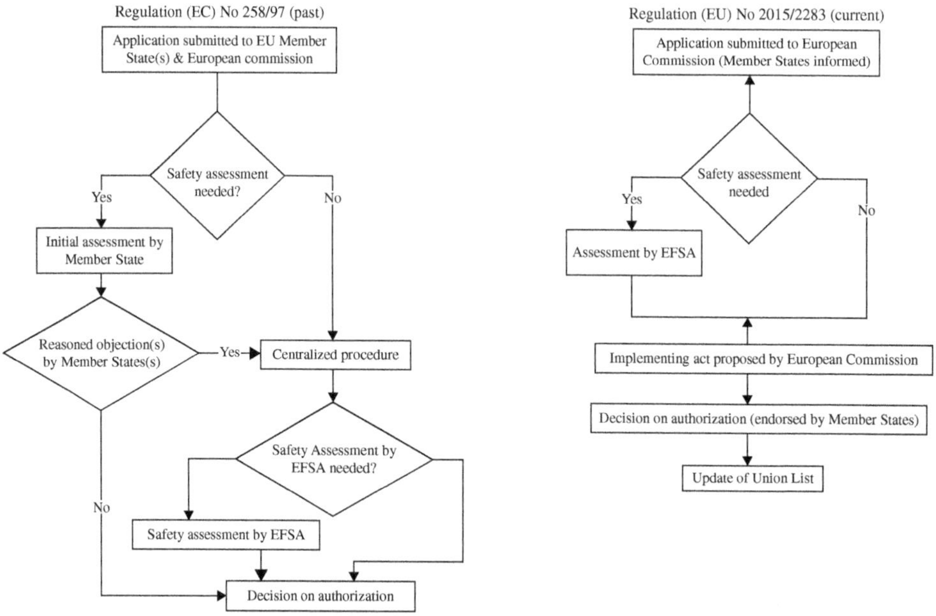

Figure 16.4 Key steps of the past and the current EU novel food regulations. *Source:* Ververis et al. (2020) / With permission of Elsevier.

conventional foods is costlier than the 3D printed foods, there might be chances for food fraud in the selling of 3D printed foods as conventional foods without proper labelling. Regardless of the well-distinguished appearance of 3D printed foods from that of conventional foods, the practice of specific labelling must be strongly recommended to avoid confusion of food choice among the consumers (Grofke et al. 2021). Further, the implementation of a certain threshold percentage of food ingredients used for 3D printing should be set as labelling cut-off. 3D printed foods are regarded as novel foods which imply the development of food from innovative food sources. Food printing involves the transformation of a wide range of ingredients into a homogeneous printing mixture with appropriate consistency conducive for pre-defined deposition of material supply in a layered fashion. With advancements in printing technology and material science, foods are manufactured directly from the chemical compounds at the molecular level to alleviate the increased demand of food supply in near future. This emphasizes the need for specific standard regulations in analyzing the fate of macro and micronutrients at the individual level in addition to the epidemiological studies among the population groups (Baiano 2020). Also, the globalization of food printing would encounter legal concerns related to intellectual property rights, copyrights, licenses, and patents.

Although consumers are doubtful about the safety of 3D printed foods, if the technology is proven to be safe then 3D printing becomes the source of food supply in the future. The conceptual framework and drivers of traceability in the food supply chain were presented in Figures 16.5 and 16.6. As stated earlier, several possibilities would mislead the consumers with insufficient labelling of 3D printed foods. The people's argument for

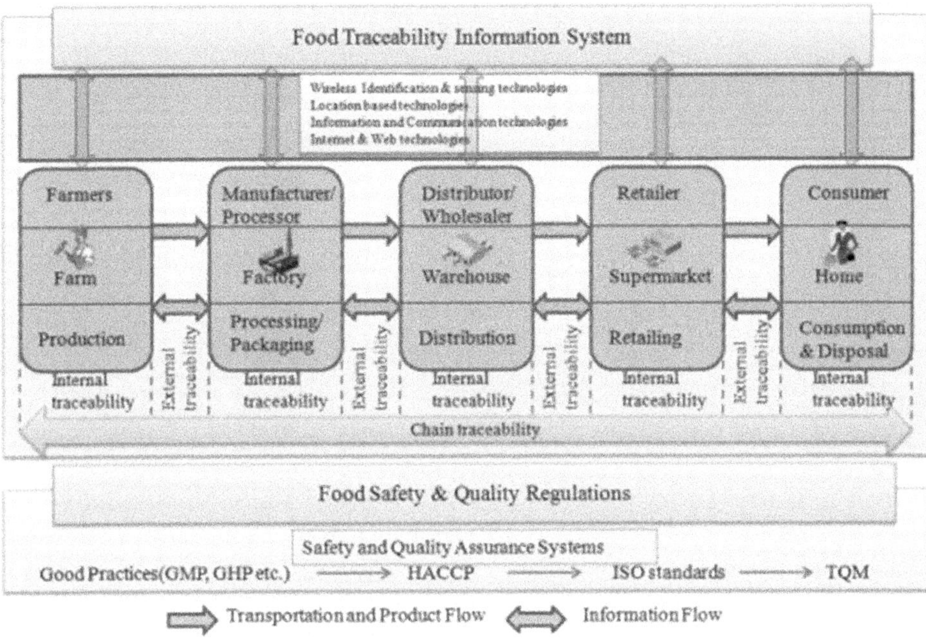

Figure 16.5 Conceptual framework of food traceability system. *Source:* Aung and Chang (2014) / With permission of Elsevier.

clear labelling is the right to know what they consume (Tran 2019). The labelling challenges encountered with 3D printed foods are more like that of the GM foods. Consumers are scared of consuming GM foods as they might possess unknown long-term effects. The labelling of 3D printed foods is applicable when the printed foods are delivered in food packages. The labelling requirement of the 3D printed foods can be demonstrated by considering different scenarios that range from non-labelling of printed foods at domestic kitchen to the labelled 3D printed foods delivered at retail stores. Sometimes, the 3D printed foods are failed to be distinguished with their appearance that enforces the need for clear labelling. If the printed foods are distinguishable then there will be no need for labelling, but this is not the case in reality that often requires standard regulations related to labelling of 3D printed foods.

16.5.1 Packed 3D Printed Foods for Mass Population

The large-scale implementation of food printing deals with big corporate food companies that print majority of foods in packages to be sold at retail stores for addressing the nutritional needs of mass population groups. This scenario is the case where labelling of 3D printed foods would be strongly recommended. The issues and concerns with labelling of 3D printed foods in this case are similar in agreement with the GM foods as the consumers have the right to know what food they consume. Another case is the 3D printing of foods with larger industrial players for addressing the special needs of people like dysphagia

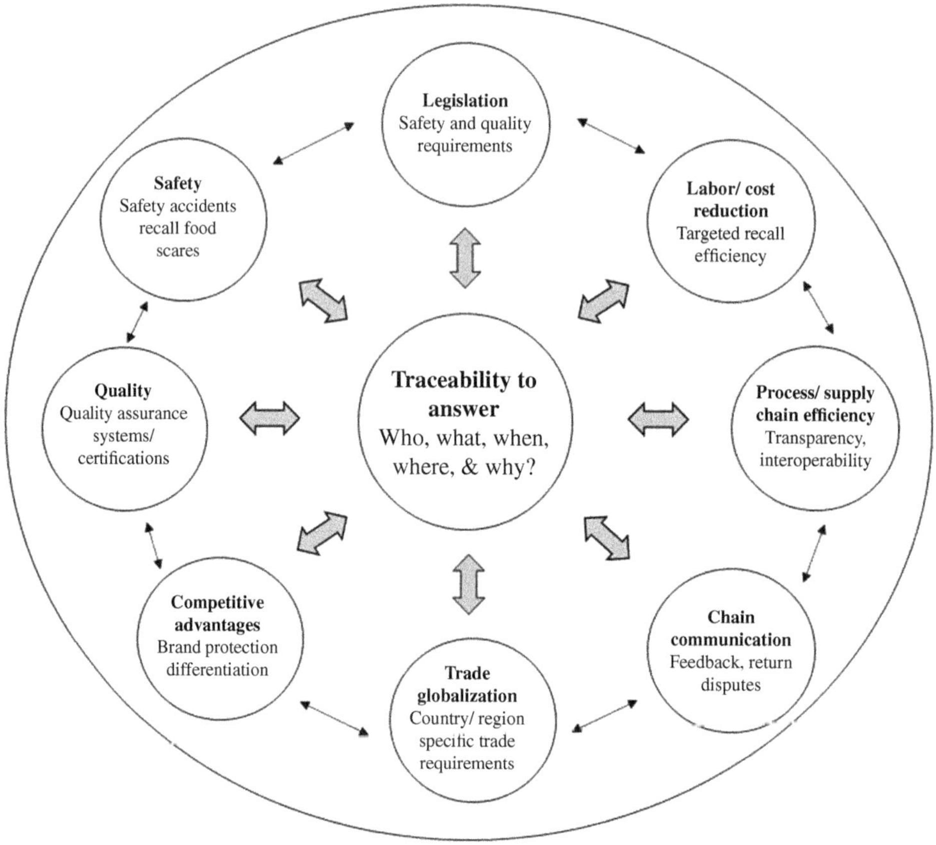

Figure 16.6 Drivers for traceability in the food supply chain. *Source:* Aung and Chang (2014) / With permission of Elsevier.

diet, cutting down of salt, sugar, and fat (Prakash et al. 2019). This case also enforces the adequacy of clear labelling. In contrast to these cases, if the targeted audience is a small group of the general public, there comes a scenario where mandatory labelling becomes questionable. Suppose if the amount of 3D printed foods (alternate ingredient) is relatively small say <1% then labelling is not necessary; however, when the portion becomes larger, the labelling becomes mandatory. This is similar to that of traditional labelling practices followed for trans-fat foods. The foods that contain trans-fat of <0.5% is often negligible with the marking of trans-fat as 0% in labelling. However, this might not be applicable always for 3D printed foods because some people who prefer only natural whole foods are more cautious with even the traces of ingredients (Thierer 2013). Furthermore, people are often prone to allergic reactions to alternative food ingredients used in 3D printing and most importantly consumers should feel safe while consuming food without any worries related to unknown food sources. In this regard, labelling of 3D printed foods is the best practice to avoid food safety concerns which must be abiding by the law. The presence of labels often allows the consumers to be selective in their choices of food to be consumed.

16.5.2 Unpacked 3D Printed Foods at Restaurants and Domestic Kitchen

This is the case where the food is 3D printed for immediate consumption without storage in selling the printed foods to the local community. The printing of foods directly in front of the consumers will not require labelling as like vending machines where consumers can select their own printing recipes with desired food ingredients. However, if the printed foods at vending shops are meant for selling after business hours, the consumers might not be aware of a situation from where the food (food sources) was 3D printed (Tran 2019). In such situations, labelling is required where the consumers lack knowledge about 3D printing. However, the majority of vending applications would occur in front of consumer's sight that favours no labelling requirements. Another scenario is the preparation of 3D printed foods using domestic kitchen 3D printers for regular meals. This is the case that clearly supports no labelling because the consumers who use 3D food printers are well aware of the process and ingredient formulation of what they are consuming.

16.6 Challenges and Research Needs

3D printing is a digitalized technology that transforms the global economy with a gradual shift of the economies of sales. The strategic foresight of food printing seems to be disruptive that opens up a dimension of food trade with innovative business policies (Cook et al. 2014). Compared to conventional food manufacturing, 3D printing is a cost-effective method for customized and personalized fabrication of food products. As a rivalry technology, 3D printing has a great potential in changing the dynamics of production in terms of process efficiency, utilization of resources, management of demand and supply chain, and government regulations. This would bridge up the gap between the small- and large-scale businesses in producing quality food products that are affordable and reachable to the people of all classes of the economy. The production of foods based on demand would reformulate the food supply chain that eases the inventory management of resources (Nachal et al. 2019). The delivery of foods at consumption points significantly reduces the warehousing, shipping, and transportation costs. This smart way of food production is often reflecting the Food Economy 4.0 that focused on localized production of foods with less added preservatives for immediate consumption. However, these benefits of food printing come along with the issues on food safety and health concerns of 3D printed foods. The 3D food printing provides a promissory result in terms of safe delivery of foods through proper sanitation but the acceptability and allergenicity of used food ingredients are questionable. Further, at large-scale implementation maintenance of hygienic sanitary conditions would be difficult until unless with the formation of standard protocols and regulations (Tran 2019). Other than food safety, the labelling of 3D printed foods remains debatable. Various concerns related to 3D printed foods are microbial safety, process safety, personnel safety, food poisoning, food contamination, food adulteration, food fraud, issues related to copyrights, piracy of 3D model designs, e-commerce trade, and patents (Baiano 2020). All these issues can be only be addressed by government agencies, statutory and regulatory bodies through the proper framework of standard practices in abide with law. This would regulate the business policies and flow of the supply chain of 3D food printing that remains to be beneficial for the major players of the food supply chain, i.e. product providers and service providers including manufacturers, suppliers, retailers, distributors, and consumers.

16.7 Conclusion

3D printing is a disruptive emerging technology that undoubtedly would revolutionize the food industry. It has a great scope to be used as a kitchen domestic appliance as well as for commercial large-scale production. The benefits of 3D printing of foods are customization, personalization, reduced complexity of food supply chain, process efficiency, less wastage, efficient utilization of resources, and sustainability. However, these advantages of 3D food printing can be fully enjoyed when food printing has become a business trade with the globalization of 3D printed foods. As a reforming technology, 3D printing aids in the decentralization and localized production of foods closer to the consumers. Despite these features, the acceptance of the developed 3D printed foods among the consumers is the most significant factor that decides the success of any technology. Some of the concerns related to 3D printed foods are safety, edibility, labelling, and piracy. These aspects cannot be overviewed by only considering the merits of 3D printing. A clear knowledge must be required in addressing these issues and requires the attention of government legislations and legal bodies in framing a protocol by regulating the food printing process. The consumer's acceptance and legal framework are two aspects that go on hand in hand and have a significant impact on behavioural change and dietary habits. Hence, it is necessary to clarify the doubts on the safety aspects of food printing before its implementation. Thus, the present chapter discussed the various implications of food printing technology on consumer's food habits, lifestyle, convenience, food supply chain, sustainability, societal, cultural, economic, and environmental aspects. Through the present work, it is clear that gaining consumer's acceptance and trust for the developed food products helps in the sustenance of 3D food printing in a long run. In this point of view, different aspects of safety considerations, legislation framework, and labelling have been highlighted for streamlining the food supply chain of 3D food printing. The intervention of government bodies in addressing issues on food safety, labelling, business policy, intellectual property, and ethics help in framing a standardized protocol for 3D food printing at the global level which would lead to the commercialization of this sunrising technology in the food sector.

References

Anukiruthika, T., Moses, J.A., and Anandharamakrishnan, C. (2020a). 3D printing of egg yolk and white with rice flour blends. *Journal of Food Engineering* **265**: 109691. https://doi.org/10.1016/j.jfoodeng.2019.109691.

Anukiruthika, T., Moses, J.A., and Anandharamakrishnan, C. (2020b). Electrohydrodynamic drying of foods: principle, applications, and prospects. *Journal of Food Engineering* **295**: 110449.

Aung, M.M. and Chang, Y.S. (2014). Traceability in a food supply chain: safety and quality perspectives. *Food Control* **39**: 172–184.

Baiano, A. (2020). 3D printed foods: a comprehensive review on technologies, nutritional value, safety, consumer attitude, regulatory framework, and economic and sustainability issues. *Food Reviews International*, pp. 1–31.

Boden, M., Dadswell, R., and Hattersley, S. (2005). Review of statutory and voluntary labelling of food allergens. *Proceedings of the Nutrition Society* **64** (4): 475–480.

Cook, C.N., Inayatullah, S., Burgman, M.A. et al. (2014). Strategic foresight: how planning for the unpredictable can improve environmental decision-making. *Trends in Ecology and Evolution* **29** (9): 531–541.

Crevel, R.W.R., Ballmer-Weber, B.K., Holzhauser, T. et al. (2008). Thresholds for food allergens and their value to different stakeholders. *Allergy* **63** (5): 597–609.

De Boeck, E., Jacxsens, L., Bollaerts, M., and Vlerick, P. (2015). Food safety climate in food processing organizations: development and validation of a self-assessment tool. *Trends in Food Science and Technology* **46** (2): 242–251.

Derossi, A., Husain, A., Caporizzi, R., and Severini, C. (2020). Manufacturing personalized food for people uniqueness. An overview from traditional to emerging technologies. *Critical Reviews in Food Science and Nutrition* **60** (7): 1141–1159.

Deshwal, G.K. and Panjagari, N.R. (2020). Review on metal packaging: materials, forms, food applications, safety and recyclability. *Journal of Food Science and Technology* **57** (7): 2377–2392.

Dick, A., Bhandari, B., Dong, X., and Prakash, S. (2020). Feasibility study of hydrocolloid incorporated 3D printed pork as dysphagia food. *Food Hydrocolloids* **107**: 105940.

Dick, A., Bhandari, B., and Prakash, S. (2019). Post-processing feasibility of composite-layer 3D printed beef. *Meat Science* **153**: 9–18. https://doi.org/10.1016/j.meatsci.2019.02.024.

Dick, A., Bhandari, B., and Prakash, S. (2021). Printability and textural assessment of modified-texture cooked beef pastes for dysphagia patients. *Future Foods* **3**: 100006.

Doppler, S. and Steffen, A. (2020). The future of food experiences. In: *Case Studies on Food Experiences in Marketing, Retail, and Events* (eds. A. Cavicchi and C. Santini), pp. 197–210. United Kingdom: Woodhead Publishing.

Feng, C., Zhang, M., Liu, Z. et al. (2020). Effect of drying method on post-processing stability and quality of 3D printed rose-yam paste. *Drying Technology* **39**: 1–9.

Forgacs, G., Marga, F., and Jakab, K.R. (2014). Engineered comestible meat. Google Patents.

Gerbens-Leenes, P.W., Mekonnen, M.M., and Hoekstra, A.Y. (2013). The water footprint of poultry, pork and beef: a comparative study in different countries and production systems. *Water Resources and Industry* **1**: 25–36.

Godoi, F.C., Prakash, S., and Bhandari, B.R. (2016). 3D printing technologies applied for food design: status and prospects. *Journal of Food Engineering* **179**: 44–54.

Grofke, N., Duplat, V., Wickert, C., and Tjemkes, B. (2021). A multi-stakeholder perspective on food labelling for environmental sustainability: attitudes, perceived barriers, and solution approaches towards the "traffic light index". *Sustainability* **13** (2): 933.

Hadde, E.K. and Chen, J. (2019). Shear and extensional rheological characterization of thickened fluid for dysphagia management. *Journal of Food Engineering* **245**: 18–23. https://doi.org/10.1016/j.jfoodeng.2018.10.007.

He, C., Zhang, M., and Fang, Z. (2019). 3D printing of food: pretreatment and post-treatment of materials. *Critical Reviews in Food Science and Nutrition* **60**: 2379–2392.

Keerthana, K., Anukiruthika, T., Moses, J.A., and Anandharamakrishnan, C. (2020). Development of fiber-enriched 3D printed snacks from alternative foods: a study on button mushroom. *Journal of Food Engineering*. https://doi.org/10.1016/j.jfoodeng.2020.110116.

Koenig, N. (2016). How 3D printing is shaking up high end dining. BBC News. https://www.bbc.com/news/business-35631265 (accessed 20 January 2021).

Krishnaraj, P., Anukiruthika, T., Choudhary, P. et al. (2019). 3D extrusion printing and post-processing of fibre-rich snack from indigenous composite flour. *Food and Bioprocess Technology* **12** (10): 1776–1786. https://doi.org/10.1007/s11947-019-02336-5.

Kumar, S., Anukiruthika, T., Dutta, S. et al. (2020). Iron deficiency anemia: a comprehensive review on iron absorption, bioavailability and emerging food fortification approaches. *Trends in Food Science and Technology* https://doi.org/10.1016/j.tifs.2020.02.021.

Liu, Z., Bhandari, B., and Zhang, M. (2020). Incorporation of probiotics (*Bifidobacterium animalis* subsp. Lactis) into 3D printed mashed potatoes: effects of variables on the viability. *Food Research International* **128**: 108795.

Liu, Z., Zhang, M., Bhandari, B., and Wang, Y. (2017). 3D printing: printing precision and application in food sector. *Trends in Food Science and Technology* **69**: 83–94.

Liu, Z., Zhang, M., and Yang, C. (2018). Dual extrusion 3D printing of mashed potatoes/strawberry juice gel. *LWT* **96**: 589–596.

Lupton, D. (2017). 'Download to delicious': promissory themes and sociotechnical imaginaries in coverage of 3D printed food in online news sources. *Futures* **93**: 44–53.

Lupton, D. and Turner, B. (2016). 'Both fascinating and disturbing': consumer responses to 3D food printing and implications for food activism. In: *Digital Food Activism* (eds. T. Schneider, K. Eli, C. Dolan and S. Ulijaszek), pp. 1–17. London: Routledge.

Lupton, D. and Turner, B. (2018a). Food of the future? Consumer responses to the idea of 3D-printed meat and insect-based foods. *Food and Foodways* **26** (4): 269–289.

Lupton, D. and Turner, B. (2018b). "I can't get past the fact that it is printed": consumer attitudes to 3D printed food. *Food, Culture and Society* **21** (3): 402–418.

Malav, O.P., Talukder, S., Gokulakrishnan, P., and Chand, S. (2015). Meat analog: a review. *Critical Reviews in Food Science and Nutrition* **55** (9): 1241–1245. https://doi.org/10.1080/10408398.2012.689381.

Manstan, T. and McSweeney, M.B. (2020). Consumers' attitudes towards and acceptance of 3D printed foods in comparison with conventional food products. *International Journal of Food Science and Technology* **55** (1): 323–331.

Marga, F.S. (2012). *Engineered Comestible Meat*. National Institute of Food and Agriculture.

Martinez-Hernandez, G.B., Amodio, M.L., and Colelli, G. (2017). Carvacrol-loaded chitosan nanoparticles maintain quality of fresh-cut carrots. *Innovative Food Science and Emerging Technologies* **41**: 56–63.

Moerman, F. (2017). Personal hygiene and good maintenance practices for the servicing of food processing equipment. In: *Food Protection and Security* (ed. S. Kennedy), pp. 267–327. United Kingdom: Woodhead Publishing.

Nachal, N., Moses, J.A., Karthik, P., and Anandharamakrishnan, C. (2019). Applications of 3D printing in food processing. *Food Engineering Reviews* **11** (3): 123–141. https://doi.org/10.1007/s12393-019-09199-8.

Nida, S., Anukiruthika, T., Moses, J.A., and Anandharamakrishnan, C. (2020). 3D printing of grinding and milling fractions of rice husk. *Waste and Biomass Valorization*, pp. 1–10. https://doi.org/10.1007/s12649-020-01000-w.

Portanguen, S., Tournayre, P., Sicard, J. et al. (2019). Toward the design of functional foods and biobased products by 3D printing: a review. *Trends in Food Science and Technology* **86**: 188–198.

Prakash, S., Bhandari, B.R., Godoi, F.C., and Zhang, M. (2019). Future outlook of 3D food printing. In: *Fundamentals of 3D Food Printing and Applications* (eds. F.C. Godoi, B.R. Bhandari, S. Prakash and M. Zhang), pp. 373–381. United Kingdom: Academic Press.

Pulatsu, E., Su, J.-W., Lin, J., and Lin, M. (2020). Factors affecting 3D printing and post-processing capacity of cookie dough. *Innovative Food Science and Emerging Technologies* **61**: 102316.

Salah, M., Tayebi, L., Moharamzadeh, K., and Naini, F.B. (2020). Three-dimensional bio-printing and bone tissue engineering: technical innovations and potential applications in maxillofacial reconstructive surgery. *Maxillofacial Plastic and Reconstructive Surgery* **42**: 1–9.

Severini, C., Derossi, A., Ricci, I. et al. (2018). Printing a blend of fruit and vegetables. New advances on critical variables and shelf life of 3D edible objects. *Journal of Food Engineering* **220**: 89–100. https://doi.org/10.1016/j.jfoodeng.2017.08.025.

Sun, J., Peng, Z., Zhou, W. et al. (2015a). A review on 3D printing for customized food fabrication. *Procedia Manufacturing* **1**: 308–319.

Sun, J., Zhou, W., Huang, D. et al. (2015b). An overview of 3D printing technologies for food fabrication. *Food and Bioprocess Technology* **8** (8): pp. 1605–1615.

Sun, J., Zhou, W., Huang, D., and Yan, L. (2018a). 3D food printing: perspectives. In: *Polymers for Food Applications* (ed. T.J. Gutierrez), 725–755. Argentina: Springer International Publishing.

Sun, J., Zhou, W., Yan, L. et al. (2018b). Extrusion-based food printing for digitalized food design and nutrition control. *Journal of Food Engineering* **220**: 1–11.

Theagarajan, R., Moses, J.A., and Anandharamakrishnan, C. (2020). 3D extrusion printability of rice starch and optimization of process variables. *Food and Bioprocess Technology*. https://doi.org/10.1007/s11947-020-02453-6.

Thierer, A. (2013). Technopanics, threat inflation, and the danger of an information technology precautionary principle. *Minnesota Journal of Law, Science and Technology* **14**: 309.

Tomasevic, I., Putnik, P., Valjak, F. et al. (2021). 3D printing as novel tool for fruit-based functional food production. *Current Opinion in Food Science* **41**: 138–145.

Tran, J.L. (2016). Press clause and 3D printing. *Northwestern Journal of Technology and Intellectual Property* **14**: 75.

Tran, J.L. (2019). Safety and labelling of 3D printed food. In: *Fundamentals of 3D Food Printing and Applications* (eds. F.C. Godoi, B.R. Bhandari, S. Prakash and M. Zhang), 355–371. United Kingdom: Academic Press.

Trenfield, S.J., Tan, H.X., Awad, A. et al. (2019). Track-and-trace: novel anti-counterfeit measures for 3D printed personalized drug products using smart material inks. *International Journal of Pharmaceutics* **567**: 118443.

Ververis, E., Ackerl, R., Azzollini, D. et al. (2020). Novel foods in the European Union: scientific requirements and challenges of the risk assessment process by the European Food Safety Authority. *Food Research International* **137**: 109515.

Vishnyak, M.N., Mashenskaya, E.A., and Melbert, A.A. (2020). Occupational risk assessment method for food industry. *IOP Conference Series: Earth and Environmental Science* **459** (2): 22038.

Wilson, A., Anukiruthika, T., Moses, J.A., and Anandharamakrishnan, C. (2020). Customized shapes for chicken meat–based products: feasibility study on 3D-printed nuggets. *Food and Bioprocess Technology* **13**: 1968–1983.

Yoha, K.S., Moses, J.A., and Anandharamakrishnan, C. (2020). Conductive hydro drying through refractance window drying – an alternative technique for drying of Lactobacillus plantarum (NCIM 2083). *Drying Technology* **38** (5–6): 610–620.

Zhang, J.Y., Pandya, J.K., McClements, D.J. et al. (2021). Advancements in 3D food printing: a comprehensive overview of properties and opportunities. *Critical Reviews in Food Science and Nutrition*, pp. 1–18.

Zhang, L., Lou, Y., and Schutyser, M.A.I. (2018). 3D printing of cereal-based food structures containing probiotics. *Food Structure* **18**: 14–22.

Index